SELECTED PAPERS

S. Chandrasekhar

SELECTED PAPERS

S. Chandrasekhar

*

VOLUME 4

Plasma Physics, Hydrodynamic and Hydromagnetic Stability, and Applications of the Tensor-Virial Theorem

THE UNIVERSITY OF CHICAGO PRESS
CHICAGO AND LONDON

The University of Chicago Press, Chicago 60637
The University of Chicago Press, Ltd., London

Printed in the United States of America

98 97 96 95 94 93 92 91 90 89 5 4 3 2 1

 Acknowledgments appear on page 586.

Library of Congress Cataloging in Publication Data

Chandrasekhar, S. (Subrahmanyan), 1910–
Plasma physics, hydrodynamic and hydromagnetic stability, and applications of the tensor-virial theorem / S. Chandrasekhar ; with a foreword by Norman R. Lebovitz.
p. cm. — (Selected papers ; v. 4)
1. Plasma (Ionized gases) 2. Hydrodynamics.
3. Magnetohydrodynamics. 4. Plasma stability. I. Title.
II. Title: Tensor-virial theorem. III. Series: Chandrasekhar, S. (Subrahmanyan), 1910– Selections. 1989; v. 4.
QC718.C38 1989 89-4906
530.4'4—dc20 CIP
ISBN 0-226-10096-0 (Vol. 4)
ISBN 0-226-10097-9 (Vol. 4, pbk.)

The paper used in this publication meets the minimum requirements of the American National Standard for Information Sciences—Permanence of Paper for Printed Library Materials, ANSI Z39.48-1984.

S. CHANDRASEKHAR is the Morton D. Hull Distinguished Service Professor Emeritus in the Department of Astronomy and Astrophysics, the Department of Physics, and the Enrico Fermi Institute at the University of Chicago.

Contents

PART THREE

Tensor Virial Theorem and Its Applications

A complete list of publications by S. Chandrasekhar will appear at the end of the final volume

Foreword

The present volume comprises papers in three subject groupings: (1) plasma physics, (2) hydrodynamic and hydromagnetic stability and (3) applications of the tensor-virial theorem, particularly in the context of self-gravitating, ellipsoidal masses. S. Chandrasekhar's efforts in these areas were largely concentrated in circumscribed time periods. The efforts devoted to areas 2 and 3 culminated in a monograph in the manner that Chandra has himself described in the preface to volume 1.

In that preface he also gives the principal criteria of selection for the present series of volumes, as follows: "first, whether the papers have not been included in any of my published books; and second, whether the papers include matters not treated in sufficient detail elsewhere and the possible historical interest they may have." These criteria leave considerable latitude for judgment. The monographs represent syntheses of Chandra's work and that of others, giving his view of the status of the subject at the time: they are not collections of reprints. No paper can be exactly reproduced in this setting if only for reasons of organization and continuity. Moreover, in most cases, there is a substantive difference between the paper and the corresponding sections of the book. The judgment lies in deciding whether material appearing in the original paper but omitted in the book is substantial enough to include the former in these selected papers. Although I have had help in arriving at decisions, the final decisions have been mine, and I hope my judgment has been good.

Part 1 on plasma physics belongs to the period of Chandra's work on hydrodynamic and hydromagnetic stability, but is not in the direct line of development of that work (which is described below). It consists of three papers on the stability of the pinch written jointly with A. N. Kaufman and K. M. Watson, and one on Chandra's own approach to the subject of adiabatic invariants, which was very active at the time and is experiencing a renewal among applied mathematicians and physicists at this writing.

The second group of papers is on hydrodynamic and hydromagnetic stability. It is a characteristic feature of this subject that a small number of model problems has done much to elucidate complex patterns of behavior in fluid dynamics. For example, the Bènard problem (of a layer of fluid heated below) is the prototype for problems of convection; the Taylor problem (of flow between rotating cylinders) is the prototype of rotational flow. Indeed, these and a small number of other, idealized model problems continue to play major roles in theoretical, experimental and computational fluid dynamics today.

As I have mentioned in the preface to volume 3, Chandra began, in the early fifties, a reconsideration of the fundamental problems of hydrodynamics from several standpoints. These were, the adequacy of their formulation for drawing inferences in natural settings, the nature of the mathematical formulation and its relation to the underlying physics, and the kinds of mathematical techniques best suited to these problems. Regarding the adequacy of formulation, whereas it had long been appreciated that rotation is pervasive and important in astrophysical and geophysical settings, the pervasive character of magnetic fields in astrophysical settings (especially on the galactic scale) had only recently been appreciated. Chandra's reconsideration generalized the fundamental problems to include magnetic fields. Regarding mathematical formulation, the novelty of Chandra's approach was the systematic exploitation of variational methods. Related matters, such as constructing appropriate basis functions and the use of the system of equations adjoint to the one under consideration, augmented the store of mathematical techniques available for solving these problems.

Chandra summarized the status of the field in his monograph *Hydrodynamic and Hydromagnetic Stability* (HHS), published in 1961, and including much of his own research in the period 1952–1961. It was about this time that serious scientific activity began in the area of nonlinear hydrodynamic stability. Inasmuch as many nonlinear analyses rest heavily on a precise understanding of the corresponding linear analysis, the basic subject matter of HHS, its appearance was timely and it has become a standard reference for all stripes of fluid dynamicists.

Since HHS is principally devoted to mathematical development, a number of papers devoted to specific scientific applications like plasma physics and astrophysics are not included there; they appear elsewhere in these *Selected Papers* (principally in part 4 of volume 3 and in part 1 of the present volume, respectively). Most of the papers published in this period and devoted to fundamental problems are adequately represented in HHS and are omitted from the present volume. Those that are included in this volume are a few papers nominally covered in HHS, but only in abridged form excluding substantial material, and one published after the appearance of HHS. Also included are three papers on mathematical technique.

The third group of papers included in this volume is devoted to applications of the tensor-virial theorem. The most thorough and systematic use of this technique was in the context of rotating, self-gravitating fluid masses of uniform density in the shape of ellipsoids. These figures include the familiar Jacobi ellipsoid and Maclaurin spheroid, which are characterized by solid-body rotation.

They also include a much wider class of ellipsoidal figures discovered by Dedekind and, especially, by Riemann, which are characterized by a motion of uniform vorticity even in a frame of reference in which the ellipsoidal surface is at rest. These Riemann ellipsoids, as they are now known, had remained in relative obscurity until Chandra investigated their properties. The monograph *Ellipsoidal Figures of Equilibrium* (EFE), published in 1969, contains a comprehensive account of the ellipsoids of Maclaurin, Jacobi, Dedekind, and Riemann, and their stability. One of the lasting effects of this work, and of the book EFE, is the restoration of Riemann's neglected work to its proper place. It is now frequently referred to in the literature of theoretical astronomy and of applied mathematics.

The tensor-virial method used to study these figures was novel. It is a moment, or Galerkin, method. It takes its name from the classical virial widely used in astrophysics, which is the trace of the tensor virial. Normally moment methods require a closure assumption and, similarly, Galerkin methods require a truncation; in either case they are approximation methods. In the context of the equilibrium and stability of the ellipsoidal figures, however, the tensor-virial method is exact. It had the advantage of being elementary in character, whereas the classical methods previously used in the subject needed to develop the elaborate machinery of ellipsoidal-harmonic analysis.

Since the monograph EFE (like HHS) is principally devoted to mathematical development, Chandra's papers addressing scientific applications are not covered there, and a number of them consequently appear in this volume. A small number of papers nominally, but not completely, represented in EFE are also reproduced here. Further included here are several papers referring to figures other than the ellipsoids, including some applying the virial method to such figures and others addressing more generally the equilibrium and the stability of self-gravitating figures, such as the brief but influential paper on the variational characterization of the oscillation frequencies of spherical masses.

NORMAN R. LEBOVITZ

PART ONE

Plasma Physics

1

Properties of an Ionized Gas of Low Density in a Magnetic Field. III[1]

S. Chandrasekhar, A. N. Kaufman, and K. M. Watson

Los Alamos Scientific Laboratory, Los Alamos, New Mexico

and

University of California Radiation Laboratory, Livermore, California

Equations are derived which describe the hydrodynamic properties of an ionized gas in a strong magnetic field and in states close to an initial stationary state. The development is based on the Boltzmann equation in which the effects of collisions between the constituents of the gas are ignored. The theory is, therefore, applicable only for following the evolution of the gas for durations which are short compared to the thermalization time. The method of solution followed is essentially one of expansion in inverse powers of the strength of the impressed magnetic field; and in the first approximation the Boltzmann equation is reducible to a one-dimensional inhomogeneous wave equation which describes the motions of the particles along the magnetic field lines.

1. INTRODUCTION

The behavior of a conducting fluid in the presence of electromagnetic fields is usually considered in terms of the so-called hydromagnetic equations (*1*). These considerations may be expected to apply to liquids and ionized gases of sufficiently high density for a local Maxwellian distribution of the particle velocities to prevail. On the other hand, when the characteristic periods of the motions are so small that a Maxwellian distribution of velocities cannot be established locally, then we should not expect the conventional hydromagnetic treatment to apply. In this paper we shall be concerned with precisely those situations when the latter is the case.

In the absence of effective uniformising collisions, a gas will in general behave as though it had more degrees of freedom than if it were a simple hydrodynamic system. However, in a strong magnetic field much of the possible complexity does not manifest itself and a quasi-hydrodynamic description becomes possible (*2*, *3*). Nevertheless, under the circumstances envisaged, the Boltzmann equation, as always, provides the only sure starting point.

In this paper the problem investigated in papers I and II will be reconsidered

[1] Work performed under the auspices of the U. S. Atomic Energy Commission.

with a view to obtaining a complete set of equations in terms of which the properties of the gas, in states infinitesimally close to an initial stationary state, can be uniquely determined. The possible stationary states themselves are the subject of a later paper.

2. THE BASIC EQUATIONS

We consider a fully ionized gas consisting of electrons and singly charged ions of masses m^- and m^+ and charges $-e$ and $+e$. Quite generally the state of such a gas can be fully described if we know the distribution functions

$$f^+(x_i\,, v_i\,, t) \quad \text{and} \quad f^-(x_i\,, v_i\,, t)$$

of the ions and electrons, respectively, in the coordinate (x_i) and velocity (v_i) spaces. Each of the functions f^+ and f^- must satisfy the appropriate Boltzmann equation; thus,

$$\frac{\partial f}{\partial t} + v_j \frac{\partial f}{\partial x_j} + \left\{ g_i + \frac{e}{m}\left[E_i + \frac{1}{c}\, \epsilon_{ijk} v_j B_k \right] \right\} \frac{\partial f}{\partial v_i} = 0 \tag{1}$$

where we have used the notation and summation convention of tensor calculus. Also, we have suppressed the superscripts + and − distinguishing the ions and electrons; we shall always do this whenever an equation of the same form applies to both ions and electrons; it must be remembered, however, that in all these cases a duplicate set of equations referring to ions and electrons, separately, must be understood.

In Eq. (1), g_i denotes the gravitational force[2] per unit mass acting on the particles and E_i and B_i are the electric and the magnetic field intensities, respectively.

It will be observed that the usual collision term is omitted in Eq. (1); this is in accordance with our assumption that the motions of interest develop in intervals of time short compared to the relaxation (i.e., thermalization) time.

The first three moments of f are

$$\begin{aligned} N &= \int f\, d\mathbf{v} = \text{electron or ion density} \\ V_i &= \frac{1}{N}\int f v_i\, d\mathbf{v} = \text{local mean velocity} \\ p_{ij} &= m \int f(v_i - V_i)(v_j - V_j)\, d\mathbf{v} = \text{pressure tensor.} \end{aligned} \tag{2}$$

There are, of course, two such sets of moments with the superscripts + and −

[2] For simplicity, we suppose $\mathbf{g}$ to be due to external sources and thus to be a given force. The extension to cases in which $\mathbf{g}$ is determined by the self-gravitational field of the gas is straightforward.

referring to the ions and electrons, respectively. In terms of these moments we can deduce the other relevant macroscopic quantities; thus,

$$\begin{aligned} &\text{charge density, } \epsilon &&= (N^+ - N^-)e \\ &\text{mass density, } \rho &&= m^+N^+ + m^-N^- \\ &\text{total pressure, } P_{ij} &&= p_{ij}^+ + p_{ij}^- \\ &\text{current density } J_i &&= e(N^+V_i^+ - N^-V_i^-). \end{aligned} \tag{3}$$

One can readily obtain equations governing these moments by taking successive moments of the Boltzman equation. For this latter purpose it is convenient to rewrite the equation in terms of the distribution function of the peculiar velocities,

$$\nu_i = v_i - V_i\,. \tag{4}$$

Let this distribution function be

$$f(x_i\,, v_i\,, t) \equiv f(x_i\,, \nu_i\,, t). \tag{5}$$

For f redefined in this manner,

$$N = \int f\,d\mathbf{v}; \qquad 0 = \int f\nu_i\,d\mathbf{v}; \qquad p_{ij} = m\int f\nu_i\nu_j\,d\mathbf{v}. \tag{6}$$

Also, when f is expressed in terms of the peculiar velocities, the dependence on time and space coordinates occurs both explicitly and implicitly (through the definition of ν_i). Thus, for the transformation (4)

$$\frac{\partial}{\partial x_j} \to \frac{\partial}{\partial x_j} - \frac{\partial V_k}{\partial x_j}\frac{\partial}{\partial \nu_k}; \qquad \frac{\partial}{\partial t} \to \frac{\partial}{\partial t} - \frac{\partial V_k}{\partial t}\frac{\partial}{\partial \nu_k}; \qquad \frac{\partial}{\partial v_j} \to \frac{\partial}{\partial \nu_j}. \tag{7}$$

With the foregoing substitutions the Boltzmann equation becomes [cf. (4)]

$$\frac{Df}{Dt} + \nu_j\frac{\partial f}{\partial x_j} + \left[g_k + \frac{e}{m}\left(E_k + \frac{1}{c}\,\epsilon_{klm}V_lB_m\right) - \frac{DV_k}{Dt}\right]\frac{\partial f}{\partial \nu_k} + \frac{e}{mc}\,\epsilon_{klm}\frac{\partial f}{\partial \nu_k}\,\nu_lB_m - \nu_j\frac{\partial V_k}{\partial x_j}\frac{\partial f}{\partial \nu_k} = 0 \tag{8}$$[3]

where

$$\frac{D}{Dt} = \frac{\partial}{\partial t} + V_j\frac{\partial}{\partial x_j}. \tag{9}$$

[3] This equation is formally valid for *any* transformation (5) whether or not **V** is the local mean velocity; for it depends only on the transformation laws (7). However, the choice of **V** as the local mean velocity is necessary if the moment equations to be derived, namely Eqs. (10) and (11), are to be valid.

On taking the zero- and the first-order moments of Eq. (7) we readily obtain (*4*)

$$\frac{DN}{Dt} + N\frac{\partial V_k}{\partial x_k} = 0 \tag{10}$$

and

$$\frac{DV_i}{Dt} = -\frac{1}{mN}\frac{\partial p_{ij}}{\partial x_j} + g_i + \frac{e}{m}\left(E_i + \frac{1}{c}\,\epsilon_{ilm}V_lB_m\right). \tag{11}$$

Equivalent forms of these equations are

$$\frac{\partial N}{\partial t} + \frac{\partial}{\partial x_j}(NV_j) = 0 \tag{12}$$

and

$$mN\left(\frac{\partial V_i}{\partial t} + V_j\frac{\partial V_i}{\partial x_j}\right) = -\frac{\partial p_{ij}}{\partial x_j} + mNg_i + Ne\left(E_i + \frac{1}{c}\,\epsilon_{ilm}V_lB_m\right) \tag{13}$$

in which forms we recognize them as the hydrodynamic equations of continuity and motion. (It should, however, be emphasized that since we do not, as yet, have an equation of state relating p_{ij} and N, Eqs. (12) and (13) do not constitute a complete set of equations; this is, indeed, one of the principal difficulties of this subject.)

By adding the equations of motion referring to the ions and electrons, respectively, we obtain [cf. Eq. (3)]

$$\sum_{+,-} mN\left(\frac{\partial V_i}{\partial t} + V_j\frac{\partial V_i}{\partial x_j}\right) = -\frac{\partial P_{ij}}{\partial x_j} + \rho g_i + \epsilon E_i + \frac{1}{c}\,\epsilon_{ilm}J_lB_m\,. \tag{14}$$

Connecting the charge (ϵ) and the current (J_i) density which occur in this equation with the field intensities E_i and B_i, we have Maxwell's equations:

$$\operatorname{div}\mathbf{E} = 4\pi\epsilon \tag{15}$$

$$\operatorname{curl}\mathbf{E} = -\frac{1}{c}\frac{\partial\mathbf{B}}{\partial t} \tag{16}$$

and

$$\operatorname{curl}\mathbf{B} = \frac{4\pi}{c}\mathbf{J} + \frac{1}{c}\frac{\partial\mathbf{E}}{\partial t}. \tag{17}$$

We shall derive certain alternative forms of Eqs. (13) and (14) which we shall find useful.

In virtue of the equation of continuity, we clearly have

$$\frac{\partial}{\partial t}(NV_i) + \frac{\partial}{\partial x_j}(NV_iV_j) = N\frac{DV_i}{Dt}. \tag{18}$$

Hence instead of Eq. (11) we may write

$$m\frac{\partial}{\partial t}(NV_i) = -\frac{\partial}{\partial x_j}(p_{ij} + mNV_iV_j) + mNg_i + eN\left(E_i + \frac{1}{c}\epsilon_{ijk}V_jB_k\right). \quad (19)$$

Now adding these equations referring to the ions and electrons, respectively, we obtain

$$\sum_{+,-} m\frac{\partial}{\partial t}(NV_i) = -\frac{\partial}{\partial x_j}\{P_{ij} + \sum_{+,-} mNV_iV_j\} + \rho g_i + \epsilon E_i + \frac{1}{c}\epsilon_{ijk}J_jB_k. \quad (20)$$

We can eliminate **E** and **B** from Eq. (20) by making use of Maxwell's equations. As we may directly verify,

$$\left(\epsilon\mathbf{E} + \frac{1}{c}\mathbf{J}\times\mathbf{B}\right)_i = \frac{1}{4\pi}\left(\mathbf{E}\operatorname{div}\mathbf{E} + \operatorname{curl}\mathbf{B}\times\mathbf{B} - \frac{1}{c}\frac{\partial\mathbf{E}}{\partial t}\times\mathbf{B}\right)_i$$
$$= \frac{1}{4\pi}\frac{\partial}{\partial x_j}\left\{E_iE_j + B_iB_j - \frac{1}{2}\delta_{ij}(|\mathbf{E}|^2 + |\mathbf{B}|^2)\right\} - \frac{1}{4\pi c}\frac{\partial}{\partial t}(\mathbf{E}\times\mathbf{B})_i. \quad (21)$$

Hence, we may rewrite Eq. (20) in the form

$$\frac{\partial}{\partial t}\left\{\sum_{+,-} mNV_i + \frac{1}{4\pi}\epsilon_{ijk}E_jB_k\right\} = \frac{\partial}{\partial x_j}(T_{ij} - \sum_{+,-} mNV_iV_j) + \rho g_i \quad (22)$$

where T_{ij} denotes the combined stress tensor,

$$T_{ij} = -P_{ij} + \frac{1}{4\pi}(E_iE_j + B_iB_j) - \frac{\delta_{ij}}{8\pi}(|\mathbf{E}|^2 + |\mathbf{B}|^2). \quad (23)$$

Finally, we may note that making use of Eq. (11) we can rewrite Eq. (8) as

$$\frac{\partial f}{\partial t} + (\nu_j + V_j)\frac{\partial f}{\partial x_j} + \left[\frac{e}{mc}\epsilon_{ijk}\nu_jB_k + \frac{1}{mN}\frac{\partial p_{ij}}{\partial x_j} - \nu_j\frac{\partial V_i}{\partial x_j}\right]\frac{\partial f}{\partial \nu_i} = 0. \quad (24)$$

3. METHOD OF SOLUTION

As we have already stated we shall be considering states of a system close to a stationary (not necessarily a static) solution of the Boltzmann equation. A superscript "0" will be normally used to distinguish quantities which refer to the stationary state, while a superscript prime will similarly distinguish nonstationary departures from the stationary state. Thus, we shall write

$$B_i = B_i^0 + B_i'; \qquad E_i = E_i^0 + E_i'; \qquad N = N^0 + N' \text{ etc.} \quad (25)$$

(A) The Boltzmann Equation for the Stationary State

The equation governing a stationary state is [see Eq. (24)]

$$(\nu_j + V_j^0)\frac{\partial f^0}{\partial x_j} + \left[\frac{e}{mc}\epsilon_{ijk}\nu_jB_k^0 + \frac{1}{mN^0}\frac{\partial p_{ij}^0}{\partial x_j} - \nu_j\frac{\partial V_i^0}{\partial x_j}\right]\frac{\partial f^0}{\partial \nu_i} = 0 \quad (26)$$

where it may be recalled that $f^0 = f^0(x_i, v_i - V_i)$ is the distribution function referred to an appropriate stationary local mean velocity $\mathbf{V}^0$; the occurrence of a nonvanishing $\mathbf{V}^0$ implies the existence of "static" flows.

(B) The Boltzmann Equation for the Nonstationary Deviations

We shall start from Eq. (1) and seek a solution of this equation of the form

$$f(x_i, v_i, t) = f^0(x_i, v_i - U_i) + f'(x_i, v_i, t) \tag{27}$$

where U_i (to be specified, presently) and f' are both considered to be small quantities. To the first order, we may write instead

$$f(x_i, v_i, t) = f^0(x_i, v_i) - U_j \frac{\partial f^0}{\partial v_j} + f'(x_i, v_i, t). \tag{28}$$

Since we shall be expressing f^0 in terms of the peculiar velocities relative to the stationary local mean velocity $\mathbf{V}^0$, the purpose of introducing $\mathbf{U}$ is to allow for the first-order change in the mean velocity caused by the perturbation as distinct and separate from the first-order change in the distribution function with respect to the proper mean velocity.

Substituting (27) in (1) and neglecting terms which are quadratic in U_i and/or f', we get

$$\mathfrak{D}f' + \frac{e}{m}\left(E_i' + \frac{1}{c}\,\epsilon_{ijk}v_jB_k'\right)\frac{\partial f^0}{\partial v_i} = \mathfrak{D}\left(U_l\,\frac{\partial f^0}{\partial v_l}\right) \tag{29}$$

where

$$\mathfrak{D} = \frac{\partial}{\partial t} + v_j\,\frac{\partial}{\partial x_j} + \left[g_i + \frac{e}{m}\left(E_i^0 + \frac{1}{c}\,\epsilon_{ijk}v_jB_k^0\right)\right]\frac{\partial}{\partial v_i}. \tag{30}$$

The right-hand side of Eq. (29) is

$$\begin{aligned}\mathfrak{D}\left(U_l\,\frac{\partial f^0}{\partial v_l}\right) = \frac{\partial f^0}{\partial v_l}\frac{\partial U_l}{\partial t} + v_jU_l\,\frac{\partial^2 f^0}{\partial x_j\partial v_l} + v_j\,\frac{\partial f^0}{\partial v_l}\frac{\partial U_l}{\partial x_j}\\ + U_l\left\{g_i + \frac{e}{m}\left(E_i^0 + \frac{1}{c}\,\epsilon_{ijk}v_jB_k^0\right)\right\}\frac{\partial^2 f^0}{\partial v_i\partial v_l}.\end{aligned} \tag{31}$$

Simplifying the last term on the right-hand side of this equation in the manner

$$\begin{aligned}U_l\,\frac{\partial}{\partial v_l}\left\{g_i + \frac{e}{m}\left(E_i^0 + \frac{1}{c}\,\epsilon_{ijk}v_jB_k^0\right)\right\}\frac{\partial f^0}{\partial v_i} - \frac{e}{mc}\,\epsilon_{ilk}\,\frac{\partial f^0}{\partial v_i}\,U_lB_k^0\\ = -U_l\,\frac{\partial}{\partial v_l}\left(v_j\,\frac{\partial f^0}{\partial x_j}\right) - \frac{e}{mc}\,\epsilon_{ijk}\,\frac{\partial f^0}{\partial v_i}\,U_jB_k^0\end{aligned} \tag{32}$$

we have

$$\mathfrak{D}\left(U_l \frac{\partial f^0}{\partial v_l}\right) = \frac{\partial f^0}{\partial v_i}\frac{\partial U_i}{\partial t} - U_j \frac{\partial f^0}{\partial x_j} + v_j \frac{\partial f^0}{\partial v_l}\frac{\partial U_l}{\partial x_j} - \frac{e}{mc}\epsilon_{ijk}\frac{\partial f^0}{\partial v_i} U_j B_k^0. \tag{33}$$

We shall now *define* **U** by

$$\frac{\partial U_i}{\partial t} = \frac{e}{m}\left(E_i' + \frac{1}{c}\epsilon_{ijk} U_j B_k^0\right). \tag{34}$$

With the right-hand side of Eq. (29) given by (33) and in virtue of this definition, Eq. (29) reduces to

$$\mathfrak{D}f' = -U_j \frac{\partial f^0}{\partial x_j} + v_j \frac{\partial f^0}{\partial v_k}\frac{\partial U_k}{\partial x_j} - \frac{e}{mc}\epsilon_{ijk} v_j B_k' \frac{\partial f^0}{\partial v_i}. \tag{35}$$

If we express f' and f^0 *both* in terms of the peculiar velocities with respect to the mean velocity $\mathbf{V}^0$ which obtains in the stationary state, Eq. (35) becomes [cf. Eqs. (7) and (24)]

$$\begin{aligned}\frac{\partial f'}{\partial t} + (\nu_j + V_j^0)\frac{\partial f'}{\partial x_j} + \left[\frac{e}{mc}\epsilon_{ijk}\nu_j B_k^0 + \frac{1}{mN^0}\frac{\partial p_{ij}^0}{\partial x_j} - \nu_j \frac{\partial V_i^0}{\partial x_j}\right]\frac{\partial f'}{\partial \nu_i}\\ = -U_j\left(\frac{\partial f^0}{\partial x_j} - \frac{\partial f^0}{\partial \nu_k}\frac{\partial V_k^0}{\partial x_j}\right) + (\nu_j + V_j^0)\frac{\partial U_k}{\partial x_j}\frac{\partial f^0}{\partial \nu_k}\\ - \frac{e}{mc}\epsilon_{ijk}\frac{\partial f^0}{\partial \nu_i}(\nu_j + V_j^0)B_k'.\end{aligned} \tag{36}$$

The solution of Eqs. (34) and (36) together with Maxwell's equations constitutes our problem. As in papers I and II we shall suppose that the strength of the impressed magnetic field, $|\mathbf{B}^0|$, is large; and we shall obtain solutions in inverse powers of $|\mathbf{B}^0|$.

4. THE INTRODUCTION OF A VARIABLE ANALOGOUS TO THE HYDRODYNAMIC LAGRANGIAN VARIABLE

We shall introduce a new variable $\boldsymbol{\xi}$ related to the velocity $\mathbf{U}^+$ (appropriate for the ions) by

$$\mathbf{U}^+ = \frac{\partial \boldsymbol{\xi}}{\partial t}. \tag{37}$$

We shall find it convenient to consider the components, $\xi_\perp$ and $\xi_\parallel$, of $\boldsymbol{\xi}$ which are perpendicular and parallel, respectively, to the direction of the static magnetic field $\mathbf{B}^0$.

Let $\mathbf{n}$ denote a unit vector in the direction of $\mathbf{B}^0$:

$$\mathbf{n} = \frac{\mathbf{B}^0}{|\mathbf{B}^0|}. \tag{38}$$

In terms of $\mathbf{n}$ we can express $\boldsymbol{\xi}_\perp$ and $\xi_\parallel$ by

$$\xi_{\perp,i} = (\delta_{ij} - n_i n_j)\xi_j \quad \text{and} \quad \xi_\parallel = \xi_j n_j\,. \tag{39}$$

This resolution of $\boldsymbol{\xi}$ into a perpendicular and a parallel component can, of course, be applied to any vector, and we shall frequently have occasion to do so.

From Eq. (34) defining $\mathbf{U}$, we find

$$\frac{\partial \mathbf{U}_\perp}{\partial t} = \frac{e}{m}\left(\mathbf{E}_\perp' + \frac{1}{c}\,\mathbf{U}_\perp \times \mathbf{B}^0\right) \tag{40}$$

and

$$\frac{\partial U_\parallel}{\partial t} = \frac{e}{m}\,\mathbf{E}'\boldsymbol{\cdot}\mathbf{n}. \tag{41}$$

A basic assumption which will underlie our entire treatment is that the characteristic times in which any of the macroscopic quantities change appreciably are very large compared to the Larmor period ($\sim mc/eB^0$). On this assumption, Eq. (40) allows us to conclude that

$$\mathbf{E}_\perp' + \frac{1}{c}\,\mathbf{U}_\perp \times \mathbf{B}^0 = 0. \tag{42}$$

This relation is valid for electrons as well as ions. Accordingly,

$$\mathbf{U}_\perp{}^+ = \mathbf{U}_\perp{}^- = \frac{\partial \boldsymbol{\xi}_\perp}{\partial t} \tag{43}$$

whereas from Eq. (41) we conclude that

$$U_\parallel{}^- = -\frac{m^+}{m^-}\,U_\parallel{}^+ = -\frac{m^+}{m^-}\frac{\partial \xi_\parallel}{\partial t}\,. \tag{44}$$

5. THE ELIMINATION OF THE ELECTROMAGNETIC VARIABLES

One of the principal advantages of the introduction of $\boldsymbol{\xi}$ as a variable, is that in terms of it we can eliminate the electromagnetic field variables. Thus, from Eqs. (34) and (37) we have

$$\mathbf{E}' = \frac{m^+}{e}\frac{\partial^2 \boldsymbol{\xi}}{\partial t^2} - \frac{1}{c}\frac{\partial \boldsymbol{\xi}_\perp}{\partial t} \times \mathbf{B}^0, \tag{45}$$

while from the linearized form of Eq. (17), namely,

$$\operatorname{curl} \mathbf{E}' = -\frac{1}{c}\frac{\partial \mathbf{B}'}{\partial t} \tag{46}$$

we obtain

$$-\frac{1}{c}\frac{\partial \mathbf{B}'}{\partial t} = \text{curl}\left\{\frac{m^+}{e}\frac{\partial^2 \boldsymbol{\xi}}{\partial t^2} - \frac{1}{c}\frac{\partial \boldsymbol{\xi}_\perp}{\partial t} \times \mathbf{B}^0\right\} \tag{47}$$

or, after integration,

$$\mathbf{B}' = \text{curl}\,(\boldsymbol{\xi}_\perp \times \mathbf{B}^0) - \frac{m^+ c}{e}\,\text{curl}\left(\frac{\partial \boldsymbol{\xi}}{\partial t}\right). \tag{48}$$

The second term in the right-hand side of this equation is of order η (see Section 6 below) relative to the first if $\xi_\parallel$ is at all comparable to $\boldsymbol{\xi}_\perp$. To a sufficient approximation, under most conditions we may write

$$\mathbf{B}' = \text{curl}\,(\boldsymbol{\xi}_\perp \times \mathbf{B}^0). \tag{49}$$

Using Eqs. (45) and (49) we may eliminate $\mathbf{E}'$ and $\mathbf{B}'$ from all our subsequent equations.

As we have already indicated, the variable $\boldsymbol{\xi}$ will play approximately the role of a hydrodynamic Lagrangian variable. It will satisfy a set of hydrodynamic equations which must be solved together with Eq. (36).

6. THE PARAMETERS CHARACTERIZING THE STATE OF THE GAS

Before we proceed to derive from the Boltzmann equation a complete set of equations for $\boldsymbol{\xi}$, it will be convenient to introduce a number of parameters characterizing the state of the gas. These are

$$\begin{aligned}
\bar{v}_\perp &= \text{mean thermal speed perpendicular to } \mathbf{B}^0\\
\omega &= eB^0/mc = \text{gyration (or Larmor) frequency}\\
r_\perp &= \bar{v}_\perp/\omega = \text{radius of gyration; or Larmor radius}\\
L &= \text{a measure of the linear dimension of the system}\\
\Omega^{-1} &= \text{``time constant'' of motion } (\sim L/\bar{v}_\perp)\\
\eta &= r_\perp/L\ (\simeq \Omega/\omega)\\
\mathbf{V}^0 &= \text{drift velocity in the stationary state.}
\end{aligned} \tag{50}$$

The assumption that B^0 $(= |\mathbf{B}^0|)$ is large will be interpreted to mean

$$\eta \ll 1 \quad \text{and} \quad \Omega/\omega \ll 1. \tag{51}$$

These conditions are basic to our treatment; it has already been used in deriving (42) from (40).

For simplicity we may also suppose that

$$|N^+ - N^-|/(N^+ + N^-) \ll 1 \quad \text{and} \quad |\mathbf{E}^0| \ll |\mathbf{B}^0|. \tag{52}$$

If in addition the motion should be subsonic then

$$| \mathbf{V}^0 | < \bar{v}_\perp \tag{53}$$

however, this last approximation will not be used in this paper.

7. THE EQUATIONS OF MOTION

Returning to the perturbed Boltzmann equation for f' [Eqs. (35) and (36)], we observe that the quantities **U** and $\mathbf{B}'$ which occur in these equations have been expressed in terms of a single Lagrangian variable $\boldsymbol{\xi}$ [Eqs. (43), (44), and (48)]. We shall now derive equations of motion for $\boldsymbol{\xi}$ in terms of the moments of f'^+ and f'^-; the solution for these latter quantities is postponed to Sections 9 and 10. The required equations of motion for ξ will be derived from Eq. (14) (or 20) together with the linearized form of Eq. (17), namely,

$$\text{curl } \mathbf{B}' = \frac{4\pi}{c} \mathbf{J}' + \frac{1}{c} \frac{\partial \mathbf{E}'}{\partial t}. \tag{54}$$

In linearizing Eq. (14) and others, we shall denote the nonstatic departures of any vector **A** from its equilibrium value $\mathbf{A}^0$ by $\mathbf{A}'$ and decompose it into a "parallel" and a perpendicular component as in Eq. (39); thus

$$\mathbf{A} = \mathbf{A}^0 + \mathbf{A}', \qquad \mathbf{A}' = \mathbf{n}A_{\|}' + \mathbf{A}_\perp' \quad (\mathbf{n}\cdot\mathbf{A}_\perp' = 0). \tag{55}$$

Similarly, for scalars we shall write

$$A = A^0 + A'. \tag{56}$$

We shall first consider the perpendicular component of Eq. (14). The right-hand side gives

$$-(\text{div } \mathbf{P}')_\perp + \rho'\mathbf{g}_\perp + \epsilon'\mathbf{E}_\perp^0 + \epsilon^0\mathbf{E}_\perp' + \frac{1}{c}(\mathbf{J}^0 \times \mathbf{B}')_\perp + \frac{1}{c}\mathbf{J}' \times \mathbf{B}^0 \tag{57}$$

where

$$(\text{div } \mathbf{P}')_\perp = (\delta_{ij} - n_i n_j) \frac{\partial P_{jk}'}{\partial x_k}. \tag{58}$$

The term in P_{ij}' can be evaluated once the Boltzmann equation for f' has been solved; we shall postpone this to Section 9. As for the other terms, we have already expressed $\mathbf{E}_\perp'$ and $\mathbf{B}'$ in terms of $\boldsymbol{\xi}$; $\mathbf{J}'$ can be found from Eq. (54) and ϵ' from

$$4\pi\epsilon' = \text{div } \mathbf{E}'. \tag{59}$$

Finally, ρ' can be found from the linearized form of the equation of continuity,

$$\frac{\partial N'}{\partial t} + \text{div } (N'\mathbf{V}^0 + N^0\mathbf{V}') = 0 \tag{60}$$

and a knowledge of $\mathbf{V}'$ (derived from the solution of Boltzmann equation for f').

Before linearizing the left-hand side of Eq. (14) consider Eq. (13) from which it was derived. Dividing Eq. (13) throughout by N and linearizing, we have

$$\left(m\frac{D\mathbf{V}}{Dt}+\frac{1}{N}\operatorname{div}\mathbf{p}\right)'=e\left(\mathbf{E}'+\frac{1}{c}\mathbf{V}^0\times\mathbf{B}'+\frac{1}{c}\mathbf{V}'\times\mathbf{B}^0\right). \tag{61}$$

Taking the vector product of this equation with $\mathbf{n}$ and remembering that [cf. Eq. (42)]

$$\mathbf{n}\times\mathbf{E}_\perp'=-\frac{B^0}{c}\mathbf{n}\times[\mathbf{U}_\perp\times\mathbf{n}]=-\frac{B^0}{c}\mathbf{U}_\perp \tag{62}$$

and

$$\mathbf{V}_\perp'=\mathbf{n}\times[\mathbf{V}'\times\mathbf{n}] \tag{63}$$

we obtain

$$\begin{aligned}\mathbf{V}_\perp'&=\mathbf{U}_\perp-\frac{1}{B^0}\mathbf{n}\times(\mathbf{V}^0\times\mathbf{B}')+\frac{mc}{eB^0}\mathbf{n}\times\left(\frac{D\mathbf{V}}{Dt}+\frac{1}{mN}\operatorname{div}\mathbf{p}\right)'\\&=\mathbf{U}_\perp+\mathbf{V}_\perp''\ \text{(say)}.\end{aligned} \tag{64}$$

It will be observed that the dominant contribution to $\mathbf{V}_\perp'$ is $\mathbf{U}_\perp$ while $\mathbf{V}_\perp''$ consists of two terms one of which is of order $\mathbf{V}^0$ (and may be of importance in some problems) and the other of which is of order η and therefore negligible.

Returning to Eq. (14) and linearizing the left-hand side and remembering that the stationary $\mathbf{V}^0$ is independent of time, we have

$$\sum_{+,-}mN^0\frac{\partial\mathbf{V}'}{\partial t}+\sum_{+,-}\{mN(\mathbf{V}\cdot\operatorname{grad})\mathbf{V}\}'. \tag{65}$$

Taking the perpendicular component of this expression, substituting for $\mathbf{V}_\perp'$ in accordance with Eq. (64), and remembering that $\mathbf{U}_\perp$ is the same for ions and electrons, we obtain

$$\rho^0\frac{\partial^2\boldsymbol{\xi}_\perp}{\partial t^2}+\boldsymbol{\Gamma} \tag{66}$$

where

$$\boldsymbol{\Gamma}=\sum_{+,-}\left[mN^0\frac{\partial\mathbf{V}_\perp''}{\partial t}+\{mN(\mathbf{V}\cdot\operatorname{grad})\mathbf{V}\}_\perp'\right]. \tag{67}$$

The quantity $\boldsymbol{\Gamma}$ is of order V^0 or η. If $\mathbf{V}_\parallel'$ is needed to evaluate $\boldsymbol{\Gamma}$ in a particular problem, it may be found from the parallel component of Eq. (61) or more directly from the solution of the Boltzmann equation for f' [see Eq. (139) below].

Finally, by equating (57) and (66), eliminating $\mathbf{J}'$ by making use of Eq. (54), (and expressing $\mathbf{E}_\perp'$ in Eq. (54) in terms of $\boldsymbol{\xi}_\perp$), we obtain

$$\rho^0\left(1 + \frac{|\mathbf{B}^0|^2}{4\pi\rho^0 c^2}\right)\frac{\partial^2\boldsymbol{\xi}_\perp}{\partial t^2} = -(\text{div } \mathbf{P}')_\perp + \epsilon^0\mathbf{E}_\perp' + \epsilon'\mathbf{E}_\perp^0 + \frac{1}{c}(\mathbf{J}^0 \times \mathbf{B}')_\perp + \frac{1}{4\pi}[\text{curl } \mathbf{B}'] \times \mathbf{B}^0 + \rho'\mathbf{g}_\perp - \boldsymbol{\Gamma}. \tag{68}$$

This is the required equation of motion for $\boldsymbol{\xi}_\perp$. In most problems, the stationary conditions are such that this equation simplifies considerably. For example, if the equilibrium plasma is uniform (i.e., $\mathbf{J}^0 = 0$, $\epsilon^0 = 0$ and $\mathbf{E}^0 = \text{constant}$) we may transform to a frame in which $\mathbf{E}^0 = 0$ and the equation becomes

$$\rho^0\left(1 + \frac{|\mathbf{B}^0|^2}{4\pi\rho^0 c^2}\right)\frac{\partial^2\boldsymbol{\xi}_\perp}{\partial t^2} = -(\text{div } \mathbf{P}')_\perp + \frac{1}{4\pi}[\text{curl } \mathbf{B}'] \times \mathbf{B}^0 + \rho'\mathbf{g}_\perp - \boldsymbol{\Gamma}. \tag{69}$$

Generally, $|\mathbf{B}^0|^2 \ll 4\pi\rho^0 c^2$ and $\boldsymbol{\Gamma}$ is frequently negligible, unless there are large static drift velocities.

To obtain an equation of motion for $\xi_\parallel$, we first take the parallel component of Eq. (54); we have

$$\begin{aligned}\frac{\partial E_\parallel'}{\partial t} &= c(\text{curl } \mathbf{B}')_\parallel - 4\pi\mathbf{J}_\parallel' \\ &= c(\text{curl } \mathbf{B}')_\parallel - 4\pi\sum_{+,-}(eN\mathbf{V})_\parallel'.\end{aligned} \tag{70}$$

Taking the left-hand side of Eq. (13) in the form given in (18), linearizing it and taking the parallel component, we have

$$\frac{\partial}{\partial t}(N\mathbf{V})_\parallel' = -\frac{1}{m}(\text{div } \mathbf{p}')_\parallel + \frac{eN^0}{m}\left(\mathbf{E}' + \frac{1}{c}\mathbf{V}^0 \times \mathbf{B}'\right)_\parallel + N'\left(\frac{e}{m}\mathbf{E}^0 + \mathbf{g}\right)_\parallel - [\text{div }(N\mathbf{V}\mathbf{V})']_\parallel. \tag{71}$$

We shall rewrite this equation in the manner,

$$\frac{\partial}{\partial t}(N\mathbf{V})_\parallel' = -\frac{1}{m}(\text{div } \mathbf{p}')_\parallel + \frac{eN^0}{mc}(\mathbf{V}^0 \times \mathbf{B}')_\parallel + \frac{eN^0}{m}E_\parallel' + \Phi \tag{72}$$

where Φ includes the remaining terms on the right-hand side of Eq. (71). There is (as always) a duplicate set of equations for ions and electrons with + and − superscripts. We can use these equations to eliminate the term in $(eN\mathbf{V})'$ on the right-hand side of Eq. (70). We thus obtain

$$\left(\frac{\partial^2}{\partial_t^2} + \omega_{p,+}^2 + \omega_{p,-}^2\right)E_\parallel' = 4\pi\sum_{+,-}\frac{e}{m}\left\{(\text{div } \mathbf{p}')_\parallel - \frac{eN^0}{c}(\mathbf{V}^0 \times \mathbf{B}')_\parallel\right\} + c\left(\text{curl }\frac{\partial\mathbf{B}'}{\partial t}\right)_\parallel - 4\pi e(\Phi^+ - \Phi^-) \tag{73}$$

where $\omega_{p,\pm}$ denotes the plasma frequency

$$\omega_{p,\pm} = \left[\frac{4\pi N^{0,\pm} e^2}{m^\pm}\right]^{1/2}. \tag{74}$$

It is again clear that the terms included in Φ will be negligible (as was the case with $\mathbf{\Gamma}$) unless there are large static drifts.

8. ON THE STATIONARY STATES DESCRIBED BY THE BOLTZMANN EQUATION

Solutions of the stationary states described by the Boltzmann equation are considered in greater detail in a later paper. Here we shall consider them briefly in a manner more appropriate to our present context.

The equation, the solutions of which we wish to consider, is Eq. (26)

$$(\nu_j + V_j)\frac{\partial f}{\partial x_j} + \left(\frac{e}{m}\epsilon_{ijk}\nu_j B_k + \frac{1}{mN}\frac{\partial p_{ij}}{\partial x_j} - \nu_j\frac{\partial V_i}{\partial x_j}\right)\frac{\partial f}{\partial \nu_i} = 0 \tag{75}$$

where, for the time being, the superscripts "0" to the various quantities (f, V_j, N, etc.,) have been suppressed. The superscripts will be restored when we proceed to the solution of the equation for f'.

We shall solve Eq. (75) by a method of approximation which is essentially one of expansion in inverse powers of B. We shall denote the first approximation by f_{I}. This satisfies the equation

$$\epsilon_{ijk}\frac{\partial f_{\mathrm{I}}}{\partial \nu_i}\nu_j n_k = 0 \tag{76}$$

where [as in Section 4, Eq. (38)] $\mathbf{n}$ is a unit vector in the direction of $\mathbf{B}$ (i.e., $\mathbf{B}^0$!). Equation (76) clearly implies that f_{I} *is a function of* ν^2 ($= \nu_i\nu_i$) *and of the component of* $\mathbf{v}$ *in the direction of* $\mathbf{n}$. We shall express this dependence of f_{I} on ν^2 and $\mathbf{v}\cdot\mathbf{n}$ as follows.

Let $\mathbf{s}$ denote the component of $\mathbf{v}$ perpendicular to $\mathbf{n}$ and q the component of $\mathbf{v}$ parallel to $\mathbf{n}$:

$$q = n_i\nu_i \quad \text{and} \quad s_i = \nu_i - qn_i. \tag{77}$$

Clearly,

$$s_i n_i = 0 \quad \text{and} \quad s^2 = \nu^2 - q^2. \tag{78}$$

We shall now write the solution of Eq. (76) as

$$f_{\mathrm{I}} \equiv f_{\mathrm{I}}(x_i, s^2, q). \tag{79}$$

Frequently, we shall assume that f_{I} is even in q.

(A) The Pressure Tensor and Its Divergence

It is known [cf. paper II, also Chew, Goldberger, and Low (*2*)] that for a distribution function of the form (79) the pressure tensor is given by

$$p_{ij} = p_{\|}n_i n_j + p_{\perp}(\delta_{ij} - n_i n_j)$$

where

$$p_{\|} = \frac{1}{2} m \iint f_{\mathrm{I}}\, dq\, ds^2; \qquad p_{\perp} = \frac{1}{4} m \iint f_{\mathrm{I}} s^2\, dq\, ds^2 \tag{80}$$

$p_{\|}$ and $p_{\perp}$ are the components of the pressure tensor, parallel and perpendicular, respectively, to the lines of force. In writing (80), we have assumed that f_{I} is so normalized that, for example,

$$N = \frac{1}{2} \iint f_{\mathrm{I}}\, dq\, ds^2.$$

As the divergence of the pressure tensor (80) plays an important role in the subsequent analysis, we shall obtain an expression for it. By direct calculation, we find

$$\frac{\partial p_{ij}}{\partial x_j} = n_i \left\{ n_j \frac{\partial p_{\|}}{\partial x_j} + (p_{\|} - p_{\perp}) \frac{\partial n_j}{\partial x_j} \right\} + \left(\frac{\partial}{\partial x_i} - n_i n_j \frac{\partial}{\partial x_j} \right) p_{\perp} + (p_{\|} - p_{\perp}) n_j \frac{\partial n_i}{\partial x_j}. \tag{81}$$

The quantities div **n** and $n_j \partial n_i / \partial x_j$ which occur in Eq. (81) have simple equivalents. Thus, from

$$0 = \operatorname{div} \mathbf{B} = \operatorname{div} B\mathbf{n} = B \operatorname{div} \mathbf{n} + (\mathbf{n}\cdot\operatorname{grad})B \tag{82}$$

we conclude that

$$\operatorname{div} \mathbf{n} = -\frac{1}{B} n_j \frac{\partial B}{\partial x_j} = -(\mathbf{n}\cdot\operatorname{grad}) \log |\mathbf{B}| = \frac{1}{D} \text{ (say)}. \tag{83}$$

Again, by elementary differential geometry,

$$n_j \frac{\partial n_i}{\partial x_j} = \frac{\mathbf{e}_2}{R_1} \tag{84}$$

where $\mathbf{e}_2$ is a unit vector, normal to **n** and in the direction of the principal normal to the line of force and R_1 is its principal radius of curvature.

It is convenient to express Eq. (81) in the form

$$\operatorname{div} \mathbf{p} = \mathbf{n}(\nabla \mathbf{p})_{\|} + (\nabla \mathbf{p})_{\perp} \tag{85}$$

where

$$(\nabla \mathbf{p})_{\|} = n_j \frac{\partial p_{\|}}{\partial x_j} + (p_{\|} - p_{\perp}) \frac{\partial n_j}{\partial x_j} = n_j \frac{\partial p_{\|}}{\partial x_j} + \frac{p_{\|} - p_{\perp}}{D} \tag{86}$$

and

$$(\nabla \mathbf{p})_{\perp} = \nabla_{\perp} p_{\perp} + (p_{\|} - p_{\perp}) n_j \frac{\partial n_i}{\partial x_j} = \nabla_{\perp} p_{\perp} + (p_{\|} - p_{\perp}) \frac{\mathbf{e}_2}{R_1} \tag{87}$$

where

$$\nabla_{\perp} = \frac{\partial}{\partial x_i} - n_i n_j \frac{\partial}{\partial x_j} = (\delta_{ij} - n_i n_j) \frac{\partial}{\partial x_j}. \tag{88}$$

(B) The Integrability Condition on f_{I}

We shall now show that there is an integrability condition on f_{I} which arises when we proceed to the next higher approximation in the solution for f.

Let

$$f = f_{\mathrm{I}}(x_i, s^2, q) + f_{\mathrm{II}}(x_i, s_i, q) \tag{89}$$

where f_{II} is of order $1/B$ relative to f_{I}. We must now substitute (89) in Eq. (75) and neglect terms of order $1/B$ or higher. To effect this substitution we need certain transformation formulas; and we shall first write them down (as they will also be needed later).

It can be readily verified that

$$\frac{\partial}{\partial \nu_i} \longrightarrow (\delta_{ij} - n_i n_j) \frac{\partial}{\partial s_j} + n_i \frac{\partial}{\partial q} \tag{90}$$

and

$$\frac{\partial}{\partial x_j} \longrightarrow \frac{\partial}{\partial x_j} - \left(n_i s_k \frac{\partial n_k}{\partial x_j} + q \frac{\partial n_i}{\partial x_j} \right) \frac{\partial}{\partial s_i} + s_k \frac{\partial n_k}{\partial x_j} \frac{\partial}{\partial q} \tag{91}$$

when the differential operations on the left-hand side are applied to an arbitrary function of x_i, s_i, and q (as f_{II}). When, however, the function depends on s_i only through s^2 (as in f_{I}) the foregoing formulas simplify to

$$\frac{\partial}{\partial \nu_i} \longrightarrow 2s_i \frac{\partial}{\partial s^2} + n_i \frac{\partial}{\partial q} \tag{92}$$

and

$$\frac{\partial}{\partial x_j} \longrightarrow \frac{\partial}{\partial x_j} + s_i \frac{\partial n_i}{\partial x_j} \left(\frac{\partial}{\partial q} - 2q \frac{\partial}{\partial s^2} \right). \tag{93}$$

Now substituting (89) in Eq. (75), neglecting terms of order $1/B$, and making use of the transformation formulas (90)–(93), we get

$$(s_j + qn_j + V_j)\left[\frac{\partial f_{\mathrm{I}}}{\partial x_j} + s_i\frac{\partial n_i}{\partial x_j}\left(\frac{\partial}{\partial q} - 2q\frac{\partial}{\partial s^2}\right)f_{\mathrm{I}}\right] + \left[\frac{1}{mN}\frac{\partial p_{ij}}{\partial x_j} - (s_j + qn_j)\frac{\partial V_i}{\partial x_j}\right]\left(2s_i\frac{\partial}{\partial s^2} + n_i\frac{\partial}{\partial q}\right)f_{\mathrm{I}} + \frac{eB}{mc}\epsilon_{ijk}\frac{\partial f_{\mathrm{II}}}{\partial s_i}s_jn_k = 0. \tag{94}$$

The last term in this equation is

$$-\frac{eB}{mc}\mathbf{n}\cdot\mathbf{s}\times\mathrm{grad}_{\mathbf{s}}\, f_{\mathrm{II}} = -\frac{eB}{mc}\frac{\partial f_{\mathrm{II}}}{\partial\phi} \tag{95}$$

where ϕ is an azimuthal angle in a plane normal to $\mathbf{n}$. Accordingly, when this term is integrated over ϕ from 0 to 2π (or, equivalently, averaged over all directions of $\mathbf{s}$) it vanishes. This implies that

$$\left\langle (s_j + qn_j + V_j)\left[\frac{\partial f}{\partial x_j} + s_i\frac{\partial n_i}{\partial x_j}\left(\frac{\partial}{\partial q} - 2q\frac{\partial}{\partial s^2}\right)f\right] + \left[\frac{1}{mN}\frac{\partial p_{ij}}{\partial x_j} - (s_j + qn_j)\frac{\partial V_i}{\partial x_j}\right]\left(2s_i\frac{\partial}{\partial s^2} + n_i\frac{\partial}{\partial q}\right)f\right\rangle = 0 \tag{96}$$

where the angular brackets (here and in the sequel) should be taken to mean that the quantities enclosed are to be averaged over all directions of $\mathbf{s}$. In Eq. (96) we have suppressed the subscript "I" to f as now no longer necessary.

The averaging over the directions of $\mathbf{s}$ indicated in (96) can be readily effected, when it is noted that

$$\langle s_i\rangle = 0 \quad\text{and}\quad \langle s_is_j\rangle = \tfrac{1}{2}s^2(\delta_{ij} - n_in_j). \tag{97}$$

In this manner, we obtain

$$(qn_j + V_j)\frac{\partial f}{\partial x_j} + s^2\frac{\partial n_j}{\partial x_j}\left(\frac{1}{2}\frac{\partial f}{\partial q} - q\frac{\partial f}{\partial s^2}\right) + \frac{1}{mN}n_i\frac{\partial p_{ij}}{\partial x_j}\frac{\partial f}{\partial q} - s^2(\delta_{ij} - n_in_j)\frac{\partial V_i}{\partial x_j}\frac{\partial f}{\partial s^2} - qn_i\frac{\partial V_i}{\partial x_j}n_j\frac{\partial f}{\partial q} = 0. \tag{98}$$

Letting [cf. Eq. (88)]

$$\nabla_{\|}\cdot\mathbf{A} = n_i\frac{\partial A_i}{\partial x_j}n_j \quad\text{and}\quad \nabla_{\perp}\cdot\mathbf{A} = \frac{\partial A_j}{\partial x_j} - n_i\frac{\partial A_i}{\partial x_j}n_j \tag{99}$$

where **A** is any vector and making use of Eqs. (83) and (85), we can rewrite Eq. (98) in the following form:

$$(qn_j + V_j)\frac{\partial f}{\partial x_j} + \frac{qs^2}{D}\left(\frac{\partial f}{\partial q^2} - \frac{\partial f}{\partial s^2}\right) + \frac{2q}{mN}(\operatorname{div}\mathbf{p})_{\|}\frac{\partial f}{\partial q^2}$$
$$- s^2(\nabla_{\perp}\cdot\mathbf{V})\frac{\partial f}{\partial s^2} - 2q^2(\nabla_{\|}\cdot\mathbf{V})\frac{\partial f}{\partial q^2} = 0. \tag{100}$$

We have derived Eq. (100) as an integrability condition on the solution obtained in a first approximation, which arises when we proceed to the solution in the next higher approximation. The physical counterpart of this mathematical process is the following. The solution (79) obtained in the first approximation is no more than the physical statement that under the conditions in which we have investigated the problem (specifically $\eta \ll 1$) the "adiabatic" or the guiding center approximation is a valid one. Granted that this approximation is legitimate, we still need a "Boltzmann equation" which will govern the distribution of the quantities (such as the energy and the magnetic moment) which are relevant in the resulting physical picture; and this is what Eq. (100) provides.

As we have stated, it is often permissible to suppose that the stationary distribution function is even in q. When this is the case, we can equate the odd and the even parts of Eq. (100) separately and obtain two equations. Thus,

$$n_j\frac{\partial f^0}{\partial x_j} + \frac{s^2}{D}\left(\frac{\partial f^0}{\partial q^2} - \right)\frac{\partial f^0}{\partial s^2} = -\frac{2}{mN^0}(\operatorname{div}\mathbf{p}^0)_{\|}\frac{\partial f^0}{\partial q^2} \tag{101}$$

and

$$V_j^0\frac{\partial f^0}{\partial x_j} - 2q^2(\nabla_{\|}\cdot\mathbf{V}^0)\frac{\partial f^0}{\partial q^2} - s^2(\nabla_{\perp}\cdot\mathbf{V}^0)\frac{\partial f^0}{\partial s^2} = 0 \tag{102}$$

where we have restored the superscripts "0" to the various quantities. We may add to these equations the pair of moment equations [Eqs. (12) and (13)] appropriate to this case; these are

$$\frac{\partial}{\partial x_j}(N^0V_j^0) = 0 \tag{103}$$

and

$$mN^0V_j^0\frac{\partial V_i^0}{\partial x_j} = -\frac{\partial p_{ij}^0}{\partial x_j} + mN^0g_i + eN^0\left(E_i^0 + \frac{1}{c}\epsilon_{ijk}V_j^0B_k^0\right). \tag{104}$$

Frequently $(\operatorname{div}\mathbf{p}^0)_{\|} = 0$; then Eq. (101) when multiplied by q^2 and integrated over all of $\mathbf{v}$ space expresses this fact. Similarly, Eq. (102) when integrated over the $\mathbf{v}$ space reduces to the equation of continuity (103.)

9. THE REDUCTION OF THE BOLTZMANN EQUATION FOR f'

The equation to be solved is Eq. (36)

$$\frac{\partial f'}{\partial t} + (\nu_j + V_j^0)\frac{\partial f'}{\partial x_j} + \left[\frac{e}{mc}\epsilon_{ijk}\nu_j B_k^0 + \frac{1}{mN^0}\frac{\partial p_{ij}^0}{\partial x_j} - \nu_j\frac{\partial V_i^0}{\partial x_j}\right]\frac{\partial f'}{\partial \nu_i}$$
$$= -U_j\left(\frac{\partial f^0}{\partial x_j} - \frac{\partial f^0}{\partial \nu_k}\frac{\partial V_k^0}{\partial x_j}\right) + (\nu_j + V_j^0)\frac{\partial U_k}{\partial x_j}\frac{\partial f^0}{\partial \nu_k}$$
$$- \frac{e}{mc}\epsilon_{ijk}\frac{\partial f^0}{\partial \nu_i}(\nu_j + V_j^0)B_k' \tag{105}$$

where f' and f^0 are both referred to the same mean velocity $\mathbf{V}^0$.

In solving Eq. (105) we follow the same procedure as in the stationary case: we first find the solution in the "first approximation" by considering the dominant terms in the equation; and then find an "integrability condition" by inserting the solution in the first approximation back into Eq. (105) and averaging with respect to all directions of $\mathbf{s}$.

The dominant terms in Eq. (105) are those which contain $\mathbf{B}^0$ or $\mathbf{B}'$ as a factor. Thus, the equation we have to consider in the first approximation would appear to be

$$\frac{eB^0}{mc}\epsilon_{ijk}\frac{\partial f'}{\partial \nu_i}\nu_j n_k = -\frac{e}{mc}\epsilon_{ijk}\frac{\partial f^0}{\partial \nu_i}(\nu_j + V_j^0)B_k'. \tag{106}$$

If as in Section 8 we express f^0 as a function of x_i, s^2, and q and f' as a function of x_i, s_i, q and t, the foregoing equation becomes [cf. Eqs. (90) and (92)]

$$\frac{eB^0}{mc}\epsilon_{ijk}\frac{\partial f'}{\partial s_i}s_j n_k = -\frac{e}{mc}\epsilon_{ijk}\left(2s_i\frac{\partial f^0}{\partial s^2} + n_i\frac{\partial f^0}{\partial q}\right)(s_j + qn_j + V_j^0)B_k'. \tag{107}$$

On further simplification, this equation reduces to

$$\frac{eB^0}{mc}\epsilon_{ijk}\frac{\partial f'}{\partial s_i}s_j n_k = \frac{e}{mc}\left\{2q\left(\frac{\partial f^0}{\partial q^2} - \frac{\partial f^0}{\partial s^2}\right)\epsilon_{ijk}B_i' s_j n_k\right.$$
$$\left. + 2\frac{\partial f^0}{\partial s^2}\epsilon_{ijk}V_i^0 s_j B_k' - 2q\frac{\partial f^0}{\partial q^2}\epsilon_{ijk}n_i V_j^0 B_k'\right\}. \tag{108}$$

We now observe that when this equation is averaged over all directions normal to $\mathbf{n}$, the only surviving term is the last one on the right-hand side, namely,

$$-\left(\frac{e}{mc}\right)2q\frac{\partial f^0}{\partial q^2}\epsilon_{ijk}n_i V_j^0 B_k' = -\left(\frac{e}{mc}\right)2q\frac{\partial f^0}{\partial q^2}\mathbf{n}\cdot(\mathbf{V}^0 \times \mathbf{B}'). \tag{109}$$

According to the ideas expressed in Section 8 (b), a term which does not vanish when averaged over all directions of $\mathbf{s}$ strictly belongs in the Boltzmann equation in the guiding center approximation. Consequently, we should have taken [instead of (106)]

$$\frac{eB^0}{mc}\epsilon_{ijk}\frac{\partial f'}{\partial s_i}s_jn_k = -\frac{e}{mc}\epsilon_{ijk}\frac{\partial f^0}{\partial v_i}(v_j + V_j^0)B_k' + \frac{2e}{mc}q\frac{\partial f^0}{\partial q^2}\mathbf{n}\cdot(\mathbf{V}^0\times\mathbf{B}') \quad (110)$$

as the equation appropriate for the first approximation; in the latter case we should have obtained the equation

$$\frac{eB^0}{mc}\epsilon_{ijk}\frac{\partial f'}{\partial s_i}s_jn_k = \frac{e}{mc}\left\{2q\left(\frac{\partial f^0}{\partial q^2} - \frac{\partial f^0}{\partial s^2}\right)\epsilon_{ijk}B_i's_jn_k + 2\frac{\partial f^0}{\partial s^2}\epsilon_{ijk}V_i^0s_jB_k'\right\} \quad (111)$$

which is self-consistent.

A particular integral of Eq. (111) can be readily written down when it is remembered that for the purposes of evaluating the left-hand side of the equation, an arbitrary function of s^2 and q can be treated as a constant when differentiating with respect to s_i. Thus, we can verify that

$$f_P' = \frac{2}{B^0}\mathbf{s}\cdot\left\{\mathbf{B}'q\left(\frac{\partial f^0}{\partial q^2} - \frac{\partial f^0}{\partial s^2}\right) + \mathbf{n}\times(\mathbf{V}^0\times\mathbf{B}')\frac{\partial f^0}{\partial s^2}\right\} \quad (112)$$

satisfies Eq. (111).

We have distinguished the solution (112) by a subscript "P" to emphasize that it is only a particular integral. The general solution can be clearly expressed in the form

$$f' = f_P' + f_L'(x_i, s^2, q, t) \quad (113)$$

where f_L' is an arbitrary function of the arguments specified.

As in Section 8 we can obtain an integrability condition on the solution (113) by substituting it in Eq. (105) and averaging over all directions of **s**. Thus, the equation to be satisfied by f_L' is

$$\begin{aligned} &\left\langle\left\{\frac{\partial}{\partial t} + (v_j + V_j^0)\frac{\partial}{\partial x_j} + \left[-v_j\frac{\partial V_i^0}{\partial x_j} + \frac{1}{mN^0}\frac{\partial p_{ij}^0}{\partial x_j}\right]\frac{\partial}{\partial v_i}\right\}f_L'\right\rangle \\ &\quad + \left\langle\left\{(v_j + V_j^0)\frac{\partial}{\partial x_j} + \left[-v_j\frac{\partial V_i^0}{\partial x_j} + \frac{1}{mN^0}\frac{\partial p_{ij}^0}{\partial x_j}\right]\frac{\partial}{\partial v_i}\right\}f_P'\right\rangle \\ &\quad = G(\mathbf{U}) - \frac{2e}{mc}q\frac{\partial f^0}{\partial q^2}\mathbf{n}\cdot(\mathbf{V}^0\times\mathbf{B}') \end{aligned} \quad (114)$$

where

$$G(\mathbf{U}) = \left\langle -U_j\left(\frac{\partial f^0}{\partial x_j} - \frac{\partial V_k^0}{\partial x_j}\frac{\partial f^0}{\partial v_k}\right) + (v_j + V_j^0)\frac{\partial U_k}{\partial x_j}\frac{\partial f^0}{\partial v_k}\right\rangle. \quad (115)$$

It is to be noted that in Eq. (114) the term $\partial f_P'/\partial t$ on the left-hand side has been omitted since this clearly vanishes on averaging; also the origin of the term in $\mathbf{n}\cdot(\mathbf{V}^0 \times \mathbf{B}')$ on the right-hand side of the equation is the circumstance that in solving for f_P' we have used Eq. (110) (as the equation appropriate for the first approximation) in which this additional term was included.

The averaging indicated in Eq. (114) can be effected by first substituting for the various derivatives in accordance with Eqs. (90)–(93) and then making use of (97). We thus find [cf. Eq. (100)]

$$\left\langle\left\{\frac{\partial}{\partial t} + (\nu_j + V_j^0)\frac{\partial}{\partial x_j} + \left[-\nu_j\frac{\partial V_i^0}{\partial x_j} + \frac{1}{mN^0}\frac{\partial p_{ij}^0}{\partial x_j}\right]\frac{\partial}{\partial \nu_i}\right\} f_L'\right\rangle$$
$$= \left\{\frac{\partial}{\partial t} + (qn_j + V_j^0)\frac{\partial}{\partial x_j} + \frac{qs^2}{D}\left(\frac{1}{2q}\frac{\partial}{\partial q} - \frac{\partial}{\partial s^2}\right) + \frac{1}{mN^0}(\operatorname{div}\mathbf{p}^0)_\parallel\frac{\partial}{\partial q} - s^2(\nabla_\perp\cdot\mathbf{V}^0)\frac{\partial}{\partial s^2} - q(\nabla_\parallel\cdot\mathbf{V}^0)\frac{\partial}{\partial q}\right\} f_L' \tag{116}$$

and

$$G(\mathbf{U}) = -U_j\frac{\partial f^0}{\partial x_j} + n_k\frac{\partial V_k^0}{\partial x_j}U_j\frac{\partial f^0}{\partial q} + (qn_j + V_j^0)n_k\frac{\partial U_k}{\partial x_j}\frac{\partial f^0}{\partial q} + \left(\frac{\partial U_j}{\partial x_j} - n_j\frac{\partial U_k}{\partial x_j}n_k\right)s^2\frac{\partial f^0}{\partial s^2}. \tag{117}$$

Considering next the term in f_P' on the left-hand side of Eq. (114) we shall first write the solution (112) for f_P' in the form

$$f_P' = qF\mathbf{s}\cdot\mathbf{B}' + \mathbf{K}\cdot\mathbf{s} \tag{118}$$

where

$$F = \frac{2}{B^0}\left(\frac{\partial f^0}{\partial q^2} - \frac{\partial f^0}{\partial s^2}\right) \quad\text{and}\quad \mathbf{K} = \frac{2}{B^0}\mathbf{n}\times(\mathbf{V}^0\times\mathbf{B}')\frac{\partial f^0}{\partial s^2}. \tag{119}$$

It is to be noted that F and the components of $\mathbf{K}$ are functions of x_i, s^2, and q; also, that

$$\mathbf{n}\cdot\mathbf{K} = 0. \tag{119'}$$

Using Eqs. (92) and (93), we verify that for f_P' given by (118)

$$\frac{\partial f_P'}{\partial \nu_i} = (\delta_{ij} - n_i n_j)B_j' qF + n_i(s_j B_j')\left(F + 2q^2\frac{\partial F}{\partial q^2}\right) + 2s_i(s_j B_j')q\frac{\partial F}{\partial s^2} + K_i + 2s_l s_i\frac{\partial K_l}{\partial s^2} + n_i s_l\frac{\partial K_l}{\partial q} \tag{120}$$

and

$$\frac{\partial f_P'}{\partial x_j} = s_k \frac{\partial n_k}{\partial x_j} \left\{ (s_l B_l') \left[F + 2q^2 \left(\frac{\partial F}{\partial q^2} - \frac{\partial F}{\partial s^2} \right) \right] - (\mathbf{n}\cdot\mathbf{B}') qF \right\}$$
$$+ q s_l \frac{\partial}{\partial x_j} (B_l' F) - \left(B_i' \frac{\partial n_i}{\partial x_j} \right) q^2 F \tag{121}$$
$$+ s_l \frac{\partial K_l}{\partial x_j} - q \frac{\partial n_i}{\partial x_j} K_i + 2 q s_k s_l \frac{\partial n_k}{\partial x_j} \left(\frac{\partial K_l}{\partial q^2} - \frac{\partial K_l}{\partial s^2} \right).$$

Making use of these expressions and (97), we find

$$\left\langle \left[\frac{1}{mN^0} \frac{\partial p_{ij}^0}{\partial x_j} - \nu_j \frac{\partial V_i^0}{\partial x_j} \right] \frac{\partial f_P'}{\partial \nu_i} \right\rangle$$
$$= \left\langle \left[\frac{1}{mN^0} \frac{\partial p_{ij}^0}{\partial x_j} - s_j \frac{\partial V_i^0}{\partial x_j} - q n_j \frac{\partial V_i^0}{\partial x_j} \right] \frac{\partial f_P'}{\partial \nu_i} \right\rangle$$
$$= \left\{ q \frac{\partial}{\partial s^2} (s^2 F) \left[-q n_j \frac{\partial V_i^0}{\partial x_j} + \frac{1}{mN^0} \frac{\partial p_{ij}^0}{\partial x_j} \right] \right. \tag{122}$$
$$\left. - \frac{1}{2} s^2 \left(F + 2q^2 \frac{\partial F}{\partial q^2} \right) n_j \frac{\partial V_j^0}{\partial x_i} \right\} (\delta_{ik} - n_i n_k) B_k'$$
$$+ \left(\frac{1}{mN^0} \frac{\partial p_{ij}^0}{\partial x_j} - q n_j \frac{\partial V_i^0}{\partial x_j} \right) \left(K_i + s^2 \frac{\partial K_i}{\partial s^2} \right) - \frac{1}{2} s^2 n_i \frac{\partial V_i^0}{\partial x_j} \frac{\partial K_j}{\partial q}$$

and

$$\left\langle (\nu_j + V_j^0) \frac{\partial f_P'}{\partial x_j} \right\rangle = \left\langle (s_j + q n_j + V_j^0) \frac{\partial f_P'}{\partial x_j} \right\rangle$$
$$= \frac{1}{2} q s^2 \left\{ (\delta_{jk} - n_j n_k) \frac{\partial}{\partial x_j} (B_k' F) - \frac{\partial n_j}{\partial x_j} (\mathbf{n}\cdot\mathbf{B}') F \right\}$$
$$+ (q n_j + V_j^0) \left\{ s^2 \left[\frac{1}{2} F + q^2 \left(\frac{\partial F}{\partial q^2} - \frac{\partial F}{\partial s^2} \right) \right] - q^2 F \right\} \left(B_k' \frac{\partial n_k}{\partial x_j} \right) \tag{123}$$
$$- q (q n_j + V_j^0) \frac{\partial n_i}{\partial x_j} \left\{ K_i - s^2 \left(\frac{\partial K_i}{\partial q^2} - \frac{\partial K_i}{\partial s^2} \right) \right\} + \frac{1}{2} s^2 (\delta_{jl} - n_j n_l) \frac{\partial K_l}{\partial x_j}$$

where it may be noted that in virtue of the orthogonality of $\mathbf{n}$ and $\mathbf{K}$

$$\frac{1}{2} s^2 (\delta_{jl} - n_j n_l) \frac{\partial K_l}{\partial x_j} = \frac{1}{2} s^2 \left(\frac{\partial K_j}{\partial x_j} + n_j \frac{\partial n_i}{\partial x_j} K_i \right). \tag{123'}$$

Combining Eqs. (116), (117), (122), and (123) we can write down the equation satisfied by f_L'. In writing this equation it is convenient to express f_L' as the sum of an even and an odd function of q in the manner

$$f_L' = A_1(x_i, s^2, q^2, t) + qA_2(x_i, s^2, q^2, t) \tag{124}$$

and separate the even and the odd parts of the various terms in Eqs. (116), (117), (122), and (123). Thus we shall write

$$G(\mathbf{U}) = G_1(\mathbf{U}) + qG_2(\mathbf{U}) \tag{125}$$

and

$$\left\langle (\nu_j + V_j^0) \frac{\partial f_P'}{\partial x_j} + \left[\frac{1}{mN^0} \frac{\partial p_{ij}^0}{\partial x_j} - \nu_j \frac{\partial V_i^0}{\partial x_j}\right] \frac{\partial f_P'}{\partial \nu_i} \right\rangle = Q_1 + qQ_2 \tag{126}$$

where

$$G_1 = -U_j \frac{\partial f^0}{\partial x_j} + 2(\nabla_\parallel \cdot \mathbf{U}) q^2 \frac{\partial f^0}{\partial q^2} + (\nabla_\perp \cdot \mathbf{U}) s^2 \frac{\partial f^0}{\partial s^2} \tag{127}$$

$$G_2 = 2\left(U_j \frac{\partial V_k^0}{\partial x_j} n_k + V_j^0 \frac{\partial U_k}{\partial x_j} n_k\right) \frac{\partial f^0}{\partial q^2} \tag{128}$$

$$\begin{aligned} Q_1 = \frac{2}{B^0} \Bigg\{ & \left(B_k' \frac{\partial n_k}{\partial x_j} V_j^0\right) \left[\left(\frac{1}{2} s^2 - q^2\right) \frac{\delta f^0}{\delta \nu^2} + q^2 s^2 \left(\frac{\delta}{\delta \nu^2}\right)^2 f^0\right] \\ & - \left[q^2 \frac{\partial}{\partial s^2}\left(s^2 \frac{\delta f^0}{\delta \nu^2}\right) n_j \frac{\partial V_i^0}{\partial x_j} + \frac{1}{2} s^2 \left(\frac{\delta f^0}{\delta \nu^2} + 2q^2 \frac{\partial}{\partial q^2} \frac{\delta f^0}{\delta \nu^2}\right) n_j \frac{\partial V_j^0}{\partial x_i}\right] \\ & \times (\delta_{ik} - n_i n_k) B_k' \Bigg\} + \frac{1}{2} s^2 \frac{\partial K_j}{\partial x_j} + n_j \frac{\partial n_i}{\partial x_j} \left\{\frac{1}{2} s^2 K_i - q^2 \left(K_i - s^2 \frac{\delta K_i}{\delta \nu^2}\right)\right\} \\ & + \frac{1}{mN^0} \frac{\partial p_{ij}^0}{\partial x_j} \left(K_i + s^2 \frac{\partial K_i}{\partial s^2}\right) \end{aligned} \tag{129}$$

and

$$\begin{aligned} Q_2 = s^2 & \left\{ (\delta_{jk} - n_j n_k) \frac{\partial}{\partial x_j} \left(\frac{B_k'}{B^0} \frac{\delta f^0}{\delta \nu^2}\right) - \frac{\partial n_j}{\partial x_j} \left(\frac{\mathbf{n} \cdot \mathbf{B}'}{B^0}\right) \frac{\delta f^0}{\delta \nu^2} \right\} \\ & + \frac{2}{B^0} \left\{ \left(B_k' \frac{\partial n_k}{\partial x_j} n_j\right) \left[\left(\frac{1}{2} s^2 - q^2\right) \frac{\delta f^0}{\delta \nu^2} + q^2 s^2 \left(\frac{\delta}{\delta \nu^2}\right)^2 f^0\right] \right. \\ & \left. + \frac{1}{mN^0} \frac{\partial}{\partial s^2} \left(s^2 \frac{\delta f^0}{\delta \nu^2}\right) \frac{\partial p_{ij}^0}{\partial x_j} (\delta_{ik} - n_i n_k) B_k' \right\} \\ & - V_j^0 \frac{\partial n_i}{\partial x_j} \left(K_i - s^2 \frac{\delta K_i}{\delta \nu^2}\right) - \frac{\partial V_i^0}{\partial x_j} \left\{ s^2 n_i \frac{\partial K_j}{\partial q^2} + n_j \left(K_i + s^2 \frac{\partial K_i}{\partial s^2}\right) \right\} \end{aligned} \tag{130}$$ [4]

[4] It can be readily verified that the first term in this expression for Q_2 has the alternative form

$$(\delta_{jk} - n_j n_k) \frac{\partial}{\partial x_j} \left[\frac{2}{B^0} \frac{\delta f^0}{\delta \nu^2} (\delta_{ik} - n_i n_k) B_i'\right].$$

where, in writing the expressions for Q_1 and Q_2 we have substituted for F in accordance with Eq. (119) and introduced the operator

$$\frac{\delta}{\delta \nu^2} = \frac{\partial}{\partial q^2} - \frac{\partial}{\partial s^2} \qquad \text{so that} \qquad F = \frac{2}{B^0}\frac{\delta f^0}{\delta \nu^2}. \tag{131}$$

The equations governing A_1 and A_2 can now be written. We have [cf. Eqs. (114), (116), (125), and (126)]

$$\begin{aligned}\frac{\partial A_1}{\partial t} + V_j^0 \frac{\partial A_1}{\partial x_j} + \frac{s^2}{D}\left(\frac{1}{2} A_2 + q^2 \frac{\delta A_2}{\delta \nu^2}\right) - 2q^2(\nabla_\parallel \cdot \mathbf{V}^0)\frac{\partial A_1}{\partial q^2} \\ - s^2(\nabla_\perp \cdot \mathbf{V}^0)\frac{\partial A_1}{\partial s^2} + q^2 n_j \frac{\partial A_2}{\partial x_j} + \frac{2}{mN^0}(\operatorname{div} \mathbf{p}^0)_\parallel \left(\frac{1}{2} A_2 + q^2 \frac{\partial A_2}{\partial q^2}\right) \\ + Q_1 = G_1(\mathbf{U})\end{aligned} \tag{132}$$

and

$$\begin{aligned}\frac{\partial A_2}{\partial t} + V_j^0 \frac{\partial A_2}{\partial x_j} + \frac{s^2}{D}\frac{\delta A_1}{\delta \nu^2} - 2(\nabla_\parallel \cdot \mathbf{V}^0)\left(\frac{1}{2} A_2 + q^2 \frac{\partial A_2}{\partial q^2}\right) \\ - s^2(\nabla_\perp \cdot \mathbf{V}^0)\frac{\partial A_2}{\partial s^2} + n_j \frac{\partial A_1}{\partial x_j} + \frac{2}{mN^0}(\operatorname{div} \mathbf{p}^0)_\parallel \frac{\partial A_1}{\partial q^2} \\ + Q_2 = G_2(\mathbf{U}) - \frac{2e}{mc}\frac{\partial f^0}{\partial q^2}\mathbf{n} \cdot (\mathbf{V}^0 \times \mathbf{B}').\end{aligned} \tag{133}$$

10. THE FIRST-ORDER CHANGES IN THE MEAN VELOCITY AND IN THE PRESSURE TENSOR CAUSED BY THE PERTURBATION

We first recall that the solution for f' has been expressed in the form

$$f'(x_i, s_i, q, t) = A_1(x_i, s^2, q^2, t) + qA_2(x_i, s^2, q^2, t) + f_P' \tag{134}$$

where

$$f_P' = \frac{2}{B^0}\left\{q \frac{\delta f^0}{\delta \nu^2} \mathbf{s} \cdot \mathbf{B}' + \mathbf{s} \cdot [\mathbf{n} \times (\mathbf{V}^0 \times \mathbf{B}')] \frac{\partial f^0}{\partial s^2}\right\}. \tag{135}$$

The change in the mean velocity caused by the perturbation is clearly given by

$$V_i' = U_i + \frac{1}{N^0}\int f' v_i \, d\mathbf{v} = U_i + \frac{1}{2N^0}\iint \langle (s_i + qn_i) f' \rangle \, dq \, ds^2. \tag{136}$$

The term in A_1 makes no contribution to the integral on the right-hand side and we are left with

$$V_i' = U_i + \frac{n_i}{2N^0}\iint A_2 q^2 \, dq \, ds^2 + \frac{1}{2N^0}\iint \langle (s_i + qn_i) f_P' \rangle \, dq \, ds^2. \tag{137}$$

The integral over f_P' on the right-hand side can be reduced as follows:

$$\frac{1}{2N^0}\iint \langle (s_i + qn_i)f_P'\rangle \, dq\, ds^2$$

$$= \frac{1}{N^0B^0}\iint \left\langle (s_i + qn_i)\left\{ q(s_jB_j')\left(\frac{\partial f^0}{\partial q^2} - \frac{\partial f^0}{\partial s^2}\right) + s_l\epsilon_{ljk}n_j[\mathbf{V}^0 \times \mathbf{B}']_k \frac{\partial f^0}{\partial s^2}\right\}\right\rangle dq\, ds^2$$

$$= \frac{1}{2N^0B^0}\iint s^2(\delta_{il} - n_in_l)\epsilon_{ljk}n_j[\mathbf{V}^0 \times \mathbf{B}']_k \frac{\partial f^0}{\partial s^2}\, dq\, ds^2 \tag{138}$$

$$= \frac{1}{2N^0B^0}\iint [\mathbf{n} \times (\mathbf{V}^0 \times \mathbf{B}')]_i s^2 \frac{\partial f^0}{\partial s^2}\, dq\, ds^2$$

$$= -\frac{1}{B^0}[\mathbf{n} \times (\mathbf{V}^0 \times \mathbf{B}')]_i.$$

Thus

$$\mathbf{V}' = \mathbf{U} + \frac{\mathbf{n}}{2N^0}\iint q^2A_2\, dq\, ds^2 - \frac{1}{B^0}\mathbf{n} \times (\mathbf{V}^0 \times \mathbf{B}'). \tag{139}$$

This expression for $\mathbf{V}'$ agrees with Eq. (64) to the first order in η.

To find the corresponding first-order change in the pressure tensor, we must evaluate

$$p_{ij}' = m\int f' v_iv_j\, d\mathbf{v}$$
$$= \frac{1}{2}m\iint \langle [s_is_j + q^2n_in_j + q(n_is_j + n_js_i)]f'\rangle\, dq\, ds^2. \tag{140}$$

The contribution to p_{ij}', from f_L' arises only from the term in A_1. Thus [cf. Eq. (80)]

$$p_{ij,L}' = (\delta_{ij} - n_in_j)p_{\perp;L}' + n_in_jp_{\parallel;L}' \tag{141}$$

where

$$p_{\parallel,L}' = \frac{1}{2}m\iint q^2A_1\, dq\, ds^2 \quad \text{and} \quad p_{\perp;L}' = \frac{1}{4}m\iint s^2A_1\, dq\, ds^2. \tag{142}$$

Similarly, the contribution to p_{ij}' from f_P' is given by

$$p'_{ij;P} = \frac{m}{2B^0} \iint dq\, ds^2 q \left(\frac{\partial f^0}{\partial q^2} - \frac{\partial f^0}{\partial s^2}\right) \times \langle [s_i s_j + q^2 n_i n_j + q\,(n_i s_j + n_j s_i)] s_k B_k' \rangle$$

$$= \frac{m}{4B^0} [n_i(\delta_{jk} - n_j n_k) B_k' + n_j(\delta_{ik} - n_i n_k) B_k'] \times \iint q^2 s^2 \left(\frac{\partial f^0}{\partial q^2} - \frac{\partial f^0}{\partial s^2}\right) dq\, ds^2 \tag{143}$$

$$= \frac{1}{B^0} [n_i(\delta_{jk} - n_j n_k) B_k' + n_j(\delta_{ik} - n_i n_k) B_k'] (p_\parallel^0 - p_\perp^0).$$

We observe that p_{ij}' (unlike p_{ij}^0) is not "diagonal" in our representation; this arises from the fact that $\mathbf{n}$ is a unit vector in the direction of the static field $\mathbf{B}^0$ and not of the total field $\mathbf{B}^0 + \mathbf{B}'$.

11. THE CASE WHEN $\mathbf{V}^0$ IS NEGLIBIGLE AND $(\text{DIV } \mathbf{p})_\parallel = 0$

In case $\mathbf{V}^0$ is negligible,

$$G_2 = Q_1 = 0 \qquad (\mathbf{V}^0 = 0) \tag{144}$$

and if further $(\text{div } \mathbf{p}^0)_\parallel = 0$, Eqs. (132) and (133) reduce to the forms

$$\frac{\partial A_1}{\partial t} + q^2 L(A_2) + \frac{s^2}{2D} A_2 = G_1(\mathbf{U}) \tag{145}$$

and

$$\frac{\partial A_2}{\partial t} + L(A_1) = -Q_2 - \frac{2e}{mc} \frac{\partial f^0}{\partial q^2} \mathbf{n} \cdot (\mathbf{V}^0 \times \mathbf{B}') \tag{146}$$ [5]

where L stands for the operator,

$$L = n_j \frac{\partial}{\partial x_j} + \frac{s^2}{D} \frac{\delta}{\delta \nu^2} = n_j \frac{\partial}{\partial x_j} + \frac{s^2}{D} \left(\frac{\partial}{\partial q^2} - \frac{\partial}{\partial s^2}\right). \tag{147}$$

By writing

$$A_2 = \frac{\partial \Lambda}{\partial t} \tag{148}$$

and noting that $G_1(\mathbf{U})$ is linear in $\mathbf{U}$ and $\mathbf{U}^+ = \partial \boldsymbol{\xi}/\partial t$, we can immediately integrate Eq. (145) to give

$$A_1^+ = -\left[\frac{s^2}{2D} \Lambda + q^2 L(\Lambda)\right] + G_1(\boldsymbol{\xi}). \tag{149}$$

[5] Note that we do not discard the term in $\mathbf{V}^0$ in this equation, since it occurs with $\mathbf{B}'$ (which is of the order $\mathbf{B}^0$) as a factor.

Equation (146) now becomes

$$\frac{\partial^2 \Lambda}{\partial t^2} + L(A_1) = -Q_2 - \frac{2e}{mc}\frac{\partial f^0}{\partial q^2}\,\mathbf{n}\cdot(\mathbf{V}^0 \times \mathbf{B}'). \tag{150}$$

By eliminating A_1 from Eqs. (149) and (150) we shall obtain a generalized one-dimensional inhomogeneous wave equation for Λ.

Equation (149) is (as we have indicated) valid only for ions; for electrons the equation should be modified to be in accord with Eqs. (43) and (44).

Finally, it may be noted that if f^0 is isotropic in the $\mathbf{v}$ space, $\delta f^0/\delta v^2 = 0$ and Q_2 also vanishes [cf. Eq. (130)].

12. THE ADIABATIC RELATIONS

From the analysis in Section 10, it is apparent that A_1 determines the even moments of f_L' while A_2 determines the odd moments. Consequently, if A_2 should be zero, the third moments (i.e., the heat-flow tensor) will vanish and there will be no exchange of heat between different volume elements. Under these circumstances we should expect to obtain the so-called adiabatic relations. We shall now verify that this is indeed the case.

If $A_2 = 0$ then $\Lambda = 0$ and Eq. (149) becomes

$$A_1 = G(\xi) = -\xi_j \frac{\partial f^0}{\partial x_j} + 2(\nabla_\parallel\cdot\xi)q^2\frac{\partial f^0}{\partial q^2} + (\nabla_\perp\cdot\xi)s^2\frac{\partial f^0}{\partial s^2}. \tag{151}$$

Using the solution (151) for A_1, we can evaluate $p_\parallel'$ and $p_\perp'$ in accordance with Eq. (142). Thus, making use of the elementary relations

$$\frac{1}{2}m\iint qs^2\frac{\partial f^0}{\partial q}\,dq\,ds^2 = -2p_\perp^0; \qquad \frac{1}{2}m\iint s^4\frac{\partial f^0}{\partial s^2}\,dq\,ds^2 = -4p_\perp^0 \tag{152}$$

$$\frac{1}{2}m\iint q^3\frac{\partial f^0}{\partial q}\,dq\,ds^2 = -3p_\parallel^0; \qquad \frac{1}{2}m\iint q^2s^2\frac{\partial f^0}{\partial s^2}\,dq\,ds^2 = -p_\parallel^0 \tag{153}$$

we find

$$\begin{aligned} p_\parallel' &= -\xi_j\frac{\partial p_\parallel^0}{\partial x_j} - (3\nabla_\parallel\cdot\xi + \nabla_\perp\cdot\xi)p_\parallel^0 \\ &= -\xi_j\frac{\partial p_\parallel^0}{\partial x_j} - \left(\frac{\partial \xi_j}{\partial x_j} + 2n_i\frac{\partial \xi_i}{\partial x_j}n_j\right)p_\parallel^0 \end{aligned} \tag{154}$$

and

$$\begin{aligned} p_\perp' &= -\xi_j\frac{\partial p_\perp^0}{\partial x_j} - (\nabla_\parallel\cdot\xi + 2\nabla_\perp\cdot\xi)p_\perp^0 \\ &= -\xi_j\frac{\partial p_\perp^0}{\partial x_j} - \left(2\frac{\partial \xi_j}{\partial x_j} - n_i\frac{\partial \xi_i}{\partial x_j}n_j\right)p_\perp^0. \end{aligned} \tag{155}$$

These laws have already been discussed in paper II [see particularly the remarks following Eq. (8)] and it has been shown that they represent simple generalizations of the usual adiabatic law for scalar pressure. It should be further pointed out that these adiabatic relations can be obtained formally by letting $\Omega^2 \to \infty$; however Ω must not be so large that the expansion in η is violated.

13. THE TIME-INDEPENDENT CASE

In case the perturbed state is independent of time, Eq. (150) for A_1 becomes (we are continuing to consider the case when $\mathbf{V}^0$ is negligible and $(\operatorname{div} \mathbf{p})_\parallel = 0$)

$$L(A_1) = -Q_2 - \frac{2e}{mc}\frac{\partial f^0}{\partial q^2}\mathbf{n}\cdot(\mathbf{V}^0 \times \mathbf{B}') \tag{156}$$

and Λ, which is not needed, is determined by Eq. (149).

Equation (156) can be formally integrated by a change of variables. Thus, let

$$A_1 \equiv A_1\left(\nu^2, \frac{s^2}{B^0}, x_i, t\right). \tag{157}$$

It can be readily verified that the effect of the operation of L on A_1 regarded as a function of x_i, s^2, q^2, and t is the same as the operation of $n_j\partial/\partial x_j$ on A_1 regarded [as in (157)] as a function of x_i, $\nu^2(= s^2 + q^2)$, s^2/B^0 and t. Accordingly, Eq. (156) is equivalent to

$$n_j \frac{\partial}{\partial x_j} A_1\left(x_i, \nu^2, \frac{s^2}{B^0}, t\right) = -\left\{Q_2 + \frac{2e}{mc}\frac{\partial f^0}{\partial q^2}\mathbf{n}\cdot(\mathbf{V}^0 \times \mathbf{B}')\right\} \tag{158}$$

an equation which can be integrated directly in the form

$$A_1 = -\int\left\{Q_2 + \frac{2e}{mc}\frac{\partial f^0}{\partial \nu^2}\mathbf{n}\cdot(\mathbf{V}^0 \times \mathbf{B}')\right\} n_j\, dx_j\,. \tag{159}$$

Once A_1 has been determined in this fashion, the pressures $p_{\parallel;L}{}'$ and $p_{\perp;L}{}'$ can be found from Eq. (142). It may be noted in this connection that if f^0 is isotropic and $\mathbf{V}^0$ vanishes to the first order, A_1 (and therefore also $\mathbf{p}'$) vanishes.

As one may expect on general grounds, it must follow from Eq. (156) that there is, in this case of time independence, a balance of pressure along the lines of force. We can verify this directly by multiplying Eq. (156) by mq^2 and integrating over all of $\mathbf{v}$ space. Thus,

$$\frac{1}{2}m\iint q^2\left\{L(A_1) + Q_2 + \frac{2e}{mc}\frac{\partial f^0}{\partial q^2}\mathbf{n}\cdot(\mathbf{V}^0 \times \mathbf{B}')\right\} dq\, ds^2 = 0. \tag{160}$$

The term in $L(A_1)$ in this equation is

$$\begin{aligned}\frac{1}{2}m\iint\left\{n_j\frac{\partial A_1}{\partial x_j} + \frac{s^2}{D}\left(\frac{\partial A_1}{\partial q^2} - \frac{\partial A_1}{\partial s^2}\right)\right\} q^2\, dq\, ds^2 \\ = n_j\frac{\partial}{\partial x_j}p_{\parallel;L}{}' + \frac{1}{D}(p_{\parallel;L}{}' - p_{\perp;L}{}') = n_j\frac{\partial}{\partial x_j}p_{ij;L}{}'\end{aligned} \tag{161}$$

where we have made use of Eqs. (85), (86), (142), (152), and (153). The term in Q_2 in Eq. (160) can be reduced by making use of Eq. (126); thus (remembering that we are putting $\mathbf{V}^0 = 0$ in these and similar expressions) we have

$$\begin{aligned} \frac{1}{2} m \iint q^2 Q_2 \, dq \, ds^2 &= m \int q \left\{ \left\langle \left(\nu_j \frac{\partial}{\partial x_j} + \frac{1}{mN^0} \frac{\partial p_{ij}{}^0}{\partial x_j} \frac{\partial}{\partial \nu_i} \right) f_P{}' \right\rangle - Q_1 \right\} dq \, ds^2 \\ &= m \int (n_k \nu_k) \left(\nu_j \frac{\partial f_P{}'}{\partial x_j} + \frac{1}{mN^0} \frac{\partial p_{ij}{}^0}{\partial x_j} \frac{\partial f_P{}'}{\partial \nu_i} \right) d\mathbf{v} \end{aligned} \tag{162}$$

since Q_1 being even in q does not contribute to the integral. By an integration by parts, we find

$$\frac{1}{N^0} \frac{\partial p_{ij}{}^0}{\partial x_j} \int \frac{\partial f_P{}'}{\partial \nu_i} (n_k \nu_k) \, d\mathbf{v} = -\frac{1}{N^0} n_i \frac{\partial p_{ij}{}^0}{\partial x_j} \int f_P{}' \, d\mathbf{v} = 0 \tag{163}$$

since we have assumed that $(\text{div } \mathbf{p})_\parallel = 0$. The surviving term in (162) is clearly

$$n_k \frac{\partial}{\partial x_j} p'_{jk;P} \,. \tag{164}$$

Finally, considering the last term in the integral in Eq. (160), we have

$$\frac{e}{c} \mathbf{n} \cdot (\mathbf{V}^0 \times \mathbf{B}') \iint q^2 \frac{\partial f^0}{\partial q^2} dq \, ds^2 = -\frac{N^0 e}{c} \mathbf{n} \cdot (\mathbf{V}^0 \times \mathbf{B}'). \tag{165}$$

Now, combining Eqs. (161), (164), and (165) we obtain

$$\mathbf{n} \cdot \left\{ \text{div } \mathbf{p}' - \frac{N^0 e}{c} (\mathbf{V}^0 \times \mathbf{B}') \right\} = 0 \tag{166}$$

which is the relation we wished to establish.

We may observe that in virtue of Eqs. (46) and (166), Eq. (73) is identically satisfied. Also, by adding Eq. (166) for ions and electrons, we obtain

$$\mathbf{n} \cdot \left\{ \text{div } \mathbf{P}' - \frac{1}{c} \mathbf{J}^0 \times \mathbf{B}' \right\} = 0. \tag{167}$$

We see that retaining the term in $\mathbf{V}^0 \times \mathbf{B}'$ in Eq. (146) (and the subsequent ones) is essential if $\mathbf{J}^0 \neq 0$.

A special case of Eq. (166) arises when $\mathbf{V}^0$ is a pure pressure drift [see a forthcoming paper (*5*); Eq. (59)] given by

$$\mathbf{V}^0 = -c\mathbf{B}^0 \times (\text{div } \mathbf{P}^0)/em \mid \mathbf{B}^0 \mid^2.$$

In this cace, Eq. (166) takes the simple form

$$\mathbf{n} \cdot [\text{div } (\mathbf{P}^0 + \mathbf{P}')] = 0. \tag{166'}$$

This means that the total pressure has no component of its divergence *along* the distorted field lines.

14. CONDITIONS AT A SURFACE OF DISCONTINUITY

In practice, the mathematical idealization of a surface of discontinuity is realized when the properties of a system change appreciably in a distance, small compared to the macroscopic dimensions of the system. Measures of the latter are the characteristic dimensions of the equilibrium state, the wavelengths of the disturbance and the extent of the displacements which the region undergoes during the disturbance. In order that the theory developed in the preceding sections may be valid, it is necessary that the dimension of the transition region is still large compared to the Larmor radii of the particles i.e., the basic assumption, $\eta \ll 1$, continues to hold.

To derive the conditions which must be satisfied at a "surface of discontinuity" we must use the full nonlinear equations since the departures from the equilibrium state will be large in the neighborhood of the surface (*7*). On either side of the surface, however, we may use the ξ formalism.

It is physically reasonable to require that in the equilibrium state the magnetic field lines and all the macroscopic velocities lie on the surface. Further, the motion of the surface, when undergoing the disturbance, must be specified in such a way that we are led to a consistent set of conditions, within the limitations of the theory, and which we may require as the proper boundary conditions to be met at a surface of discontinuity.

We shall define the motion of the surface, when disturbed, as that of the magnetic field lines which lie on the surface in the equilibrium state. The velocity field, $\boldsymbol{\alpha}$, which specifies the motion of the field lines is not unique [cf. (*6*)]; but it must satisfy the equation

$$\frac{\partial \mathbf{B}}{\partial t} = \text{curl}\ (\boldsymbol{\alpha} \times \mathbf{B}) \tag{168}$$

or, equivalently

$$\frac{\partial B_i}{\partial t} + \alpha_j \frac{\partial B_i}{\partial x_j} = B_j \frac{\partial \alpha_i}{\partial x_j} - B_i \frac{\partial \alpha_j}{\partial x_j}. \tag{169}$$

From a comparison of Eq. (168) with Maxwell's equation

$$-\frac{1}{c}\frac{\partial \mathbf{B}}{\partial t} = \text{curl}\ \mathbf{E} \tag{170}$$

we conclude that

$$\mathbf{E} = -\frac{1}{c}\boldsymbol{\alpha} \times \mathbf{B} + \text{grad}\ \chi \tag{171}$$

where χ is a scalar function. We shall find it convenient to let

$$\boldsymbol{\alpha}^0 = 0 \tag{172}$$

in the equilibrium state. Then according to Eq. (171)

$$\mathbf{E}^0 = \text{grad } \chi^0 \tag{173}$$

and in the disturbed state,

$$\mathbf{E}' = -\frac{1}{c}\boldsymbol{\alpha}' \times \mathbf{B}^0 + \text{grad } \chi'. \tag{174}$$

As it involves no loss of generality, we shall further suppose that

$$\boldsymbol{\alpha}' \cdot \mathbf{B}^0 = 0 \tag{175}$$

in which case the solution of Eq. (174) for $\boldsymbol{\alpha}'$ is

$$\boldsymbol{\alpha}' = \frac{c}{|\mathbf{B}^0|^2}\mathbf{E}' \times \mathbf{B}^0 + \frac{c}{B^0}\mathbf{n} \times \text{grad } \chi' \tag{176}$$

also

$$\mathbf{n} \cdot \text{grad } \chi' = \mathbf{n} \cdot \mathbf{E}' = E_{\parallel}'. \tag{177}$$

Let $\mathbf{N}$ denote the unit normal to the surface. Given that the element of the surface moves with the field lines with the velocity $\boldsymbol{\alpha}$, it follows that the "equation of motion" governing $\mathbf{N}$ is [cf. (7)]

$$\frac{\partial N_i}{\partial t} + \alpha_j \frac{\partial N_i}{\partial x_j} = -N_k \frac{\partial \alpha_k}{\partial x_i} + N_i N_k N_j \frac{\partial \alpha_k}{\partial x_j}. \tag{178}$$

Multiplying Eqs. (169) and (178) by N_i and B_i respectively and adding, we get

$$\frac{\partial}{\partial t}(\mathbf{N} \cdot \mathbf{B}) + \alpha_j \frac{\partial}{\partial x_j}(\mathbf{N} \cdot \mathbf{B}) = (\mathbf{N} \cdot \mathbf{B})\left(N_k \frac{\partial \alpha_k}{\partial x_j} N_j - \frac{\partial \alpha_j}{\partial x_j}\right). \tag{179}$$

From this equation it follows that if $\mathbf{N} \cdot \mathbf{B}$ vanishes everywhere at any time, then it must continue to vanish at all times. This is, of course, to be expected from our definition of $\boldsymbol{\alpha}$.

The linearized form of Eq. (178) is [cf. Eq. (172)]

$$\frac{\partial N_i'}{\partial t} + \alpha_j' \frac{\partial N_i^0}{\partial x_j} = -N_k^0 \frac{\partial \alpha_k'}{\partial x_i} + N_i^0 N_k^0 N_j^0 \frac{\partial \alpha_k'}{\partial x_j}. \tag{180}$$

where N_i' is the change in N_i caused by the disturbance.

A condition we may impose at a surface of discontinuity is that there be no flux of the plasma across the surface; this requires that

$$\mathbf{N} \cdot (\mathbf{V} - \boldsymbol{\alpha}) = 0 \tag{181}$$

where **V**, as usual, denotes the mean fluid velocity. The linearized form of this condition is

$$\boldsymbol{\alpha}'\cdot\mathbf{N}^0 = \mathbf{V}'\cdot\mathbf{N}^0 + \mathbf{V}^0\cdot\mathbf{N}'. \tag{182}$$

This last condition, as we shall presently see, determines the term in grad χ' in Eq. (174).

Taking the scalar product of Eq. (139) (giving $\mathbf{V}'$) with $\mathbf{N}^0$, we get

$$\begin{aligned}\mathbf{V}'\cdot\mathbf{N}^0 &= \frac{\partial}{\partial t}(\boldsymbol{\xi}\cdot\mathbf{N}^0) - \frac{1}{B^0}\mathbf{N}^0\cdot[\mathbf{n}\times(\mathbf{V}^0\times\mathbf{B}')] \\ &= \frac{\partial}{\partial t}(\boldsymbol{\xi}\cdot\mathbf{N}^0) - \frac{1}{B^0}\mathbf{N}^0\cdot[(\mathbf{n}\cdot\mathbf{B}')\mathbf{V}^0 - (\mathbf{n}\cdot\mathbf{V}^0)\mathbf{B}'].\end{aligned} \tag{183}$$

The second term on the right-hand side of Eq. (183) can be simplified by making use of the relations

$$\mathbf{N}^0\cdot\mathbf{V}^0 = 0 \quad\text{and}\quad \mathbf{N}\cdot\mathbf{B} = \mathbf{N}^0\cdot\mathbf{B}' + \mathbf{N}'\cdot\mathbf{B}^0 = 0. \tag{184}$$

We find

$$\mathbf{V}'\cdot\mathbf{N}^0 = \frac{\partial}{\partial t}(\boldsymbol{\xi}\cdot\mathbf{N}^0) - (\mathbf{n}\cdot\mathbf{V}^0)(\mathbf{n}\cdot\mathbf{N}'). \tag{185}$$

Using this result in Eq. (182), we have

$$\alpha_j'N_j^0 = \frac{\partial}{\partial t}(\xi_j N_j^0) - N_j'(\delta_{ij} - n_i n_j)V_i^0. \tag{186}$$

On the other hand, according to Eqs. (42) and (176)

$$\alpha_j'N_j^0 = \frac{\partial}{\partial t}(\xi_j N_j^0) + \frac{c}{B^0}\mathbf{N}^0\cdot\mathbf{n}\times\text{grad}\,\chi'. \tag{187}$$

We may, therefore, identify the second terms on the right-hand sides of Eqs. (186) and (187) and thus determine grad χ'.

Since the motion of the surface must be the same as measured on either side of the surface of discontinuity, it is clear that $\boldsymbol{\alpha}'\cdot\mathbf{N}^0$ *as given by Eq.* (186) *must be continuous across the interface.* Since to the lowest order in η, $\mathbf{V}^0$ is the same for ions and electrons, Eq. (186) applies (to this order) to either constituent.

A further condition can be obtained from the equation of motion as expressed in Eq. (22). We integrate this equation over a small cylindrical volume enclosing an element of the surface of discontinuity, and *in a coordinate system moving with the volume.* Keeping the generators of the cylinder normal to the surface of discontinuity, we let the height of the cylinder approach zero. By this process, contributions to the integral arise only from the discontinuities. In general, if

A is a vector which suffers a discontinuity at the interface, the process described leads to

$$\int d\tau \operatorname{div} \mathbf{A} = \int \mathbf{A}\cdot d\mathbf{S} \rightarrow N_j\Delta[A_j]\, dS \tag{188}$$

and

$$\begin{aligned}\int d\tau \frac{\partial A_i}{\partial t} \rightarrow -\alpha_j \int d\tau \frac{\partial A_i}{\partial x_j} &= -\alpha_j\, dS_j\Delta[A_i] \\ &= -\alpha_j N_j\Delta[A_i]\, dS\end{aligned} \tag{189}$$

where $d\mathbf{S}$ is the element of the surface of discontinuity enclosed by the small cylinder and $\Delta[A_i]$ is the extent of the discontinuity which A_i suffers at the interface.

Applying the foregoing process to Eq. (22), we obtain

$$\begin{aligned}-\alpha_j N_j\Delta[\sum_{+,-} mNV_i] - \frac{\alpha_j N_j}{4\pi c}\Delta[(\mathbf{E}\times\mathbf{B})_i]& \\ = N_j\Delta[T_{ij}] - N_j\Delta[\sum_{+,-} mNV_iV_j]&.\end{aligned} \tag{190}$$

Rearranging this equation and remembering that N_j is continuous across the interface, we get

$$N_j\Delta[T_{ij}] + \frac{\alpha_j N_j}{4\pi c}\Delta[(\mathbf{E}\times\mathbf{B})_i] = N_j\Delta[\sum_{+,-} mNV_i(V_j - \alpha_j].\tag{191}$$

But according to condition (181) which must be met at a surface of discontinuity, the right-hand side of Eq. (191) vanishes and we are left with

$$N_j\Delta[T_{ij}] + \frac{\alpha_j N_j}{4\pi c}\Delta[(\mathbf{E}\times\mathbf{B})_i] = 0. \tag{192}$$

It may be recalled here that [cf. Eq. (23)]

$$T_{ij} = -P_{ij} + \frac{1}{4\pi}(E_iE_j + B_iB_j) - \frac{\delta_{ij}}{8\pi}(|\mathbf{E}|^2 + |\mathbf{B}|^2). \tag{193}$$

In the stationary case when $\boldsymbol{\alpha}^0 = 0$, Eq. (192) leads to the condition

$$N_j^0\Delta[T_{ij}^0] = 0 \tag{194}$$

i.e., the normal component of the stress tensor must be continuous across a surface of discontinuity.

We can apply the process we have described in connection with Eq. (22) to

Maxwell's equations (15) and (16) and obtain

$$N_j \Delta[B_j] = 0 \tag{195}$$

and

$$\frac{1}{c} N_j \alpha_j \Delta[B_i] = \epsilon_{ijk} N_j \Delta[E_k]. \tag{195$'$}$$

Equation (195) does not provide any new information, since it is automatically satisfied by the requirement that $\mathbf{N}\cdot\mathbf{B} = 0$ on a surface of discontinuity. If $\Delta[\mathbf{n}] = 0$, then by taking the vector product of Eq. (195′) with $\mathbf{n}$ we obtain the condition

$$\mathbf{n} \times (\mathbf{N} \times \Delta[\mathbf{E}]) = 0. \tag{196}$$

When applying the condition (192) to a perturbed state, one must remember that the various quantities are to be evaluated on the displaced surface. Since the element of surface will be displaced by an amount

$$\zeta_N = \mathbf{N}^0 \cdot \boldsymbol{\alpha}' \, dt \tag{197}$$

in the direction of the normal, we should allow for the contribution,

$$\Delta \left[\zeta_N N_k^{\,0} \frac{\partial}{\partial x_k} N_j^{\,0} T_{ij}^{\,0} \right] \tag{198}$$

arising from this cause when we linearize Eq. (192). Thus, the condition appropriate for the perturbed problem is

$$\Delta \left[\zeta_N N_k^{\,0} \frac{\partial}{\partial x_k} N_j^{\,0} T_{ij}^{\,0} \right] + N_j^{\,0} \Delta[T_{ij}'] + N_j' \Delta[T_{ij}^{\,0}] + \frac{\alpha_j' N_j^{\,0}}{4\pi c} \Delta[(\mathbf{E}^0 \times \mathbf{B}^0)_i] = 0. \tag{199}$$

We shall now obtain expressions for the various terms in Eq. (199). According to the definition of T_{ij} [Eq. (193)]

$$N_j T_{ij} = - N_j P_{ij} + \frac{1}{4\pi} (N_j E_j) E_i - \frac{N_i}{8\pi} (|\mathbf{E}|^2 + |\mathbf{B}|^2) \tag{200}$$

since $N_j B_j = 0$. By using the form for the equilibrium pressure tensor given by Eq. (80), we find

$$N_j^{\,0} T_{ij}^{\,0} = - N_i^{\,0} P_\perp^{\,0} + \frac{1}{4\pi} (N_j^{\,0} E_j^{\,0}) E_i^{\,0} - \frac{N_i^{\,0}}{8\pi} (|\mathbf{E}^0|^2 + |\mathbf{B}^0|^2). \tag{201}$$

If we neglect terms of order $|\mathbf{E}^0|^2$, then

$$N_j^0 T_{ij}^0 = -N_i^0 \left(P_\perp^0 + \frac{|\mathbf{B}^0|^2}{8\pi}\right) \tag{202}$$

and the continuity of the normal component of the stress tensor across an interface in the stationary state requires now that $(P_\perp^0 + |\mathbf{B}^0|^2/8\pi)$ be continuous. To the same accuracy as Eq. (202) [cf. Eq. (184)]

$$\begin{aligned} N_j' T_{ij}^0 &= -N_j' P_{ij}^0 + \frac{N_j'}{4\pi}(B_i^0 B_j^0) - \frac{N_i'}{8\pi}|\mathbf{B}^0|^2 \\ &= -N_j' P_{ij}^0 - \frac{1}{4\pi}(N_j^0 B_j')B_i^0 - \frac{N_i'}{8\pi}|\mathbf{B}^0|^2. \end{aligned} \tag{203}$$

In this equation [again making use of Eq. (80)],

$$N_j' P_{ij}^0 = n_i(n_j N_j')(P_\parallel^0 - P_\perp^0) + N_i' P_\perp^0. \tag{204}$$

Similarly,

$$\begin{aligned} N_j^0 T_{ij}' = -N_j^0 P_{ij}' + \frac{1}{4\pi}\{(N_j^0 E_j^0)E_i' + (N_j^0 E_j')E_i^0 + (N_j^0 B_j')B_i^0\} \\ - \frac{N_i^0}{4\pi}(E_j^0 E_j' + B_j^0 B_j'). \end{aligned} \tag{205}$$

Now making use of the solution for P_{ij}' given in Section 10 [Eqs. (141) and (143)], we find

$$N_j^0 P_{ij}' = -n_i(n_j N_j')(P_\parallel^0 - P_\perp^0) + N_i^0 P_\perp'. \tag{206}$$

Using the foregoing expressions in Eq. (199) and taking the scalar product of the resulting equation with $\mathbf{N}^0$, we get

$$\begin{aligned} \Delta\Bigg[\zeta_N N_k^0 \frac{\partial}{\partial x_k}\left(P_\perp^0 + \frac{|\mathbf{B}^0|^2}{8\pi}\right) + P_\perp' + \frac{1}{4\pi}(E_j^0 E_j' + B_j^0 B_j') \\ - \frac{1}{2\pi}(N_j^0 E_j^0)(N_i^0 E_i')\Bigg] = 0 \end{aligned} \tag{207}$$

the terms in $\Delta[T_{ij}^0]$ and $(\mathbf{E}^0 \times \mathbf{B}^0)$ not making any contributions in view of the condition (194) and the fact that $\Delta[\mathbf{N}^0 \times \mathbf{E}^0 \boldsymbol{\cdot} \mathbf{B}^0] = 0$.

Next taking the vector product of Eq. (199) with $\mathbf{N}^0$ we similarly find

$$\begin{aligned} \Delta\Bigg[\frac{1}{4\pi}(\mathbf{N}^0\boldsymbol{\cdot}\mathbf{E}')(\mathbf{E}^0 \times \mathbf{N}^0) + \frac{1}{4\pi}(\mathbf{N}^0\boldsymbol{\cdot}\mathbf{E}^0)(\mathbf{E}' \times \mathbf{N}^0) + \frac{\mathbf{N}^0\boldsymbol{\cdot}\boldsymbol{\alpha}'}{4\pi c}(\mathbf{N}^0\boldsymbol{\cdot}\mathbf{E}^0)\mathbf{B}^0 \\ + \left(P_\perp^0 + \frac{|\mathbf{B}^0|^2}{8\pi}\right)\left(\mathbf{N}' - \zeta_N N_k \frac{\partial \mathbf{N}'}{\partial x_k}\right) \times \mathbf{N}^0\Bigg] = 0. \end{aligned} \tag{208}$$

The last term clearly vanishes in virtue of Eqs. (194) and(202), while the second and the third terms cancel on account of Eq. (196). We are thus left with

$$(\mathbf{E}^0 \times \mathbf{N}^0)\Delta[\mathbf{N}^0 \cdot \boldsymbol{\xi}_\perp \times \mathbf{B}^0] = 0 \tag{209}$$

where we have substituted for $\mathbf{E}_\perp'$ from Eq. (42). The condition (209) is generally satisfied by $\mathbf{E}^0 \times \mathbf{N}^0 = 0$. However, Eq. (207) is of importance in providing a condition for $\boldsymbol{\xi}$ when $P_\perp'$, $\mathbf{B}'$ and $\mathbf{E}'$ are expressed in terms of $\boldsymbol{\xi}$.

15. CONCLUDING REMARKS

The Boltzmann equation for the perturbed distribution function f' as reduced in Section 9 together with the equations of motion of Section 7 define the behavior of the ionized gas in states close to an initial stationary state. The equations which we have derived can be used as a basis for the study of the oscillations of plasmas and of their stability when confined by magnetic fields. Examples illustrative of the use of our equations in such problems will be published separately.

In concluding we may draw attention to two parameters which play an important role in the applications of theories such as we are describing [cf. (*8*)]. These are

$$\kappa = 1 + \frac{4\pi m^+ N^+ c^2}{|\mathbf{B}^0|^2} \quad \text{and} \quad \beta = \frac{P^0}{|\mathbf{B}^0|^2/8\pi}. \tag{210}$$

The parameter κ is of the nature of a dielectric constant. In most cases of interest $\kappa \gg 1$ and the methods of the present paper will apply; if, on the other hand, $\kappa \sim 1$, then it would appear that the description should be based on the individual orbits of the particles in the prescribed field. The parameter β is a measure of the partition of the available energy between the energy of motion of the particles and the energy of the magnetic field. When $\beta \ll 1$, the prescribed magnetic field will tend to limit the possible low-frequency modes of motion; for, the plasma does not contain sufficient energy to influence the prevailing magnetic field appreciably. As will be illustrated in our discussion of applications (*9*) the assumption, $\beta \ll 1$, considerably simplifies the theory of plasma motion.

Received: July 11, 1957

REFERENCES

1. H. Alfvén, "Cosmical Electrodynamics," Clarendon Press, Oxford, 1950; L. Spitzer, "Physics of Fully Ionized Gases," Interscience, New York, 1956.

2. G. Chew, M. Goldberger, and F. Low, *Proc. Roy. Soc.* (*London*) **A236,** 112 (1956).

3. K. M. Watson, *Phys. Rev.* **102,** 12 (1956) and K. A. Brueckner and K. M. Watson, *Phys. Rev.* **102,** 19 (1956); these papers are referred to as papers I and II in the text. However, the development in the present paper is self-contained although occasional references will be made to papers I and II.

4. S. Chapman and T. G. Cowling, "Mathematical Theory of Nonuniform Gases," Cambridge Univ. Press, London and New York, 1952,
5. S. Chandrasekhar, A. N. Kaufman, and K. M. Watson, *Annals of Physics.*, (in press).
6. W. A. Newcomb, "The Motion of Magnetic Lines of Force," Princeton University Observatory, Technical Report No. 1 (1955).
7. T. G. Northrop, *Phys. Rev.* **103,** 1150 (1956); this paper contains a useful description of the boundary conditions for ordinary hydromagnetics.
8. M. Rosenbluth and C. Longmire, *Annals of Physics* **1,** 120 (1957).
9. Unpublished.

2

Properties of an Ionized Gas of Low Density in a Magnetic Field

Part IV*

S. Chandrasekhar, A. N. Kaufman, and K. M. Watson

Los Alamos Scientific Laboratory, Los Alamos, New Mexico, and University of California Radiation Laboratory, Livermore, California

A systematic method is described for solving the Boltzmann equation for the steady states of an ionized gas of low density in a strong magnetic field. On the assumption that the term representing the Lorentz force in the Boltzmann equation dominates all others, the solution is developed as a series in inverse powers of the gyration frequency ω. The solution is explicitly carried out to the first order in ω^{-1}. Expressions for the drifts which arise in the first order are also obtained.

1. INTRODUCTION

This paper is devoted to the steady state solutions of the Boltzmann equation for an ionized gas of low density in a strong magnetic field; specifically, it deals with the problem only briefly considered in Section 8 of paper III (*1*; an important correction to this paper will be found in Appendix III) of this series. While the method of solution to be described is derived, largely, from paper II (*2*), it has many similarities with the treatments given by Chew, Goldberger, and Low (*3*, *4*) in their papers.

2. THE OUTLINE OF THE METHOD

Consider a fully ionized gas of low density consisting of electrons and singly charged ions of masses m^- and m^+ and charges $-e$ and $+e$. In a stationary state we can describe the gas by giving the distribution functions

$$f^+(x_i, v_i) \quad \text{and} \quad f^-(x_i, v_i) \tag{1}$$

of the ions and electrons in the coordinate (x_i) and velocity (v_i) spaces. On the assumption that the collisions between the constituents can be ignored, the Boltzmann equation is

$$v_j \frac{\partial f}{\partial x_j} + A_j \frac{\partial f}{\partial v_j} + \frac{e}{mc} \epsilon_{ijk} \frac{\partial f}{\partial v_i} v_j B_k = 0, \tag{2}$$

* Work performed under the auspices of the U. S. Atomic Energy Commission.

where we have used the notation and the summation convention of tensor calculus. In writing Eq. (2), we have suppressed the superscripts + and − distinguishing the ions and the electrons; we shall always do this whenever an equation of the same form applies to both; but in all these cases a duplicate set of equations referring to the ions and electrons, separately, must be understood.

In Eq. (2), mA_j denotes an external force (an electric field and/or gravity) and B_i is the intensity of the magnetic field.

In seeking solutions of Eq. (2) we shall suppose that the term representing the Lorentz force dominates all others in the equation.

Let

$$\mathbf{B} = \bar{B}\lambda(x_i)\mathbf{n}, \tag{3}$$

where $\mathbf{n}$ is a unit vector in the direction of $\mathbf{B}$ and $\bar{B}\lambda(x_i)$ is the magnitude of $\mathbf{B}$ expressed as a product of a function $\lambda(x_i)$ (allowing for the variation of the field strength with position) and a certain constant field strength $\bar{B}$. We may take $\bar{B}$ to be an "average" field so that $\lambda(x_i)$ (which is dimensionless) is, numerically, of order unity.

We now write the Boltzmann equation in the form

$$\mathfrak{D}f + \omega\lambda(x_i)\mathfrak{L}f = 0, \tag{4}$$

where

$$\mathfrak{D} = v_j \frac{\partial}{\partial x_j} + A_j \frac{\partial}{\partial v_j} \quad \text{and} \quad \mathfrak{L} = \epsilon_{ijk} v_j n_k \frac{\partial}{\partial v_i}, \tag{5}$$

and

$$\omega = e\bar{B}/mc \tag{6}$$

is the gyration frequency appropriate for the average field $\bar{B}$.

An alternative expression for the operator $\mathfrak{L}$ is

$$\mathfrak{L} = -\frac{\partial}{\partial \varphi} \tag{7}$$

where φ is an azimuthal angle in the velocity space, normal to $\mathbf{n}$. In view of this, it is clear that if we average Eq. (4) over all angles φ (i.e., all directions in the velocity space normal to the direction of the prevailing field $\mathbf{B}$) we shall obtain

$$\langle \mathfrak{D}f \rangle = 0. \tag{8}$$

In Eq. (8) we have used the angular brackets to signify that the quantity enclosed has been averaged over all φ; in this paper the angular brackets will always be used with this meaning.

We may regard Eq. (8) as an *integrability condition* on the solutions.

As we have already stated, in the applications of the theory we have in view,

$$|\omega| \gg 1. \tag{9}$$

Accordingly, we shall seek solutions of Eq. (4) which are expressible as series in inverse powers of ω. Thus, assuming

$$f = \sum_{n=0}^{\infty} \omega^{-n} f_n , \tag{10}$$

we obtain from Eqs. (4) and (8) the set of equations

$$\begin{aligned} \mathfrak{L} f_0 &= 0 \\ \mathfrak{D} f_{n-1} + \lambda \mathfrak{L} f_n &= 0 \end{aligned} \qquad (n = 1, 2, \cdots) \tag{11}$$

and

$$\langle \mathfrak{D} f_n \rangle = 0 \qquad (n = 0, 1, 2, \cdots). \tag{12}$$

The zero-order approximation to f, namely, f_0, is determined by the equations

$$\mathfrak{L} f_0 = 0 \quad \text{and} \quad \langle \mathfrak{D} f_0 \rangle = 0. \tag{13}$$

The higher order corrections are then determined successively by the equations

$$-\lambda \mathfrak{L} f_n = \mathfrak{D} f_{n-1} \tag{14}$$

and

$$\langle \mathfrak{D} f_n \rangle = 0. \tag{15}$$

Making use of Eq. (12), we can rewrite Eq. (14) in the form

$$-\lambda \mathfrak{L} f_n = \mathfrak{D} f_{n-1} - \langle \mathfrak{D} f_{n-1} \rangle, \tag{16}$$

which makes it manifest that the average over φ of both sides of the equation vanishes.

We shall presently see that it is always possible to write down, by inspection, a particular integral of the equation

$$-\mathfrak{L} \phi = \mathfrak{D} f_{n-1} - \langle \mathfrak{D} f_{n-1} \rangle. \tag{17}$$

Denoting by ϕ_n the particular integral, we can write the general solution of Eq. (16) as

$$f_n = \frac{1}{\lambda} \phi_n + \psi_n, \tag{18}$$

where

$$\mathfrak{L} \psi_n = 0. \tag{19}$$

Next, substituting for f_n in accordance with Eq. (18) in the integrability condition (15), we obtain

$$\langle \mathfrak{D}\psi_n \rangle + \frac{1}{\lambda} \langle \mathfrak{D}\phi_n \rangle + \left\langle \phi_n \mathfrak{D}\left(\frac{1}{\lambda}\right) \right\rangle = 0, \tag{20}$$

an equation which makes ψ_n determinate.

It is clear that by following the procedure outlined, we shall, in principle, be able to obtain the solution to all orders in an entirely consistent manner.

3. THE SOLUTION IN THE CASE WHEN THERE ARE NO EXTERNAL FORCES

We shall illustrate the procedure described in Section 2 by explicitly carrying out the solution to the first order in the case when $A_j = 0$; the modifications in the solution when $A_j \neq 0$ will be considered in Section 4.

The general solution of the equation

$$\mathfrak{L} f_0 = \epsilon_{ijk} \frac{\partial f_0}{\partial v_i} v_j n_k = 0 \tag{21}$$

is clearly,

$$f_0 \equiv f_0(s^2, q, x_i), \tag{22}$$

where

$$s_i = v_i - q n_i \quad \text{and} \quad q = n_i v_i \tag{23}$$

are the components of the velocity perpendicular to **n** and parallel to **n**, respectively. [It may be noticed here that according to Eq. (19) the complimentary function in the solution for f_n has the same functional form as f_0.]

(A) Transformation to the Variables s_i and q

In view of the special role which the components of the velocity parallel and perpendicular to **n** play, it is convenient to express the distribution function f in terms of the variables s_i, q, and x_i:

$$f \equiv f(s_i, q, x_i). \tag{24}$$

Expressed in the same variables, the operators $\mathfrak{D}$ and $\mathfrak{L}$ are:

$$\mathfrak{D} = (s_j + q n_j) \left\{ \frac{\partial}{\partial x_j} - \left(n_i s_k \frac{\partial n_k}{\partial x_j} + q \frac{\partial n_i}{\partial x_j} \right) \frac{\partial}{\partial s_i} + s_k \frac{\partial n_k}{\partial x_j} \frac{\partial}{\partial q} \right\} \tag{25}$$

and

$$\mathfrak{L} = \epsilon_{ijk} s_j n_k \frac{\partial}{\partial s_i}. \tag{26}$$

When applied to a function

$$\psi \equiv \psi(s^2, q, x_i) \tag{27}$$

the operators $\mathfrak{D}$ and $\mathfrak{L}$ have the results:

$$\mathfrak{D}\psi = (s_j + qn_j)\left(\frac{\partial}{\partial x_j} + 2qs_i \frac{\partial n_i}{\partial x_j}\frac{\delta}{\delta \nu^2}\right)\psi, \tag{28}$$

and

$$\mathfrak{L}\psi = 0, \tag{29}$$

where in Eq. (28) we have introduced the operator

$$\frac{\delta}{\delta \nu^2} = \frac{1}{2q}\frac{\partial}{\partial q} - \frac{\partial}{\partial s^2}. \tag{30}$$

(B) The Solution in the Zero-Order Approximation

Making use of Eq. (28) we can write the intergrability condition on f_0 as:

$$\left\langle (s_j + qn_j)\left(\frac{\partial f_0}{\partial x_j} + 2qs_i \frac{\partial n_i}{\partial x_j}\frac{\delta f_0}{\delta \nu^2}\right)\right\rangle = 0. \tag{31}$$

Since

$$\langle s_i \rangle = 0 \quad \text{and} \quad \langle s_i s_j \rangle = \tfrac{1}{2}s^2(\delta_{ij} - n_i n_j), \tag{32}$$

Eq. (31) reduces to

$$n_j \frac{\partial f_0}{\partial x_j} + s^2 \frac{\partial n_j}{\partial x_j}\frac{\delta f_0}{\delta \nu^2} = 0. \tag{33}$$

The divergence of $\mathbf{n}$ which occurs in Eq. (33) is related to the gradient of B in the direction of $\mathbf{n}$. We have [see III, Eq. (83); also, Appendix II, Eq. $(A_2, 9)$]

$$\frac{\partial n_j}{\partial x_j} = -\frac{n_j}{B}\frac{\partial B}{\partial x_j} = -n_j \frac{\partial \log \lambda}{\partial x_j} = \frac{1}{D} \text{ (say)}. \tag{34}$$

Accordingly, we have

$$n_j \frac{\partial f_0}{\partial x_j} - s^2 \frac{n_j}{B}\frac{\partial B}{\partial x_j}\left(\frac{\partial f_0}{2q\partial q} - \frac{\partial f_0}{\partial s^2}\right) = 0. \tag{35}$$

The Lagrangian subsidiary equations associated with Eq. (35) are:

$$\frac{dx_j}{dt} = n_j; \qquad \frac{dq^2}{dt} = -\frac{s^2}{B} n_j \frac{\partial B}{\partial x_j} \quad \text{and} \quad \frac{ds^2}{dt} = \frac{s^2}{B} n_j \frac{\partial B}{\partial x_j}. \tag{36}$$

These equations clearly admit the integrals

$$q^2 + s^2 = 2H = \text{constant}; \qquad s^2/H\lambda = 2\mu = \text{constant} \tag{37}$$

and

$$\int \epsilon_{ijk} n_j \, dx_k = \text{constant vector.} \tag{38}$$

The integrals (37) express the conservation of the kinetic energy ($H = w_\parallel + w_\perp$) and the adiabatic invariance of $w_\perp/B$.

To make the meaning of the integral (38) clearer, we shall first rewrite it in the form

$$\int_{\mathbf{r}_0}^{\mathbf{r}} d\mathbf{r} \times \frac{\mathbf{B}}{B} = \text{constant vector.} \tag{39}$$

Under a variation of **r**, this integral remains constant only if **r** remains on the same line of force. Thus the integral represents a possible (nonunique) labelling of the lines of force. The fact that it is an integral of Eq. (35) shows that, to this order, a particle remains on the same line of force.

In terms of the integrals (37) and (39) of the equations of motion (36) we can write down the general solution for f_0. We have

$$f_0 \equiv f_0\left(H, \mu, \int_{\mathbf{r}_0}^{\mathbf{r}} d\mathbf{r} \times \frac{\mathbf{B}}{B}\right). \tag{40}$$

Eq. (40) expresses the maximum information we can obtain on f_0 from the Boltzmann equation.

In the natural system of coordinates described in Appendix II, the solution for f_0 can be written as

$$f_0 \equiv f_0(H, \mu, x_2, x_3). \tag{41}$$

The "simplicity" of (41) [in contrast to (40)] is more apparent than real: for the true meanings of x_2 and x_3 cannot be defined except via the integral (39).

It is important to observe that in writing the solution for f_0 in the form (40) or (41) we are not implying that f_0 is necessarily even in q. What can be said, however, about the dependence of f_0 on q is the following: Since the differential operator acting on f_0 in Eq. (35) is even in q, it is clear that if $f_0(+q)$ is a solution then so is $f_0(-q)$. Consequently the solutions of Eq. (35) can be classified into odd and even solutions; and any linear combination of them will, of course, also be a solution. The essential physical distinction here is one of the particles being either "trapped" (as in a "mirror machine") or "untrapped" (as in a linear pinch). In the former case we should expect f_0 to be even in q; but in the latter case any linear superposition of an odd and an even solution is allowed.

(C) THE SOLUTION FOR f_1

According to the method described in Section 2, the solution for f_1 proceeds in two steps. First, we seek a particular integral of the equation

$$-\epsilon_{ijk}\frac{\partial\phi}{\partial s_i}s_jn_k = \mathfrak{D}f_0 - \langle\mathfrak{D}f_0\rangle; \tag{42}$$

then determine the arbitrary function of s^2 and q which we can add to the particular integral by making use of the integrability condition (20).

From Eqs. (28) and (33) we find that the quantity on the right-hand side of Eq. (42) is

$$\begin{aligned}(s_j + qn_j)\left(\frac{\partial f_0}{\partial x_j} + 2qs_i\frac{\partial n_i}{\partial x_j}\frac{\delta f_0}{\delta v^2}\right) - q\left(n_j\frac{\partial f_0}{\partial x_j} + 2\langle s_is_j\rangle\frac{\partial n_i}{\partial x_j}\frac{\delta f_0}{\delta v^2}\right)\\ = s_j\left(\frac{\partial f_0}{\partial x_j} + 2q^2n_i\frac{\partial n_j}{\partial x_i}\frac{\delta f_0}{\delta v^2}\right) + 2q(s_is_j - \langle s_is_j\rangle)\frac{\partial n_i}{\partial x_j}\frac{\delta f_0}{\delta v^2}.\end{aligned} \tag{43}$$

In view of the symmetry of $(s_is_j - \langle s_is_j\rangle)$ in its indices, the term which occurs with this factor in (43) can also be written as

$$q\left(\frac{\partial n_i}{\partial x_j} + \frac{\partial n_j}{\partial x_i}\right)(s_is_j - \langle s_is_j\rangle). \tag{44}$$

Equation (42) is, therefore, expressible in the form

$$-\epsilon_{ijk}\frac{\partial\phi}{\partial s_i}s_jn_k = s_j\Phi_j + (s_is_j - \langle s_is_j\rangle)\Phi_{ij}, \tag{45}$$

where

$$\Phi_j = \frac{\partial f_0}{\partial x_j} + 2q^2n_i\frac{\partial n_j}{\partial x_i}\frac{\delta f_0}{\delta v^2} \tag{46}$$

and

$$\Phi_{ij} = q\frac{\delta f_0}{\delta v^2}\left(\frac{\partial n_i}{\partial x_j} + \frac{\partial n_j}{\partial x_i}\right). \tag{47}$$

Observe that Φ_{ij} is symmetrical in its indices.

Making use of the result given in Appendix I, we can readily write down the required particular integral of Eq. (45). We have

$$\phi_1 = \epsilon_{ijk}(\Phi_i + \tfrac{1}{2}\Phi_{il}s_l)s_jn_k\,. \tag{48}$$

The solution for f_1 is therefore given by

$$f_1 = \psi_1(s^2, q, x_i) + \frac{1}{\lambda}\epsilon_{ijk}\left(\Phi_i + \frac{1}{2}\Phi_{il}s_l\right)s_jn_k\,, \tag{49}$$

where ψ_1 is determined by

$$q\left(n_j \frac{\partial \psi_1}{\partial x_j} + \frac{s^2}{D}\frac{\delta \psi_1}{\delta \nu^2}\right) + \frac{1}{\lambda}\langle \mathfrak{D}\phi_1 \rangle + \left\langle \phi_1 \mathfrak{D}\left(\frac{1}{\lambda}\right)\right\rangle = 0. \tag{50}$$

The evaluation of $\langle \mathfrak{D}\phi_1 \rangle$ with ϕ_1 given by Eq. (48) is lengthy but straight forward. We find that the term in Φ_{il} makes no contribution and that

$$\langle \mathfrak{D}\phi_1 \rangle = \epsilon_{ijk}\left\{\frac{1}{2}s^2 \frac{\partial}{\partial x_j}(\Phi_i n_k) + n_l \frac{\partial n_j}{\partial x_l} n_k \left[\left(\frac{1}{2}s^2 - q^2\right)\Phi_i + q^2 s^2 \frac{\delta}{\delta \nu^2}\Phi_i\right]\right\}. \tag{51}$$

On substituting for Φ_i in accordance with Eq. (46) in the foregoing equation, we find after some further reductions that

$$\begin{aligned}\langle \mathfrak{D}\phi_1 \rangle = \epsilon_{ijk}\Bigg\{&\frac{\partial f_0}{\partial x_i}\left[\frac{1}{2}s^2 \frac{\partial n_k}{\partial x_j} + \left(\frac{1}{2}s^2 - q^2\right) n_l \frac{\partial n_j}{\partial x_l} n_k\right] \\ &+ q^2 s^2 \frac{\delta f_0}{\delta \nu^2}\frac{\partial}{\partial x_j}\left(n_l \frac{\partial n_i}{\partial x_l} n_k\right)\Bigg\}.\end{aligned} \tag{52}$$

Similarly, we find

$$\left\langle \phi_1 \mathfrak{D}\left(\frac{1}{\lambda}\right)\right\rangle = -\frac{1}{2}s^2 \epsilon_{ijk}\left[\frac{\partial f_0}{\partial x_i} + 2q^2 n_l \frac{\partial n_i}{\partial x_l}\frac{\delta f_0}{\delta \nu^2}\right]\left(\frac{1}{\lambda^2}\frac{\partial \lambda}{\partial x_j}\right) n_k\,. \tag{53}$$

Finally, combining Eqs. (50), (52), and (53), we find

$$\begin{aligned}&q\left(n_j \frac{\partial \psi_1}{\partial x_j} + \frac{s^2}{D}\frac{\partial \psi_1}{\delta \nu^2}\right) \\ &\quad + \frac{1}{\lambda}\epsilon_{ijk}\Bigg\{\frac{\partial f_0}{\partial x_i}\left[\frac{1}{2}s^2 \frac{\partial n_k}{\partial x_j} + \left(\frac{1}{2}s^2 - q^2\right) n_l \frac{\partial n_j}{\partial x_l} n_k - \frac{1}{2}s^2 \frac{\partial \log \lambda}{\partial x_j} n_k\right] \\ &\qquad + q^2 s^2 \frac{\delta f_0}{\delta \nu^2}\left[\frac{\partial}{\partial x_j}\left(n_l \frac{\partial n_i}{\partial x_l} n_k\right) - \left(n_l \frac{\partial n_i}{\partial x_l} n_k\right)\frac{\partial \log \lambda}{\partial x_j}\right]\Bigg\} = 0.\end{aligned} \tag{54}$$

An immediate consequence of this equation is that ψ_1 is odd in q *if f_0 is assumed to be even in q*; therefore, under these circumstances, drifts parallel to $\mathbf{n}$ occur only in the first approximation: they do not occur in the zeroth approximation.

(D) Discussion of Eq. (54)

Upon changing the velocity variables from s^2 and q to $H = \frac{1}{2}(q^2 + s^2)$ and $\mu = \frac{1}{2}s^2/H\lambda$, Eq. (54) simplifies somewhat to

$$\begin{aligned}&q\frac{\partial \psi_1}{\partial x_1} + \frac{1}{\lambda}\operatorname{grad} f_0 \cdot \mathbf{n} \times \left(q^2 \partial \mathbf{n}/\partial x_1 + \frac{1}{2}s^2 \operatorname{grad}\log \lambda\right) \\ &\qquad + \frac{\mu}{\lambda}\frac{\partial f_0}{\partial \mu}\left[q^2 \operatorname{div}(\mathbf{nn}\cdot \operatorname{curl}\mathbf{n}) + \frac{1}{2}s^2(\operatorname{div}\mathbf{n})(\mathbf{n}\cdot\operatorname{curl}\mathbf{n})\right] = 0.\end{aligned} \tag{55}$$

Note that in these variables, it follows from Eq. (35) or (41) that $\partial f_0/\partial x_1 = 0$. We identify the coefficient of grad f_0 as the guiding-center drift due to the field gradient and the curvature of the lines of force (*5*):

$$\mathbf{v}_d = \frac{1}{\omega\lambda}\,\mathbf{n} \times (q^2\partial\mathbf{n}/\partial x_1 + \tfrac{1}{2}\, s^2 \text{ grad} \log \lambda). \tag{56}$$

The coefficient of $\partial f_0/\partial\mu$ is also related to $\mathbf{v}_d$:

$$\nabla_\perp\cdot(\omega\lambda\mathbf{v}_d) = -[q^2 \text{ div } (\mathbf{n}\mathbf{n}\cdot\text{curl } \mathbf{n}) + \tfrac{1}{2}s^2(\text{div } \mathbf{n})(\mathbf{n}\cdot\text{curl } \mathbf{n})], \tag{57}$$

where

$$\nabla_\perp\cdot\mathbf{A} \equiv \text{div } \mathbf{A} - \mathbf{n}\cdot(\nabla\mathbf{A})\cdot\mathbf{n}. \tag{58}$$

Thus Eq. (55) can be written as

$$\frac{q}{\omega}\frac{\partial\psi_1}{\partial x_1} + \mathbf{v}_d\cdot\text{grad } f_0 - \frac{1}{\lambda}\nabla_\perp\cdot(\lambda\mathbf{v}_d)\mu\frac{\partial f_0}{\partial\mu} = 0. \tag{59}$$

(*a*) If the zero-order distribution is isotropic, so that $\partial f_0/\partial\mu = 0$, Eq. (59) reduces to

$$\frac{q}{\omega}\frac{\partial\psi_1}{\partial x_1} + \mathbf{v}_d\cdot\text{grad } f_0 = 0; \tag{60}$$

and if, in addition, f_0 does not vary in the direction of the drift, we obtain

$$\frac{\partial\psi_1}{\partial x_1} = 0. \tag{61}$$

We conclude that

$$\psi_1 = 0, \tag{62}$$

since the general solution of Eq. (61) can be incorporated into f_0. For an isotropic f_0, Eq. (46) reduces to

$$\Phi_j = \frac{\partial f_0}{\partial x_j} \tag{63}$$

and therefore the distribution function is, to first order,

$$f = f_0 + \frac{1}{\lambda\omega}\,\epsilon_{ijk}\,\frac{\partial f_0}{\partial x_i}\,s_j n_k\,. \tag{64}$$

(*b*) If the magnetic field is (approximately) that appropriate in a vacuum, then (see Appendix II)

$$\mathbf{n}\cdot\text{curl } \mathbf{n} = 0, \qquad \mathbf{n} \times \text{grad} \log \lambda = \mathbf{n} \times \partial\mathbf{n}/\partial x_1 = \rho_1\mathbf{e}_3\,.$$

Eq. (55) reduces to

$$q\frac{\partial\psi_1}{\partial x_1} + \rho_1\frac{q^2 + \frac{1}{2}s^2}{\lambda}\frac{\partial f_0}{\partial x_3} = 0. \tag{65}$$

If, as in (a), f_0 is again uniform in the direction of the drift

$$\mathbf{v}_d = \frac{1}{\omega\lambda}\left(q^2 + \frac{1}{2}s^2\right)\rho_1\mathbf{e}_3\,, \tag{66}$$

we again find $\psi_1 = 0$.

(c) Consider an axisymmetric field $\mathbf{B} = (B_r\,, 0, B_z)$, so that (see Appendix II)

$$\mathbf{n}\cdot\text{curl}\,\mathbf{n} = 0, \qquad \frac{\partial\mathbf{n}}{\partial x_1} = \rho_1\mathbf{e}_2\,.$$

Eq. (59) now reads

$$\frac{q}{\omega}\frac{\partial\psi_1}{\partial x_1} + \frac{v_d}{r}\frac{\partial f_0}{\partial\theta} = 0, \tag{67}$$

where

$$v_d = \frac{1}{\omega\lambda}\left(q^2\rho_1 + \frac{1}{2}s^2\frac{\partial\log\lambda}{\partial x_2}\right)$$

is in the θ-direction. Divide Eq. (67) by q and, for a trapped particle, integrate between the turning points. Assuming that $\psi_1 = 0$ at the end points, we obtain

$$\frac{\partial f_0}{\partial\theta}\int\frac{dx_1}{q}\frac{v_d}{r} = 0. \tag{68}$$

Now v_d/r is $\dot\theta$, the drift in θ, and the integral is proportioned to the time-average, $\bar{\dot\theta}$. It is physically evident that for a particle (type) for which $\bar{\dot\theta} \neq 0$, f_0 must be independent of θ in order to be time-independent.

(d) Consider a field such that a line of force ergodically generates a "magnetic surface", rather than closing on itself. Since $\partial f_0/\partial x_1 = 0$, f_0 can now be a function only of H, μ, and ϕ (by which we label the magnetic surfaces). Equation (59) now reads

$$\frac{q}{\omega}\frac{\partial\psi_1}{\partial x_1} + \dot\phi\frac{\partial f_0}{\partial\phi} - \frac{1}{\lambda}\nabla_\perp\cdot(\lambda\mathbf{v}_d)_\mu\frac{\partial f_0}{\partial\mu} = 0,$$

where

$$\dot\phi \equiv \mathbf{v}_d\cdot\text{grad}\,\phi.$$

Upon dividing by q, and integrating over the line of force (for an untrapped particle), we obtain

$$\bar{\dot{\phi}} \frac{\partial f_0}{\partial \phi} - \left\langle \frac{1}{\lambda} \nabla_\perp \cdot (\lambda \mathbf{v}_d) \right\rangle_{\text{Av}} \mu \frac{\partial f_0}{\partial \mu} = 0, \tag{69}$$

where

$$\langle A \rangle_{\text{Av}} = \bar{A} \equiv \int \frac{dx_1}{q} A$$

is proportional to the time-average experienced by a particle of given H, μ on ϕ. In order that f_0 may be an arbitrary function of ϕ and μ, we must require of the magnetic surfaces that

$$\bar{\dot{\phi}} = 0, \qquad \left\langle \frac{1}{\lambda} \nabla_\perp \cdot (\lambda \mathbf{v}_d) \right\rangle_{\text{Av}} = 0 \tag{70}$$

for all μ and ϕ. (There is no dependence on H; see below.) These requirements are evidently that a particle does not drift off the surface (on the average), and that its motion is, in some sense, divergence-free. In terms of the constants of the motion and the curvature coefficients, the requirements are

$$\int \frac{dx_1}{\lambda(1 - \mu\lambda)^{1/2}} \left\{ \frac{\partial \phi}{\partial x_3} (1 - \mu\lambda)\rho_1 + \frac{1}{2} \mu \left(\frac{\partial \phi}{\partial x_3} \frac{\partial \lambda}{\partial x_2} - \frac{\partial \phi}{\partial x_2} \frac{\partial \lambda}{\partial x_3} \right) \right\} = 0$$

and

$$\int \frac{dx_1}{\lambda(1 - \mu\lambda)^{1/2}} \left\{ \left(1 - \frac{1}{2}\mu\lambda\right) \frac{\tau_2 + \tau_3}{D} + (1 - \mu\lambda) \frac{\partial}{\partial x_1} (\tau_2 + \tau_3) \right\} = 0.$$

(E) The First-Order Drift

The entire contribution to the mean velocity (the drift) normal to **B**, in the first order, is made by the particular integral in the solution for f_1. This drift, **V**, is given by

$$\begin{aligned} N^0 V_p &= \frac{1}{2\lambda\omega} \iint \left\langle (s_p + q n_p) \epsilon_{ijk} \left(\Phi_i + \frac{1}{2} \Phi_{il} s_l \right) s_j n_k \right\rangle dq\, ds^2 \\ &= \frac{1}{4\lambda\omega} \iint \epsilon_{ijk} s^2 \left[\Phi_i (\delta_{jp} - n_j n_p) + \frac{1}{2} q n_p \Phi_{il} (\delta_{lj} - n_l n_j) \right] n_k\, dq\, ds^2 \qquad (71)^1 \\ &= \frac{1}{4\lambda\omega} \iint \epsilon_{ipk} \Phi_i n_k s^2\, dq\, ds^2. \end{aligned}$$

[1] We are assuming that f_0 is so normalized that, for example, $N^0 = \frac{1}{2} \int \int f_0\, dq\, ds^2$.

On substituting for Φ_i in accordance with Eq. (46) we have

$$N^0 V_p = -\frac{1}{4\lambda\omega}\iint \epsilon_{pik}\left(\frac{\partial f_0}{\partial x_i} + 2q^2 n_l \frac{\partial n_i}{\partial x_l}\frac{\delta f_0}{\delta v^2}\right) n_k s^2\, dq\, ds^2$$
$$= -\frac{1}{m\lambda\omega}\epsilon_{pik}\left\{\frac{\partial p_\perp{}^0}{\partial x_i} + (p_\parallel{}^0 - p_\perp{}^0) n_l \frac{\partial n_i}{\partial x_l}\right\} n_k\,. \tag{72}$$

In the notation introduced in III, Eqs. (85)–(88), the drift given by the foregoing formula is:

$$N^0 \mathbf{V} = \frac{1}{m\lambda\omega}\mathbf{n} \times (\nabla \mathbf{p}^0)_\perp\,; \tag{73}$$

this is the usual pressure-drift.

In addition to the transverse drift given by Eq. (73) there will, in general, be also a drift, $\bar{q}$, parallel to $\mathbf{n}$ given by the first moment of ψ_1; this is finite since, as we have seen, ψ_1 is an odd function of q. An equation relating $\bar{q}$ to the other macroscopic quantities can be obtained by integrating Eq. (55) over all velocity space. We find:

$$\frac{\partial \bar{q}}{\partial x_1} + \frac{\bar{q}}{D} + \frac{1}{\lambda}\left\{-(\tau_2+\tau_3)\frac{\partial p_\perp{}^0}{\partial x_1} + \left(\rho_1 \frac{\partial p_\parallel{}^0}{\partial x_3} + \frac{\partial \log\lambda}{\partial x_2}\frac{\partial p_\perp{}^0}{\partial x_3}\right)\right.$$
$$\left. - \frac{\partial \log\lambda}{\partial x_2}\frac{\partial p_\perp{}^0}{\partial x_3} + (p_\parallel{}^0 - p^0{}_\perp)\left(\frac{\partial \rho_1}{\partial x_3} - \rho_1\sigma_2 - \rho_1 \frac{\partial \log\lambda}{\partial x_3}\right)\right\} = 0. \tag{74}$$

There are no additional contributions to the pressure tensor in the first approximation.

4. THE SOLUTION TO THE FIRST APPROXIMATION IN THE CASE WHEN A FIRST-ORDER EXTERNAL FORCE IS PRESENT

With some more algebra the analysis of the preceding section can be generalized to include the term in $\mathbf{A}$ in $\mathfrak{D}$. We shall briefly indicate the necessary generalizations.

First, we observe that the inclusion of the term $A_j\partial/\partial v_j$ in the Boltzmann operator does not affect the solution (22). However, it modifies the operator $\mathfrak{D}$ and therefore all the subsequent analysis. Thus, instead of Eq. (25) we now have

$$\mathfrak{D} = (s_j + qn_j)\left\{\frac{\partial}{\partial x_j} - \left(n_i s_k \frac{\partial n_k}{\partial x_j} + q\frac{\partial n_i}{\partial x_j}\right)\frac{\partial}{\partial s_i} + s_k \frac{\partial n_k}{\partial x_j}\frac{\partial}{\partial q}\right\}$$
$$+ A_j\left\{(\delta_{jk} - n_j n_k)\frac{\partial}{\partial s_k} + n_j \frac{\partial}{\partial q}\right\}. \tag{75}$$

The expression for $\mathfrak{L}$ is unchanged.

When applied to a function of s^2, q, and x_i [such as ψ defined in Eq. (27)] the operator $\mathfrak{D}$ has the effect

$$\mathfrak{D}\psi = (s_j + qn_j)\left(\frac{\partial}{\partial x_j} + 2qs_i \frac{\partial n_i}{\partial x_j}\frac{\delta}{\delta v^2}\right)\psi + A_j\left(2s_j \frac{\partial}{\partial s^2} + n_j \frac{\partial}{\partial q}\right)\psi. \quad (76)$$

The additional term in $\mathfrak{D}$ modifies the integrability condition on f_0. Thus, instead of (35) we now have

$$n_j \frac{\partial f_0}{\partial x_j} - s^2 \frac{n_j}{B}\frac{\partial B}{\partial x_j}\left(\frac{\partial f_0}{\partial q^2} - \frac{\partial f_0}{\partial s^2}\right) + 2A_j n_j \frac{\partial f_0}{\partial q^2} = 0. \quad (77)$$

If A_j should be derivable from a potential χ, or more generally of the form

$$A_i = -\frac{\partial \chi}{\partial x_i} + \epsilon_{ijk}F_j n_k , \quad (78)$$

then the Lagrangian subsidiary equations associated with Eq. (77) again admit three integrals; they are:

$$\left.\begin{aligned} &s^2 + q^2 + 2\chi = \text{constant}; \qquad s^2/B = \text{constant} \\ &\text{and} \\ &\int_{\mathbf{r}_0}^{\mathbf{r}} d\mathbf{r} \times \mathbf{B}/B = \text{constant vector.} \end{aligned}\right\} \quad (79)$$

Thus, in place of Eq. (41) we shall now have

$$f_0 \equiv f_0(s^2 + q^2 + 2\chi, s^2/B, x_2, x_3). \quad (80)$$

Proceeding to the solution of the next term in the expansion of f, we find that the equation for ϕ [Eq. (42)] can be reduced to the same form as Eq. (45) with a redefinition only of Φ_j. The definition which is now valid is

$$\Phi_j = \frac{\partial f_0}{\partial x_j} + 2q^2 n_i \frac{\partial n_j}{\partial x_i}\frac{\delta f_0}{\delta v^2} + 2A_j \frac{\partial f_0}{\partial s^2}. \quad (81)$$

With this definition of Φ_j the solution expressed by Eqs. (48) and (49) continue to hold. Similarly, Eq. (51) holds; however, the additional term in Φ_j results in the addition of the terms [cf. Eq. (51)]

$$\epsilon_{ijk}\left\{s^2 \frac{\partial}{\partial x_j}\left(A_i \frac{\partial f_0}{\partial s^2} n_k\right) + 2n_l \frac{\partial n_j}{\partial x_l} n_k \left[\left(\frac{1}{2}s^2 - q^2\right)\frac{\partial f_0}{\partial s^2} + q^2 s^2 \frac{\delta}{\delta v^2}\left(\frac{\partial f_0}{\partial s^2}\right)\right]\right\}, \quad (82)$$

to the right-hand side of Eq. (52). Similarly, the additional terms on the right-hand side of Eq. (53) are

$$- s^2 \frac{\partial f_0}{\partial s^2}\epsilon_{ijk}A_i\left(\frac{1}{\lambda^2}\frac{\partial \lambda}{\partial x_j}\right) n_k . \quad (83)$$

With these new terms, Eq. (54) becomes

$$q\left(n_j \frac{\partial \psi_1}{\partial x_j} + \frac{s^2}{D}\frac{\delta \psi_1}{\delta \nu^2} + 2A_j n_j \frac{\partial \psi_1}{\partial q^2}\right)$$
$$+ \frac{1}{\lambda}\epsilon_{ijk}\left\{\left(\frac{\partial f_0}{\partial x_i} + 2A_i \frac{\partial f_0}{\partial s^2}\right)\left[\left(\frac{1}{2}s^2 - q^2\right) n_l \frac{\partial n_j}{\partial x_l} n_k\right.\right.$$
$$\left.\left. + \frac{1}{2}s^2\left(\frac{\partial n_k}{\partial x_j} - \frac{\partial \log \lambda}{\partial x_j} n_k\right)\right] + q^2 s^2 \frac{\delta f_0}{\delta \nu^2}\left[\frac{\partial}{\partial x_j}\left(n_l \frac{\partial n_i}{\partial x_j} n_k\right)\right. \tag{84}$$
$$\left. - \left(n_l \frac{\partial n_i}{\partial x_l} n_k\right)\frac{\partial \log \lambda}{\partial x_j}\right] + 2q^2 s^2 \left(\frac{\delta}{\delta \nu^2}\frac{\partial f_0}{\partial s^2}\right) A_i n_l \frac{\partial n_j}{\partial x_l} n_k$$
$$\left. + s^2 A_i \left(\frac{\partial}{\partial x_j}\frac{\partial f_0}{\partial s^2}\right) n_k + s^2 \frac{\partial f_0}{\partial s^2}\frac{\partial A_i}{\partial x_j} n_k\right\} = 0.$$

If A_i is derivable from a potential, the last term in this equation in $\partial A_i/\partial x_j$ clearly vanishes.

In the system of coordinates described in Appendix II, Eq. (84) becomes

$$q\left(\frac{\partial \psi_1}{\partial x_1} + \frac{s^2}{D}\frac{\delta \psi_1}{\delta \nu^2} + 2A_1 \frac{\partial \psi_1}{\partial q^2}\right) + \frac{1}{\lambda}\left\{-\frac{1}{2}s^2(\tau_2 + \tau_3)\left(\frac{\partial f_0}{\partial x_1} + 2A_1 \frac{\partial f_0}{\partial s^2}\right)\right.$$
$$- \frac{1}{2}s^2 \frac{\partial \log \lambda}{\partial x_3}\left(\frac{\partial f_0}{\partial x_2} + 2A_2 \frac{\partial f_0}{\partial s^2}\right)$$
$$+ \left(\rho_1 q^2 + \frac{1}{2}s^2 \frac{\partial \log \lambda}{\partial x_2}\right)\left(\frac{\partial f_0}{\partial x_3} + 2A_3 \frac{\partial f_0}{\partial s^2}\right) \tag{85}$$
$$+ q^2 s^2\left[\frac{\delta f_0}{\delta \nu^2}\left(\frac{\partial \rho_1}{\partial x_3} - \rho_1\sigma_2 - \rho_1 \frac{\partial \log \lambda}{\partial x_3}\right) - 2A_3\rho_1 \frac{\delta}{\delta \nu^2}\left(\frac{\partial f_0}{\partial s^2}\right)\right]$$
$$\left. + s^2\left(A_2 \frac{\partial}{\partial x_3} - A_3 \frac{\partial}{\partial x_2}\right)\frac{\partial f_0}{\partial s^2} + s^2 \frac{\partial f_0}{\partial s^2}\mathbf{e}_1 \cdot \operatorname{curl} \mathbf{A}\right\} = 0.$$

In the presence of an external force there is, as one may directly verify, an additional transverse drift of amount $(-\mathbf{n} \times \mathbf{A})/(\lambda\omega)$.

5. THE SOLUTION FOR THE CASE WHEN THERE ARE ZERO-ORDER TRANSVERSE DRIFTS

In solving the Boltzmann equation in the manner described in the preceeding sections, we have supposed that the term in the Lorentz force dominates all others in the equation including that representing the external force (**A**). There are, however, circumstances when this need not be the case: they arise when forces are present which cause large drifts normal to the direction of **B**. Thus when an electric field, $\mathbf{E}$, is present there will be a drift normal to $\mathbf{B}$ given by

$$\text{electric drift} = c\,\frac{\mathbf{E}\times\mathbf{B}}{B^2}. \tag{86}$$

Suppose then that an electric field, $\mathbf{E}$, in addition to an external force $m\mathbf{X}$ is present. The Boltzmann equation we have to consider then is

$$v_j\,\frac{\partial f}{\partial x_j} + \left(X_i + \frac{e}{m}\,E_i + \frac{e}{mc}\,\epsilon_{ijk}v_jB_k\right)\frac{\partial f}{\partial v_i} = 0. \tag{87}$$

Resolving $\mathbf{E}$ and $\mathbf{X}$ parallel and perpendicular to $\mathbf{B}$ and defining the drift

$$\boldsymbol{\alpha} = c\,\frac{\mathbf{E}\times\mathbf{B}}{B^2} + \frac{mc}{e}\,\frac{\mathbf{X}\times\mathbf{B}}{B^2}, \tag{88}$$

we can rewrite Eq. (87) in the manner

$$v_j\,\frac{\partial f}{\partial x_j} + \left[An_i + \frac{e}{mc}\,\epsilon_{ijk}(v_j - \alpha_j)B_k\right]\frac{\partial f}{\partial v_i} = 0, \tag{89}$$

where

$$A = X_{\parallel} + \frac{e}{m}\,E_{\parallel}. \tag{90}$$

In terms of the gyration frequency ω and the function $\lambda(x_i)$ [see Eq. (3)], Eq. (90) is

$$v_j\,\frac{\partial f}{\partial x_j} + [An_i + \omega\lambda\epsilon_{ijk}(v_j - \alpha_j)n_k]\,\frac{\partial f}{\partial v_i} = 0. \tag{91}$$

Let

$$\mathbf{s} = \mathbf{v} - q\mathbf{n} - \boldsymbol{\alpha} \quad \text{and} \quad q = (\mathbf{v} - \boldsymbol{\alpha})\cdot\mathbf{n} = \mathbf{v}\cdot\mathbf{n}; \tag{92}$$

$\mathbf{s}$ denotes, then, the transverse velocity in excess of the drift $\boldsymbol{\alpha}$. Also, express the distribution function f in terms of the variables s_i, q, and x_i:

$$f(x_i, v_i) \equiv f(s_i, q, x_i). \tag{93}$$

The Boltzmann equation then takes the form

$$\mathfrak{D}f + \lambda\omega\mathfrak{L}f = 0, \tag{94}$$

where

$$\begin{aligned}\mathfrak{D} = (s_j + qn_j + \alpha_j)&\left\{\frac{\partial}{\partial x_j} + (s_k + \alpha_k)\,\frac{\partial n_k}{\partial x_j}\,\frac{\partial}{\partial q}\right.\\ &\left.- \left[n_i(s_k + \alpha_k)\,\frac{\partial n_k}{\partial x_j} + q\,\frac{\partial n_i}{\partial x_j} + \frac{\partial \alpha_i}{\partial x_j}\right]\frac{\partial}{\partial s_i}\right\} + A\,\frac{\partial}{\partial q}\end{aligned} \tag{95}$$

and

$$\mathfrak{L} = \epsilon_{ijk} s_j n_k \frac{\partial}{\partial s_i}. \tag{96}$$

Once the Boltzmann equation has been reduced to the form (94), its solution in a series in inverse powers of ω can be accomplished in the manner described in Section 2.

(A) THE ZERO-ORDER APPROXIMATION

The solution for f_0 in the zeroth approximation can be readily obtained. The solution of the equation $\mathfrak{L} f_0 = 0$ again leads to

$$f_0 \equiv f_0(s^2, q, x_i). \tag{97}$$

For f_0 of this form

$$\begin{aligned}\mathfrak{D} f_0 = (s_j + q n_j + \alpha_j) \left\{ \frac{\partial f_0}{\partial x_j} + (s_k + \alpha_k) \frac{\partial n_k}{\partial x_j} \frac{\partial f_0}{\partial q} \right. \\ \left. - 2 s_i \left(q \frac{\partial n_i}{\partial x_j} + \frac{\partial \alpha_i}{\partial x_j} \right) \frac{\partial f_0}{\partial s^2} \right\} + A \frac{\partial f_0}{\partial q} \\ = s_i s_j \Phi_{ij} + s_j \Phi_j + \Phi_0 \text{ (say)},\end{aligned} \tag{98}$$

where

$$\Phi_{ij} = q \left(\frac{\partial n_i}{\partial x_j} + \frac{\partial n_j}{\partial x_i} \right) \frac{\delta f_0}{\delta \nu^2} - \left(\frac{\partial \alpha_i}{\partial x_j} + \frac{\partial \alpha_j}{\partial x_i} \right) \frac{\partial f_0}{\partial s^2}, \tag{99}$$

$$\Phi_j = \frac{\partial f_0}{\partial x_j} + 2(q n_i + \alpha_i) \left(q \frac{\partial n_j}{\partial x_i} \frac{\delta f_0}{\delta \nu^2} - \frac{\partial \alpha_j}{\partial x_i} \frac{\partial f_0}{\partial s^2} \right) + \alpha_i \frac{\partial n_i}{\partial x_j} \frac{\partial f_0}{\partial q}, \tag{100}$$

$$\Phi_0 = (q n_j + \alpha_j) \left(\frac{\partial f_0}{\partial x_j} + \alpha_i \frac{\partial n_i}{\partial x_j} \frac{\partial f_0}{\partial q} \right) + A \frac{\partial f_0}{\partial q}. \tag{101}$$

The integrability condition on f_0 is then

$$\langle \mathfrak{D} f_0 \rangle = \tfrac{1}{2} s^2 (\delta_{ij} - n_i n_j) \Phi_{ij} + \Phi_0 = 0. \tag{102}$$

Making use of the definition of Φ_{ij}, we find

$$\begin{aligned}(\delta_{ij} - n_i n_j) \Phi_{ij} &= \Phi_{jj} - n_i n_j \Phi_{ij} \\ &= \frac{2q}{D} \frac{\delta f_0}{\delta \nu^2} - 2(\nabla_\perp \cdot \boldsymbol{\alpha}) \frac{\partial f_0}{\partial s^2},\end{aligned} \tag{103}$$

where [see III, Eq. (99)]

$$\nabla_\perp \cdot \boldsymbol{\alpha} = \frac{\partial \alpha_j}{\partial x_j} - n_i \frac{\partial \alpha_j}{\partial x_i} n_j, \tag{104}$$

and D has the same meaning as in Eq. (34). Therefore the equation for f_0 is

$$q\,\frac{s^2}{D}\,\frac{\delta f_0}{\delta v^2} - s^2(\nabla_\perp\cdot\boldsymbol{\alpha})\,\frac{\partial f_0}{\partial s^2} + A\,\frac{\partial f_0}{\partial q} + (qn_j + \alpha_j)\left(\frac{\partial f_0}{\partial x_j} + \alpha_i\,\frac{\partial n_i}{\partial x_j}\,\frac{\partial f_0}{\partial q}\right) = 0. \tag{105}$$

Rearranging this equation, we have

$$\begin{aligned} q\left\{n_j\,\frac{\partial f_0}{\partial x_j} + \frac{s^2}{D}\left(\frac{\partial f_0}{\partial q^2} - \frac{\partial f_0}{\partial s^2}\right) + 2\left(A + \alpha_i\,\frac{\partial n_i}{\partial x_j}\,\alpha_j\right)\frac{\partial f_0}{\partial q^2}\right\} \\ - s^2(\nabla_\perp\cdot\boldsymbol{\alpha})\,\frac{\partial f_0}{\partial s^2} + \alpha_j\,\frac{\partial f_0}{\partial x_j} + 2\alpha_i\,\frac{\partial n_i}{\partial x_j}\,n_j q^2\,\frac{\partial f_0}{\partial q^2} = 0. \end{aligned} \tag{106}$$

This is the general equation which f_0 must satisfy.

Comparing Eq. (106) with the equations for f_0 obtained in Sections 3 and 4 [Eqs. (35) and (77)], we observe one important difference: In Eq. (106) the differential operator acting on f_0 is not even in q as it was in the other cases. Therefore, Eq. (106) will not in general allow us to express f_0 as a superposition of an odd and an even solution [see the remarks following Eq. (41)].

(B) The Case When $\boldsymbol{\alpha}$ Is a Pure Electric Drift

Equation (106) simplifies, somewhat, in case $\boldsymbol{\alpha}$ is purely an electric drift. In this case, Maxwell's equation

$$\frac{\partial \mathbf{B}}{\partial t} = \operatorname{curl}\,(\boldsymbol{\alpha}\times\mathbf{B}) \tag{107}$$

can be split into an equation for the absolute value (B) of $\mathbf{B}$ and an equation for the direction ($\mathbf{n}$) of $\mathbf{B}$. Thus

$$\frac{\partial B}{\partial t} + \alpha_j\,\frac{\partial B}{\partial x_j} = -\,B(\nabla_\perp\cdot\boldsymbol{\alpha}) \tag{108}$$

and

$$\frac{\partial n_i}{\partial t} = n_j\,\frac{\partial \alpha_i}{\partial x_j} - \alpha_j\,\frac{\partial n_i}{\partial x_j} - n_i n_j n_k\,\frac{\partial \alpha_k}{\partial x_j}. \tag{109}$$

Multiplying Eq. (109) by α_i we obtain

$$\alpha_i\,\frac{\partial n_i}{\partial t} = \frac{1}{2}\,n_j\,\frac{\partial}{\partial x_j}\,\alpha^2 - \alpha_i\,\frac{\partial n_i}{\partial x_j}\,\alpha_j. \tag{110}$$

In a stationary state, Eqs. (108) and (110) give

$$(\nabla_\perp\cdot\boldsymbol{\alpha}) = -\frac{\alpha_j}{B}\,\frac{\partial B}{\partial x_j} = -\alpha_j\,\frac{\partial \log \lambda}{\partial x_j} \tag{111}$$

and

$$\alpha_i \frac{\partial n_i}{\partial x_j} \alpha_j = \frac{1}{2} n_j \frac{\partial}{\partial x_j} \alpha^2. \tag{112}$$

Substituting from Eqs. (111) and (112) in Eq. (106) we obtain

$$\begin{aligned} q \left\{ n_j \frac{\partial f_0}{\partial x_j} + \left(\frac{s^2}{D} + n_j \frac{\partial}{\partial x_j} \alpha^2 + 2A \right) \frac{\partial f_0}{\partial q^2} - \frac{s^2}{D} \frac{\partial f_0}{\partial s^2} \right\} \\ + \alpha_j \left(\frac{\partial f_0}{\partial x_j} + 2q^2 \frac{\partial f_0}{\partial q^2} n_i \frac{\partial n_j}{\partial x_i} + s^2 \frac{\partial f_0}{\partial s^2} \frac{\partial \log \lambda}{\partial x_j} \right) = 0. \end{aligned} \tag{113}$$

If there is no drift in the direction $\mathbf{e}_2$ and none of the quantities vary in the direction $\mathbf{e}_3$ the terms in Eq. (113) which are contracted with respect to α_j vanish and we will be left with

$$n_j \frac{\partial f_0}{\partial x_j} + \left(\frac{s^2}{D} + n_j \frac{\partial}{\partial x_j} \alpha^2 + 2A \right) \frac{\partial f_0}{\partial q^2} - \frac{s^2}{D} \frac{\partial f_0}{\partial s^2} = 0. \tag{114}$$

We are led to this same equation for f_0 if we *assume* that f_0 is even in q. In this case, we must *require* that

$$\alpha_j \left(\frac{\partial f_0}{\partial x_j} + 2q^2 \frac{\partial f_0}{\partial q^2} n_i \frac{\partial n_j}{\partial x_i} + s^2 \frac{\partial f_0}{\partial s^2} \frac{\partial \log \lambda}{\partial x_j} \right) = 0; \tag{115}$$

and we can *deduce* from this equation that

$$\partial f_0 / \partial x_3 = 0, \tag{116}$$

if there is no drift in the $\mathbf{e}_2$-direction and none of the *field* quantities vary in the $\mathbf{e}_3$-direction.

If the external forces (besides $\mathbf{E}$) included in $\mathbf{A}$ should be derivable from a potential (χ) then Eq. (114) allows the general solution [see Eq. (80)]

$$f_0 \equiv f_0(s^2 + q^2 - \alpha^2 + 2\chi, s^2/B, x_2, x_3). \tag{117}$$

Equation (117) implies that

$$H = \tfrac{1}{2}(s^2 + q^2 - \alpha^2) + \chi \tag{118}$$

is an integral of the equations of motion under the conditions considered. It is not difficult to verify the existence of this integral directly: the term $-\tfrac{1}{2}\alpha^2$ is in fact the centrifugal potential energy (per unit mass).

(C) THE PARTICULAR INTEGRAL FOR THE SOLUTION IN THE FIRST APPROXIMATION

We shall not complete the solution in the first approximation since the various formulae become excessively complicated. However, the particular integral which will suffice to determine the additional transverse drifts can be readily written down.

According to Eq. (16) the required particular integral is a solution of

$$-\lambda \mathfrak{L} f_1 = \Phi_{ij}(s_i s_j - \langle s_i s_j \rangle) + s_j \Phi_j ; \tag{119}$$

and making use of the result given in Appendix I, we have

$$f_1 = \frac{1}{\lambda} \epsilon_{ijk} \left(\Phi_i + \frac{1}{2} \Phi_{il} s_l \right) s_j n_k , \tag{120}$$

where Φ_i and Φ_{il} are defined as in Eqs. (99) and (100).

From the solution (120) we find that the transverse drift arising from the first-order correction to the distribution function is

$$\frac{\mathbf{n}}{\lambda \omega} \times \left[\frac{1}{N^0 m} (\nabla \mathbf{p}^0)_\perp + (\boldsymbol{\alpha} \cdot \operatorname{grad}) \boldsymbol{\alpha} \right], \tag{121}$$

where $(\nabla \mathbf{p}^0)_\perp$ has the same meaning as in Eq. (73). This result can also be obtained from the equation of motion:

$$Nm(\mathbf{V} \cdot \operatorname{grad})\mathbf{V} = - \operatorname{div} \mathbf{p} + N \left(m\mathbf{A} + \frac{e}{c} \mathbf{V} \times \mathbf{B} \right)$$

by solving for $\mathbf{V}$ to first order.

APPENDIX I

In developments relating to the solution of the Boltzmann equation in plasma physics, one often requires to find a particular integral of an equation of the form

$$- \epsilon_{ijk} \frac{\partial \phi}{\partial s_i} s_j n_k = \Phi_i s_i + \Phi_{ij}(s_i s_j - \langle s_i s_j \rangle) + \Phi_{ijk} s_i s_j s_k + \Phi_{ijkl}(s_i s_j s_k s_l - \langle s_i s_j s_k s_l \rangle), \tag{A$_1$, 1}$$

where ϕ is a function of $\mathbf{s} = (s_1, s_2, s_3)$; $\mathbf{n}$ is a unit vector; $\mathbf{s} \cdot \mathbf{n} = 0$; Φ_i a vector independent of the direction of $\mathbf{s}$; Φ_{ij}, Φ_{ijk}, and Φ_{ijkl} are constant tensors which are symmetric in their indices; and

$$\langle s_i s_j \rangle = \tfrac{1}{2} s^2 (\delta_{ij} - n_i n_j), \tag{A_1, 2}$$

$$\begin{aligned} \langle s_i s_j s_k s_l \rangle = \tfrac{1}{8} s^4 [\delta_{ij}\delta_{kl} &+ \delta_{ik}\delta_{jl} + \delta_{il}\delta_{jk} \\ &- (\delta_{ij} n_k n_l + \delta_{ik} n_j n_l + \delta_{il} n_j n_k \\ &+ n_i n_j \delta_{kl} + n_i n_k \delta_{jl} + n_i n_l \delta_{jk}) \\ &+ 3 n_i n_j n_k n_l], \end{aligned} \tag{A_1, 3}$$

are the averages of the quantities enclosed in the angular brackets over all directions transverse to **n**.

One may verify by direct substitution that a particular integral of Eq. (A_1, 1) is given by

$$\begin{aligned} \phi = \epsilon_{ijk}[\Phi_i + \tfrac{1}{2}\Phi_{il} s_l + \tfrac{1}{3}\Phi_{ilm}(s_l s_m + 4\langle s_l s_m \rangle) \\ + \tfrac{1}{4}\Phi_{ilmp} s_l (s_m s_p + 3\langle s_m s_p \rangle)] s_j n_k . \end{aligned} \tag{A_1, 4}$$

The general solution is obtained by adding an arbitrary function of s^2 and $s_i n_i$ to the foregoing.

An identity useful in the analysis may be noted here:

$$\begin{aligned} \epsilon_{\alpha\beta\gamma}\epsilon_{abc} = \delta_{\alpha a}\delta_{\beta b}\delta_{\gamma c} &- \delta_{\alpha a}\delta_{\beta c}\delta_{\gamma b} \\ + \delta_{\alpha b}\delta_{\beta c}\delta_{\gamma a} &- \delta_{\alpha b}\delta_{\beta a}\delta_{\gamma c} \\ + \delta_{\alpha c}\delta_{\beta a}\delta_{\gamma b} &- \delta_{\alpha c}\delta_{\beta b}\delta_{\gamma a} . \end{aligned} \tag{A_1, 5}$$

APPENDIX II

In this appendix we shall describe a coordinate system which is often quite useful in interpreting the formulas of plasma physics. (The coordinate system was chosen in paper I.)

Let **B** denote the intensity of the magnetic field; and $\mathbf{e}_1$, $\mathbf{e}_2$, and $\mathbf{e}_3$ be three mutually perpendicular unit vectors chosen at each point in such a way that $\mathbf{e}_1$ is along the direction of **B** (it is, therefore, the same as **n**) and $\mathbf{e}_2$ is along the principal normal to the line of force. Further, let dx_1, dx_2, and dx_3 denote elements of arcs along the directions of $\mathbf{e}_1$, $\mathbf{e}_2$, and $\mathbf{e}_3$, respectively.

The results of differentiating the vectors $\mathbf{e}_1$, $\mathbf{e}_2$, and $\mathbf{e}_3$ along the x_1-direction (i.e., along the line of force) are given by the Serret-Frenet formulas:

$$\begin{aligned} \frac{\partial \mathbf{e}_1}{\partial x_1} &= \rho_1 \mathbf{e}_2 , \\ \frac{\partial \mathbf{e}_2}{\partial x_1} &= -(\tau_1 \mathbf{e}_3 + \rho_1 \mathbf{e}_1) \\ \frac{\partial \mathbf{e}_3}{\partial x_1} &= \tau_1 \mathbf{e}_2 , \end{aligned} \tag{A_2, 1}$$

where ρ_1 is the reciprocal of the principal radius of curvature and τ_1 is the *torsion*. In addition to the foregoing formulas we shall also need the results of differentiating the unit vectors along the x_2- and the x_3-directions. Compatible with the orthogonality $\mathbf{e}_i \cdot \mathbf{e}_j = \delta_{ij}$ we can write these further derivatives in the forms

$$\begin{aligned} \frac{\partial \mathbf{e}_1}{\partial x_2} &= \rho_2 \mathbf{e}_2 - \tau_2 \mathbf{e}_3 \,; & \frac{\partial \mathbf{e}_1}{\partial x_3} &= \tau_3 \mathbf{e}_2 + \rho_3 \mathbf{e}_3 \,, \\ \frac{\partial \mathbf{e}_2}{\partial x_2} &= \sigma_2 \mathbf{e}_3 - \rho_2 \mathbf{e}_1 \,; & \frac{\partial \mathbf{e}_2}{\partial x_3} &= \sigma_3 \mathbf{e}_3 - \tau_3 \mathbf{e}_1 \,, \\ \frac{\partial \mathbf{e}_3}{\partial x_2} &= \tau_2 \mathbf{e}_1 - \sigma_2 \mathbf{e}_2 \,; & \frac{\partial \mathbf{e}_3}{\partial x_3} &= -\rho_3 \mathbf{e}_1 - \sigma_3 \mathbf{e}_2 \,, \end{aligned} \tag{A$_2$, 2}$$

where ρ_2, ρ_3, τ_2, τ_3, σ_2 and σ_3 are further curvature coefficients.

The gradient operator in the chosen coordinate system is

$$\nabla = \mathbf{e}_1 \frac{\partial}{\partial x_1} + \mathbf{e}_2 \frac{\partial}{\partial x_2} + \mathbf{e}_3 \frac{\partial}{\partial x_3}. \tag{A$_2$, 3}$$

Since x_1, x_2, and x_3 do not represent a true curvilinear system of coordinates in which the coordinates can be treated as independent variables satisfying conditions such as $\partial^2/\partial x_i \partial x_j = \partial^2/\partial x_j \partial x_i$, some care should be exercised in the use of the operator ∇. Thus, if $\mathbf{A}$ is an arbitrary vector its curl and its divergence should be taken to mean

$$\begin{aligned} \operatorname{curl} \mathbf{A} &= \mathbf{e}_j \times \frac{\partial}{\partial x_j} (A_k \mathbf{e}_k) \\ &= (\mathbf{e}_j \times \mathbf{e}_k) \frac{\partial A_k}{\partial x_j} + A_k \left(\mathbf{e}_j \times \frac{\partial \mathbf{e}_k}{\partial x_j} \right), \end{aligned} \tag{A$_2$, 4}$$

and

$$\begin{aligned} \operatorname{div} \mathbf{A} &= \mathbf{e}_j \cdot \frac{\partial}{\partial x_j} (A_k \mathbf{e}_k) \\ &= \frac{\partial A_j}{\partial x_j} + A_k \mathbf{e}_j \cdot \frac{\partial \mathbf{e}_k}{\partial x_j}, \end{aligned} \tag{A$_2$, 5}$$

where summation over repeated indices is to be understood and the terms in $\mathbf{e}_j \times \partial \mathbf{e}_k / \partial x_j$ and $\mathbf{e}_j \cdot \partial \mathbf{e}_k / \partial x_j$ are to be evaluated in accordance with Eqs. (A$_2$, 1) and (A$_2$, 2). As particular examples of Eqs. (A$_2$, 4) and (A$_2$, 5), we may note that

$$\operatorname{div} \mathbf{e}_1 = \mathbf{e}_j \cdot \frac{\partial \mathbf{e}_1}{\partial x_j} = \rho_2 + \rho_3$$

and

$$\operatorname{curl} \mathbf{e}_1 = \mathbf{e}_j \times \frac{\partial \mathbf{e}_1}{\partial x_j} = \rho_1 \mathbf{e}_3 - (\tau_2 + \tau_3) \mathbf{e}_1. \tag{A$_2$, 6}$$

A relation between the curvature coefficients and the variation of **B** along the field lines follows from the requirement

$$\operatorname{div} \mathbf{B} = 0. \qquad (A_2, 7)$$

According to Eqs. $(A_2, 1)$, $(A_2, 2)$, $(A_2, 5)$, and $(A_2, 6)$.

$$\operatorname{div} \mathbf{B} = (\rho_2 + \rho_3)B + \frac{\partial B}{\partial x_1} = 0. \qquad (A_2, 8)$$

Hence

$$-\frac{1}{B}\frac{\partial B}{\partial x_1} = \rho_2 + \rho_3 = \frac{1}{D} \text{ (say)}. \qquad (A_2, 9)$$

Further relations follow from imposing on **B** a condition of the form

$$\operatorname{curl} \mathbf{B} = a\mathbf{B} \qquad (A_2, 10)$$

where $a = a(x_i)$ is some function of position. When $a = 0$, the condition is that appropriate for a vacuum field; and when $a \neq 0$, the condition is that appropriate for a *force-free* field. In the latter case it is clear that

$$\frac{\partial a}{\partial x_1} = 0. \qquad (A_2, 11)$$

From Eqs. $(A_2, 1)$, $(A_2, 2)$, $(A_2, 4)$, and $(A_2, 6)$ we find that

$$\operatorname{curl} B\mathbf{e}_1 = \mathbf{e}_2 \frac{\partial B}{\partial x_3} - \mathbf{e}_3 \frac{\partial B}{\partial x_2} + B[\rho_1 \mathbf{e}_3 - (\tau_2 + \tau_3)\mathbf{e}_1]. \qquad (A_2, 12)$$

Hence

$$\frac{\partial B}{\partial x_3} = 0; \qquad \frac{1}{B}\frac{\partial B}{\partial x_2} = \rho_1 \quad \text{and} \quad \tau_2 + \tau_3 = -a; \qquad (A_2, 13)$$

in particular, for a vacuum field

$$\tau_2 + \tau_3 = 0. \qquad (A_2, 14)$$

APPENDIX III

In the reduction of the Boltzmann equation for f' in III, Section 9, Eq. (105), the solution for f^0 obtained in the first approximation in III, Section 8 (and denoted there by f_{I}) was used. However, it is clear that in a consistent approximation we must include terms of order $1/B^0$ in f^0 in the evaluation of

$$-\frac{e}{mc}\,\epsilon_{ijk}\,\frac{\partial f^0}{\partial v_i}\,(v_j + V_j^0)B_k' \qquad (A_3, 1)$$

on the right-hand side of III, Eq. (105). Unfortunately, this was overlooked in

III, Section 9; and we are grateful to Dr. R. Kulsrud for drawing our attention to this. In this appendix we shall make the necessary corrections to the analysis of paper III.

First, we need the solution for $f_{\rm II}$. We can obtain this in the manner described in the present paper. Subtracting III, Eq. (96), from III, Eq. (94), we obtain an equation which we can write in the form

$$-\frac{eB^0}{mc}\epsilon_{ijk}\frac{\partial f_{\rm II}}{\partial s_i}s_j n_k = \Phi_i s_i + \Phi_{ij}(s_i s_j - \langle s_i s_j\rangle), \qquad (\mathrm{A}_3, 2)$$

where

$$\Phi_i = \frac{\partial f_{\rm I}}{\partial x_i} + 2q(qn_j + V_j^0)\frac{\partial n_i}{\partial x_j}\frac{\delta f_{\rm I}}{\delta v^2} + \frac{2}{mN}\frac{\partial p_{ij}}{\partial x_j}\frac{\partial f_{\rm I}}{\partial s^2}$$

$$-2qn_j\left(\frac{\partial V_i^0}{\partial x_j}\frac{\partial f_{\rm I}}{\partial s^2} + \frac{\partial V_j^0}{\partial x_1}\frac{\partial f_{\rm I}}{\partial q^2}\right), \qquad (\mathrm{A}_3, 3)$$

$$\Phi_{ij} = q\left(\frac{\partial n_i}{\partial x_j} + \frac{\partial n_j}{\partial x_i}\right)\frac{\delta f_{\rm I}}{\delta v^2} - \left(\frac{\partial V_i^0}{\partial x_j} + \frac{\partial V_j^0}{\partial x_i}\right)\frac{\partial f_{\rm I}}{\partial s^2}. \qquad (\mathrm{A}_3, 4)$$

The general solution of Eq. $(\mathrm{A}_3, 2)$ is

$$f_{\rm II} = \frac{mc}{eB^0}\{\epsilon_{ijk}(\Phi_i + \tfrac{1}{2}\Phi_{il}s_l)s_j n_k + \psi(s^2, q, x_i)\}, \qquad (\mathrm{A}_3, 5)$$

where ψ is an arbitrary function of the arguments specified. The equation which determines ψ can be obtained by considering the integrability condition for the equation which governs the next higher term $f_{\rm III}$ in the solution for f^0. It will appear that for our present purposes it is not necessary to know ψ besides the nature of its dependence on the arguments.

Returning to III, Eq. (105), we must replace the last term on the right-hand side of this equation by

$$-\frac{e}{mc}\epsilon_{ijk}\frac{\partial}{\partial v_i}(f_{\rm I} + f_{\rm II})(v_j + V_j^0)B_k'. \qquad (\mathrm{A}_3, 6)$$

Additional terms in III, Eq. (105) resulting from the inclusion of $f_{\rm II}$ in $(A_3, 6)$ are of two kinds: those which are derived from ψ in the solution $(A_3, 5)$ for $f_{\rm II}$ and those which are derived from the particular integral involving Φ_i and Φ_{il}. Whereas the former terms vanish when averaged over the different directions of **s**, the latter, clearly, do not. Consequently, these latter terms must be included in obtaining the equation for f_L' defined in III, Eq. (113). It is also clear that the arguments leading to the solution III, Eq. (112), for f_P' is unaffected by the presence of these additional terms in $f_{\rm II}$.

The final result, then, is that we must *add* to the left-hand side of III, Eq. (114), the term

$$\frac{1}{B^0}\left\langle \epsilon_{ijk}\frac{\partial}{\partial \nu_i}[\epsilon_{\alpha\beta\gamma}(\Phi_\alpha + \tfrac{1}{2}\Phi_{\alpha\delta}s_\delta)s_\beta n_\gamma](s_j + qn_j + V_j^0)B_k'\right\rangle, \qquad (\mathrm{A}_3\,,7)$$

where the partial derivative with respect to ν_i must be carried out in accordance with III, Eq. (91).

On evaluating $(A_3\,,7)$ we find that the term $\Phi_{\alpha\delta}$ makes no contribution and we are left with

$$\frac{1}{B^0}\left\{\frac{\partial}{\partial s^2}(s^2\Phi_i)[n_kV_k^0)B_i' - (n_kB_k')V_i^0] - q\frac{\delta}{\delta\nu^2}(s^2\Phi_i)(\delta_{ik} - n_in_k)B_k'\right\}. \qquad (\mathrm{A}_3\,,8)$$

Letting

$$\mathbf{X} = \mathbf{n}\times(\mathbf{V}^0\times\mathbf{B}')/B^0, \qquad (A_3\,,9)$$

(so that $\mathbf{K}$ defined in III, Eq. (119) is $2\mathbf{X}\partial f^0/\partial s^2$), we can rewrite $(A_3\,,8)$ in the form

$$-\frac{\partial}{\partial s^2}(s^2\Phi_i)X_i - \frac{1}{B^0}q\frac{\delta}{\delta\nu^2}(s^2\Phi_i)(\mathbf{B}_\perp')_i. \qquad (\mathrm{A}_3\,,10)$$

This last expression must be added to $Q_1 + qQ_2$ as defined in III, Eq. (126). Separating the even and the odd (with respect to q) parts of $(A_3\,,10)$ we can redefine Q_1 and Q_2 so that Eqs. (132) and (133) for A_1 and A_2 continue to be valid. We find

$$\begin{aligned}Q_1 = \frac{1}{B^0}\left(s^2\frac{\partial f^0}{\partial s^2} - 2q^2\frac{\partial f^0}{\partial q^2}\right)\left(\frac{\partial V_i^0}{\partial x_j} + \frac{\partial V_j^0}{\partial x_i}\right)(\mathbf{B}_\perp')_i n_j + s^2\frac{\partial X_i}{\partial x_i}\frac{\partial f^0}{\partial s^2}\\ - X_i\left\{\frac{\partial f^0}{\partial x_i} - n_j\frac{\partial n_i}{\partial x_j}\left(s^2\frac{\partial f^0}{\partial s^2} - 2q^2\frac{\partial f^0}{\partial q^2}\right)\right\}\end{aligned} \qquad (\mathrm{A}_3\,,11)$$

and

$$\begin{aligned}Q_2 = s^2\left[\nabla_\perp\cdot\left(\frac{\mathbf{B}'}{B^0}\right) - \frac{\mathbf{n}\cdot\mathbf{B}'}{B^0}\frac{\partial n_j}{\partial x_j} - n_j\frac{\partial n_i}{\partial x_j}\frac{(\mathbf{B}_\perp')_i}{B^0}\right]\frac{\delta f^0}{\delta\nu^2}\\ + \frac{1}{B^0}\left(\frac{\partial f^0}{\partial x_i} + \frac{2}{mN^0}\frac{\partial p_{ij}^0}{\partial x_j}\frac{\partial f^0}{\partial q^2}\right)(\mathbf{B}_\perp')_i\\ + 2X_i\left(n_j\frac{\partial V_j^0}{\partial x_i} - V_j^0\frac{\partial n_i}{\partial x_j}\right)\frac{\partial f^0}{\partial q^2};\end{aligned} \qquad (\mathrm{A}_3\,,12)$$

these expressions are obtained by simply adding the even and the odd parts of $(A_3\,,10)$ to Q_1 and qQ_2, respectively, as defined in III, Eqs. (129) and (130). In Eqs. $(A_3\,,11)$ and $(A_3\,,12)$, f^0 is what has hitherto been denoted as f_{I}.

With the foregoing redefinitions of Q_1 and Q_2, the analysis of III, Sections 10, 11, 12, and 14 remain unaffected. However, the analysis of III, Section 13, requires revision even apart from the changes in the meanings of Q_1 and Q_2; these arise from the fact that $\partial A_2/\partial t$ does not necessarily tend to zero in the time-independent case (cf. *6*). These revisions will be considered in Paper V of this series which will include applications of the theory developed in these papers; however, we may note here that by the transformation to the variables ν^2 and s^2/B^0 considered in III, Section 13, the foregoing expressions for Q_1 and Q_2 are considerably simplified.

Received: May 26, 1958

REFERENCES

1. S. Chandrasekhar, A. N. Kaufman, and K. M. Watson, *Annals of Physics* **2,** 435 (1957); referred to in the text as III.
2. K. M. Watson and K. A. Brueckner, *Phys. Rev.* **102,** 19 (1956); this is paper II of the series. Paper I is by K. M. Watson, *Phys. Rev.* **102,** 12 (1956).
3. G. Chew, M. Goldberger, and F. Low, *Proc. Roy. Soc.* **A236,** 112 (1956).
4. G. Chew, M. Goldberger, and F. Low (in press).
5. See, for example, H. Alfvén, "Cosmical Electrodynamics." Oxford Univ. Press, London and New York, 1950;
L. Spitzer, "Physics of Fully Ionized Gases." Interscience, New York, 1956.
6. S. Chandrasekhar, A. N. Kaufman, and K. M. Watson, *Proc. Roy. Soc.* **A245,** 435 (1958).

3

The stability of the pinch

By S. Chandrasekhar, F.R.S., A. N. Kaufman and K. M. Watson
Los Alamos Scientific Laboratory, Los Alamos, New Mexico

(*Received* 9 *December* 1957)

The stability of a cylindrical plasma with an axial magnetic field and confined between conducting walls is investigated by solving, for small oscillations about equilibrium, the linearized Boltzmann and Maxwell equations. A criterion for marginal stability is derived; this differs slightly from the one derived by Rosenbluth from an analysis of the particle orbits. However, Rosenbluth's principal results on the possibility of stabilizing the pinch under suitable external conditions are confirmed. In the appendix a dispersion relation appropriate for plane hydromagnetic waves in an infinite medium is obtained; this relation discloses under the simplest conditions certain types of instabilities which may occur in plasma physics.

1. Introduction

The stability of a cylindrical plasma (the 'pinch') with an axial magnetic field has recently been investigated by Kruskal & Tuck (1958), Rosenbluth (1957), Tayler (1957), Shrafranov (1957), and others. Rosenbluth has, in particular, shown that when the plasma is confined between conducting walls, the presence of an axial magnetic field can, under suitable circumstances, stabilize the pinch. Moreover, Rosenbluth has treated the problem not only from the standpoint of conventional hydromagnetics (with the usual assumptions of scalar pressure and adiabatic changes of state), but also from the physically more important standpoint of the orbits described by the ions and electrons in the external magnetic field and under conditions when collisions between particles play no role. The importance of Rosenbluth's treatment from the latter standpoint (along the general lines described by Longmire & Rosenbluth 1957) arises from the fact that under the conditions pinches are usually realized in the laboratory, collisions between ions and electrons do not, indeed, play any significant role.

In this paper we shall re-examine the problem treated by Rosenbluth by going directly to the Boltzmann equation appropriate under the conditions. This method has certain advantages over Rosenbluth's in that certain assumptions justified by him on physical grounds can now be examined for their validity. Also, we are able to treat the general time-dependent problem without being, necessarily, restricted to the marginal case distinguishing stability from instability. Further, the method provides an illustration of a general theory we have recently developed (Chandrasekhar, Kaufman & Watson 1957; this paper will be referred to hereafter as III) for a rigorous treatment for problems of this kind.

2. The method of treatment and the basic equations

Consider a uniform cylindrical plasma of radius r_0 with a constant magnetic field, B_P^0, along the z-axis. The plasma is confined in a cylinder of radius R_0 with conducting walls. In the space between r_0 and R_0 there is a vacuum field with both z- and θ-components; thus,

$$\mathbf{B}_V^0 = B_V^0 \mathbf{1}_z + B_\theta \frac{r_0}{r} \mathbf{1}_\theta \quad (r_0 \leqslant r \leqslant R_0), \tag{1}$$

where $\mathbf{1}_z$ and $\mathbf{1}_\theta$ are unit vectors in the z- and the θ-directions. We are interested in the stability of the plasma under these conditions.

A general perturbation theory for treating small departures from stationary solutions of the Boltzmann equation (in which the collision term is neglected) has been given in III. We shall begin by briefly describing this theory and quoting the basic equations under the simplified conditions of the problem on hand. The essential simplifications are that there are no static drifts (denoted by $\mathbf{V}^0$ in III); also the unperturbed field lines in the plasma are straight.

In treating the perturbation problem, a variable $\boldsymbol{\xi}$ which plays the same role as the Lagrangian displacement in the usual hydromagnetic treatments (cf. Bernstein, Frieman, Kruskal & Kulsrud 1958) is first introduced. In the present theory, $\boldsymbol{\xi}$ is related to the perturbations in the electric ($\mathbf{E}'$) and the magnetic ($\mathbf{B}'$) fields. In the case when the dependence on time of all quantities describing departures from equilibrium is given by

$$e^{\Omega t},$$

these relations are as follows: resolving $\boldsymbol{\xi}$ into two components, $\xi_\parallel$ and $\boldsymbol{\xi}_\perp$, respectively parallel and perpendicular to the direction of the unperturbed magnetic field $\mathbf{B}^0$ (we are suppressing the subscript, P, for the present) we have (cf. III, equations (41), (42) and (46) and note that $\mathbf{U} = \Omega\boldsymbol{\xi}$)

$$E'_\parallel = \frac{m^+\Omega^2}{e}\xi_\parallel, \quad \mathbf{E}'_\perp = -\frac{\Omega}{c}\boldsymbol{\xi}_\perp \times \mathbf{B}^0 \tag{2}$$

and

$$\mathbf{B}' = \operatorname{curl}\left(\boldsymbol{\xi}_\perp \times \mathbf{B}^0 - \frac{m^+c}{e}\Omega\xi_\parallel \mathbf{n}\right), \tag{3}$$

where $\mathbf{n}$ denotes a unit vector in the direction of $\mathbf{B}^0$—the z-axis in the problem on hand; further, m^+ denotes the mass of the ion, e the charge on the ion and c the velocity of light.

Before proceeding further, we may make some general remarks on the notation we shall adopt. The subscripts $\parallel$ and $\perp$ will indicate that the components of the particular vector parallel and perpendicular, respectively, to $\mathbf{n}$ are meant. Similarly, the superscripts $+$ and $-$ will distinguish the quantities referring to the ions and the electrons; when, however, an equation (or a quantity) is to be understood as applying to both ions and electrons, these superscripts will in general be omitted. Finally, superscripts '0' will be used to denote the equilibrium values of the respective quantities while primes will denote the corresponding perturbations.

From equation (2) it follows that

$$\frac{E'_\parallel}{|\mathbf{E}_\perp|} = \frac{m^+c}{eB^0}\Omega \sim \frac{\Omega}{\omega_{\text{Larmor}}}. \tag{2'}$$

Since a basic assumption underlying the present treatment of plasmas (cf. III, § 6) is that the changes which the system undergoes take place in times which are long compared with the Larmor periods of the particle orbits, it is clear that in calculating $\mathbf{B}'$ in accordance with equation (3), we may neglect the term in $\xi_\parallel$ if $\xi_\parallel$ and $|\boldsymbol{\xi}_\perp|$ should be of comparable magnitudes. And even if the term in $\xi_\parallel$ makes no contribution to $\mathbf{B}'$, we shall find that $\xi_\parallel$ does contribute a term to the perturbation in the pressure tensor which cannot be ignored.

The equations of motion governing $\xi_\parallel$ and $\boldsymbol{\xi}_\perp$ are (III, equations (68) and (73))

$$\rho^0\left(1+\frac{|B^0|^2}{4\pi\rho^0c^2}\right)\Omega^2\boldsymbol{\xi}_\perp = -(\operatorname{div}\mathbf{P}')_\perp+\frac{1}{4\pi}(\operatorname{curl}\mathbf{B}')\times\mathbf{B}^0 \tag{4}$$

and

$$(\Omega^2+\sum_{+,-}\omega_p^2)\,E'_\parallel = c\Omega\mathbf{n}.\operatorname{curl}\mathbf{B}'+4\pi\sum_{+,-}\left\{\frac{e}{m}(\operatorname{div}\mathbf{p}')_\parallel\right\}, \tag{5}$$

where $\mathbf{P}'$ $(\equiv P'_{ij})$ denotes the perturbation in the total pressure tensor† and the summations in equation (5) are over the terms referring to the ions and electrons; also ρ^0 denotes the (unperturbed) density and $\omega_p^\pm$ the plasma frequency,

$$\omega_p^\pm = (4\pi N^{0,\pm}e^2/m^\pm)^{\frac{1}{2}}, \tag{6}$$

and $N^{0,\pm}$ the (unperturbed) concentration of the ions and electrons.

It will be observed that equations (4) and (5) involve the perturbations in the pressure tensor, $\mathbf{p}'^{,\pm}$. We, therefore, need equations for determining $\mathbf{p}'$; and these are provided by the appropriately linearized form of the Boltzmann equation.

Let $f^0(q^2, s^2, \mathbf{r})$ denote the function governing the distribution of the velocities q and $\mathbf{s}$ parallel and perpendicular, respectively, to the direction of $\mathbf{n}$, in the stationary equilibrium state. Then, by expressing the perturbation in the distribution function, $f'(q, \mathbf{s}, \mathbf{r}, t)$ in the manner

$$f'(q,\mathbf{s},\mathbf{r},t) = A_1(q^2,s^2,\mathbf{r},t)+qA_2(q^2,s^2,\mathbf{r},t)+f'_P(q,\mathbf{s},\mathbf{r},t), \tag{7}$$

where f'_P is such that it vanishes when averaged over all directions of $\mathbf{s}$, equations for A_1 and A_2 and an explicit expression for f'_P were derived in III (equations (112), (145), (148) and (150)). Under the simpler conditions of the problem on hand, these equations are

$$f'_P = \frac{2q}{B^0}\left(\frac{\partial f^0}{\partial q^2}-\frac{\partial f^0}{\partial s^2}\right)\mathbf{s}.\mathbf{B}', \tag{8}$$

$$\left.\begin{aligned} A_2 = \Omega\Lambda, \quad A_1 &= -q^2\frac{\partial\Lambda}{\partial z}+G_1 \\ \text{and} \qquad \Omega^2\Lambda+\frac{\partial A_1}{\partial z} &= -Q_2, \end{aligned}\right\} \tag{9}$$

where (III, equations (127) and (130))

$$G_1 = 2q^2\frac{\partial\xi_\parallel}{\partial z}\frac{\partial f^0}{\partial q^2}+s^2(\boldsymbol{\nabla}_\perp.\boldsymbol{\xi}_\perp)\frac{\partial f^0}{\partial s^2} \tag{10}$$

and

$$Q_2 = s^2\left(\frac{\partial f^0}{\partial q^2}-\frac{\partial f^0}{\partial s^2}\right)\frac{\boldsymbol{\nabla}_\perp.\mathbf{B}'_\perp}{B^0}. \tag{11}$$

† Note that P^0_{ij} and P'_{ij} refer to the total pressure (due to the electrons and the ions) while p^0_{ij} and p'_{ij} with superscripts + and − refer to the pressures due to the ions and electrons, separately; thus $P^0_{ij} = \sum_{+,-} p^0_{ij}$ and $P'_{ij} = \sum_{+,-} p'_{ij}$.

In equations (10) and (11), $\nabla_\perp$ denotes the projection of the usual gradient operator on a plane perpendicular to **n**:

$$(\nabla_\perp)_i = \frac{\partial}{\partial x_i} - n_i n_j \frac{\partial}{\partial x_j}. \tag{12}$$

In terms of the solution (7) for f', the required perturbation in the pressure tensor, p'_{ij}, is given by

$$p'_{ij} = p'_{\parallel;L} n_i n_j + p'_{\perp;L}(\delta_{ij} - n_i n_j) + p'_{ij;P}, \tag{13}$$

where

$$p'_{\parallel;L} = \tfrac{1}{2}m \iint q^2 A_1 \, dq \, ds^2; \quad p'_{\perp;L} = \tfrac{1}{4}m \iint s^2 A_1 \, dq \, ds^2 \tag{14†}$$

and

$$p'_{ij;P} = (n_i \mathbf{B}'_{\perp,j} + n_j \mathbf{B}'_{\perp,i}) \frac{p^0_\parallel - p^0_\perp}{B^0}. \tag{15}$$

3. The solution of the perturbation equations for the plasma

We shall now show how the various equations governing the perturbed plasma can be solved. In solving these equations we shall suppose that the disturbance has been analyzed into normal modes and that the dependence on z and θ of all the perturbed quantities is given by

$$e^{i(kz+m\theta)}, \tag{16}$$

where k is the wave number of the disturbance and m is an integer (positive, zero or negative). Apart from the factor (16), the various quantities are functions only of r, the distance from the z-axis. For solutions having the dependence on z and θ given by (16),

$$\frac{\partial}{\partial z} = ik \quad \text{and} \quad \frac{\partial}{\partial \theta} = im; \tag{17}$$

also,

$$\nabla_\perp = \mathbf{1}_r \frac{\partial}{\partial r} + \mathbf{1}_\theta \frac{im}{r}. \tag{18}$$

(a) *The perturbations in the electric and the magnetic fields*

Let

$$F' = \frac{c}{\Omega} E'_\parallel \quad \text{so that} \quad \xi_\parallel = \frac{e}{m^+ c\Omega} F'. \tag{19}$$

In terms of F' the expression for $\mathbf{B}'$ is (cf. equation (3))

$$\mathbf{B}' = \operatorname{curl}(\boldsymbol{\xi}_\perp \times \mathbf{B}^0) + \mathbf{n} \times \operatorname{grad} F'. \tag{20}$$

When $\mathbf{B}^0$ is in the direction of the z-axis, the foregoing becomes

$$\mathbf{B}' = B^0\{ik\boldsymbol{\xi}_\perp - (\nabla_\perp . \boldsymbol{\xi}_\perp)\,\mathbf{n}\} + (\mathbf{n} \times \nabla_\perp)\,F', \tag{21}$$

where we have made use of the fact that when **n** is a constant vector,

$$\nabla . \mathbf{A}_\perp \equiv \nabla_\perp . \mathbf{A} \equiv \nabla_\perp . \mathbf{A}_\perp,$$

where **A** is an arbitrary vector. Accordingly,

$$B'_\parallel = -B^0 \nabla_\perp . \boldsymbol{\xi}_\perp \quad \text{and} \quad \mathbf{B}'_\perp = ikB^0 \boldsymbol{\xi}_\perp + (\mathbf{n} \times \nabla_\perp)\,F'. \tag{22}$$

† In writing (14) we have assumed that f^0 is so normalized that, for example,

$$N^0 = \frac{1}{2}\int_0^\infty \int_0^\infty f^0 \, dq \, ds^2.$$

We may note here (for later use) that for $\mathbf{B}'$ given by equation (21)

$$\operatorname{curl}\mathbf{B}' = B^0\left\{ik\operatorname{curl}\boldsymbol{\xi}_\perp + (\mathbf{n}\times\boldsymbol{\nabla}_\perp)(\boldsymbol{\nabla}_\perp.\boldsymbol{\xi}_\perp) + \frac{1}{B^0}(\mathbf{n}\nabla^2 - ik\boldsymbol{\nabla})F'\right\}, \tag{23}$$

so that
$$\mathbf{n}\times\operatorname{curl}\mathbf{B}' = B^0\left\{ik\mathbf{n}\times\operatorname{curl}\boldsymbol{\xi}_\perp - \boldsymbol{\nabla}_\perp(\boldsymbol{\nabla}_\perp.\boldsymbol{\xi}_\perp) - \frac{ik}{B^0}(\mathbf{n}\times\boldsymbol{\nabla}_\perp)F'\right\} \tag{24}$$

and
$$\mathbf{n}.\operatorname{curl}\mathbf{B}' = B^0\left\{ik\mathbf{n}.\operatorname{curl}\boldsymbol{\xi}_\perp + \frac{1}{B^0}\nabla_\perp^2 F'\right\}. \tag{25}†$$

(b) *The solution for* A_1

According to equations (9) and (17)

$$A_1 = -ikq^2\Lambda + G_1 \tag{26}$$

and
$$\Omega^2\Lambda + ikA_1 = -Q_2. \tag{27}$$

From these equations, we find

$$A_1 = \frac{\Omega^2}{\Omega^2+k^2q^2}G_1 + \frac{ikq^2}{\Omega^2+k^2q^2}Q_2; \tag{28}$$

and substituting for G_1 and Q_2 from equations (10) and (11), we have

$$A_1 = \frac{\Omega^2}{\Omega^2+k^2q^2}\left\{2q^2\frac{\partial f^0}{\partial q^2}(ik\xi_\parallel) + s^2\frac{\partial f^0}{\partial s^2}(\boldsymbol{\nabla}_\perp.\boldsymbol{\xi}_\perp)\right\} + \frac{ikq^2s^2}{\Omega^2+k^2q^2}\left(\frac{\partial f^0}{\partial q^2} - \frac{\partial f^0}{\partial s^2}\right)\frac{\boldsymbol{\nabla}_\perp.\mathbf{B}'_\perp}{B^0}. \tag{29}$$

(c) *The pressure tensor* p'_{ij} *and its divergence*

On inserting for A_1 from (29) in equations (14), we observe that $p'_{\parallel;L}$ and $p'_{\perp;L}$ can be expressed in the forms

$$p'_{\parallel;L} = ik\xi_\parallel I_1 + (\boldsymbol{\nabla}_\perp.\boldsymbol{\xi}_\perp)I_2 + ik\frac{\boldsymbol{\nabla}_\perp.\mathbf{B}'_\perp}{B^0}I_3 \tag{30}$$

and
$$p'_{\perp;L} = ik\xi_\parallel J_1 + (\boldsymbol{\nabla}_\perp.\boldsymbol{\xi}_\perp)J_2 + ik\frac{\boldsymbol{\nabla}_\perp.\mathbf{B}'_\perp}{B^0}J_3, \tag{31}$$

where
$$\begin{pmatrix}I_1\\J_1\end{pmatrix} = \begin{pmatrix}\frac{1}{2}\\\frac{1}{4}\end{pmatrix}m\iint\frac{2\Omega^2q^2}{\Omega^2+k^2q^2}\begin{pmatrix}q^2\\s^2\end{pmatrix}\frac{\partial f^0}{\partial q^2}dq\,ds^2, \tag{32}$$

$$\begin{pmatrix}I_2\\J_2\end{pmatrix} = \begin{pmatrix}\frac{1}{2}\\\frac{1}{4}\end{pmatrix}m\iint\frac{\Omega^2s^2}{\Omega^2+k^2q^2}\begin{pmatrix}q^2\\s^2\end{pmatrix}\frac{\partial f^0}{\partial s^2}dq\,ds^2 \tag{33}$$

and
$$\begin{pmatrix}I_3\\J_3\end{pmatrix} = \begin{pmatrix}\frac{1}{2}\\\frac{1}{4}\end{pmatrix}m\iint\frac{q^2s^2}{\Omega^2+k^2q^2}\begin{pmatrix}q^2\\s^2\end{pmatrix}\left(\frac{\partial f^0}{\partial q^2} - \frac{\partial f^0}{\partial s^2}\right)dq\,ds^2. \tag{34}$$

Now substituting for $\xi_\parallel$ and $\mathbf{B}'_\perp$ from equations (19) and (22) in the expressions for $p'_{\parallel;L}$ and $p'_{\perp;L}$, we obtain

$$p'_{\parallel;L} = (I_2 - k^2I_3)\boldsymbol{\nabla}_\perp.\boldsymbol{\xi}_\perp + ik\frac{eI_1}{mc\Omega}F' \tag{35}$$

and
$$p'_{\perp;L} = (J_2 - k^2J_3)\boldsymbol{\nabla}_\perp.\boldsymbol{\xi}_\perp + ik\frac{eJ_1}{mc\Omega}F'. \tag{36}$$

† In deriving this, use has been made of the relation $\nabla^2 - ik\mathbf{n}.\boldsymbol{\nabla} = \nabla^2 + k^2 = \nabla_\perp^2$.

Using the definitions of I_2 and I_3, we readily find that

$$I_2 - k^2 I_3 = \tfrac{1}{2}m \iint q^2 s^2 \left\{\frac{\partial f^0}{\partial s^2} - \frac{\partial f^0}{\partial q^2} + \frac{\Omega^2}{\Omega^2 + k^2 q^2}\frac{\partial f^0}{\partial q^2}\right\} \mathrm{d}q\, \mathrm{d}s^2$$
$$= (p_\perp^0 - p_\parallel^0) + J_1. \tag{37}$$

Similarly,

$$J_2 - k^2 J_3 = -\tfrac{1}{4}m \iint s^4 \left(\frac{\partial f^0}{\partial q^2} - \frac{\partial f^0}{\partial s^2}\right) \mathrm{d}q\, \mathrm{d}s^2 + \tfrac{1}{4}m\Omega^2 \iint \frac{s^4}{\Omega^2 + k^2 q^2}\frac{\partial f^0}{\partial q^2} \mathrm{d}q\, \mathrm{d}s^2; \tag{38}$$

or, letting

$$S \iint s^2 f^0 \mathrm{d}q\, \mathrm{d}s^2 = \iint s^4 \left(\frac{\partial f^0}{\partial q^2} - \frac{\partial f^0}{\partial s^2}\right) \mathrm{d}q\, \mathrm{d}s^2 - \Omega^2 \iint \frac{s^4}{\Omega^2 + k^2 q^2}\frac{\partial f^0}{\partial q^2} \mathrm{d}q\, \mathrm{d}s^2, \tag{39}$$

we can write

$$J_2 - k^2 J_3 = -S p_\perp^0. \tag{40}$$

Thus,

$$p'_{\parallel;L} = (J_1 + p_\perp^0 - p_\parallel^0)\,\nabla_\perp . \boldsymbol{\xi}_\perp + \mathrm{i}k\frac{eI_1}{mc\Omega}F' \tag{41}$$

and

$$p'_{\perp;L} = -S p_\perp^0 (\nabla_\perp . \boldsymbol{\xi}_\perp) + \mathrm{i}k\frac{eJ_1}{mc\Omega}F'. \tag{42}$$

For later use we shall obtain here the divergence of the tensor p'_{ij}. For p'_{ij} given by equation (13) we may write (cf. III, equations (85) to (88))

$$\operatorname{div} \mathbf{p}' = \nabla_\perp p'_{\perp;L} + \mathrm{i}k p'_{\parallel;L}\mathbf{n} + \operatorname{div} p'_P. \tag{43}$$

By making use of equation (15), we can reduce the last term in equation (43) as follows:

$$\frac{\partial}{\partial x_j} p'_{ij;P} = \left[n_i \frac{\partial}{\partial x_j} B'_{\perp,j} + n_j \frac{\partial}{\partial x_j} B'_{\perp,i}\right] \frac{p_\parallel^0 - p_\perp^0}{B^0}$$
$$= \left[\mathrm{i}k B'_{\perp,i} + n_i \frac{\partial}{\partial x_j}(B'_j - n_j n_l B'_l)\right] \frac{p_\parallel^0 - p_\perp^0}{B^0}$$
$$= \mathrm{i}k(B'_{\perp,i} - n_i B'_\parallel)\frac{p_\parallel^0 - p_\perp^0}{B^0}; \tag{44}$$

and substituting for $\mathbf{B}'_\perp$ and $B'_\parallel$ from (22), we obtain

$$\operatorname{div} \mathbf{p}'_P = \mathrm{i}k\left\{\mathrm{i}k\boldsymbol{\xi}_\perp + \frac{1}{B^0}(\mathbf{n} \times \nabla_\perp)\,F' + (\nabla_\perp . \boldsymbol{\xi}_\perp)\,\mathbf{n}\right\}(p_\parallel^0 - p_\perp^0). \tag{45}$$

Finally, combining equations (41), (42), (43) and (45), we have

$$\operatorname{div} \mathbf{p}' = \nabla_\perp\left\{-S p_\perp^0 (\nabla_\perp . \boldsymbol{\xi}_\perp) + \mathrm{i}k\frac{eJ_1}{mc\Omega}F'\right\} + \mathrm{i}k\left\{[J_1 + (p_\perp^0 - p_\parallel^0)]\,(\nabla_\perp . \boldsymbol{\xi}_\perp) + \mathrm{i}k\frac{eI_1}{mc\Omega}F'\right\}\mathbf{n}$$
$$+ \mathrm{i}k\left\{\mathrm{i}k\boldsymbol{\xi}_\perp + \frac{1}{B^0}(\mathbf{n} \times \nabla_\perp)\,F' + \mathbf{n}(\nabla_\perp . \boldsymbol{\xi}_\perp)\right\}(p_\parallel^0 - p_\perp^0). \tag{46}$$

Hence

$$(\operatorname{div} \mathbf{p}')_\parallel = \mathrm{i}k\left\{J_1(\nabla_\perp . \boldsymbol{\xi}_\perp) + \mathrm{i}k\frac{eI_1}{mc\Omega}F'\right\} \tag{47}$$

and

$$(\operatorname{div} \mathbf{p}')_\perp = \nabla_\perp\left\{-S p_\perp^0 (\nabla_\perp . \boldsymbol{\xi}_\perp) + \mathrm{i}k\frac{eJ_1}{mc\Omega}F'\right\} + \left\{k^2\boldsymbol{\xi}_\perp - \frac{\mathrm{i}k}{B^0}(\mathbf{n} \times \nabla_\perp)\,F'\right\}(p_\perp^0 - p_\parallel^0). \tag{48}$$

(d) The reduction of the equations governing F' and $\boldsymbol{\xi}_\perp$

With the pressure tensor $\mathbf{p}'$ determined, we can now proceed to the solution of the equations governing $\xi_\parallel$ and $\boldsymbol{\xi}_\perp$.

Considering first equation (4) for $\boldsymbol{\xi}_\perp$, and making use of equations (24) and (48), we have

$$\rho^*\Omega^2\boldsymbol{\xi}_\perp = (\sum_{+,-} Sp_\perp^0)\boldsymbol{\nabla}_\perp(\boldsymbol{\nabla}_\perp.\boldsymbol{\xi}_\perp) - ik\left(\sum_{+,-}\frac{eJ_1}{mc\Omega}\right)\boldsymbol{\nabla}_\perp F'$$
$$-\left\{k^2\boldsymbol{\xi}_\perp - \frac{ik}{B^0}(\mathbf{n}\times\boldsymbol{\nabla}_\perp)F'\right\}(P_\perp^0 - P_\parallel^0)$$
$$-\frac{|B^0|^2}{4\pi}\left\{ik\mathbf{n}\times\operatorname{curl}\boldsymbol{\xi}_\perp - \boldsymbol{\nabla}_\perp(\boldsymbol{\nabla}_\perp.\boldsymbol{\xi}_\perp) - \frac{ik}{B^0}(\mathbf{n}\times\boldsymbol{\nabla}_\perp)F'\right\}, \quad (49)$$

where
$$\rho^* = \rho^0\left(1 + \frac{|B^0|^2}{4\pi\rho^0c^2}\right). \quad (50)$$

Letting
$$\sum_{+,-} Sp_\perp^0 = S^+p_\perp^{0,+} + S^-p_\perp^{0,-} = \bar{S}p_\perp^0 \quad \text{(say)}, \quad (51)$$

and noting that $\mathbf{n}\times\operatorname{curl}\boldsymbol{\xi}_\perp = -ik\boldsymbol{\xi}_\perp$, we find after some rearranging that

$$\left(\bar{S}P_\perp^0 + \frac{|B^0|^2}{4\pi}\right)\boldsymbol{\nabla}_\perp(\boldsymbol{\nabla}_\perp.\boldsymbol{\xi}_\perp) - k^2\left(P_\perp^0 - P_\parallel^0 + \frac{|B^0|^2}{4\pi} + \rho^*\frac{\Omega^2}{k^2}\right)\boldsymbol{\xi}_\perp$$
$$= -\left(P_\perp^0 - P_\parallel^0 + \frac{|B^0|^2}{4\pi}\right)\frac{ik}{B^0}(\mathbf{n}\times\boldsymbol{\nabla}_\perp)F' + ik\left(\sum_{+,-}\frac{eJ_1}{mc\Omega}\right)\boldsymbol{\nabla}_\perp F'. \quad (52)$$

Similarly, by combining equations (5), (19), (25) and (47), we obtain

$$(\Omega^2 + \sum_{+,-}\omega_p^2)\frac{\Omega}{c}F' = c\Omega\nabla_\perp^2 F' + ikcB^0\Omega\mathbf{n}.\operatorname{curl}\boldsymbol{\xi}_\perp$$
$$+ 4\pi ik\left(\sum_{+,-}\frac{e}{m}J_1\right)\boldsymbol{\nabla}_\perp.\boldsymbol{\xi}_\perp - k^2\left(\sum_{+,-}\frac{4\pi e^2}{m^2c\Omega}I_1\right)F', \quad (53)$$

or, equivalently (cf. equation (6))

$$\left\{\Omega^2 + \sum_{+,}\omega_p^2\left(1 + \frac{k^2}{mN^0}\frac{I_1}{\Omega^2}\right)\right\}\frac{\Omega}{c}F'$$
$$= c\Omega\nabla_\perp^2 F' + ikcB^0\Omega\mathbf{n}.\operatorname{curl}\boldsymbol{\xi}_\perp + 4\pi ik\left(\sum_{+,-}\frac{e}{m}J_1\right)\boldsymbol{\nabla}_\perp.\boldsymbol{\xi}_\perp. \quad (54)$$

Making use of the definition of I_1 (equation (32)) we find,

$$1 + \frac{k^2I_1}{mN^0\Omega^2} = 1 + \frac{1}{N^0}\iint\frac{k^2q^4}{\Omega^2 + k^2q^2}\frac{\partial f^0}{\partial q^2}\,dq\,ds^2$$
$$= 1 + \frac{1}{N^0}\iint q^2\frac{\partial f^0}{\partial q^2}\,dq\,ds^2 - \frac{1}{N^0}\iint\frac{\Omega^2q^2}{\Omega^2 + k^2q^2}\frac{\partial f^0}{\partial q^2}\,dq\,ds^2$$
$$= -\frac{\Omega^2}{N^0}\iint\frac{q^2}{\Omega^2 + k^2q^2}\frac{\partial f^0}{\partial q^2}\,dq\,ds^2 = -\Omega^2K \quad \text{(say)}. \quad (55)$$

Equation (54) may, therefore, be rewritten as

$$c\Omega\left\{\frac{\Omega^2}{c^2}(1 - \sum_{+,-}\omega_p^2K) - \nabla_\perp^2\right\}F' = ik\left\{c\Omega B^0\mathbf{n}.\operatorname{curl}\boldsymbol{\xi}_\perp + 4\pi\left(\sum_{+,-}\frac{e}{m}J_1\right)\boldsymbol{\nabla}_\perp.\boldsymbol{\xi}_\perp\right\}. \quad (56)$$

(*e*) *The solution of the equations for* F' *and* $\boldsymbol{\xi}_\perp$ *in the limit* $kc/\Omega \to \infty$

Under most conditions of physical interest, the fluid velocities are very small compared with the velocity of light and $kc/\Omega \gg 1$. In the limit $kc/\Omega \to \infty$, equations (52) and (56) simplify somewhat and we are left with

$$a\boldsymbol{\nabla}_\perp(\boldsymbol{\nabla}_\perp . \boldsymbol{\xi}_\perp) = k^2\left(b+\rho\frac{\Omega^2}{k^2}\right)\boldsymbol{\xi}_\perp - ik\frac{b}{B^0}(\mathbf{n}\times\boldsymbol{\nabla}_\perp)F' + ik\left(\sum_{+,-}\frac{eJ_1}{mc\Omega}\right)\boldsymbol{\nabla}_\perp F' \tag{57}$$

and $$\nabla_\perp^2 F' = -ik\left\{B^0\mathbf{n}.\operatorname{curl}\boldsymbol{\xi}_\perp + 4\pi\left(\sum_{+,-}\frac{eJ_1}{mc\Omega}\right)\boldsymbol{\nabla}_\perp . \boldsymbol{\xi}_\perp\right\} - \frac{\Omega^2}{c^2}\left(\sum_{+,-}\omega_p^2 K\right)F', \tag{58}$$

where we have introduced the abbreviations

$$a = \bar{S}P_\perp^0 + \frac{|B^0|^2}{4\pi} \quad \text{and} \quad b = P_\perp^0 - P_\parallel^0 + \frac{|B^0|^2}{4\pi}. \tag{59}$$

Letting $$\frac{ik}{B^0}F' = \phi, \quad \boldsymbol{\nabla}_\perp . \boldsymbol{\xi}_\perp = \chi, \tag{60}$$

$$\Sigma_1 = B^0 \sum_{+,-}\frac{eJ_1}{mc\Omega}, \quad \Sigma_2 = \frac{\Omega^2}{c^2}\left(\sum_{+,-}\omega_p^2 K\right) \tag{61}$$

and $$\gamma^2 = \frac{1}{a}\left(b+\rho\frac{\Omega^2}{k^2}\right), \tag{62}$$

we can rewrite equations (57) and (58) more conveniently in the forms

$$\boldsymbol{\nabla}_\perp\chi = k^2\gamma^2\boldsymbol{\xi}_\perp - \frac{b}{a}(\mathbf{n}\times\boldsymbol{\nabla}_\perp)\phi + \frac{\Sigma_1}{a}\boldsymbol{\nabla}_\perp\phi \tag{63}$$

and $$\nabla_\perp^2\phi = k^2\left\{\mathbf{n}.\operatorname{curl}\boldsymbol{\xi}_\perp + \frac{4\pi}{|B^0|^2}\Sigma_1\chi\right\} - \Sigma_2\phi. \tag{64}$$

Taking the vector product of equation (63) with $\boldsymbol{\nabla}_\perp$, we get

$$k^2\gamma^2\boldsymbol{\nabla}_\perp\times\boldsymbol{\xi}_\perp = \frac{b}{a}\mathbf{n}\nabla_\perp^2\phi. \tag{65}$$

Multiplying this equation scalarly by $\mathbf{n}$ and making use of equation (64), we obtain

$$k^2\gamma^2\mathbf{n}.\operatorname{curl}\boldsymbol{\xi}_\perp = \frac{b}{a}k^2\left\{\mathbf{n}.\operatorname{curl}\boldsymbol{\xi}_\perp + \frac{4\pi}{|B^0|^2}\Sigma_1\chi\right\} - \frac{b}{a}\Sigma_2\phi. \tag{66}$$

On further simplification (in which use is made of equation (62)) equation (66) reduces to

$$\rho\Omega^2\mathbf{n}.\operatorname{curl}\boldsymbol{\xi}_\perp = b\left\{\frac{4\pi k^2\Sigma_1}{|B^0|^2}\chi - \Sigma_2\phi\right\}. \tag{67}$$

The terms in $\boldsymbol{\xi}_\perp$ in this equation are of relative order (cf. equations (32), (59) and (61))

$$\frac{\rho\Omega^2}{k^2\Sigma_1}\frac{b}{|B^0|^2} = O\left(\frac{\rho\Omega^2}{k^2 J_1}\frac{mc\Omega}{eB^0}\right) = O\left(\frac{mc\Omega}{eB^0}\right) = O\left(\frac{\Omega}{\omega_{\text{Larmor}}}\right); \tag{68}$$

and quantities of this order are neglected in our present treatment (cf. the remarks following equation (2′)). Accordingly, we may equate the quantity on the right-hand side of equation (67) to zero and obtain

$$\phi = \frac{4\pi k^2}{|B^0|^2}\frac{\Sigma_1}{\Sigma_2}\chi. \tag{69}$$

Returning to equation (63), we observe that of the two terms in ϕ on the right-hand side of this equation, the first can be neglected since it is of order $(\Omega/\omega)(\omega_p^2/k^2c^2)$ relative to the second. Thus, we can write

$$\nabla_\perp \chi = k^2\gamma^2 \boldsymbol{\xi}_\perp + \frac{\Sigma_1}{a}\nabla_\perp \phi. \tag{70}$$

Eliminating ϕ from this equation by making use of equation (69), we obtain

$$\left(1 - \frac{4\pi k^2}{|B^0|^2}\frac{\Sigma_1^2}{a\Sigma_2}\right)\nabla_\perp \chi = k^2\gamma^2 \boldsymbol{\xi}_\perp. \tag{71}$$

On substituting for Σ_1, Σ_2 and ω_p in accordance with equations (6) and (61), we find that the quantity in parentheses on the left-hand of equation (71) simplifies to

$$1 - \frac{k^2}{a\Omega^4}\frac{\{\sum_{+,-}(eJ_1/m)\}^2}{\sum_{+,-} e^2(N^0K/m)}. \tag{72}$$

Letting

$$\gamma^2 = \Gamma^2\left\{1 - \frac{k^2}{a\Omega^4}\frac{\{\sum_{+,-}(eJ_1/m)\}^2}{\sum_{+,-} e^2(N^0K/m)}\right\}, \tag{73}$$

we can rewrite equation (71) in the form

$$\nabla_\perp \chi = k^2\Gamma^2 \boldsymbol{\xi}_\perp. \tag{74}$$

Taking the scalar product of this equation with $\nabla_\perp$ and remembering that $\nabla_\perp . \boldsymbol{\xi}_\perp = \chi$, we obtain

$$\nabla_\perp^2 \chi = k^2\Gamma^2\chi. \tag{75}$$

The general solution of equation (75) which has no singularity at $r = 0$ is a multiple of

$$I_m(\Gamma kr)\, e^{im\theta} = X(r)\, e^{im\theta} \quad \text{(say)}, \tag{76}$$

where $I_m(x)$ is the Bessel function of order m for a purely imaginary argument. (In this paper we shall adopt Watson's (1952) notation regarding Bessel functions.) With the solution for X known (apart from an arbitrary constant factor which we shall ignore), we can, in accordance with equation (74), write

$$\boldsymbol{\xi}_\perp = \nabla_\perp (X\, e^{im\theta}). \tag{77}$$

Thus

$$\xi_r = X' = k\Gamma I_m'(\Gamma kr) \tag{78}$$

and

$$\xi_\theta = \frac{im}{r}X = \frac{im}{r}I_m(\Gamma kr), \tag{79}$$

where (as in all other cases) we have suppressed the (common) factor $e^{i(kz+m\theta)}$ in the expressions for ξ_r and ξ_θ.

(*f*) *The perturbation in the magnetic field*

In terms of the solution for $\boldsymbol{\xi}_\perp$ obtained in the preceding subsection, the perturbation in the magnetic field $\mathbf{B}'_P$ (given by equation (21)) becomes

$$\mathbf{B}'_P = ikB^0_P\left(X'\mathbf{1}_r + \frac{im}{r}X\mathbf{1}_\theta\right) - k^2\Gamma^2 B^0_P X\mathbf{1}_z, \tag{80}$$

where we have restored the subscript P to indicate that this solution refers to the interior of the plasma. (Note that to the order of accuracy of the present treatment, the term in F' in equation (21) does not make any contribution to $\mathbf{B}'_P$ (cf. the remarks following equation (2')).)

(*g*) *The perturbation in the pressure tensor*

The perturbation in the transverse component of the pressure tensor can now be found. According to equations (42), (60) and (69), we have

$$p'_{\perp;L} = \left\{-Sp^0_\perp + \left(\frac{eB^0}{mc\Omega}J_1\right)\frac{4\pi k^2}{|B^0|^2}\frac{\Sigma_1}{\Sigma_2}\right\}\nabla_\perp.\boldsymbol{\xi}_\perp. \tag{81}$$

On substituting for Σ_1 and Σ_2 from (61), we find that the second term in braces on the right-hand side of equation (81) can be simplified to the form

$$\frac{k^2}{\Omega^4}\frac{(eJ_1/m)\sum\limits_{+,-}(eJ_1/m)}{\sum\limits_{+,-}(e^2N^0K/m)} = -Rp^0_\perp \quad \text{(say)}. \tag{82}$$

Thus, $$p'_{\perp;L} = -(S+R)\,p^0_\perp\nabla_\perp.\boldsymbol{\xi}_\perp = -k^2\Gamma^2(S+R)\,p^0_\perp X. \tag{83}$$

From equations (82) and (83) it follows that

$$P'_{\perp;L} = -k^2\Gamma^2(\bar{S}+\bar{R})\,P^0_\perp X, \tag{84}$$

where (cf. equations (62) and (73))

$$\bar{R}P^0_\perp = -\frac{k^2}{\Omega^4}\frac{\{\sum\limits_{+,-}(eJ_1/m)\}^2}{\sum\limits_{+,-}(e^2N^0K/m)} = \left(\frac{\gamma^2}{\Gamma^2}-1\right)a, \tag{85}$$

and $\bar{S}$ has the same meaning as in equation (51).

From equation (85) we can derive an alternative formulae for Γ^2. We have

$$(\bar{S}+\bar{R})\,P^0_\perp\,\Gamma^2 = (\bar{S}P^0_\perp - a)\,\Gamma^2 + \gamma^2 a; \tag{86}$$

or making use of equations (59), we have

$$(\bar{S}+\bar{R})\,P^0_\perp\,\Gamma^2 = -\frac{|B^0|^2}{4\pi}\Gamma^2 + P^0_\perp - P^0_\parallel + \rho\frac{\Omega^2}{k^2} + \frac{|B^0|^2}{4\pi}. \tag{87}$$

Thus $$\Gamma^2 = \frac{P^0_\perp - P^0_\parallel + |B^0|^2/4\pi + \rho\Omega^2/k^2}{(\bar{S}+\bar{R})\,P^0_\perp + |B^0|^2/4\pi}. \tag{88}$$

4. The solution for the perturbed field in the vacuum

The unperturbed field outside the plasma is given by equation (1). Let $\mathbf{B}'_V$ denote the perturbation in the field. Since no currents can flow in a vacuum, we can derive $\mathbf{B}'_V$ from a scalar potential Ψ; thus

$$\mathbf{B}'_V = \boldsymbol{\nabla}\Psi, \quad \text{where} \quad \nabla^2\Psi = 0. \tag{89}$$

For solutions which have the $e^{i(kz+m\theta)}$ dependence on z and θ, the appropriate form for Ψ is

$$\Psi = B_\theta r_0[C_1 I_m(kr) + C_2 K_m(kr)]\, e^{i(kz+m\theta)}, \tag{90}$$

where C_1 and C_2 are constants to be determined, and I_m and K_m are the Bessel functions of order m and of the two kinds for a purely imaginary argument (cf. Watson 1952). The constant factor $B_\theta r_0$ has been introduced in (90) for later convenience. The required solution for $\mathbf{B}'_V$ can, therefore, be expressed in the form

$$\mathbf{B}'_V = B_\theta r_0\left(\psi'\mathbf{1}_r + \frac{im}{r}\psi\mathbf{1}_\theta + ik\psi\mathbf{1}_z\right), \tag{91}$$

where

$$\psi(r) = C_1 I_m(kr) + C_2 K_m(kr), \tag{92}$$

and the factor $e^{i(kz+m\theta)}$ has again been suppressed.

At the outer boundary, $r = R_0$, the radial component of $\mathbf{B}'_V$ must vanish. This leads to the relation

$$C_1 I'_m(kR_0) + C_2 K'_m(kR_0) = 0. \tag{93}$$

A further relation arises from applying the boundary conditions at the surface of the plasma. These latter conditions are considered in the following section.

5. The characteristic equation for Ω^2

The solutions of the equations governing the departures from equilibrium of the plasma have been obtained in §§ 3 and 4. It remains to satisfy the boundary conditions which must be met at the surface of the plasma. As we shall presently see, these conditions will lead to an equation for Ω^2 and thus to the criterion for stability.

The boundary conditions which must in general be satisfied at a surface of discontinuity have been formulated in III, § 14. The principal requirements are that the normal components of the magnetic field and the stress tensor, T_{ij}, are continuous across the boundary. Thus, if $\mathbf{N}$ denotes the unit outward normal on a surface of discontinuity, then

$$N_j \Delta[B_j] = 0 \tag{94}$$

and

$$N_j \Delta[T_{ij}] = 0, \tag{95}$$

where $\Delta[X]$ is the jump experienced by a quantity X at the surface. The conditions (94) and (95) must, of course, be satisfied for both the perturbed and the unperturbed problems.

(a) *The continuity of the normal component of* $\mathbf{B}$

Consider first the condition (94). Since $\mathbf{N}^0 = \mathbf{1}_r$, the unperturbed fields

$$\mathbf{B}^0_P = B^0_P\mathbf{1}_z = \alpha_P B_\theta \mathbf{1}_z \quad (r \leqslant r_0) \tag{96}$$

and

$$\mathbf{B}^0_V = B_\theta\left(\alpha_V\mathbf{1}_z + \frac{r_0}{r}\mathbf{1}_\theta\right) \quad (r_0 \leqslant r \leqslant R_0), \tag{97}$$

(where $\alpha_P = B_P^0/B_\theta$ and $\alpha_V = B_V^0/B_\theta$) clearly satisfy the required condition. That the condition be satisfied for the perturbed problem, as well, requires that

$$\mathbf{1}_r.\Delta[\mathbf{B}'] + \delta\mathbf{N}.\Delta[\mathbf{B}^0] = 0, \tag{98}$$

where $\delta\mathbf{N}$ is the change in the direction of the outward normal caused by the perturbation. This latter change can be inferred from the equation of motion relating $\mathbf{N}$ to the velocity of displacement, $\boldsymbol{\alpha}$, of the surface, namely (III, equation (178))

$$\frac{\partial \mathbf{N}}{\partial t} + (\boldsymbol{\alpha}.\boldsymbol{\nabla})\,\mathbf{N} = \mathbf{N}\times[\mathbf{N}\times\{(\boldsymbol{\nabla}\boldsymbol{\alpha}).\mathbf{N}\}]. \tag{99}$$

Since there are no static drifts in the problem under consideration, the appropriate, linearized form of equation (99) for the perturbed problem is

$$\Omega\delta\mathbf{N} = \mathbf{1}_r\times[\mathbf{1}_r\times\{(\boldsymbol{\nabla}\boldsymbol{\alpha}).\mathbf{1}_r\}]. \tag{100}$$

The velocity of displacement of the surface of the plasma is clearly given by

$$\boldsymbol{\alpha} = \Omega\boldsymbol{\xi}_\perp. \tag{101}$$

From equations (100) and (101) we deduce that

$$\delta\mathbf{N} = -\frac{im}{r_0}\xi_r(r_0)\,\mathbf{1}_\theta - ik\xi_r(r_0)\,\mathbf{1}_z. \tag{102}$$

Now substituting for $\mathbf{B}'$, $\mathbf{B}^0$ and $\delta\mathbf{N}$ in accordance with equations (80), (91), (96), (97) and (102) in equation (98), we find, after some reductions, that

$$i\xi_r(r_0)\,(\alpha_V y + m) = r_0^2\psi'(r_0), \tag{103}$$

where

$$y = kr_0 \tag{104}$$

is the wave number of the disturbance in the unit $1/r_0$. The explicit form of equation (103) is (cf. equations (78) and (92))

$$ik\Gamma I_m'(\Gamma y)\,(\alpha_V y + m) = r_0^2 k[C_1 I_m'(y) + C_2 K_m'(y)]. \tag{105}$$

Equations (92) and (105) now determine the constants C_1 and C_2. We find:

$$\left.\begin{aligned} C_1 &= -\frac{i}{r_0^2}\frac{\Gamma I_m'(\Gamma y)}{I_m'(y)}\frac{G_{m,\zeta}(y)}{1-G_{m,\zeta}(y)}(\alpha_V y + m),\\ C_2 &= +\frac{i}{r_0^2}\frac{\Gamma I_m'(\Gamma y)}{K_m'(y)}\frac{1}{1-G_{m,\zeta}(y)}(\alpha_V y + m), \end{aligned}\right\} \tag{106}$$

where

$$G_{m,\zeta}(y) = \frac{I_m'(y)\,K_m'(R_0 y/r_0)}{K_m'(y)\,I_m'(R_0 y/r_0)} = \frac{I_m'(y)\,K_m'(\zeta y)}{K_m'(y)\,I_m'(\zeta y)}, \tag{107}$$

where $\zeta = R_0/r_0$.

(*b*) *The continuity of the normal component of T_{ij}*

Consider next the requirement of the continuity of the normal component of T_{ij}. Since

$$T_{ij} = -P_{ij} + \frac{1}{4\pi}(B_iB_j + E_iE_j) - \frac{\delta_{ij}}{8\pi}(|\,\mathbf{E}\,|^2 + |\,\mathbf{B}\,|^2), \tag{108}$$

the condition for the unperturbed problem is

$$\Delta\left[P_\perp^0+\frac{|\mathbf{B}|^2}{8\pi}\right]=0. \tag{109}$$

According to equations (96) and (97) this condition requires that

$$P_\perp^0+\alpha_P^2\frac{B_\theta^2}{8\pi}=\frac{1}{8\pi}(\alpha_V^2B_\theta^2+B_\theta^2) \tag{110}$$

or

$$\frac{4\pi}{B_\theta^2}P_\perp^0=\tfrac{1}{2}(1+\alpha_V^2-\alpha_P^2); \tag{111}$$

this is a condition for the equilibrium of the stationary plasma.

The continuity of N_jT_{ij} for the perturbed problem on the displaced surface of the plasma leads to the single condition (cf. III, equation (207))

$$\Delta\left[\xi_r\frac{\partial}{\partial r}\left(P_\perp^0+\frac{|\mathbf{B}^0|^2}{8\pi}\right)+P'_\perp+\frac{1}{4\pi}\mathbf{B}^0.\mathbf{B}'\right]=0. \tag{112}$$

According to equations (96) and (97)

$$\begin{aligned}\Delta\left[\xi_r\frac{\partial}{\partial r}\left(P_\perp^0+\frac{|\mathbf{B}^0|^2}{8\pi}\right)\right]&=-\frac{B_\theta^2}{4\pi}\left[\xi_r\frac{\partial}{\partial r}\left(\frac{r_0}{r}\right)\right]_{r=r_0}\\&=\frac{B_\theta^2}{4\pi}\left(\frac{\xi_r}{r}\right)_{r=r_0}.\end{aligned} \tag{113}$$

Similarly, from equations (80), (91), (96) and (97) we find

$$\begin{aligned}\Delta[\mathbf{B}^0.\mathbf{B}']&=\mathbf{B}_P^0.\mathbf{B}'_P-\mathbf{B}_V^0.\mathbf{B}'_V\\&=-B_\theta^2[\alpha_P^2\nabla_\perp.\boldsymbol{\xi}_\perp+\mathrm{i}\psi(\alpha_Vy+m)]_{r=r_0};\end{aligned} \tag{114}$$

also (cf. equation (84)),

$$\Delta[P'_\perp]=-(\bar{S}+\bar{R})P_\perp^0\nabla_\perp.\boldsymbol{\xi}_\perp. \tag{115}$$

Substituting for the various terms in equation (112) in accordance with the foregoing equations, we obtain

$$-(\bar{S}+\bar{R})P_\perp^0\nabla_\perp.\boldsymbol{\xi}_\perp=\frac{B_\theta^2}{4\pi}\left[\alpha_P^2\nabla_\perp.\boldsymbol{\xi}_\perp+\mathrm{i}\psi(\alpha_Vy+m)-\frac{\xi_r}{r_0}\right]_{r=r_0}, \tag{116}$$

or, more explicitly,

$$\begin{aligned}&\left[\frac{\alpha_P^2B_\theta^2}{4\pi}+(\bar{S}+\bar{R})P_\perp^0\right]k^2\Gamma^2I_m(\Gamma y)\\&\quad=-\mathrm{i}\frac{B_\theta^2}{4\pi}(\alpha_Vy+m)[C_1I_m(y)+C_2K_m(y)]+\frac{B_\theta^2k\Gamma}{4\pi\,r_0}I'_m(\Gamma y).\end{aligned} \tag{117}$$

Finally, substituting for C_1 and C_2 from equations (106), we obtain

$$\begin{aligned}&\left[\frac{\alpha_P^2B_\theta^2}{4\pi}+(\bar{S}+\bar{R})P_\perp^0\right]k^2\Gamma^2I_m(\Gamma y)\\&\quad=-\frac{B_\theta^2}{4\pi}(\alpha_Vy+m)^2\frac{\Gamma}{r_0^2}\frac{I'_m(\Gamma y)}{1-G_{m,\zeta}(y)}\left\{G_{m,\zeta}(y)\frac{I_m(y)}{I'_m(y)}-\frac{K_m(y)}{K'_m(y)}\right\}+\frac{B_\theta^2k\Gamma}{4\pi\,r_0}I'_m(\Gamma y).\end{aligned} \tag{118}$$

Letting
$$P_m(x) = \frac{I_m(x)}{xI'_m(x)} \quad \text{and} \quad Q_m(x) = \frac{K_m(x)}{xK'_m(x)}, \tag{119}†$$

and making use of equation (88) we find that equation (118) can be reduced to the form
$$\left\{1 + 4\pi\frac{P^0_\perp - P^0_\parallel}{\alpha_P^2 B_\theta^2} + \frac{4\pi}{k^2}\frac{\rho^0\Omega^2}{\alpha_P^2 B_\theta^2}\right\}\alpha_P^2 y^2 P_m(\Gamma y)$$
$$+(\alpha_V y \pm m)^2 \frac{G_{m,\zeta}(y)\,P_m(y) - Q_m(y)}{1 - G_{m,\zeta}(y)} = 1, \tag{120}$$

where we have replaced m by $\pm m$ to emphasize that m can be a positive or a negative integer (including, of course, the value zero).

Equation (120) is the required equation for determining Ω^2.

6. The marginal state $\Omega^2 = 0$

If one assumes that the principle of the exchange of stabilities is valid, then the marginal state separating the domains of stability and instability will be characterized by $\Omega^2 = 0$; and when $\Omega^2 = 0$, the characteristic equation (120) reduces to
$$\left\{1 + 4\pi\frac{P^0_\perp - P^0_\parallel}{\alpha_P^2 B_\theta^2}\right\}\alpha_P^2 y^2 P_m(\Gamma y) + (\alpha_V y \pm m)^2 \frac{G_{m,\zeta}(y)\,P_m(y) - Q_m(y)}{1 - G_{m,\zeta}(y)} = 1, \tag{121}$$

where it may be recalled that now (cf. equation (88))
$$\Gamma^2 = \frac{P^0_\perp - P^0_\parallel + \alpha_P^2 B_\theta^2/4\pi}{(\bar{S} + \bar{R})\,P^0_\perp + \alpha_P^2 B_\theta^2/4\pi}, \tag{122}$$
$$\bar{S}P^0_\perp = \sum_{+,-} Sp^0_\perp,$$
$$\bar{R}P^0_\perp = \lim_{\Omega\to 0} -\frac{k^2}{\Omega^4}\frac{\{\sum_{+,-}(eJ_1/m)\}^2}{\sum_{+,-}(e^2N^0K/m)} \tag{123}$$

and
$$S\iint f^0 s^2\,dq\,ds^2 = \iint s^4\left(\frac{\partial f^0}{\partial q^2} - \frac{\partial f^0}{\partial s^2}\right)dq\,ds^2. \tag{124}$$

Except for the appearance of the term in $\bar{R}$, these equations are the same as those derived by Rosenbluth from an analysis of the particle orbits. The term in $\bar{R}$ arises from electric fields induced in the direction of $\mathbf{B}^0_P$; and the possibility of such electric fields are not allowed for in Rosenbluth's treatment. As we shall presently see, only under very special conditions does $\bar{R}$ vanish.

If the distributions of q and $\mathbf{s}$ are both Gaussian but with different dispersions ('temperatures') then it can be readily shown from the foregoing definitions that
$$S = 2(1-\eta) \quad \text{and} \quad Rp^0_\perp = \frac{eN^0\eta \sum_{+,-} eN^0\eta}{\sum_{+,-}[e^2(N^0)^2\eta/p^0_\perp]}, \tag{125}$$

where
$$\eta = p^0_\perp/p^0_\parallel. \tag{126}$$

† These are the functions denoted by L_m and K_m by Kruskal & Tuck (1958).

If, in addition, the ratio of the temperatures in the longitudinal and the transverse directions are the same for ions and electrons, then

$$\bar{S} = S = 2(1-\eta) \quad \text{and} \quad \bar{R} = R = 0, \tag{127}$$

since $N^{0,+} = N^{0,-}$. In our further discussion we shall assume the validity of equations (127).

In discussing the implications of equation (121) (under the circumstances leading to (127)) it is convenient to introduce the parameter

$$\beta = \frac{4\pi}{B_\theta^2} P_\perp^0; \tag{128}$$

defined in this manner, it has (apart from a factor) the usual meaning of 'β'. In terms of η and β

$$\Gamma^2 = \frac{\alpha_P^2 + \beta(1-\eta^{-1})}{\alpha_P^2 + 2\beta(1-\eta)} \tag{129}$$

and equation (121) becomes

$$\left\{\alpha_P^2 + \beta\left(1 - \frac{1}{\eta}\right)\right\} y^2 P_m(\Gamma y) + (\alpha_V y \pm m)^2 \frac{G_{m,\zeta}(y)\, P_m(y) - Q_m(y)}{1 - G_{m,\zeta}(y)} = 1. \tag{130}$$

In equation (130), α_P^2, α_V^2 and β are not all independent; for, according to equation (111),

$$2\beta = 1 + \alpha_V^2 - \alpha_P^2; \tag{131}$$

in particular,

$$\alpha_P^2 \leqslant (1 + \alpha_V^2), \tag{132}$$

and the maximum value of α_P^2 (namely, $1+\alpha_V^2$) corresponds to $\beta = 0$ and, therefore, to a vanishing plasma.

The case $\eta = 1$ has been discussed quite completely by Rosenbluth. The case $\eta \neq 1$ can be discussed in a very similar manner, though there is one important limitation which does not arise when $\eta = 1$. For, if η is sufficiently large (or small) Γ^2 can be negative; and one can convince oneself that if Γ^2 should be negative then the pinch would be definitely unstable for certain values of y. This arises from the fact that if $\Gamma^2 < 0$, then $P_m(\Gamma y)$ becomes

$$P_m(|\Gamma| y) = \frac{J_m(|\Gamma| y)}{|\Gamma| y J_m'(|\Gamma| y)}, \tag{133}$$

where $J_m(x)$ is the ordinary Bessel function. The function $P_m(|\Gamma| y)$ thus defined has poles for infinitely many values of y; in the neighbourhood of these poles, the right-hand side of equation (130) will become negatively infinite; and this clearly implies instability. Therefore, a supplementary condition for stability is

$$\left.\begin{aligned} &\alpha_P^2 + 2\beta(1-\eta) > 0 \quad \text{when} \quad \eta > 1 \\ \text{and} \quad &\alpha_P^2 + \beta(1-\eta^{-1}) > 0 \quad \text{when} \quad \eta < 1. \end{aligned}\right\} \tag{134}$$

On eliminating β by making use of equation (131), we find that these conditions are

$$\left.\begin{aligned} &\alpha_P^2 > (1+\alpha_V^2)(1-\eta^{-1}) \quad \text{when} \quad \eta > 1 \\ \text{and} \quad &\alpha_P^2 > (1+\alpha_V^2)(1-\eta)/(1+\eta) \quad \text{when} \quad \eta < 1. \end{aligned}\right\} \tag{135}$$

(Note that by definition η is positive.) For stability, the possible range of α_P^2 is, therefore, limited by

$$\left.\begin{aligned} &(1+\alpha_V^2) \geqslant \alpha_P^2 > (1+\alpha_V^2)(1-\eta^{-1}) && (\eta > 1) \\ \text{and}\quad &(1+\alpha_V^2) \geqslant \alpha_P^2 > (1+\alpha_V^2)(1-\eta)/(1+\eta) && (\eta < 1). \end{aligned}\right\} \tag{136}$$

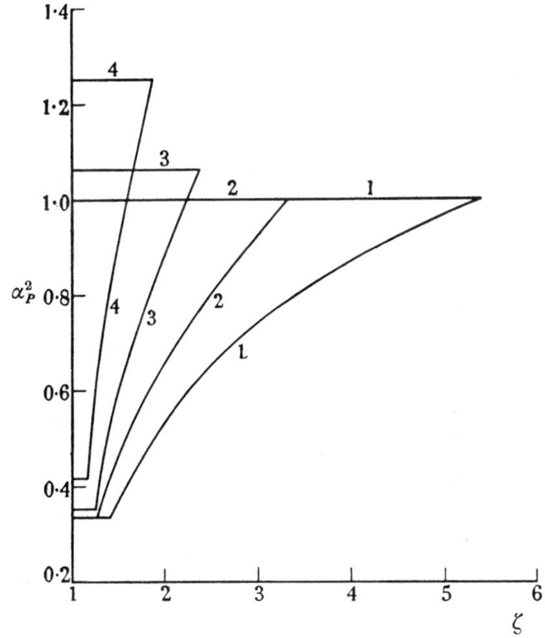

FIGURE 1

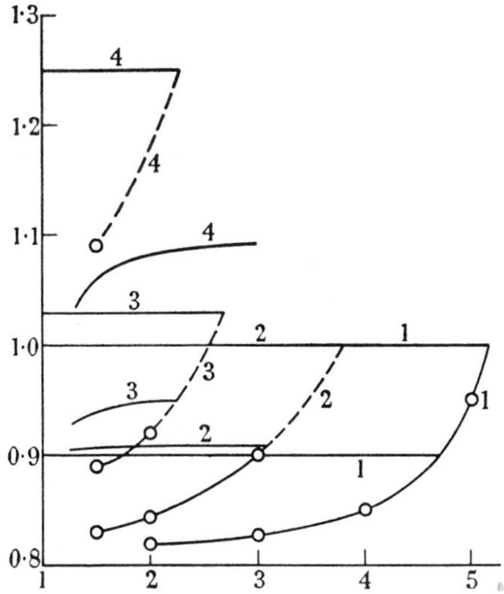

FIGURE 2

FIGURE 1. The regions of stability in the (α_P^2, ζ)-plane for the case $\eta = \frac{1}{2}$ and $\alpha_V = 0$, 0·1, 0·25 and 0·5 (distinguished by the numbers 1, 2, 3 and 4, respectively). For a given α_V, a stable pinch occurs in a region bounded on the left by the α_P^2-axis, on the right by the marginal curve for $m = -1$, above the equilibrium condition $\alpha_P^2 < (1+\alpha_V^2)$ and below by the condition $\alpha_P^2 > \frac{1}{3}(1+\alpha_V^2)$ (see (136)).

FIGURE 2. The regions of stability in the (α_P^2, ζ)-plane for the case $\eta = 5$ and $\alpha_V = 0$, 0·1, 0·25 and 0·5 (distinguished by the numbers 1, 2, 3 and 4, respectively). For a given α_V, a stable pinch occurs in a region bounded on the left by the α_P^2-axis, on the right by the marginal curve for $m = -1$, below by the condition (138) and above by the equilibrium condition $\alpha_P^2 < (1+\alpha_V^2)$. The circles are the calculated points on the marginal curve for $m = -1$; and the extrapolated parts of these curves are shown dashed.

Returning to equations (130) and (131) and considering first the case $m = 0$, we can show, as in Rosenbluth's discussion for the case $\eta = 1$, that a sufficient condition for the stability of the mode $m = 0$, for all y, is that it obtains in the limit $y = 0$. This yields the condition

$$\alpha_P^2 + 2\beta(1-\eta) + \frac{\alpha_V^2}{\zeta^2 - 1} > \tfrac{1}{2}; \tag{137}$$

or, eliminating β by means of equation (131), we obtain

$$\eta\alpha_P^2 + (1-\eta)(1+\alpha_V^2) + \frac{\alpha_V^2}{\zeta^2 - 1} > \tfrac{1}{2}. \tag{138}$$

This condition is always satisfied for $\eta < \frac{1}{2}$; and for $\eta > \frac{1}{2}$, equation (138) defines, for each prescribed η, a locus in the (α_P^2, ζ)-plane which, together with (136), delimits a region of stability (for the mode $m = 0$) in this plane. These regions for different values of η and α_V^2 are shown in figures 1 and 2.

When $|m| = 1$, a similar locus can be defined in the (α_P^2, ζ)-plane. However, the specification of this locus requires a careful numerical examination of equations (130) and (131). (It is sufficient to consider $m = -1$ as being more unfavourable for stability.) Precisely, it requires the determination (by trial and error) of the maximum value of α_P^2 which will solve equation (130) (for some y) for assigned values of ζ, η and α_V. Mrs Josephine Powers has carried out the necessary calculations and her results are included in figures 1 and 2.

Finally, considering the modes $|m| \geqslant 2$, we can again show, following Rosenbluth, that the pinch will be stable for all y and $|m| > 2$ if it is stable for $|m| = 2$, all y and $\zeta = \infty$. The case $|m| = 2$ and $\zeta = \infty$ needs a separate discussion. The most unfavourable circumstances for stability, in case $|m| = 2$, arise from

$$\{\alpha_P^2 + \beta(1-\eta^{-1})\}\, y^2 P_2(\Gamma y) + (\alpha_V y - 2)^2 Q_2(y) = 1. \tag{139}$$

For a given α_V^2 this equation determines a maximum value for α_P^2 for which it allows a solution for some y. These values are listed in table 1. From an examination of this table it appears that if the modes $m = 0$ and -1 are stable and the supplementary conditions (136) are satisfied then all the remaining modes are stable.

TABLE 1. THE MAXIMUM VALUE OF α_P^2 FOR WHICH EQUATION (139) ALLOWS A SOLUTION FOR ASSIGNED VALUES OF α_V AND η

$\eta = 0\cdot5$			$\eta = 5\cdot0$		
α_V	α_P^2	y	α_V	α_P^2	y
0·10	0·383	3·52	0·10	0·307	5·00
0·25	0·464	2·20	0·25	0·348	3·55
0·50	0·688	1·50	0·50	0·422	1·95
1·00	1·482	1·00	1·00	0·970	1·00

7. THE TIME-DEPENDENT CASE

When $\Omega^2 \neq 0$, we must go back to equation (120) in which the definition of Γ^2 also involves Ω^2; thus,

$$\Gamma^2 = \frac{P_\perp^0 - P_\parallel^0 + \alpha_P^2 B_\theta^2/4\pi + \rho\Omega^2/k^2}{(\bar{S}+\bar{R})\,P_\perp^0 + \alpha_P^2 B_\theta^2/4\pi}, \tag{140}$$

where the 'shape factor' $(\bar{S}+\bar{R})$ also depends on Ω^2.

If as in § 6 we consider Gaussian distributions of q and $\mathbf{s}$ (but with different temperatures) then it can be shown that (cf. equation (125))

$$\bar{S}P_\perp^0 = 2\sum_{+,-}(1-\tilde{\eta})\,p_\perp^0, \quad \bar{R}P_\perp^0 = \frac{\{\sum_{+,-}(eN\tilde{\eta})\}^2}{\sum_{+,-}(e^2N^2\tilde{\eta}/p_\perp^0)}, \tag{141}$$

where
$$\tilde{\eta} = \eta[1-H(\sigma)] = \frac{p_\perp^0}{p_\parallel^0}[1-H(\sigma)], \tag{142}$$

$$\sigma = \frac{\Omega}{k}\left(\frac{m}{2\boldsymbol{k}T_\parallel}\right)^{\frac{1}{2}} \quad (\boldsymbol{k} = \text{Boltzmann constant}) \tag{143}$$

and
$$H(x) = \pi^{\frac{1}{2}} x\, e^{x^2}\left[1 - \frac{2}{\sqrt{\pi}}\int_0^x e^{-y^2}\,dy\right]. \tag{144}$$

From the definition of $H(x)$ it follows that
$$H(x) \to 0 \text{ as } x \to 0 \quad\text{and}\quad H(x) \to 1 \text{ as } x \to \infty. \tag{145}$$

If $\Omega^2 > 0$, then the characteristic equation can be discussed without any difficulty of principle. If, however, we should be interested in stable oscillations, then we should exercise some care since the integrals such as I_1, J_1, etc., which we have to define, are divergent. Under these circumstances, we may suppose that we are solving the Boltzmann equation by considering its Laplace transform; thus
$$f(\mathbf{r}, \Omega) = \int_0^\infty f(\mathbf{r}, t)\, e^{-\Omega t}\,dt. \tag{146}$$

The solution we have found is formally $f(\mathbf{r}, \Omega)$; and the required solution satisfying suitable boundary conditions can then be obtained by inversion. This will require us to go into the complex Ω-plane; and the solution of the problem can be completed as has been done in the case of plasma oscillations by Landau (1946; see also van Kampen 1955). We hope to return to these matters in greater detail on a later occasion.

NOTES ADDED IN PROOF 10 MAY 1958

(a) *The non-occurrence of overstability in the pinch*

In discussing the stability of the pinch in § 6, we have assumed that $\Omega^2 = 0$ separates the domains of stability and instability. To complete the discussion, it is clearly necessary to examine whether overstability can occur. It can be shown that this is not possible. We had constructed a proof for this; at the same time Dr Marshall Rosenbluth communicated to us a somewhat different but very elegant proof. The following is an outline of Dr Rosenbluth's proof; we are grateful to him for allowing us to include it in this paper.

The basic equations are (107), (119), (120) and (140) to (144). These equations must be considered in the complex Ω-plane.

It is convenient to rewrite equation (120) in the form
$$\Phi(\Omega) = \left\{1 + \frac{4\pi}{\alpha_P^2 B_\theta^2}\left(P_\perp^0 - P_\parallel^0 + \frac{\rho\Omega^2}{k^2}\right)\right\}\alpha_P^2 y^2 P_m(\Gamma y) - L = 0.$$

Here L is independent of Ω and all quantities except Γ^2 and Ω are real.

Considering first $H(\Omega)$ (as defined in equations (143) and (144)), we observe that this is analytic on the right half of the complex plane and on the imaginary axis; further the sign of its imaginary part is the same as that of Ω; and the imaginary part vanishes only for real Ω. From these properties of $H(\Omega)$ we deduce that: $\Gamma^2(\Omega)$ is analytic on the right half plane; it has zeros only at $\pm i\Omega_0$ where
$$\Omega_0 = k\sqrt{(P_\perp^0 - P_\parallel^0 + \alpha_P^2 B_\theta^2/4\pi)/\sigma};$$

it has a non-vanishing imaginary part except on the real axis and on the curves $L-1$ and $L-2$ passing through $\pm i\Omega_0$; and on these latter lines $\Gamma^2 > 0$.

Considering next $P_m(\Gamma y)$ (as defined in equation (119)) for complex Γ, we infer from the non-existence of complex zeros of $AJ_m(z)+BzJ'_m(z)$ for all real A and B (see Watson 1952, p. 482), that the imaginary part of $P_m(\Gamma y)$ is non-vanishing and that it has no poles except when Γ^2 is real and negative. Therefore $\Phi(\Omega)$ has no poles on the right half of the complex Ω-plane and is in fact analytic. Consequently, we may determine the number of zeros of $\Phi(\Omega)$ on this half of the complex plane by the principle of the argument, namely by evaluating the integral,

$$C = \frac{1}{2\pi i}\int \frac{1}{\Phi}\frac{d\Phi}{d\Omega}d\Omega,$$

along a contour consisting of a semi-circular arc of sufficiently large radius and the intercepted part of the imaginary axis; C is the total change in the argument of Φ around the contour. From the fact that $\Phi(\Omega)$ becomes proportional to Ω as $\Omega \to \infty$, it is clear that the contribution to C from the semi-circular arc is $\frac{1}{2}$. On the imaginary axis, the imaginary part of Φ is positive during the whole contour and the change in the argument is $\pm\frac{1}{2}\pi$ depending on whether $\Phi(0)$ is positive or negative. Hence $C = 1$ if $\Phi(0) < 0$ and $C = 0$ if $\Phi(0) > 0$. The conclusion then is that if $\Gamma^2(0) > 0$ and $\Phi(0) > 0$, the pinch is stable; otherwise it is unstable. Moreover, in the latter case, since on the real axis $\Phi(\Omega)$ increases from a negative number to infinity, the instability evidently occurs with a real frequency. In the case of a stable pinch no purely imaginary roots are possible, i.e. the oscillations are Landau-damped.

(b) *The induced electric field parallel to* $\mathbf{B}$

By combining equations (19), (60) and (69) we clearly obtain an expression for $E'_\parallel$. This latter expression can be derived more directly as follows: If the displacement current is ignored

$$\operatorname{curl}\mathbf{B} = 4\pi\mathbf{j}/c. \tag{i}$$

Taking the divergence of this equation, we clearly obtain

$$\text{charge density} = \sum_{+,-} eN' = 0. \tag{ii}$$

On the other hand according to equation (7)

$$N' = \frac{1}{2}\iint A_1 \,dq\,ds^2; \tag{iii}$$

and in limit $\Omega^2 = 0$ we have (cf. equations (19), (22) and (29))

$$A_1 = \frac{B'_\parallel}{B^0}s^2\left(\frac{\partial f^0}{\partial q^2}-\frac{\partial f^0}{\partial s^2}\right)-\frac{2e}{imk}\frac{\partial f^0}{\partial q^2}E'_\parallel. \tag{iv}$$

We thus and

$$2E'_\parallel\left\{\sum_{+,-}\frac{e^2}{m}\iint dq\,ds^2\frac{\partial f^0}{\partial q^2}\right\} = ik\frac{B'_\parallel}{B^0}\sum_{+,-}\iint dq\,ds^2\left(\frac{\partial f^0}{\partial q^2}-\frac{\partial f^0}{\partial s^2}\right)s^2, \tag{v}$$

in agreement with the result given in the paper.

(c) *The case* $\eta = 1$

As we have stated, Rosenbluth (1958) has carried out the discussion of the stability for the case $\eta = 1$. Since his paper may not be readily available we reproduce, with his permission, his results in figure 3.

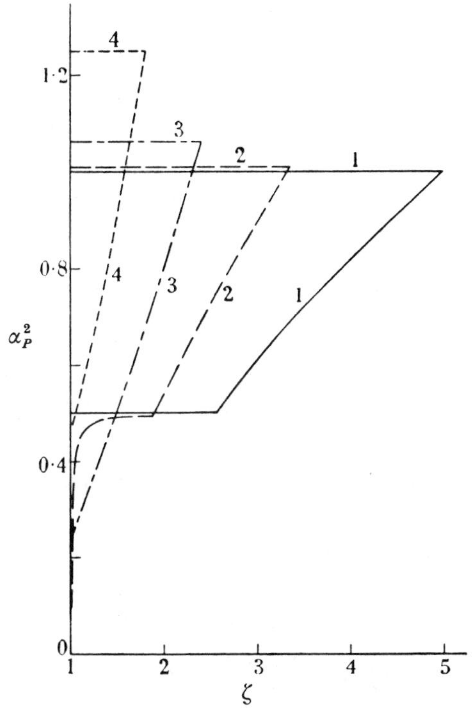

FIGURE 3. The regions of stability in the (α_P^2, ζ)-plane for the case $\eta = 1$ and $\alpha_V = 0$, 0·1, 0·25 and 0·5 (distinguished by the numbers 1, 2, 3 and 4, respectively). For a given α_V, a stable pinch occurs in a region bounded on the left by the α_P^2-axis, on the right by the marginal curve for $m = -1$, above by the equilibrium condition $\alpha_P^2 < (1 + \alpha_V^2)$ and below by the marginal curve for $m = 0$.

This investigation was carried out under the auspices of the United States Atomic Energy Commission.

REFERENCES

Bernstein, I. B., Frieman, E. A., Kruskal, M. D. & Kulsrud, R. M. 1958 *Proc. Roy. Soc.* A, **244**, 17.
Chandrasekhar, S., Kaufman, A. N. & Watson, K. M. 1957 *Ann. Phys.* (New York) **2**, 435.
Kruskal, M. & Tuck, J. L. 1958 *Proc. Roy. Soc.* A, **245**, 222.
Landau, L. D. 1946 *J. Phys. U.S.S.R.* **10**, 25.
Rosenbluth, M. 1957 *Stability of the Pinch.* Los Alamos Scientific Laboratory Report No. 2030 (written April 1956; distributed April 1957).
Rosenbluth, M. & Longmire, C. 1957 *Ann. Phys.* (New York) **1**, 120.
Shrafranov, V. D. 1957 *J. Nucl. Energy*, **2**, 86.
Tayler, R. J. 1957 *Proc. Phys. Soc.* B, **70**, 1049.
Van Kampen, N. G. 1955 *Physica*, **21**, 949.
Watson, G. N. 1952 *Theory of Bessel functions.* Cambridge University Press.

APPENDIX. THE PROPAGATION OF PLANE HYDROMAGNETIC WAVES IN AN INFINITE MEDIUM

In deriving the equation (§ 3, equation (75))

$$\nabla_\perp^2 \chi = k^2 \Gamma^2 \chi, \tag{A 1}$$

no assumptions (in addition to those underlying the general perturbation theory) were made except that there is a uniform field, $\mathbf{B}^0$, in the z-direction and that all quantities describing the perturbation have the dependence

$$e^{\Omega t + \mathrm{i}kz} \tag{A 2}$$

on t and z. Accordingly, if the plasma should be of infinite extent, we may seek solutions of (A 1) whose dependence on x and y is also periodic with a (total) wave number $k_\perp$. For such plane waves (A 1) provides the dispersion relation

$$k_\perp^2 + k_\parallel^2 \Gamma^2 = 0, \tag{A 3}$$

where we have written $k_\parallel$ in place of k to emphasize that this represents the wave number of the disturbance in a direction parallel to $\mathbf{B}^0$. Substituting for Γ^2 from equation (88) in (A 3), we obtain

$$k_\perp^2 \left\{ \frac{|B^0|^2}{4\pi} + (\bar{S} + \bar{R}) P_\perp^0 \right\} + k_\parallel^2 \left\{ P_\perp^0 - P_\parallel^0 + \frac{|B^0|^2}{4\pi} - \rho \frac{\omega^2}{k_\parallel^2} \right\} = 0, \tag{A 4}$$

where we have further written $\mathrm{i}\omega$ in place of Ω; ω denotes, therefore, the frequency of the wave.

Letting (cf. equations (126) and (128))

$$\beta = \frac{4\pi}{|B^0|^2} P_\perp^0, \quad \eta = \frac{P_\perp^0}{P_\parallel^0} \tag{A 5}$$

and

$$k_\perp^2 = k^2 \cos^2 \vartheta, \quad k_\parallel^2 = k^2 \sin^2 \vartheta \quad (k^2 = k_\parallel^2 + k_\perp^2), \tag{A 6}$$

we can rewrite (A 4) in the form

$$\beta(\bar{S} + \bar{R}) \cos^2 \vartheta + \beta \left(1 - \frac{1}{\eta}\right) \sin^2 \vartheta + 1 = \frac{4\pi\rho\omega^2}{|B^0|^2 k^2}. \tag{A 7}$$

If the distributions of q and $\mathbf{s}$ are Gaussian (but with different dispersions) then $\bar{S}$ and $\bar{R}$ in (A 7) have the values given in § 7, equations (141) to (144).

From (A 7) it follows that we shall have *instability* if

$$\beta(\bar{S} + \bar{R}) \cos^2 \vartheta + \beta(1 - 1/\eta) \sin^2 \vartheta + 1 < 0, \tag{A 8}$$

where $\bar{S}$ and $\bar{R}$ are now to be evaluated for the limit $\omega = 0$.

If the particular assumptions of § 6 leading to equation (127) are made, then (A 8) becomes

$$2\beta(1 - \eta) \cos^2 \vartheta + \beta(1 - 1/\eta) \sin^2 \vartheta + 1 < 0. \tag{A 9}$$

A case of special interest is when the waves are propagated in the direction of $\mathbf{B}^0$ Then $k_\perp = 0$ and the dispersion relation is (cf. (A 4))

$$P_\perp^0 - P_\parallel^0 + \frac{|B^0|^2}{4\pi} = \frac{\rho\omega^2}{k^2} \quad (k_\perp = 0;\ k = k_\parallel). \tag{A 10}$$

When $P_\perp^0 = P_\parallel^0$ this represents the usual Alfvén wave. The waves described by (A 8) represent unstable modes if

$$P_\perp^0 - P_\parallel^0 + \frac{|B^0|^2}{4\pi} < 0. \tag{A 11}$$

This special case of (A 9) has been stated by Longmire (1956, private communication); and it has been discussed more recently by Parker (1957, private communication). Also, it may be noted that (A 11) is equivalent to one of the conditions for stability considered in § 6 (cf. (134)).

ADIABATIC INVARIANTS IN THE MOTIONS OF CHARGED PARTICLES

S. CHANDRASEKHAR*

I. INTRODUCTION

In current treatments of plasma physics one often considers the motions of charged particles in varying magnetic fields in a so-called guiding center approximation. In this approximation one separates the spiraling motion of the charged particles about the lines of force, from the motion along the lines of force. This separation of the motion into the two parts is possible only so long as the magnetic field remains sensibly constant, spatially, over several Larmor radii and, temporally, over several Larmor periods. When these latter conditions are fulfilled, one generally supposes that the transverse kinetic energy ($w_\perp$) of the spiraling motion divided by the strength of the magnetic field (B) remains constant during the motion. This constancy of

$$\mu = \frac{w_\perp}{B} = \frac{m v_\perp{}^2}{2B} \tag{1}$$

is not strictly an integral of the equations of motion; it is an *adiabatic invariant* in the sense that it is a constant in the limit of infinitely slow variation of the field.

If in virtue of the constancy of μ, the particle should be trapped between two regions of relatively strong field,† then one supposes that the integral

$$\oint \mathbf{v} \cdot \frac{d\mathbf{B}}{B} \tag{2}$$

* The Enrico Fermi Institute for Nuclear Studies, University of Chicago.

† If W $(= w_\perp + w_{\|})$ denotes the total kinetic energy of the particle, then the points between which the particle will be trapped will be determined by $B_{\text{max}} = W/\mu$.

of the component of the velocity parallel to the field taken over a complete cycle is a further adiabatic invariant.*

In this paper we shall examine the precise meaning which must be attached to the notion of adiabatic invariance with a view to clarifying the limitations in its use in plasma physics.

II. THE NOTION OF ADIABATIC INVARIANCE

The notion of adiabatic invariance played an important role in the early developments of the quantum theory in the context of formulating the general rules of quantization. Historically, it arose from a question proposed by Lorentz at the first Solvay Congress in 1911. Lorentz's question was: How does a simple pendulum behave when the length of the suspending thread is gradually shortened? The relevance of this question for the quantum theory of the time was the following: If an oscillator has originally the correct energy appropriate to an elementary quantum ($h\nu$), would the energy suffice to make up a quantum at the end of a process

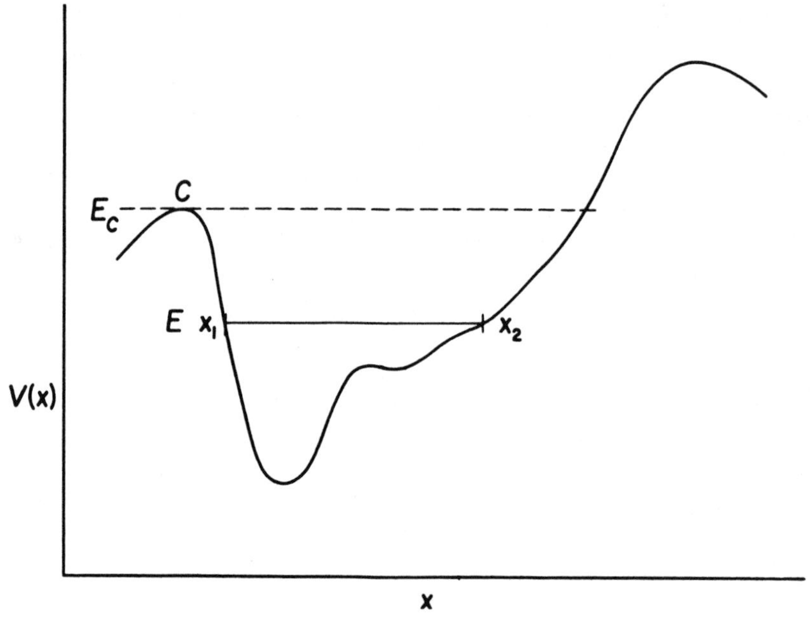

FIG. 1. Potential field with a barrier

* The conditions which must prevail in order that Eq. (2) may be an adiabatic invariant are very much more stringent than those which must obtain for $w_\perp/B$ to be an invariant (see Section XII).

(such as shortening the length of a pendulum) in which the frequency has been increased? To Lorentz's question, Einstein furnished the correct answer by saying that if the suspending thread is shortened infinitely slowly, then the energy, E, will increase proportionately to the "instantaneous" frequency, ν, so that a quantum of energy remains a quantum of energy; in other words, E/ν *is an adiabatic invariant.* We shall presently return to a somewhat more precise formulation of what this adiabatic invariance of E/ν really means; but it may be noted here, parenthetically, that the invariance of E/ν to slow changes was extended by Ehrenfest into a general "adiabatic hypothesis" applicable to multiply periodic systems. In this paper we shall, however, be concerned with adiabatic invariance as a concept in classical mechanics and, indeed, only for one-dimensional systems.

III. THE ADIABATIC INVARIANCE OF THE ACTION INTEGRAL

Consider the motion of a particle (of unit mass) in a potential field $V(x)$ (see Fig. 1). The equation governing its motion is

$$\frac{d^2x}{dt^2} + \frac{dV(x)}{dx} = 0. \tag{3}$$

This equation allows the energy integral

$$\frac{1}{2}\left(\frac{dx}{dt}\right)^2 + V(x) = E = \text{constant}; \tag{4}$$

from this we deduce that

$$t = \int \frac{dx}{\{2[E - V(x)]\}^{\frac{1}{2}}}. \tag{5}$$

If two points x_1 and x_2 exist such that (see Fig. 1)

$$E = V(x_1) = V(x_2) \qquad (x_2 > x_1), \tag{6}$$

then the particle will oscillate between x_1 and x_2; and the period of the oscillation will be given by

$$P = 2\int_{x_1}^{x_2} \frac{dx}{\{2[E - V(x)]\}^{\frac{1}{2}}} = \oint \frac{dx}{\{2[E - V(x)]\}^{\frac{1}{2}}}. \tag{7}$$

Associated with such periodic motions one defines the *action integral*

$$J = \oint dx\{2[E - V(x)]\}^{\frac{1}{2}} = \oint dx\left(\frac{dx}{dt}\right). \tag{8}$$

Suppose that $V(x)$ depends (apart from x) on a set of parameters a_j $(j=1,\ldots,n)$ such that

$$V(x)=V(x;a_1,\ldots,a_n). \tag{9}$$

Now let the a_j's instead of being constants be slowly varying functions of time. "Slowly varying" in this connection means that the particle executes several oscillations (appropriate to a set of a_j's which occur) before any of the a_j's change appreciably; in other words, we require:

$$\left|\frac{1}{a_j}\frac{da_j}{dt}\right| << \frac{1}{P(a_1,\ldots,a_n)} \qquad (j=1,\ldots,n). \tag{10}$$

When this is the case, we may characterize the motion by a certain energy, $\overline{E}$, appropriate for a single oscillation, namely,

$$\begin{aligned}\overline{E}&=\oint\frac{dx}{\dot{x}}\left[\frac{1}{2}\left(\frac{dx}{dt}\right)^2+V(x;a_1,\ldots,a_n)\right]\Big/\oint\frac{dx}{\dot{x}}\\ &=\oint dt\left[\frac{1}{2}\left(\frac{dx}{dt}\right)^2+V(x;a_1,\ldots,a_n)\right]\Big/\oint dt.\end{aligned} \tag{11}$$

It will be observed that in defining $\overline{E}$ in this way, we are averaging the energy during an oscillation, weighting each element of the orbit by the fraction of the time the particle spends in that element.

The theorem on adiabatic invariance is to the effect that *the action integral,*

$$J=\oint dx\,\{2[\overline{E}-V(x;a_1,\ldots,a_n)]\}^{\frac{1}{2}}, \tag{12}$$

defined in terms of $\overline{E}$ is a constant for infinitely slow variations of the a_j's. The arguments by which one attempts to establish this constancy of J are, essentially, as follows:

We have

$$\frac{dJ}{dt}=\frac{\partial J}{\partial\overline{E}}\frac{d\overline{E}}{dt}+\sum_{j=1}^{n}\frac{\partial J}{\partial a_j}\frac{da_j}{dt}. \tag{13}$$

On the other hand, from the definition of J we have

$$\left.\begin{aligned} \frac{\partial J}{\partial a_j} &= -\oint \frac{\partial V}{\partial a_j} \frac{dx}{\{2[\bar{E}-V]\}^{\frac{1}{2}}} \\ \text{and} \qquad & \\ \frac{\partial J}{\partial \bar{E}} &= \oint \frac{dx}{\{2[\bar{E}-V]\}^{\frac{1}{2}}} = \oint \frac{dx}{\dot{x}}; \end{aligned}\right\} \tag{14}$$

while from the definition of $\bar{E}$ we have

$$\frac{d\bar{E}}{dt} = \oint dt \left[\frac{dx}{dt}\frac{d^2x}{dt^2} + \frac{\partial V}{\partial x}\frac{dx}{dt} + \sum_{j=1}^{n} \frac{\partial V}{\partial a_j}\frac{da_j}{dt} \right] \Big/ \oint dt. \tag{15}$$

Making use of the equation of motion governing x, we can reduce the foregoing equation to the form

$$\begin{aligned} \frac{d\bar{E}}{dt} &= \sum_{j=1}^{n} \frac{da_j}{dt} \oint \frac{\partial V}{\partial a_j} dt \Big/ \oint dt \\ &= \sum_{j=1}^{n} \frac{da_j}{dt} \oint \frac{\partial V}{\partial a_j} \frac{dx}{\{2[\bar{E}-V]\}^{\frac{1}{2}}} \Big/ \oint \frac{dx}{\dot{x}}. \end{aligned} \tag{16}$$

Combining Eqs. (13), (14), and (16), we have

$$\begin{aligned} \frac{dJ}{dt} &= \sum_{j=1}^{n} \frac{da_j}{dt} \oint \frac{\partial V}{\partial a_j} \frac{dx}{\{2[\bar{E}-V]\}^{\frac{1}{2}}} \\ &\quad - \sum_{j=1}^{n} \frac{da_j}{dt} \oint \frac{\partial V}{\partial a_j} \frac{dx}{\{2[\bar{E}-V]\}^{\frac{1}{2}}} = 0, \end{aligned} \tag{17}$$

which establishes the constancy of J.

IV. ADIABATIC INVARIANCE IN THE SMALL AND IN THE LARGE

The proof of the constancy of the action integral given in Section III, for slow changes in the parameters which occur in the potential function, contains many heuristic elements. For example, the proof at no stage involved any explicit estimate of the error involved in the concept of $\bar{E}$ and of the action integral defined in terms of it. Indeed, the notion of adiabatic invariance was taken as equivalent to a justification of $\overline{E}$ and J as defined. This is not altogether satisfactory; and we shall accordingly try to formulate more explicitly the conceptual prerequisites underlying the notion of adiabatic invariance. In order that we may clarify the essential physical concepts without any of the formal complexities which a completely general formulation will require, we shall consider the relevant questions explicitly in the context of Lorentz's original problem and leave the generalizations to the reader.

Consider then the equation of motion of a simple pendulum. We have

$$\frac{d^2x}{dt^2} + \omega^2 x = 0, \tag{18}$$

where ω denotes the circular frequency. If ω is a constant, the general solution of this equation can be written as

$$x = A \cos(\omega t + \varepsilon), \tag{19}$$

where A and ε are constants. From this solution it follows that

$$\langle \dot{x}^2 \rangle = \tfrac{1}{2} A^2\omega^2; \quad \langle x^2 \rangle = \tfrac{1}{2} A^2; \tag{20}$$

and, further, that

$$\overline{E} = \tfrac{1}{2}\{\langle \dot{x}^2 \rangle + \omega^2 \langle x^2 \rangle\} = \tfrac{1}{2} A^2\omega^2. \tag{21}$$

The adiabatic invariance of the action integral as applied to this problem states the following:

Let ω, instead of being a constant, be a slowly varying function of time such that

$$\left| \frac{1}{\omega}\frac{d\omega}{dt} \right| << \omega. \tag{22}$$

For such slow variations

$$\overline{E}/\omega \text{ is a constant.} \tag{23}$$

If one wants to formulate the principle of adiabatic invariance without any of the heuristic elements implied in the forgeoing statement, then it would appear that the physical premises should be restated somewhat differently.

Clearly, the concepts of energy and period taken over from the case when ω is a constant cannot strictly apply to a dynamical system in which ω is some function of time no matter how slowly varying. They can apply only if certain limiting conditions are satisfied. Thus, let ω (t) be a function of time such that

$$\omega(t) \to \omega_1 \text{ as } t \to -\infty,$$

and (24)

$$\omega(t) \to \omega_2 \text{ as } t \to +\infty;$$

and also that

$$\frac{d^n\omega}{dt^n} \to 0 \quad \text{as } t \to +\infty \quad \text{and } t \to -\infty \quad \text{for all } n \geqslant 1. \tag{25}$$

When these conditions are satisfied, the system is a simple harmonic oscillator, in the strict sense, both when $t \to -\infty$ and when $t \to +\infty$. Accordingly, $\langle x^2 \rangle$ can be defined uniquely for both these limits and we may consider the ratio

$$\lambda = \frac{\omega_2 \langle x^2 \rangle_{t \to +\infty}}{\omega_1 \langle x^2 \rangle_{t \to -\infty}}. \tag{26}$$

This will clearly depend on the manner in which ω varies between ω_1 and ω_2. Let $d\omega/dt$ be bounded and

$$\text{Maximum of } \left| \frac{1}{\omega} \frac{d\omega}{dt} \right| = \frac{1}{T} \qquad (-\infty < t < +\infty). \tag{27}$$

A precise statement of the theorem on adiabatic invariance would be the assertion:

$$\lambda \to 1 \quad \text{as} \quad T \to \infty. \tag{28}$$

The remarkable aspect of this assertion is that it is not restricted by any limitation on ω_2/ω_1. Nevertheless, it will be convenient to distinguish two cases: the case when

$$\delta = \tfrac{1}{2} \left| \frac{\omega_2 - \omega_1}{\omega_2 + \omega_1} \right| << 1; \tag{29}$$

and when no such restriction applies. If, in the former case, we can show that

$$\lambda = 1 + O(\delta^2) \quad \text{as} \quad T \to \infty, \tag{30}$$

then we shall say that we have *adiabatic invariance in the small.* If on the other hand, Eq. (28) holds with no restriction on ω_2/ω_1, we shall say that we have *adiabatic invariance in the large.*

The principle of adiabatic invariance in the large can be formulated somewhat differently as follows: Let

$$\omega \equiv \omega(\alpha t) \tag{31}$$

represent a one-parameter family of time variations satisfying the requirements (24) and (25). For such a family of time variations, λ defined as in Eq. (26) will depend on α; and adiabatic invariance in the large implies that

$$\lambda(\omega_1, \omega_2; \alpha) \to 1 \quad \text{as} \quad \alpha \to 0, \tag{32}$$

independently of ω_1 and ω_2.

It is clear that the foregoing ideas formulated in the context of Lorentz's original problem can be extended to include the more general problem considered in Section III.

V. THE RELATION OF THE ADIABATIC INVARIANCE OF $w_\perp/B$ TO THE INVARIANCE OF $\omega \langle |x|^2 \rangle$ IN THE PENDULUM PROBLEM

Before discussing in some detail the adiabatic invariance of $\omega \langle |x|^2 \rangle$ in the problem of the simple pendulum, it will be useful to relate this problem to the invariance of $w_\perp/B$ in the motion of a charged particle in a varying magnetic field.

With the substitution

$$\zeta = x \exp\left(-i \int \omega \, dt\right), \tag{33}$$

the equation

$$\frac{d^2 x}{dt^2} + \omega^2 x = 0 \tag{34}$$

becomes

$$\frac{d^2\zeta}{dt^2} + 2i\omega \frac{d\zeta}{dt} + i \frac{d\omega}{dt} \zeta = 0; \tag{35}$$

and this is the equation of motion of a charged particle in a spatially uniform but a temporally varying magnetic field, if we identify 2ω as the Larmor frequency ($= eB/mc$) and the real and the imaginary parts of ζ (regarded as a complex variable) as the Cartesian coordinates of the particle in a plane normal to the lines of force. The transverse kinetic energy divided by the instantaneous strength of the field is, apart from constant factors of proportionality, given by

$$\frac{1}{\omega} \left| \frac{d\zeta}{dt} \right|^2 = \frac{1}{\omega} \left| \frac{dx}{dt} - i\omega x \right|^2. \tag{36}$$

The adiabatic invariance of $w_{\perp}/B$ in the motion of a charged particle in a uniform but time-dependent magnetic field is, therefore, very directly related to the adiabatic invariance of $\bar{E}/\omega$ in the problem of the simple pendulum; they represent, in fact, the same problem.

VI. THE ADIABATIC INVARIANCE OF $\omega\langle|x|^2\rangle$ IN THE SMALL

The proof of the adiabatic invariance of $\omega\langle|x|^2\rangle$ in the small can be accomplished very simply.

By letting

$$t_1 = \int \omega dt \quad \text{and} \quad dt_1 = \omega dt, \tag{37}$$

we find that Eq. (34) becomes

$$\frac{d^2x}{dt_1{}^2} + x = -\left(\frac{1}{\omega}\frac{d\omega}{dt_1}\right)\frac{dx}{dt_1}. \tag{38}$$

We solve this equation by an iteration procedure, i.e., by evaluating the right-hand side of the equation in terms of the solution obtained when this side is ignored; and solving the resulting equation as a non-homogeneous equation for x. Thus, if

$$x = e^{it_1} \tag{39}$$

is the solution appropriate for $t_1 \to -\infty$, the equation we have to solve is

$$\frac{d^2x}{dt_1{}^2} + x = -i\left(\frac{1}{\omega}\frac{d\omega}{dt_1}\right)e^{it_1}. \tag{40}$$

The solution of Eq. (40) which tends to (39) as $t_1 \to -\infty$ is readily found to be

$$x = e^{it_1}\left(1 - \tfrac{1}{2}\log\frac{\omega}{\omega_1}\right) + \tfrac{1}{2}e^{-it_1}\int_{-\infty}^{t_1}\frac{1}{\omega}\frac{d\omega}{dt_1'}e^{2it_1'}dt_1'. \tag{41}$$

From this solution it follows that

$$\lim_{t_1\to+\infty}\left\{\omega\langle|x|^2\rangle\right\} = \omega_2\left(1 - \tfrac{1}{2}\log\frac{\omega_2}{\omega_1}\right)^2 + \omega_2|P|^2, \tag{42}^*$$

where

$$P = \tfrac{1}{2}\int_{-\infty}^{+\infty}\frac{1}{\omega}\frac{d\omega}{dt_1}e^{2it_1}dt_1. \tag{43}$$

* In writing this equation we have further averaged over all initial phases.

Accordingly [cf. Eq. (26)]

$$\lambda = \frac{\omega_2}{\omega_1}\left(1 - \tfrac{1}{2}\log\frac{\omega_2}{\omega_1}\right)^2 + \frac{\omega_2}{\omega_1}|P|^2. \tag{44}$$

If we now suppose that

$$\omega_2 = \omega_1 + \delta\omega, \tag{45}$$

as is permissible when considering adiabatic invariance in the small, then we deduce from Eq. (44) that

$$\lambda = 1 + O(\delta\omega^2) + \frac{\omega_2}{\omega_1}|P|^2. \tag{46}$$

It is evident that in the limit of infinitely slow variation,

$$|P|^2 \to 0. \tag{47}$$

The adiabatic invariance in the small is therefore established.

VII. THE ADIABATIC INVARIANCE OF $\omega\langle|x|^2\rangle$ IN THE LARGE

The adiabatic invariance of $\omega\langle|x|^2\rangle$ in the large is more difficult to establish. It can, however, be made to "look obvious" by a further transformation of Eq. (38). Thus, by introducing the variable

$$x_1 = \sqrt{\omega}\, x \tag{48}$$

in place of x, we find that Eq. (38) becomes

$$\frac{d^2x_1}{dt_1^2} + x_1 + \Omega x_1 = 0 \tag{49}$$

where

$$\Omega = \tfrac{1}{4}\left(\frac{1}{\omega}\frac{d\omega}{dt_1}\right)^2 - \frac{1}{2\omega}\frac{d^2\omega}{dt_1^2}. \tag{50}$$

It will be observed that in contrast to Eq. (38), the "correction terms" included in Ω in Eq. (49) are of the second order in $d\omega/dt$. If we can ignore the term in Ω in Eq. (49), then it is clearly a consequence of this equation that $\langle|x_1|^2\rangle$ is indeed a constant; and the constancy, in this approximation, of $\omega\langle|x|^2\rangle$ directly follows. More precisely, we may try to solve Eq. (49) by an iteration procedure by evaluating the term Ωx_1 in terms of the solution obtained by ignoring this term; and solving the resulting equation as an inhomogeneous equation for x_1. In this manner we find that the solution of Eq. (49), which tends to e^{it_1} as $t_1 \to -\infty$, is given by

$$x_1 = e^{it_1} + \tfrac{1}{2} i \left\{ e^{it_1} \int_{-\infty}^{t_1} \Omega(t_1')\, dt_1' - e^{-it_1} \int_{-\infty}^{t_1} \Omega(t_1')\, e^{2it_1'}\, dt_1' \right\}. \quad (51)$$

If we now suppose that

$$\omega(t) \equiv \omega(\alpha t) = \omega(\tau) \qquad (\tau = \alpha t), \quad (52)$$

as in Eq. (31), we may conclude from Eq. (51) that

$$x_1 \to e^{it_1} + i\alpha\,(e^{it_1} X - e^{-it_1} Y) \qquad (t_1 \to +\infty), \quad (53)$$

where

$$X = \tfrac{1}{2} \int_{-\infty}^{+\infty} \Omega(\tau_1) d\tau_1; \quad Y = \tfrac{1}{2} \int_{-\infty}^{+\infty} \Omega(\tau_1) e^{2i\tau_1/\alpha} d\tau_1, \quad (54)$$

$$\Omega(\tau_1) = \tfrac{1}{4} \left(\frac{1}{\omega} \frac{d\omega}{d\tau_1} \right)^2 - \frac{1}{2\omega} \frac{d^2\omega}{d\tau_1{}^2} \qquad [\omega \equiv \omega(\tau_1) \text{ and } \tau_1 = \alpha t_1]. \quad (55)$$

If the integral defining X exists, then it clearly follows from Eq. (53) that

$$x \to e^{it} \text{ as } \alpha \to 0 \qquad (\text{and } t_1 \to +\infty). \quad (56)$$

From this, the adiabatic invariance of $\omega \langle |x|^2 \rangle$ in the large follows.

While the foregoing suffices to establish the adiabatic invariance of $\omega \langle |x|^2 \rangle$ in the large, it is not sufficient to yield the error term with exactitude for $\alpha \to 0$. This arises from the fact that the iteration scheme by which we obtained the solution (51) is not uniformly convergent for the entire (infinite) range of t_1. A method which appears sufficient to establish the true order of the error term as $\alpha \to 0$ was discovered by Hertweck and Schlüter.[1] In this method, we start from Eq. (40) and make the following sequence of transformations. First, we let

$$x = \exp\left(\int \gamma\, dt_1\right), \quad (57)$$

when Eq. (40) becomes

$$\frac{d\gamma}{dt_1} + \gamma^2 + 1 + \left(\frac{1}{\omega} \frac{d\omega}{dt_1} \right) \gamma = 0. \quad (58)$$

Next, we let

$$y = \frac{\gamma + i}{\gamma - i}, \quad (59)$$

and obtain

$$\frac{dy}{dt_1} - 2iy + \frac{1}{2\omega}\frac{d\omega}{dt_1}(1 - y^2) = 0. \tag{60}$$

Hertweck and Schlüter now show that by neglecting y^2 in comparison with 1 in this equation and solving the equation consistently in this approximation, one obtains

$$\lambda = 1 + |P|^2 \tag{61}$$

where

$$P = \tfrac{1}{2}\int_{-\infty}^{+\infty} \frac{1}{\omega}\frac{d\omega}{dt_1} e^{-2it_1}\, dt_1. \tag{62}$$

[Note that this definition of P agrees with that given in Eq. (43).] In this way, Hertweck and Schlüter establish the adiabatic invariance of $\omega\langle|x|^2\rangle$ in the large and obtain at the same time the correct order of the error term as $\alpha \to 0$.

VIII. THE HIERARCHY OF INVARIANTS

Starting from the equation

$$\frac{d^2x}{dt^2} + \omega^2 x = 0, \tag{63}$$

where ω is a function of time, we have found that the substitutions

$$dt_1 = \omega\, dt \quad \text{and} \quad x_1 = \sqrt{\omega}\, x \tag{64}$$

lead to an equation of the same form as the original, namely [cf. Eqs. (49) and (50)]

$$\frac{d^2x_1}{dt_1^2} + \omega_1^2 x_1 = 0, \tag{65}$$

where

$$\begin{aligned}\omega_1^2 = 1 + \Omega &= 1 + \tfrac{1}{4}\left(\frac{1}{\omega}\frac{d\omega}{dt_1}\right)^2 - \frac{1}{2\omega}\frac{d^2\omega}{dt_1^2} \\ &= 1 + \frac{1}{\omega^{3/2}}\frac{d^2}{dt^2}\frac{1}{\sqrt{\omega}}.\end{aligned} \tag{66}$$

Consequently, by the further transformations

$$dt_2 = \omega_1\, dt_1 \quad \text{and} \quad x_2 = \sqrt{\omega_1}\, x_1, \tag{67}$$

Eq. (65) can be brought once again to the form

$$\frac{d^2 x_2}{dt_2^{\,2}} + \omega_2^{\,2}\, x_2 = 0, \tag{68}$$

where

$$\omega_2^{\,2} = 1 + \frac{1}{\omega_1^{\,3/2}} \frac{d^2}{dt_1^{\,2}} \frac{1}{\sqrt{\omega_1}}. \tag{69}$$

More generally, by defining the transformations

$$dt_n = \omega_{n-1}\, dt_{n-1}, \quad x_n = \sqrt{\omega_{n-1}}\, x_{n-1}, \tag{70}$$

by induction, we can derive the sequence of equations

$$\frac{d^2 x_n}{dt_n^{\,2}} + \omega_n^{\,2}\, x_n = 0 \tag{71}$$

where

$$\omega_n^{\,2} = 1 + \frac{1}{\omega_{n-1}^{\,3/2}} \frac{d^2}{dt_{n-1}^{\,2}} \frac{1}{\sqrt{\omega_{n-1}}}. \tag{72}$$

We have seen that whenever we have an equation of the form (63) in which ω^2 is a slowly varying function of time,

$$\omega \, \langle |x|^2 \rangle \quad \text{is an adiabatic invariant.} \tag{73}$$

By applying this principle to Eq. (71), we obtain the hierarchy of invariants

$$\begin{aligned} \mu_n &= \omega_n \langle |x_n|^2 \rangle = \omega_n\, \omega_{n-1} \langle |x_{n-1}|^2 \rangle \\ &= \ldots = \omega_n \omega_{n-1} \ldots \omega_1 \omega \langle |x|^2 \rangle. \end{aligned} \tag{74}$$

IX. THE HIGHER-ORDER INVARIANTS FOR THE MOTION OF CHARGED PARTICLES; AN EXAMPLE

We shall now consider some applications of the higher-order invariants to the motion of charged particles in varying magnetic fields.

We have seen in Section V that the equation

$$\frac{d^2\zeta}{dt^2} + 2i\omega \frac{d\zeta}{dt} + i \frac{d\omega}{dt} \zeta = 0, \tag{75}$$

representing the motion of a charged particle in a uniform but a time-dependent magnetic field, becomes

$$\frac{d^2x}{dt^2} + \omega^2 x = 0, \tag{76}$$

by the transformation

$$x = \zeta \exp\,(i\textstyle\int \omega dt) = \zeta\, e^{it_1}. \tag{77}$$

Since by the succession of transformations [cf. Eq. (70)]

$$(x, t) \to (x_1, t_1) \to (x_2, t_2) \to \ldots \to (x_n, t_n) \to \ldots, \tag{78}$$

Eq. (76) can be repeatedly brought to the same form, it is clear that by a corresponding sequence of transformations

$$(\zeta, t) \to (\zeta_1, t_1) \to (\zeta_2, t_2) \to \ldots \to (\zeta_n, t_n) \to \ldots, \tag{79}$$

Eq. (75) can be similarly brought, repeatedly, to the same form. According to Eqs. (70) and (77), the required transformations (79) are, inductively, defined by

$$\left.\begin{aligned} &\zeta_n = x_n e^{-it_{n+1}} = \sqrt{\omega_{n-1}}\, x_{n-1}\, e^{-it_{n+1}} = \sqrt{\omega_{n-1}}\, e^{i(t_n - t_{n+1})}\, \zeta_{n-1}, \\ &\text{and} \\ &dt_n = \omega_{n-1}\, dt_{n-1} \quad (n = 1, 2, \ldots,), \end{aligned}\right\} \tag{80}$$

where the ω_n's are given, as before, by Eq. (72).

As an example we may note that the transformations

$$\zeta_1 = \sqrt{\omega}\, e^{i(t_1 - t_2)}\, \zeta, \quad dt_1 = \omega\, dt, \tag{81}$$

will lead to an equation of the form

$$\frac{d^2\zeta_1}{dt_1^{\,2}} + 2i\omega_1 \frac{d\zeta_1}{dt_1} + i\,\frac{d\omega_1}{dt_1}\,\zeta_1 = 0, \tag{82}$$

where ω_1 is given by Eq. (66).

We have seen that whenever an equation of the form (75) obtains in which ω is a slowly varying function of time, the quantity

$$\mu = \frac{1}{\omega}\left|\frac{d\zeta}{dt}\right|^2 \tag{83}$$

is an adiabatic invariant. Since ζ_n satisfies, with respect to t_n, a differential equation of the same form as (75), we may conclude that

$$\mu_n = \frac{1}{\omega_n}\left|\frac{d\zeta_n}{dt_n}\right|^2 \tag{84}$$

represents a hierarchy of adiabatic invariants for the motion of a charged particle in a time-dependent magnetic field.

The relation in which the various higher-order invariants stand with respect to each other and to the basic invariant (83) can be seen by considering the first of these higher-order ones, namely,

$$\mu_1 = \frac{1}{\omega_1} \left| \frac{d\zeta_1}{dt_1} \right|^2 . \tag{85}$$

Reverting to the original variables by means of Eqs. (81), we readily find:

$$\mu_1 = \frac{1}{\omega\omega_1} \left| \frac{d\zeta}{dt} + \zeta \left\{ \tfrac{1}{2} \frac{d}{dt} \log \omega + i\,\omega\,(1 - \omega_1) \right\} \right|^2 . \tag{86}$$

If we set in Eq. (86) the derivatives of ω equal to zero, we recover the basic invariant (83).

By averaging over all relative phases between ζ and $d\zeta/dt$, we obtain from Eq. (86), the invariant

$$\mu_1 = \frac{1}{\omega\omega_1} \left[\left| \frac{d\zeta}{dt} \right|^2 + |\zeta|^2 \left\{ \omega^2 (1 - \omega_1)^2 + \tfrac{1}{4} \left(\frac{d}{dt} \log \omega \right)^2 \right\} \right] . \tag{87}$$

The (relative) constancy of μ_1 under circumstances in which μ may not be treated as a constant, may have some practical applications.

X. THE CHANGE IN THE INVARIANT WHEN THERE IS A DISCONTINUITY IN $B^{\cdot}$ OR A HIGHER DERIVATIVE OF B

By rewriting Eq. (75) in the manner

$$\frac{d}{dt} \left(\frac{d\zeta}{dt} + 2i\omega\zeta \right) - i \frac{d\omega}{dt} \zeta = 0, \tag{88}$$

and integrating over t, we get

$$\frac{d\zeta}{dt} + 2i\omega\zeta - i \int^t \frac{d\omega}{dt} \zeta dt = C, \tag{89}$$

where C is a constant. If ω should suffer a discontinuity at $t = 0$ (say), at which time it jumps discontinuously from a value ω^- to a value ω^+, we may deduce from Eq. (89) that

$$\frac{d\zeta}{dt} + 2i\omega^-\zeta = C \quad (t < 0), \tag{90}$$

and

$$\frac{d\zeta}{dt} + 2i\omega^+\zeta - i\,(\omega^+ - \omega^-)\,\zeta_0 = C \quad (t > 0), \tag{91}$$

where ζ_0 is the value of ζ at $t = 0$. From the foregoing equations we readily find that

$$\frac{\mu^+}{\mu^-} = \frac{1}{\omega^+}\left|\frac{d\zeta}{dt}\right|^2_{t\to+0} \Bigg/ \frac{1}{\omega^-}\left|\frac{d\zeta}{dt}\right|^2_{t\to-0} = \frac{(\omega^+ + \omega^-)^2}{4\omega^+\omega^-}, \tag{92}$$

if ζ is assumed to be of the form $\rho e^{-2i\omega^- t}$ (ρ real) for $t < 0$.

If instead of ω a derivative of ω should suffer a discontinuity at $t = 0$, then we can obtain a result similar to (92) for one of the higher invariants μ_n. Thus, if ω_n is the first of the ω_n's which experience a discontinuity at $t = 0$, then by treating the equation satisfied by ζ_n in a similar manner, we shall obtain the result

$$\frac{\mu_n^+}{\mu_n^-} = \frac{(\omega_n^+ + \omega_n^-)^2}{4\,\omega_n^+\,\omega_n^-}. \tag{93}$$

XI. THE CHANGE IN $w_\perp/B$ FOR A SLOW BUT A FINITE RATE OF VARIATION OF B

We have seen that $w_\perp/B$ is strictly a constant only in the limit of infinitely slow variation of B. For any finite rate of variation, $w_\perp/B$ will change by calculable amounts; and the asymptotic dependence of this change on the rate of variation of B, as it tends to zero, can be ascertained by Hertweck and Schlüter's method. In this method one starts from the equation of motion of a simple harmonic oscillator (with ω^2 a function of time) and obtains by the sequence of transformations [cf. Eqs. (57) and (59)]

$$x \to \gamma \to y \quad \text{and} \quad t \to t_1, \tag{94}$$

the differential equation [Eq. (60)]

$$\frac{dy}{dt_1} - 2iy + \frac{1}{2\omega}\frac{d\omega}{dt_1}(1 - y^2) = 0. \tag{95}$$

By neglecting the term in y^2 in this equation and solving the equation

$$\frac{dy}{dt_1} - 2iy + \frac{1}{2\omega}\frac{d\omega}{dt_1} = 0 \tag{96}$$

consistently in this approximation, Hertweck and Schlüter show that for sufficiently slow variations of B

$$\frac{\mu^{+\infty}}{\mu^{-\infty}} - 1 \simeq |P|^2 \tag{97}$$

where

$$P = \tfrac{1}{2} \int_{-\infty}^{+\infty} \frac{1}{\omega} \frac{d\omega}{dt_1} e^{-2it_1} dt_1. \tag{98}$$

It is clear that we may improve upon Hertweck and Schlüter's method by starting from Eq. (71) for x_n and treating this equation in a similar fashion. We shall then obtain

$$\frac{\mu^{+\infty}}{\mu^{-\infty}} - 1 \simeq |P_n|^2 \tag{99}$$

where

$$P_n = \tfrac{1}{2} \int_{-\infty}^{+\infty} \frac{1}{\omega_n} \frac{d\omega_n}{dt_{n+1}} e^{-2it_{n+1}} dt_{n+1}. \tag{100}$$

Integrating the expression on the right-hand side of this equation by parts and remembering that $d\omega_n/dt_{n+1}$ tends to zero at both limits of integration for all $n > 1$, we obtain

$$P_n = i \int_{-\infty}^{+\infty} (\log \omega_n)\, e^{-2it_{n+1}} dt_{n+1}, \tag{101}$$

or [cf. Eq. (70)]

$$P_n = i \int_{-\infty}^{+\infty} (\omega_n \log \omega_n)\, e^{-2it_{n+1}} dt_n. \tag{102}$$

We shall illustrate the use of these higher-order approximations for $(\mu^{+\infty}/\mu^{-\infty}) - 1$ by considering the case $n = 1$. In this case Eq. (102) takes the form

$$P_1 = \tfrac{1}{2} i \int_{-\infty}^{+\infty} [(1 + \Omega)^{\frac{1}{2}} \log (1 + \Omega)] \exp \{-2i\textstyle\int (1 + \Omega)^{\frac{1}{2}} dt_1\} dt_1, \tag{103}$$

where we have substituted for ω_1 in accordance with Eq. (66). When $\Omega \to 0$, we can write

$$P_1 = \tfrac{1}{2} i \int_{-\infty}^{+\infty} \Omega (t_1)\, e^{-2it_1} dt_1. \tag{104}$$

In Hertweck and Schlüter's paper the case when $\omega(t)$ has the form

$$\omega(t) = \omega_0 \left(\tfrac{3}{2} + \tfrac{1}{2} \tanh \alpha t\right), \tag{105}$$

where ω_0 and α are constants, is treated in some detail. For this form of $\omega\,(t)$ it can be readily shown that

$$\Omega = \left(\frac{\alpha}{\omega_0}\right)^2 \frac{2Q^{3/2} - 5Q + 3}{4Q^2} \tag{106}$$

where

$$Q = 1 + 8\, e^{2y} \text{ and } y = \alpha t_1/\omega_0. \tag{107}$$

The integral defining P_1 [Eq. (104)] can be evaluated explicitly and we find

$$|\, P_1 \,|^2 = \frac{\pi^2 q^2}{16\left(\sinh \dfrac{\pi}{2q}\right)^2} \left| 1 - \frac{3i}{4q} - \frac{1}{\sqrt{\pi}} \frac{\Gamma\left(\dfrac{1}{2} + \dfrac{i}{2q}\right)}{\Gamma\left(1 + \dfrac{i}{2q}\right)} \right|^2, \tag{108}$$

where

$$q = \tfrac{1}{2}\,(\alpha/\omega_0). \tag{109}$$

Equation (108) may be contrasted with

$$|\, P \,|^2 = \frac{\pi^2}{16\left(\sinh \dfrac{\pi}{2q}\right)^2} \left| 1 - \frac{1}{\sqrt{\pi}} \frac{\Gamma\left(\dfrac{1}{2} + \dfrac{i}{2q}\right)}{\Gamma\left(1 + \dfrac{i}{2q}\right)} \right|^2, \tag{110}$$

obtained by Hertweck and Schlüter. The values of $|\, P \,|^2$ and $|\, P_1 \,|^2$ given by the foregoing formulae for some values of q are listed in Table 1, and are further compared with the results of exact numerical integrations.* It will be observed that $|\, P_1 \,|^2$ gives for intermediate values of q a better representation of the exact results than does $|\, P \,|^2$.

XII. THE LONGITUDINAL INVARIANT

As we have stated in Section I, in treating the motions of charged particles in the guiding center approximation, we separate the motions parallel and perpendicular to the field. The variation of $w_\perp$ along the lines of force is determined from the (assumed) constancy of $w_\perp/B$ as an adiabatic invariant. The variation of the velocity ($v_\parallel$) parallel to the lines of

* I am indebted to Dr. R. Lüst for supplying me with these.

force is, then, determined from the equation of motion [cf. Spitzer[2] or Rosenbluth and Longmire[3]]

$$\frac{dv_{\parallel}}{dt} = -\frac{w_{\perp}}{B}\frac{\partial B}{\partial s}, \tag{111}$$

where ds denotes the element of arc along $\boldsymbol{B}$. The content of Eq. (111) is that the motion along the line of force is governed by the potential function

$$V(s) = \frac{w_{\perp}}{B} B(s). \tag{112}$$

If the particle should describe a periodic orbit in this potential function, then the considerations of Section III apply and we may infer the adiabatic invariance of the action integral

$$J = \oint v_{\parallel}\, ds = \oint \mathbf{v}\cdot\frac{d\boldsymbol{B}}{B}. \tag{113}$$

This is the longitudinal invariant of Chew, Goldberger, and Low.[4] It is, however, clear that this invariant can be used only if the variation of B with time takes place very slowly in the time scale of the period of oscillation of the particle in the potential field (112). Also, it is necessary that during the variation of B, particles which are initially "trapped" do not escape over the potential barrier (as at C in Fig. 1); and, conversely, particles, once "free," are not trapped. For these reasons the conditions

TABLE 1

THE CHANGE IN $w_{\perp}/B$ FOR FINITE RATES OF VARIATION OF B IN ACCORDANCE WITH EQ. (105)

q	$\lvert P\rvert^2$	$\lvert P_1\rvert^2$	Exact
0.20	2.223×10^{-7}	1.914×10^{-7}	
0.25	4.823×10^{-6}	4.291×10^{-6}	
0.3333	1.013×10^{-4}	9.384×10^{-5}	1.13×10^{-4}
0.50	1.956×10^{-3}	1.903×10^{-3}	2.04×10^{-3}
0.625	6.066×10^{-3}	6.028×10^{-3}	
0.8333...	1.762×10^{-2}	1.788×10^{-2}	
1.00	2.884×10^{-2}	2.958×10^{-2}	2.99×10^{-2}
1.25	4.527×10^{-2}	4.694×10^{-2}	
2.00	7.919×10^{-2}	8.411×10^{-2}	8.26×10^{-2}
2.50	9.141×10^{-2}	9.869×10^{-2}	
∞	...	...	1.25×10^{-1}

which must obtain for a valid use of the longitudinal invariant are very much more stringent than for the use of $w_{\perp}^2/B$ as an invariant during the motion.

REFERENCES

1. F. Hertweck and A. Schlüter, "Die 'adiabatische Invarianz' des magnetischen Bahnmomentes geladener Teilchen," Z. Naturforsch. **12A,** 844 (1957); see also R. Kulsrud, "Adiabatic Invariant of Harmonic Oscillator," Phys. Rev. **106,** 205 (1957).
2. L. Spitzer, Physics of Fully Ionized Gases. Interscience Publishers, New York (1956).
3. M. N. Rosenbluth and C. L. Longmire, "Stability of Plasma Confined by Magnetic Fields," Ann. Phys. **1,** 120 (1957).
4. G. F. Chew, M. L. Goldberger, and F. E. Low, "An Adiabatic Invariant for Motion Along the Magnetic Lines of Force" (September 1955), Los Alamos Report LA-2055 T-767.

PART TWO

Hydrodynamic and Hydromagnetic Stability

5

The Stability of Viscous Flow between Rotating Cylinders in the Presence of a Radial Temperature Gradient

S. CHANDRASEKHAR

Yerkes Observatory, University of Chicago

CONTENTS

The research reported in this paper has in part been supported by the Geophysics Research Directorate of the Air Force Cambridge Research Center, Air Research and Development Command, under Contract AF 19(604)–299 with the University of Chicago.

107

1. Introduction. Recent experimental studies by D. Fultz [**1**] and R. Hide [**2**] on the patterns of fluid motion which occur in the space between two rotating co-axial cylinders in the presence of a radial temperature gradient have disclosed remarkable analogies with the phenomenon of the jet stream in the upper atmosphere. It is the opinion of Fultz and Hide that under the conditions of their experiments the effect of gravity was at least as important as that of the Coriolis force. Nevertheless, in this paper we shall ignore the effect of gravity and consider the simpler problem of the stability of the two-dimensional viscous flow between rotating cylinders when a radial temperature gradient is present. This simpler problem is of some interest in itself: it discloses a type of rotationally induced thermal instability which appears to be novel; and as we shall see this new type of thermal instability has several features in common with the more familiar Bénard type of gravitationally induced thermal instability.

It may be stated here that the present paper is one of a series which is devoted to a systematic study of the various problems of stability in hydrodynamics and hydromagnetics (Chandrasekhar [**3**]–[**9**]; also Chandrasekhar & Fermi [**10**]).

2. The equations of the problem. As we have stated we shall consider the problem as two-dimensional. The equations of motion and heat conduction appropriate to the problem on hand are in cylindrical polar co-ordinates (r, θ),

$$(1)\qquad \rho\left(\frac{\partial}{\partial t}+u_r\frac{\partial}{\partial r}+\frac{u_\theta}{r}\frac{\partial}{\partial \theta}\right)u_r-\rho\frac{u_\theta^2}{r}=-\frac{\partial p}{\partial r}+\rho\nu\left(\nabla^2 u_r-\frac{2}{r^2}\frac{\partial u_\theta}{\partial \theta}-\frac{u_r}{r^2}\right),$$

$$(2)\qquad \rho\left(\frac{\partial}{\partial t}+u_r\frac{\partial}{\partial r}+\frac{u_\theta}{r}\frac{\partial}{\partial \theta}\right)u_\theta+\rho\frac{u_\theta u_r}{r}=-\frac{1}{r}\frac{\partial p}{\partial \theta}+\rho\nu\left(\nabla^2 u_\theta+\frac{2}{r^2}\frac{\partial u_r}{\partial \theta}-\frac{u_\theta}{r^2}\right)$$

and

$$(3)\qquad \left(\frac{\partial}{\partial t}+u_r\frac{\partial}{\partial r}+\frac{u_\theta}{r}\frac{\partial}{\partial \theta}\right)T=\kappa\nabla^2 T,$$

where ρ, p and T denote the density, pressure and temperature, u_r and u_θ the components of the velocity in the radial and the transverse (θ) directions and ν and κ are the coefficients of kinematic viscosity and thermometric conductivity, respectively. Also, we have the "equation of state"

$$(4)\qquad \rho=\rho_0(1-\alpha\Delta T),\qquad \Delta T=T-T_0,$$

where α denotes the coefficient of volume expansion, ρ_0 the density corresponding to a mean temperature T_0 and ΔT is the deviation of the local temperature from T_0.

It may be readily verified that equations (1) and (2) admit the stationary solution

$$(5)\qquad u_r=0,\qquad u_\theta=V(r)=Ar+\frac{B}{r}\qquad \text{and}\qquad p=p_0(r),$$

where A and B are constants (related to the angular velocities of rotation, Ω_1 and Ω_2 of the inner and the outer cylinders of radii R_1 and R_2, $R_2 > R_1$) and $p_0(r)$ is determined by

$$\frac{\partial p_0}{\partial r} = \rho_0\{1 - \alpha\Delta T_0(r)\}\frac{V^2}{r}. \tag{6}$$

In equation (6), $\Delta T_0(r)$ represents a stationary solution of equation (3). Regarding this latter, we shall make two assumptions:

$$\text{case I:} \qquad \kappa\nabla^2\{\Delta T_0(r)\} = \epsilon,$$

(7) and

$$\text{case II:} \qquad \kappa\nabla^2\{\Delta T_0(r)\} = 0.$$

Case I will be appropriate to a situation when there is a uniform distribution of heat sources such that in the absence of conduction the temperature at each point will rise at the rate ϵ; while case II will be appropriate to a situation when the two cylinders confining the liquid are maintained at constant, different, temperatures. The stationary temperature gradients maintained in the two cases will be

$$\text{case I:} \qquad \frac{d}{dr}\Delta T_0(r) = \beta r \qquad \left(\beta = \frac{\epsilon}{2\kappa} = \text{constant}\right) \tag{8}$$

and

$$\text{case II:} \qquad \frac{d}{dr}\Delta T_0(r) = \frac{\beta}{r} \qquad (\beta = \text{constant}). \tag{9}$$

In general we shall write

$$\frac{d}{dr}\Delta T_0(r) = \beta(r). \tag{10}$$

We shall now suppose that the solution represented by equations (5) and (6) is slightly perturbed. Let this perturbed motion be characterized by

$$u_r = u, \qquad u_\theta = V + v, \qquad \Delta T = \Delta T_0(r) + \tau \qquad \text{and} \qquad p = p_0 + \delta p, \tag{11}$$

where u and v are small compared to V while τ and δp are small compared to ΔT_0 and p_0, respectively. With these assumptions the linearized form of equations (1) and (2) are

$$\rho_0\left(\frac{\partial u}{\partial t} + \frac{V}{r}\frac{\partial u}{\partial \theta}\right) + \rho_0\,\alpha\tau\frac{V^2}{r} - \rho_0\frac{2Vv}{r} = -\frac{\partial}{\partial r}\delta p + \rho_0\,\nu\left(\nabla^2 u - \frac{2}{r^2}\frac{\partial v}{\partial \theta} - \frac{u}{r^2}\right) \tag{12}$$

and

$$\rho_0\left(\frac{\partial v}{\partial t} + u\frac{\partial V}{\partial r} + \frac{V}{r}\frac{\partial v}{\partial \theta}\right) + \rho_0\frac{Vu}{r} = -\frac{1}{r}\frac{\partial}{\partial \theta}\delta p + \rho_0\,\nu\left(\nabla^2 v + \frac{2}{r^2}\frac{\partial u}{\partial \theta} - \frac{v}{r^2}\right), \tag{13}$$

where it should be noted that in obtaining these equations the further assumption has been made that α, except when multiplied by V^2, leads to a quantity of the second order of smallness. The assumption in particular that $\alpha\tau V^2$ is a quantity of the first order of smallness is a necessary one: on it depends the onset of instability when the temperature gradient which is maintained exceeds a certain critical value. It may be recalled in this connection that a similar assumption is made in the theory of the gravitationally induced thermal instability (Rayleigh [**11**], Jeffreys [**12**]) when the effect of the variation in density (due to thermal expansion) is ignored in all terms in the equations of motion except the one in gravity.

In the framework of the approximations leading to equations (12) and (13) the equation of continuity reduces simply to the statement that the velocity is a solenoidal vector. Thus,

$$\frac{\partial u}{\partial r} + \frac{u}{r} + \frac{1}{r}\frac{\partial v}{\partial \theta} = 0. \tag{14}$$

On making use of the particular form of V (equation (5)) we find that equations (12) and (13) can be reduced to the forms

$$\frac{\partial \varpi}{\partial r} = -\frac{\partial u}{\partial t} - \frac{V}{r}\frac{\partial u}{\partial \theta} - \alpha\tau\frac{V^2}{r} + 2\frac{Vv}{r} + \nu\left(\nabla^2 u - \frac{2}{r^2}\frac{\partial v}{\partial \theta} - \frac{u}{r^2}\right) \tag{15}$$

and

$$\frac{1}{r}\frac{\partial \varpi}{\partial \theta} = -\frac{\partial v}{\partial t} - 2Au - \frac{V}{r}\frac{\partial v}{\partial \theta} + \nu\left(\nabla^2 v + \frac{2}{r^2}\frac{\partial u}{\partial \theta} - \frac{v}{r^2}\right), \tag{16}$$

where for the sake of brevity we have written

$$\varpi = \delta p/\rho_0. \tag{17}$$

Eliminating ϖ between equations (15) and (16) we obtain

$$\begin{aligned}\frac{\partial}{\partial \theta}\left\{-\frac{\partial u}{\partial t} - \frac{V}{r}\frac{\partial u}{\partial \theta} - \alpha\tau\frac{V^2}{r} + 2\frac{Vv}{r} + \nu\left(\nabla^2 u - \frac{2}{r^2}\frac{\partial v}{\partial \theta} - \frac{u}{r^2}\right)\right\}\\ = \frac{\partial}{\partial r} r\left\{-\frac{\partial v}{\partial t} - 2Au - \frac{V}{r}\frac{\partial v}{\partial \theta} + \nu\left(\nabla^2 v + \frac{2}{r^2}\frac{\partial u}{\partial \theta} - \frac{v}{r^2}\right)\right\}.\end{aligned} \tag{18}$$

Returning to the equation of heat conduction (3), we find that the corresponding linearized form of this equation is

$$\frac{\partial \tau}{\partial t} + \beta(r)u + \frac{V}{r}\frac{\partial \tau}{\partial \theta} = \kappa\nabla^2 \tau \tag{19}$$

We shall now seek solutions of equations (14), (18) and (19) which are of the forms

$$u = u(r)e^{i(n\theta+\sigma t)}, \qquad v = v(r)e^{i(n\theta+\sigma t)}, \qquad \tau = \tau(r)e^{i(n\theta+\sigma t)}, \tag{20}$$

where n is an integer and σ is a constant unspecified for the present. For solutions of the form (20) equations (14), (18) and (19) become

$$rD^* u = D(ru) = -inv, \tag{21}$$

$$in\left\{-i\sigma u - in\frac{Vu}{r} - \alpha\tau\frac{V^2}{r} + 2\frac{Vv}{r} + \nu\left[\left(DD^* - \frac{n^2}{r^2}\right)u - 2in\frac{v}{r^2}\right]\right\}$$
$$= rD^*\left\{-i\sigma v - 2Au - in\frac{Vv}{r} + \nu\left[\left(DD^* - \frac{n^2}{r^2}\right)v + 2in\frac{u}{r^2}\right]\right\} \tag{22}$$

and

$$i\sigma\tau + \beta(r)u + in\frac{V\tau}{r} = \kappa\left(D^* D - \frac{n^2}{r^2}\right)\tau, \tag{23}$$

where

$$D = \frac{d}{dr} \quad \text{and} \quad D^* = D + \frac{1}{r}. \tag{24}$$

The operators D and D^* satisfy the commutation relation

$$DD^* = D^* D - \frac{1}{r^2} = \frac{d^2}{dr^2} + \frac{1}{r}\frac{d}{dr} - \frac{1}{r^2}; \tag{25}$$

this relation is useful in the subsequent reductions.

Equation (22) can be rearranged in the manner

$$rD^*\left[\nu\left(DD^* - \frac{n^2}{r^2}\right) - in\Omega - i\sigma\right]v - in\left[\nu\left(DD^* - \frac{n^2}{r^2}\right) - in\Omega - i\sigma\right]u$$
$$- 2\left[rD^*\left(A - in\frac{\nu}{r^2}\right)u + in\left(\frac{V}{r} - in\frac{\nu}{r^2}\right)v\right] = -in\alpha\frac{V^2}{r}\tau, \tag{26}$$

where

$$\Omega = \frac{V}{r} = A + \frac{B}{r^2}, \tag{27}$$

denotes the angular velocity. We can now eliminate v from equation (26) by making use of equation (21). We find

$$\left[rD^*\left\{\nu\left(DD^* - \frac{n^2}{r^2}\right) - in\Omega - i\sigma\right\}rD^* - n^2\left\{\nu\left(DD^* - \frac{n^2}{r^2}\right) - in\Omega - i\sigma\right\}\right.$$
$$\left. - 2in\frac{B}{r}D^* - \frac{4n^2\nu}{r^2}\right]u = -(n^2\Omega^2\alpha)r\tau; \tag{28}$$

while equation (23) can be written in the form

$$\left\{\kappa\left(D^* D - \frac{n^2}{r^2}\right) - in\Omega - i\sigma\right\}\tau = \beta(r)u. \tag{29}$$

Equations (28) and (29) must be solved together with the boundary conditions which state that at $r = R_1$ and $r = R_2$ the fluctuations in the velocity and in the temperature must vanish, *i.e.*,

$$u = v = 0 \quad \text{and} \quad \tau = 0 \quad \text{at} \quad r = R_1 \quad \text{and} \quad r = R_2. \tag{30}$$

3. The case Ω = constant; the equations governing marginal stability. In the remaining part of this paper the assumption will be made that the two cylinders are rotated at the same constant rate. As may be expected on general grounds and as is also apparent from equations (28) and (29) this assumption will lead to considerable simplification.

When

$$\Omega = A = \text{constant and } B = 0, \tag{31}$$

it is permissible to introduce a *phase angular velocity*

$$\omega = n\Omega + \sigma. \tag{32}$$

In terms of ω equations (28) and (29) take the simpler forms

$$\begin{aligned}\Bigg[rD^*\left\{\nu\left(DD^* - \frac{n^2}{r^2}\right) - i\omega\right\}rD^* \\ - n^2\left\{\nu\left(DD^* - \frac{n^2}{r^2}\right) - i\omega\right\} - \frac{4n^2\nu}{r^2}\Bigg]u = -(n^2\Omega^2\alpha)r\tau\end{aligned} \tag{33}$$

and

$$\left[\kappa\left(D^*D - \frac{n^2}{r^2}\right) - i\omega\right]\tau = \beta(r)u, \tag{34}$$

while the form of the solutions sought is (*cf.* equations (20))

$$u = u(r)e^{in(\theta-\Omega t)+i\omega t} \text{ and } \tau = \tau(r)e^{in(\theta-\Omega t)+i\omega t} \tag{35}$$

Accordingly, in a frame of reference rotating with the angular velocity Ω, what distinguishes stability from instability of a pattern of fluid motion with $2n$ vortices, is the real part, $\Re(\omega)$, of ω. Now, it is known that in several related problems (Meksyn [**13**]; also Pellew & Southwell [**14**]) the situation in marginal stability (*i.e.*, one on the verge of stability) is characterized by $\omega = 0$ (and not merely by $\Re(\omega) = 0$); in other words, the principle of the exchange of stabilities is valid. In this paper we shall assume that this principle is valid also for the problem on hand. In a later paper we shall return to a justification of the principle.

On the assumption then that the principle of the exchange of stabilities is valid, the equations governing marginal stability are

$$\text{(36)}\qquad \left\{rD^*\left(DD^* - \frac{n^2}{r^2}\right)rD^* - n^2\left(DD^* - \frac{n^2}{r^2}\right) - \frac{4n^2}{r^2}\right\}u = -\frac{n^2\,\Omega^2\,\alpha}{\nu}\,r\tau$$

and

$$\text{(37)}\qquad \kappa\left(D^*\,D - \frac{n^2}{r^2}\right)\tau = \beta(r)u.$$

On expanding the differential operator on the left-hand side of equation (36), we find that the equation is in fact equivalent to

$$\text{(38)}\qquad \left(D^*\,D - \frac{n^2}{r^2}\right)^2(ru) = -\frac{n^2\,\Omega^2\,\alpha}{\nu}\,\tau.$$

From equations (37) and (38) it now follows that

$$\text{(39)}\qquad \left(D^*\,D - \frac{n^2}{r^2}\right)^3(ru) = -\frac{n^2\,\Omega^2\,\alpha}{\kappa\nu}\,\beta(r)u,$$

where it may be noted that (*cf*. equation (25))

$$\text{(40)}\qquad D^*\,D - \frac{n^2}{r^2} = \frac{d^2}{dr^2} + \frac{1}{r}\frac{d}{dr} - \frac{n^2}{r^2}.$$

According to equations (21) and (38) the boundary conditions with respect to which we must solve equation (39) are

$$\text{(41)}\qquad \begin{gathered} u = 0, \quad D(ru) = 0 \quad \text{and} \quad \left(D^*\,D - \frac{n^2}{r^2}\right)^2(ru) = 0 \\ \text{at} \quad r = R_1 \quad \text{and} \quad r = R_2. \end{gathered}$$

It is evident that solving equation (39) together with the six boundary conditions (41), three at each of the two boundaries, is equivalent to a characteristic value problem. In the following sections it will be shown how we can solve this problem for the two cases (8) and (9). It is the manner of solving this characteristic value problem that constitutes the essential mathematical content of this paper.

4. A variational procedure for solving the characteristic value problem in case $\beta(r) = \beta r$ **(case I).** In this case, letting

$$\text{(42)}\qquad W = ru$$

and measuring r in units of the radius, R_2, of the outer cylinder, we may first observe that the characteristic value problem we have to solve is one of determining

$$S_n = \frac{\Omega^2 \alpha\beta}{\kappa\nu} R_2^6, \tag{43}$$

such that the equation

$$\left(\frac{d^2}{dr^2} + \frac{1}{r}\frac{d}{dr} - \frac{n^2}{r^2}\right)^3 W = -n^2 S_n W, \tag{44}$$

may have a non-trivial solution which satisfies the boundary conditions

$$W = 0, \quad DW = 0 \quad \text{and} \quad \left(\frac{d^2}{dr^2} + \frac{1}{r}\frac{d}{dr} - \frac{n^2}{r^2}\right)^2 W = 0 \tag{45}$$
$$\text{at} \quad r = 1 \quad \text{and} \quad r = R_1/R_2 = \eta \text{ (say)}.$$

Reduced in this manner, the problem is seen to be very similar to the one encountered in the theory of the thermal instability of fluid spheres and spherical shells (*cf.* Chandrasekhar [**4**], [**5**] and [**6**]). And as in this latter theory the solution for the lowest characteristic number, S_n, can be effected by the application of a variational principle which we shall now proceed to formulate.

Letting

$$\begin{aligned} G &= \left(\frac{d^2}{dr^2} + \frac{1}{r}\frac{d}{dr} - \frac{n^2}{r^2}\right) W = \frac{1}{r}\frac{d}{dr}\left(r\frac{dW}{dr}\right) - \frac{n^2 W}{r^2}, \\ F &= \left(\frac{d^2}{dr^2} + \frac{1}{r}\frac{d}{dr} - \frac{n^2}{r^2}\right)^2 W = \frac{1}{r}\frac{d}{dr}\left(r\frac{dG}{dr}\right) - \frac{n^2 G}{r^2}, \end{aligned} \tag{46}$$

we can rewrite the differential equation governing W in the form

$$\frac{1}{r}\frac{d}{dr}\left(r\frac{dF}{dr}\right) - \frac{n^2 F}{r^2} = -n^2 S_n W. \tag{47}$$

The boundary conditions (45) now require that

$$F = W = DW = 0 \quad \text{at} \quad r = 1 \quad \text{and} \quad r = \eta. \tag{48}$$

Multiply equation (47) by rF and integrate over the range of r. The left-hand side of the equation gives

$$\int_\eta^1 F\frac{d}{dr}\left(r\frac{dF}{dr}\right) r\, d - n^2 \int_\eta^1 F^2 \frac{dr}{r}. \tag{49}$$

By integrating by parts the first of the two integrals in (49) and remembering that F vanishes at both limits, we obtain

$$-\int_\eta^1 \left\{ r\left(\frac{dF}{dr}\right)^2 + \frac{n^2 F^2}{r} \right\} dr. \tag{50}$$

Turning next to the right-hand side of equation (47) we have (*cf.* equation (46))

$$\int_\eta^1 rWF\,dr = \int_\eta^1 W\frac{d}{dr}\left(r\frac{dG}{dr}\right)dr - n^2\int_\eta^1 WG\frac{dr}{r}. \tag{51}$$

After two integrations by parts the foregoing becomes

$$\int_\eta^1 rWF\,dr = \left(rW\frac{dG}{dr} - rG\frac{dW}{dr}\right)_\eta^1 + \int_\eta^1 rG\left\{\frac{1}{r}\frac{d}{dr}\left(r\frac{dW}{dr}\right) - \frac{n^2}{r^2}W\right\}dr. \tag{52}$$

The integrated parts vanish on account of the boundary conditions (*cf.* equation (48)) and we are left with

$$\int_\eta^1 rWF\,dr = \int_\eta^1 rG^2\,dr. \tag{53}$$

The result of multiplying equation (47) by rF and integrating is, therefore,

$$n^2 S_n = \frac{\int_\eta^1 r\{(dF/dr)^2 + n^2F^2/r^2\}\,dr}{\int_\eta^1 rG^2\,dr}. \tag{54}$$

This formula expresses S_n as the ratio of two positive definite integrals.

If we now consider the effect on S_n of a variation δW in W compatible with the boundary conditions, we readily find that

$$n^2\,\delta S_n = -\frac{2}{\int_\eta^1 rG^2\,dr}\int_\eta^1 r\left\{\frac{d^2F}{dr^2} + \frac{1}{r}\frac{dF}{dr} - \frac{n^2}{r^2}F + n^2 S_n W\right\}\delta F\,dr. \tag{55}$$

Hence to the first order, $\delta S_n \equiv 0$ *for all small arbitrary variations* δF. Further, it follows from (55) that the true solution of the problem gives a minimal value for S_n. This last fact enables us to formulate the following variational procedure of solving equation (47) and satisfying the boundary conditions of the problem:

Assume for F an expression involving one or more parameters A_j which vanishes at $r = 1$ and $r = \eta$. With the chosen form of F determine W as a solution of the equation

$$\left(\frac{d^2}{dr^2} + \frac{1}{r}\frac{d}{dr} - \frac{n^2}{r^2}\right)^2 W = F, \tag{56}$$

which satisfies the boundary conditions

(57) $$W = 0 \text{ and } \frac{dW}{dr} = 0 \text{ at } r = 1 \text{ and } r = \eta.$$

Then evaluate S_n according to formula (54) and minimize it with respect to the parameters A_j. In this way we shall obtain the "best" value of S_n for the chosen form of F.

5. The variational solution of the characteristic value problem for case I. One confining cylinder. We shall first consider the case $\eta = 0$ *i.e.*, when there is only one confining cylinder. In this case the continuity of the solutions at $r = 0$ and the form of the equation to be solved for W suggest that we assume for F the trial function

(58) $$F = \Sigma_j A_j J_n(\alpha_j r),$$

where J_n denotes the Bessel function of order n, the α_j's ($j = 1, 2, \cdots$) are its zeros and the A_j's are the variational parameters. With this choice, F vanishes at $r = 1$ as required. It may be recalled here that the functions $J_n(\alpha_j r)$ (for a given n) satisfy the orthogonality relations

(59) $$\int_0^1 rJ_n(\alpha_j r)J_n(\alpha_k r)\, dr = \tfrac{1}{2}[J'_n(\alpha_j)]^2\, \delta_{jk},$$

where primes denote differentiation with respect to the argument of the Bessel function and δ_{jk} is the usual Kronecker symbol.

With F given by equation (58) the equation to be solved for W is

(60) $$\left(\frac{d^2}{dr^2} + \frac{1}{r}\frac{d}{dr} - \frac{n^2}{r^2}\right)^2 W = \Sigma_j A_j J_n(\alpha_j r).$$

The general solution of this equation which has no singularity at the origin is

(61) $$W = \Sigma_j(A_j/\alpha_j^4)J_n(\alpha_j r) + Br^n + Cr^{n+2},$$

where B and C are constants of integration. The condition $W = 0$ at $r = 1$ requires $B = -C$ and we have

(62) $$W = \Sigma_j(A_j/\alpha_j^4)J_n(\alpha_j r) + B(r^n - r^{n+2}).$$

The constant B is determined by the remaining condition at $r = 1$, namely that here dW/dr must vanish. This leads to

(63) $$B = \tfrac{1}{2}\Sigma_j(A_j/\alpha_j^3)J'_n(\alpha_j).$$

Turning next to the evaluation of S_n according to formula (54), we find that for F and W given by equations (58) and (62)

$$\int_0^1 r\left\{\left(\frac{dF}{dr}\right)^2 + \frac{n^2 F^2}{r^2}\right\} dr$$

(64)
$$= -\int_0^1 rF\left\{\frac{d^2 F}{dr^2} + \frac{1}{r}\frac{dF}{dr} - \frac{n^2 F}{r^2}\right\} dr$$
$$= \int_0^1 r\Sigma_j A_j J_n(\alpha_j r)\Sigma_k A_k \alpha_k^2 J_n(\alpha_k r)\, dr$$
$$= \tfrac{1}{2}\Sigma_j A_j^2 \alpha_j^2 [J'_n(\alpha_j)]^2$$

and

(65)*
$$\int_0^1 rG^2\, dr = \int_0^1 rWF\, dr$$
$$= \tfrac{1}{2}\Sigma_j (A_j^2/\alpha_j^4)[J'_n(\alpha_j)]^2 + B\Sigma_j A_j \int_0^1 (r^{n+1} - r^{n+3}) J_n(\alpha_j r)\, dr$$
$$= \tfrac{1}{2}\Sigma_j (A_j^2/\alpha_j^4)[J'_n(\alpha_j)]^2 + 2B\Sigma_j (A_j/\alpha_j^2) J_{n+2}(\alpha_j).$$

The resulting expression for S_n is, therefore

(66)
$$n^2 S_n = \frac{\Sigma_j A_j^2 \alpha_j^2 [J'_n(\alpha_j)]^2}{\Sigma_j (A_j^2/\alpha_j^4)[J'_n(\alpha_j)]^2 + 2\Sigma_j (A_j/\alpha_j^2) J_{n+2}(\alpha_j)\Sigma_k (A_k/\alpha_k^3) J'_n(\alpha_j)},$$

where we have substituted for B from equation (63). By minimizing this last expression with respect to the A_j's we shall obtain the "best" value for the lowest characteristic number S_n for the chosen form of F.

The simplest trial function for F of the chosen form is

(67)
$$F = J_n(\alpha_1 r),$$

where α_1 is the first zero of J_n. For this choice of F there is no variational parameter with respect to which we have to minimize and equation (66) directly gives

(68)
$$n^2 S_n = \frac{\alpha_1^7 J'_n(\alpha_j)}{\alpha_1 J'_n(\alpha_1) + 2J_{n+2}(\alpha_1)}.$$

Values of S_n obtained with the aid of this formula are listed in Table 1.

The values given by (68) can be improved by including a second term in F. Thus, with the assumption

(69)
$$F = J_n(\alpha_1 r) + AJ_n(\alpha_2 r),$$

* The transformations used in going from (49) to (50) and similarly from (51) to (53) can be used in the reverse fashion since these depend (apart from definitions) only on the boundary conditions of the problem.

TABLE 1

The Lowest Characteristic Numbers S_n for Case I and One Confining Cylinder

n	FIRST APPROXIMATION	A	SECOND APPROXIMATION
1	6954	0.0811	6873
2	8416	0.0944	8297
3	1.235×10^4	0.1017	1.216×10^4
4	1.828×10^4	0.1059	1.800×10^4
5	2.648×10^4	0.1083	2.607×10^4
6	3.731×10^4	0.1097	3.675×10^4
7	5.123×10^4	0.1104	5.048×10^4
8	6.871×10^4	0.1107	6.773×10^4
9	9.027×10^4	0.1107	8.902×10^4
10	1.165×10^5	0.1105	1.149×10^5
11	1.478×10^5	0.1101	1.459×10^5
12	1.851×10^5	0.1097	1.828×10^5
13	2.287×10^5	0.1092	2.260×10^5
14	2.795×10^5	0.1086	2.762×10^5
15	3.381×10^5	0.1080	3.343×10^5

where α_2 denotes the second zero of J_n and A is a variational parameter, equation (66) gives

$$n^2 S_n = \alpha_1^7\{[J'_n(\alpha_1)]^2 + A^2(\alpha_2/\alpha_1)^2[J'_n(\alpha_2)]^2\} \times [J'_n(\alpha_1)\{\alpha_1 J'_n(\alpha_1) + 2J_{n+2}(\alpha_1)\} \tag{70}$$
$$+ 4A(\alpha_1/\alpha_2)^3 J_{n+2}(\alpha_1)J'_n(\alpha_2) + A^2(\alpha_1/\alpha_2)^5 J'_n(\alpha_2)\{\alpha_2 J'_n(\alpha_2) + 2J_{n+2}(\alpha_2)\}]^{-1}.$$

The values of S_n obtained after minimizing the foregoing expression with respect to A are listed in Table 1 together with the values of A which give the minimum values.

From a comparison of the results obtained in the first and the second approximations it would appear that the second approximation gives values which are probably correct to one part in 10^4.

6. The variational solution of the characteristic value problem for case I. Two confining cylinders. Turning next to the solution of the characteristic value problem for case I when there are two confining cylinders, we shall assume as a trial function for F a linear combination of the Bessel functions, $J_n(\alpha r)$ and $Y_n(\alpha r)$, of the two kinds which vanishes at $r = 1$ and $r = \eta$. For this latter purpose we first define the cylinder function (of order ν)

$$\mathcal{C}_{n,\nu}(z) = Y_n(\alpha\eta)J_\nu(z) - J_n(\alpha\eta)Y_\nu(z), \tag{71}$$

where α is a constant which we shall leave unspecified for the present. Then,

$$\mathcal{C}_{n,n}(\alpha r) = Y_n(\alpha\eta)J_n(\alpha r) - J_n(\alpha\eta)Y_n(\alpha r), \tag{72}$$

clearly vanishes for $r = \eta$; it will also vanish for $r = 1$ provided

(73) $$Y_n(\alpha\eta)J_n(\alpha) - J_n(\alpha\eta)Y_n(\alpha) = 0.$$

It is known (*cf.* Gray & Mathews **[15]** p. 82, theorem X) that equation (73) admits an infinite number of roots all of which are real and simple; and that if $\alpha_j (j = 1, 2, \cdots)$ are the distinct roots of the equation, the functions $\mathcal{C}_{n,n}(\alpha_j r)$ $(j = 1, 2, \cdots)$ form an orthogonal set with the integral property

(74) $$\int_\eta^1 r\, \mathcal{C}_{n,n}(\alpha_j r)\, \mathcal{C}_{n,n}(\alpha_k r)\, dr = N_{j,n}\, \delta_{jk},$$

where

(75) $$N_{j,n} = \frac{2}{\pi^2 \alpha_j^2}\left\{\frac{J_n^2(\alpha_j \eta)}{J_n^2(\alpha_j)} - 1\right\}.$$

For later use we may note here that the derivatives of $\mathcal{C}_{n,n}(\alpha_j r)$ at $r = 1$ and $r = \eta$ (which we shall denote by $\mathcal{C}_n'(\alpha_j)$ and $\mathcal{C}_n'(\alpha_j \eta)$, respectively) are given by

(76) $$\mathcal{C}_n'(\alpha_j) = \left[\frac{d}{dr}\, \mathcal{C}_{n,n}(\alpha_j r)\right]_{r=1} = -\frac{2}{\pi}\frac{J_n(\alpha_j \eta)}{J_n(\alpha_j)}$$

and

(77) $$\mathcal{C}_n'(\alpha_j \eta) = \left[\frac{d}{dr}\, \mathcal{C}_{n,n}(\alpha_j r)\right]_{r=\eta} = -\frac{2}{\eta\pi}.$$

Also, since $\mathcal{C}_{n,\nu}(z)$ is a cylinder function of ν it satisfies (with respect to ν) the same recurrence relations as the Bessel functions J_ν and Y_ν.

Returning to the variational solution of the characteristic value problem we assume for F the trial function

(78) $$F(r) = \Sigma_j\, A_j\, \mathcal{C}_{n,n}(\alpha_j r),$$

where the A_j's are the variational parameters. With this choice, F vanishes at $r = 1$ and $r = \eta$. With F given by (78) the equation governing W (equation (56)) can be explicitly solved and for $n > 1$ we have (*cf.* equation (61))

(79) $$W = \Sigma_j(A_j/\alpha_j^4)\mathcal{C}_{n,n}(\alpha_j r) + B_1 r^n + B_2 r^{n+2} + B_3 r^{-n} + B_4 r^{-n+2},$$

where B_1, B_2, B_3 and B_4 are constants of integration to be determined by the boundary conditions (57). (It may be explicitly noted here that the solution given by (79) does not apply for $n = 1$; we shall return to this case presently.) The boundary conditions (57) lead to the equations

$$(80) \qquad B_1 + B_2 + B_3 + B_4 = 0,$$

$$(81) \qquad B_1 \eta^n + B_2 \eta^{n+2} + B_3 \eta^{-n} + B_4 \eta^{-n+2} = 0,$$

$$(82) \qquad nB_1 + (n+2)B_2 - nB_3 - (n-2)B_4 = -\Sigma_j (A_j/\alpha_j^4)\mathcal{C}'_n(\alpha_j),$$

$$(83) \qquad nB_1 \eta^{n-1} + (n+2)B_2 \eta^{n+1} - nB_3 \eta^{-n-1} - (n-2)B_4 \eta^{-n+1} = -\Sigma_j (A_j/\alpha_j^4)\mathcal{C}'_n(\alpha_j \eta),$$

where $\mathcal{C}'_n(\alpha_j)$ and $\mathcal{C}'_n(\alpha_j \eta)$ are defined as in equations (76) and (77). On solving the foregoing equations we find that the solution can be expressed in the form

$$(84) \qquad B_i = \Sigma_j (A_j/\alpha_j^4) b_{ij} \qquad (i = 1, 2, 3 \text{ and } 4),$$

where the constants b_{1j}, b_{2j}, b_{3j} and b_{4j} depend only on α_j. Thus

$$(85) \qquad b_{2j} = K_n \{(n-1)(1-\eta^2)\Delta_n(\alpha_j) + (1-\eta^{-2n+2})\delta_n(\alpha_j)\}$$

and

$$(86) \qquad b_{4j} = K_n \{(1-\eta^{2n+2})\Delta_n(\alpha_j) - (n+1)(1-\eta^2)\delta_n(\alpha_j)\},$$

where

$$(87) \qquad \Delta_n(\alpha_j) = -\tfrac{1}{2}\{\mathcal{C}'_n(\alpha_j) - \eta^{-n+1}\mathcal{C}'_n(\alpha_j \eta)\},$$

$$(88) \qquad \delta_n(\alpha_j) = -\tfrac{1}{2}\{\mathcal{C}'_n(\alpha_j) - \eta^{n+1}\mathcal{C}'_n(\alpha_j \eta)\}$$

and

$$(89) \qquad K_n = [(n^2-1)(1-\eta^2)^2 + (1-\eta^{2n+2})(1-\eta^{-2n+2})]^{-1}.$$

With b_{2j} and b_{4j} given by equations (85) and (86), b_{1j} and b_{3j} follow from the equations:

$$(90) \qquad b_{1j} = -\frac{1}{n}\{\tfrac{1}{2}\mathcal{C}'_n(\alpha_j) + (n+1)b_{2j} + b_{4j}\}$$

and

$$(91) \qquad b_{3j} = +\frac{1}{n}\{\tfrac{1}{2}\mathcal{C}'_n(\alpha_j) + b_{2j} - (n-1)b_{4j}\}.$$

Turning next to the evaluation of S_n we find that for F and W given by equations (78) and (79) the integrals occurring in (54) can be reduced by steps similar to those indicated in (64) and (65). Thus, we now find

$$(92) \qquad \int_\eta^1 r\left\{\left(\frac{dF}{dr}\right)^2 + \frac{n^2 F^2}{r^2}\right\} dr = \Sigma_j A_j^2 \alpha_j^2 N_{j,n}$$

and

$$\int_\eta^1 rG^2\,dr = \Sigma_j(A_j^2/\alpha_j^4)N_{j,n} \tag{93}$$
$$+ \Sigma_j A_j \int_\eta^1 \mathcal{C}_{n,n}(\alpha_j r)[B_1 r^{n+1} + B_2 r^{n+3} + B_3 r^{-n+1} + B_4 r^{-n+3}]\,dr.$$

The remaining integrals on the right-hand side of (93) can be evaluated if proper use is made of the various recurrence relations satisfied by the cylinder functions $\mathcal{C}_{n,\nu}$ and also of equations (80) and (81); and we find after some lengthy reductions that

$$\int_\eta^1 rG^2\,dr = \Sigma_j(A_j^2/\alpha_j^4)N_{j,n} - 2B_2\,\Sigma_j(A_j/\alpha_j^2)\mathfrak{M}_{n,n+2}^{(j)} \tag{94}$$
$$- 2B_4\,\Sigma_j(A_j/\alpha_j^2)\mathfrak{N}_{n,n-2}^{(j)},$$

where

$$\mathfrak{M}_{n,\nu}^{(j)} = \mathcal{C}_{n,\nu}(\alpha_j) - \eta^\nu \mathcal{C}_{n,\nu}(\alpha_j\eta) \tag{95}$$

and

$$\mathfrak{N}_{n,\nu}^{(j)} = \mathcal{C}_{n,\nu}(\alpha_j) - \eta^{-\nu}\mathcal{C}_{n,\nu}(\alpha_j\eta). \tag{96}$$

The resulting expression for S_n is, therefore,

$$n^2 S_n = \frac{\Sigma_j A_j^2\alpha_j^2 N_{j,n}}{\Sigma_j\{(A_j^2/\alpha_j^4)N_{j,n} - 2B_2(A_j/\alpha_j^2)\mathfrak{M}_{n,n+2}^{(j)} - 2B_4(A_j/\alpha_j^2)\mathfrak{N}_{n,n-2}^{(j)}\}} \tag{97}$$

where B_2 and B_4 are given by equations (84) to (89).

As we have already stated the foregoing solution does not apply for the case $n = 1$. The reason why we must distinguish this case is that the complementary function in the solution (79) is no longer the general one as the terms in B_1 and B_4 become identical when $n = 1$. However, it is readily found that in the case $n = 1$ the general solution for W is

$$W = \Sigma_j(A_j/\alpha_j^4)\mathcal{C}_{1,1}(\alpha_j r) + B_1 r + B_2 r^3 + B_3 r^{-1} + B_4 r\log r, \tag{98}$$

and the equations determining the constants of integration are:

$$\begin{aligned} B_1 + B_2 + B_3 &= 0,\\ B_1\eta + B_2\eta^3 + B_3\eta^{-1} + B_4\eta\log\eta &= 0,\\ B_1 + 3B_2 - B_3 + B_4 &= -\Sigma_j(A_j/\alpha_j^4)\mathcal{C}_1'(\alpha_j),\\ B_1 + 3B_2\eta^2 - B_3\eta^{-2} + B_4(1+\log\eta) &= -\Sigma_j(A_j/\alpha_j^4)\mathcal{C}_1'(\alpha_j\eta). \end{aligned} \tag{99}$$

On solving these equations we find that the solutions for B_1, B_2, B_3 and B_4 can be expressed in the same way as before (*i.e.* by equations of the form (84)); but equations (85), (86), (89), (90) and (91) are now replaced by

$$b_{2j} = K_1\{\tfrac{1}{2}(1 - \eta^2)\Delta_1(\alpha_j) + (\log \eta)\delta_1(\alpha_j)\},$$

$$b_{4j} = K_1\{-(1 - \eta^4)\Delta_1(\alpha_j) + 2(1 - \eta^2)\delta_1(\alpha_j)\},$$

$$K_1 = [(1 - \eta^2)^2 + (1 - \eta^4)\log \eta]^{-1}, \tag{100}$$

$$b_{1j} = -[\tfrac{1}{2}\mathcal{C}_1'(\alpha_j) + 2b_{2j} + \tfrac{1}{2}b_{4j}]$$

and

$$b_{3j} = +[\tfrac{1}{2}\mathcal{C}_1'(\alpha_j) + b_{2j} + \tfrac{1}{2}b_{4j}].$$

The reduction of the formula for S_1 proceeds somewhat differently but essentially along the same lines as for $n > 1$. We finally obtain

$$S_1 = \frac{\Sigma_j A_j^2 \alpha_j^2 N_{j,1}}{\Sigma_j\{(A_j/\alpha_j^4)N_{j,1} - 2B_2(A_j/\alpha_j^2)\mathfrak{M}_{1,3}^{(j)} + 2B_4(A_j/\alpha_j^3)\mathfrak{M}_{1,0}^{(j)}\}}. \tag{101}$$

(In equations (100) and (101) the general definitions of $\Delta_n(\alpha_j)$, $\delta_n(\alpha_j)$, $\mathfrak{M}_{n,\nu}^{(j)}$ and $\mathfrak{N}_{n,\nu}^{(j)}$ are retained without modifications.)

The simplest trial function for F of the chosen form is

$$F = \mathcal{C}_{n,n}(\alpha_1 r), \tag{102}$$

where α_1 is the first zero of the expression on the left-hand side of equation (73). For this choice of F there is no variational parameter with respect to which we have to minimize and equations (97) and (101) directly give (after substituting for B_2 and B_4):

$$\begin{aligned} S_n = \frac{\alpha_1^8}{n^2} N_{1,n}\{\alpha_1^2 N_{1,n} & - 2K_n[(n-1)(1-\eta^2)\Delta_n(\alpha_1) + (1 - \eta^{-2n+2})\delta_n(\alpha_1)]\mathfrak{M}_{n,n+2}^{(1)} \\ & - 2K_n[(1-\eta^{2n+2})\Delta_n(\alpha_1) - (n+1)(1-\eta^2)\delta_n(\alpha_1)]\mathfrak{N}_{n,n-2}^{(1)}\}^{-1} \end{aligned} \tag{103}$$

$$(n > 1)$$

and

$$\begin{aligned} S_1 = \alpha_1^8 N_{1,1}\{\alpha_1^2 N_{1,1} & - 2K_1[\tfrac{1}{2}(1-\eta^2)\Delta_1(\alpha_1) + (\log\eta)\delta_1(\alpha_1)]\mathfrak{M}_{1,3}^{(1)} \\ & + 2K_1[-(1-\eta^4)\Delta_1(\alpha_1) + 2(1-\eta^2)\delta_1(\alpha_1)]\mathfrak{M}_{1,0}^{(1)}/\alpha_1\}^{-1} \quad (n = 1). \end{aligned} \tag{104}$$

In Table 2 are listed the values of S_n obtained with the aid of the foregoing formulae and the recent tabulation of Chandrasekhar & Donna Elbert **[16]**

TABLE 2

The Lowest Characteristic Numbers S_n for Case I and Two Confining Cylinders: η denotes the Ratio of the Radius of the Inner to that of the Outer Cylinder

n	$\eta = 0.2$	$\eta = 0.3$	$\eta = 0.4$	$\eta = 0.5$	$\eta = 0.6$	$\eta = 0.8$
1	2.399×10^4	4.945×10^4	1.184×10^5	3.418×10^5	1.275×10^6	7.974×10^7
2	1.225×10^4	2.006×10^4	4.057×10^4	1.038×10^5	3.553×10^5	2.038×10^7
3	1.343×10^4	1.736×10^4	2.856×10^4	6.204×10^4	1.886×10^5	9.395×10^6
4	1.855×10^4	2.052×10^4	2.786×10^4	5.065×10^4	1.332×10^5	5.557×10^6
5	2.649×10^4	2.725×10^4	3.223×10^4	4.960×10^4	1.118×10^5	3.789×10^6
6		3.765×10^4	4.069×10^4	5.436×10^4	1.053×10^5	2.837×10^6
7					1.075×10^5	2.274×10^6
8					1.160×10^5	1.918×10^6
9						1.686×10^6
10						1.532×10^6
11						1.445×10^6
12						1.368×10^6

of the roots of equation (73). The dependence of S_n on n is further illustrated in Figure 1. From Table 2 and Figure 1 it is apparent how the pattern of convection which first appears at marginal stability shifts progressively to systems with a large number of vortices as η approaches unity.

The values given by (103) and (104) could be improved by including a second term in F. However, in view of the fact that for $\eta = 0$ the first approximation already gives values accurate to a fraction of a per cent (*cf.* Table 1) it was not considered necessary to carry out this improvement.

7. The solution of the characteristic value problem for case II. One confining cylinder. In case II the confining cylinders are maintained at constant, but different, temperatures. From a laboratory standpoint this is therefore the more important of the two cases considered.

In case II (*cf.* equation (9))

$$\beta(r) = \beta/r \qquad (\beta = \text{a constant}), \tag{105}$$

and equation (39) becomes

$$\left(D^* D - \frac{n^2}{r^2}\right)^3 W = -\frac{n^2 \Omega^2 \alpha\beta}{\kappa\nu}\frac{W}{r^2}, \tag{106}$$

where, as before, $W = ru$. Again, if we measure r in units of the radius R_2 of the outer cylinder the problem reduces to one of determining

$$S_n = \frac{\Omega^2 \alpha\beta}{\kappa\nu} R_2^{\,4}, \tag{107}$$

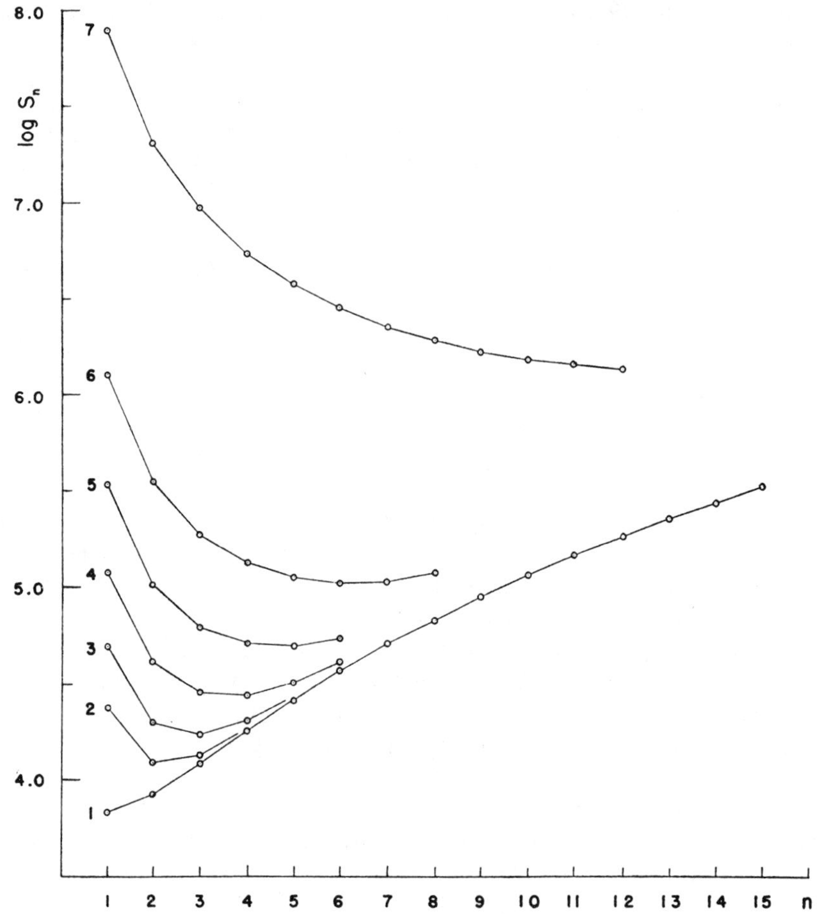

FIGURE 1. The criterion (S_n) for the onset of instability with a convection pattern with $2n$ vortices in the viscous flow between two rotating cylinders in the presence of a radial temperature gradient, for various values of the ratio (η) of the radii of the two cylinders and for case I. The curves labelled 1, 2, $\cdots$, 7 refer to $\eta = 0, 0.2, 0.3, 0.4, 0.5, 0.6$ and 0.8, respectively.

such that the differential equation

$$\left(\frac{d^2}{dr^2} + \frac{1}{r}\frac{d}{dr} - \frac{n^2}{r^2}\right)^3 W = -n^2 S_n \frac{W}{r^2}, \tag{108}$$

allows a non-trivial solution which satisfies the boundary conditions (45).

Now, it can be readily verified that the occurrence of $1/r^2$ on the right-hand side of equation (108) prevents the formulation of a variational procedure for solving the underlying characteristic value problem. We shall therefore describe a different method of solution which, it will appear, is quite satisfactory for the purposes of determining the lowest characteristic number S_n. We shall illustrate the method by considering first the special case when there is only one confining cylinder. Strictly, the consideration of this special case under the conditions postulated is improper: For, the admission of a temperature gradient which varies as $1/r$ implies a logarithmic singularity in the temperature distribution at the origin if r should take the value zero. However, in spite of this singularity at the origin, the characteristic value problem itself has a meaning and the consideration of this case as a limit of permissible situations may be allowed.

As in §5 we shall let

$$\left(\frac{d^2}{dr^2} + \frac{1}{r}\frac{d}{dr} - \frac{n^2}{r^2}\right)^2 W = F, \tag{109}$$

and rewrite equation (108) in the form

$$\left(\frac{d^2}{dr^2} + \frac{1}{r}\frac{d}{dr} - \frac{n^2}{r^2}\right) F = -n^2 S_n \frac{W}{r^2}. \tag{110}$$

We shall now assume that in the interval (0, 1), F can be expanded as a series in the form

$$F(r) = \Sigma_j A_j J_n(\alpha_j r), \tag{111}$$

where (*cf*. equation (59))

$$A_j = 2[J'_n(\alpha_j)]^{-2} \int_0^1 rF(r)J_n(\alpha_j r)\, dr. \tag{112}$$

With F given by (111) the solution of equation (109) which is continuous at the origin and satisfies the boundary conditions $W = dW/dr = 0$ at $r = 1$, is the same as that given in §5. We have (*cf*. equations (62) and (63))

$$W = \Sigma_j (A_j/\alpha_j^4) J_n(\alpha_j r) + B(r^n - r^{n+2}), \tag{113}$$

where

$$B = \tfrac{1}{2}\Sigma_j (A_j/\alpha_j^3) J'_n(\alpha_j). \tag{114}$$

Substituting (111) and (113) in equation (110), we are left with

$$\Sigma_j A_j \alpha_j^2 J_n(\alpha_j r) = \frac{n^2 S_n}{r^2}\{\Sigma_j (A_j/\alpha_j^4) J_n(\alpha_j r) + B(r^n - r^{n+2})\}. \tag{115}$$

Now multiplying equation (115) by $rJ_n(\alpha_k r)$ and integrating over the range r, we obtain

$$\tfrac{1}{2}A_k \alpha_k^2 [J'_n(\alpha_k)]^2 = n^2 S_n \int_0^1 \frac{1}{r}\left\{\Sigma_j \frac{A_j}{\alpha_j^4} J_n(\alpha_j r) + B(r^n - r^{n+2})\right\} J_n(\alpha_k r)\, dr. \tag{116}$$

Letting

$$(117)\qquad \left(j\left|\frac{1}{r^2}\right|k\right) = \int_0^1 J_n(\alpha_j r)\,\frac{1}{r}\,J_n(\alpha_k r)\,dr$$

and

$$(118)\qquad D_k = \int_0^1 (r^{n-1} - r^{n+1})J_n(\alpha_k r)\,dr,$$

we can rewrite equation (116) in the form

$$(119)\qquad \tfrac{1}{2}A_k\,\alpha_k^2[J'_n(\alpha_k)]^2\,Q = \Sigma_j(A_j/\alpha_j^4)\,(j\,|\,r^{-2}\,|\,k) + BD_k,$$

where

$$(120)\qquad Q = \frac{1}{n^2}\,S_n.$$

Finally, substituting for B from equation (114) we obtain the following infinite system of homogeneous equations for the A_j's:

$$(121)\qquad \Sigma_j\,\frac{A_j}{\alpha_j^4}\left\{\left(j\left|\frac{1}{r^2}\right|k\right) + \tfrac{1}{2}\alpha_j\,J'_n(\alpha_j)D_k - \tfrac{1}{2}\alpha_j^6\,[J'_n(\alpha_j)]^2\,Q\delta_{jk}\right\} = 0.$$

Hence Q is to be determined as a root of the infinite determinantal equation

$$(122)\qquad \left\|\left(j\left|\frac{1}{r^2}\right|k\right) + \tfrac{1}{2}\alpha_j J'_n(\alpha_j)D_k - \tfrac{1}{2}\alpha_j^6\,[J'_n(\alpha_j)]^2\,\delta Q_{jk}\right\| = 0.$$

By using the known properties of the Bessel functions, we can show that

$$(123)\qquad \begin{aligned} D_k = \frac{1}{\alpha_k^n}\,\{&2^{n-1}(n-1)! - 2(n-1)\alpha_k^{n-2}\,J_{n-2}(\alpha_k) \\ &- 2^2(n-1)(n-2)\alpha_k^{n-3}\,J_{n-3}(\alpha_k) \\ &- 2^3(n-1)(n-2)(n-3)\alpha_k^{n-4}\,J_{n-4}(\alpha_k) - \cdots\cdots - 2^{n-1}(n-1)!J_0(\alpha_k)\}.\end{aligned}$$

Also the diagonal elements of the matrix $(j|\,r^{-2}\,|k)$ can be explicitly evaluated. We find (*cf.* Watson [**17**] p. 137) that

$$(124)\qquad \left(j\left|\frac{1}{r^2}\right|j\right) = \frac{1}{2n}\left\{1 - 2\sum_{m=1}^{n-1} J_m^2(\alpha_j) - J_0^2(\alpha_j)\right\}.$$

But it does not appear that the non-diagonal elements can be similarly evaluated.

Now a method of solving equation (122) of infinite order would be to set the determinant formed by the first j rows and columns equal to zero and let j take increasingly larger values. In practise the success of this method will depend on how rapidly the lowest root of the equation of order j tends to its limit as $j \to \infty$. It will appear that for the problem on hand the process converges sufficiently rapidly.

On the method of solution outlined in the preceding paragraph the *first approximation* for the lowest characteristic number S_n will be given by setting the $(1,1)$ element of the matrix equal to zero. Thus

$$\tfrac{1}{2}\alpha_1^2 [J'_n(\alpha_1)]^2 Q = \frac{1}{\alpha_1^4}\left(1 \left| \frac{1}{r^2} \right| 1\right) + \frac{J'_n(\alpha_1)}{2\alpha_1^3} D_1, \tag{125}$$

where α_1 denotes the first zero of $J_n(z)$. Substituting for $(1| r^{-2} |1)$ and D_1 in accordance with equations (123) and (124), we obtain

$$\begin{aligned} n^2 S_n = \alpha_1^6 [J'_n(\alpha_1)]^2 \Bigg[\frac{1}{n} \Bigg\{ 1 - 2 \sum_{m=1}^{n-1} J_m^2(\alpha_1) - J_0^2(\alpha_1) \Bigg\} \\ + \frac{J'_n(\alpha_1)}{\alpha_1^{n-1}} \{ 2^{n-1}(n-1)! - 2(n-1)\alpha_k^{n-2} J_{n-2}(\alpha_1) \\ - 2^2(n-1)(n-2)\alpha_1^{n-3} J_{n-3}(\alpha_1) - \cdots - 2^{n-1}(n-1)! J_0(\alpha_1) \} \Bigg]^{-1}. \end{aligned} \tag{126}$$

The values of S_n obtained with the aid of the foregoing formula for $n = 1, 2, \cdots, 6$ are given in Table 3. For these same values of n a second approximation was also carried out by setting the determinant formed by the first two rows and columns of (122) equal to zero; and for $n = 1$ a third approximation was carried out as well. In carrying out these higher approximations the non-diagonal elements $(j| r^{-2} |k) (j \neq k)$ had to be evaluated numerically. This part of the work was enormously lightened by tables of the functions

$$J_n(\alpha_j) \text{ for } j = 1(1)5, \qquad n = 1(1)6 \text{ and } r = 0(0.01)1.00, \tag{127}$$

calculated for the purpose by the Electronic Computer Project of the Institute for Advanced Study at Princeton. (A fuller acknowledgement is made at the end of the paper.)

A comparison of the values of S_n given in Table 3 shows that on going from the first to the second approximation we reduce the value of S_n by a nearly

TABLE 3

The Lowest Characteristic Numbers S_n for Case II and One Confining Cylinder

n	FIRST APPROXIMATION	SECOND APPROXIMATION	THIRD APPROXIMATION
1	1180	1075	1077
2	2340	2128	
3	4390	4000	
4	7560	6910	
5	12180	11170	
6	18570	17100	

constant factor: in fact, the factor varies only between 1.100 and 1.086 in the tabulated range of n. Also, in the one case for which the result of a third approximation is available, the value of S_n was changed from that given by the second approximation by only one part in 500. From these facts we may conclude that already the second approximation provides the required characteristic numbers with errors not probably exceeding one per cent.

8. The solution of the characteristic value problem for case II. Two confining cylinders. Turning finally to the solution for case II when there are two confining cylinders, we shall assume that F can be expanded as a series in the form (*cf.* equations (74) and (75))

$$F = \Sigma_j A_j \mathcal{C}_{n,n}(\alpha_j r), \tag{128}$$

where

$$A_j = \frac{1}{N_{j,n}} \int_\eta^1 rF(r)\mathcal{C}_{n,n}(\alpha_j r)\, dr. \tag{129}$$

With F given by (128) the solution of equation (109) which satisfies the boundary conditions (57) is evidently the same as that given in §6. Combining equations (79) and (84) (and similarly (98) and (84)) we can write the required solution for W in the form

$$\begin{aligned} W = \Sigma_j(A_j/\alpha_j^4)\,\{\mathcal{C}_{n,n}(\alpha_j r) + b_{1j}r^n + b_{2j}r^{n+2} + b_{3j}r^{-n} \\ + b_{4j}r^{-n+2} \text{ (or } r\log r \text{ in case } n = 1)\}, \end{aligned} \tag{130}$$

where the coefficients b_{1j}, *etc.*, have the same meanings as in §6 (equations (85) to (91) and (100)).

Now substituting for F and W according to equations (128) and (130) in equation (109) we obtain

$$\begin{aligned} \Sigma_j A_j \alpha_j^2 \mathcal{C}_{n,n}(\alpha_j r) = \frac{n^2 S_n}{r^2} \Sigma_j(A_j/\alpha_j^4)\{\mathcal{C}_{n,n}(\alpha_j r) \\ + b_{1j}r^n + b_{2j}r^{n+2} + b_{3j}r^{-n} + b_{4j}r^{-n+2} \quad \text{(or } r\log r \text{ in case } n = 1)\}. \end{aligned} \tag{131}$$

Next multiplying this equation by $r\mathcal{C}_{n,n}(\alpha_k r)$ and integrating over the range of r we obtain (*cf.* equations (74) and (75))

$$\begin{aligned} A_k \alpha_k^2 N_{k,n} Q = \Sigma_j(A_j/\alpha_j^4) \\ \times \{(j|\,r^{-2}\,|k) + b_{1j}\langle r^{n-1}\rangle_k + b_{2j}\langle r^{n+1}\rangle_k + b_{3j}\langle r^{-n-1}\rangle_k \\ + b_{4j}\langle r^{-n+1}\rangle_k \text{ (or } \langle \log r\rangle_k \text{ in case } n = 1)\}, \end{aligned} \tag{132}$$

where $Q = 1/n^2 S_n$,

$$\left(j \left| \frac{1}{r^2} \right| k\right) = \int_\eta^1 \mathcal{C}_{n,n}(\alpha_j r)\, \frac{1}{r}\, \mathcal{C}_{n,n}(\alpha_k r)\, dr \tag{133}$$

and

(134) $$\langle X\rangle_k = \int_\eta^1 X\mathfrak{C}_{n,n}(\alpha_k r)\, dr.$$

Equation (132) represents the following infinite system of homogeneous equations for the constants A_j:

(135) $$\Sigma_j(A_j/\alpha_j^4)\,\{(j|\,r^{-2}\,|k) + b_{1j}\langle r^{n-1}\rangle_k + b_{2j}\langle r^{n+1}\rangle_k + b_{3j}\langle r^{-n-1}\rangle_k + b_{4j}\langle r^{-n+1}\rangle_k \text{ (or } \langle\log r\rangle_k \text{ in case } n = 1) - Q\alpha_j^6 N_{j,n}\,\delta_{jk}\} = 0.$$

Hence Q is to be determined as a root of the infinite determinantal equation

(136) $$\|\,(j|\,r^{-2}\,|k) + b_{1j}\langle r^{n-1}\rangle_k + b_{2j}\langle r^{n+1}\rangle_k + b_{3j}\langle r^{-n-1}\rangle_k + b_{4j}\langle r^{-n+1}\rangle_k \text{ (or } \langle\log r\rangle_k \text{ in case } n = 1) - Q\alpha_j^6 N_{j,n}\,\delta_{jk}\,\| = 0.$$

By using the known properties of cylinder functions, we can show that

(137) $$\langle r^{n-1}\rangle_k = -\frac{1}{\alpha_k^n}\{\alpha_k^{n-1}\,\mathfrak{M}_{n,n-1}^{(k)} + 2(n-1)\alpha_k^{n-2}\,\mathfrak{M}_{n,n-2}^{(k)} + 2^2(n-1)(n-2)\alpha_k^{n-3}\,\mathfrak{M}_{n,n-3}^{(k)} + \cdots + 2^{n-1}(n-1)!\,\mathfrak{M}_{n,0}^{(k)}\},$$

(138) $$\langle r^{n+1}\rangle_k = \mathfrak{M}_{n,n+1}^{(k)}/\alpha_k,$$

(139) $$\langle r^{-n+1}\rangle_k = -\mathfrak{N}_{n,n-1}^{(k)}/\alpha_k$$

and

(140) $$\left(k\left|\frac{1}{r^2}\right|k\right) = -\frac{1}{2n}\left\{[\mathfrak{C}_{n,0}^2(\alpha_k) - \mathfrak{C}_{n,0}^2(\alpha_k\eta)] + 2\sum_{m=1}^{n-1}[\mathfrak{C}_{n,m}^2(\alpha_k) - \mathfrak{C}_{n,m}^2(\alpha_k\eta)]\right\}.$$

But it does not appear that

(141) $$\langle r^{-n-1}\rangle_k = \int_\eta^1 \frac{dr}{r^{n+1}}\,\mathfrak{C}_{n,n}(\alpha_k r),$$
$$\langle\log r\rangle_k = \int_\eta^1 \log r\,\mathfrak{C}_{1,1}(\alpha_k r)\, dr \qquad (n = 1),$$

or the non-diagonal elements of the matrix (133) can be expressed similarly in terms of known quantities; in these cases numerical integration would seem unavoidable.

As in §7 a first approximation to the value of the lowest characteristic number S_n can be obtained by setting the (1, 1) element of the matrix on the left-hand

TABLE 4

The Lowest Characteristic Numbers S_n for Case II and Two Confining Cylinders: η denotes the Ratio of the Radius of the Inner to the Outer Cylinder

n	FIRST APPROXIMATION				SECOND APPROXIMATION
	$\eta = 0.2$	$\eta = 0.3$	$\eta = 0.4$	$\eta = 0.5$	$\eta = 0.5$
1	7478			1.855×10^5	1.897×10^5
2	4109	7936		5.652×10^4	5.813×10^4
3	5013	7188	1.356×10^4	3.399×10^4	3.505×10^4
4	7750	9025	1.299×10^4	2.799×10^4	2.893×10^4
5			1.620×10^4	2.772×10^4	2.868×10^4
6				3.081×10^4	3.171×10^4

side of equation (136) equal to zero. The values of S_n obtained in such a first approximation are given in Table 4 for various values of n and η.

We already know (§7) that for $\eta = 0$, the error of the first approximation is to give values of S_n which are too *high* by ten per cent. To estimate the errors in the present calculations, a second approximation was carried out for $\eta = 0.5$. The values obtained in the second approximation (by setting the determinant formed by the first two rows and columns of the secular matrix equal to zero) are also listed in Table 4. And a comparison of the values of $S_n(n = 1, 2, \cdots, 6)$ obtained in the first and in the second approximations indicates that now the error of the first approximation is to give values of S_n which are too *low* by three per cent. From these facts we may conclude that no entry in Table 4 is likely to be in error by more than ten per cent; indeed, for $\eta = 0.3$, 0.4 and 0.5 the errors may be very much less.

The results of the calculations of this as well as of the preceding section are illustrated in Figure 2. The remarkable similarity of these results with those obtained for case I (compare particularly Figures 1 and 2) is especially noteworthy.

9. An illustrative example. As an example illustrative of the theory developed in this paper we shall consider the case of water (at room temperatures) enclosed between two long cylinders maintained at constant different temperatures and of radii 5.0 cm and 2.5 cm, respectively. Let the entire system be rotated at a constant angular velocity of 1 radian per second. Then according to the results of Table 4, the lowest mode of instability is one with ten vortices ($n = 5$) and occurs when

$$S = \frac{\Omega^2 \alpha\beta}{\kappa\nu} R_2^4 = 2.87 \times 10^4. \tag{142}$$

(Actually, the values of S at which convection patterns with eight or twelve vortices can occur are not substantially higher.) Inserting the numerical values

$$\Omega = 1 \text{ sec}^{-1}; \qquad R_2 = 5 \text{ cm}; \qquad \alpha = 2.1 \times 10^{-4} \text{ degree}^{-1},$$

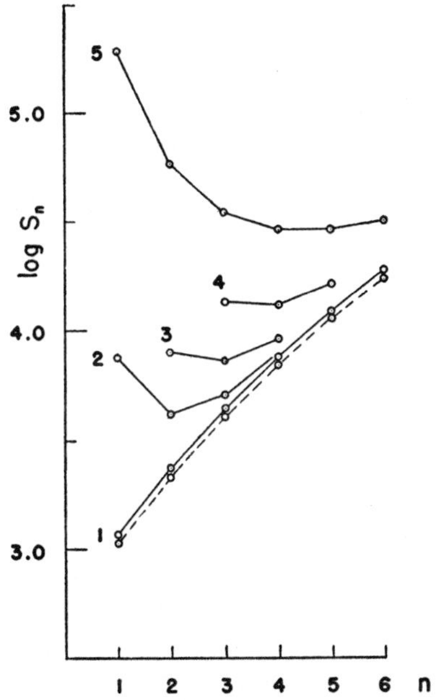

FIGURE 2. Same as Figure 1 but for case II. The curves labelled 1, 2, 3, 4 and 5 refer to $\eta = 0, 0.2, 0.3, 0.4$ and 0.5, respectively; of the two curves labelled 1 the dotted one refers to the results of the second approximation while the other to the results of the first approximation. For $\eta = 0.5$ (curve 5) the results of the second approximation are plotted; the remaining curves refer to the results of the first approximation.

(143)

$$\kappa = 1.4 \times 10^{-3}\ \text{cm}^2/\text{sec} \quad \text{and} \quad \nu = 1.0 \times 10^{-2}\ \text{cm}^2/\text{sec},$$

we find from (142) that the critical value of β at which instability will set in is

(144) $$\beta_{\text{critical}} = 3.1\ \text{degree/cm}.$$

Under the circumstances considered (namely that of case II)

(145) $$\frac{dT}{dr} = \frac{\beta}{r};$$

accordingly, β is related to the difference in temperature of the two cylinders by

(146) $$\Delta T = T_2 - T_1 = \beta \log (R_2/R_1) = -\beta \log \eta.$$

Hence (144) implies a difference in temperature of

(147) $$\Delta T = 3.1 \log_e 2 = 2.1°.$$

In other words, for water enclosed between two cylinders of radii 5.0 and 2.5 cm and rotated at the rate 1 radian/sec, instability of the type considered will arise when the difference in temperature between the two cylinders increases above 2.1°. On account of the occurrence of R_2^4 in the definition of S, this difference in temperature predicted for $R_2 = 5.0$ cm and $R_2 = 2.5$ cm will be reduced by a factor 16 if the scale of the apparatus is increased by a factor 2 and the rate of rotation is unaltered *i.e.*, $\Delta T = 0.13°$ under these latter circumstances. Now the experiments of Fultz and Hide seem to have been performed with cylinders of radii 10 cm or more and rates of rotation much larger than 1 radian/sec; moreover, the patterns of motion they have studied are for differences of temperature generally in excess of one degree. Consequently, in their experiments factors not included in this theory must have been operative. Nevertheless, it would be of interest to repeat their experiments with cylinders of radii 5 cm or less (or with smaller rates of rotation) to see if the effects predicted here can be observed.

10. Acknowledgements. As has already been acknowledged the calculations for case II when there is only one confining cylinder (§7) were greatly facilitated by the Electronic Computer Project of the Institute for Advanced Study at Princeton having provided tables of the functions

(148) $J_n(\alpha_j r)$ for $n = 1(1)6$, $j = 1(1)5$ and $r = 0(0.01)1.00$.

Similarly the calculations presented in §8 for two confining cylinders were facilitated by the Project having provided tables of the functions

(149) $\mathcal{C}_{n,n}(\alpha_j r)$ for $\eta = 0.2(0.1)0.6$ and 0.8, $n = 1(1)6$ and $j = 1$

and, also

(150) $\mathcal{C}_{n,n}(\alpha_j r)$ for $\eta = 0.5$, $n = 1(1)6$ and $j = 2$ and 3.

All these functions have been tabulated at intervals of 0.01 to eight significant figures of which at least seven are believed to be reliable.

It must be evident that without the tables provided by the Electronic Computer Project of the Institute for Advanced Study the task of evaluating numerically the various integrals needed in the calculation of S_n (for case II) would have been too formidable to have undertaken. It is therefore a particular pleasure to acknowledge here my indebtedness to Professor J. von Neumann and Dr. Herman H. Goldstine of the Institute for Advanced Study for their most generous co-operation. My thanks are also due to Miss Donna Elbert who gave valuable assistance in all the remaining calculations.

References

[1] Fultz, D., "Experimental analogies to atmospheric motion", *Compendium of Meteorology*, Amer. Met. Soc. (1951) 1235–1248.

[2] Hide, R., (unpublished) Reported in Geophysical Discussion of the Royal Astronomical Society, *Observatory* **72** (1952) 231–232.

[3] Chandrasekhar, S., "On the inhibition of convection by a magnetic field", *Phil. Mag.* (7) **43** (1952) 501–532.

[4] ———, "The thermal instability of a fluid sphere heated within", *Phil. Mag.* (7) **43** (1952) 1317–1329.

[5] ———, "The onset of convection by thermal instability in spherical shells", *Phil. Mag.* (7) **44** (1953) 233–241.

[6] ———, "The onset of convection by thermal instability in spherical shells (a correction)", *Phil. Mag.* (7) **44** (1953) 1129–1130.

[7] ———, "The stability of viscous flow between rotating cylinders in the presence of a magnetic field", *Proc. Roy. Soc.* A, **216** (1953) 293–309.

[8] ———, "The instability of a layer of fluid heated below and subject to Coriolis forces", *Proc. Roy. Soc.* A, **217** (1953) 306–327.

[9] ———, "The gravitational instability of an infinite homogeneous medium when Coriolis force is acting and a magnetic field is present", *Ap. J.* (in press).

[10] Chandrasekhar, S. & Fermi, E., "Problems of gravitational stability in the presence of a magnetic field", *Ap. J.* 118 (1953) 116–141.

[11] Lord Rayleigh, "On convection currents in a horizontal layer of fluid, when the higher temperature is on the under side", *Scientific Papers*, Vol. 6, Cambridge University Press, London (1920) 432–446.

[12] Jeffreys, H., "The stability of a layer of fluid heated below", *Phil. Mag.* (7) **2** (1926) 833–844.

[13] Meksyn, D., "Stability of viscous flow between rotating cylinders. I", *Proc. Roy. Soc.* A, **187** (1946) 115–128.

[14] Pellew, A. & Southwell, R. V., "On maintained convective motion in a fluid heated from below", *Proc. Roy. Soc.* A, **176** (1940) 312–343.

[15] Gray, A. & Mathews, G. B., *A treatise on Bessel functions*, 2nd ed. Macmillan, London (1922).

[16] Chandrasekhar, S. & Elbert, D., "The roots of $Y_n(\lambda\eta)J_n(\lambda) - J_n(\lambda\eta)Y_n(\lambda) = 0$", *Camb. Phil. Soc.* (in press).

[17] Watson, G. N., *A treatise on the theory of Bessel functions*, 2nd ed., Cambridge (1944) 137.

6

The stability of viscous flow between rotating cylinders. II

By S. Chandrasekhar, F.R.S. and Donna D. Elbert
University of Chicago

(*Received* 31 *January* 1962)

In this paper a new method is described for solving the characteristic value problem underlying the theory of the stability of viscous flow between rotating cylinders. The method depends on solving a simpler adjoint system of equations. It is shown that the present method is superior to the one described in an earlier paper (Chandrasekhar 1958). Numerical results are given for the case when the ratio of the radii of the two cylinders is one-half.

1. Introduction

In an earlier paper (Chandrasekhar 1958; this paper will be referred to hereafter as I; see also Chandrasekhar 1961 *a*, § 73), the mathematical problem of the stability of viscous flow between two rotating co-axial cylinders was solved without making the customary approximation of a narrow gap. In this paper we shall present an alternative method of solution of the same problem which leads to results of higher precision. The method exploits the notion of adjointness of differential systems in the theory of hydrodynamic stability which has recently been developed (Roberts 1960; Chandrasekhar 1961 *a*, *b*).

2. The characteristic value problem and the proposed method of solution

By considering symmetric perturbations of the solution,

$$V(r) = Ar + B/r, \tag{1}$$

which the Navier–Stokes equations allow for the distribution of the rotational velocity $V(r)$ between two rotating cylinders (where A and B are constants related to the angular velocities, Ω_1 and Ω_2, with which the inner and the outer cylinders are rotated), one finds that the equations governing the state of marginal stability, for disturbances which are periodic in the axial z-direction with a wave number k, can be reduced to the pair of equations (cf. Chandrasekhar 1961 *a*, p. 295, equations (164) and (165))

$$(\mathrm{DD}_* - a^2)^2 u = a^2 \frac{2B}{\nu}\left(\frac{1}{r^2} - \kappa\right) v \tag{2}$$

and

$$(\mathrm{DD}_* - a^2)\, v = \frac{2A}{\nu} R_2^2 u. \tag{3}$$

In the foregoing equations r (the radial distance from the axis) is measured in units of the radius R_2 of the outer cylinder;

$$\mathrm{D} = \mathrm{d}/\mathrm{d}r, \quad \mathrm{D}_* = \mathrm{D} + 1/r, \quad a = kR_2, \quad \kappa = -AR_2^2/B; \tag{4}$$

and u and v are the amplitudes of the periodically varying velocities in the radial and in the transverse directions, respectively.

The boundary conditions with respect to which equations (2) and (3) must be solved are

$$u = \mathrm{D}u = v = 0 \quad \text{for} \quad r = 1 \quad \text{and} \quad r = \eta = R_1/R_2. \tag{5}$$

By the transformation

$$u \to a^2(2B/\nu)\, u, \tag{6}$$

equations (2) and (3) take the more convenient forms

$$(\mathrm{DD}_* - a^2)^2 u = (1/r^2 - \kappa)\, v \tag{7}$$

and

$$(\mathrm{DD}_* - a^2)\, v = -Ta^2 u, \tag{8}$$

where

$$T = -(4AB/\nu^2)\, R_2^4 \tag{9}$$

is the Taylor number. The boundary conditions (5) are, of course, unaffected by the transformation (6).

It may be noted here that, if A and B are expressed in terms of Ω_1 and Ω_2,

$$T = \frac{4\Omega_1^2 R_1^4}{\nu^2} \frac{(1-\mu)(1-\mu/\eta^2)}{(1-\eta^2)^2} \quad \text{and} \quad \kappa = \frac{1-\mu/\eta^2}{1-\mu}, \tag{10}$$

where

$$\mu = \Omega_2/\Omega_1 \quad \text{and} \quad \eta = R_1/R_2. \tag{11}$$

From our experience with the solution of the present problem in the framework of the narrow-gap approximation (Chandrasekhar 1954*a*, 1961*a*, § 71*a*,*b*) the method which would naturally occur to one to apply to the solution of equations (7) and (8) is the following.

Since v must vanish at $r = 1$ and η we should want to expand it in terms of a set of orthogonal functions (defined in the interval $\eta \leqslant r \leqslant 1$) which vanish at the same points. For the problem on hand the most suitable functions are those defined by

$$\mathscr{C}_1(\alpha_j r) = Y_1(\alpha_j \eta)\, J_1(\alpha_j r) - J_1(\alpha_j \eta)\, Y_1(\alpha_j r), \tag{12}$$

where J_1 and Y_1 are the Bessel functions of the two kinds of order 1 and α_j is a root of the equation

$$Y_1(\alpha\eta)\, J_1(\alpha) - J_1(\alpha\eta)\, Y_1(\alpha) = 0. \tag{13}$$

The functions $\mathscr{C}_1(\alpha_j r)$ satisfy the orthogonality relation

$$\int_\eta^1 \mathscr{C}_1(\alpha_j r)\, \mathscr{C}_1(\alpha_k r)\, r\, \mathrm{d}r = N_j \delta_{jk}, \tag{14}$$

where N_j is a normalization constant. We expand, then, v in the form

$$v = \sum_j A_j \mathscr{C}_1(\alpha_j r), \tag{15}$$

and express u as the sum

$$u = \sum_j A_j u_j, \tag{16}$$

where u_j is the unique solution of the equation

$$(\mathrm{DD}_* - a^2)^2 u_j = (1/r^2 - \kappa)\, \mathscr{C}_1(\alpha_j r) \tag{17}$$

which satisfies the boundary conditions

$$u_j = \mathrm{D}u_j = 0 \quad \text{for} \quad r = 1 \quad \text{and} \quad \eta. \tag{18}$$

Having determined u_j in this fashion, we insert the expansions (15) and (16) in equation (8) to obtain the equation

$$\sum_j A_j(\alpha_j^2+a^2)\mathscr{C}_1(\alpha_j r) = Ta^2 \sum_j A_j u_j. \tag{19}$$

By multiplying this equation by $r\mathscr{C}_1(\alpha_k r)$ and integrating over the range of r, we obtain an infinite set of linear homogeneous equations for the A_j's which in turn leads to the secular equation

$$\|(1/Ta^2)(\alpha_j^2+a^2)N_j\delta_{jk}-(j|k)\| = 0, \tag{20}$$

where
$$(j|k) = \int_\eta^1 u_j\mathscr{C}_1(\alpha_k r)\,r\,\mathrm{d}r. \tag{21}$$

The foregoing method of solution was *not* adopted in I as it was considered 'impracticable' on the grounds that the solution of equation (17) cannot be explicitly written down in terms of known (or tabulated) functions; that the solution will involve many indefinite integrals over the functions $\mathscr{C}_1(\alpha_j r)$; and that in consequence the evaluation of the matrix element $(j|k)$ will require as many double quadratures. On these grounds, equations (7) and (8) were considered in their equivalent forms

$$(\mathrm{DD}_*-a^2)\,v = u \tag{22}$$

and
$$(\mathrm{DD}_*-a^2)^2\,u = -Ta^2(1/r^2-\kappa)\,v; \tag{23}$$

and u (instead of v) was expanded in a series of orthogonal functions which together with their derivatives vanish at $r = 1$ and η.

However, the 'difficulties' which led to the abandonment of the method based on the expansion of v are more apparent than real: for, the system *adjoint* to equations (7) and (8) and the boundary conditions (5), namely,

$$(\mathrm{DD}_*-a^2)^2\,u^\dagger = v^\dagger, \tag{24}$$

$$(\mathrm{DD}_*-a^2)\,v^\dagger = -Ta^2(1/r^2-\kappa)\,u^\dagger, \tag{25}$$

and
$$u^\dagger = \mathrm{D}u^\dagger = v^\dagger = 0 \quad \text{for} \quad r = 1 \quad \text{and} \quad \eta, \tag{26}$$

define the same set of characteristic values for T as the original system (cf. Chandrasekhar 1961 *b*); and as we shall presently see, the solution of this adjoint system by the same method encounters no such 'difficulties'.

3. The solution of the adjoint system

In solving the adjoint system of equations (24) to (26), we express $v^\dagger$ and $u^\dagger$ as series in the forms

$$v^\dagger = \sum_j B_j\mathscr{C}_1(\alpha_j r) \quad \text{and} \quad u^\dagger = \sum_j B_j u_j^\dagger, \tag{27}$$

where $u_j^\dagger$ is the unique solution of the equation

$$(\mathrm{DD}_*-a^2)^2\,u_j^\dagger = \mathscr{C}_1(\alpha_j r), \tag{28}$$

which satisfies the boundary conditions

$$u_j^\dagger = \mathrm{D}u_j^\dagger = 0 \quad \text{for} \quad r = 1 \quad \text{and} \quad \eta. \tag{29}$$

The required solution for $u_j^\dagger$ is readily found. It can be expressed in the form

$$u_j^\dagger = E_j I_1(ar) + F_j r I_0(ar) + G_j K_1(ar) + H_j r K_0(ar) + \frac{\mathscr{C}_1(\alpha_j r)}{(\alpha_j^2+a^2)^2}, \tag{30}$$

where I_1, I_0, K_1, and K_0 are the Bessel functions, for purely imaginary arguments, of the two kinds and of orders one and zero; and E_j, F_j, G_j, and H_j are constants of integration to be determined by the boundary conditions (29); and the application of the boundary conditions leads to the equations

$$\left.\begin{aligned} E_j I_1(a) + F_j I_0(a) + G_j K_1(a) + H_j K_0(a) &= 0,\\ E_j I_1(a\eta) + F_j \eta I_0(a\eta) + G_j K_1(a\eta) + H_j \eta K_0(a\eta) &= 0,\\ E_j I_2(a) + F_j I_1(a) - G_j K_2(a) - H_j K_1(a) &= -\frac{\alpha_j \mathscr{C}_0(\alpha_j)}{a(\alpha_j^2+a^2)^2},\\ E_j I_2(a\eta) + F_j \eta I_1(a\eta) - G_j K_2(a\eta) - H_j \eta K_1(a\eta) &= -\frac{\alpha_j \mathscr{C}_0(\alpha_j \eta)}{a(\alpha_j^2+a^2)^2}. \end{aligned}\right\} \tag{31}$$

The solution of the foregoing equations for the case $\eta = \frac{1}{2}$ and for the values of a which are relevant for the critical Taylor numbers to be deduced in §3 are listed in table 1.

Having determined $u_j^\dagger$, we insert the expansions for $u^\dagger$ and $v^\dagger$ in equation (25) and obtain in the usual manner the secular equation

$$\|(1/Ta^2)(\alpha_j^2+a^2) N_j \delta_{jk} - (j|k)^\dagger\| = 0, \tag{32}$$

where, now,

$$(j|k)^\dagger = \int_\eta^1 u^\dagger{}_j \left(\frac{1}{r^2} - \kappa\right) \mathscr{C}_1(\alpha_k r)\, r\, \mathrm{d}r. \tag{33}$$

It has been shown (Chandrasekhar 1961 *b*) that this matrix, $(j|k)^\dagger$, is the transposed of the matrix, $(j|k)$, defined by equation (21); it is, in fact, on this account that the characteristic values of the original and the adjoint systems are the same.

Returning to the matrix element $(j|k)^\dagger$, we can write in accordance with equation (30),

$$\begin{aligned} (j|k)^\dagger = {} & E_j[I_k^{(-1)}(a) - \kappa I_k^{(+1)}(a)] + F_j[L_k^{(-1)}(a) - \kappa L_k^{(+1)}(a)]\\ & + G_j[K_k^{(-1)}(a) - \kappa K_k^{(+1)}(a)] + H_j[M_k^{(-1)}(a) - \kappa M_k^{(+1)}(a)]\\ & + \frac{1}{(\alpha_j^2+a^2)^2}[M_{jk} - \kappa N_j \delta_{jk}], \end{aligned} \tag{34}$$

where we have used the abbreviations

$$\left.\begin{aligned} I_k^{(\pm1)}(a) &= \int_\eta^1 \mathscr{C}_1(\alpha_k r)\, I_1(ar)\, r^{\pm1} \mathrm{d}r,\\ K_k^{(\pm1)}(a) &= \int_\eta^1 \mathscr{C}_1(\alpha_k r)\, K_1(ar)\, r^{\pm1} \mathrm{d}r,\\ L_k^{(\pm1)}(a) &= \int_\eta^1 \mathscr{C}_1(\alpha_k r)\, r I_0(ar)\, r^{\pm1} \mathrm{d}r,\\ M_k^{(\pm1)}(a) &= \int_\eta^1 \mathscr{C}_1(\alpha_k r)\, r K_0(ar)\, r^{\pm1} \mathrm{d}r, \end{aligned}\right\} \tag{35}$$

and

$$M_{jk} = \int_\eta^1 \mathscr{C}_1(\alpha_j r)\, \mathscr{C}_1(\alpha_k r) \frac{\mathrm{d}r}{r}. \tag{36}$$

The diagonal elements of M_{jk} can be explicitly evaluated. We find (cf. Watson 1944, p. 137)

$$M_{jj} = \tfrac{1}{2}[\mathscr{C}_0^2(\alpha\eta) - \mathscr{C}_0^2(\alpha)]. \tag{37}‡$$

However, the non-diagonal elements require numerical evaluation.

TABLE 1. THE SOLUTION OF EQUATIONS (31) FOR $\eta = \frac{1}{2}$ AND $j = 1, 2,$ AND 3 FOR VARIOUS VALUES OF a

a	j	E_j	F_j	G_j	H_j
6·2	1	$+1{\cdot}7574\times10^{-6}$	$-1{\cdot}6390\times10^{-6}$	$-2{\cdot}1693\times10^{-3}$	$4{\cdot}7835\times10^{-3}$
	2	$-4{\cdot}6387\times10^{-7}$	$+4{\cdot}1618\times10^{-7}$	$-5{\cdot}0687\times10^{-4}$	$1{\cdot}2255\times10^{-3}$
	3	$+6{\cdot}9762\times10^{-8}$	$-6{\cdot}5086\times10^{-8}$	$-8{\cdot}7570\times10^{-5}$	$1{\cdot}9325\times10^{-4}$
6·4	1	$+1{\cdot}3803\times10^{-6}$	$-1{\cdot}2891\times10^{-6}$	$-2{\cdot}3149\times10^{-3}$	$5{\cdot}1038\times10^{-3}$
	2	$-3{\cdot}6570\times10^{-7}$	$+3{\cdot}2972\times10^{-7}$	$-5{\cdot}4970\times10^{-4}$	$1{\cdot}3175\times10^{-3}$
	3	$+5{\cdot}7612\times10^{-8}$	$-5{\cdot}3823\times10^{-8}$	$-9{\cdot}8208\times10^{-5}$	$2{\cdot}1667\times10^{-4}$
6·8	1	$+8{\cdot}5068\times10^{-7}$	$-7{\cdot}9656\times10^{-7}$	$-2{\cdot}6314\times10^{-3}$	$5{\cdot}7984\times10^{-3}$
	2	$-2{\cdot}2803\times10^{-7}$	$+2{\cdot}0740\times10^{-7}$	$-6{\cdot}4594\times10^{-4}$	$1{\cdot}5245\times10^{-3}$
	3	$+3{\cdot}9209\times10^{-8}$	$-3{\cdot}6723\times10^{-8}$	$-1{\cdot}2317\times10^{-4}$	$2{\cdot}7156\times10^{-4}$
7·2	1	$+5{\cdot}2359\times10^{-7}$	$-4{\cdot}9146\times10^{-7}$	$-2{\cdot}9847\times10^{-3}$	$6{\cdot}5714\times10^{-3}$
	2	$-1{\cdot}4270\times10^{-7}$	$+1{\cdot}3077\times10^{-7}$	$-7{\cdot}5822\times10^{-4}$	$1{\cdot}7661\times10^{-3}$
	3	$+2{\cdot}6603\times10^{-8}$	$-2{\cdot}4975\times10^{-8}$	$-1{\cdot}5391\times10^{-4}$	$3{\cdot}3900\times10^{-4}$
7·8	1	$+2{\cdot}5230\times10^{-7}$	$-2{\cdot}3761\times10^{-7}$	$-3{\cdot}5937\times10^{-3}$	$7{\cdot}8980\times10^{-3}$
	2	$-7{\cdot}1022\times10^{-8}$	$+6{\cdot}5695\times10^{-8}$	$-9{\cdot}6262\times10^{-4}$	$2{\cdot}2061\times10^{-3}$
	3	$+1{\cdot}4778\times10^{-8}$	$-1{\cdot}3919\times10^{-8}$	$-2{\cdot}1345\times10^{-4}$	$4{\cdot}6924\times10^{-4}$
8·2	1	$+1{\cdot}5490\times10^{-7}$	$-1{\cdot}4618\times10^{-7}$	$-4{\cdot}0599\times10^{-3}$	$8{\cdot}9092\times10^{-3}$
	2	$-4{\cdot}4726\times10^{-8}$	$+4{\cdot}1589\times10^{-8}$	$-1{\cdot}1275\times10^{-3}$	$2{\cdot}5607\times10^{-3}$
	3	$+9{\cdot}9465\times10^{-9}$	$-9{\cdot}3876\times10^{-9}$	$-2{\cdot}6423\times10^{-4}$	$5{\cdot}7997\times10^{-4}$
8·8	1	$+7{\cdot}4432\times10^{-8}$	$-7{\cdot}0448\times10^{-8}$	$-4{\cdot}8650\times10^{-3}$	$1{\cdot}0649\times10^{-2}$
	2	$-2{\cdot}2421\times10^{-8}$	$+2{\cdot}0988\times10^{-8}$	$-1{\cdot}4274\times10^{-3}$	$3{\cdot}2050\times10^{-3}$
	3	$+5{\cdot}4600\times10^{-9}$	$-5{\cdot}1681\times10^{-9}$	$-3{\cdot}6150\times10^{-4}$	$7{\cdot}9141\times10^{-4}$
9·2	1	$+4{\cdot}5644\times10^{-8}$	$-4{\cdot}3280\times10^{-8}$	$-5{\cdot}4832\times10^{-3}$	$1{\cdot}1980\times10^{-2}$
	2	$-1{\cdot}4172\times10^{-8}$	$+1{\cdot}3317\times10^{-8}$	$-1{\cdot}6690\times10^{-3}$	$3{\cdot}7235\times10^{-3}$
	3	$+3{\cdot}6468\times10^{-9}$	$-3{\cdot}4582\times10^{-9}$	$-4{\cdot}4364\times10^{-4}$	$9{\cdot}6941\times10^{-4}$
9·6	1	$+2{\cdot}7987\times10^{-8}$	$-2{\cdot}6584\times10^{-8}$	$-6{\cdot}1768\times10^{-3}$	$1{\cdot}3469\times10^{-2}$
	2	$-8{\cdot}9672\times10^{-9}$	$+8{\cdot}4545\times10^{-9}$	$-1{\cdot}9504\times10^{-3}$	$4{\cdot}3266\times10^{-3}$
	3	$+2{\cdot}4290\times10^{-9}$	$-2{\cdot}3074\times10^{-9}$	$-5{\cdot}4273\times10^{-4}$	$1{\cdot}1836\times10^{-3}$

By making use of the relations

$$(\mathrm{DD}_* - a^2)\times\begin{Bmatrix} I_0(ar)\,r \\ K_0(ar)\,r \\ I_0(ar)/r \\ K_0(ar)/r \\ \mathscr{C}_1(\alpha_k r) \end{Bmatrix} = \begin{Bmatrix} +2aI_1(ar) \\ -2aK_1(ar) \\ -2aI_1(ar)/r^2 \\ +2aK_1(ar)/r^2 \\ -(\alpha_k^2+a^2)\,\mathscr{C}_1(\alpha_k r) \end{Bmatrix}, \tag{38}$$

and remembering that

$$\left.\begin{aligned} \mathrm{D}_*\mathscr{C}_1(\alpha_j r) &= \alpha_j\mathscr{C}_0(\alpha_j) \quad \text{at} \quad r = 1 \quad \text{and} \\ &= \alpha_j\mathscr{C}_0(\alpha_j\eta) \quad \text{at} \quad r = \eta, \end{aligned}\right\} \tag{39}$$

‡ The function $\mathscr{C}_\nu(\alpha_j r)$ is defined by (cf. equation (12))

$$\mathscr{C}_\nu(\alpha_j r) = Y_1(\alpha_j\eta)\,J_\nu(\alpha_j r) - J_1(\alpha_j\eta)\,Y_\nu(\alpha_j r).$$

we can reduce the various integrals listed in (37) to give

$$\left.\begin{aligned} I_k^{(+1)}(a) &= -\frac{\alpha_k}{\alpha_k^2+a^2}[\mathscr{C}_0(\alpha_k)\,I_1(a)-\eta\mathscr{C}_0(\alpha_k\eta)\,I_1(a\eta)],\\ K_k^{(+1)}(a) &= -\frac{\alpha_k}{\alpha_k^2+a^2}[\mathscr{C}_0(\alpha_k)\,K_1(a)-\eta\mathscr{C}_0(\alpha_k\eta)\,K_1(a\eta)],\\ L_k^{(+1)}(a) &= -\frac{\alpha_k}{\alpha_k^2+a^2}[\mathscr{C}_0(\alpha_k)\,I_0(a)-\eta^2\mathscr{C}_0(\alpha_k\eta)\,I_0(a\eta)]-\frac{2a}{\alpha_k^2+a^2}I_k^{(+1)}(a),\\ M_k^{(+1)}(a) &= -\frac{\alpha_k}{\alpha_k^2+a^2}[\mathscr{C}_0(\alpha_k)\,K_0(a)-\eta^2\mathscr{C}_0(\alpha_k\eta)\,K_0(a\eta)]+\frac{2a}{\alpha_k^2+a^2}K_k^{(+1)}(a),\\ L_k^{(-1)}(a) &= -\frac{\alpha_k}{\alpha_k^2+a^2}[\mathscr{C}_0(\alpha_k)\,I_0(a)-\mathscr{C}_0(\alpha_k\eta)\,I_0(a\eta)]+\frac{2a}{\alpha_k^2+a^2}I_k^{(-1)}(a),\\ M_k^{(-1)}(a) &= -\frac{\alpha_k}{\alpha_k^2+a^2}[\mathscr{C}_0(\alpha_k)\,K_0(a)-\mathscr{C}_0(\alpha_k\eta)\,K_0(a\eta)]-\frac{2a}{\alpha_k^2+a^2}K_k^{(-1)}(a), \end{aligned}\right\} \tag{40}$$

where

$$I_k^{(-1)}(a) = \int_\eta^1 \mathscr{C}_1(\alpha_k r)\,I_1(ar)\frac{\mathrm{d}r}{r} \quad\text{and}\quad K_k^{(-1)}(a) = \int_\eta^1 \mathscr{C}_1(\alpha_k r)\,K_1(ar)\frac{\mathrm{d}r}{r}. \tag{41}$$

Thus to set up the secular matrix explicitly, the only integrals (dependent on a) which need to be evaluated numerically are $I_k^{(-1)}(a)$ and $K_k^{(-1)}(a)$; and this is not a formidable problem.

4. Numerical results for the case $\eta = \frac{1}{2}$

For the purposes of evaluating the secular equation, a knowledge of the function $\mathscr{C}_1(\alpha_j r)$ (at least of the lower orders) must be presupposed. Fortunately, we have in our possession tables of the required functions for $\eta = \frac{1}{2}$ and $j = 1, 2,$ and 3 which the late Professor J. von Neumann had provided us in another connexion (see Chandrasekhar 1954*b*). Making use of these functions, we have evaluated the integrals and the matrix elements required for the setting up of the secular matrix up to the third order. And for various assigned values of κ and a, the secular equation was solved in the second and in the third approximations.‡ In practice, for each value of κ, values of a were considered in the range in which the minimum of the lowest characteristic value of T was expected. The results of the calculations are summarized in tables 2 and 3.

In table 2 we have included for comparison the results obtained in I. The clear superiority of the present method over the one used in I is evident; and this superiority was, of course, to be expected.

In table 3 we have listed the coefficients (B_j/B_1) in the expansion of $v^\dagger$ obtained in the third approximation. Since the matrix $(j|k)^\dagger$ is the transposed of the matrix $(j|k)$, the coefficients (A_j/A_1) in the expansion of the dual function v can also be determined; and they are included in the table. Finally, in figure 1 we illustrate the functions v and $v^\dagger$ for three typical cases.

‡ The *order* of an approximation is the same as the number of terms retained in the expansion of $v^\dagger$ (or v).

TABLE 2. TAYLOR NUMBERS FOR VARIOUS ASSIGNED VALUES OF κ AND a ($\eta = \frac{1}{2}$)

		T obtained by present method		T obtained by earlier method‡	
κ	a	2nd approx.	3rd approx.	2nd approx.	3rd approx.
0	6·0	$1·5406 \times 10^4$	$1·5362 \times 10^4$	$1·5470 \times 10^4$	$1·5370 \times 10^4$
	6·2	$1·5363 \times 10^4$	$1·5320 \times 10^4$	$1·5434 \times 10^4$	$1·5328 \times 10^4$
	6·4	$1·5366 \times 10^4$	$1·5322 \times 10^4$	$1·5444 \times 10^4$	$1·5332 \times 10^4$
0·4	6·0	$1·9632 \times 10^4$	$1·9584 \times 10^4$	$1·9728 \times 10^4$	$1·9594 \times 10^4$
	6·2	$1·9575 \times 10^4$	$1·9527 \times 10^4$	$1·9680 \times 10^4$	$1·9539 \times 10^4$
	6·4	$1·9577 \times 10^4$	$1·9529 \times 10^4$	$1·9692 \times 10^4$	$1·9542 \times 10^4$
0·6	6·0	$2·2744 \times 10^4$	$2·2696 \times 10^4$		
	6·2	$2·2677 \times 10^4$	$2·2628 \times 10^4$	$2·2811 \times 10^4$	$2·2642 \times 10^4$
	6·4	$2·2677 \times 10^4$	$2·2629 \times 10^4$	$2·2823 \times 10^4$	$2·2644 \times 10^4$
1·0	6·2	$3·3124 \times 10^4$	$3·3088 \times 10^4$	$3·3386 \times 10^4$	$3·3110 \times 10^4$
	6·4	$3·3111 \times 10^4$	$3·3076 \times 10^4$	$3·3393 \times 10^4$	$3·3100 \times 10^4$
	6·6	$3·3190 \times 10^4$	$3·3157 \times 10^4$	$3·3492 \times 10^4$	$3·3182 \times 10^4$
$1·3\dot{3}$	6·2	$5·3249 \times 10^4$	$5·3307 \times 10^4$	$5·389 \times 10^4$	$5·3352 \times 10^4$
	6·4	$5·3169 \times 10^4$	$5·3231 \times 10^4$	$5·386 \times 10^4$	$5·3280 \times 10^4$
	6·6	$5·3234 \times 10^4$	$5·3300 \times 10^4$	$5·397 \times 10^4$	$5·3354 \times 10^4$
1·6	6·4	—	—	$1·007 \times 10^5$	$9·9072 \times 10^4$
	6·6	$9·8063 \times 10^4$	$9·8647 \times 10^4$	$1·005 \times 10^5$	$9·8831 \times 10^4$
	6·8	$9·8032 \times 10^4$	$9·8631 \times 10^4$	$1·006 \times 10^5$	$9·8832 \times 10^4$
	7·0	$9·8225 \times 10^4$	$9·8837 \times 10^4$	—	—
1·7	7·0	$1·3547 \times 10^5$	$1·3663 \times 10^5$		
	7·2	$1·3547 \times 10^5$	$1·3663 \times 10^5$		($1·377 \times 10^5$)
	7·4	$1·3573 \times 10^5$	$1·3689 \times 10^5$		
1·8	7·6	$1·9682 \times 10^5$	$1·9815 \times 10^5$	$2·0840 \times 10^5$	$1·9967 \times 10^5$
	7·8	$1·9669 \times 10^5$	$1·9794 \times 10^5$	$2·0862 \times 10^5$	$1·9954 \times 10^5$
	8·0	$1·9688 \times 10^5$	$1·9806 \times 10^5$	$2·0917 \times 10^5$	$1·9972 \times 10^5$
1·85	8·2	$2·3938 \times 10^5$	$2·3936 \times 10^5$		($2·419 \times 10^5$)
	8·4	$2·3958 \times 10^5$	$2·3939 \times 10^5$		
1·9	8·4	—	—	$2·024 \times 10^5$	$2·939 \times 10^5$
	8·6	$2·9239 \times 10^5$	$2·8873 \times 10^5$	$2·042 \times 10^5$	$2·9363 \times 10^5$
	8·8	$2·9260 \times 10^5$	$2·8863 \times 10^5$	$2·065 \times 10^5$	$2·9365 \times 10^5$
	9·0	$2·9323 \times 10^5$	$2·8893 \times 10^5$	—	—
1·95	9·0	$3·5901 \times 10^5$	$3·4957 \times 10^5$		
	9·2	$3·5930 \times 10^5$	$3·4943 \times 10^5$		($3·556 \times 10^5$)
	9·4	$3·6010 \times 10^5$	$3·4974 \times 10^5$		
2·0	9·4	$4·4523 \times 10^5$	$4·1655 \times 10^5$	$5·02 \times 10^5$	$4·29 \times 10^5$
	9·6	$4·4574 \times 10^5$	$4·1624 \times 10^5$	$5·046 \times 10^5$	$4·286 \times 10^5$
	9·8	—	—	$5·092 \times 10^5$	$4·305 \times 10^5$

‡ Values in parentheses were obtained in I by interpolation.

TABLE 3. THE COEFFICIENTS IN THE EXPANSION FOR v AND $v^\dagger$

κ	a	A_2/A_1	A_3/A_1	B_2/B_1	B_3/B_1
0	6·2	0·03635	−0·08636	0·17961	−0·06116
0·4	6·2	0·04597	−0·08585	0·22869	−0·05295
0·6	6·2	0·05307	−0·08547	0·26489	−0·04667
1·0	6·4	0·08008	−0·08558	0·39576	−0·02363
$1·3\dot{3}$	6·4	0·12845	−0·08230	0·63719	+0·02831
1·6	6·8	0·25560	−0·07413	1·2218	+0·1939
1·7	7·2	0·37630	−0·06229	1·7230	+0·3922
1·8	7·8	0·58280	−0·03056	2·4800	+0·8192
1·85	8·2	0·72443	+0·00045	2·9239	+1·1702
1·9	8·8	0·89601	+0·05016	3·3464	+1·6384
1·95	9·2	1·0658	+0·0913	3·7865	+1·7516
2·0	9·6	1·2421	+0·2021	4·0829	+2·8407

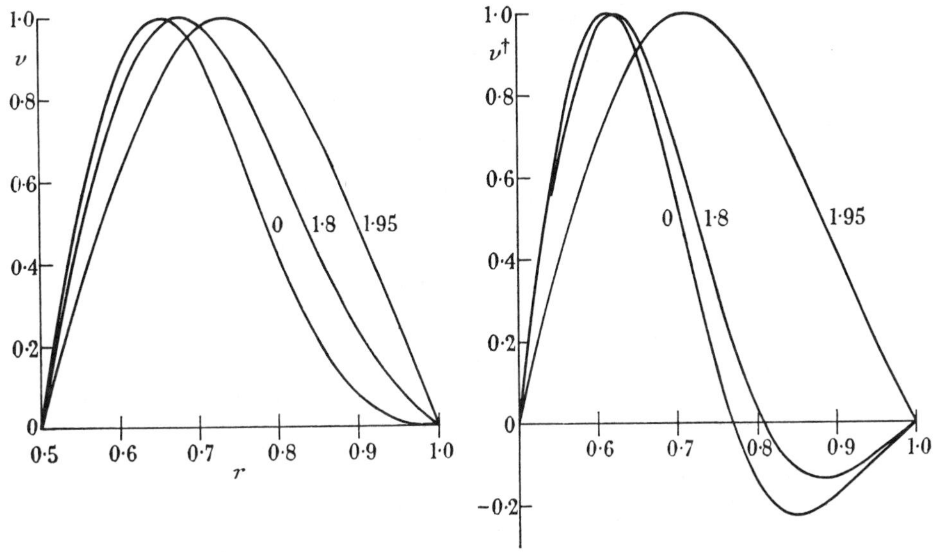

FIGURE 1. The dual functions v and $v^\dagger$ (normalized to unit amplitude at maxima) for different values of κ; the curves are labelled by the values of κ to which they belong. The reduced amplitude of v, in the outer half of the gap, for increasing κ, should be noted.

In conclusion we may merely state that a knowledge of the solutions of the original system as well as of the adjoint system are needed for a theory (along the lines of Stuart 1958) of the finite amplitudes of the flow which prevail just beyond marginal instability.

The research reported in this paper has in part been supported by the Geophysics Research Directorate of the Air Force Cambridge Research Center, Air Research and Development Command, under Contract AF 19(604)–2046 with the University of Chicago.

REFERENCES

Chandrasekhar, S. 1954*a* *Mathematika*, **1**, 5.
Chandrasekhar, S. 1954*b* *J. Rat. Mech. Anal.* **3**, 181.
Chandrasekhar, S. 1958 *Proc. Roy. Soc.* A, **246**, 301.
Chandrasekhar, S. 1961*a* *Hydrodynamic and hydromagnetic stability*. Oxford: Clarendon Press.
Chandrasekhar, S. 1961*b* *J. Math. Mech.* **10**, 683.
Roberts, P. H. 1960 *J. Math. Anal. Applic.* **1**, 195.
Stuart, J. T. 1958 *J. Fluid Mech.* **4**, 1.
Watson, G. N. 1944 *A treatise on the theory of Bessel functions*, 2nd edition. Cambridge University Press.

7

THE CHARACTER OF THE EQUILIBRIUM OF AN INCOMPRESSIBLE FLUID SPHERE OF VARIABLE DENSITY AND VISCOSITY SUBJECT TO RADIAL ACCELERATION

By S. CHANDRASEKHAR

(*Yerkes Observatory and Institute for Nuclear Studies, University of Chicago*)

[Received 13 April 1954]

SUMMARY

This paper is devoted to a consideration of the following problem: An incompressible fluid sphere, in which the density and the viscosity are functions of the distance r from the centre only, is subject to a radial acceleration $-\gamma r$, where γ is a function of r: to determine the manner of initial development of an infinitesimal disturbance. By analysing the disturbance in spherical harmonics, the mathematical problem is reduced to one in characteristic values in a fourth-order differential equation and a variational principle characterizing the solution is enunciated. The particular case of a sphere of radius R and density ρ_1 embedded in a medium of a different density ρ_2 (but of the same kinematic viscosity ν) is considered in some detail; and it is shown that the character of the equilibrium depends on the sign of $\gamma_R(\rho_2-\rho_1)$ and the magnitude of $\mathfrak{G} = \gamma_R R^4/\nu^2$. If $\gamma_R(\rho_2-\rho_1) > 0$, the situation is unstable and the mode of maximum instability is $l = 1$ for all $\mathfrak{G} < 230$; for larger values of $\mathfrak{G}$ it shifts progressively to higher harmonics. In the case $\gamma_R(\rho_2-\rho_1) > 0$ the results of both an exact calculation and an approximate calculation (based on the variational principle) are given and contrasted. In the case $\gamma_R(\rho_2-\rho_1) < 0$ when the situation is stable, the manner of decay of the disturbance is briefly discussed in terms of an approximate theory only.

1. Introduction

THE character of the equilibrium of an incompressible heavy fluid of variable density stratified in parallel planes has been the subject of investigations by Rayleigh (**1**), Harrison (**2**), Lamb (**3**), Chandrasekhar (**4**), and Hide (**5**). But the related problem of the equilibrium of an incompressible fluid sphere of variable density and viscosity does not seem to have attracted any attention, though as an example of a problem in hydrodynamic stability it has a definite interest. The problem recently arose in connexion with certain geophysical and astrophysical questions; and the solution (on a certain approximation) is presented here with a view to those applications.

2. The equations of the problem

A static state in which an incompressible fluid subject to radial acceleration is arranged in concentric shells and the pressure p and the density ρ are functions of the distance r from the centre only is clearly a kinematically

Reprinted from the *Quarterly Journal of Mechanics and Applied Mathematics* 8, pt. 1 (1955): 1–21, by permission of Oxford University Press.

realizable one. The character of the equilibrium of this static state can be analysed by supposing that there is a slight disturbance and then following its further evolution. Let the actual density at any point x_i ($i = 1, 2, 3$) following the disturbance be $\rho+\delta\rho$, where $\delta\rho$ is a function of x_i and the time t. Let δp be the corresponding increment in the pressure; δp will also be a function of x_i and t. Finally, let u_i ($i = 1, 2, 3$) denote the components of the velocity; u_i (like $\delta\rho$ and δp) will be considered as a small quantity of the first order.

In formulating the relevant equations of motion we shall make the simplifying assumption that the changes in the prevailing field of force caused by the disturbance can be ignored. In the case of a fluid sphere in equilibrium under its own gravitation this assumption will be justified only if the variations in the initial density distribution over the relevant distances are small compared with the mean density: in other words, the theory we shall present will provide the asymptotically correct solution *either* when the variations in density become very small *or* when the order of the spherical harmonic disturbance considered becomes very large (cf. Cowling (**6**) and Ledoux (**7**)). Alternatively, the theory may also be considered as applying when the prevailing field of force is of 'external origin' such as a source of intense radiation at the centre. Equally, the theory will also apply to a spherical shell of fluid subject to an acceleration (caused by an 'explosion' (say) at the centre) which impinges on the surrounding medium.

On the assumption made in the foregoing paragraph, the equations of motion and continuity in Cartesian tensor notation are

$$\rho\frac{\partial u_i}{\partial t} = -\frac{\partial}{\partial x_i}\,\delta p+\frac{\partial}{\partial x_k}p_{ik}-g_i\,\delta\rho \tag{1}$$

and

$$\frac{\partial u_i}{\partial x_i} = 0, \tag{2}$$

where (in accordance with the assumption) the acceleration g_i (unaffected by the disturbance) is of the form

$$g_i = \gamma(r)x_i; \tag{3}$$

$\gamma(r)$, as the notation implies, is a function of r only. Also, in equation (1), p_{ik} is the viscous stress tensor given by

$$p_{ik} = \mu\left(\frac{\partial u_k}{\partial x_i}+\frac{\partial u_i}{\partial x_k}\right), \tag{4}$$

where μ (which is a function of r) denotes the coefficient of viscosity. In addition to equations (1) and (2) we have the equation

$$\frac{\partial}{\partial t}\,\delta\rho+u_i\frac{\partial\rho}{\partial x_i} = 0, \tag{5}$$

which ensures that the density of every element of liquid remains unchanged.

Inserting equation (4) in equation (1) and making use of the solenoidal character of u_i, we find

$$\rho\frac{\partial u_i}{\partial t} = -\frac{\partial}{\partial x_i}\delta p+\mu\nabla^2 u_i+\left(\frac{\partial u_k}{\partial x_i}+\frac{\partial u_i}{\partial x_k}\right)\frac{\partial \mu}{\partial x_k}-\gamma\,\delta\rho\, x_i; \tag{6}$$

or, since

$$\frac{\partial \mu}{\partial x_k} = \frac{x_k}{r}\frac{d\mu}{dr}, \tag{7}$$

we have

$$\rho\frac{\partial u_i}{\partial t} = -\frac{\partial}{\partial x_i}\delta p+\mu\nabla^2 u_i+\frac{x_k}{r}\left(\frac{\partial u_i}{\partial x_k}+\frac{\partial u_k}{\partial x_i}\right)\frac{d\mu}{dr}-\gamma x_i\,\delta\rho. \tag{8}$$

Similarly, equation (5) can be rewritten in the form

$$\frac{\partial}{\partial t}\delta\rho+u_r\frac{d\rho}{dr} = 0, \tag{9}$$

where

$$u_r = u_i x_i/r \tag{10}$$

is the radial component of the velocity.

We shall seek solutions of equations (8) and (9) whose dependence on t is given by the factor

$$e^{nt} \tag{11}$$

where n is a constant. For solutions having this dependence on time, equations (8) and (9) become

$$\frac{\partial}{\partial x_i}\delta p = -(n\rho u_i+\gamma x_i\,\delta\rho)+\mu\nabla^2 u_i+\frac{x_k}{r}\left(\frac{\partial u_i}{\partial x_k}+\frac{\partial u_k}{\partial x_i}\right)\frac{d\mu}{dr}, \tag{12}$$

and

$$\delta\rho = -\frac{u_r}{n}\frac{d\rho}{dr}. \tag{13}$$

Multiplying equation (12) by x_i/r and remembering that

$$x_i\frac{\partial}{\partial x_i} = r\frac{\partial}{\partial r}, \tag{14}$$

we get

$$\frac{\partial}{\partial r}\delta p = -(n\rho u_r+\gamma r\,\delta\rho)+\frac{\mu}{r}x_i\,\nabla^2 u_i+2\frac{d\mu}{dr}\frac{x_i x_k}{r^2}\frac{\partial u_i}{\partial x_k}; \tag{15}$$

or, since

$$\frac{x_i x_k}{r^2}\frac{\partial u_i}{\partial x_k} = \frac{1}{r^2}\left(x_k\frac{\partial}{\partial x_k}u_i x_i-u_i x_i\right) = \frac{1}{r}\left\{\frac{\partial}{\partial r}(ru_r)-u_r\right\} = \frac{\partial u_r}{\partial r}, \tag{16}$$

and the operations of x_i and ∇^2 are permutable when applied to a solenoidal vector, we have

$$\frac{\partial}{\partial r}\delta p = -(n\rho u_r+\gamma r\,\delta\rho)+\frac{\mu}{r}\nabla^2(ru_r)+2\frac{d\mu}{dr}\frac{\partial u_r}{\partial r}. \tag{17}$$

Next, taking the divergence of equation (12), we get

$$\nabla^2\,\delta p = -\left\{n u_r \frac{d\rho}{dr} + 3\gamma\,\delta\rho + r\frac{\partial}{\partial r}(\gamma\,\delta\rho)\right\} + \frac{1}{r}\frac{d\mu}{dr}\nabla^2(r u_r) + \\ + \frac{\partial}{\partial x_i}\left\{\frac{x_k}{r}\left(\frac{\partial u_i}{\partial x_k} + \frac{\partial u_k}{\partial x_i}\right)\frac{d\mu}{dr}\right\}. \tag{18}$$

Again, using the solenoidal property of u_i, we can reduce the last term on the right-hand side of equation (18) in the manner (cf. equation (16))

$$\frac{\partial}{\partial x_i}\left\{\frac{1}{r}\frac{d\mu}{dr}x_k\left(\frac{\partial u_i}{\partial x_k} + \frac{\partial u_k}{\partial x_i}\right)\right\}$$

$$= \frac{2}{r}\frac{d}{dr}\left(\frac{1}{r}\frac{d\mu}{dr}\right)x_i x_k \frac{\partial u_i}{\partial x_k} + \frac{1}{r}\frac{d\mu}{dr}\left\{\delta_{ik}\left(\frac{\partial u_i}{\partial x_k} + \frac{\partial u_k}{\partial x_i}\right) + x_k\,\nabla^2 u_k\right\}$$

$$= 2r\frac{d}{dr}\left(\frac{1}{r}\frac{d\mu}{dr}\right)\frac{\partial u_r}{\partial r} + \frac{1}{r}\frac{d\mu}{dr}\nabla^2(r u_r). \tag{19}$$

Hence

$$\nabla^2\,\delta p = -\left\{n u_r \frac{d\rho}{dr} + 3\gamma\,\delta\rho + r\frac{\partial}{\partial r}(\gamma\,\delta\rho)\right\} + \frac{2}{r}\frac{d\mu}{dr}\nabla^2(r u_r) + 2r\frac{d}{dr}\left(\frac{1}{r}\frac{d\mu}{dr}\right)\frac{\partial u_r}{\partial r}. \tag{20}$$

Equations (13), (17), and (20) are the basic equations of this theory.

3. The equation for determining the rate of growth of a spherical harmonic disturbance of order l

We shall seek solutions of equations (13), (17), and (20) which are of the forms

$$\delta p = \varpi Y_l(\vartheta, \varphi), \qquad \delta\rho = \sigma Y_l(\vartheta, \varphi), \quad \text{and} \quad u_r = w Y_l(\vartheta, \varphi), \tag{21}$$

where $Y_l(\vartheta, \varphi)$ is a spherical harmonic of order l and ϖ, σ, and w are functions of r only. For solutions of this form, the equations reduce to

$$\left\{\frac{d^2}{dr^2} + \frac{2}{r}\frac{d}{dr} - \frac{l(l+1)}{r^2}\right\}\varpi = -\left\{n w \frac{d\rho}{dr} + 3\gamma\sigma + r\frac{d}{dr}(\gamma\sigma)\right\} + \\ + \frac{2}{r}\frac{d\mu}{dr}\left\{\frac{d^2}{dr^2} + \frac{2}{r}\frac{d}{dr} - \frac{l(l+1)}{r^2}\right\}(rw) + 2r\frac{d}{dr}\left(\frac{1}{r}\frac{d\mu}{dr}\right)\frac{dw}{dr}, \tag{22}$$

$$\frac{d\varpi}{dr} = -(n\rho w + \gamma r\sigma) + \frac{\mu}{r}\left\{\frac{d^2}{dr^2} + \frac{2}{r}\frac{d}{dr} - \frac{l(l+1)}{r^2}\right\}(rw) + 2\frac{d\mu}{dr}\frac{dw}{dr} \tag{23}$$

and

$$\sigma = -\frac{w}{n}\frac{d\rho}{dr}. \tag{24}$$

Letting

$$F = \left\{\frac{d^2}{dr^2} + \frac{2}{r}\frac{d}{dr} - \frac{l(l+1)}{r^2}\right\}(rw), \tag{25}$$

we shall rewrite equations (22) and (23) as

$$\frac{d}{dr}\left(r^2\frac{d\varpi}{dr}\right)-l(l+1)\varpi = -\left\{nr^2w\frac{d\rho}{dr}+\frac{d}{dr}(r^3\gamma\sigma)\right\}+2r\frac{d\mu}{dr}F+2r^3\frac{d}{dr}\left(\frac{1}{r}\frac{d\mu}{dr}\right)\frac{dw}{dr} \tag{26}$$

and

$$\frac{d\varpi}{dr} = -(n\rho w+\gamma r\sigma)+\frac{\mu}{r}F+2\frac{d\mu}{dr}\frac{dw}{dr}. \tag{27}$$

By differentiating equation (26) with respect to r and eliminating $d\varpi/dr$ between the resulting equation and equation (27), we obtain

$$\frac{d^2}{dr^2}\left\{r^2\left[-(n\rho w+\gamma r\sigma)+\frac{\mu}{r}F+2\frac{d\mu}{dr}\frac{dw}{dr}\right]\right\}-$$

$$-l(l+1)\left\{-(n\rho w+\gamma r\sigma)+\frac{\mu}{r}F+2\frac{d\mu}{dr}\frac{dw}{dr}\right\}+$$

$$+\frac{d}{dr}\left\{nr^2w\frac{d\rho}{dr}+\frac{d}{dr}(r^3\gamma\sigma)-2r\frac{d\mu}{dr}F-2\left(r^2\frac{d^2\mu}{dr^2}-r\frac{d\mu}{dr}\right)\frac{dw}{dr}\right\} = 0. \tag{28}$$

After some further reductions equation (28) can be brought to the form

$$-n\frac{d}{dr}\left\{\rho\frac{d}{dr}(r^2w)\right\}+\frac{d}{dr}\left\{\mu\frac{d}{dr}(rF)\right\}+\frac{d}{dr}\left\{rF\frac{d\mu}{dr}\right\}-$$

$$-2\frac{d}{dr}\left\{\frac{d\mu}{dr}\left[r\frac{dw}{dr}+2w-l(l+1)w\right]\right\}+$$

$$+l(l+1)\left\{n\rho w-\frac{\mu}{r}F-2\frac{d\mu}{dr}\frac{dw}{dr}\right\} = -l(l+1)\gamma r\sigma. \tag{29}$$

Now it can be readily verified that

$$rF-2\left\{r\frac{dw}{dr}+2w-l(l+1)w\right\} = r\left\{\frac{d^2W}{dr^2}+(l+2)(l-1)\frac{W}{r^2}\right\}, \tag{30}$$

where

$$W = rw. \tag{31}$$

Substituting from equations (24) and (30) in equation (29) we finally obtain, after some rearrangement of the terms:

$$\frac{d}{dr}\left\{\rho\frac{d}{dr}(rW)-\frac{\mu}{n}\frac{d}{dr}(rF)\right\}-$$

$$-\frac{1}{n}\frac{d}{dr}\left\{r\frac{d\mu}{dr}\left[\frac{d^2W}{dr^2}+(l+2)(l-1)\frac{W}{r^2}\right]\right\}-\frac{l(l+1)}{r}\left(\rho W-\frac{\mu}{n}F\right)$$

$$= -\frac{l(l+1)}{n^2}\left\{\gamma W\frac{d\rho}{dr}+2n\frac{d\mu}{dr}\frac{dw}{dr}\right\}, \tag{32}$$

where it may be recalled that

$$F = \left\{\frac{d^2}{dr^2}+\frac{2}{r}\frac{d}{dr}-\frac{l(l+1)}{r^2}\right\}W. \tag{33}$$

Equations (32) and (33) together represent a fourth-order differential equation for W. In seeking a solution of these equations we must satisfy certain conditions at the centre and on the bounding sphere. At the centre we must clearly require that none of the physical quantities has any singularity; this requires in particular that (cf. equation (31))

$$W \to 0 \quad \text{as } r \to 0. \tag{34}$$

And on the bounding sphere we must have

$$W = 0 \text{ and either } \frac{dW}{dr} \text{ or } \frac{d^2W}{dr^2} = 0 \text{ (on the bounding sphere)}, \tag{35}$$

depending on whether the bounding surface is rigid or free† (cf. Jeffreys and Bland (**8**); and Chandrasekhar (**9**)). The requirement that a solution of equations (32) and (33) satisfies the foregoing boundary conditions will determine a sequence of possible values for n; and the sign and the magnitude of the real part of n will decide whether or not the initial state is stable for a spherical harmonic disturbance of order l and will determine the rate of decay (or growth) of the disturbance.

4. A fluid sphere of constant density and viscosity in an infinite medium of different density and viscosity

We shall first apply the equation derived in the preceding section to the case when a fluid sphere of radius R, density ρ_1, and viscosity μ_1 is embedded in an infinite medium of density ρ_2 and viscosity μ_2. In each of the two regions of constant ρ and μ, equation (32) reduces to

$$\rho\frac{d^2}{dr^2}(rW)-\frac{\mu}{n}\frac{d^2}{dr^2}(rF)-\frac{l(l+1)}{r}\left(\rho W-\frac{\mu}{n}F\right) = 0, \tag{36}$$

where, for the present, we are suppressing the subscripts distinguishing the two regions.

Equation (36) can be written alternatively in the form

$$\left\{\frac{d^2}{dr^2}+\frac{2}{r}\frac{d}{dr}-\frac{l(l+1)}{r^2}\right\}\left(W-\frac{\nu}{n}F\right) = 0, \tag{37}$$

where $\nu = \mu/\rho$ denotes the coefficient of kinematic viscosity. The general solution of this equation is

$$W-\frac{\nu}{n}F = A_1 r^l+A_2 r^{-(l+1)}, \tag{38}$$

† In the present connexion a free bounding surface is equivalent to an interface with a frictionless surface.

where A_1 and A_2 are constants of integration. Using this solution in equation (33) we have

$$\left\{\frac{d^2}{dr^2}+\frac{2}{r}\frac{d}{dr}-\frac{l(l+1)}{r^2}\right\}W = \frac{n}{\nu}\{W-A_1 r^l-A_2 r^{-(l+1)}\}. \tag{39}$$

A particular integral of this equation is clearly

$$W = A_1 r^l+A_2 r^{-(l+1)}, \tag{40}$$

while the complementary function is a linear combination of the integrals

$$\frac{1}{\sqrt{r}}I_{l+\frac{1}{2}}\{r\sqrt{(n/\nu)}\} \quad \text{and} \quad \frac{1}{\sqrt{r}}K_{l+\frac{1}{2}}\{r\sqrt{(n/\nu)}\}, \tag{41}$$

where $I_{l+\frac{1}{2}}$ and $K_{l+\frac{1}{2}}$ are the Bessel functions of the two kinds for a purely imaginary argument. (Strictly speaking, we should, without 'prejudice', express the complementary function as a linear combination of $J_{l+\frac{1}{2}}$ and $J_{-(l+\frac{1}{2})}$ with the argument $r\sqrt{(-n/\nu)}$ and allow n to be complex; but we are expressing the solution as a linear combination of $I_{l+\frac{1}{2}}$ and $K_{l+\frac{1}{2}}$ with the argument $r\sqrt{(n/\nu)}$ since we shall be particularly interested in the unstable case when n is real and positive.) The general solution of equation (37) is therefore

$$W = A_1 r^l+A_2 r^{-(l+1)}+\frac{B_1}{\sqrt{r}}I_{l+\frac{1}{2}}\{r\sqrt{(n/\nu)}\}+\frac{B_2}{\sqrt{r}}K_{l+\frac{1}{2}}\{r\sqrt{(n/\nu)}\}, \tag{42}$$

where B_1 and B_2 are further constants of integration.

Since W must vanish both when $r \to 0$ (in the 'core') and when $r \to \infty$ (in the 'mantle') we can write

$$W_1 = A_1 r^l+\frac{B_1}{\sqrt{r}}I_{l+\frac{1}{2}}\{r\sqrt{(n/\nu_1)}\} \quad (r \leqslant R) \tag{43}$$

and

$$W_2 = A_2 r^{-(l+1)}+\frac{B_2}{\sqrt{r}}K_{l+\frac{1}{2}}\{r\sqrt{(n/\nu_2)}\} \quad (r \geqslant R), \tag{44}$$

as the solutions appropriate for the core and the mantle respectively.

On the interface ($r = R$) the components of the velocity as well as the tangential viscous stresses should be continuous. These conditions require (cf. Jeffreys and Bland (**8**)) that

$$W, \quad \frac{dW}{dr}, \quad \text{and} \quad \mu\left\{r^2\frac{d^2W}{dr^2}+(l+2)(l-1)W\right\} \tag{45}$$

are continuous on the interface $r = R$.

A further boundary condition is obtained by integrating equation (32) across the interface $r = R$ between $r = R+\epsilon$ and $r = R-\epsilon$ ($\epsilon > 0$)

and then letting $\epsilon \to 0$. In view of the continuity conditions (45) this limiting process leads to

$$\rho_2\left[\frac{d}{dr}\left\{r\left(W_2-\frac{\nu_2}{n}F_2\right)\right\}\right]_{r=R}-\rho_1\left[\frac{d}{dr}\left\{r\left(W_1-\frac{\nu_1}{n}F_1\right)\right\}\right]_{r=R}$$
$$=-\frac{l(l+1)}{n^2}\left\{\gamma_R(\rho_2-\rho_1)[W]_R+2n(\mu_2-\mu_1)\left[\frac{dw}{dr}\right]_R\right\}, \quad (46)$$

where $[W]_R$ and $[dw/dr]_R$ are the common values of W_1 and W_2 and similarly of dw_1/dr and dw_2/dr at $r=R$.

Since (cf. equation (38))

$$\frac{d}{dr}\left\{r\left(W-\frac{\nu}{n}F\right)\right\}=(l+1)A_1 r^l-lA_2 r^{-(l+1)}, \quad (47)$$

an equivalent form of the boundary condition (46) as applied to the solutions (43) and (44) is

$$(l+1)\rho_1 A_1 R^l+l\rho_2 A_2 R^{-(l+1)}=\frac{l(l+1)}{n^2}\left\{\gamma_R(\rho_2-\rho_1)[W]_R+2n(\mu_2-\mu_1)\left[\frac{dw}{dr}\right]_R\right\}. \quad (48)$$

In applying the boundary conditions (45) and (48) to the solutions (43) and (44), it is convenient to measure r in units of R. Writing

$$q_1=\sqrt{(nR^2/\nu_1)} \quad \text{and} \quad q_2=\sqrt{(nR^2/\nu_2)}, \quad (49)$$

we can express the solutions for W_1 and W_2 in the forms

$$W_1=A_1 r^l+\frac{B_1}{\sqrt{r}}I_{l+\frac{1}{2}}(q_1 r) \quad (r\leqslant 1) \quad (50)$$

and

$$W_2=A_2 r^{-(l+1)}+\frac{B_2}{\sqrt{r}}K_{l+\frac{1}{2}}(q_2 r) \quad (r\geqslant 1). \quad (51)$$

Now applying the boundary conditions (45) and (48) to the foregoing solutions we find

$$A_1+B_1 I_{l+\frac{1}{2}}(q_1)=A_2+B_2 K_{l+\frac{1}{2}}(q_2) \quad (=[W]_1), \quad (52)$$

$$lA_1+B_1\{q_1 I'_{l+\frac{1}{2}}(q_1)-\tfrac{1}{2}I_{l+\frac{1}{2}}(q_1)\}=-(l+1)A_2+B_2\{q_2 K'_{l+\frac{1}{2}}(q_2)-\tfrac{1}{2}K_{l+\frac{1}{2}}(q_2)\}, \quad (53)$$

$$\mu_1[2(l^2-1)A_1+B_1\{-2q_1 I'_{l+\frac{1}{2}}(q_1)+(q_1^2+2l^2+2l-1)I_{l+\frac{1}{2}}(q_1)\}]$$
$$=\mu_2[2l(l+2)A_2+B_2\{-2q_2 K'_{l+\frac{1}{2}}(q_2)+(q_2^2+2l^2+2l-1)K_{l+\frac{1}{2}}(q_2)\}], \quad (54)$$

and (cf. equations (48), (52), and (53))

$$(l+1)\rho_1 A_1+l\rho_2 A_2$$
$$=\frac{l(l+1)}{2n^2}\gamma_R(\rho_2-\rho_1)\{A_1+B_1 I_{l+\frac{1}{2}}(q_1)+A_2+B_2 K_{l+\frac{1}{2}}(q_2)\}+$$
$$+\frac{l(l+1)}{nR^2}(\mu_2-\mu_1)\{(l-1)A_1+B_1[q_1 I'_{l+\frac{1}{2}}(q_1)-\tfrac{3}{2}I_{l+\frac{1}{2}}(q_1)]-$$
$$-(l+2)A_2+B_2[q_2 K'_{l+\frac{1}{2}}(q_2)-\tfrac{3}{2}K_{l+\frac{1}{2}}(q_2)]\}, \quad (55)$$

where primes denote differentiation with respect to the argument of the Bessel functions. Introducing the abbreviations

$$\alpha_1 = \frac{\rho_1}{\rho_1+\rho_2}, \qquad \alpha_2 = \frac{\rho_2}{\rho_1+\rho_2} \quad (\alpha_1+\alpha_2 = 1), \tag{56}$$

$$\left.\begin{aligned} \mathfrak{R} &= \frac{l(l+1)}{n^2}\frac{\rho_1-\rho_2}{\rho_1+\rho_2}\gamma_R = \frac{1}{n^2}\gamma_R\, l(l+1)(\alpha_1-\alpha_2) \\ \text{and}\qquad C &= \frac{l(l+1)}{nR^2}\frac{\mu_1-\mu_2}{\rho_1+\rho_2} = \frac{1}{nR^2}l(l+1)(\alpha_1\nu_1-\alpha_2\nu_2) \end{aligned}\right\} \tag{57}$$

we can rewrite equations (52)–(55) in the form

$$A_1+I_{l+\frac{1}{2}}(q_1)B_1-A_2-K_{l+\frac{1}{2}}(q_2)B_2 = 0, \tag{58}$$

$$lA_1+[q_1 I'_{l+\frac{1}{2}}(q_1)-\tfrac{1}{2}I_{l+\frac{1}{2}}(q_1)]B_1+(l+1)A_2+ \\ +[-q_2 K'_{l+\frac{1}{2}}(q_2)+\tfrac{1}{2}K_{l+\frac{1}{2}}(q_2)]B_2 = 0, \tag{59}$$

$$2\alpha_1\nu_1(l^2-1)A_1+\alpha_1\nu_1\{-2q_1 I'_{l+\frac{1}{2}}(q_1)+(q_1^2+2l^2+2l-1)I_{l+\frac{1}{2}}(q_1)\}B_1- \\ -2\alpha_2\nu_2 l(l+2)A_2-\alpha_2\nu_2\{-2q_2 K'_{l+\frac{1}{2}}(q_2)+(q_2^2+2l^2+2l-1)K_{l+\frac{1}{2}}(q_2)\}B_2 = 0 \tag{60}$$

and

$$[(l+1)\alpha_1+\tfrac{1}{2}\mathfrak{R}+(l-1)C]A_1+\{\tfrac{1}{2}\mathfrak{R}I_{l+\frac{1}{2}}(q_1)+C[q_1 I'_{l+\frac{1}{2}}(q_1)-\tfrac{3}{2}I_{l+\frac{1}{2}}(q_1)]\}B_1+ \\ +[l\alpha_2+\tfrac{1}{2}\mathfrak{R}-(l+2)C]A_2+\{\tfrac{1}{2}\mathfrak{R}K_{l+\frac{1}{2}}(q_2)+C[q_2 K'_{l+\frac{1}{2}}(q_2)-\tfrac{3}{2}K_{l+\frac{1}{2}}(q_2)]\}B_2 = 0. \tag{61}$$

The foregoing equations represent a system of linear homogeneous equations for the constants A_1, B_1, A_2, and B_2; and the determinant of the system must vanish if we are to have a non-trivial solution. By equating the determinant of equations (58)–(61) to zero we shall obtain the required characteristic equation for determining n.

5. The case $\nu_1 = \nu_2$

In the further discussion of equations (58)–(61) we shall restrict ourselves to the case when the kinematic viscosities of the two fluids are the same, i.e. when

$$\nu_1 = \nu_2. \tag{62}$$

This assumption simplifies the characteristic equation considerably; but one would not expect the essential features of the problem to be obscured by this simplifying assumption (cf. Chandrasekhar (**4**) and Hide (**5**) for the parallel discussion of the plane problem).

When $\nu_1 = \nu_2$,

$$q_1 = q_2 = \sqrt{(nR^2/\nu)} = q \text{ (say)}, \tag{63}$$

and we find that the determinant of equations (58)–(61) can be reduced to one of third order by adding or subtracting suitable multiples of the different rows (and columns) from other rows (and columns). Thus we find

$$\begin{vmatrix} qI_{l+\frac{3}{2}} & 2l+1 & qK_{l-\frac{1}{2}} \\ \alpha_1 q(-2I_{l+\frac{3}{2}}+qI_{l+\frac{1}{2}}) & 2[\alpha_1(l^2-1)-\alpha_2 l(l+2)] & -\alpha_2 q(2K_{l-\frac{1}{2}}+qK_{l+\frac{1}{2}}) \\ -(l+1)\alpha_1 I_{l+\frac{1}{2}}+qCI_{l+\frac{3}{2}} & (l+1)\alpha_1+l\alpha_2+\Re-3C & -l\alpha_2 K_{l+\frac{1}{2}}-qCK_{l-\frac{1}{2}} \end{vmatrix} = 0, \tag{64}$$

where the argument of all the Bessel functions is now q. In reducing the determinant of equations (58)–(61) to the foregoing, one must make use of the various recurrence relations satisfied by the Bessel functions.

On expanding the determinant (64) we find after some straightforward but lengthy reductions that

$$l(l+1)(\alpha_2-\alpha_1)\frac{\mathfrak{G}}{q^4} = (l+1)\alpha_1+l\alpha_2+$$

$$+\frac{1}{q^2}[\![2(\alpha_2-\alpha_1)(2l+1)q\{\alpha_2 lI_{l+\frac{3}{2}}K_{l+\frac{1}{2}}-\alpha_1(l^2-1)I_{l+\frac{1}{2}}K_{l-\frac{1}{2}}\}-$$

$$-4(\alpha_2-\alpha_1)^2 l(l^2-1)(l+2)I_{l+\frac{3}{2}}K_{l-\frac{1}{2}}+\alpha_1\alpha_2(2l+1)^2q^2 I_{l+\frac{1}{2}}K_{l+\frac{1}{2}}]\!]\times$$

$$\times\{q(\alpha_1 I_{l+\frac{1}{2}}K_{l-\frac{1}{2}}+\alpha_2 I_{l+\frac{3}{2}}K_{l+\frac{1}{2}})+2(\alpha_2-\alpha_1)I_{l+\frac{3}{2}}K_{l-\frac{1}{2}}\}^{-1}, \tag{65}$$

where
$$\mathfrak{G} = \gamma_R R^4/\nu^2 \tag{66}$$

is a 'Grashoff' number.

Equation (65) determines $\mathfrak{G}$ as a function of q for each specified l and given α_1 and α_2 $(= 1-\alpha_1)$. On the other hand, according to equation (63)

$$\frac{n}{\sqrt{\gamma_R}} = q^2\frac{\nu}{R^2\sqrt{\gamma_R}} = \frac{q^2}{\sqrt{\mathfrak{G}}}; \tag{67}$$

thus $q^2/\sqrt{\mathfrak{G}}$ gives n in units of $\sqrt{\gamma_R}$. Hence, the $(q, \mathfrak{G}; l)$-relation which is directly given by equation (65) can be transformed into an $(n, \mathfrak{G}; l)$-relation; and using this last relation we can determine the dependence of n on l for given $\mathfrak{G}$ and thus complete the solution of the problem.

6. The mode of maximum instability for the case $\nu_1 = \nu_2$ and $\rho_2 > \rho_1$: an illustrative example

When $\rho_2 > \rho_1$ and $\alpha_2 > \alpha_1$, equation (65) determines a $(q, \mathfrak{G})$-relation which for q real and positive is monotonic increasing; thus as q increases from zero to infinity, $\mathfrak{G}$ also increases from zero to infinity. According to equation (67), $n > 0$ and the situation considered is unstable, so we should expect that for a given $\mathfrak{G}$ there will be a mode of maximum instability.

The modes of maximum instability for various assigned values of $\mathfrak{G}$ and $\alpha_2-\alpha_1 = 0{\cdot}1$ were determined in the following manner.

By using the recent tabulation by the Royal Society (**10**) of the spherical Bessel functions for purely imaginary arguments, the $(q, \mathfrak{G})$-relations were determined for $l = 1, 2, \ldots, 9$ and for q in the range (0, 10). Then, by interpolating in these relations, the values of q (and thence the values of n) were determined (for each l) at which $\mathfrak{G}$ had assigned values. The results of such calculations are summarized in Table 1 and are further illustrated in Fig. 1.

TABLE I

The $(n/\sqrt{\gamma_R}, l)$-relation for $\alpha_2 - \alpha_1 = 0{\cdot}1$ and for various values of $\mathfrak{G}$. The mode of maximum instability is in bold type in each case

l \ $\mathfrak{G}$	10	100	200	400	600	800	1000	2000	3000
1	**0·106**	**0·195**	**0·221**	0·245	0·258	0·266	0·272	0·290	0·298
2	0·073	0·178	0·218	**0·258**	**0·280**	**0·295**	**0·306**	0·337	0·354
3	0·052	0·146	0·190	0·238	0·267	0·288	0·303	**0·348**	**0·372**
4	0·040	0·119	0·161	0·211	0·243	0·266	0·285	0·341	0·372
5	0·032	0·100	0·137	0·184	0·217	0·241	0·261	0·324	0·361
6	0·027	0·085	0·118	0·162	0·193	0·217	0·236	0·303	0·343
7	0·023	0·074	0·103	0·143	0·172	0·195	0·214	0·280	0·323
8	0·021	0·065	0·092	0·128	0·154	0·176	0·194	0·259	0·302
9	0·018	0·059	0·082	0·115	0·140	0·160	0·177	0·239	0·282

l \ $\mathfrak{G}$	4000	5000	6000	8000	10000	15000	20000	25000	30000
1	0·304	0·308	0·311	0·316	0·319	0·324	0·328	0·331	0·333
2	0·364	0·372	0·378	0·387	0·394	0·404	0·411	0·416	0·420
3	0·388	0·400	0·409	0·423	0·433	0·450	0·460	0·468	0·474
4	**0·393**	**0·409**	**0·421**	0·440	0·453	0·476	0·491	0·501	0·510
5	0·386	0·405	0·420	**0·444**	**0·460**	0·490	0·508	0·522	0·533
6	0·372	0·394	0·412	0·439	0·459	**0·494**	0·517	0·534	0·547
7	0·354	0·378	0·397	0·428	0·451	0·492	**0·518**	**0·538**	0·554
8	0·334	0·359	0·380	0·413	0·439	0·484	0·515	0·538	**0·555**
9	0·314	0·340	0·362	0·397	0·424	0·473	0·507	0·532	0·552

From the results given in Table 1 it is apparent that for $\mathfrak{G}$ less than a certain critical value in the neighbourhood of 200 the mode of maximum instability always occurs for $l = 1$. It is found that

$$n/\sqrt{\gamma_R} = 0{\cdot}226 \quad \text{for } l = 1 \text{ and } l = 2 \quad \text{when } \mathfrak{G} = 230; \tag{68}$$

the critical value of $\mathfrak{G}$ is therefore more nearly 230. For $\mathfrak{G} > 230$ the mode of maximum instability shifts to higher l's.

7. A variational principle

The discussion in the preceding sections of the character of the instability of a fluid sphere of constant density embedded in a medium of a different density but the same kinematic viscosity makes it clear that an investigation of the problem under even slightly more general conditions is likely to be very troublesome. However, it appears that as in the case of the plane problem (cf. Hide (**5**)) the essential features of the physical problem in the present case can be equally inferred from an approximate theory based on a variational principle. We shall now deduce this principle.

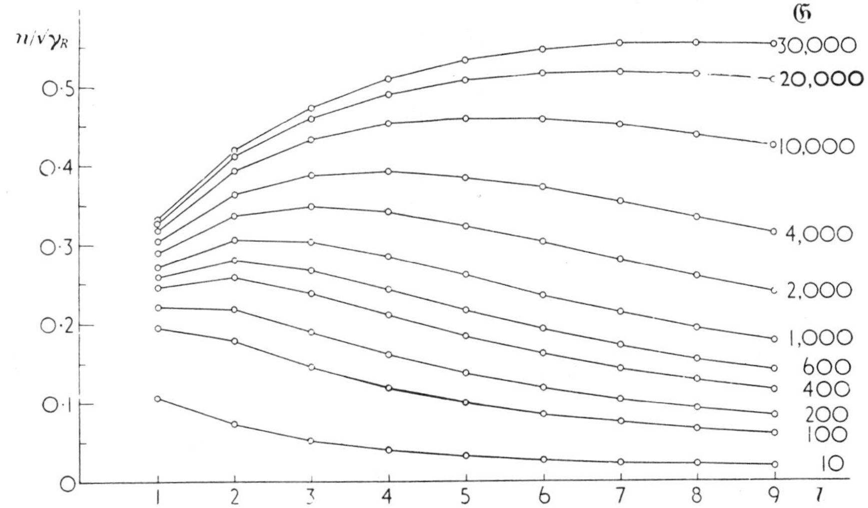

FIG. 1. The (n, l)-relation for various values of $\mathfrak{G}$ $(= \gamma_R R^4/\nu^2)$. The values of $\mathfrak{G}$ to which the various curves refer are indicated.

The equation to be solved is (cf. equations (24) and (29))

$$-n\frac{d}{dr}\left\{\rho\frac{d}{dr}(rW)\right\}+\frac{d^2}{dr^2}(\mu rF)-2\frac{d}{dr}\left\{\frac{1}{r}\frac{d\mu}{dr}\frac{d}{dr}(rW)\right\}+$$

$$+2l(l+1)\frac{d}{dr}\left(\frac{d\mu}{dr}\frac{W}{r}\right)+l(l+1)\left\{n\rho\frac{W}{r}-\frac{\mu}{r}F-2\frac{d\mu}{dr}\frac{d}{dr}\left(\frac{W}{r}\right)\right\}$$

$$=\frac{l(l+1)}{n}\gamma\frac{d\rho}{dr}W. \qquad (69)$$

This can be further simplified to the form

$$-n\frac{d}{dr}\left\{\rho\frac{d}{dr}(rW)\right\}+l(l+1)n\rho\frac{W}{r}-\frac{l(l+1)}{n}\gamma\frac{d\rho}{dr}W+$$

$$+\frac{d^2}{dr^2}(\mu rF)-2\frac{d}{dr}\left\{\frac{1}{r}\frac{d\mu}{dr}\frac{d}{dr}(rW)\right\}+2l(l+1)\frac{1}{r}\frac{d^2\mu}{dr^2}W-l(l+1)\frac{\mu}{r}F=0, \qquad (70)$$

where it may be recalled that F is given by equation (33). And the boundary conditions with respect to which equation (70) must be solved are that

$$W \text{ and } \textit{either } \frac{dW}{dr} \textit{ or } \frac{d^2W}{dr^2} \text{ vanish on the bounding surfaces.} \qquad (71)$$

As we have already stated, the requirement that a solution of equation (70) satisfies the conditions (71) on the bounding surfaces will lead to a determinate sequence of possible values for n. Let n_i and n_j denote two of these characteristic values; and let the solutions belonging to these characteristic values be distinguished by subscripts i and j respectively.

Now consider equation (70) for the characteristic value n_i and after multiplying it by rW_j (belonging to n_j) integrate over the range of r (which we shall assume to be $R_1 \leqslant r \leqslant R_2$). We obtain after one or more integrations by parts:

$$n_i \int_{R_1}^{R_2} \rho\left\{\frac{d}{dr}(rW_i)\frac{d}{dr}(rW_j)+l(l+1)W_i W_j\right\} dr - \frac{l(l+1)}{n_i}\int_{R_1}^{R_2} \frac{d\rho}{dr} r\gamma W_i W_j\, dr$$

$$= -2\int_{R_1}^{R_2} \frac{1}{r}\frac{d\mu}{dr}\frac{d}{dr}(rW_i)\frac{d}{dr}(rW_j)\, dr - 2l(l+1)\int_{R_1}^{R_2} \frac{d^2\mu}{dr^2} W_i W_j\, dr -$$

$$-\int_{R_1}^{R_2} rW_j \frac{d^2}{dr^2}(\mu r F_i)\, dr + l(l+1)\int_{R_1}^{R_2} \mu F_i W_j\, dr. \quad (72)$$

After two successive integrations by parts we find that

$$\int_{R_1}^{R_2} rW_j \frac{d^2}{dr^2}(\mu r F_i)\, dr = -\left[\mu r F_i \frac{d}{dr}(rW_j)\right]_{R_1}^{R_2} + \int_{R_1}^{R_2} \mu r F_i \frac{d^2}{dr^2}(rW_j)\, dr. \quad (73)$$

Now $$\frac{d}{dr}(rW) = r\frac{dW}{dr} \quad \text{on a bounding surface.} \quad (74)$$

If the bounding surface is rigid, this vanishes. On the other hand, if the bounding surface is free, $d^2W/dr^2 = 0$ and (cf. equation (33))

$$F = \frac{2}{r}\frac{dW}{dr} \quad \text{on a free bounding surface.} \quad (75)$$

Hence in all cases we may write

$$\int_{R_1}^{R_2} rW_j \frac{d^2}{dr^2}(\mu r F_i)\, dr = -\left[2\mu r \frac{dW_i}{dr}\frac{dW_j}{dr}\right]_{R_1}^{R_2} + \int_{R_1}^{R_2} \mu r F_i \frac{d^2}{dr^2}(rW_j)\, dr. \quad (76)$$

Using the foregoing result in equation (72) and substituting further for F_i

in accordance with equation (33), we obtain after some rearrangement of the terms:

$$n_i \int_{R_1}^{R_2} \rho \left\{ \frac{d}{dr}(rW_i)\frac{d}{dr}(rW_j)+l(l+1)W_i W_j \right\} dr - \frac{l(l+1)}{n_i} \int_{R_1}^{R_2} \frac{d\rho}{dr} r\gamma W_i W_j \, dr$$

$$= -2 \int_{R_1}^{R_2} \frac{1}{r}\frac{d\mu}{dr}\frac{d}{dr}(rW_i)\frac{d}{dr}(rW_j)\, dr - 2l(l+1) \int_{R_1}^{R_2} \frac{d^2\mu}{dr^2} W_i W_j \, dr -$$

$$-l^2(l+1)^2 \int_{R_1}^{R_2} \frac{\mu}{r^2} W_i W_j \, dr + \left[2\mu r \frac{dW_i}{dr}\frac{dW_j}{dr} \right]_{R_1}^{R_2} -$$

$$- \int_{R_1}^{R_2} \mu \frac{d^2}{dr^2}(rW_i)\frac{d^2}{dr^2}(rW_j)\, dr + l(l+1) \int_{R_1}^{R_2} \frac{\mu}{r} \left\{ W_i \frac{d^2}{dr^2}(rW_j) + W_j \frac{d^2}{dr^2}(rW_i) \right\} dr. \tag{77}$$

Rewriting the last integral on the right-hand side of equation (77) in the manner

$$\int_{R_1}^{R_2} \mu \left\{ \frac{d^2}{dr^2}(W_i W_j) - 2\frac{dW_i}{dr}\frac{dW_j}{dr} + \frac{2}{r}\frac{d}{dr}(W_i W_j) \right\} dr, \tag{78}$$

we can transform it by further integration by parts into

$$\int_{R_1}^{R_2} \frac{d^2\mu}{dr^2} W_i W_j \, dr - 2 \int_{R_1}^{R_2} \mu \frac{dW_i}{dr}\frac{dW_j}{dr}\, dr - 2 \int_{R_1}^{R_2} W_i W_j \left(\frac{1}{r}\frac{d\mu}{dr} - \frac{\mu}{r^2} \right) dr. \tag{79}$$

Thus, we finally obtain

$$n_i \int_{R_1}^{R_2} \rho \left\{ \frac{d}{dr}(rW_i)\frac{d}{dr}(rW_j)+l(l+1)W_i W_j \right\} dr - \frac{l(l+1)}{n_i} \int_{R_1}^{R_2} \frac{d\rho}{dr} r\gamma W_i W_j \, dr$$

$$= \left[2\mu r \frac{dW_i}{dr}\frac{dW_j}{dr} \right]_{R_1}^{R_2} -$$

$$- \int_{R_1}^{R_2} \mu \left\{ \frac{d^2}{dr^2}(rW_i)\frac{d^2}{dr^2}(rW_j) + 2l(l+1)\frac{dW_i}{dr}\frac{dW_j}{dr} + l(l^2-1)(l+2)\frac{W_i W_j}{r^2} \right\} dr -$$

$$-2 \int_{R_1}^{R_2} \frac{1}{r}\frac{d\mu}{dr} \left\{ \frac{d}{dr}(rW_i)\frac{d}{dr}(rW_j) + l(l+1)W_i W_j \right\} dr - l(l+1) \int_{R_1}^{R_2} \frac{d^2\mu}{dr^2} W_i W_j \, dr. \tag{80}$$

Interchanging i and j in equation (80) and subtracting the resulting equation from it, we obtain

$$(n_i-n_j)\left[\int\limits_{R_1}^{R_2} \rho\left\{\frac{d}{dr}(rW_i)\frac{d}{dr}(rW_j)+l(l+1)W_i W_j\right\} dr + \frac{1}{n_i n_j}\int\limits_{R_1}^{R_2} \frac{d\rho}{dr} r\gamma W_i W_j\, dr\right] = 0. \tag{81}$$

Hence. if $n_i \neq n_j$.

$$\int\limits_{R_1}^{R_2} \rho\left\{\frac{d}{dr}(rW_i)\frac{d}{dr}(rW_j)+l(l+1)W_i W_j\right\} dr + \frac{1}{n_i n_j}\int\limits_{R_1}^{R_2} \frac{d\rho}{dr} r\gamma W_i W_j\, dr = 0 \quad (i \neq j). \tag{82}$$

If n_i should be complex, we can suppose that n_i and n_j are complex conjugates and we deduce from equation (82) that

$$\int\limits_{R_1}^{R_2} \rho\left\{\left|\frac{d}{dr}(rW)\right|^2+l(l+1)|W|^2\right\} dr + \frac{1}{|n|^2}\int\limits_{R_1}^{R_2} \frac{d\rho}{dr} r\gamma |W|^2\, dr = 0, \tag{83}$$

a relation which cannot be true if $\gamma\, d\rho/dr$ is everywhere positive.

Returning to equation (80) and setting $i = j$, we get (on further suppressing the subscripts)

$$n\int\limits_{R_1}^{R_2} \rho\left\{\left[\frac{d}{dr}(rW)\right]^2+l(l+1)W^2\right\} dr - \frac{l(l+1)}{n}\int\limits_{R_1}^{R_2} \frac{d\rho}{dr} r\gamma W^2\, dr$$

$$= -\int\limits_{R_1}^{R_2} \mu\left\{\left[\frac{d^2}{dr^2}(rW)\right]^2+2l(l+1)\left(\frac{dW}{dr}\right)^2+l(l^2-1)(l+2)\frac{W^2}{r^2}\right\} dr -$$

$$-2\int\limits_{R_1}^{R_2} \frac{1}{r}\frac{d\mu}{dr}\left\{\left[\frac{d}{dr}(rW)\right]^2+l(l+1)W^2\right\} dr -$$

$$-l(l+1)\int\limits_{R_1}^{R_2} \frac{d^2\mu}{dr^2} W^2\, dr + 2\left[\mu r\left(\frac{dW}{dr}\right)^2\right]_{R_1}^{R_2}. \tag{84}$$

This last equation provides the basis for a convenient variational procedure for determining n. For, by considering the effect on n (determined in accordance with equation (84)) of an arbitrary variation δW in W compatible only with the boundary conditions on W, we find after some straightforward but lengthy reductions that

$$-2\left[\left[\int\limits_{R_1}^{R_2}\rho\left\{\left[\frac{d}{dr}(rW)\right]^2+l(l+1)W^2\right\}dr+\frac{l(l+1)}{n^2}\int\limits_{R_1}^{R_2}\frac{d\rho}{dr}r\gamma W^2\,dr\right]\right]\delta n$$

$$=\int\limits_{R_1}^{R_2}r\,\delta W\left\{-n\frac{d}{dr}\left[\rho\frac{d}{dr}(rW)\right]+l(l+1)n\rho\frac{W}{r}-\frac{l(l+1)}{n}\gamma\frac{d\rho}{dr}W+\right.$$

$$\left.+\frac{d^2}{dr^2}(\mu rF)-2\frac{d}{dr}\left[\frac{1}{r}\frac{d\mu}{dr}\frac{d}{dr}(rW)\right]+2l(l+1)\frac{1}{r}\frac{d^2\mu}{dr^2}W-l(l+1)\frac{\mu}{r}F\right\}dr. \quad (85)$$

It will be observed that the quantity which appears as a factor of δW under the integral sign on the right-hand side of equation (85) vanishes if equation (70) governing W is satisfied. Hence a necessary and sufficient condition for δn to be zero to the first order for all small arbitrary variations in W which are compatible with the boundary conditions is that W be a solution of the characteristic-value problem. Accordingly, equation (84) provides the basis for the determination of n by a variational procedure.

8. Illustration of the use of the variational principle

We shall illustrate the use of the variational principle derived in section 7 by reconsidering on its basis the problem which has been exactly solved in sections 4–6. However, in applying equation (84) to this case, in which discontinuities in ρ and μ occur at $r = R$, we must be careful to make allowance for them by dividing the range of integration into three intervals, 0 to $R-\epsilon$, $R-\epsilon$ to $R+\epsilon$, and finally $R+\epsilon$ to ∞, where $\epsilon > 0$, and then passing to the limit $\epsilon = 0$. By this limiting process we readily derive from equation (84) that for the case considered in section 4,

$$n\left[\left[\rho_1\int\limits_0^1\left\{\left[\frac{d}{dr}(rW_1)\right]^2+l(l+1)W_1^2\right\}dr+\rho_2\int\limits_1^\infty\left\{\left[\frac{d}{dr}(rW_2)\right]^2+l(l+1)W_2^2\right\}dr\right]\right]+$$

$$+\frac{1}{R^2}\left[\left[\mu_1\int\limits_0^1\left\{\left[\frac{d^2}{dr^2}(rW_1)\right]^2+2l(l+1)\left(\frac{dW_1}{dr}\right)^2+l(l^2-1)(l+2)\frac{W_1^2}{r^2}\right\}dr+\right.\right.$$

$$\left.\left.+\mu_2\int\limits_1^\infty\left\{\left[\frac{d^2}{dr^2}(rW_2)\right]^2+2l(l+1)\left(\frac{dW_2}{dr}\right)^2+l(l^2-1)(l+2)\frac{W_2^2}{r^2}\right\}dr\right]\right]-$$

$$-\frac{l(l+1)}{n}\gamma_R(\rho_2-\rho_1)[W^2]_1+\frac{2}{R^2}(\mu_2-\mu_1)\left[\left\{\frac{d}{dr}(rW)\right\}^2+l(l+1)W^2\right]_1=0, \quad (86)$$

where r is measured in units of R. In equation (86) the symbol $[\;]_1$ denotes the common value, at $r = 1$, of the quantity in the brackets (which is continuous at the interface).

In the absence of viscosity the solution for W appropriate to the problem on hand is (cf. equation (38))

$$W = W_1 = r^l \quad (r \leqslant 1)$$

and $$W = W_2 = r^{-(l+1)} \quad (r \geqslant 1). \tag{87}$$

This solution does not have a continuous derivative at $r = 1$. Nevertheless, from Hide's experience with the plane problem (**5**) we can expect that the use of this solution (valid only in the absence of viscosity) as a 'trial' function in the variational expression for n will lead to the correct dependence of n on the parameters of the problem. However, in the present instance (in contrast to the plane problem) the discontinuity in the derivative of W affects one of the terms in equation (86), namely the term which occurs as the factor of $(\mu_2-\mu_1)$. Thus,

$$\left[\left\{\frac{d}{dr}(rW_1)\right\}^2 + l(l+1)W_1^2\right]_1 = (l+1)(2l+1), \tag{88}$$

while $$\left[\left\{\frac{d}{dr}(rW_2)\right\}^2 + l(l+1)W_2^2\right]_1 = l(2l+1). \tag{89}$$

In using equation (86) in conjunction with (87) we shall take the average of the two foregoing values for the term in question. (This is admittedly a crude procedure, but it will presently appear that this term is only of secondary importance in the equation which we shall derive (equation (90) below) for n.) In this manner we find from equation (86) that

$$[(l+1)\alpha_1+l\alpha_2]n^2+\frac{1}{R^2}\{2l(l+1)[(l+1)\alpha_1\nu_1+l\alpha_2\nu_2]+(2l+1)^2(\alpha_2\nu_2-\alpha_1\nu_1)\}n-$$
$$-l(l+1)\gamma_R(\alpha_2-\alpha_1) = 0, \tag{90}$$

where α_1 and α_2 have the meanings given in (56).

The two terms in the quantity in braces in (90) are approximately in the ratio $l(l+1):(2l+1)(\alpha_2-\alpha_1)$; the second term can therefore be neglected for $|\alpha_2-\alpha_1| \leqslant 0{\cdot}1$, and in all cases for large values of l.

(i) *The case* $\nu_1 = \nu_2$. To see how much reliance we may place on deductions based on (90) (and therefore on results derived from the variational procedure quite generally) we shall consider the case of equal kinematic viscosities for which we have the results of an exact calculation (Table 1).

For $\nu_1 = \nu_2$ equation (90) becomes

$$n^2+\frac{\nu}{R^2}\left\{2l(l+1)+(\alpha_2-\alpha_1)\frac{(2l+1)^2}{l+\alpha_1}\right\}n-(\alpha_2-\alpha_1)\gamma_R\frac{l(l+1)}{l+\alpha_1} = 0, \tag{91}$$

or, alternatively,

$$\left(\frac{n}{\sqrt{\gamma_R}}\right)^2+\frac{1}{\sqrt{\mathfrak{G}}}\left\{2l(l+1)+(\alpha_2-\alpha_1)\frac{(2l+1)^2}{l+\alpha_1}\right\}\frac{n}{\sqrt{\gamma_R}}-(\alpha_2-\alpha_1)\frac{l(l+1)}{l+\alpha_1}=0. \tag{92}$$

Hence

$$\frac{n}{\sqrt{\gamma_R}}=-\frac{1}{\sqrt{\mathfrak{G}}}\left\{l(l+1)+(\alpha_2-\alpha_1)\frac{(2l+1)^2}{2(l+\alpha_1)}\right\}\pm$$
$$\pm\left[\frac{1}{\mathfrak{G}}\left\{l(l+1)+(\alpha_2-\alpha_1)\frac{(2l+1)^2}{2(l+\alpha_1)}\right\}^2+(\alpha_2-\alpha_1)\frac{l(l+1)}{l+\alpha_1}\right]^{\frac{1}{2}}. \tag{93}$$

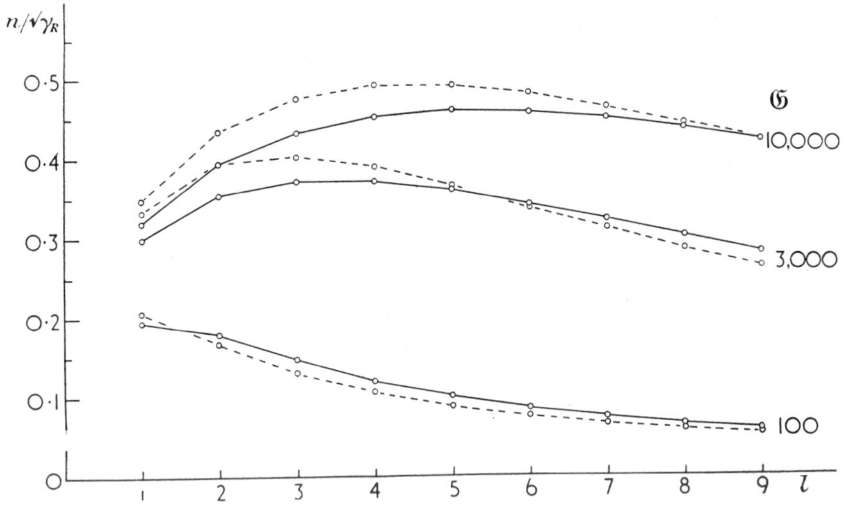

FIG. 2. A comparison of the (n, l)-relations given by the exact theory (the points joined by the full line) and an approximate theory (the points joined by the dashed line) based on the variational principle.

(ii) *The mode of maximum instability for the case $\nu_1 = \nu_2$ and $\rho_2 > \rho_1$.* When $\rho_2 > \rho_1$ and $\alpha_2 > \alpha_1$, the two roots given by (93) are both real; one of these is positive and is the only one which is physically significant. Thus

$$\frac{n}{\sqrt{\gamma_R}}=\left[\frac{1}{\mathfrak{G}}\left\{l(l+1)+(\alpha_2-\alpha_1)\frac{(2l+1)^2}{2(l+\alpha_1)}\right\}^2+(\alpha_2-\alpha_1)\frac{l(l+1)}{l+\alpha_1}\right]^{\frac{1}{2}}-$$
$$-\frac{1}{\sqrt{\mathfrak{G}}}\left\{l(l+1)+(\alpha_2-\alpha_1)\frac{(2l+1)^2}{2(l+\alpha_1)}\right\}; \tag{94}$$

and since n is positive the situation considered is an unstable one.

The values of n derived from (94) for various values of l and $\mathfrak{G}$ are plotted in Fig. 2 (the points joined by the dashed lines) along with the results of the exact calculations (the points joined by the full lines). From a comparison of the two sets of curves it is apparent that the approximate theory predicts qualitatively the correct dependence of n on l and $\mathfrak{G}$; and the quantitative agreement is not unsatisfactory.

From equation (94) it follows in particular that, for sufficiently large l,

$$\frac{n}{\sqrt{\gamma_R}} \to \frac{\alpha_2-\alpha_1}{2l}\sqrt{\mathfrak{G}} \quad (l\to\infty); \tag{95}$$

thus $n \to 0$ as $l \to \infty$ and this is in agreement with what one should expect on physical grounds.

(iii) *The manner of decay in the case* $\nu_1 = \nu_2$ *and* $\rho_2 < \rho_1$. For $\rho_2 < \rho_1$ and $\alpha_2 < \alpha_1$ both the roots given by (93) have negative real parts: the situation considered is therefore a stable one. However, depending on the value of $\mathfrak{G}$, n can be complex for the lower modes. To see this clearly we shall neglect the terms in $(2l+1)^2(\alpha_2-\alpha_1)/2(l+\alpha_1)$ in equation (93): this, as we have seen, is permissible so long as we do not consider $|\alpha_2-\alpha_1| > 0{\cdot}1$. With this additional approximation we can now write

$$\frac{n}{\sqrt{\gamma_R}} = -\frac{l(l+1)}{\sqrt{\mathfrak{G}}} \pm \left[\frac{l^2(l+1)^2}{\mathfrak{G}} - (\alpha_1-\alpha_2)\frac{l(l+1)}{l+\alpha_1}\right]^{\frac{1}{2}}. \tag{96}$$

The condition for complex roots is therefore

$$\frac{l^2(l+1)^2}{\mathfrak{G}} < \frac{l(l+1)}{l+\alpha_1}(\alpha_1-\alpha_2), \tag{97}$$

or

$$\mathfrak{G} > \frac{l(l+1)(l+\alpha_1)}{\alpha_1-\alpha_2}. \tag{98}$$

From (98) it follows that for a given $\mathfrak{G}$ all modes beyond a certain l (say l_*) will decay aperiodically; and this aperiodic decay can take place at one of two alternative rates. And all modes with $l < l_*$ will be damped periodically. However, if $\mathfrak{G} < 2(1+\alpha_1)/(\alpha_1-\alpha_2)$, then all modes (including $l = 1$) will decay aperiodically. Thus for $\alpha_1-\alpha_2 = 0{\cdot}1$ all modes will decay aperiodically for $\mathfrak{G} < 29$; while for $\mathfrak{G} > 29$ one or more of the lower modes will decay periodically: e.g. for $\mathfrak{G} = 1000$ the modes $l = 1, 2, 3$, and 4 will be damped periodically and all the higher modes will be damped aperiodically. This behaviour of the solution in the stable case has certain similarities to the plane problem (cf. Chandrasekhar (**4**) and Hide (**5**)).

9. Concluding remarks

While this is clearly not the place to go into any detailed geophysical or other applications of the theory presented here, a brief comparison of the manner in which instability arises under the circumstances considered in this paper and under the analogous circumstances of thermal instability might be of some interest. For a fluid sphere heated within, it is known (cf. Jeffreys and Bland (**8**) and Chandrasekhar (**9**)) that the mode of instability which is easiest to excite is $l = 1$; in spherical shells this shifts to harmonics of higher orders. On the other hand, if the instability is that caused by an overlying material of higher density, then the order of the harmonic in which the instability will manifest itself will depend on the value of $\mathfrak{G}$. But one general result can be stated: as long as $\mathfrak{G} < 230$ the mode of maximum instability will always be $l = 1$. This condition on $\mathfrak{G}$ is essentially a condition for a sufficiently high viscosity. For the condition $\mathfrak{G} < 230$ is equivalent to (cf. equation (66))

$$\nu > R^2\sqrt{(\gamma_R/230)}. \tag{99}$$

If the acceleration on the heavier material causing the instability is that due to the gravitational attraction of the mass interior to R, then,

$$\gamma_R = \tfrac{4}{3}\pi\rho_1 G, \tag{100}$$

where G denotes the constant of gravitation. Hence in this case

$$\nu > (\tfrac{4}{3}\pi\rho_1 G/230)^{\frac{1}{2}} R^2. \tag{101}$$

Choosing $\rho_1 = 10$ gm./cm.^{2} and $R = 5\times 10^8$ cm. (102)

as 'typical' values, we find

$$\nu > 2{\cdot}8\times 10^{13} \text{ cm.}^2/\text{sec.} \tag{103}$$

Now it has been variously estimated that the viscosity of the material of the earth's mantle is 10^{22}–10^{23}. This value is larger than (103) by a very large margin. And one might conclude from this that if the viscosity of the earth's mantle should at some time (in the process of cooling?) have 'passed' through the value (103) before reaching its present high value and *if* an instability of the kind considered in this paper should have arisen, then the mode $l = 1$ should be exhibited as the last surviving feature. But before one can be certain of such conclusions it is important that the basic theory be extended and generalized along several directions: and this it is hoped to do in the near future.

In conclusion I should like to acknowledge my indebtedness to Miss Donna Elbert, who carried out all the numerical calculations pertaining to this paper.

The research reported in this paper has in part been supported by the Geophysics Research Directorate of the Air Force Cambridge Research Center, Air Research and Development Command, under Contract AF 19(604)–299 with the University of Chicago.

REFERENCES

1. Lord Rayleigh, *Proc. Lond. Math. Soc.* **14** (1883), 170–7 (also *Scientific Papers*, **2**, 200–7).
2. W. J. Harrison, *Proc. Lond. Math. Soc.* **6** (1908), 396–405.
3. H. Lamb, *Hydrodynamics*, 6th edn. (Cambridge, 1931); see particularly pp. 370–9.
4. S. Chandrasekhar, *Proc. Camb. Phil. Soc.* (in the press).
5. R. Hide, ibid. (in the press).
6. T. G. Cowling, *Month. Not. R. Astr. Soc.* **101** (1941), 367–75.
7. P. Ledoux, *Astrophys. J.* **114** (1951), 373–84.
8. H. Jeffreys and M. E. M. Bland, *Month. Not. R. Astr. Soc. Geophys. Suppl.* **6** (1951), 148–58.
9. S. Chandrasekhar, *Phil. Mag.* (7) **43** (1953), 1317–29; also *Phil. Mag.* (7) **44** (1953), 233–41 and 1129–30.
10. C. W. Jones, *A Short Table for the Bessel Functions* $I_{n+\frac{1}{2}}(x)$ *and* $(2/\pi)K_{n+\frac{1}{2}}(x)$ (Cambridge, 1952). (Prepared on behalf of the Mathematical Tables Committee of the Royal Society.)

8

Rumford Medal Lecture 1957

Thermal Convection

S. CHANDRASEKHAR

I. INTRODUCTION

THE SIMPLEST example of thermally induced convection arises when a horizontal layer of fluid is heated from below and an adverse temperature gradient is maintained. The adjective "adverse" is used to qualify the prevailing temperature gradient, since, on account of thermal expansion, the fluid at the bottom becomes lighter than the fluid at the top; and this is a top-heavy arrangement which is potentially unstable. Under these circumstances the fluid will try to redistribute itself to redress this weakness in its arrangement. This is how thermal convection originates: It represents the efforts of the fluid to restore to itself some degree of stability. These basic facts concerning thermally induced convection were discovered by Count Rumford when he observed that currents were set up in the bore of a large thermometer which he had been using in an experiment. I am indebted to Professor Sanborn Brown for the following extract from Count Rumford's writings, in which his discovery is announced:

> I saw the whole mass of the liquid in the tube in a most rapid motion running swiftly in two opposite directions, up and down at the same time. The bulb of the thermometer, which is of copper, had been made two years before I found leisure to begin my experiments, and having been left unfilled without being closed with a stopple, some fine particles of dust had found their way into it and these particles which were intimately mixed with the spirits of wine, on their being illuminated by the sun's beam, became perfectly visible . . . and by their motions discovered the violent motions by which the spirits of wine in the tube of the thermometer was agitated. . . . On examining the motion of the spirits of wine with a lens, I found that the ascending current occupied the axis of the tube and that it descended by the sides of the tube. On inclining the tube a little, the rising current moved out of the axis and occupied the side of the tube which was uppermost, while the descending currents occupied the whole of the lower side of it.

Reprinted by permission of *Daedalus*, Journal of the American Academy of Arts and Sciences, vol. 86, no. 4, October 1957, pp. 323–339.

However, as everyone has known since King Alfred's time, one cannot always depend on thermally induced circulation to prevent burning at the bottom! The reason is, the natural tendency of the fluid to react instantly to its unstable arrangement is inhibited by its own viscosity, and the more viscous the fluid, the less agile is it to react to its potential instability.

The first quantitative experiments to establish the extent to which viscosity inhibits the onset of instability, and to determine the precise manner in which instability does set in, are those of Bénard at the turn of the century. Bénard worked with very thin layers of liquid (only about 1 mm. deep) standing on a leveled metallic plate which was maintained at a uniform temperature. The upper surface was usually free and, being in contact with air, was at a lower temperature. Various liquids were employed – and some, indeed, which would have been solid under ordinary conditions. Bénard's experiments established the following two fundamental facts: *First*, a certain critical temperature gradient has to be exceeded before instability can set in and, *second*, the motions which ensue on surpassing of the critical temperature gradient have a *cellular pattern*. What actually happens on the onset of instability is that the layer of liquid rapidly resolves itself into a number of cells which after a while become equal and regular and align themselves to form a beautiful hexagonal pattern. Figure 1 is a reproduction of an early photograph of Bénard's; Figure 2, which illustrates the same phenomenon by a different experimental arrangement, is taken from a paper by Schmidt and Milverton.

The correct interpretation of Bénard's experiments was given by Lord Rayleigh in 1916. Rayleigh showed that what decides the stability or otherwise of a fluid heated on the "underside" – as he expressed it – is the numerical value of the nondimensional parameter,

$$R = \frac{g\alpha\beta}{\kappa\nu} d^4 \qquad (1)$$

– now called the Rayleigh number – where g denotes the acceleration due to gravity, d the depth of the layer, $\beta = | dT / dZ |$ the constant adverse temperature gradient which is maintained, and α, κ, and ν are the coefficients of volume expansion, thermometric conductivity, and kinematic viscosity, respectively. Rayleigh showed that instability must set in when R exceeds a certain determinate critical value R_c, and that when $R = R_c$ a stationary pattern of motions will come to prevail.

If β_c denotes the temperature gradient which must be exceeded for the occurrence of convection, then

$$\beta_c = R_c \frac{\kappa\nu}{g\alpha} d^{-4} \;; \qquad (2)$$

and, as we observe, this is higher the higher the viscosity.

In this lecture I want to consider especially the effects of rotation and magnetic field, separately and jointly, on the onset of thermal instability. However, to understand fully the meaning of the predicted effects, it is necessary that we have some appreciation of the manner in which instability under the circumstances envisaged can manifest itself, and how, indeed, one can calculate the critical Rayleigh number for the onset of instability. I shall, therefore, consider these aspects of the problem first.

II. ON THE TWO WAYS IN WHICH INSTABILITY CAN SET IN: AS CONVECTION AND AS OVERSTABILITY

Suppose, then, that an initial *static* state in which a certain adverse temperature gradient is maintained is slightly disturbed. We ask: Will the disturbance gradually die down and the original state be eventually restored? Or, will the disturbance grow in amplitude in such a way that the system progressively departs from the initial state and never reverts to it? If the latter should be the case, then the state we started from is clearly an unstable one. On the other hand, we cannot positively conclude stability if the former obtains, for, in order that we may consider the initial state as stable, it is necessary that not merely a particular disturbance, but all conceivable disturbances to which the state may be subject, be damped. The criterion for instability is, then, that there exist at least *one* mode of disturbance for which the system is unstable; and the criterion for stability is that there exist *no* mode of disturbance for which it is unstable. If all initial states are classified as stable or unstable according to these criteria, then the boundary (in the set theoretical sense) between these two classes of states in the manifold of all states will define a certain *marginal state*. By this definition, a marginal state is a state of *neutral stability*. The isolation and characterization of the marginal state is clearly the prime object of an investigation on stability.

It is not necessary for our present purposes to go into the details of the mathematical processes by which one specifies the marginal state, but it is important to recognize that marginal states can be one of two distinct kinds. The two kinds correspond to the two different ways in which the amplitude of a disturbance can grow or be damped (see figure 3). Thus if $A(t)$ denotes the amplitude of a disturbance, then its dependence on time can be either of the two kinds:

$$A(t) = A_0 \, e^{qt} \, , \tag{3}$$

or

$$A(t) = A_0 \, e^{qt} \cos pt \, , \tag{4}$$

where p and q are real. In either case, we shall have stability or instability according as q is negative or positive; in the former case (equation 3) the amplitude of the disturbance will be aperiodically damped or amplified; in the latter case (equation 4) the amplitudes of certain characteristic oscillations will be damped or amplified. In both cases the marginal state will be distinguished by $q = 0$, but with this essential difference: In case 3 the marginal state will exhibit a stationary pattern of motions, while in case 4 it will exhibit oscillatory motions with a certain characteristic frequency.

Quite generally, if at the onset of instability a stationary pattern of motions prevails, then one says that the *principle of the exchange of stabilities* is valid and that instability sets in as convection. On the other hand, if instability sets in via a marginal state of purely oscillatory motions, then one says (following Eddington) that one has a case of *overstability*. The use of the term overstability in this connection is not entirely common. So it may be worth recalling Eddington's own definition of that term: "In the usual kinds of *instability* a slight displacement provokes restoring forces tending away from equilibrium; in *overstability* it provokes restoring forces so strong as to overshoot the corresponding position on the other side of equilibrium."

One final general remark: Since stability means stability with respect to all possible disturbances, it is clear that for an investigation of the stability of a system to be complete, it is necessary that the reaction of the system to all possible disturbances be examined. In practice one accomplishes this by expressing an arbitrary disturbance as a superposition of certain basic possible modes and examining the stability of the system with respect to each of these modes. Thus, in the problem of the stability of a layer of fluid heated from below, an arbitrary disturbance is expressed (in accordance with Fourier's theorem) as a superposition of two-dimensional periodic waves, and the stability is then investigated with respect to each of these waves.

III. THE SOLUTION OF THE CLASSICAL BÉNARD PROBLEM

As I have already stated, Rayleigh first gave the correct interpretation of Bénard's experiments. However, in solving the underlying mathematical problem, Rayleigh did not attempt to satisfy the correct (physical) boundary conditions. Later investigations by Jeffreys, Low, and Pellew and Southwell have corrected this deficiency in Rayleigh's original solution, so that we may now consider the problem as solved. I shall briefly describe the outcome of these investigations.

Rayleigh and, more generally, Pellew and Southwell have proved that for this problem the principle of the exchange of stabilities is valid and that, in agreement with the experiments, the onset of instability must manifest itself as a stationary pattern of motions. And the critical Rayleigh number for the onset of instability has been determined as follows:

Considering a horizontal two-dimensional periodic disturbance of an assigned wave number a ($= 2\pi/\lambda$ where λ denotes the wavelength of the disturbance) one asks: What is the lowest Rayleigh number $R(a)$ at which a mode of disturbance with the wave number a, when excited, does not get damped? On solving this problem, one finds that the resulting function $R(a)$ has a single minimum at $a = a_c$ (say) where $R = R_c$. It is clear that R_c specifies the required critical Rayleigh number for the onset of thermal convection, for if $R < R_c$, then all disturbances (expressible, as they are, as superpositions of two-dimensional waves) will be damped; and when $R = R_c$ all disturbances will again be damped *except* for a periodic disturbance with precisely the wave number a_c; this is, therefore, the disturbance which will manifest itself at marginal stability. In this manner the critical Rayleigh number for the onset of instability has been determined for the two cases of interest, namely, when the layer of fluid is confined between two rigid planes and when the layer of fluid is supported by a rigid plane and the top surface is left free. The results of the mathematical analysis are:

$$\begin{array}{lll} \text{Both surfaces rigid} & R_c = 1708, & a_c = 3.13/d\,, \\ \left.\begin{array}{l}\text{Bottom surface rigid and}\\ \text{the top surface free}\end{array}\right\} & R_c = 1100, & a_c = 2.68/d\,. \end{array} \qquad (5)$$

Several experiments have been performed to verify whether or not instability does set in at the predicted Rayleigh numbers. I shall refer to only one such set of experiments, by Schmidt and Milverton, since the principle underlying their experiments is an important one and has provided the basis for other experiments to which I shall presently refer.

In the experiments of Schmidt and Milverton the layer of fluid was confined between two rigid planes which were maintained at constant temperatures, and heat was supplied (by an electrical coil) to the bottom plate at a constant rate. The experiments consisted in determining the difference in temperature, $(T_2 - T_1)$, between the two plates for varying rates of heating of the bottom plate; a measure of the latter is provided by the square of the heating current, C^2. Experiments of this kind generally give a plot similar to the one illustrated in figure 4. It will be seen that this plot shows a distinct break for a particular $T_2 - T_1 = \Delta T_c$ (say); this determines the critical temperature gradient, $\beta = \Delta T_c/d$ for the onset of instability. For, when $T_2 - T_1 < \Delta T_c$, the relation $(T_2 - T_1, C^2)$ is linear, with a constant slope corresponding to what must be (and is verified to be) the conductive temperature gradient; at $T_2 - T_1 = \Delta T_c$, the slope of the $(T_2 - T_1, C^2)$-relation suddenly decreases, indicating that after instability a new mechanism of heat transport – namely, convective heat transport – has begun to be operative. By such experiments Schmidt and Milverton were able to confirm that the critical Rayleigh number for the onset of instability for their experimental arrangement is 1770 ± 140; this is in good agreement with the theoretical value 1708 (cf. equation 5).

IV. THE EFFECT OF ROTATION

We shall now pass on to the consideration of the case when the layer of fluid which is being heated from below is set in rotation with a constant angular velocity Ω about the vertical. The effect of the rotation is to subject the fluid to Coriolis acceleration in addition to that of gravity. And Coriolis acceleration can have a decisive effect on the onset of instability, as can be seen from the following argument:

In the absence of viscosity and temperature gradients, the equations governing the motions of the fluid are:

$$\frac{\partial \boldsymbol{u}}{\partial t} + (\boldsymbol{u}. \, grad)\boldsymbol{u} = - \, grad \, P + 2\boldsymbol{u} \times \Omega \qquad (6)$$

and

$$div \, \boldsymbol{u} = 0, \qquad (7)$$

where

$$P = -\tfrac{1}{2} |\Omega \times \boldsymbol{r}|^2 + p/\rho \qquad (8)$$

In the foregoing equations $\boldsymbol{u}$ denotes the velocity, p the pressure and ρ the density.

If the state of motions is stationary, and the velocities are sufficiently small for the nonlinear terms in equation 6 to be negligible, then

$$grad \, P = 2\boldsymbol{u} \times \Omega \qquad (9)$$

From this equation it follows that

$$curl \, (\boldsymbol{u} \times \Omega) = 0 \, . \qquad (10)$$

When Ω is assumed to be in the direction of the z-axis, the three components of the single vector equation (10) are

$$\frac{\partial u_x}{\partial z} = 0, \ \frac{\partial u_y}{\partial z} = 0 \text{ and } \frac{\partial u_x}{\partial x} + \frac{\partial u_y}{\partial y} = - \frac{\partial u_z}{\partial z} = 0 \, . \qquad (11)$$

Hence, in the absence of viscosity, for sufficiently slow motions, the velocity components cannot depend on z. This result is a special case of a general theorem due to J. Proudman and G. I. Taylor that *all slow motions in a rotating inviscid fluid are necessarily two dimensional.* This theorem has an important bearing on the ensuing of thermal convection in a rotating fluid.* For, convection implies that the motions have a three-dimensional character; and this the Taylor-Proudman theorem forbids for an inviscid fluid so long as the nonlinear terms in the equations of motion are neglected. Accordingly, in contrast to the case of a nonrotating fluid, an inviscid fluid in rotation is thermally stable for *all* adverse temperature gradients.

*I am indebted to Dr. Raymond Hide for pointing out to me the relevance of the Taylor-Proudman theorem for these considerations.

Indeed, only in the presence of viscosity can instability arise, for only then can the Taylor-Proudman theorem be violated.

It is clear, then, that the effect of rotation will be to inhibit the onset of convection. More precisely, it follows from a theoretical analysis of this problem that the extent of the inhibition depends on Ω through the nondimensional parameter

$$T = \frac{4\Omega^2}{\nu^2} d^4 ; \qquad (12)$$

this has now come to be called the Taylor number.

A further important difference between the problem with and without rotation is the following: While in the problem without rotation the principle of the exchange of stabilities is always valid, this is no longer the case when Coriolis forces are acting. The discriminating parameter in this connection is the ratio of the kinematic viscosity (ν) to the thermometric conductivity (κ); this is sometimes called the Prandtl number and denoted by

$$\omega = \nu/\kappa . \qquad (13)$$

From a theoretical analysis of the problem it follows that if ω is less than a certain critical value ω^* (say), then for all values of T greater than a certain determinate T^* (depending on ω), the mode of instability which should set in first is overstability and not convection. But if $\omega > \omega^*$, the principle of the exchange of stabilities obtains and instability should set in as ordinary cellular convection. The type of results to be expected under these circumstances is shown in figure 5. The precise value of ω^* depends on the boundary conditions that have to be satisfied on the confining boundaries (such as whether they are rigid or free). For the (nonphysical) case of two free boundaries $\omega^* = 0.677$; the exact value of ω^* for other more realistic boundary conditions has not been determined; but it is known that in all cases it is not very different from unity.

It is remarkable that the Prandtl number should play this decisive role in determining the manner of the onset of thermal instability in rotating fluids; and that overstability should be the rule for $\omega \ll 1$, as is the case with metallic liquids such as mercury.

(*a*) *Theoretical Predictions*

The results of the theoretical calculations on the dependence of R_c on T are shown in figure 6. The cases when instability sets in as a stationary pattern of convection as well as when it sets in via a state of oscillatory motions are both illustrated; in the latter case the curves refer to a value of $\omega = 0.025$ which is appropriate for mercury at room temperatures. The asymptotic behaviors of these dependences for $T \to \infty$ may be noted.

In case the instability sets in as convection, the dependence of the critical Rayleigh number for the onset of instability on the Taylor number T has the asymptotic behavior,

$$R_c \to \text{constant} \times T^{\frac{2}{3}} \qquad (T \to \infty); \qquad (14)$$

while the wave number a_c (in units of $1/d$) of the disturbance which manifests itself at marginal stability increases with T according to the law

$$a_c \to \text{constant} \times T^{\frac{1}{6}} \qquad (T \to \infty). \qquad (15)$$

These relations apply so long as ω is greater than a certain critical value ω^*. If the Prandtl number is less than this critical value, then for all T greater than a certain determinate value, instability will set in as overstability; and for sufficiently large T, we have the asymptotic relations:

$$\left.\begin{aligned} R_c &\to \text{constant} \times \frac{\omega^{\frac{4}{3}}}{(1+\omega)^{\frac{1}{3}}}\, T^{\frac{2}{3}}, \\ a_c &\to \text{constant} \times \left(\frac{\omega}{1+\omega}\right)^{\frac{1}{3}} T^{\frac{1}{6}}. \end{aligned}\right\} (T \to \infty). \qquad (16)$$

Also, the frequency, p, of the characteristic oscillations at marginal stability has the behavior,

$$p/\Omega \to \text{constant} \times T^{-\frac{1}{6}} \qquad (T \to \infty). \qquad (17)$$

(*b*) *Experimental Verifications*

The various predictions described in the preceding paragraph and exhibited in figure 6 have been fully confirmed by some very beautiful experiments of Nakagawa and Frenzen and of Fultz and Nakagawa at the University of Chicago.

First, the experiments of Nakagawa and Frenzen with water (which has a Prandtl number $\omega = 6.0$) fully confirm that in this case, thermal instability does indeed set in as cellular convection (see figure 7) and that the $T^{\frac{2}{3}}$-law is obeyed (see figure 8). On the other hand, the experiments of Fultz and Nakagawa with mercury (which has a Prandtl number $\omega = 0.025$) show that in this instance, in agreement with theoretical prediction, instability does set in as overstability. This is shown in a particularly striking manner by the temperature records (see figures 9 and 10 in which the temperature records obtained in the experiments with water and mercury are contrasted). Further, the experimentally determined (R_c, T)-dependence is in accord with the theoretical relation (see figure 8). And, finally, the frequencies of the characteristic oscillations at marginal stability (as determined from temperature records such as those shown in figure 10) are also in very good agreement with the predicted values (see figure 11).

V. THE EFFECT OF A MAGNETIC FIELD

We now turn to the effect of an impressed magnetic field on thermal instability. We suppose that the fluid under consideration is an electrical conductor (such as mercury) and that an external magnetic field ($\boldsymbol{H}$) is impressed in a direction parallel to gravity, $\boldsymbol{g}$. (The case when the directions of $\boldsymbol{H}$ and $\boldsymbol{g}$ are not parallel has also been treated; but we shall not consider it here.)

On general grounds we may expect that the effect of a magnetic field will also be to inhibit the onset of convection and that the inhibiting effect will be the greater the stronger the magnetic field (H), and the higher the electrical conductivity (σ): for, when the field is strong (or the conductivity high) the lines of magnetic force tend to be glued to the material and this will make motions at right angles to the field difficult. In this latter respect the inhibiting effect of a magnetic field has a different origin from that of rotation: for, while in the absence of viscosity motions parallel to the axis of rotation are forbidden (in accordance with the Taylor-Proudman theorem), in the absence of electrical resistivity motions perpendicular to the field are forbidden.

As to whether instability will set in as convection or as overstability, it can be shown that with liquid metals such as mercury overstability cannot arise. Moreover, when cellular convection sets in, we must in accordance with what we have stated, expect that the cells become progressively elongated as the strength of the magnetic field is increased. And in the limit of infinite electrical conductivity (or, infinite field strength) when the fluid elements are

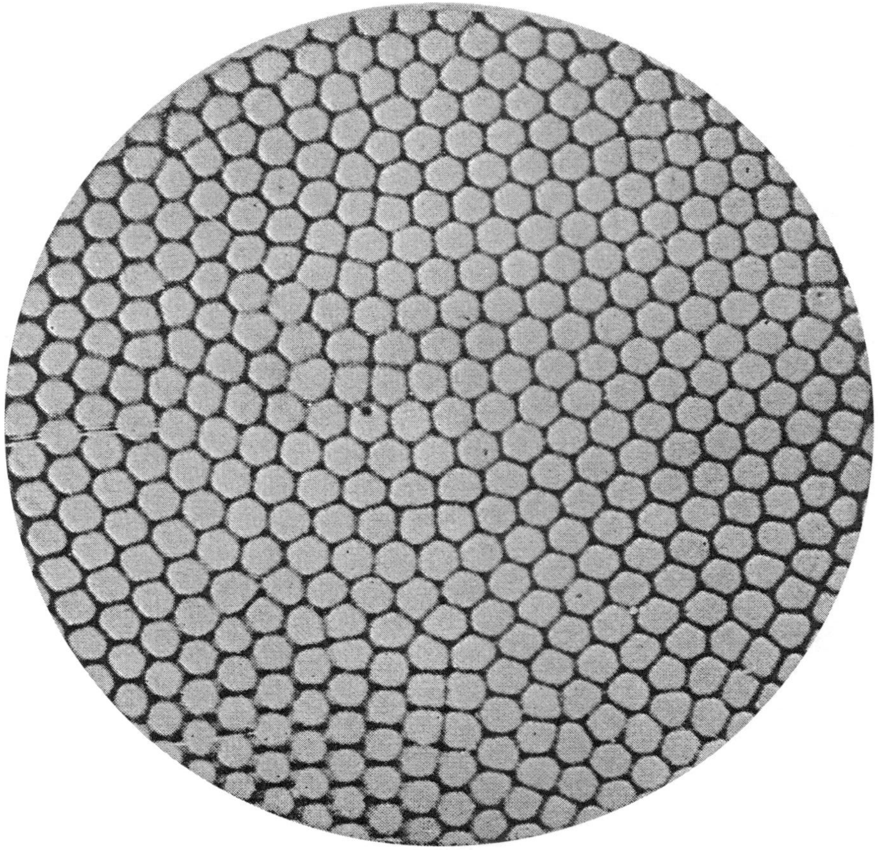

FIGURE 1. Bénard cells in spermaceti. A reproduction of one of Bénard's original photographs.

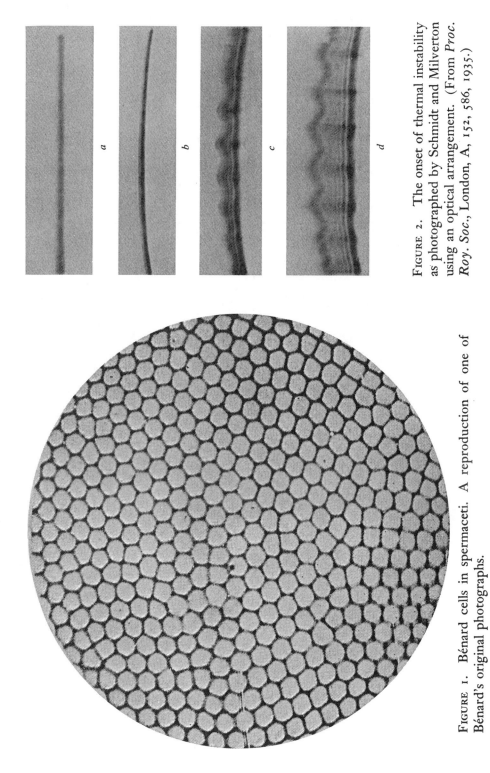

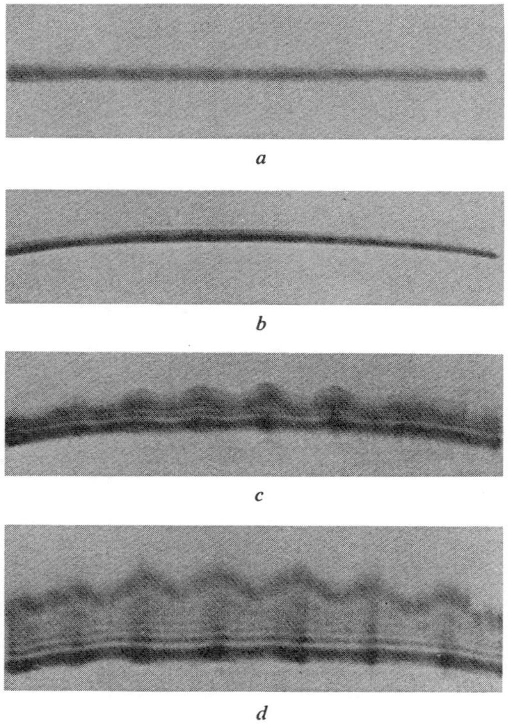

FIGURE 2. The onset of thermal instability as photographed by Schmidt and Milverton using an optical arrangement. (From *Proc. Roy. Soc.*, London, A, 152, 586, 1935.)

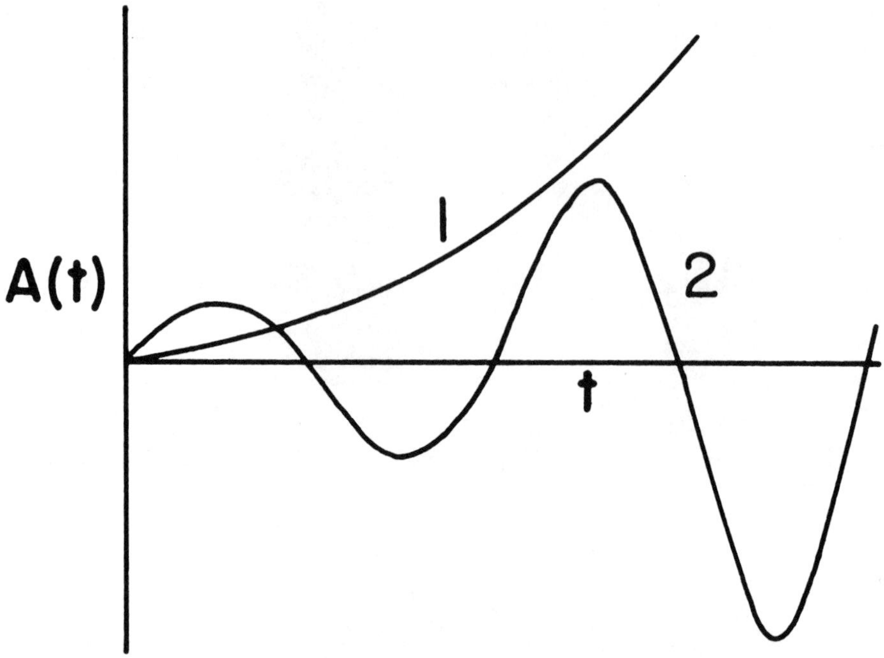

FIGURE 3. Aperiodic instability (curve 1) and overstability (curve 2).

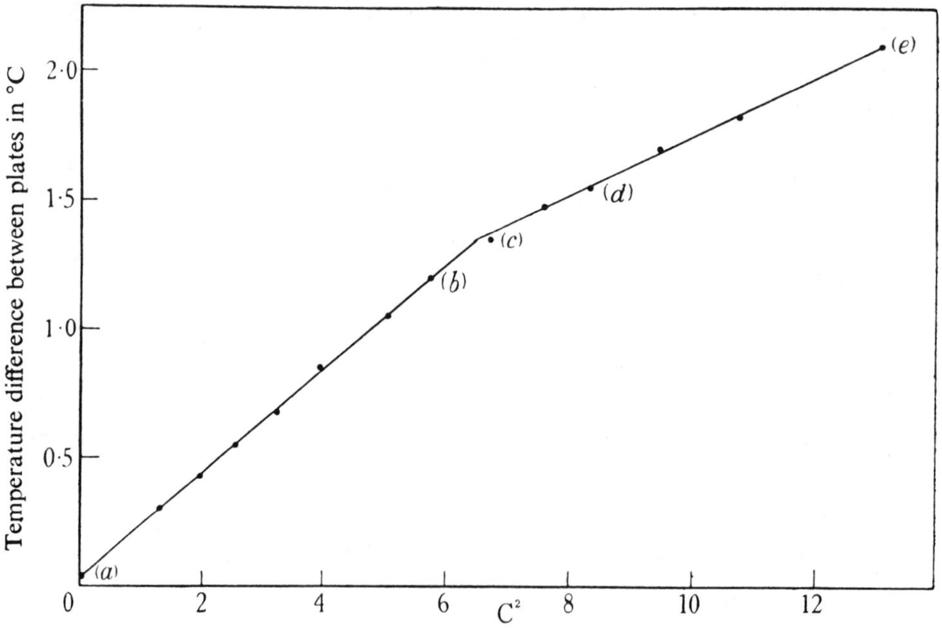

FIGURE 4. Curve showing the discontinuity in the rate of heat transfer in a layer of fluid between two horizontal plates when heated from below. (R. J. Schmidt and S. W. Milverton, *Proc. Roy. Soc.*, London, A, 152, 586, 1935.)

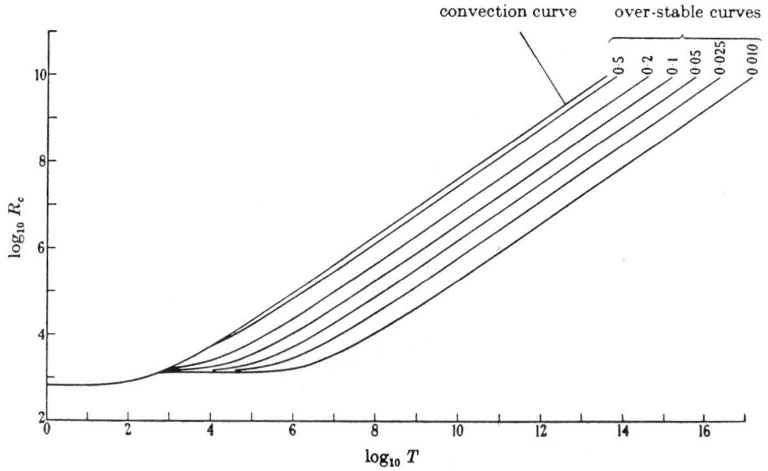

FIGURE 5. The (R_c,T)-relations for a rotating horizontal layer of fluid heated below. The curves have been derived for the case when both bounding surfaces are free. The curve labelled " convection curve " is the (R_c,T)-relation for the onset of ordinary cellular convection. The remaining curves are the corresponding relations for the onset of overstability. The value of ω to which the various curves refer are shown at the top of each curve. It will be seen that for each value of $\omega < 0.677$, the instability sets in as ordinary cellular convection for T less than a certain T^* while it sets in as overstability for $T > T^*$. (From *Proc. Roy. Soc.*, London, A, 231, 198, 1955.)

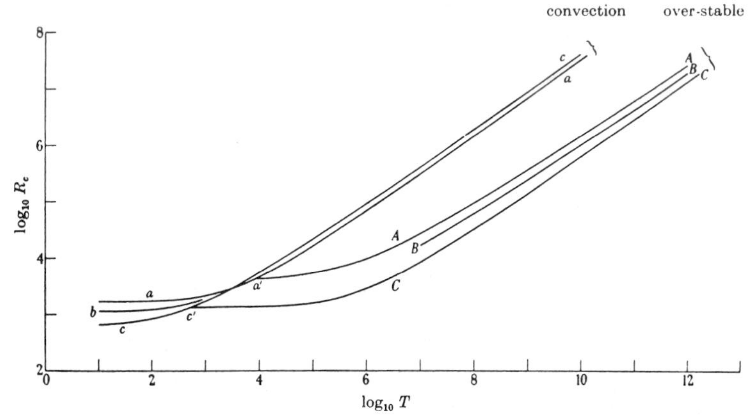

FIGURE 6. The (R_c,T)-relations for the three cases (a) both bounding surfaces rigid, (b) one bounding surface rigid and the other free, and (c) both bounding surfaces free. The curves labelled *aa*, *b*, and *cc* are the relations for the onset of ordinary cellular convection for the three cases, respectively. The curves labelled *a'AA*, *BB*, and *c'CC* are the corresponding relations for the onset of overstability for $\omega = 0.025$. At *a'* (respectively *c'*) we have a change from one type of instability to another as T increases. (From *Proc. Roy. Soc.*, London, A, 231, 198, 1955.)

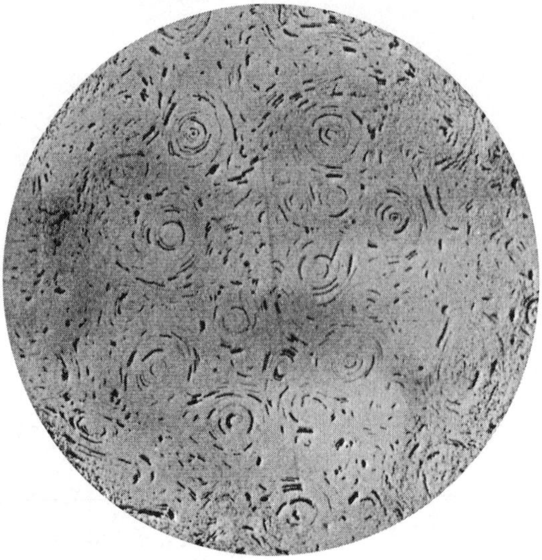

FIGURE 7. Convection cells which appear in water when in rotation and heated from below: depth 18 cm.; difference in temperature 0.7°; rate of rotation 5.0 rpm; Taylor number 1.2×10^9. (Y. Nakagawa and P. Frenzen, *Tellus*, 7, 1, 1955.)

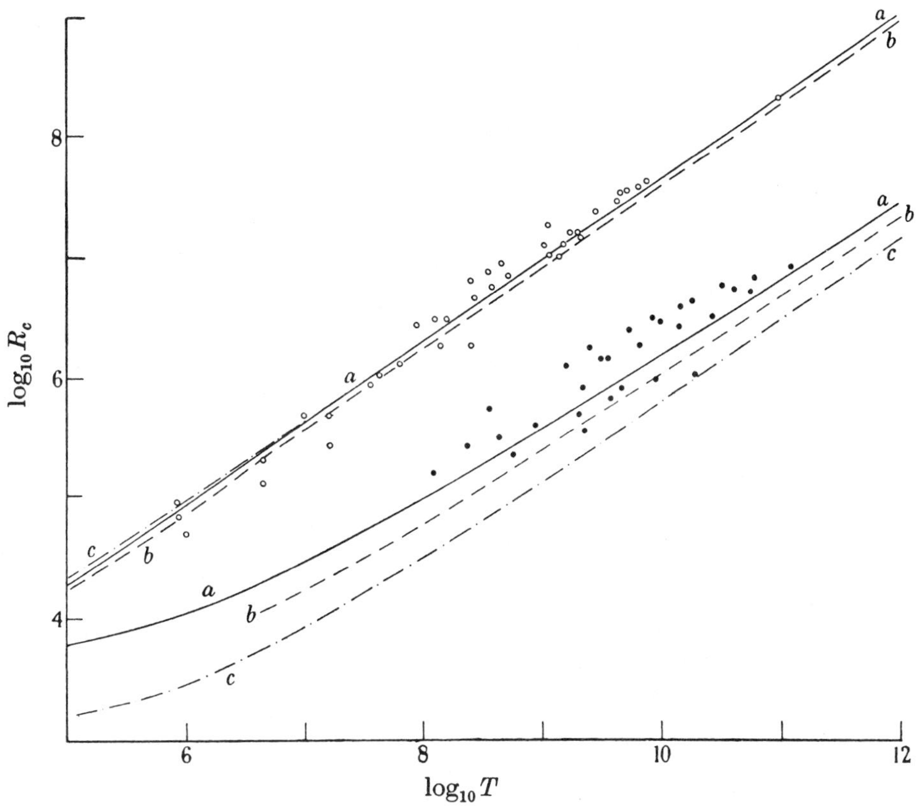

FIGURE 8. A summary of the experimental and the theoretical results. The curves *aa*, *bb*, and *cc* are the theoretical (R_c, T)-relations derived for the three cases (a) both bounding surfaces rigid, (b) one bounding surface rigid and the other free, and (c) both bounding surfaces free. The upper group of curves are for the onset of instability as ordinary cellular convection. The lower group of curves are for the onset of overstability for a value of the Prandtl number $\omega = 0.025$. The open circles are the experimentally determined points for water ($\omega = 6$). The solid circles are the experimentally determined points for mercury ($\omega = 0.025$). (D. Fultz and Y. Nakagawa, *Proc. Roy. Soc.*, London, A, 231, 211, 1955.)

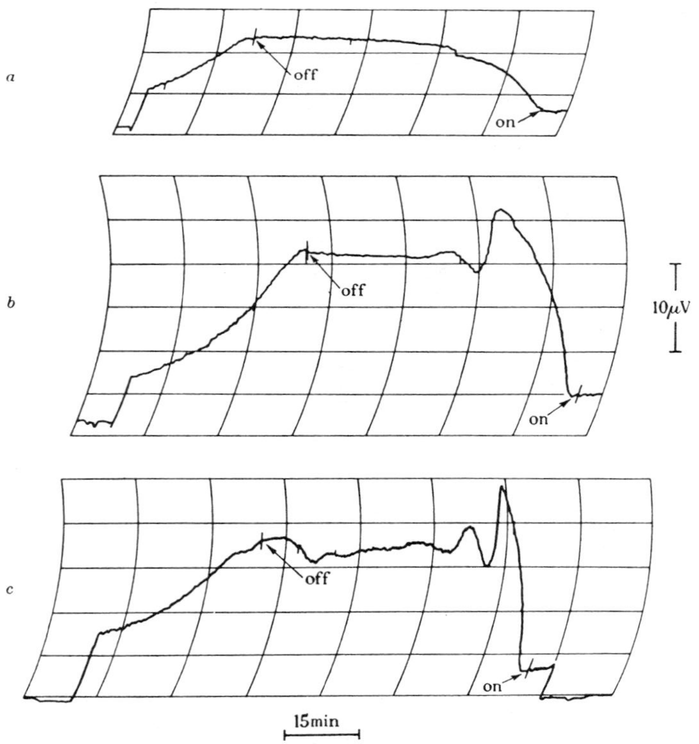

FIGURE 9. A time record of the adverse temperature gradient for water for three different rates of heating; $d = 3$ cm., $\Omega = 10$ rpm. (D. Fultz and Y. Nakagawa, *Proc. Roy. Soc.*, London, A, 231, 211, 1955.)

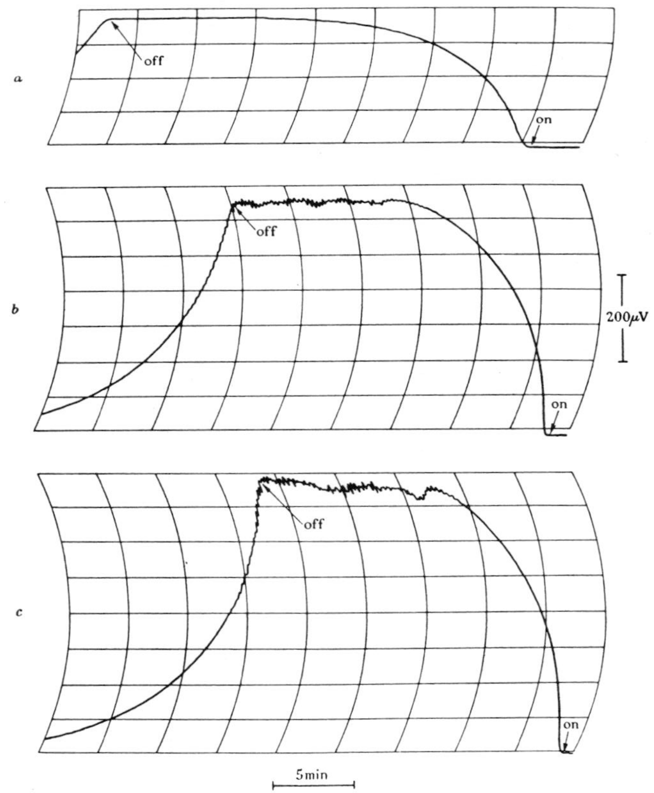

FIGURE 10. A time record of the adverse temperature gradient for mercury for three different rates of heating; $d = 6$ cm., $\Omega = 15$ rpm. (D. Fultz and Y. Nakagawa, *Proc. Roy. Soc.*, London, A, 231, 211, 1955.)

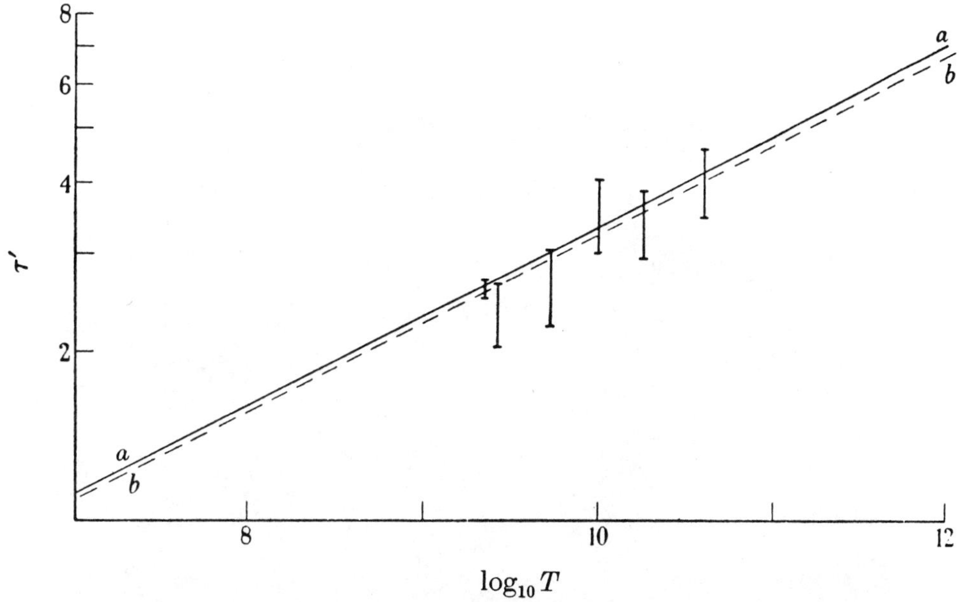

FIGURE 11. A comparison of the observed periods of oscillation at marginal stability with the theoretical periods. The ordinate (τ') gives the period in units of $2\pi/\Omega$. The curves *aa* and *bb* are the theoretical relations derived for $\omega = 0.025$ and for the case of two bounding surfaces rigid (*aa*) and one bounding surface rigid and the other free (*bb*). (D. Fultz and Y. Nakagawa, *Proc. Roy. Soc.*, London, A, 231, 211, 1955.)

FIGURE 12. The hydromagnetic laboratory at the Enrico Fermi Institute for Nuclear Studies of the University of Chicago.

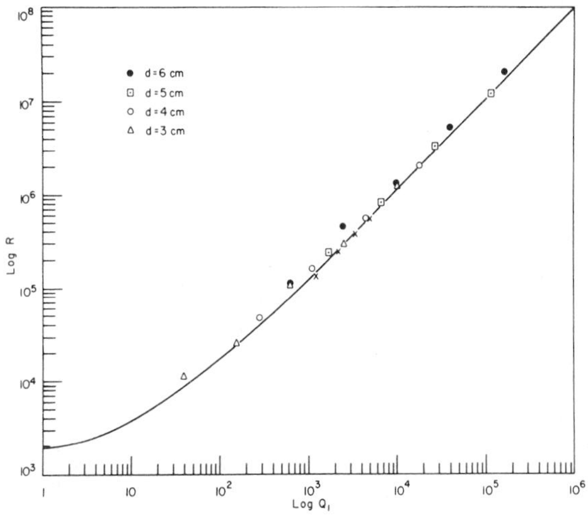

FIGURE 13. A comparison of the experimental and the theoretical results. The theoretical (R_c, Q_1)-relation is shown by the full line curve. The solid circles, squares, open circles, and triangles are the experimentally determined points for $d = 6$, 5, 4 and 3cm., respectively, with the 36½ inch magnet; the triangles represent the results with a smaller magnet with $H = 1500$ gauss and $d = 6$, 5, 4 and 3 cm. (Y. Nakagawa, *Proc. Roy. Soc.*, London, A, in press.)

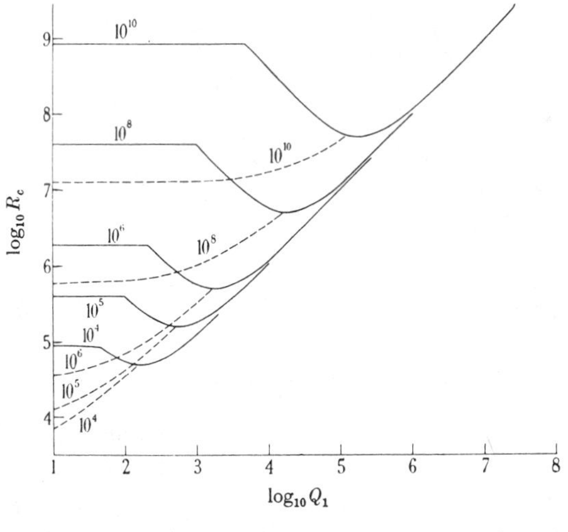

FIGURE 14. The critical Rayleigh number (R_c) for the onset of ordinary cellular convection (solid line) and overstability (for $\omega = 0.025$) (broken line) as a function of Q_1 $(= \mu^2 d^2 H^2 \sigma / \pi^2 \rho \nu)$ for various assigned values of T_1 $(= 4\Omega^2 d^4 / \pi^4 \nu^2)$. The curves are labelled by the values of T_1 to which they refer. For a given value of T_1, instability will set in as overstability for all values of Q_1 less than that at the point of intersection of the corresponding full line and dashed curves; for all larger values of Q_1, it will set in as ordinary convection. (*Proc. Roy. Soc.*, London, A, 237, 476, 1956.)

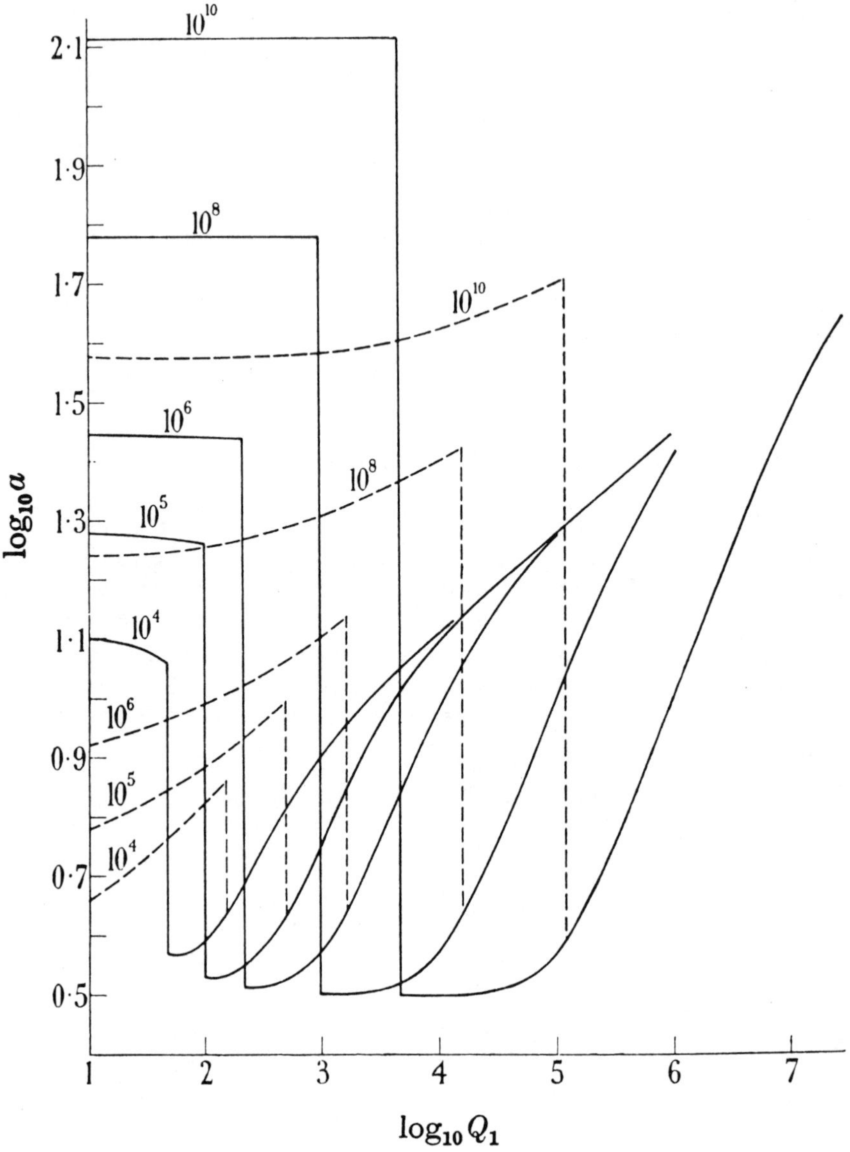

FIGURE 15. The dependence on Q_1 (for various assigned values of T_1) of the wave number a (in the unit $1/d$) of the disturbance at which instability first sets in as convection (solid line) and as overstability (for $\omega = 0.025$) (broken line). It will be observed that a discontinuous change in a occurs when (for increasing Q_1) the manner of instability changes from overstability to cellular convection. (*Proc. Roy. Soc.*, London, A, 237, 476, 1956.)

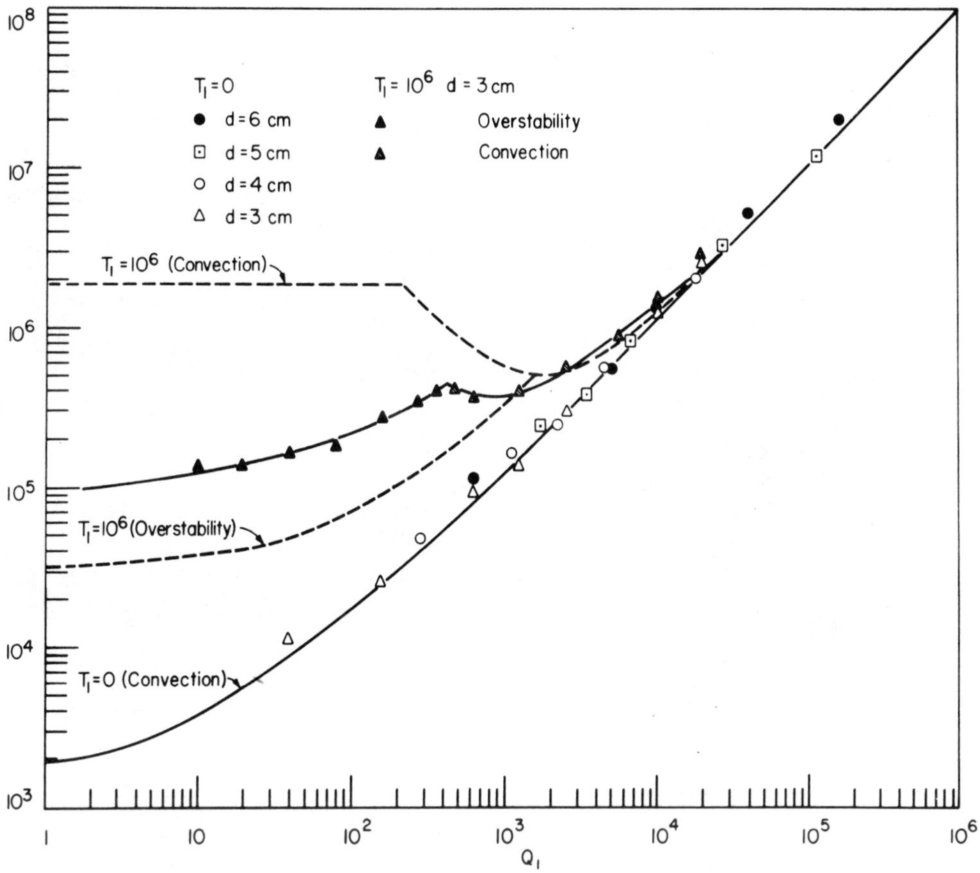

FIGURE 16. A summary of the experimental and the theoretical results. The curves labelled $T_1 = 10^6$ (convection), $T_1 = 10^6$ (overstability) and $T_1 = 0$ (convection) represent the theoretically derived relations. The value of T_1 appropriate for the experimental points represented by the solid and shaded triangles is 7.75×10^5.

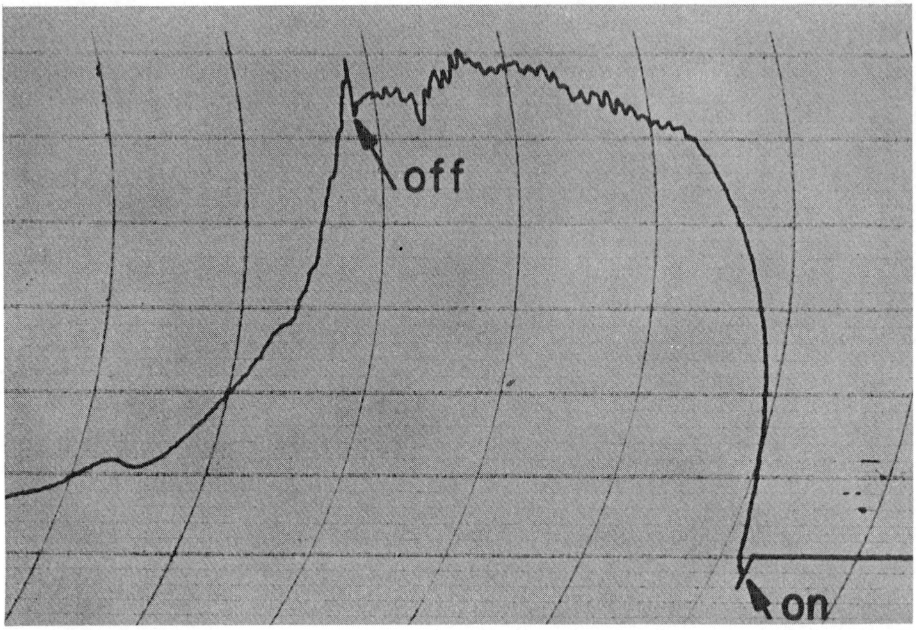

FIGURE 17. A time record of the adverse temperature gradient for mercury: $d = 3$ cm., $\Omega = 5$ rpm, $H = 125$ gauss, $Q_1 = 1.01 \times 10^1$, $T_1 = 7.90 \times 10^5$.

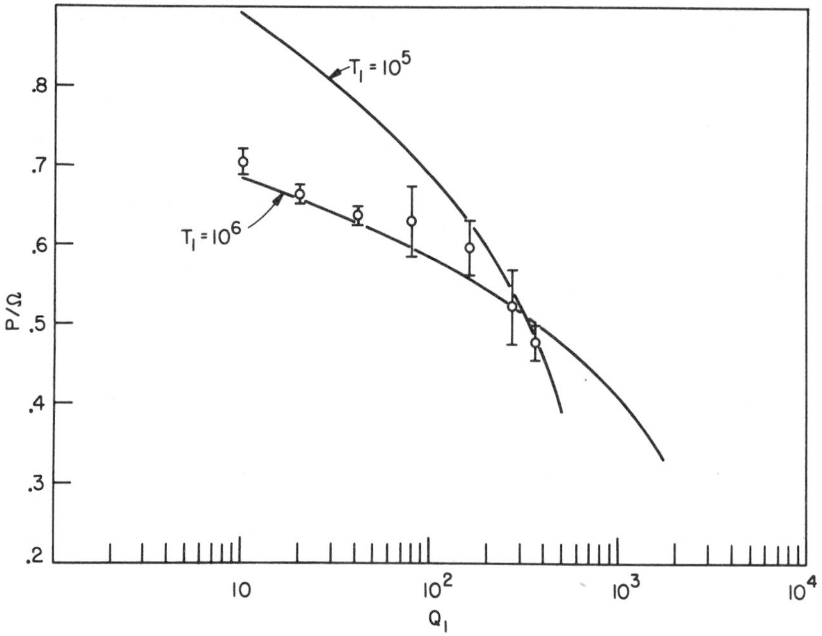

FIGURE 18. A comparison of the observed periods of overstable oscillations with the theoretical values (the full line curves). The value of T_1 appropriate for the experimental results is $(8.05 \pm 0.07) \times 10^5$.

permanently attached to the lines of force and motions parallel to the field are the only ones permissible, convection in the usual sense cannot occur.

(a) Theoretical Predictions

A detailed theoretical treatment of the inhibition of thermal convection by a magnetic field confirms one's general expectations. Precisely, it is found that the critical Rayleigh number, R_c, for the onset of instability depends on the strength of the magnetic field (H) and the electrical conductivity (σ) through the nondimensional parameter

$$Q = \frac{\mu^2 H^2 \sigma}{\rho \nu} d^2, \tag{18}$$

where μ denotes the magnetic permeability (see figure 13). In particular it is found that the dependence of R_c on Q has the asymptotic behavior

$$R_c \to \pi^2 Q \qquad (Q \to \infty), \tag{19}$$

independently of the boundary conditions; and further that the wave number a_c (in units of $1/d$) of the disturbance which manifests itself at marginal stability has the behavior

$$a_c \to \pi^{\frac{2}{3}} (\tfrac{1}{2} Q)^{\frac{1}{6}} \qquad (Q \to \infty), \tag{20}$$

again, independently of the boundary conditions.

If we insert in the asymptotic relation (19) the expressions for R_c and Q given by equations 2 and 18, we obtain

$$\beta_c = \pi^2 \frac{\mu^2 H^2 \sigma \kappa}{g \alpha} d^{-2} \qquad (H^2 \sigma \to \infty): \tag{21}$$

a formula for the critical temperature gradient in which the viscosity no longer enters. The physical meaning of this independence on viscosity is simply that as the strength of the impressed magnetic field is increased, ohmic dissipation—rather than viscous dissipation—becomes the principal factor in arresting incipient convection. Indeed, if we consider an *inviscid* fluid with a finite electrical conductivity, then we shall find that in the presence of an external magnetic field, the critical temperature gradient for the onset of instability is given precisely by equation 21. The presence of a magnetic field, therefore, imparts to the fluid characteristics which we normally associate with viscosity; we may, if we like, even define an effective kinematic viscosity

$$\nu_{eff} = \frac{\mu^2 H^2 \sigma}{\rho} d^2, \tag{22}$$

for motions perpendicular to the field.

(b) Experimental Verifications

Experiments on the inhibition of thermal convection by a magnetic field have been carried out by Nakagawa at the University of Chicago. For these and similar hydromagnetic experiments, the electromagnet of a (discarded!) 36½ inch cyclotron has been reconditioned at the Enrico Fermi Institute for Nuclear Studies at the university (see figure 12). This magnet provides a uniform magnetic field in a cylindrical volume 78 cm. in diameter and 22 cm. in height; and the strength of the field can be varied up to a maximum of 13,000 gauss. By using layers of mercury of depth 3 to 6 cm. and magnetic fields of strength 500 to 8,000 gauss, Nakagawa has determined the dependence of the critical Rayleigh number for the onset of instability on the parameter Q. The results of his experiments together with the theoretically predicted (R_c, Q)-relation are shown in figure 13. It will be seen that the experiments fully confirm the theoretical predictions.

(c) The Magnetic Inhibition of Convection in Sunspots

It is now generally believed that the lower temperatures of the sunspots relative to the surrounding photosphere is due to the absence of a convection zone in the sunspots at depths where convection is operative in the photospheric layers; and that this absence of convection in the sunspots is due to its having been suppressed by the prevailing magnetic fields. This suggestion was originally made by Walén and Biermann.

VI. THE EFFECT OF ROTATION AND MAGNETIC FIELD

We now consider the effect on thermal instability of a magnetic field and rotation acting simultaneously. From the physical interpretation which we have given of the effects which arise when rotation alone or magnetic field alone is present, it is clear that when they are both present, the fluid must be subject to conflicting tendencies.

For, while both have inhibitive effects on the onset of instability, the origin in the two cases is different. As we have explained, the inhibitive effect of rotation arises from the fact that in the absence of viscosity motions parallel to the axis of rotation are forbidden; while the inhibitive effect of a magnetic field arises from the fact that in the absence of electrical resistivity motions perpendicular to the field are forbidden. Consequently, when both rotation and magnetic field are present, motions compatible with the one are incompatible with the other. The fluid has therefore to obey two masters with contrary inclinations; and we may conclude that the fluid is in danger of a nervous breakdown! To express the same thing less anthropomorphically: The behavior of the fluid must be characterized by discontinuities indicating which of the two laws – the conservation of vorticity or the conservation of magnetic flux – is violated at the expense of the other (by virtue of viscous or ohmic dissipation).

An alternative way of looking at the problem is the following: We have seen that in the presence of rotation viscosity facilitates the onset of instability; we have also seen that a magnetic field imparts to the fluid characteristics associated with viscosity. Consequently, even though each acting separately inhibits the onset of convection, they will not necessarily conspire toward the same ends when they act together; indeed, we may expect that under certain circumstances the effect of a magnetic field on a rotating fluid may actually be one of accelerating the onset of convection.

We may look at the problem in still another way: In the absence of viscosity we have the theorem of Helmholtz and Kelvin that the normal flux of the vorticity ($\boldsymbol{\omega} = \text{curl}\, \boldsymbol{u}$) across any element of surface is an integral of the equations of motion; i.e.

$$\frac{d}{dt}\int_S \boldsymbol{\omega} \,.\, d\boldsymbol{S} = 0\,, \tag{23}$$

where the integral is over an element of surface S and we follow the element with the motion. At the same time, in the absence of electrical resistivity we have the theorem of Cowling that the normal flux of the magnetic field ($\boldsymbol{H}$) across any element of surface is also an integral of the equations of motion; i.e.

$$\frac{d}{dt}\int_S \boldsymbol{H} .\, d\boldsymbol{S} = 0\,. \tag{24}$$

It is the existence of these two integrals simultaneously that is at the base of the complex behavior of a real fluid (i.e. one with finite viscosity and finite electrical resistivity) when circulating and in the presence of a magnetic field.

(*a*) *Theoretical Predictions*

A complete theoretical discussion of the thermal instability of a layer of fluid such as mercury, in rotation and in the presence of an impressed magnetic field, is further complicated by the fact that the onset of instability in the presence of rotation is generally as overstability while it is as convection in the presence of a magnetic field. While the general solution of this complex problem taking correctly into account all the boundary conditions has not been carried out, the analogue of the case of two free boundaries (first considered by Rayleigh in his solution of the classical problem) has been fully analyzed and the results of the analysis are summarized in figures 14 and 15.

Considering first the case when instability sets in as convection, we observe that the following sequence of events will occur if we gradually increase the strength of the magnetic field keeping the angular velocity of rotation constant. Suppose for example that T_1 $(= T/\pi^4) = 10^5$ and we gradually increase the magnetic field; i.e., Q_1 $(= Q/\pi^2)$. When $Q_1 = 0$ (and $T_1 = 10^5$) the cells which appear at marginal stability will be elongated ($a/d = 18.9$); and as Q_1 is increased, at first, the critical Rayleigh number and the dimension of the marginal cells will be hardly affected. However, when the magnetic field has increased to a value corresponding to $Q_1 = 100$, cells of two very different sizes will suddenly appear simultaneously: one set, which will be highly elongated and the other set, which will be relatively quite wide; in fact, in this particular case the ratio of the two cell diameters is about 6. As the magnetic field increases beyond this value, the critical Rayleigh number will start decreasing and pass through a minimum; and eventually the effect of the magnetic field will predominate.

The sequence of events we have described will occur if the Prandtl number ($\omega = \nu/\kappa$) exceeds a certain value of the order of one. But if one considers a fluid such as mercury with a low Prandtl number, the occurrence of overstability (when the effects of rotation dominate) alters the situation. The sequence of events we may then observe can be directly read from figures 14 and 15. It is seen for example from figure 14 that the transition from overstability to convection occurs at about the place where the convection curve passes through its minimum; and from figure 15 it is seen that this transition is accompanied by a substantial discontinuity in the wave number of the disturbance which manifests itself at marginal stability. This latter discontinuity is in the sense that the convection cells (for increasing Q_1) suddenly become very much widened; under experimentally realizable conditions this widening can be by a factor as large as 10.

(*b*) *Experimental Verifications*

Again at the Fermi Institute, Nakagawa has carried out experiments on the thermal instability of rotating layers of mercury in a homogeneous external magnetic field. The results of his experiments are presented in figures 16 to 18.

Nakagawa was able to distinguish whether the onset of instability is as overstability or as convection by examining the temperature records. Figure 16 is one of these temperature records; it shows in a very striking manner the onset of overstability. The experimentally determined critical Rayleigh numbers for $T_1 = 7.75 \times 10^5$ and varying values of Q_1 are exhibited in figure 17 together with the theoretically expected relations for $T_1 = 0$ and $T_1 = 10^6$. It is seen that the transition from overstability to convection does occur, as predicted, at a certain critical field strength. However, the quantitative agreement between theory and experiment in this case is not as good as in the others; but this was not to be expected, since, in the solution of the theoretical problem in this case, the correct boundary conditions were not satisfied. Nevertheless, the experiments do confirm the broad features of the predicted variations; the agreement, in particular, between the observed frequencies of the oscillations at marginal stability and the predicted values is very satisfactory.

* * * * *

That the nonlinearity of the hydrodynamical equations leads to a complex and often unexpected behavior of a fluid is well known. If the fluid should in addition be an electrical conductor and magnetic fields are present, then its behavior contains even greater elements for surprise. It is this unexpectedness which makes hydromagnetics a fascinating field for study; but it also calls for a disciplined and a systematic examination.

In our efforts to understand why a fluid behaves in a particular way under particular conditions the study of hydrodynamic and hydromagnetic stability is likely to be useful, for it may disclose the relative importance and sometimes the conflicting tendencies of the different forces and constraints which are kept dormant in the undisturbed state. Thus, the study of thermal stability in the context of the simplest problem already reveals several phases of fluid behavior which, in these connections, go to make for its complexity. A similar examination of other situations may give us further insight. To the critic who rightly points out that in these studies we are very remote from the conditions under which Nature presents her most extravagant displays, let me only say:

> There is a square. There is an oblong. The players take the square and place it upon the oblong. They place it very accurately. They make a perfect dwelling place. The structure is now visible. What was inchoate is here stated. We are not so various or so mean. We have made oblongs and stood them upon squares. This is our triumph. This is our consolation.
>
> —Virginia Woolf

References

The following list, while very incomplete as a bibliography, represents, however, the principal source for the lecture.

On the Classical Problem

1. Rumford, *Complete Works* (Boston: American Academy of Arts and Sciences, 1870) I, 237, et. seq.
2. Bénard, H., 1900, *Rev. gen. Sci. por. appl.* 12, 1261, 1309.
3. Rayleigh, 1916, *Phil. Mag.* 6, 32, 529.
4. Jeffreys, H., 1926, *Phil. Mag.* 7, 2, 833.
5. Low, A. R., 1929, *Proc. Roy. Soc.* A, 125, 180.
6. Pellew, A., and Southwell, R. V., 1940 *Proc. Roy. Soc.* A, 176, 312.
7. Schmidt, R. J., and Milverton, S. W., 1935, *Proc. Roy. Soc.* A, 152, 586.

On the Effect of Rotation

1. Chandrasekhar, S., 1953, *Proc. Roy. Soc.* A, 217, 306.
2. Chandrasekhar, S., and Elbert, D. D., 1955, *Proc. Roy. Soc.* A, 231, 198.
3. Nakagawa, Y., and Frenzen, P., 1955, *Tellus,* 7, 1.
4. Fultz, D., and Nakagawa, Y., 1955, *Proc. Roy. Soc.* A, 231, 211.

On the Effect of a Magnetic Field

1. Chandrasekhar, S., 1952, *Phil. Mag.* 7, 43, 501.
2. Chandrasekhar, S., 1954, *Phil. Mag.* 7, 45, 1177.

3. Thompson, W. B., 1951, *Phil. Mag.* 7, 42, 1417.
4. Nakagawa, Y., 1957, *Proc. Roy. Soc.* A, 240, 108.

On the Effect of Rotation and Magnetic Field

1. Chandrasekhar, S., 1954, *Proc. Roy. Soc.* A, 225, 173.
2. Chandrasekhar, S., 1956, *Proc. Roy. Soc.* A, 237, 476.
3. Nakagawa, Y., 1957, *Proc. Roy. Soc.* A (in press).

9

THE THERMODYNAMICS OF THERMAL INSTABILITY IN LIQUIDS

by S. Chandrasekhar

The Enrico Fermi Institute for Nuclear Studies
The University of Chicago
Chicago

"*Every natural process involves in greater or less degree friction or conduction of heat. But in the domain of irreversible processes the principle of least action is no longer sufficient; for the principle of the increase of entropy brings into the system of physics a wholly new element foreign to the action principle and which demands special mathematical treatment.*" [1]

MAX PLANCK (1909)

ABSTRACT

In this paper the thermodynamics of thermal instability in liquids is considered and the following theorem is proved: Thermal instability as cellular convection will set in at the minimum (adverse) temperature gradient which is necessary to maintain a balance between the rate of dissipation of energy by all irreversible processes present and the rate of liberation of the thermodynamically available energy by the buoyancy force acting on the fluid. Likewise, the onset of thermal instability will be as overstable oscillations if it is possible (at a lower adverse temperature gradient) to balance in a synchronous manner the periodically varying amounts of kinetic and other forms of energy with similarly varying rates of dissipation and liberation of energy.

I

INTRODUCTION

It is known that when a horizontal layer of liquid is heated from below, instability with ensuing convection will set in if the prevailing adverse temperature gradient (β) exceeds a certain critical value. As RAYLEIGH [1] first showed, the stability of the liquid under these circumstances depends on the numerical value of the non-dimensional parameter,

$$R = \frac{g\,\alpha\,\beta}{\varkappa\,\nu}\,d^4, \tag{1}$$

– called the RAYLEIGH number – where d denotes the depth of the layer, g the acceleration due to gravity, and α, $\varkappa$ and ν are the coefficients of volume expansion, thermometric conductivity and kinematic viscosity, respectively. The condition for instability is that R exceeds a certain determinate critical value, R_c.

During the past few years, the effects of rotation and magnetic field, separately and jointly, on the onset of thermal instability have been investigated both experimentally and theoretically (for a general account of these investigations see CHANDRASEKHAR [2]).

[1] Translated from M. PLANCK, Leidener Vortrag, 9.12.1909: „Die Einheit des physikalischen Weltbildes".

The theoretical determination of the critical RAYLEIGH number, R_c, for the onset of instability depends, always, on the solution of a characteristic value problem which gives the value of R at which a disturbance periodic in the horizontal plane with a specified wave number, a, first becomes unstable; the required value of R_c is the minimum of R, determined in this fashion, as a function of a. In all cases which have been investigated so far, it has been found that the underlying characteristic value problem can be solved by a variational method based on a formula for R which expresses it as the ratio of two integrals which are, in most instances, positive-definite. For the classical RAYLEIGH problem, it was shown by MALKUS [3] (see also [4]) that the formula which provides the basis for the variational treatment (in case the instability sets in as a stationary pattern of motions) equates, simply, the rate at which energy is dissipated by viscosity to the rate at which the thermodynamically available energy is released by the buoyancy force acting on the fluid. In this paper we shall generalize this principle to include the effects of rotation and magnetic field and allow also for the possibility of convection setting in as oscillations of increasing amplitude (i.e., as overstability).

II

THE CLASSICAL PROBLEM

As a preliminary to the consideration of the more general cases in the following sections, it will be convenient to have a parallel treatment of the classical problem. As we have stated, this problem has, effectively, been considered by MALKUS [3]; however our point of view is slightly different and the presentation will be adapted to the discussion which is to follow.

The equations which govern the perturbation of an initial static state in which the temperature gradient $-\beta$ prevails, are:

$$\frac{\partial u_i}{\partial t} = -\frac{\partial \varpi}{\partial x_i} + \gamma \theta \lambda_i + \nu \nabla^2 u_i \tag{2}$$

and

$$\frac{\partial \theta}{\partial t} = \beta w + \varkappa \nabla^2 \theta, \tag{3}$$

where u_i denotes the component of the velocity $\mathfrak{u}$ in the direction i ($i = 1, 2, 3$) of a rectangular system of reference x_i; θ the perturbation in the temperature; $\vec{\lambda}$ a unit vector in the direction of the vertical (i.e., of the acceleration of gravity $\mathfrak{g}$); $w (= \lambda_j u_j)$ is the vertical component of the velocity; $\gamma = g\alpha$; and $\varrho \varpi$ (where ϱ is the density) is related to the departures from the static pressure distribution caused by the perturbation.

By taking the curl of equation (2), we can eliminate ϖ and obtain

$$\frac{\partial \omega_i}{\partial t} = \gamma \varepsilon_{ijk} \frac{\partial \theta}{\partial x_j} \lambda_k + \nu \nabla^2 \omega_i, \tag{4}$$

where

$$\vec{\omega} = \operatorname{curl} \mathfrak{u} \tag{5}$$

denotes the vorticity. ε_{ijk} is a tensor with the following properties: $\varepsilon_{ijk} = -\varepsilon_{ikj}$, $\varepsilon_{iik} = 0$, $\varepsilon_{ijk} \cdot \varepsilon_{ljk} = \delta_{il}$, where $\delta_{il} = 1$ for $i = l$ and $\delta_{il} = 0$ for $i \neq l$. Taking once again the curl of equation (4) and making use of the solenoidal character of u, we obtain

$$-\frac{\partial}{\partial t} \nabla^2 u_i = \gamma \left(\lambda_j \frac{\partial^2 \theta}{\partial x_j \partial x_i} - \lambda_i \nabla^2 \theta \right) - \nu \nabla^4 u_i . \tag{6}$$

Multiplying equation (4) by λ_i, we get

$$\frac{\partial \zeta}{\partial t} = \nu \nabla^2 \zeta , \tag{7}$$

where

$$\zeta = \lambda_j \omega_j = \frac{\partial v}{\partial x} - \frac{\partial u}{\partial y} \tag{8}$$

is the z-component of the vorticity and u and v are the horizontal components of the velocity in the rectangular directions x and y respectively. Similarly, the z-component of equation (6) is

$$\frac{\partial}{\partial t} \nabla^2 w = \gamma \left(\frac{\partial^2 \theta}{\partial x^2} + \frac{\partial^2 \theta}{\partial y^2} \right) + \nu \nabla^4 w. \tag{9}$$

In treating the perturbation problem characterized by the foregoing equations, we analyze the disturbance into normal modes and consider each mode separately. In the present problem, the normal modes are disturbances, periodic in the (x, y)-plane and of assigned wave numbers. And a disturbance belonging to a particular wave number, a, will be such that all relevant quantities (q) describing it will satisfy the equations

$$\left(\frac{\partial^2}{\partial x^2} + \frac{\partial^2}{\partial y^2} \right) q = -a^2 q \quad \text{and} \quad \nabla^2 q = \left(\frac{\partial^2}{\partial z^2} - a^2 \right) q . \tag{10}$$

Considering then a disturbance belonging to a wave number a and assuming that the marginal state is a stationary one, we can, after eliminating θ between equations (3) and (9), reduce the problem to one in characteristic values, for R (as dependent on a). We shall not carry out this reduction as it is available in the literature in several places. But it is important for our purposes to show how, once equations (3) and (9) have been solved, we can complete the solution by determining the velocities in the horizontal plane as well. We can accomplish this, quite generally, as follows:

Resolving the horizontal component of the velocity into a rotational and an irrotational part in the manner

$$u = \frac{\partial \varphi}{\partial x} - \frac{\partial \psi}{\partial y} \quad \text{and} \quad v = \frac{\partial \varphi}{\partial y} + \frac{\partial \psi}{\partial x} , \tag{11}$$

where φ and ψ are two scalar functions, we have

$$-\frac{\partial w}{\partial z} = \frac{\partial u}{\partial x} + \frac{\partial v}{\partial y} = \frac{\partial^2 \varphi}{\partial x^2} + \frac{\partial^2 \varphi}{\partial y^2} \tag{12}$$

and

$$\zeta = \frac{\partial v}{\partial x} - \frac{\partial u}{\partial y} = \frac{\partial^2 \psi}{\partial x^2} + \frac{\partial^2 \psi}{\partial y^2} . \tag{13}$$

If φ and ψ are FOURIER-analyzed in the same way as all the other quantities, and we are considering a disturbance belonging to a wave number a, then it follows from equations (12) and (13) that

$$\varphi = \frac{1}{a^2}\frac{\partial w}{\partial z} \quad \text{and} \quad \psi = -\frac{1}{a^2}\zeta. \tag{14}$$

Hence,

$$u = \frac{1}{a^2}\left(\frac{\partial^2 w}{\partial x\,\partial z} + \frac{\partial \zeta}{\partial y}\right) \quad \text{and} \quad v = \frac{1}{a^2}\left(\frac{\partial^2 w}{\partial y\,\partial z} - \frac{\partial \zeta}{\partial x}\right). \tag{15}$$

These equations express u and v in terms of the z-components of the velocity and the vorticity; they are of general applicability.

Returning to equations (3), (7) and (9), we shall consider them under conditions of a stationary state. Then from equation (7), it follows directly that

$$\zeta \equiv 0, \tag{16}$$

and that, therefore,

$$u = \frac{1}{a^2}\frac{\partial^2 w}{\partial x\,\partial z} \quad \text{and} \quad v = \frac{1}{a^2}\frac{\partial^2 w}{\partial y\,\partial z}. \tag{17}$$

Consider, now, the average rate of dissipation of energy by viscosity in a unit column of the fluid (i.e., in a vertical column of unit cross section). This is given by[1]

$$\varepsilon_\nu = -\nu \int_0^d \langle u_i \nabla^2 u_i \rangle \, dz, \tag{18}$$

where the angular brackets signify that the quantity enclosed is to be averaged over the entire (x, y)-plane. (In this paper the angular brackets will always be used in this sense.)

Making use of equations (10) and (17), we have

$$\langle u_i \nabla^2 u_i \rangle = \left\langle \frac{1}{a^4}\frac{\partial^2 w}{\partial x\,\partial z}\left(\frac{\partial^2}{\partial z^2} - a^2\right)\frac{\partial^2 w}{\partial x\,\partial z} + \frac{1}{a^4}\frac{\partial^2 w}{\partial y\,\partial z}\left(\frac{\partial^2}{\partial z^2} - a^2\right)\frac{\partial^2 w}{\partial y\,\partial z} + w\left(\frac{\partial^2}{\partial z^2} - a^2\right) w \right\rangle. \tag{19}$$

Without loss of generality, we may suppose that w is of the form

$$w = W(z) \sin a_x x \sin a_y y, \tag{20}$$

where $W(z)$ is a function of z only, and

$$a_x^2 + a_y^2 = a^2. \tag{21}$$

With w given by equation (20), equation (19) becomes

$$\langle u_i \nabla^2 u_i \rangle = \frac{1}{4}\left\{ W(D^2 - a^2)\, W + \frac{1}{a^2} D W\, (D^2 - a^2)\, D W \right\}, \tag{22}$$

where

$$D = \frac{d}{dz}. \tag{23}$$

Inserting this expression for $\langle u_i \nabla^2 u_i \rangle$ in equation (18), we have

$$\varepsilon_\nu = -\frac{\nu}{4a^2}\int_0^d \left\{ a^2 W(D^2 - a^2)\, W + D W\, (D^2 - a^2)\, D W \right\} dz. \tag{24}$$

[1] This is per unit mass. We will not repeat this every time: it must be understood.

By an integration by parts, we find that

$$\int_0^d DW(D^2 - a^2)\,DW\,dz = -\int_0^d D^2 W(D^2 - a^2)\,W\,dz, \tag{25}$$

the integrated part vanishes on account of the boundary conditions[1] on W. Hence

$$\varepsilon_\nu = \frac{\nu}{4a^2}\int_0^d [(D^2 - a^2)\,W]^2\,dz, \tag{26}$$

which expresses ε_ν as a positive-definite integral.

Consider next, the average rate of liberation of energy, ε_g, in a unit column of the fluid by the buoyancy force acting on it. This is given by

$$\varepsilon_g = \gamma\int_0^d \langle u_i\,\lambda_i\,\theta\rangle\,dz = \gamma\int_0^d \langle\theta\,w\rangle\,dz. \tag{27}$$

On the other hand, according to equation (3), in a steady state,

$$w = -\frac{\varkappa}{\beta}\,\nabla^2\theta = -\frac{\varkappa}{\beta}\left(\frac{\partial^2}{\partial z^2} - a^2\right)\theta. \tag{28}$$

Substituting for w in accordance with this equation in the expression for ε_g and writing θ in the form (cf. equation (20))

$$\theta = \Theta(z)\sin a_x x \sin a_y y, \tag{29}$$

we obtain

$$\varepsilon_g = -\frac{\varkappa\gamma}{4\beta}\int_0^d \Theta\,(D^2 - a^2)\,\Theta\,dz. \tag{30}$$

By an integration by parts this can be brought to the positive-definite form

$$\varepsilon_g = \frac{\varkappa\gamma}{4\beta}\int_0^d [(D\Theta)^2 + a^2\,\Theta^2]\,dz. \tag{31}$$

(The integrated part again vanishes since $\theta = 0$ on the boundaries.)

If we now demand that

$$\varepsilon_\nu = \varepsilon_g, \tag{32}$$

then

$$\frac{\varkappa\gamma}{\beta}\int_0^d [(D\,\Theta)^2 + a^2\,\Theta^2]\,dz = \frac{\nu}{a^2}\int_0^d [(D^2 - a^2)\,W]^2\,dz. \tag{33}$$

To verify that this is exactly the relation for β which is minimized in Pellew and Southwell's variational treatment of the problem [5], we first observe that according to equation (9), in a steady state,

$$\gamma\,a^2\,\Theta = \nu\,(D^2 - a^2)^2\,W. \tag{34}$$

[1] These are $W = 0$; and either $DW = 0$ or $D^2 W = 0$ depending on whether the bounding surface is rigid or free.

Hence, we may write

$$\Theta = \frac{\nu}{\gamma a^2} F \quad \text{where} \quad F = (D^2 - a^2)^2 W. \tag{35}$$

With Θ expressed in terms of W in this manner, equation (33) becomes

$$\frac{\varkappa \nu^2}{\beta \gamma a^4} \int_0^d [(DF)^2 + a^2 F^2]\, dz = \frac{\nu}{a^2} \int_0^d [(D^2 - a^2) W]^2\, dz, \tag{36}$$

or,

$$\frac{\beta \gamma}{\varkappa \nu} = \frac{\int_0^d [(DF)^2 + a^2 F^2]\, dz}{a^2 \int_0^d [(D^2 - a^2) F]^2\, dz}. \tag{37}$$

Equation (37) clearly expresses the RAYLEIGH number (for the specified a) as the ratio of two positive definite integrals. And it can now be verified that in the variational treatment of the problem the expression which is minimized is indeed the quantity on the right-hand side of equation (37). Therefore, *the condition for instability is, simply, that the adverse temperature gradient be greater than the minimum necessary to maintain, steadily, the equality between ε_ν and ε_g.*

III

THE EFFECT OF A MAGNETIC FIELD

Suppose now that the liquid which is being investigated is an electrical conductor and that a uniform external magnetic field of intensity H is impressed on it in the direction of $\mathfrak{g}$. (The generalization to the case when the directions of $\mathfrak{H}$ and $\mathfrak{g}$ do not coincide is not difficult, but since it does not add anything essential to the physics of the problem, we shall not consider it here.) Under these circumstances, equation (2) is replaced by

$$\frac{\partial u_i}{\partial t} = -\frac{\partial \varpi}{\partial x_i} + \gamma \theta \lambda_i + \nu \nabla^2 u_i + \frac{\mu H}{4\pi \varrho} \frac{\partial h_i}{\partial z}, \tag{38}$$

where μ denotes the magnetic permeability and h_i is the perturbation in the magnetic field; the latter is governed by the equations

$$\frac{\partial h_i}{\partial t} = H \frac{\partial u_i}{\partial z} + \eta \nabla^2 h_i, \tag{39}$$

and

$$\frac{\partial h_i}{\partial x_i} = 0. \tag{40}$$

In equation (39)

$$\eta = \frac{1}{4\pi \mu \sigma}, \tag{41}$$

where σ is the coefficient of electrical conductivity.

In a stationary state, equation (39) gives

$$\eta \nabla^2 h_i = \eta \left(\frac{\partial^2}{\partial z^2} - a^2\right) h_i = -H \frac{\partial u_i}{\partial z}. \tag{42}$$

It can be shown that in a steady state the presence of a magnetic field will not induce any vorticity in the z-direction; therefore, the horizontal components of the velocity will still be given by equation (17) of the last section.

Since the liquid considered is an electrical conductor, currents will be generated by motions across the lines of force; and these currents will give rise to JOULE heating. Thus, there is in this case an additional source of irreversible dissipation of energy; and we must take this into account.

Let ε_σ denote, then, the average rate of dissipation of energy in a unit column of the fluid by JOULE heating. This is given by

$$\varepsilon_\sigma = -\frac{\mu \eta}{4 \pi \varrho} \int_0^d \langle h_i \nabla^2 h_i \rangle \, dz, \tag{43}$$

or, making use of equations (17) and (42), we have

$$\varepsilon_\sigma = \frac{\mu H}{4 \pi \varrho} \int_0^d \left\langle h_i \frac{\partial u_i}{dz} \right\rangle dz = \frac{\mu H}{4 \pi \varrho} \int_0^d \left\langle \frac{1}{a^2} \left(h_x \frac{\partial^3 w}{\partial x \, \partial z^2} + h_y \frac{\partial^3 w}{\partial y \, \partial z^2} \right) + h_z \frac{\partial w}{\partial z} \right\rangle dz. \tag{44}$$

Since averaging over the horizontal plane is equivalent to integrating over a square with a side equal to the wave length, $\lambda\!\!\!^-$, of the disturbance, we can write, for example,

$$\left\langle h_x \frac{\partial^3 w}{\partial x \, \partial z^2} \right\rangle = C \iint h_x \frac{\partial}{\partial x} \left(\frac{\partial^2 w}{\partial z^2} \right) dx \, dy, \tag{45}$$

where the integration is over a square of area $\lambda\!\!\!^-{}^2$ and C is an appropriate constant of proportionality. By an integration by parts, we obtain

$$\left\langle h_x \frac{\partial^3 w}{\partial x \, \partial z^2} \right\rangle = -C \iint \frac{\partial h_x}{\partial x} \frac{\partial^2 w}{\partial z^2} \, dx \, dy = -C \left\langle \frac{\partial h_x}{\partial x} \frac{\partial^2 w}{} \right\rangle. \tag{46}$$

The terms in h_x and h_y in the integral defining ε_σ thus combine to give

$$-\frac{1}{a^2} \int_0^d \left\langle \left(\frac{\partial h_x}{\partial x} + \frac{\partial h_y}{\partial y} \right) \frac{\partial^2 w}{\partial z^2} \right\rangle dz = \frac{1}{a^2} \int_0^d \left\langle \frac{\partial h_z}{\partial z} \frac{\partial^2 w}{\partial z^2} \right\rangle dz. \tag{47}$$

Hence,

$$\varepsilon_\sigma = \frac{\mu H}{4 \pi \varrho} \int_0^d \left\langle h_z \frac{\partial w}{\partial z} + \frac{1}{a^2} \frac{\partial h_z}{\partial z} \frac{\partial^2 w}{\partial z^2} \right\rangle dz. \tag{48}$$

A further integration by parts (this time with respect to z) of the term in $\partial h_z / \partial z$ n (48), transforms this expression into

$$\varepsilon_\sigma = -\frac{\mu H}{4 \pi \varrho a^2} \int_0^d \left\langle \frac{\partial w}{\partial z} \left(\frac{\partial^2}{\partial z^2} - a^2 \right) h_z \right\rangle dz. \tag{49}$$

But, according to equation (42),

$$\left(\frac{\partial^2}{\partial z^2} - a^2\right) h_z = -\frac{H}{\eta}\frac{\partial w}{\partial z}. \tag{50}$$

Hence

$$\varepsilon_\sigma = \frac{\mu H^2}{4\pi \varrho a^2 \eta}\int_0^d \left\langle \left(\frac{\partial w}{\partial z}\right)^2 \right\rangle dz = \frac{\mu H^2}{16\pi \varrho a^2 \eta}\int_0^d (D W)^2\, dz. \tag{51}$$

On substituting for η from equation (41), the expression for ε_σ finally becomes

$$\varepsilon_\sigma = \frac{\mu^2 H^2 \sigma}{4 \varrho a^2}\int_0^d (D W)^2\, dz. \tag{52}$$

Combining equation (52) with equations (27) for ε_ν, we obtain, for the total average rate of irreversible dissipation of energy in a unit column of the fluid, the expression

$$\varepsilon_\nu + \varepsilon_\sigma = \frac{\nu}{4a^2}\int_0^d \left\{[(D^2 - a^2) W]^2 + \frac{\mu^2 H^2 \sigma}{\varrho \nu}(D W)^2\right\} dz. \tag{53}$$

The expression for the average rate of liberation of energy, ε_g, by the buoyancy force is unchanged. We have (cf. equation (31)):

$$\varepsilon_g = \frac{\varkappa \gamma}{4\beta}\int_0^d [(D\Theta)^2 + a^2 \Theta^2]\, dz. \tag{54}$$

(However, the further equation (35) relating Θ with W is not valid in the present problem.)

By demanding

$$\varepsilon_\sigma + \varepsilon_\nu = \varepsilon_g, \tag{55}$$

we obtain a relation which expresses β as the ratio of two positive-definite integrals; and it can be verified that it is indeed this expression which is minimized in the variational treatment of the problem [6].

IV

THE EFFECT OF ROTATION

We shall now consider a case in which $\zeta \neq 0$ and the motion in the (x, y)-plane is not irrotational. Such a case arises when the fluid is subject to a rotation and we have to allow for the resulting Coriolis acceleration in the equation of motion.

To illustrate the problem in its simplest context, we shall suppose that the axis of rotation coincides with the vertical. Equation (2) is then replaced by

$$\frac{\partial u_i}{\partial t} = -\frac{\partial \varpi}{\partial x_i} + \gamma \theta \lambda_i + \nu \nabla^2 u_i + 2\Omega \varepsilon_{ijk} u_j \lambda_k, \tag{56}$$

where Ω denotes the angular velocity of rotation about $\vec{\lambda}$; further, in place of equations (4), (7) and (9) we now have

$$\frac{\partial \zeta}{\partial t} = \nu \nabla^2 \zeta + 2\Omega \frac{\partial w}{\partial z}, \tag{57}$$

and

$$\frac{\partial}{\partial t}(\nabla^2 w) = \gamma \left(\frac{\partial^2 \theta}{\partial x^2} + \frac{\partial^2 \theta}{\partial y^2}\right) + \nu \nabla^4 w - 2\Omega \frac{\partial \zeta}{\partial z}. \tag{58}$$

In a stationary state, these equations become

$$\nu \nabla^2 \zeta = -2\Omega \frac{\partial w}{\partial z} \tag{59}$$

and

$$\nu \nabla^4 w - 2\Omega \frac{\partial \zeta}{\partial z} = -\gamma \left(\frac{\partial^2 \theta}{\partial x^2} + \frac{\partial^2 \theta}{\partial y^2}\right) = \gamma a^2 \theta. \tag{60}$$

From equation (59) it is apparent that in this problem $\zeta \neq 0$. We must, therefore, use the general solution (15) for the horizontal components of the velocity.

Consider again the average rate of dissipation of energy, ε_ν, by viscosity in a unit column of the fluid. The contribution to ε_ν by the horizontal component, u, of the velocity is

$$-\nu \int_0^d \langle u \nabla^2 u \rangle \, dz = -\frac{\nu}{a^4} \int_0^d \left\langle \left(\frac{\partial^2 w}{\partial x \partial z} + \frac{\partial \zeta}{\partial y}\right) \nabla^2 \left(\frac{\partial^2 w}{\partial x \partial z} + \frac{\partial \zeta}{\partial y}\right) \right\rangle dz. \tag{61}$$

It is clear from equation (59) (and indeed on general grounds) that the phases of the waves associated with w and ζ must agree. Therefore, $\partial^2 w/\partial x \partial z$ and $\partial \zeta/\partial y$ are out of phase and the average of their product must vanish. The cross terms in (61) do not, therefore, contribute and we are left with

$$\begin{aligned} -\nu \int_0^d \langle u \nabla^2 u \rangle \, dz &= -\frac{\nu}{a^4} \int_0^d \left\langle \frac{\partial^2 w}{\partial x \partial z} \nabla^2 \frac{\partial^2 w}{\partial x \partial z} + \frac{\partial \zeta}{\partial y} \nabla^2 \frac{\partial \zeta}{\partial y} \right\rangle dz \\ &= -\frac{\nu}{a^4} \int_0^d \left\{ a_x^2 D W (D^2 - a^2) D W + a_y^2 Z (D^2 - a^2) Z \right\} dz, \end{aligned} \tag{62}$$

where we have supposed (in agreement with equations (20) and (59)) that

$$\zeta = Z(z) \sin a_x x \sin a_y y. \tag{63}$$

We have a similar contribution from the terms in v. Thus, all together, we have

$$\varepsilon_\nu = -\frac{\nu}{4a^2} \int_0^d \left\{ a^2 W (D^2 - a^2) W + D W (D^2 - a^2) D W + Z (D^2 - a^2) Z \right\} dz. \tag{64}$$

The terms in W clearly combine to give the same expression as that found in II. The terms in Z can also be transformed into a positive-definite form by an integration by parts.[1] Thus, we finally obtain

$$\varepsilon_\nu = \frac{\nu}{4a^2} \int_0^d \left\{ [(D^2 - a^2) W]^2 + (DZ)^2 + a^2 Z^2 \right\} dz. \tag{65}$$

[1] The integrated part vanishes on account of the boundary conditions on Z. These are: $Z = 0$ on a rigid surface and $DZ = 0$ on a free surface.

The expression for ε_g is unchanged:

$$\varepsilon_g = \frac{\varkappa \gamma}{4 \beta} \int_0^d [(D \Theta)^2 + a^2 \Theta^2] \, dz. \tag{66}$$

By equating these expressions for ε_ν and ε_g we again obtain an expression for β as the ratio of two positive-definite integrals which is identical with the quantity which is minimized in the variational treatment of this problem [7].

It should be remarked that the expression for ε_ν given by equation (65) is entirely general[1]; it is in no way restricted to the particular problem under discussion: for, we have not used in its derivation equation (59) relating w and ζ which distinguishes this problem. (However, in the variational treatment this relation must be used.)

V

THE CASE WHEN OVERSTABILITY OCCURS

So far we have considered the marginal state as a stationary one. But it is known that instability can set in also via a state of oscillatory motions. Indeed, in the case of a layer of liquid heated from below and subject to rotation, instability can arise either, as cellular convection, or as overstable oscillations, depending on the value of the Prandtl number, $\nu/\varkappa$; the theoretical predictions [8] relative to this problem have been fully confirmed by experiments [9]. Examples of overstability also occur when rotation and magnetic field are simultaneously present [10], [11], [12]. The question, therefore, arises: How should one modify, under these circumstances, the principle which equates the rate of irreversible dissipation of energy with the rate of liberation of thermodynamically available energy by the buoyancy force, as a criterion for stability. It would appear natural that we generalize the principle along the following lines:

If the motion is periodic, then the kinetic energy of the fluid, as well as the rate of release of thermodynamic energy by the buoyancy force, will be subject to similar variations. Suppose that all quantities which describe the perturbation (such as w, θ, etc.,) vary with a circular frequency p so that all the amplitudes have a time dependent factor e^{ipt}. Then

$$u_i \frac{\partial u_i}{\partial t} = i p \, u_i^2. \tag{67}$$

We must allow for this change in the kinetic energy (per unit mass) in writing an equation of energy balance. If there are no sources of dissipation besides viscosity, then it would appear that we must equate

$$\varepsilon_\nu + i p \int_0^d \langle u_i^2 \rangle \, dz \tag{68}$$

[1] The assumption that the waves associated with $\partial^2 w / \partial x \, \partial z$ and $\partial \zeta / \partial y$ are out of phase is, however, essential: There are situations in which this need not be the case, e.g. when the directions of the axis of rotation and of $\mathfrak{g}$ do not coincide.

to the rate of release of energy

$$\varepsilon_g = \gamma \int_0^d \langle \theta\, w \rangle\, dz, \tag{69}$$

and allow for the fact that θ and w are no longer stationary.

On first sight, the appearance of the imaginary i, and complex numbers generally, in the equation for the energy balance may strike one as very odd. Its origin must be traced to the fact that the oscillations in the velocity and in the acceleration are out of phase. Consequently, the excess (or defect) of energy dissipated during one phase of the cycle must be exactly compensated by a similar excess (or defect) of energy liberated in a synchronous manner. It is this need for synchronism which determines the period of oscillation, as well as the RAYLEIGH number, as characteristics of the marginal state.[1]

We shall now verify that the thermodynamic principle as reformulated above is in agreement with the variational treatment of the problem considered in IV, for the case of overstability.

When the state is not stationary, the vertical component of the vorticity is necessarily non-vanishing and we must allow for this in the solution for the horizontal components of the velocity. The expression for ε_ν is, therefore, formally the same as that found in IV.

We must now evaluate the additional term in $\langle u_i^2 \rangle$ in (68). Considering the contribution to the integral by the x-component of the velocity, we have

$$\int_0^d \langle u^2 \rangle\, dz = \frac{1}{a^4} \int_0^d \left\langle \left(\frac{\partial^2 w}{\partial x\, \partial z} + \frac{\partial \zeta}{\partial y} \right)^2 \right\rangle dz. \tag{70}$$

Remembering that in the case under consideration the waves in the horizontal plane associated with $\partial^2 w / \partial x\, \partial z$ and $\partial \zeta / \partial y$ are out of phase and that, therefore, the average of their product vanishes, we have

$$\int_0^d \langle u^2 \rangle\, dz = \frac{1}{4 a^4} \int_0^d [a_x^2 (D\,W)^2 + a_y^2 Z^2]\, dz. \tag{71}$$

We have a similar contribution from $\langle v^2 \rangle$. Thus, all together, we have

$$i p \int_0^d \langle u_i^2 \rangle\, dz = \frac{i p}{4 a^2} \int_0^d [a^2 W^2 + (D\,W)^2 + Z^2]\, dz. \tag{72}$$

This must be added to the expression (65) for ε_ν.

Turning to the evaluation of ε_g, we must now use

$$\beta w = -\varkappa \nabla^2 \theta + i p\, \theta, \tag{73}$$

[1] The mathematical counterpart of this is that the determination of R and p is through the solution of a *double* characteristic value problem [8].

instead of (28) which we have used up to the present, to eliminate w. Accordingly we now have

$$\varepsilon_g = -\frac{\varkappa\gamma}{\beta}\int_0^d \left\langle \theta\left\{\left(\frac{\partial^2}{\partial z^2} - a^2\right)\theta - \frac{ip}{\varkappa}\,\theta\right\}\right\rangle dz = \frac{\varkappa\gamma}{4\beta}\int_0^d \left[(D\Theta)^2 + a^2\Theta^2 + \frac{ip}{\varkappa}\Theta^2\right] dz. \tag{74}$$

Hence, the equation which, under these circumstances, will determine a marginal state which is oscillatory is:

$$\frac{\nu}{a^2}\int_0^d \left\{[(D^2 - a^2)W]^2 + (DZ)^2 + a^2Z^2 + \frac{ip}{\nu}\,[a^2W^2 + (DW)^2 + Z^2]\right\} dz$$
$$= \frac{\varkappa\gamma}{\beta}\int_0^d \left\{(D\Theta)^2 + a^2\Theta^2 + \frac{ip}{\varkappa}\,\Theta^2\right\} dz; \tag{75}$$

and this is indeed the expression for β which is used in the variational treatment of the problem [8].

VI

CONCLUDING REMARKS

The analysis of the foregoing sections allows us to formulate the following general principle:

Thermal instability as cellular convection will set in at the minimum (adverse) temperature gradient which is necessary to maintain a balance between the rate of dissipation of energy by all irreversible processes present and the rate of liberation of the thermodynamically available energy by the buoyancy force acting on the fluid. Likewise, the onset of thermal instability will be as overstable oscillations if it is possible (at a lower adverse temperature gradient) to balance in a synchronous manner the periodically varying amounts of kinetic and other forms of energy with similarly varying rates of dissipation and liberation of energy.

REFERENCES

[1] Rayleigh, Phil. Mag. [6], *32*, 529, 1916.
[2] Chandrasekhar, S., Daedalus *86*, 323, 1957.
[3] Malkus, W.V.R., Proc. Roy. Soc. A, *225*, 185, 196, 1954.
[4] Jeffreys, H., Quart. Jour. of Mech. and Applied Math. *9*, 1, 1956.
[5] Pellew, A., and R. Southwell, Proc. Roy. Soc. A, *152*, 586, 1940.
[6] Chandrasekhar, S., Phil. Mag. [7], *43*, 501, 1952.
[7] Chandrasekhar, S., Proc. Roy. Soc. A, *217*, 306, 1953.
[8] Chandrasekhar, S., and D.D. Elbert, Proc. Roy. Soc. A, *231*, 198, 1955.
[9] Fultz, D., and Y. Nakagawa, Proc. Roy. Soc. A, *23*, 211, 1955.
[10] Chandrasekhar, S., Proc. Roy. Soc. A, *237*, 476, 1956.
[11] Nakagawa, Y., Proc. Roy. Soc. A, *242*, 81, 1957.
[12] Nakagawa, Y., Proc. Roy. Soc. A, (in press) 1958.

10

VARIATIONAL METHODS IN HYDRODYNAMICS

BY

S. CHANDRASEKHAR

1. Introduction. Studies in hydrodynamic and hydromagnetic stability have led to characteristic-value problems in differential equations of high order, and it has been possible to solve several of them by variational methods which, at least in the manner of their applications, appear novel. In this paper we shall briefly describe two such examples; a list of further examples will be found in Chandrasekhar [1].

2. First example. In the study of the instability of a layer of fluid heated from below and subject to Coriolis forces resulting from rotation with an angular velocity Ω about the vertical, one is led to the following problem: To solve

$$(D^2 - a^2 - i\sigma)Z = -\frac{2\Omega}{\nu} d\, DW, \tag{1}$$

$$(D^2 - a^2)(D^2 - a^2 - i\sigma)W - \frac{2\Omega}{\nu} d^3\, DZ = F, \tag{2}$$

and

$$(D^2 - a^2 - i\tilde{\omega}\sigma)F = -Ra^2W, \tag{3}$$

together with the boundary conditions

$$W = F = 0 \text{ for } z = \pm\tfrac{1}{2}, \tag{4}$$

and

$$\begin{aligned} &\textit{Either}\ DW = Z = 0 \text{ on } z = \pm\tfrac{1}{2},\\ &\textit{Or}\quad D^2W = DZ = 0 \text{ on } z = \pm\tfrac{1}{2},\\ &\textit{Or}\quad DW = Z = 0 \text{ on } z = +\tfrac{1}{2} \quad \text{and} \quad D^2W = DZ = 0 \text{ on } z = -\tfrac{1}{2}, \end{aligned} \tag{5}$$

where $D = d/dz$, a, Ω, ν, and $\tilde{\omega}$ are assigned constants, and σ is a parameter to be determined by the condition that the characteristic value R is real. The solution of the physical problem requires the minimum (with respect to a^2) of these real characteristic values of R for various assigned values of T $(= 4\Omega^2 d^4/\nu^2)$ and $\tilde{\omega}$.

A consideration of the foregoing *double* characteristic-value problem (double since both σ and R are to be determined) leads to the following variational principle:

From equations (1) to (3) it follows that

$$R = \frac{\int_{-\frac{1}{2}}^{+\frac{1}{2}} [(DF)^2 + (a^2 + i\tilde{\omega}\sigma)F^2]\, dz}{a^2 \int_{-\frac{1}{2}}^{+\frac{1}{2}} \{[(D^2 - a^2)W]^2 + d^2[(DZ)^2 + a^2Z^2] + i\sigma[(DW)^2 + a^2W^2 + d^2Z^2]\}\, dz}. \tag{6}$$

It can now be readily verified that the variation δR in R given by equation (6) due to variations δW and δZ in W and Z compatible only with the boundary conditions on W, Z, and F, is given by

$$\text{(7)} \qquad \delta R = -\frac{2}{a^2 I_2} \int_{-\frac{1}{2}}^{+\frac{1}{2}} \delta F[(D^2 - a^2 - i\tilde{\omega}\sigma)F + Ra^2 W]\, dz,$$

where I_2 stands for the integral in the denominator of the expression on the right-hand side of (6). Accordingly, $\delta R \equiv 0$ for all small arbitrary variations δF, provided that

$$\text{(8)} \qquad (D^2 - a^2 - i\tilde{\omega}\sigma)F + Ra^2 W = 0,$$

i.e., if the differential equation governing W is satisfied. On this account, formula (6) provides the basis for a variational procedure for solving equations (1) to (3) (for any assigned a^2 and σ) and satisfying the boundary conditions of the problem.

It should be noted that formula (6) does not express R as the ratio of two positive-definite integrals; indeed, for an arbitrarily assigned σ, R will be complex. Nevertheless, it appears that the simplest trial function for F, namely, $F = \cos \pi z$, already leads to surprisingly accurate determinations for the characteristic values provided that, for the chosen form of F, the functions W and Z are determined as solutions of equations (1) and (2). (For the details of the solution and for the comparison with experimental results, see Chandrasekhar and Elbert [2] and Fultz and Nakagawa [3].)

3. Second example. The stability of an incompressible, heavy, viscous fluid of variable density leads to the following characteristic-value problem (Chandrasekhar [4]): To solve

$$\text{(9)} \quad D\left\{\left[\rho - \frac{\mu}{n}(D^2 - k^2)\right] Dw - \frac{1}{n}(D\mu)(D^2 + k^2)w\right\}$$
$$= k^2 \left\{-\frac{g}{n^2}(D\rho)w + \left[\rho - \frac{\mu}{n}(D^2 - k^2)\right] w - \frac{2}{n}(D\mu)(Dw)\right\},$$

together with the boundary conditions

$$\text{(10)} \qquad w = 0 \text{ for } z = 0 \text{ and } l,$$

and

$$\text{(11)} \quad \begin{array}{llll} \textit{Either } Dw = 0 \text{ for } z = 0 & \text{and} & l, & \\ \textit{Or } \quad D^2w = 0 \text{ for } z = 0 & \text{and} & l, & \\ \textit{Or } \quad Dw = 0 \text{ for } z = 0 & \text{and} & D^2w = 0 \text{ for } z = l, \\ \textit{Or } \quad D^2w = 0 \text{ for } z = 0 & \text{and} & Dw = 0 \text{ for } z = l, \end{array}$$

where $\rho = \rho(z)$ and $\mu = \mu(z)$ are given functions of z, k is an assigned (real) constant, and n is the characteristic-value parameter. (Note that n can be complex.)

One can deduce from equation (9) that

$$(12)\quad n\int_0^l \rho\left\{w^2+\frac{1}{k^2}(Dw)^2\right\}dz-\frac{g}{n}\int_0^l (D\rho)w^2\,dz$$
$$=-\int_0^l\left\{\mu\left[k^2w^2+2(Dw)^2+\frac{1}{k^2}(D^2w)^2\right]+(D^2\mu)w^2\right\}dz;$$

and again this last equation provides the basis for a convenient variational procedure for determining n. For, considering the effect on n [determined in accordance with equation (12)] of an arbitrary variation δw in w compatible only with the boundary conditions on w, we find that

$$(13)\quad -\frac{1}{2}k^2\left(I_1+\frac{g}{n^2}I_2\right)\frac{\delta n}{n}$$
$$=\int_0^l \delta w\left\{k^2\left[\rho w-\frac{\mu}{n}(D^2-k^2)w-\frac{g}{n^2}(D\rho)w-\frac{2}{n}(D\mu)(Dw)\right]\right.$$
$$\left.-D\left[\rho\,Dw-\frac{\mu}{n}(D^2-k^2)\,Dw-\frac{1}{n}(D\mu)(D^2+k^2)w\right]\right\}dz,$$

where

$$(14)\qquad I_1=\int_0^l \rho\left\{w^2+\frac{1}{k^2}(Dw)^2\right\}dz$$

and

$$(15)\qquad I_2=\int_0^l (D\rho)w^2\,dz.$$

It will be noticed that the variational procedure in this instance involves the solution of a quadratic equation none of whose coefficients are positive-definite. Nevertheless, as Hide [5] has shown, simple trial functions for w satisfying the boundary conditions enable the complicated dependence of n on the various parameters of the problem to be determined.

Bibliography

1. S. Chandrasekhar, *On characteristic value problems in high order differential equations which arise in studies on hydrodynamic and hydromagnetic stability*, Amer. Math. Monthly vol. 61 (1954) pp. 32–45.

2. ——— and D. Elbert, *The instability of a layer of fluid heated below and subject to Coriolis forces, II*, Proc. Roy. Soc. London Ser. A vol. 231 (1955) pp. 198–210.

3. D. Fultz and Y. Nakagawa, *Experiments on over-stable thermal convection in mercury*, Proc. Roy. Soc. London Ser. A vol. 231 (1955) pp. 211–225.

4. S. Chandrasekhar, *The character of the equilibrium of an incompressible heavy viscous fluid of variable density*, Proc. Cambridge Philos. Soc. vol. 51 (1955) pp. 162–178.

5. R. Hide, *The character of the equilibrium of an incompressible heavy viscous fluid of variable density: an approximate theory*, Proc. Cambridge Philos. Soc. vol. 51 (1955) pp. 179–201.

University of Chicago,
Chicago, Ill.

11

ON CHARACTERISTIC VALUE PROBLEMS IN HIGH ORDER DIFFERENTIAL EQUATIONS WHICH ARISE IN STUDIES ON HYDRODYNAMIC AND HYDROMAGNETIC STABILITY

S. CHANDRASEKHAR, University of Chicago

1. Introduction. Recent studies in hydrodynamic and hydromagnetic stability have disclosed the existence of a class of characteristic value problems in differential equations of high order—orders as high as twenty-four have been encountered—which appear to have a genuine mathematical interest. A partial list of these problems will be found in the Appendix. These problems have been solved, singly, as they have arisen in the physical connections. But one feels that there must be a general theory which embraces them all though such a theory is lacking at the present time. It is the object of this paper to bring these problems and the methods which have been developed for their solution to the attention of the mathematicians in the hope of stimulating their interest.

In the solution of the problems listed in the Appendix, methods of two different kinds have been found useful. In the first of these methods a variational procedure is developed which in the manner of its application is different from the usual ones. The second of these methods is based on expansion in orthogonal functions but again applied in an unusual manner. We shall illustrate the principles underlying these methods by considering one of the simpler problems listed in the Appendix.

2. A typical problem. G. I. Taylor's [1] investigation in 1923 on the stability of viscous flow between two concentric rotating cylinders provided the first example of a case of hydrodynamic instability for which the criterion was theoretically predicted and experimentally verified. The mathematical problem underlying this classic investigation in hydrodynamic stability is the following:

With certain simplifications suitable to the circumstances under which the experiments were performed, the problem requires the solution of the sixth order equation

$$(D^2 - a^2)^3 v = - a^2 T(1 + \alpha z)v, \tag{1}$$

with the boundary conditions

$$v = (D^2 - a^2)v = D(D^2 - a^2)v = 0 \tag{2}$$

for $z=0$ and 1, where T is the characteristic value parameter, $D=d/dz$ and a and α are assigned (real) constants. In the physical problem α is negative and one is interested in the range $0 \geqq \alpha \geqq -3.0$.

Equation (2) provides six boundary conditions (three at $z=0$ and three at $z=1$) and the requirement that a non-trivial solution of equation (1) satisfy these conditions will lead to a determinate sequence of possible values for T (for given a^2 and α). Among these possible values of T (for given α and varying a^2) there will be a smallest (positive) value; and in the physical problem particu-

Reprinted from *The American Mathematical Monthly*, (1954), vol. 61, pages 32–45, by permission of the Mathematical Association of America.

lar interest is attached to the minimum of these smallest positive values of T as a function of a^2 (for fixed α).

In discussing the characteristic value problem presented by equations (1) and (2), we shall distinguish the cases $\alpha=0$ and $\alpha\neq 0$. The problem has basically different characters in the two cases: in the former case ($\alpha=0$) it is "self-adjoint" in some sense (yet to be defined!) while in the latter case it is not. Because of this difference the solution of the problem, while it can be made to depend on an extremal principle in the case $\alpha=0$, it cannot be so accomplished in the case $\alpha\neq 0$. As far as the general solution of equations (1) and (2) is concerned, a separate detailed discussion of the case $\alpha=0$ would hardly seem justified; but it does happen that the case $\alpha=0$ is the simplest proto-type of a wide class of problems to which similar methods of solution can be applied. For this reason we shall treat the case $\alpha=0$ at some length.

3. The case $\alpha=0$: example of the variational procedure. When $\alpha=0$, the solution of the characteristic value problem can, in principle, be achieved very simply. For, in this case, the solution of the equation

$$(D^2 - a^2)^3 v = - a^2 T v \tag{3}$$

must clearly be of the form

$$v_i = \sum_{i=1}^{6} A_i e^{q_i z}, \tag{4}$$

where the q_i's, occurring in pairs, are the roots of the characteristic equation

$$(q^2 - a^2)^3 = - a^2 T, \tag{5}$$

and the A_i's are constants of integration. The requirement that the solution represented by equation (4) satisfy the boundary conditions (2) will lead to a system of six linear homogeneous equations; and the determinant of this system must vanish if we are not to have the trivial solution $A_i=0$, $i=1, \cdots, 6$. And the condition that the determinant vanish will provide an equation for determining T. There is, of course, no difficulty of principle in carrying out this procedure. Indeed, it has been carried out by Pellew and Southwell [2] who find for example that

$$T = 1707.8 \qquad \text{for } a = 3.12. \tag{6}$$

(This is the value of a at which T, as a function of a, attains its minimum.)

For the case on hand the carrying out of the direct method of solution described in the preceding paragraph is particularly simple as the roots of equation (5) can be explicitly written down in terms of the known cube roots of -1; thus

$$q^2 = a^2 + \omega \sqrt[3]{(a^2 T)}, \qquad (\omega^3 = -1). \tag{7}$$

But this simplification does not arise in other problems which have to be considered; also, the characteristic equations which have to be solved are of higher order. It would therefore be useful if we can devise a method of solution which will avoid the necessity of solving complicated algebraic equations of high order by laborious methods of trial and error. This would be specially important if, as it often happens, we are required to establish the dependence of the characteristic values on two or more parameters rather than on a single parameter, such as a, in the present problem ($\alpha=0$). And it does appear that in many of the cases which arise, convenient variational methods can be devised which we shall now illustrate by considering the problem presented by equations (2) and (3).

Letting

$$G = (D^2 - a^2)v \quad \text{and} \quad F = (D^2 - a^2)G = (D^2 - a^2)^2 v, \tag{8}$$

we can rewrite equation (3) in the form

$$(D^2 - a^2)F = - a^2 T v, \tag{9}$$

while the boundary conditions (2) require that

$$v = G = DG = 0 \qquad \text{for } z = 0 \text{ and } 1. \tag{10}$$

Let T_i denote a particular characteristic value and let the various functions derived from the solution v_i belonging to T_i be distinguished by a subscript i. Multiplying the equation governing v_i by G_j belonging to a different characteristic value T_j and integrating the resulting equation over the range of z, we obtain

$$\int_0^1 G_j(D^2 - a^2)F_i dz = - a^2 T_i \int_0^1 v_i(D^2 - a^2)v_j dz. \tag{11}$$

The integral occurring on the left-hand side of this equation can be reduced by two successive integration by parts; thus,

$$\int_0^1 G_j(D^2 - a^2)F_i dz = \left[G_j DF_i - (DG_j)F_i\right]_0^1 + \int_0^1 F_i(D^2 - a^2)G_j dz. \tag{12}$$

The integrated parts vanish in virtue of the boundary conditions (*cf.* equations (10)) and we are left with

$$\int_0^1 G_j(D^2 - a^2)F_i dz = \int_0^1 F_i F_j dz. \tag{13}$$

Similarly, after an integration by parts the right-hand side of equation (11) becomes

$$a^2 T_i \int_0^1 [(Dv_i)(Dv_j) + a^2 v_i v_j] dz. \tag{14}$$

Thus,

$$\int_0^1 F_iF_jdz = a^2T_i\int_0^1 [(Dv_i)(Dv_j) + a^2v_iv_j]dz. \tag{15}$$

Noticing that interchanging i and j in this equation replaces T_i by T_j but leaves the equation otherwise unaffected, we conclude that

$$\int_0^1 F_iF_jdz = 0 \qquad i \neq j; \tag{16}$$

and, further, that when $i=j$ the corresponding characteristic value, T, can be expressed as the ratio of two positive definite integrals in the form:

$$T = \frac{\displaystyle\int_0^1 F^2dz}{a^2\displaystyle\int_0^1 [(Dv)^2 + a^2v^2]dz}. \tag{17}$$

These facts clearly imply the existence of an extremal principle which can be made the basis of a variational method of solving the underlying characteristic value problem. For comparison with the method we shall describe in §4 for the case $\alpha\neq 0$, we shall formulate the extremal principle in the context of a slightly transformed equation.

Operating on equation (3) by (D^2-a^2) we get

$$(D^2 - a^2)^3W = -a^2TW \tag{18}$$

as the equation governing

$$W = (D^2 - a^2)v. \tag{19}$$

According to equations (2) and (3) the corresponding boundary conditions on W are:

$$W = DW = 0 \quad F \equiv (D^2 - a^2)G \equiv (D^2 - a^2)^2W = 0 \qquad \text{for } z = 0 \text{ and } 1. \tag{20}$$

Again rewriting equation (18) in the form

$$(D^2 - a^2)F = -a^2TW, \tag{21}$$

and multiplying the equation governing W_i (belonging to T_i) by F_j (belonging to T_j) we obtain

$$\int_0^1 F_j(D^2 - a^2)F_idz = -a^2T_i\int_0^1 W_i(D^2 - a^2)G_jdz. \tag{22}$$

Making use of the boundary conditions (20) we can reduce equation (22) by one or more integrations by parts to the form

$$\int_0^1 [(DF_i)(DF_j) + a^2F_iF_j]dz = a^2T_i \int_0^1 G_iG_jdz. \tag{23}$$

From this equation it follows that

$$\int_0^1 G_iG_jdz = 0, \qquad i \neq j; \tag{24}$$

and that when $i=j$ we can express T as:

$$T = \frac{\int_0^1 [(DF)^2 + a^2F^2]dz}{a^2 \int_0^1 G^2dz}. \tag{25}$$

This last formula, like (17), expresses T as the ratio of two positive definite integrals.

Consider now the effect on T (evaluated according to equation (25)) of an arbitrary variation, δW, in W compatible only with the boundary conditions on W. We find in a straightforward manner that

$$\delta T = -\frac{2}{a^2 \int_0^1 G^2dz} \int_0^1 \delta F\{(D^2 - a^2)F + a^2TW\}dz, \tag{26}$$

where it should be noted that in the reductions leading to equation (26) the relations implied in the definitions of F and G (*cf.* equation (20)) have been used; in particular

$$\delta F = (D^2 - a^2)^2\delta W. \tag{27}$$

Hence, to the first order, $\delta T \equiv 0$ for all small arbitrary variations in W which satisfy the boundary conditions, provided

$$(D^2 - a^2)F = -a^2TW, \tag{28}$$

i.e., if the differential equation governing W is satisfied. It is evident that the converse of this proposition is also true. Further, it follows from (25) that the true solution of the problem (belonging to the lowest characteristic value T) leads to the minimum value (in the sense of the calculus of variations) for T when evaluated according to (25).

While the minimal principle formulated as above is valid for all small arbitrary variations δW compatible only with the boundary conditions on W, it is true also for all small arbitrary variations δF, the variation, δW, in W being determined in terms of δF by the equation

$$(D^2 - a^2)^2\delta W = \delta F, \tag{29}$$

and the boundary conditions

(30) $$\delta W = D\delta W = 0 \qquad \text{for } z = 0 \text{ and } 1.$$

The validity of this minimal principle in this more restricted form actually provides a more effective basis for a variational procedure for determining T. For on this latter basis the procedure would be the following:

Assume for F an expression involving one or more parameters, A_k, and which vanishes at $z=0$ and 1. With the chosen form of F solve the equation

(31) $$(D^2 - a^2)^2 W = F$$

for W and arrange that the solution satisfies the boundary conditions $W=DW=0$ for $z=0$ and 1; since equation (31) is of the fourth order there will be just enough constants of integration to do this. With W determined in this fashion evaluate T according to (25) and minimize it with respect to the parameters A_k. In this way we shall obtain the "best" value of T for the chosen form of F.

In practice it is found that even with the simplest trial functions, the variational procedure in the foregoing form gives surprisingly high accuracy in the deduced values of T. Thus, with the trial function, $F=\sin \pi z$, with no variational parameter, the method gives $T=1715.1$ for $a=3.12$; while the function, $F=\sin \pi z + A \sin 3\pi z$, with one variational parameter leads to $T=1707.9$ for the same value of a; these values obtained in the "first" and the "second" approximations should be compared with the value $T=1707.8$ obtained from an "exact" solution of the problem. The origin of this high precision in the deduced values of T must clearly be traced to the fact that in satisfying four of the six boundary conditions of the problem (namely $W=DW=0$ for $z=0$ and 1) we have *exactly* satisfied the second of the pair of differential equations,

(32) $$(D^2 - a^2)F = -a^2 TW \quad \text{and} \quad (D^2 - a^2)^2 W = F,$$

which governs the problem.

[It may be noticed here that had we used equation (17) (instead of (25)) as the basis of the variational method, we should have had to assume a form for DG such that not only does it vanish at $z=0$ and 1, but also G, obtained after integration, vanishes at the same points; further, W will have to be obtained as the solution of $(D^2-a^2)W=G$ which vanishes at $z=0$ and 1. Apart from these differences in detail the application of the variational method based on (17) proceeds along essentially the same lines. But the method based on (25) is preferable since it avoids the restrictions implied by the requirement that both G and DG vanish at $z=0$ and 1.]

4. The case $\alpha \neq 0$: example of a method based on expansion in orthogonal functions when no extremal principle exists. Returning to equation (1) and the general case $\alpha \neq 0$, we can readily verify that propositions similar to those embodied in equations (15), (16) and (17), for the case $\alpha=0$, cannot be established now. Thus, if equation (1) and the boundary conditions (2) together represent a system which might be described as "self-adjoint" when $\alpha=0$, it cannot be so described when $\alpha\neq 0$. It is important to note that this conclusion in no way depends on the sign of α; in particular it is independent of the circumstance that when $\alpha<-1$ the operator whose characteristic values we are seeking becomes singular on account of $(1+\alpha z)$ having a zero in the range $0\leqq z\leqq 1$.

Since equations (1) and (2), in general, do not allow the formulation of an extremal principle on which a variational procedure might be based and since the general solution of equation (1) cannot also be written down in any convenient form, it would appear that the only remaining course is to expand the unknown functions in Fourier series. In Taylor's original investigation this was the method which was adopted: Taylor expanded v in a sine series of the form

$$v = \sum_{n=1}^{\infty} V_n \sin n\pi z \tag{33}$$

and obtained T as the characteristic root of an infinite matrix. But the process of determining the root was not a very convergent one; and he succeeded (with considerable effort) in determining the characteristic root for only one value of $\alpha<-1$. However, we may expect that a more rapidly convergent process will be obtained if in using the method of expansion in Fourier series, we incorporate in the method the same basic idea which led to the high precision of the variational method for the case $\alpha=0$. We shall now indicate how this can be accomplished (see Chandrasekhar [3]).

First we transform equation (1) by rewriting it in the form

$$\frac{1}{1+\alpha z}(D^2 - a^2)^3 v = -a^2 T v, \tag{34}$$

and operating on it by (D^2-a^2); in this way, we obtain the differential equation

$$(D^2 - a^2)\left\{\frac{1}{1+\alpha z}(D^2 - a^2)^2 W\right\} = -a^2 T W, \tag{35}$$

for

$$W = (D^2 - a^2)v. \tag{36}$$

The corresponding boundary conditions on W are the same as those given in (20).

Equation (35) is equivalent to the pair of equations

$$(D^2 - a^2)^2 W = (1+\alpha z)\psi \tag{37}$$

and

$$(D^2 - a^2)\psi = -a^2 T W, \tag{38}$$

while the boundary conditions (20) can be expressed alternatively in the forms

(39) $$W = DW = 0 \quad \text{and} \quad \psi = 0 \qquad \text{for } z = 0 \text{ and } 1.$$

Since ψ has to vanish at $z=0$ and 1, we can expand it in a sine series of the form

(40) $$\psi = \sum_{n=1}^{\infty} C_n \sin n\pi z.$$

Having chosen ψ in this manner, we next *solve* the equation

(41) $$(D^2 - a^2)^2 W = (1 + \alpha z) \sum_{n=1}^{\infty} C_n \sin n\pi z,$$

obtained by inserting (40) in (37), and arrange that the solution satisfies the four remaining conditions on W; since equation (41) is of the fourth order there will be just enough constants of integration to do this. With W determined in this fashion and ψ given by (40), equation (38) will lead to an infinite determinant which must be zero if all the C_n's are not to vanish. In this way we shall obtain a characteristic equation for determining T.

When the details of the method described in the preceding paragraphs are carried out, one finds that the process of solving the infinite order characteristic equation for T, by setting the determinant formed by the first n rows and columns equal to zero and letting n take increasingly larger values, converges very rapidly indeed. Thus for $\alpha = -2.5$ and $a = 5.00$, 5.05 and 5.10 the values of T obtained in the third and the fourth approximations (the "order" of the approximation being the order of the determinant which is set equal to zero in the determination of T) are:

(42) $$a = \begin{cases} 5.00 \\ 5.05 \\ 5.10 \end{cases} \quad T \text{ (3rd app.)} = \begin{cases} 4.607 \times 10^4 \\ 4.600 \times 10^4 \\ 4.604 \times 10^4 \end{cases} \quad T \text{ (4th app.)} = \begin{cases} 4.626 \times 10^4 \\ 4.619 \times 10^4. \\ 4.623 \times 10^4 \end{cases}$$

It is seen that the values of T given in the third and the fourth approximations differ by only four parts in a thousand. The origin of this rapid convergence clearly lies in the splitting of the original equation of order six into a pair of order two and four respectively and satisfying the equation of order four exactly. This basic idea underlying the method is capable of extension and application to a wide class of problems.

Appendix

We shall list here some of the more important characteristic value problems in high order differential equations which have occurred in recent studies on hydrodynamic and hydromagnetic stability. The particular physical connections in which they arise are indicated. The references are to papers in which the solutions of the problems will be found.

I. *The inhibition of convection by a magnetic field* [**4, 5**].

1) To solve

$$(D^2 - a^2)[(D^2 - a^2)^2 - QD^2]W = - a^2RW,$$

together with the boundary conditions

$$W = [(D^2 - a^2)^2 - QD^2]W = 0 \quad \text{on} \quad z = \pm \tfrac{1}{2},$$

and

either $DW = 0$ on $z = \pm \tfrac{1}{2}$,

or $D^2W = 0$ on $z = \pm \tfrac{1}{2}$,

or $DW = 0$ on $z = + \tfrac{1}{2}$ and $D^2W = 0$ on $z = - \tfrac{1}{2}$,

where a and Q are assigned positive constants and R is the characteristic value parameter.

The physical problem requires the minimum (with respect to a) of the lowest characteristic value R for various assigned values of Q.

2) To solve

$$(D^2 - a^2)[(D^2 - a^2)^2 - Q(D + i\bar{\omega})^2]W = - a^2RW,$$

together with the boundary conditions

$$|W| = |(D^2 - a^2)^2W - Q(D + i\bar{\omega})^2W| = 0 \quad \text{on} \quad z = \pm \tfrac{1}{2},$$

and

either $|DW| = 0$ on $z = \pm \tfrac{1}{2}$,

or $|D^2W| = 0$ on $z = \pm \tfrac{1}{2}$,

or $|DW| = 0$ on $z = + \tfrac{1}{2}$ and $|D^2W| = 0$ on $z = - \tfrac{1}{2}$,

where a, $\bar{\omega}$ and Q are assigned constants and R is the characteristic value parameter.

The physical problem requires the minimum (with respect to a and $\bar{\omega}$) of the lowest characteristic value R for various assigned values of Q. Since W is complex, we have here a genuine problem in an equation of order twelve.

II. *The instability of a layer of fluid heated below and subject to Coriolis acceleration* [**6**].

3) To solve

$$(D^2 - a^2)Z = - \left(\frac{2\Omega}{\nu} d\right) DW$$

and

$$(D^2 - a^2)\left[(D^2 - a^2)^2W - \left(\frac{2\Omega}{\nu} d^3\right) DZ\right]$$
$$= [(D^2 - a^2)^3 + TD^2]W = - a^2RW,$$

together with the boundary conditions

$$W = (D^2 - a^2)^2 W - \left(\frac{2\Omega}{\nu} d^3\right) DZ = 0 \quad \text{on} \quad z = \pm \tfrac{1}{2},$$

and

$$\begin{aligned} &\textit{either} \quad DW = Z = 0 \quad \text{on} \quad z = \pm \tfrac{1}{2}, \\ &\textit{or} \quad D^2W = DZ = 0 \quad \text{on} \quad z = \pm \tfrac{1}{2}, \\ &\textit{or} \quad DW = Z = 0 \quad \text{on} \quad z = +\tfrac{1}{2} \quad \text{and} \quad D^2W = DZ = 0 \quad \text{on} \quad z = -\tfrac{1}{2}, \end{aligned}$$

where a, ν, d and Ω are assigned constants, $T = 4\Omega^2 d^4/\nu^2$ and R is the characteristic value parameter.

The physical problem requires the minimum (with respect to a) of the lowest characteristic value R for various assigned values of T.

3a) To solve

$$(D^2 - a^2 - i\sigma)Z = -\left(\frac{2\Omega}{\nu} d\right) DW,$$

$$(D^2 - a^2)(D^2 - a^2 - i\sigma)W - \left(\frac{2\Omega}{\nu} d^3\right) DZ = F$$

and

$$(D^2 - a^2 - i\tilde{\omega}\sigma)F = -Ra^2 W,$$

together with the boundary conditions

$$|F| = |W| = 0 \quad \text{on} \quad z = \pm \tfrac{1}{2},$$

and

$$\begin{aligned} &\textit{either} \quad DW = Z = 0 \quad \text{on} \quad z = \pm \tfrac{1}{2}, \\ &\textit{or} \quad D^2W = DZ = 0 \quad \text{on} \quad z = \pm \tfrac{1}{2}, \\ &\textit{or} \quad DW = Z = 0 \quad \text{on} \quad z = +\tfrac{1}{2} \quad \text{and} \quad D^2W = DZ = 0 \quad \text{on} \quad z = -\tfrac{1}{2}, \end{aligned}$$

where $\tilde{\omega}$ is a further assigned constant and σ (which is real) is a parameter to be determined (for given $T = 4\Omega^2 d^4/\nu^2$ and a^2) by the condition that R is real; and the physical problem requires the minimum (with respect to a^2) of these real values of R for various assigned values of T and $\tilde{\omega}$.

III. *The instability of a layer of fluid heated below and subject simultaneously to a magnetic field and Coriolis acceleration* [7].

4) To solve

$$(D^2 - a^2)\left[\{(D^2 - a^2)^2 - QD^2\}^2 + TD^2(D^2 - a^2)\right]W = -Ra^2\left[(D^2 - a^2)^2 - QD^2\right]W$$

together with (several different) five pairs of boundary conditions on W at

$z = \pm\frac{1}{2}$, where a, Q and T are assigned constants and R is the characteristic value parameter.

IV. *The stability of viscous flow between rotating cylinders* [**1, 3, 8, 9**].

5) To solve

$$(D^2 - a^2)^3 v = - a^2 T(1 + \alpha z)v,$$

together with the boundary conditions

$$v = D^2 v = D(D^2 - a^2)v = 0 \quad \text{on} \quad z = 0 \quad \text{and} \quad 1,$$

where a and α (<0) are assigned constants and T is the characteristic value parameter. The physical problem requires the minimum (with respect to a) of the lowest (positive) characteristic value T for various assigned values of $\alpha(<0)$.

6) To solve

$$(DD^* - \lambda^2)^3 v = \frac{4A\lambda^2}{\nu^2}\left(A + \frac{B}{r^2}\right)v,$$

where $D = d/dr$ and $D^* = D + 1/r$, together with the boundary conditions

$$v = (DD^* - \lambda^2)v = D^*(DD^* - \lambda^2)v = 0 \qquad \text{for } r = R_1 \text{ and } R_2;$$

λ, ν and B are assigned constants and A is the characteristic value parameter.

V. *The stability of viscous flow between rotating cylinders in the presence of a magnetic field* [**10**].

7) To solve

$$[(D^2 - a^2)^2 + Qa^2]^2\psi = - Ta^2(D^2 - a^2)\psi,$$

together with the boundary conditions

$$D\psi = (D^2 - a^2)\psi = [(D^2 - a^2)^2 + Qa^2]\psi = D[(D^2 - a^2)^2 + Qa^2]\psi = 0 \text{ on } z = \pm\tfrac{1}{2},$$

where a and Q are assigned constants and T is the characteristic value parameter. The physical problem requires the minimum (with respect to a) of the lowest characteristic value T for various assigned values of Q.

VI. *The stability of viscous flow between rotating cylinders in the presence of a radial temperature gradient* [**11**].

8) To solve

$$\left(\frac{d^2}{dr^2} + \frac{1}{r}\frac{d}{dr} - \frac{n^2}{r^2}\right)^3 W = - n^2 S_n W,$$

together with the boundary conditions

$$W = \frac{dW}{dr} = \left(\frac{d^2}{dr^2} + \frac{1}{r}\frac{d}{dr} - \frac{n^2}{r^2}\right)^2 W = 0 \qquad \text{for } r = 1 \text{ and } \eta(<1),$$

where n is an integer and S_n is the characteristic value parameter.

9) To solve

$$\left(\frac{d^2}{dr^2}+\frac{1}{r}\frac{d}{dr}-\frac{n^2}{r}\right)^3 W = -n^2 S_n \frac{W}{r^2},$$

together with the same boundary conditions as in (8) above.

VII. *The thermal instability of fluid spheres and spherical shells* [**12, 13, 14**].

10) To solve

$$\left(\frac{d^2}{dr^2}+\frac{2}{r}\frac{d}{dr}-\frac{l(l+1)}{r^2}\right)^3 W = -l(l+1)C_l W,$$

together with the boundary conditions

(a) $$W = \left(\frac{d^2}{dr^2}+\frac{2}{r}\frac{d}{dr}-\frac{l(l+1)}{r^2}\right)^2 W = 0$$

and

$$\textit{either}\quad \frac{dW}{dr}\quad \textit{or}\quad \frac{d^2W}{dr^2} = 0 \qquad \text{for } r = 1,$$

and

$$W = O(r^l) \qquad \text{as } r \to 0;$$

(b) $$W = \frac{d^2W}{dr^2} = \left[\frac{d^2}{dr^2}+\frac{2}{r}\frac{d}{dr}-\frac{l(l+1)}{r^2}\right]^2 W = 0 \qquad \text{for } r = 1 \text{ and } \eta(<1),$$

where l is an integer and C_l is the characteristic value parameter.

VIII. *The stability of superposed fluids (plane problem)* [**15, 16, 17, 18, 19**].

11) To solve

$$D\left\{\left[\rho + \frac{\mu}{\sigma}(-k^2 + D^2)\right]Dw + \frac{1}{\sigma}(D\mu)(k^2 + D^2)w\right\}$$

$$= k^2\left\{\left[-\frac{g}{\sigma^2}D\rho + \rho + \frac{\mu}{\sigma}(-k^2 + D^2)\right]w + \frac{2}{\sigma}(D\mu)(Dw)\right\},$$

together with the boundary conditions

$$w = 0 \quad \text{for} \quad z = 0 \quad \text{and} \quad l$$

and

either $Dw = 0$ for $z = 0$ and l,

or $D^2w = 0$ for $z = 0$ and l,

or $Dw = 0$ for $z = 0$ and $D^2w = 0$ for $z = l$,

or $D^2w = 0$ for $z = 0$ and $Dw = 0$ for $z = l$,

where $\rho=\rho(z)$ and $\mu=\mu(z)$ are given functions of z, k is an assigned (real) constant and σ is the characteristic value parameter. (Note that σ can be complex).

IX. *The stability of superposed fluids (spherical problem)* [20].

12) To solve

$$\frac{d}{dr}\left\{\rho\frac{d}{dr}(rW)+\frac{\mu}{\sigma}\frac{d}{dr}(rF)\right\}+\frac{1}{\sigma}\frac{d}{dr}\left\{r\frac{d\mu}{dr}\left[\frac{d^2W}{dr^2}+\frac{(l+2)(l-1)}{r^2}W\right]\right\}$$
$$-\frac{l(l+1)}{r}\left(\rho W+\frac{\mu}{\sigma}F\right)=-\frac{l(l+1)}{\sigma^2}\left\{\gamma\frac{d\rho}{dr}W-2\sigma\frac{d\mu}{dr}\frac{d}{dr}\left(\frac{W}{r}\right)\right\},$$

where

$$F=\left[\frac{d^2}{dr^2}+\frac{2}{r}\frac{d}{dr}-\frac{l(l+1)}{r^2}\right]W,$$

together with the boundary conditions

$$W=0 \quad \text{and} \quad \textit{either}\ \frac{dW}{dr}\ \textit{or}\ \frac{d^2W}{dr^2}=0 \quad \text{on} \quad r=R,$$

and

$$W=O(r^l) \qquad \text{as } r\to 0.$$

In the foregoing ρ, μ and γ are given functions of r, l is an integer and σ (which may be complex) is the characteristic value parameter.

The case when

$$\mu=\mu_1=\text{constant}, \qquad \rho=\rho_1=\text{constant for } 0\leqq r<r_i$$

and

$$\mu=\mu_2=\text{constant}, \qquad \rho=\rho_2=\text{constant for } r_i<r\leqq R$$

is of particular importance. In this case the equation reduces to the form

$$\frac{d^2}{dr^2}\left\{r\left(W+\frac{\mu}{\rho\sigma}F\right)\right\}-\frac{l(l+1)}{r}\left(W+\frac{\mu}{\rho\sigma}F\right)=0$$

in each of the two regions of constant ρ and μ. And solutions are sought which satisfy the boundary conditions

$$W=0 \quad \text{and} \quad \frac{d^2W}{dr^2}=0 \qquad \text{for } r=R$$

$$W=O(r^l) \qquad \text{for } r\to 0$$

and

$$W, \quad \frac{dW}{dr} \quad \text{and} \quad \mu\left\{r^2\frac{d^2W}{dr^2} + (l+2)(l-1)W\right\}$$

are continuous on $r=r_i$.

References

1. G. I. Taylor, Stability of a viscous liquid contained between two rotating cylinders, Philos. Trans. Roy. Soc. London Ser., A, vol. 223, 1923, pp. 289–343.
2. A. Pellew and R. V. Southwell, On maintained convective motion in a fluid heated from below, Proc. Roy. Soc. London Ser. A, vol. 176, 1940, pp. 312–343.
3. S. Chandrasekhar, The stability of viscous flow between rotating cylinders, Mathematika (in press).
4. S. Chandrasekhar, On the inhibition of convection by a magnetic field, Philos. Mag. Ser. 7, vol. 43, 1952, pp. 501–532.
5. S. Chandrasekhar, On the inhibition of convection by a magnetic field. II, Philos. Mag. (in press).
6. S. Chandrasekhar, The instability of a layer of fluid heated below and subject to Coriolis forces, Proc. Roy. Soc. London Ser. A, vol. 217, 1953, pp. 306–327; also a further paper in press.
7. S. Chandrasekhar, The instability of a layer of fluid heated below and subject to the simultaneous action of a magnetic field and rotation, Proc. Roy. Soc. London Ser. A, (in press).
8. D. Meksyn, Stability of viscous flow between rotating cylinders. I, II, III, Proc. Roy. Soc. London Ser. A, vol. 187, 1946, pp. 115–128, 480–491, 492–504.
9. S. Chandrasekhar, (in press).
10. S. Chandrasekhar, The stability of viscous flow between rotating cylinders in the presence of a magnetic field, Proc. Roy. Soc. London Ser. A, vol. 216, 1953, pp. 293–309.
11. S. Chandrasekhar, The stability of viscous flow between rotating cylinders in the presence of a radial temperature gradient, Journal of Rational Mechanics and Analysis, vol. 3, 1954, pp. 181–207.
12. H. Jeffreys and M. E. M. Bland, The instability of a fluid sphere heated within, Mon. Not. R. Astr. Soc. Geophys. Sup. vol. 6, 1951, pp. 148–158.
13. S. Chandrasekhar, The thermal instability of a fluid sphere heated within, Philos. Mag. Ser. 7, vol. 43, 1952, pp. 1317–1329.
14. S. Chandrasekhar, The onset of convection by thermal instability in spherical shells, Philos. Mag. Ser. 7, vol. 44, 1953, pp. 233–241, 1129–1130.
15. Lord Rayleigh, Investigation of the character of the equilibrium of an incompressible heavy fluid of variable density, Scientific Papers, vol. 2, Cambridge University Press, Cambridge, 1900, pp. 200–207.
16. G. I. Taylor, The instability of liquid surfaces when accelerated in a direction perpendicular to their planes. I, Proc. Roy. Soc. London Ser. A, vol. 201, 1950, pp. 192–196; also D. J. Lewis, The instability of liquid surfaces when accelerated in a direction perpendicular to their planes. II, Proc. Roy. Soc. London Ser. A, vol. 202, 1950, pp. 81–96.
17. S. Chandrasekhar, (in press).
18. R. Hide, (in press).
19. Tsung-Ying Teng Fan, (in press).
20. S. Chandrasekhar, The character of the equilibrium of an incompressible fluid sphere of variable density and viscosity subject to radial acceleration, Quart. Jour. of Mechanics and Applied Mathematics, (in press).

Adjoint Differential Systems in the Theory of Hydrodynamic Stability

S. CHANDRASEKHAR

1. Introduction. In the theory of hydrodynamic stability, one often encounters characteristic value problems in differential systems of relatively high order which do not allow variational formulations in terms of expressions for the characteristic values as ratios of positive-definite integrals. However, ROBERTS [3] has recently suggested that in many of these instances one may be able to construct systems which are in some sense *adjoint* to the ones considered and in terms of which one may, indeed, obtain the solutions to the basic problems by variational methods. In this paper we shall show the practical usefulness of this concept of adjointness in a concrete case while developing at the same time ROBERT'S ideas somewhat differently.

2. The example. The characteristic value problem we shall consider is the one which occurs in the theory of the hydrodynamic stability of viscous flow between two rotating co-axial cylinders. The problem is to solve (*cf.* CHANDRASEKHAR [1]) the equations

$$(DD_* - a^2)^2 u = \left(\frac{1}{r^2} - \kappa\right) v \tag{1}$$

and

$$(DD_* - a^2) v = -\lambda u, \tag{2}$$

together with the boundary conditions

$$u = Du = v = 0 \quad \text{for} \quad r = 1 \quad \text{and} \quad \eta \qquad (< 1), \tag{3}$$

where a^2, κ, and η are assigned constants, λ is the characteristic value parameter, and

$$D = \frac{d}{dr} \quad \text{and} \quad D_* = D + \frac{1}{r}. \tag{4}$$

3. A method of solution. First we shall outline a "natural" method of solving the problem we have outlined.

Since v is required to vanish at $r = 1$ and η, we can expand it in terms of the cylinder functions of order 1 which vanish at $r = 1$ and η. These latter functions can be expressed in the manner

$$\mathcal{C}_1(\alpha_j r) = Y_1(\alpha_j \eta) J_1(\alpha_j r) - J_1(\alpha_j \eta) Y_1(\alpha_j r) \tag{5}$$

where J_1 and Y_1 are the Bessel functions of the two kinds of order 1 and α_j is a root of the equation

$$Y_1(\alpha\eta) J_1(\alpha) - J_1(\alpha\eta) Y_1(\alpha) = 0. \tag{6}$$

(The roots of this equation for various assigned values of η have been tabulated by CHANDRASEKHAR & ELBERT [2].) The functions satisfy the orthogonality relation

$$\int_\eta^1 \mathcal{C}_1(\alpha_j r)\mathcal{C}_1(\alpha_k r) r\, dr = N_j\ \delta_{jk}\ , \tag{7}$$

where N_j is the normalization constant. Associated with the functions $\mathcal{C}_1(\alpha_j r)$ are the general cylinder functions,

$$\mathcal{C}_\nu(\alpha_j r) = Y_1(\alpha_j \eta) J_\nu(\alpha_j r) - J_1(\alpha_j \eta) Y_\nu(\alpha_j r), \tag{8}$$

which satisfy the same recurrence relations as the Bessel functions.

We assume, then, that v can be expanded in the form

$$v = \sum_j A_j \mathcal{C}_1(\alpha_j r), \tag{9}$$

and express u as the sum

$$u = \sum_j A_j u_j\ , \tag{10}$$

where u_j is the unique solution of the equation

$$(DD_* - a^2)^2 u_j = \left(\frac{1}{r^2} - \kappa\right)\mathcal{C}_1(\alpha_j r) \tag{11}$$

which satisfies the boundary conditions

$$u_j = Du_j = 0 \quad \text{for} \quad r = 1 \quad \text{and} \quad \eta. \tag{12}$$

Having determined u_j in this fashion, we insert the expansions (9) and (10) in equation (2) to obtain

$$\sum_j A_j(\alpha_j^2 + a^2)\mathcal{C}_1(\alpha_j r) = \lambda \sum_j A_j u_j\ . \tag{13}$$

Multiplying this equation by $r\mathcal{C}_1(\alpha_k r)$ and integrating over the range of r, we obtain an infinite set of linear homogeneous equations for the A_j's, which in turn leads to the secular equation

$$\left\|\frac{1}{\lambda}(\alpha_j^2 + a^2)N_j\ \delta_{jk} - (j \mid k)\right\| = 0, \tag{14}$$

where

$$(j \mid k) = \int_\eta^1 u_j \mathcal{C}_1(\alpha_k r) r \, dr. \tag{15}$$

4. The adjoint system. Now consider the slightly different system

$$(DD_* - a^2)^2 u^\dagger = v^\dagger, \tag{16}$$

$$(DD_* - a^2) v^\dagger = -\lambda^\dagger \left(\frac{1}{r^2} - \kappa\right) u^\dagger, \tag{17}$$

and

$$u^\dagger = Du^\dagger = v^\dagger = 0 \quad \text{for} \quad r = 1 \quad \text{and} \quad \eta. \tag{18}$$

Let us seek the solution of this system by the same method which we used for the solution of the system (1)–(3). Accordingly, we express $v^\dagger$ and $u^\dagger$ as series in the forms

$$v^\dagger = \sum_j B_j \mathcal{C}_1(\alpha_j r) \quad \text{and} \quad u^\dagger = \sum_j B_j u_j^\dagger \tag{19}$$

where $u_j^\dagger$ is now the unique solution of the equation

$$(DD_* - a^2)^2 u_j^\dagger = \mathcal{C}_1(\alpha_j r), \tag{20}$$

which satisfies the boundary conditions

$$u_j^\dagger = Du_j^\dagger = 0 \quad \text{for} \quad r = 1 \quad \text{and} \quad \eta. \tag{21}$$

Equation (17) then leads to the secular equation

$$\left\| \frac{1}{\lambda^\dagger} (\alpha_j^2 + a^2) N_j \, \delta_{jk} - (j \mid k)^\dagger \right\| = 0, \tag{22}$$

where

$$(j \mid k)^\dagger = \int_\eta^1 u_j^\dagger \left(\frac{1}{r^2} - \kappa\right) \mathcal{C}_1(\alpha_k r) r \, dr. \tag{23}$$

We shall now show that *the matrices* $(j|k)$ *and* $(j|k)^\dagger$ *are the transposed forms of one another*. To show this substitute for $(1/r^2 - \kappa)\mathcal{C}_1(\alpha_k r)$ in the integrand for $(j|k)^\dagger$ according to equation (11); we then obtain

$$(j \mid k)^\dagger = \int_\eta^1 r u_j^\dagger (DD_* - a^2)^2 u_k \, dr. \tag{24}$$

After two integrations by parts equation (24) can be brought to the form

$$(j \mid k)^\dagger = \int_\eta^1 [(DD_* - a^2) u_j^\dagger][(DD_* - a^2) u_k] r \, dr. \tag{25}$$

Similarly, by substituting for $\mathcal{C}_1(\alpha_k r)$ in the integrand for $(j|k)$ in accordance with equation (20), we obtain

$$(j \mid k) = \int_\eta^1 r u_j (DD_* - a^2)^2 u_k^\dagger \, dr, \tag{26}$$

which after two integrations by parts becomes

$$(j \mid k) = \int_\eta^1 r[(DD_* - a^2)u_j][(DD_* - a^2)u_k^\dagger] r \, dr. \tag{27}$$

A comparison of equations (25) and (27) shows that

$$(j \mid k) = (k \mid j)^\dagger. \tag{28}$$

A consequence of this last relation is that the secular equations (14) and (22) are the transposed forms of one another. Therefore *the characteristic equations for* λ *and* $\lambda^\dagger$ *are identical.*

We have thus shown that the systems (1)–(3) and (16)–(18) determine the same set of characteristic values. For this reason we shall call the two systems the *adjoints of one another*.

5. The dual relationship and the variational principle. From the identity of the characteristic values of the systems (1)–(3) and (16)–(18) we can derive a certain *relationship of duality* between the two systems. To elucidate the nature of this duality, consider the proper solutions u_j and v_j belonging to a particular characteristic value λ_j and the solutions $u_k^\dagger$ and $v_k^\dagger$ belonging to a different characteristic value λ_k. The equations satisfied by these solutions are

$$(DD_* - a^2)^2 u_j = \left(\frac{1}{r^2} - \kappa\right) v_j \, , \tag{29}$$

$$(DD_* - a^2) v_j = -\lambda_j u_j \, , \tag{30}$$

and

$$(DD_* - a^2)^2 u_k^\dagger = v_k^\dagger \, , \tag{31}$$

$$(DD_* - a^2) v_k^\dagger = -\lambda_k \left(\frac{1}{r^2} - \kappa\right) u_k^\dagger \, . \tag{32}$$

Now consider

$$\lambda_j \int_\eta^1 r u_j v_k^\dagger \, dr = -\int_\eta^1 r v_k^\dagger (DD_* - a^2) v_j \, dr. \tag{33}$$

By carrying out two integrations by parts in succession, we find

$$\begin{aligned} \lambda_j \int_\eta^1 r u_j v_k^\dagger \, dr &= \int_\eta^1 r[(D_* v_j)(D_* v_k^\dagger) + a^2 v_j v_k^\dagger] \, dr \\ &= -\int_\eta^1 r v_j (DD_* - a^2) v_k^\dagger \, dr. \end{aligned} \tag{34}$$

On the other hand, by substituting for $v_k^\dagger$ from equation (31), we obtain

$$\begin{aligned}\lambda_j \int_\eta^1 r u_j v_k^\dagger \, dr &= \lambda_j \int_\eta^1 r u_j (DD_* - a^2)^2 u_k^\dagger \, dr \\ &= \lambda_j \int_\eta^1 r[(DD_* - a^2)u_j][(DD_* - a^2)u_k^\dagger] \, dr.\end{aligned} \tag{35}$$

Therefore, defining

$$G_j = (DD_* - a^2)u_j \quad \text{and} \quad G_k^\dagger = (DD_* - a^2)u_k^\dagger \, , \tag{36}$$

we have the relation

$$\lambda_j \int_\eta^1 G_j G_k^\dagger r \, dr = \int_\eta^1 r[(D_* v_j)(D_* v_k^\dagger) + a^2 v_j v_k^\dagger] \, dr. \tag{37}$$

Considering the second alternative form of the integral on the right hand side of equation (37) (given in (34)) and substituting for $(DD_* - a^2)v_k^\dagger$ from equation (32), we obtain

$$\int_\eta^1 r[(D_* v_j)(D_* v_k^\dagger) + a^2 v_j v_k^\dagger] \, dr = \lambda_k \int_\eta^1 v_j \left(\frac{1}{r^2} - \kappa\right) u_k^\dagger r \, dr. \tag{38}$$

On the right hand side of this equation we can replace $(1/r^2 - \kappa)v_j$ by $(DD_* - a^2)^2 u_j$ (in accordance with equation (29)); we thus find

$$\begin{aligned}\int_\eta^1 r[(D_* v_j)(D_* v_k^\dagger) + a^2 v_j v_k^\dagger] \, dr &= \lambda_k \int_\eta^1 r u_k^\dagger (DD_* - a^2)^2 u_j \, dr \\ &= \lambda_k \int_\eta^1 [(DD_* - a^2)u_j][(DD_* - a^2)u_k^\dagger] r \, dr = \lambda_k \int_\eta^1 G_j G_k^\dagger r \, dr.\end{aligned} \tag{39}$$

Now combining the results expressed by equations (37) and (39), we have the relation

$$(\lambda_j - \lambda_k) \int_\eta^1 r G_j G_k^\dagger \, dr = 0. \tag{40}$$

Hence

$$\int_\eta^1 G_j G_k^\dagger r \, dr = 0 \quad \text{if} \quad j \neq k. \tag{41}$$

The proper solutions G_j and $G_k^\dagger$ are, therefore, in a dual relationship of orthogonality.

When $j = k$, equations (37) and (39) yield the same relation

$$\lambda = \frac{\int_\eta^1 [(D_* v)(D_* v^\dagger) + a^2 v v^\dagger] r \, dr}{\int_\eta^1 r[(DD_* - a^2)u][(DD_* - a^2)u^\dagger] \, dr} = \frac{I_1}{I_2} \quad \text{(say)}, \tag{42}$$

where the distinguishing subscripts have been suppressed.

We shall now show that equation (42) provides the basis for a variational formulation of the underlying characteristic value problem.

Consider then the effect on λ (evaluated in accordance with equation (42)) due to infinitesimal variations δv and $\delta v^\dagger$ (in v and $v^\dagger$) which are arbitrary except for the requirement that they vanish at $r = 1$ and η. The variations δu and $\delta u^\dagger$ (in u and $u^\dagger$) consequent to the variations δv and $\delta v^\dagger$ (in v and $v^\dagger$) are to be determined as the unique solutions of the equations

$$(DD_* - a^2)^2\, \delta u = \left(\frac{1}{r^2} - \kappa\right) \delta v \tag{43}$$

and

$$(DD_* - a^2)^2\, \delta u^\dagger = \delta v^\dagger, \tag{44}$$

which satisfy the boundary conditions

$$\delta u = D\,\delta u = \delta u^\dagger = D\,\delta u^\dagger = 0 \quad \text{for} \quad r = 1 \quad \text{and} \quad \eta. \tag{45}$$

In other words, the variations are to be carried out subject to equations (1) and (16) as *constraints.*

Denoting by $\delta\lambda$ the first-order change in λ resulting from the variation, we have

$$\delta\lambda = \frac{1}{I_2}(\delta I_1 - \lambda\,\delta I_2), \tag{46}$$

where

$$\delta I_1 = \int_\eta^1 \{(D_*\,\delta v)(D_* v^\dagger) + (D_* v)(D_*\,\delta v^\dagger) + a^2(v\,\delta v^\dagger + v^\dagger\,\delta v)\} r\,dr \tag{47}$$

and

$$\begin{aligned}\delta I_2 = \int_\eta^1 \{&[(DD_* - a^2)\,\delta u][(DD_* - a^2)u^\dagger] \\ &+ [(DD_* - a^2)u][(DD_* - a^2)\,\delta u^\dagger]\} r\,dr\end{aligned} \tag{48}$$

are the corresponding variations in I_1 and I_2. Making use of the boundary conditions imposed on the various increments, we can reduce the expressions for δI_1 and δI_2 by integrations by parts to obtain

$$\delta I_1 = -\int_\eta^1 \{\delta v(DD_* - a^2)v^\dagger + \delta v^\dagger(DD_* - a^2)v\} r\,dr \tag{49}$$

and

$$\delta I_2 = \int_\eta^1 \{u^\dagger(DD_* - a^2)^2\,\delta u + u(DD_* - a^2)^2\,\delta u^\dagger\} r\,dr. \tag{50}$$

Since δu and $\delta u^\dagger$ are subject to equations (43) and (44), an equivalent expression for δI_2 is

$$\delta I_2 = \int_\eta^1 \left\{ \delta v \left(\frac{1}{r^2} - \kappa \right) u^\dagger + \delta v^\dagger u \right\} r\, dr. \tag{51}$$

With δI_1 and δI_2 given by equations (49) and (51), the required first-order change in λ is given by

$$\delta\lambda = -\frac{1}{I_2} \int_\eta^1 \left\{ \delta v \left[(DD_* - a^2) v^\dagger + \lambda \left(\frac{1}{r^2} - \kappa \right) u^\dagger \right] + \delta v^\dagger [(DD_* - a^2) v + \lambda u] \right\} r\, dr. \tag{52}$$

From this equation it follows that $\delta\lambda = 0$ *to the first order for all small variations* δv *and* $\delta v^\dagger$ *(which vanish at* $r = 1$ *and* η*) provided*

$$(DD_* - a^2) v^\dagger + \lambda \left(\frac{1}{r^2} - \kappa \right) u^\dagger = 0 \quad \textit{and} \quad (DD_* - a^2) v + \lambda u = 0; \tag{53}$$

i.e. provided equations (2) and (17) governing u and v and $u^\dagger$ and $v^\dagger$ are simultaneously satisfied. It is evident that the converse of this statement is also true. Therefore, equation (42) does provide the basis for a variational treatment of the problem.

6. The variational character of the method of solution of §§3 and 4. We shall now show that the method described in §§3 and 4 for the solution of equations (1)–(3) and (16)–(18) is equivalent to a variational procedure (based on equation (42)) in which the coefficients A_j and B_j in the expansions for v and $v^\dagger$ are treated as variational parameters. Thus, with the chosen forms for v and $v^\dagger$, the expression for λ becomes

$$\lambda = \frac{\sum_j A_j B_j N_j (\alpha_j^2 + a^2)}{\sum_j \sum_k A_j B_k \int_\eta^1 r[(DD_* - a^2) u_j][(DD_* - a^2) u_k^\dagger]\, dr}. \tag{54}$$

The matrix element which occurs in the denominator of this expression is, by equation (27), none other than $(j|k)$. Accordingly,

$$\lambda = \frac{\sum_j A_j B_j N_j (\alpha_j^2 + a^2)}{\sum_j \sum_k A_j (j \mid k) B_k^\dagger}. \tag{55}$$

The condition that this expression for λ is an extremum with respect to variations of the parameters A_j and B_k leads precisely to the same secular equations (14) and (22). This establishes the equivalence of the procedure outlined in §§3 and 4 with a variational treatment based on equation (42).

7. The secular determinant. We shall now briefly consider the secular determinant (14) (or (22)) from the point of view of its explicit evaluation.

When the method of solution of equations (1)–(3), outlined in §2, occurred to the writer several years ago, it was abandoned as "impracticable" (see CHANDRASEKHAR [1]) on the grounds that the solution of equation (11) cannot be written down explicitly in terms of known functions and involves many indefinite integrals over the functions $\mathcal{C}_1(\alpha_k r)$, and that, in consequence, the evaluation of the matrix elements (15) would require as many double quadratures. For these reasons equations (1) and (2) were considered in their equivalent forms

$$(DD_* - a^2)v = u \tag{56}$$

and

$$(DD_* - a^2)^2 u = -\lambda\left(\frac{1}{r^2} - \kappa\right)v, \tag{57}$$

and u was expanded in a series of orthogonal functions which, together with their derivatives, vanish at $r = 1$ and η. However, it is now clear that the "difficulty" mentioned in connection with the solution of equation (11) does not exist with respect to equation (20); and the evaluation of the matrix elements (23) can, at worst, involve only single quadratures. Actually, it appears that the matter is even simpler. For the general solution of equation (20) is

$$u_j^\dagger = A_j I_1(ar) + B_j r I_0(ar) + C_j K_1(ar) + D_j r K_0(ar) + \frac{\mathcal{C}_1(\alpha_j r)}{(\alpha_j^2 + a^2)^2}, \tag{58}$$

where the constants A_j, B_j, *etc.*, are determined by the boundary conditions (21); and it can be readily shown that most of the integrals which occur in the matrix elements $(j|k)^\dagger$ (as defined in equation (23)) can be explicitly evaluated and that numerical quadratures are needed only for the two integrals

$$\int_\eta^1 \frac{dr}{r} I_1(ar)\mathcal{C}_1(\alpha_k r) \quad \text{and} \quad \int_\eta^1 \frac{dr}{r} K_1(ar)\mathcal{C}_1(\alpha_k r). \tag{59}$$

The explicit solution of the characteristic value problem carried out along these lines will be presented in another paper.

The research reported in this paper has in part been supported by the Geophysics Research Directorate of the Air Force Cambridge Research Center, Air Research and Development Command, under Contract AF 19(604)-2046 with the University of Chicago.

BIBLIOGRAPHY

1. S. CHANDRASEKHAR, *The stability of viscous flow between rotating cylinders,* Proc. Roy. Soc. (London) A, **246**, 301–11, 1958.
2. S. CHANDRASEKHAR & DONNA ELBERT, *The roots of* $J_n(\alpha\eta)Y_n(\alpha) - Y_n(\alpha\eta)J_n(\alpha) = 0$, Proc. Camb. Phil. Soc., **50**, 266–8, 1954.
3. P. H. ROBERTS, *Characteristic value problems posed by differential equations arising in hydrodynamics and hydromagnetics,* J. Math. Analysis and Applications, **1**, 195–214, 1960.

University of Chicago
Chicago, Illinois

PART THREE

Tensor Virial Theorem and Its Applications

The Virial Theorem in Hydromagnetics

S. CHANDRASEKHAR

University of Chicago, Chicago, Illinois

A tensor form of the virial theorem appropriate for configurations in hydromagnetic equilibrium is obtained. Various elementary consequences of this theorem are derived such as the impossibility of spherical symmetry for static configurations with prevailing magnetic fields. A variational form of the virial theorem governing small departures from an initial static state is obtained; and the usefulness of this variational form for estimating the characteristic periods of oscillation of a hydromagnetic system is illustrated by considering a special case.

I. INTRODUCTION

The usual form of the virial theorem was extended to hydromagnetics by Chandrasekhar and Fermi [1]. In this paper we shall consider a generalization of this theorem.

II. THE VIRIAL THEOREM

Consider an inviscid fluid of zero electrical resistivity in which a magnetic field $\mathbf{H}(\mathbf{x})$ prevails. Suppose that the fluid is a perfect gas and that the ratio of the specific heats is γ. Suppose further that apart from the prevailing magnetic field and gas pressure, the only force acting on the medium is that derived from its own gravitation. Under these circumstances the equation of motion governing the fluid velocities is

$$\rho \frac{du_i}{dt} = -\frac{\partial}{\partial x_i}\left(p + \frac{|\mathbf{H}|^2}{8\pi}\right) + \rho \frac{\partial \mathscr{V}}{\partial x_i} + \frac{1}{4\pi}\frac{\partial}{\partial x_j} H_i H_j, \tag{1}$$

where

$$\frac{d}{dt} = \frac{\partial}{\partial t} + u_j \frac{\partial}{\partial x_j} \tag{2}$$

is the total time derivative. In Eq. (1), $\mathscr{V}$ denotes the gravitational potential, and the rest of the symbols have their usual meanings. (We are setting $\mu = 1$ in the present analysis.)

A. Some Definitions and Relations

Since we have supposed that the gravitational potential, $\mathscr{V}(\mathbf{x})$, is derived from the distribution of the matter present,

$$\mathscr{V}(\mathbf{x}) = G \int_V \frac{\rho(\mathbf{x}')}{|\mathbf{x} - \mathbf{x}'|} \, d\mathbf{x}' \tag{3}$$

where for the sake of brevity we have written

$$d\mathbf{x}' = dx_1' \, dx_2' \, dx_3' \tag{4}$$

and abridged three integral signs into one. In (3), the integration is effected over the entire volume V occupied by the fluid.

We shall find it convenient to define the symmetric tensor,

$$\mathscr{V}_{ik}(\mathbf{x}) = G \int_V \rho(\mathbf{x})' \, \frac{(x_i - x_i')(x_j - x_j')}{|\mathbf{x} - \mathbf{x}'|^3} \, d\mathbf{x}' \tag{5}$$

which represents a generalization of (3) and to which it reduces on contraction:

$$\mathscr{V}_{ii} = \mathscr{V}. \tag{6}$$

Similarly, in generalization of the usual definition of the gravitational potential energy, we shall define

$$\mathscr{W}_{ik} = -\tfrac{1}{2} G \int_V \int_V \rho(\mathbf{x}) \rho(\mathbf{x}') \, \frac{(x_i - x_i')(x_k - x_k')}{|\mathbf{x} - \mathbf{x}'|^3} \, d\mathbf{x} \, d\mathbf{x}'; \tag{7}$$

and the contraction of this tensor gives the gravitational potential energy;

$$\mathscr{W}_{ii} = \mathscr{W} = -\tfrac{1}{2} G \int_V \int_V \frac{\rho(\mathbf{x}) \rho(\mathbf{x}')}{|\mathbf{x} - \mathbf{x}'|} \, d\mathbf{x} \, d\mathbf{x}'. \tag{8}$$

An identity which follows from the foregoing definitions is

$$\mathscr{W}_{ik} = \int_V \rho(\mathbf{x}) x_i \frac{\partial \mathscr{V}}{\partial x_k} \, d\mathbf{x} = \int_V \rho(\mathbf{x}) x_k \frac{\partial \mathscr{V}}{\partial x_i} \, d\mathbf{x}. \tag{9}$$

This can be established as follows: by definition,

$$\int_V d\mathbf{x}\, \rho(\mathbf{x}) x_i \frac{\partial \mathscr{V}}{\partial x_k} = G \int_V d\mathbf{x}\, \rho(\mathbf{x}) x_i \frac{\partial}{\partial x_k} \int_V d\mathbf{x}' \frac{\rho(\mathbf{x}')}{|\mathbf{x} - \mathbf{x}'|} \tag{10}$$

$$= -G \int_V \int_V d\mathbf{x}\, d\mathbf{x}'\, \rho(\mathbf{x}) \rho(\mathbf{x}') \frac{x_i(x_k - x_k')}{|\mathbf{x} - \mathbf{x}'|^3}.$$

By transposing the primed and the unprimed variables of integrations in (10) and taking the average of the two equivalent expressions, we verify that the double integral on the last line of (10) is the same as defined in Eq. (7).

In addition to $\mathscr{V}_{ik}$ and $\mathscr{W}_{ik}$, we shall find it useful to define the further tensors

$$\mathscr{T}_{ik} = \tfrac{1}{2} \int_V \rho u_i u_k \, d\mathbf{x}, \tag{11}$$

and[1]

$$\mathscr{M}_{ik} = \frac{1}{8\pi} \int_V H_i H_k \, d\mathbf{x}; \qquad (12)^1$$

and the contraction of these tensors gives the kinetic and the magnetic energies of the system:

$$\mathscr{T} = \mathscr{T}_{ii} = \tfrac{1}{2} \int_V \rho |\mathbf{u}|^2 \, d\mathbf{x}, \tag{13}$$

and

$$\mathscr{M} = \mathscr{M}_{ii} = \frac{1}{8\pi} \int_V |\mathbf{H}|^2 \, d\mathbf{x}. \tag{14}$$

The analogous expression for the internal heat energy of the system is

$$\mathscr{U} = \frac{1}{\gamma - 1} \int_V p \, d\mathbf{x}. \tag{15}$$

[1] The specification of the volume V over which this integration is effected requires some care; we return to this matter in Section I, B below.

Even as $\mathscr{W}_{ik}$, $\mathscr{T}_{ik}$, $\mathscr{M}_{ik}$, and $\mathscr{U}$ characterize, in terms of a few parameters, the distribution of the different forms of energy in the configuration so does the *inertia-tensor*,

$$I_{ik} = \int_V \rho x_i x_k \, d\mathbf{x}, \tag{16}$$

similarly characterize the distribution of the density in the configuration. The contraction of I_{ik} gives the moment of inertia:

$$I = I_{ii} = \int_V \rho |\mathbf{x}|^2 \, d\mathbf{x}. \tag{17}$$

B. The General Form of the Virial Theorem

Returning to Eq. (1), multiply it by x_k and integrate it over the *entire* volume V in which the fluid *and* the field pervade. The left-hand side of the equation can be reduced in the manner:

$$\int_V \rho x_k \frac{du_i}{dt} \, d\mathbf{x} = \int_V \rho x_k \frac{d^2 x_i}{dt^2} \, d\mathbf{x} = \int_V \rho \frac{d}{dt}\left(x_k \frac{dx_i}{dt}\right) d\mathbf{x} - \int_V \rho \frac{dx_k}{dt}\frac{dx_i}{dt} \, d\mathbf{x}$$

$$= \int_V \rho \frac{d}{dt}\left(x_k \frac{dx_i}{dt}\right) d\mathbf{x} - 2\mathscr{T}_{ik}. \tag{18}$$

The terms on the right-hand side, similarly, give

$$-\int_V x_k \frac{\partial}{\partial x_i}\left(p + \frac{|\mathbf{H}|^2}{8\pi}\right) d\mathbf{x} = -\int_S \left(p + \frac{|\mathbf{H}|^2}{8\pi}\right) x_k \, dS_i + \delta_{ik} \int_V \left(p + \frac{|\mathbf{H}|^2}{8\pi}\right) d\mathbf{x}$$

$$= -\int_S \left(p + \frac{|\mathbf{H}|^2}{8\pi}\right) x_k \, dS_i + \delta_{ik} \{(\gamma - 1)\mathscr{U} + \mathscr{M}\}, \tag{19}$$

$$\int_V \rho x_k \frac{\partial \mathscr{V}}{\partial x_i} \, d\mathbf{x} = \mathscr{W}_{ik}, \tag{20}$$

and

$$\frac{1}{4\pi}\int_V x_k \frac{\partial}{\partial x_j} H_i H_j \, d\mathbf{x} = \frac{1}{4\pi}\int_S x_k H_i H_j \, dS_j - \frac{1}{4\pi}\int_V H_i H_k \, d\mathbf{x}$$

$$= \frac{1}{4\pi}\int_S x_k H_i H_j \, dS_j - 2\mathscr{M}_{ik}, \tag{21}$$

where in Eqs. (19) and (21) the surface integrals are extended over the surface S, bounding V.

Combining Eqs. (18)—(21), we obtain

$$\int_V \rho \frac{d}{dt}\left(x_k \frac{dx_i}{dt}\right) d\mathbf{x} = 2\mathscr{T}_{ik} + \delta_{ik}\{(\gamma-1)\mathscr{U} + \mathscr{M}\} \tag{22}$$

$$+ \mathscr{W}_{ik} - 2\mathscr{M}_{ik} + \frac{1}{8\pi}\int_S x_k(2H_i H_j\, dS_j - |\mathbf{H}|^2\, dS_i) - \int_S p x_k\, dS_i.$$

We shall suppose that the volume V over which the integrations are extended includes the *whole* system, so that all the variables p, ρ, and $\mathbf{H}$ may be assumed to vanish on S. It is important to note that this definition of V may require us to include in it volumes which one may normally consider as external to the "natural" boundary of the system, namely, the surface on which the density ρ and the material pressure p vanish. The assumption that $\mathbf{H}$ vanishes on S may, indeed, require us to place S at infinity. This latter possibility arises because magnetic fields can extend far beyond the conventional limits of a material object; but to the extent the object is the seat of the magnetic field, there is justification in including all portions of space into which the field extends as parts of the system. And since the field of an object isolated in space must decrease, at least as rapidly as that of a dipole (i.e., as r^{-3}), the surface integrals over the components of $\mathbf{H}$ in Eq. (22) will vanish under these circumstances when $S \to \infty$. However, it may sometimes be convenient to let S coincide with the natural boundary, in which case the surface integrals over S in Eq. (22) must be retained; and, moreover, $\mathscr{M}_{ik}$ will then refer to only that part of the field which is interior to S.

In the present analysis we shall suppose that V includes (as we have already remarked) all parts of space in which the fluid *and* the field pervade. With this explicit understanding, Eq. (22) becomes

$$\int_V \rho \frac{d}{dt}\left(x_k \frac{dx_i}{dt}\right) d\mathbf{x} = 2\mathscr{T}_{ik} - 2\mathscr{M}_{ik} + \mathscr{W}_{ik} + \delta_{ik}\{(\gamma-1)\mathscr{U} + \mathscr{M}\}. \tag{23}$$

Since all the tensors on the right-hand of this equation are symmetric, the tensor on the left-hand side of the equation must also be symmetric. Therefore,

$$\int_V \rho \frac{d}{dt}\left(x_k \frac{dx_i}{dt}\right) d\mathbf{x} = \int_V \rho \frac{d}{dt}\left(x_i \frac{dx_k}{dt}\right) d\mathbf{x}. \tag{24}$$

An immediate consequence of this result is

$$\int_V \rho \frac{d}{dt}\left(x_k \frac{dx_i}{dt} - x_i \frac{dx_k}{dt}\right) d\mathbf{x} = \frac{d}{dt} \int_V \rho \left(x_k \frac{dx_i}{dt} - x_i \frac{dx_k}{dt}\right) d\mathbf{x}, \tag{25}$$

where in taking d/dt outside of the integral sign, we have made use of the constancy of $\rho\, d\mathbf{x}$ assured by the equation of continuity. *Equation* (25) *expresses the constancy of the total angular momentum of the system.* It is worth noting that the existence of this integral of the equations of motion has not been affected by the presence of the magnetic field.

A further consequence of the identity (24) is that the quantity on the left-hand side of Eq. (23) can be replaced by

$$\tfrac{1}{2} \int_V \rho \frac{d}{dt}\left(x_k \frac{dx_i}{dt} + x_i \frac{dx_k}{dt}\right) d\mathbf{x} = \tfrac{1}{2} \frac{d^2}{dt^2} \int_V \rho x_i\, x_k\, d\mathbf{x} = \tfrac{1}{2} \frac{d^2 I_{ik}}{dt^2}. \tag{26}$$

With this replacement, Eq. (23) gives

$$\tfrac{1}{2} \frac{d^2 I_{ik}}{dt^2} = 2\mathcal{T}_{ik} - 2\mathcal{M}_{ik} + \mathcal{W}_{ik} + \delta_{ik}\{(\gamma - 1)\mathcal{U} + \mathcal{M}\}. \tag{27}$$

This equation represents the complete statement of *the virial theorem for hydromagnetics.*

When $i \neq k$, Eq. (27) gives

$$\tfrac{1}{2} \frac{d^2 I_{ik}}{dt^2} = 2\mathcal{T}_{ik} - 2\mathcal{M}_{ik} + \mathcal{W}_{ik} \qquad (i \neq k); \tag{28}$$

and by contracting the indices in Eq. (27), we obtain

$$\tfrac{1}{2} \frac{d^2 I}{dt^2} = 2\mathcal{T} + \mathcal{M} + \mathcal{W} + 3(\gamma - 1)\mathcal{U}. \tag{29}$$

C. The Virial Theorem for Equilibrium Configurations

For configurations in equilibrium and in a steady state, Eqs. (27) and (29) give

$$2\mathcal{T}_{ik} - 2\mathcal{M}_{ik} + \mathcal{W}_{ik} + \delta_{ik}\{(\gamma - 1)\mathcal{U} + \mathcal{M}\} = 0, \tag{30}$$

and

$$2\mathcal{T} + \mathcal{M} + \mathcal{W} + 3(\gamma - 1)\mathcal{U} = 0. \tag{31}$$

An alternative way of stating the result expressed by Eq. (30) is that *the tensor,* $2\,(\mathcal{T}_{ik} - \mathcal{M}_{ik}) + \mathcal{W}_{ik}$, *is isotropic.*

Consider first the case when there are no fluid motions or magnetic fields. In this case, Eq. (30) gives

$$\mathscr{W}_{ik} = \int_V \rho(\mathbf{x}) x_i \frac{\partial \mathscr{V}}{\partial x_k} d\mathbf{x} = -(\gamma - 1)\mathscr{U}\delta_{ik}. \tag{32}$$

This relation is compatible with a spherical symmetry of the underlying density distribution. It would be of interest to know if, *conversely*, the spherical symmetry can be *deduced* from Eq. (32).

When $\mathscr{T}_{ik}$ and $\mathscr{M}_{ik}$ are not zero, $\mathscr{W}_{ik}$ will not, in general, be diagonal. Consequently, *a spherical symmetry of the configuration is, in general, incompatible with the presence of fluid motions and magnetic fields.* An exception is possible if $\mathscr{T}_{ik} \equiv \mathscr{M}_{ik}$ — an identity which is satisfied, for example, by the equipartition solution (2), $u_i = \pm H_i/\sqrt{(4\pi\rho)}$; for, when $\mathscr{T}_{ik} \equiv \mathscr{M}_{ik}$, Eq. (30) gives

$$\mathscr{W}_{ik} = \int_V \rho(\mathbf{x}) x_i \frac{\partial \mathscr{V}}{\partial x_k} d\mathbf{x} = -\{(\gamma - 1)\mathscr{U} + \mathscr{M}\}\,\delta_{ik}; \tag{33}$$

and this equation is not incompatible with spherical symmetry.

In the special case when only magnetic fields are present,

$$\mathscr{W}_{ik} = 2\mathscr{M}_{ik} - \delta_{ik}\{(\gamma - 1)\mathscr{U} + \mathscr{M}\}; \tag{34}$$

and the impossibility of spherical symmetry, in general, follows from the relation

$$\mathscr{W}_{ik} = 2\mathscr{M}_{ik} \neq 0 \qquad \text{for} \qquad i \neq k. \tag{35}$$

Consider next the contracted form, (31), of the general relation. If $\mathscr{E}$ denotes the total energy of the configuration,

$$\mathscr{E} = \mathscr{T} + \mathscr{U} + \mathscr{M} + \mathscr{W}. \tag{36}$$

From Eqs. (31) and (36) it would appear that as far as these scalar equations go, the magnetic energy can be considered together with the gravitational potential energy. Thus, by eliminating $\mathscr{M} + \mathscr{W}$ between the two equations, we obtain the relation,

$$\mathscr{E} = -\mathscr{T} - (3\gamma - 4)\mathscr{U}, \tag{37}$$

which is independent of $\mathscr{M}$. Alternatively, by eliminating $\mathscr{U}$, we obtain

$$\mathscr{E} = \frac{1}{3(\gamma - 1)}\{(3\gamma - 4)(\mathscr{M} + \mathscr{W}) + (3\gamma - 5)\mathscr{T}\}. \tag{38}$$

In case $\mathscr{T} = 0$, Eqs. (37) and (38) give

$$\mathscr{E} = -(3\gamma - 4)\mathscr{U}, \tag{39}$$

and

$$\mathscr{E} = \frac{3\gamma - 4}{3(\gamma - 1)}(\mathscr{M} + \mathscr{W}). \tag{40}$$

III. The Virial Theorem for Small Oscillations about Equilibrium

We shall now consider small oscillations about equilibrium of a configuration in which magnetic fields are present. We shall, however, suppose that in the stationary state there are no fluid motions; and we shall seek the form which the virial theorem takes under these circumstances.

Considering periodic oscillations with a gyration frequency σ, let $\boldsymbol{\xi} e^{i\sigma t}$ denote the Lagrangian displacement of an element of mass, $dm = \rho d\mathbf{x}$, from its equilibrium position at $\mathbf{x}$. Let $\delta p e^{i\sigma t}$, $\delta \rho e^{i\sigma t}$, and $\delta \mathbf{H} e^{i\sigma t}$ denote the corresponding changes in the other physical variables as we follow the fluid element with its motion. The equation of continuity ensures the constancy of dm for such Lagrangian displacements; and if we suppose that the oscillations take place adiabatically, then

$$\frac{\delta p}{p} = \gamma \frac{\delta \rho}{\rho}, \tag{41}$$

where

$$\frac{\delta \rho}{\rho} = -\operatorname{div} \boldsymbol{\xi}. \tag{42}$$

If $\delta I_{ik} e^{i\sigma t}$, $\delta \mathscr{M}_{ik} e^{i\sigma t}$, etc., denote the first order changes in the various integrals representing these quantities in the stationary state, then Eq. (27) gives

$$-\tfrac{1}{2}\sigma^2 \delta I_{ik} = \delta \mathscr{W}_{ik} + \delta_{ik}\{(\gamma - 1)\,\delta\mathscr{U} + \delta\mathscr{M}\} - 2\delta\mathscr{M}_{ik}, \tag{43}$$

the term in $\mathscr{T}_{ik}$ makes no contribution in this order since we have supposed that there are no zero-order fluid motions.

We shall now consider, in turn, the various quantities occurring in Eq. (43). Clearly,

$$\delta I_{ik} = \int_V \rho(\xi_i x_k + x_i \xi_k)\, d\mathbf{x}. \tag{44}$$

Also, according to the definitions of $\mathscr{V}_{ik}$ and $\mathscr{W}_{ik}$ given in Eqs. (5) and (7),

$$\delta\mathscr{W}_{ik} = -G\int_V\int_V dm dm' \xi_j \frac{\partial}{\partial x_j}\frac{(x_i - x_i')(x_k - x_k')}{|\mathbf{x} - \mathbf{x}'|^3}$$

$$= -G\int_V d\mathbf{x}\,\rho(\mathbf{x})\xi_j \frac{\partial}{\partial x_j}\int_V d\mathbf{x}'\rho(\mathbf{x}')\frac{(x_i - x_i')(x_k - x_k')}{|\mathbf{x} - \mathbf{x}'|^3}$$

$$= -\int_V d\mathbf{x}\,\rho(\mathbf{x})\xi_j \frac{\partial\mathscr{V}_{ik}}{\partial x_j}. \tag{45}$$

This last relation is a generalization of the familiar formula,

$$\delta\mathscr{W} = -\int_V d\mathbf{x}\,\rho(\mathbf{x})\xi_j \frac{\partial\mathscr{V}}{\partial x_j}, \tag{46}$$

for the first order change in the gravitational potential energy consequent to a slight redistribution of the matter in the system.

Turning next to the change in the internal energy, we have

$$(\gamma - 1)\delta\mathscr{U} = \int_V \delta\left(\frac{p}{\rho}\right)\rho\, d\mathbf{x}, \tag{47}$$

or making use of the relations (41) and (42), we have

$$(\gamma - 1)\,\delta\mathscr{U} = (\gamma - 1)\int \frac{p}{\rho}\frac{\delta\rho}{\rho}\rho\, d\mathbf{x} = -(\gamma - 1)\int_V p\frac{\partial\xi_j}{\partial x_j}d\mathbf{x}. \tag{48}$$

Letting

$$\varPi = p + \frac{|\mathbf{H}|^2}{8\pi}, \tag{49}$$

we can write

$$\delta\mathscr{U} = -\int_V \varPi\frac{\partial\xi_j}{\partial x_j}d\mathbf{x} + \frac{1}{8\pi}\int_V |\mathbf{H}|^2\frac{\partial\xi_j}{\partial x_j}d\mathbf{x}. \tag{50}$$

Integrating by parts the first of the two integrals on the right-hand side, we obtain

$$\int_V \varPi\frac{\partial\xi_j}{\partial x_j}d\mathbf{x} = \int_S \varPi\,\xi_j\,dS_j - \int_V \xi_j\frac{\partial\varPi}{\partial x_j}d\mathbf{x}. \tag{51}$$

We shall suppose that S is so placed that all the physical variables and their perturbations vanish on S. (This may require that S be placed at infinity.) Then

$$\int_V \Pi \frac{\partial \xi_j}{\partial x_j} d\mathbf{x} = -\int_V \xi_j \frac{\partial \Pi}{\partial x_j} d\mathbf{x}; \tag{52}$$

or, making use of the relation,

$$\frac{\partial \Pi}{\partial x_j} = \rho \frac{\partial \mathscr{V}}{\partial x_j} + \frac{1}{4\pi} \frac{\partial}{\partial x_l} H_l H_j, \tag{53}$$

which obtains in equilibrium, we have

$$\int_V \Pi \frac{\partial \xi_j}{\partial x_j} d\mathbf{x} = -\int_V \rho \xi_j \frac{\partial \mathscr{V}}{\partial x_j} d\mathbf{x} - \frac{1}{4\pi} \int_V \xi_j \frac{\partial}{\partial x_l} H_l H_j d\mathbf{x}. \tag{54}$$

Integrating by parts the second of the two integrals on the right-hand side of this equation, we obtain

$$\int_V \Pi \frac{\partial \xi_j}{\partial x_j} d\mathbf{x} = -\int_V \rho \xi_j \frac{\partial \mathscr{V}}{\partial x_j} d\mathbf{x} + \frac{1}{4\pi} \int_V H_l H_j \frac{\partial \xi_j}{\partial x_l} d\mathbf{x}, \tag{55}$$

the integrated part, again, making no contribution. Now combining Eqs. (50) and (55), we have

$$\delta \mathscr{U} = \int_V \rho \xi_j \frac{\partial \mathscr{V}}{\partial x_j} d\mathbf{x} + \frac{1}{8\pi} \int_V \left(|\mathbf{H}|^2 \frac{\partial \xi_j}{\partial x_j} - 2 H_l H_j \frac{\partial \xi_l}{\partial x_j} \right) d\mathbf{x}. \tag{56}$$

Considering, finally, $\delta \mathscr{M}_{ik}$, we have

$$\delta \mathscr{M}_{ik} = \frac{1}{8\pi} \int_V (H_k \delta H_i + H_i \delta H_k) \, d\mathbf{x} + \frac{1}{8\pi} \int_V H_i H_k \frac{\partial \xi_j}{\partial x_j} d\mathbf{x}, \tag{57}$$

where the second term arises from allowing for the variation of the volume element, $d\mathbf{x}$, following the motion in accordance with Eq. (42) and the constancy of $dm = \rho \, d\mathbf{x}$.

Now the change in the magnetic field, $\delta \mathbf{H}$, as we follow the motion of the fluid element, is given by

$$\delta \mathbf{H} = \text{Curl} \, (\boldsymbol{\xi} \times \mathbf{H}) + (\boldsymbol{\xi} \cdot \text{grad}) \, \mathbf{H}. \tag{58}$$

The first term on the right-hand side gives the Eulerian change in $\mathbf{H}$ at a *fixed point* caused by the redistribution of the matter, while the second term is the allowance for the fact that, as we follow the fluid element, it finds itself in a slightly displaced location. In the notation of Cartesian tensors, Eq. (58) becomes

$$\delta H_i = H_j \frac{\partial \xi_i}{\partial x_j} - H_i \frac{\partial \xi_j}{\partial x_j}. \tag{59}$$

Making use of this relation, we find that Eq. (57) reduces to

$$\delta \mathscr{M}_{ik} = \frac{1}{8\pi} \int_V \left\{ H_j \left(H_k \frac{\partial \xi_i}{\partial x_j} + H_i \frac{\partial \xi_k}{\partial x_j} \right) - H_i H_k \frac{\partial \xi_j}{\partial x_j} \right\} d\mathbf{x}; \tag{60}$$

and by contracting, we find

$$\delta \mathscr{M} = -\frac{1}{8\pi} \int_V \left(|\mathbf{H}|^2 \frac{\partial \xi_j}{\partial x_j} - 2H_j H_l \frac{\partial \xi_l}{\partial x_j} \right) d\mathbf{x}. \tag{61}$$

Now inserting in Eq. (43) for δI_{ik}, $\delta \mathscr{W}_{ik}$, etc., the expressions we have derived for them, we find:

$$\begin{aligned} -\tfrac{1}{2}\sigma^2 \int_V \rho(\xi_i x_k + x_i \xi_k)\, d\mathbf{x} = &- \int_V \rho \xi_j \frac{\partial \mathscr{V}_{ik}}{\partial x_j} d\mathbf{x} \\ &+ \delta_{ik} \left\{ (\gamma - 1) \int_V \rho \xi_j \frac{\partial \mathscr{V}}{\partial x_j} d\mathbf{x} + \frac{\gamma - 2}{8\pi} \int_V \left(|\mathbf{H}|^2 \frac{\partial \xi_j}{\partial x_j} - 2H_j H_l \frac{\partial \xi_l}{\partial x_j} \right) d\mathbf{x} \right. \\ &\left. + \frac{1}{4\pi} \int_V \left\{ H_i H_k \frac{\partial \xi_j}{\partial x_j} - H_j \left(H_k \frac{\partial \xi_i}{\partial x_j} + H_i \frac{\partial \xi_k}{\partial x_j} \right) \right\} d\mathbf{x}. \right. \end{aligned} \tag{62}$$

By contracting this equation, we obtain

$$\begin{aligned} -\sigma^2 \int_V \rho \xi_i x_i\, d\mathbf{x} \\ = (3\gamma - 4) \left\{ \int_V \rho \xi_j \frac{\partial \mathscr{V}}{\partial x_j} d\mathbf{x} + \frac{1}{8\pi} \int_V \left(|\mathbf{H}|^2 \frac{\partial \xi_j}{\partial x_j} - 2H_j H_l \frac{\partial \xi_l}{\partial x_j} \right) d\mathbf{x}. \right. \end{aligned} \tag{63}$$

A. A Characteristic Equation for Determining the Periods of Oscillation

Equations (62) and (63) can be used to obtain estimates for σ^2 by inserting in them suitable "trial" functions for $\boldsymbol{\xi}$. Thus, the simplest assumption,

$$\boldsymbol{\xi} = \text{constant } \mathbf{x} \tag{64}$$

together with Eq. (63), yields the formula (3)

$$\sigma^2 = -(3\gamma - 4)\frac{\mathscr{W} + \mathscr{M}}{I}. \tag{65}$$

However, since a configuration with a prevalent magnetic field cannot be spherically symmetric, the use of a trial function which corresponds to a uniform radial expansion would appear unsuitable; indeed, the substitution of (64) in the tensor equation (62) will lead to gross inconsistencies unless the departures from spherical symmetry of the equilibrium configuration are small. If the latter should not be the case, a more consistent procedure would appear to be the following.

Assume that

$$\boldsymbol{\xi} = \boldsymbol{X}\mathbf{x} \tag{66}$$

where $\boldsymbol{X}$ is a symmetric matrix:

$$\mathscr{X}_{ij} = \mathscr{X}_{ji}. \tag{67}$$

Further, define the super-matrix

$$\mathscr{W}_{lj;ik} = \int_V \rho(\mathbf{x})\, x_l \frac{\partial}{\partial x_j} \mathscr{V}_{ik}\, d\mathbf{x}. \tag{68}$$

By contracting this super-matrix with respect to i and k, we obtain

$$\mathscr{W}_{lj;ii} = \int_V \rho(\mathbf{x}) x_l \frac{\partial \mathscr{V}}{\partial x_j}\, d\mathbf{x} = \mathscr{W}_{lj}. \tag{69}$$

Also, let

$$\boldsymbol{M} = (\mathscr{M}_{ik}), \qquad \boldsymbol{W} = (W_{ik}) \qquad \text{and} \qquad \boldsymbol{I} = (I_{ik}). \tag{70}$$

Clearly,

$$\mathscr{M} = \mathscr{M}_{ii} = \mathrm{Tr}\,(\boldsymbol{M}) \qquad \text{and} \qquad \mathscr{W} = \mathscr{W}_{ii} = \mathrm{Tr}\,(\boldsymbol{W}), \tag{71}$$

where Tr stands for trace (meaning diagonal sum). Similarly,

$$\frac{\partial \xi_j}{\partial x_j} = \mathrm{Tr}\,(\boldsymbol{X}). \tag{72}$$

With the foregoing definitions, the substitution of the trial function (66) in Eq. (62) leads to the result

$$\begin{aligned} -\tfrac{1}{2}\sigma^2 (\mathbf{I}\boldsymbol{X} + \boldsymbol{X}\mathbf{I})_{ik} = &-\mathscr{X}_{jl}\mathscr{W}_{lj;ik} \\ &+ \delta_{ik}\{(\gamma - 1)\,\mathrm{Tr}\,(\boldsymbol{X}\boldsymbol{W}) + (\gamma - 2)\,[\mathrm{Tr}\,(\boldsymbol{X})\,\mathrm{Tr}\,(\boldsymbol{M}) - 2\,\mathrm{Tr}\,(\boldsymbol{X}\boldsymbol{M})]\} \\ &+ 2\,\{\mathrm{Tr}\,(\boldsymbol{X})\mathscr{M}_{ik} - [\boldsymbol{X}\boldsymbol{M} + \boldsymbol{M}\boldsymbol{X}]_{ik}\}. \end{aligned} \tag{73}$$

Equation (73) represents a system of linear homogeneous equations for the six coefficients of the (assumed) symmetric linear transformation, $\boldsymbol{X}$. The determinant of the system must, therefore, vanish; and the resulting characteristic equation will not only determine σ^2 but also the transformation, $\boldsymbol{X}$ (apart from a constant of proportionality).

Finally, we may note that by contracting the tensor equation (73), we obtain the scalar equation,

$$-\sigma^2 \operatorname{Tr}(\mathbf{I}\boldsymbol{X}) = (3\gamma - 4)\{\operatorname{Tr}(\boldsymbol{X}\boldsymbol{W}) + \operatorname{Tr}(\boldsymbol{X})\operatorname{Tr}(\boldsymbol{M}) - 2\operatorname{Tr}(\boldsymbol{X}\boldsymbol{M})\}. \quad (74)$$

which is a generalization of (65).

References

1. Chandrasekhar, S., and Fermi, E. *Astrophys. J.* **118**, 116–141 (1953).
2. Chandrasekhar, S. *Proc. Natl. Acad. Sci. U.S.* **42**, 273–276 (1956).
3. Chandrasekhar, S., and Nelson Limber, D. *Astrophys.* J. **119**, 10–13 (1954).

14

THE HIGHER ORDER VIRIAL EQUATIONS AND THEIR APPLICATIONS TO THE EQUILIBRIUM AND STABILITY OF ROTATING CONFIGURATIONS *

S. Chandrasekhar
Professor
University of Chicago

CHAPTER I

The Virial Equations of the Various Orders

1. Introduction

A standard technique for treating the integro-differential equations of mathematical physics is to take the moments of the equations concerned and consider suitably truncated sets of the resulting equations. An advantage in considering such moment equations is that the equations of the lowest orders have, often, simple physical interpretations; and, moreover, in many instances, their solutions, with suitable assumptions of "closure", suggest methods of obtaining approximate solutions of the exact equations in a systematic way. Such methods have been used successfully in statistical mechanics (in the treatment of the Boltzmann equation) and in the theory of radiative transfer. But the method has not been used, except recently, in the treatment of the equations which govern self-gravitating systems even though, from one point of view, it is simplest to consider them as integro-differential equations by expressing the Newtonian gravitational potential as an integral over the "unknown" density distribution by Poisson's formula. The method of the virial is essentially the method of the moments as applied to systems in which gravitation determines equilibrium. In describing and developing this method of the virial in these lectures, the attempt will not be to present the method in its most general framework; the emphasis will rather be on illustrating the usefulness of the method by applying it to the solution of certain concrete problems.

2. The integrals of the n-body problem

As the simplest dynamical system in which the gravitational interactions play the determining role, we shall consider a set of n discrete mass points $m^{(\alpha)}$ ($\alpha = 1, 2, \ldots, n$). Let $x_i^{(\alpha)}$ and $u_i^{(\alpha)}$($i = 1, 2, 3$;

* Presented at the THEORETICAL PHYSICS INSTITUTE, University of Colorado, Summer 1963.

$\alpha = 1, \ldots, n$) denote the co-ordinates and the velocities of the mass points in a chosen inertial frame of reference. Note that the inferior Latin index refers to the Cartesian components while the superior Greek index distinguishes the mass points.

We define the gravitational potential $\mathfrak{V}^{(\alpha)}$ (appropriate to the mass-point α) by

$$\mathfrak{V}^{(\alpha)} = G \sum_{\beta \neq \alpha} \frac{m^{(\beta)}}{|\vec{x}^{(\alpha)} - \vec{x}^{(\beta)}|}, \tag{1}$$

where G denotes the constant of gravitation and $\vec{x}^{(\alpha)}$ and $\vec{x}^{(\beta)}$ are the instantaneous positions of the mass points α and β; and the summation in (1) is over all mass points β, exclusive of α.

The equations of motion governing the mass points are

$$\begin{aligned} m^{(\alpha)} \frac{d^2 x_i^{(\alpha)}}{dt^2} = m^{(\alpha)} \frac{du_i^{(\alpha)}}{dt} &= m^{(\alpha)} \frac{\partial \mathfrak{V}^{(\alpha)}}{\partial x_i^{(\alpha)}} \\ &= -Gm^{(\alpha)} \sum_{\beta \neq \alpha} m^{(\beta)} \frac{x_i^{(\alpha)} - x_i^{(\beta)}}{|\vec{x}^{(\alpha)} - \vec{x}^{(\beta)}|^3}. \end{aligned} \tag{2}$$

These are the equations of the classical n-body problem. The standard integrals of this problem are no more than the moments of these equations.

a) <u>The uniform motion of the center of mass</u>

Summing equation (2) over all α, we obtain

$$\sum_\alpha m^{(\alpha)} \frac{du_i^{(\alpha)}}{dt} = -G \sum_\alpha \sum_{\beta \neq \alpha} m^{(\alpha)} m^{(\beta)} \frac{x_i^{(\alpha)} - x_i^{(\beta)}}{|\vec{x}^{(\alpha)} - \vec{x}^{(\beta)}|^3} = 0, \tag{3}$$

since the summand on the right-hand side is anti-symmetrical in α and β. Hence the center of mass, as defined by the equation

$$X_i = \frac{\sum_\alpha m^{(\alpha)} x_i^{(\alpha)}}{\sum_\alpha m^{(\alpha)}}, \tag{4}$$

has the motion

$$\frac{dX_i}{dt} = V_i = \frac{\sum_\alpha m^{(\alpha)} u_i^{(\alpha)}}{\sum_\alpha m^{(\alpha)}} = \text{constant;} \tag{5}$$

therefore,

$$X_i = V_i t + X_i(o) \qquad (X_i(o) = \text{constant}). \tag{6}$$

b) The energy integral

By multiplying equation (2) by $u_i^{(\alpha)}$ and summing over α (and also over i, as a repeated index) we have

$$\sum_\alpha m^{(\alpha)} u_i^{(\alpha)} \frac{du_i^{(\alpha)}}{dt} = -G \sum_\alpha \sum_{\beta \neq \alpha} m^{(\alpha)} m^{(\beta)} \frac{u_i^{(\alpha)}\left(x_i^{(\alpha)} - x_i^{(\beta)}\right)}{|\vec{x}^{(\alpha)} - \vec{x}^{(\beta)}|^3}. \tag{7}$$

By taking the average of equation (7) and the equation derived from it by interchanging α and β in the double summation on the right-hand side, we obtain

$$\begin{aligned} \frac{1}{2} \sum_\alpha m^{(\alpha)} \frac{d}{dt} |\vec{u}^{(\alpha)}|^2 &= -\frac{1}{2} G \sum_\alpha \sum_{\beta \neq \alpha} m^{(\alpha)} m^{(\beta)} \frac{\left(x_i^{(\alpha)} - x_i^{(\beta)}\right)\left(u_i^{(\alpha)} - u_i^{(\beta)}\right)}{|\vec{x}^{(\alpha)} - \vec{x}^{(\beta)}|^3} \\ &= \frac{1}{2} G \frac{d}{dt} \sum_\alpha \sum_{\beta \neq \alpha} \frac{m^{(\alpha)} m^{(\beta)}}{|\vec{x}^{(\alpha)} - \vec{x}^{(\beta)}|}. \end{aligned} \tag{8}$$

Letting

$$\mathfrak{T} = \frac{1}{2} \sum_\alpha m^{(\alpha)} |\vec{u}^{(\alpha)}|^2 \tag{9}$$

and

$$\mathfrak{W} = -\frac{1}{2} \sum_\alpha m^{(\alpha)} \mathfrak{B}^{(\alpha)} = -\frac{1}{2} G \sum_\alpha \sum_{\beta \neq \alpha} \frac{m^{(\alpha)} m^{(\beta)}}{|\vec{x}^{(\alpha)} - \vec{x}^{(\beta)}|} \tag{10}$$

denote the kinetic and the potential energies of the system, we can write equation (8) in the form

$$\frac{d}{dt}(\mathfrak{T} + \mathfrak{W}) = 0, \tag{11}$$

or

$$\mathfrak{T} + \mathfrak{W} = \text{constant}. \tag{12}$$

This is the energy integral.

c) The angular-momentum integral and the second-order virial theorem

Multiply equation (2) by $x_j^{(\alpha)}$ and sum over all α. We obtain

$$\sum_\alpha m^{(\alpha)} x_j^{(\alpha)} \frac{d^2 x_i^{(\alpha)}}{dt^2} = -G \sum_\alpha \sum_{\beta \neq \alpha} m^{(\alpha)} m^{(\beta)} \frac{x_j^{(\alpha)}\left(x_i^{(\alpha)} - x_i^{(\beta)}\right)}{|\vec{x}^{(\alpha)} - \vec{x}^{(\beta)}|^3}. \tag{13}$$

The left-hand side of this equation can be rewritten as

$$\sum_\alpha m^{(\alpha)} \frac{d}{dt}\left(x_j^{(\alpha)} u_i^{(\alpha)}\right) - \sum_\alpha m^{(\alpha)} u_i^{(\alpha)} u_j^{(\alpha)}, \tag{14}$$

while the right-hand side can be reduced in the manner

$$-G \sum_\alpha \sum_{\beta \neq \alpha} m^{(\alpha)} m^{(\beta)} \frac{x_j^{(\alpha)}\left(x_i^{(\alpha)} - x_i^{(\beta)}\right)}{|\vec{x}^{(\alpha)} - \vec{x}^{(\beta)}|^3}$$

$$= +G \sum_\alpha \sum_{\beta \neq \alpha} m^{(\alpha)} m^{(\beta)} \frac{x_j^{(\beta)}\left(x_i^{(\alpha)} - x_i^{(\beta)}\right)}{|\vec{x}^{(\alpha)} - \vec{x}^{(\beta)}|^3} \tag{15}$$

$$= -\frac{1}{2} G \sum_\alpha \sum_{\beta \neq \alpha} m^{(\alpha)} m^{(\beta)} \frac{\left(x_j^{(\alpha)} - x_j^{(\beta)}\right)\left(x_i^{(\alpha)} - x_i^{(\beta)}\right)}{|\vec{x}^{(\alpha)} - \vec{x}^{(\beta)}|^3}.$$

The foregoing reductions of the two sides of equation (13) suggest that we define the tensors

$$\mathfrak{T}_{ij} = \frac{1}{2} \sum_\alpha m^{(\alpha)} u_i^{(\alpha)} u_j^{(\alpha)}, \tag{16}$$

$$\mathfrak{B}_{ij}^{(\alpha)} = G \sum_{\beta \neq \alpha} m^{(\beta)} \frac{\left(x_i^{(\alpha)} - x_i^{(\beta)}\right)\left(x_j^{(\alpha)} - x_j^{(\beta)}\right)}{|\vec{x}^{(\alpha)} - \vec{x}^{(\beta)}|^3}, \tag{17}$$

and

$$\mathfrak{W}_{ij} = -\frac{1}{2}\sum_{\alpha} m^{(\alpha)}\mathfrak{B}_{ij}^{(\alpha)} . \tag{18}$$

These tensors represent tensorial generalizations of the kinetic energy $\mathfrak{T}$, the potential $\mathfrak{B}$ and the potential energy $\mathfrak{W}$, as usually defined; and they reduce to the corresponding scalars on contraction:

$$\mathfrak{T}_{ii} = \mathfrak{T}, \quad \mathfrak{B}_{ii}^{(\alpha)} = \mathfrak{B}^{(\alpha)} \quad \text{and} \quad \mathfrak{W}_{ii} = \mathfrak{W}. \tag{19}$$

The tensors $\mathfrak{T}_{ij}$, $\mathfrak{B}_{ij}^{(\alpha)}$, and $\mathfrak{W}_{ij}$ are manifestly symmetric in i and j.

In terms of the tensors we have defined, equation (13), in virtue of the reductions (14) and (15), takes the form

$$\frac{d}{dt}\sum_{\alpha} m^{(\alpha)}u_i^{(\alpha)}x_j^{(\alpha)} = 2\mathfrak{T}_{ij} + \mathfrak{W}_{ij}. \tag{20}$$

The right-hand side of equation (20) is clearly symmetric in i and j. The anti-symmetric part of the tensor on the left-hand side must therefore vanish. Thus

$$\frac{d}{dt}\sum_{\alpha} m^{(\alpha)}\left(u_i^{(\alpha)}x_j^{(\alpha)} - u_j^{(\alpha)}x_i^{(\alpha)}\right) = 0, \tag{21}$$

or

$$\sum_{\alpha} m^{(\alpha)}\left(u_i^{(\alpha)}x_j^{(\alpha)} - u_j^{(\alpha)}x_i^{(\alpha)}\right) = \text{constant.} \tag{22}$$

Equation (22) expresses the conservation of the angular momentum of the system.

The symmetric part of equation (20) leads to the virial theorem:

$$\frac{1}{2}\frac{d}{dt}\sum_{\alpha} m^{(\alpha)}\left(u_i^{(\alpha)}x_j^{(\alpha)} + u_j^{(\alpha)}x_i^{(\alpha)}\right) = \frac{1}{2}\frac{d^2}{dt^2}\sum_{\alpha} m^{(\alpha)}x_i^{(\alpha)}x_j^{(\alpha)}$$
$$= 2\mathfrak{T}_{ij} + \mathfrak{W}_{ij}. \tag{23}$$

Thus the conservation of the angular momentum and the virial theorem have their common origin in equation (20).

Defining the moment of inertia tensor

$$I_{ij} = \sum_{\alpha} m^{(\alpha)} x_i^{(\alpha)} x_j^{(\alpha)}, \tag{24}$$

we can rewrite equation (23) in the manifestly symmetric form

$$\frac{1}{2}\frac{d^2 I_{ij}}{dt^2} = 2\mathfrak{T}_{ij} + \mathfrak{W}_{ij}. \tag{25}$$

On contracting this equation we obtain

$$\frac{1}{2}\frac{d^2 I}{dt^2} = 2\mathfrak{T} + \mathfrak{W}; \tag{26}$$

this is the usual scalar form of the virial theorem; and in this form it was first derived by Jacobi (though for the special case of the three-body problem it was known already to Lagrange).

If the system considered is multiply periodic (or if all the particles remain for all time in the finite part of the phase space) then by averaging equation (20) over a sufficiently long time T, we obtain

$$2\langle \mathfrak{T}_{ij}\rangle + \langle \mathfrak{W}_{ij}\rangle = \lim_{T\to\infty} \frac{1}{T}\left(\sum_{\alpha} m^{(\alpha)} u_i^{(\alpha)} x_j^{(\alpha)}\right)\Bigg|_0^T = 0, \tag{27}$$

where the angular brackets signify that the quantity enclosed has been averaged over a sufficiently long time. The contracted version of equation (27) is the relation (usually known as <u>the</u> virial theorem)

$$2\langle \mathfrak{T}\rangle + \langle \mathfrak{W}\rangle = 0. \tag{28}$$

3. <u>The tensor potential $\mathfrak{B}_{ij}^{(\alpha)}$ and the superpotential $\chi^{(\alpha)}$</u>

The tensor potential $\mathfrak{B}_{ij}^{(\alpha)}$, which we have defined in §2, can be derived from the scalar <u>superpotential</u>

$$\chi^{(\alpha)} = -G \sum_{\beta\neq\alpha} m^{(\beta)} |\vec{x}^{(\alpha)} - \vec{x}^{(\beta)}|. \tag{29}$$

Thus

$$\frac{\partial^2 \chi^{(\alpha)}}{\partial x_i^{(\alpha)} \partial x_j^{(\alpha)}} = -G \frac{\partial}{\partial x_i^{(\alpha)}} \sum_{\beta \neq \alpha} m^{(\beta)} \frac{x_j^{(\alpha)} - x_j^{(\beta)}}{|\vec{x}^{(\alpha)} - \vec{x}^{(\beta)}|}$$

$$= G \sum_{\beta \neq \alpha} m^{(\beta)} \frac{\left(x_i^{(\alpha)} - x_i^{(\beta)}\right)\left(x_j^{(\alpha)} - x_j^{(\beta)}\right)}{|\vec{x}^{(\alpha)} - \vec{x}^{(\beta)}|^3} - \delta_{ij} G \sum_{\beta \neq \alpha} \frac{m^{(\beta)}}{|\vec{x}^{(\alpha)} - \vec{x}^{(\beta)}|}$$

$$= \mathfrak{B}_{ij}^{(\alpha)} - \delta_{ij} \mathfrak{B}^{(\alpha)}, \tag{30}$$

or

$$\mathfrak{B}_{ij}^{(\alpha)} = \frac{\partial^2 \chi^{(\alpha)}}{\partial x_i^{(\alpha)} \partial x_j^{(\alpha)}} + \delta_{ij} \mathfrak{B}^{(\alpha)}. \tag{31}$$

By contracting equation (31), we obtain

$$\nabla_{(\alpha)}^2 \chi^{(\alpha)} = -2 \mathfrak{B}^{(\alpha)}. \tag{32}$$

An alternative expression for $\mathfrak{B}_{ij}^{(\alpha)}$ in terms of the potential,

$$\mathfrak{D}_i^{(\alpha)} = G \sum_{\beta \neq \alpha} m^{(\beta)} \frac{x_i^{(\beta)}}{|\vec{x}^{(\alpha)} - \vec{x}^{(\beta)}|}, \tag{33}$$

is useful in the further developments of the theory. We have

$$\frac{\partial \mathfrak{D}_i^{(\alpha)}}{\partial x_j^{(\alpha)}} = -G \sum_{\beta \neq \alpha} m^{(\beta)} \frac{x_i^{(\beta)}\left(x_j^{(\alpha)} - x_j^{(\beta)}\right)}{|\vec{x}^{(\alpha)} - \vec{x}^{(\beta)}|^3}$$

$$= G \sum_{\beta \neq \alpha} m^{(\beta)} \frac{\left(x_i^{(\alpha)} - x_i^{(\beta)}\right)\left(x_j^{(\alpha)} - x_j^{(\beta)}\right)}{|\vec{x}^{(\alpha)} - \vec{x}^{(\beta)}|^3} - G x_i^{(\alpha)} \sum_{\beta \neq \alpha} m^{(\beta)} \frac{x_j^{(\alpha)} - x_j^{(\beta)}}{|\vec{x}^{(\alpha)} - \vec{x}^{(\beta)}|^3}$$

$$= \mathfrak{B}_{ij}^{(\alpha)} + x_i^{(\alpha)} \frac{\partial \mathfrak{B}^{(\alpha)}}{\partial x_j^{(\alpha)}}, \tag{34}$$

or

$$\mathfrak{B}_{ij}^{(\alpha)} = \frac{\partial \mathfrak{D}_i^{(\alpha)}}{\partial x_j^{(\alpha)}} - x_i^{(\alpha)} \frac{\partial \mathfrak{B}^{(\alpha)}}{\partial x_j^{(\alpha)}} . \tag{35}$$

4. The second-order virial equations for a continuous distribution of matter with hydrostatic pressure

The generalizations of the results of §§1-3 for continuous distributions of matter can be readily accomplished by replacing the sums over α (and/or β) by integrals in the relevant equations by the correspondence

$$\sum_\alpha m^{(\alpha)} \ldots \rightarrow \int_V d\vec{x}' \, \rho(\vec{x}') \ldots , \tag{36}$$

where $\rho(\vec{x}')$ denotes the density at $\vec{x}'$ and the integral is extended over the entire volume V occupied by the fluid. Thus the equation

$$\mathfrak{B}^{(\alpha)} = G \sum_{\beta \neq \alpha} \frac{m^{(\beta)}}{|\vec{x}^{(\alpha)} - \vec{x}^{(\beta)}|} \tag{37}$$

becomes

$$\mathfrak{B}(\vec{x}) = G \int_V d\vec{x}' \frac{\rho(\vec{x}')}{|\vec{x} - \vec{x}'|} ; \tag{38}$$

and similarly for the others. However, in writing the expressions for the components of the energy-momentum tensor $\mathfrak{T}_{ij}$, we must be careful to distinguish the contributions arising from the macroscopic fluid motions, represented by the hydrodynamic velocity $u_i(\vec{x})$ of a fluid element, and the microscopic molecular motions which are responsible for the hydrostatic pressure (p) and the viscous stresses (p_{ij}). While the respective contributions to the energy-momentum tensor can be written down without much difficulty, it is more convenient to obtain the relevant equations, ab initio, from the hydrodynamic equations; in this way, some formal questions concerning the averaging and the limiting processes can be avoided.

Restricting ourselves to the case when the fluid is subject to an (isotropic) hydrostatic pressure p, we have the hydrodynamic equation

$$\rho \frac{du_i}{dt} = -\frac{\partial p}{\partial x_i} + \rho \frac{\partial \mathfrak{B}}{\partial x_i} , \tag{39}$$

where ρ is the density, p is the pressure, u_i is the hydrodynamic fluid velocity, and $\mathfrak{B}$ is the gravitational potential defined as in equation (38).

By integrating equation (39) over the volume V occupied by the fluid, we obtain

$$\int_V \rho(\vec{x}) \frac{du_i}{dt} d\vec{x} = 0. \tag{40}$$

The integral over $\partial p/\partial x_i$ vanishes since p must be assumed to vanish on the boundary of V; and the integral over $\partial \mathfrak{B}/\partial x_i$ also vanishes since the integrand of the double integral which it represents is anti-symmetric in the two variables of integration:

$$\int_V d\vec{x}\, \rho(\vec{x}) \frac{\partial \mathfrak{B}}{\partial x_i} = -G \int_V \int_V d\vec{x}\, d\vec{x}'\, \rho(\vec{x})\, \rho(\vec{x}') \frac{x_i - x_i'}{|\vec{x} - \vec{x}'|^3} = 0. \tag{41}$$

Equation (40) is equivalent to

$$\frac{d}{dt} \int_V \rho(\vec{x})\, u_i\, d\vec{x} = 0; \tag{42}$$

and this equation expresses no more than the uniform motion of the center of mass (in the present context). [Note that we can always write

$$\int_V d\vec{x}\, \rho(\vec{x}) \frac{d}{dt} Q(\vec{x}) = \frac{d}{dt} \int_V d\vec{x}\, \rho(\vec{x})\, Q(\vec{x}), \tag{43}$$

where $Q(\vec{x})$ is an arbitrary function, since the conservation of mass ensures

$$\frac{d}{dt} \int_V d\vec{x}\, \rho(\vec{x}) = 0. \qquad (44)]$$

In the same way, multiplying equation (39) by x_j and integrating over V, we obtain

$$\frac{d}{dt} \int_V \rho\, u_i x_j\, d\vec{x} = 2\mathfrak{T}_{ij} + \mathfrak{W}_{ij} + \Pi \delta_{ij}, \tag{45}$$

where

$$\mathfrak{T}_{ij} = \frac{1}{2}\int_V \rho u_i u_j d\vec{x}, \quad \Pi = \int_V p d\vec{x}, \quad \mathfrak{W}_{ij} = -\frac{1}{2}\int_V \rho \mathfrak{B}_{ij} d\vec{x}, \tag{46}$$

and

$$\mathfrak{B}_{ij}(\vec{x}) = G\int_V d\vec{x}'\rho(\vec{x}')\frac{(x_i - x_i')(x_j - x_j')}{|\vec{x} - \vec{x}'|^3}. \tag{47}$$

In equation (45), the term

$$\Pi\delta_{ij} = -\int_V x_j \frac{\partial p}{\partial x_i} d\vec{x} = \delta_{ij}\int_V p(\vec{x})d\vec{x}, \tag{48}$$

represents the contribution of the microscopic molecular motions to the complete energy momentum tensor.

The anti-symmetric part of the tensor on the left-hand side of equation (45) must clearly vanish. Thus

$$\frac{d}{dt}\int_V \rho(u_i x_j - u_j x_i)d\vec{x} = 0; \tag{49}$$

and this equation expresses the conservation of the angular momentum of the system. And the symmetric part of equation (45) gives

$$\frac{1}{2}\frac{d^2 I_{ij}}{dt^2} = 2\mathfrak{T}_{ij} + \mathfrak{W}_{ij} + \Pi\delta_{ij}, \tag{50}$$

where

$$I_{ij} = \int_V \rho(\vec{x}) x_i x_j d\vec{x}. \tag{51}$$

In a steady state, equation (50) gives

$$2\mathfrak{T}_{ij} + \mathfrak{W}_{ij} + \Pi\delta_{ij} = 0; \tag{52}$$

and if there should be no fluid motion and a state of hydrostatic equilibrium prevails, then

$$\mathfrak{W}_{ij} = -\Pi\delta_{ij}. \tag{53}$$

Finally, we may note that for a continuous distribution of matter, the superpotential χ is defined by the equation

$$\chi(\vec{x}) = -G\int_V \rho(\vec{x}')\,|\vec{x}-\vec{x}'|\,d\vec{x}'; \tag{54}$$

it satisfies the equation

$$\nabla^2\chi = -2\mathfrak{B}. \tag{55}$$

In terms of χ, the tensor potential $\mathfrak{B}_{ij}$ is given by

$$\mathfrak{B}_{ij} = \frac{\partial^2\chi}{\partial x_i \partial x_j} + \mathfrak{B}\delta_{ij}\,. \tag{56}$$

And if we define the Newtonian potential $\mathfrak{D}_i(\vec{x})$ due to the "fictitious density" ρx_i, by

$$\mathfrak{D}_i(\vec{x}) = G\int_V \rho(\vec{x}')\,\frac{x_i'}{|\vec{x}-\vec{x}'|}\,d\vec{x}'\,, \tag{57}$$

then

$$\mathfrak{B}_{ij} = \frac{\partial\mathfrak{D}_i}{\partial x_j} - x_i\frac{\partial\mathfrak{B}}{\partial x_j}\,. \tag{58}$$

Also, it can be readily verified that

$$\mathfrak{D}_i = x_i\mathfrak{B} + \frac{\partial\chi}{\partial x_i}\,. \tag{59}$$

5. <u>The second-order virial theorem in hydromagnetics</u>

The extension of the results of the preceding section to hydromagnetics is straightforward. Equation (39) is now replaced by

$$\rho\frac{du_i}{dt} = -\frac{\partial}{\partial x_i}\left(p + \mu\frac{|\vec{H}|^2}{8\pi}\right) + \frac{\mu}{4\pi}\frac{\partial}{\partial x_k}H_iH_k + \rho\frac{\partial\mathfrak{B}}{\partial x_i}\,, \tag{60}$$

where the additional terms in the magnetic field $\vec{H}$ arise from the Lorentz force,

$$\vec{\mathfrak{L}} = \mu \vec{J} \times \vec{H} = \frac{\mu}{4\pi} \operatorname{Curl} \vec{H} \times \vec{H}, \tag{61}$$

acting on the medium. (In eqs. [60] and [61], μ denotes the magnetic permeability and $\vec{J}$ the current density.)

In integrating equation (60), after multiplication by suitable factors, we must allow for the fact that the prevailing magnetic field $\vec{H}$ may not vanish over the "material" boundary of the configuration where the density ρ vanishes. In such cases, we may extend the integration over all regions where the density and/or the field is non-vanishing; or, alternatively, we may extend the integration over only those regions where the density is non-vanishing and allow for the fact that $\vec{H}$ may not vanish on the boundary of the volume of integration. Normally, the latter procedure will require the inclusion of surface integrals in the integral relations which embody the virial theorem. If the former procedure (which avoids the inclusion of surface integrals) is adopted, the terms in $\vec{H}$ in equation (60) will contribute the following additional terms to the second-order virial equation:

$$-\int x_j \left\{ \frac{\partial}{\partial x_i}\left(\mu \frac{|\vec{H}|^2}{8\pi}\right) - \frac{\mu}{4\pi} \frac{\partial}{\partial x_k} H_i H_k \right\} d\vec{x}$$

$$= \delta_{ij} \int \mu \frac{|\vec{H}|^2}{8\pi} d\vec{x} - \int \mu \frac{H_i H_j}{4\pi} d\vec{x} \tag{62}$$

$$= (\mathfrak{M}\delta_{ij} - 2\mathfrak{M}_{ij})$$

where

$$\mathfrak{M}_{ij} = \frac{1}{8\pi} \int \mu H_i H_j d\vec{x}. \tag{63}$$

And the second-order virial equation takes the form

$$\frac{d}{dt} \int \rho u_i x_j d\vec{x} = 2(\mathfrak{T}_{ij} - \mathfrak{M}_{ij}) + \delta_{ij}(\Pi + \mathfrak{M}) + \mathfrak{W}_{ij}. \tag{64}$$

The conservation of angular momentum obtains as in the hydrodynamic case; and equation (64) can be written in the manifestly symmetric form

$$\frac{1}{2} \frac{d^2 I_{ij}}{dt^2} = 2(\mathfrak{T}_{ij} - \mathfrak{M}_{ij}) + \delta_{ij}(\Pi + \mathfrak{M}) + \mathfrak{W}_{ij}. \tag{65}$$

In a steady state, equation (65) gives

$$2(\mathfrak{T}_{ij} - \mathfrak{M}_{ij}) + \delta_{ij}(\Pi + \mathfrak{M}) + \mathfrak{W}_{ij} = 0; \tag{66}$$

and the contracted version of this equation is

$$2\mathfrak{T} + \mathfrak{M} + \mathfrak{W} + 3\Pi = 0. \tag{67}$$

An important consequence of this last equation is

$$\mathfrak{M} < |\mathfrak{W}|; \tag{68}$$

and this inequality limits the total magnetic energy in a self-gravitating system to be less than $|\mathfrak{W}|$.

6. The third-order virial equations

In obtaining the virial equations of the second order, we multiplied the equation of motion by x_j and integrated over the volume occupied by the fluid. This procedure can clearly be extended to obtain moment equations of higher order. Thus, by multiplying the equation of motion by $x_j x_k$ and integrating over V we obtain the virial equations of the third order. Thus, restricting ourselves to the framework of hydrodynamics, we have

$$\int_V \rho x_j x_k \frac{du_i}{dt}\, d\vec{x} = -\int_V x_j x_k \frac{\partial p}{\partial x_i}\, d\vec{x} + \int_V \rho x_j x_k \frac{\partial \mathfrak{B}}{\partial x_i}\, d\vec{x}. \tag{69}$$

Considering in turn the different terms in this equation, we have

$$\int_V \rho x_j x_k \frac{du_i}{dt}\, d\vec{x} = \int_V \rho \frac{d}{dt}(u_i x_j x_k)\, d\vec{x} - \int_V \rho u_i (u_j x_k + u_k x_j)\, d\vec{x}, \tag{70}$$

$$-\int_V x_j x_k \frac{\partial p}{\partial x_i}\, d\vec{x} = \int_V p(x_k \delta_{ij} + x_j \delta_{ik})\, d\vec{x}, \tag{71}$$

and

$$\int_V \rho x_j x_k \frac{\partial \mathfrak{B}}{\partial x_i} d\vec{x} = -G \int_V \int_V d\vec{x}\, d\vec{x}'\, \rho(\vec{x}) \rho(\vec{x}') \frac{(x_i - x_i')x_j x_k}{|\vec{x} - \vec{x}'|^3}$$

$$= -\frac{1}{2} G \int_V \int_V d\vec{x}\, d\vec{x}'\, \rho(\vec{x}) \rho(\vec{x}') \frac{(x_i - x_i')(x_j x_k - x_j' x_k')}{|\vec{x} - \vec{x}'|^3}$$

$$= -\frac{1}{2} G \int_V \int_V d\vec{x}\, d\vec{x}'\, \rho(\vec{x}) \rho(\vec{x}') \frac{(x_i - x_i')[(x_k - x_k')x_j + (x_j - x_j')x_k']}{|\vec{x} - \vec{x}'|^3}$$

$$= -\frac{1}{2} G \int_V d\vec{x}\, \rho(\vec{x}) x_j \int_V d\vec{x}'\, \rho(\vec{x}') \frac{(x_i - x_i')(x_k - x_k')}{|\vec{x} - \vec{x}'|^3} \tag{72}$$

$$-\frac{1}{2} G \int_V d\vec{x}'\, \rho(\vec{x}') x_k' \int_V d\vec{x}\, \rho(\vec{x}) \frac{(x_i' - x_i)(x_j' - x_j)}{|\vec{x} - \vec{x}'|^3}$$

$$= -\frac{1}{2} \int_V \rho(\vec{x}) x_j \mathfrak{B}_{ik}\, d\vec{x} - \frac{1}{2} \int_V \rho(\vec{x}) x_k \mathfrak{B}_{ij}\, d\vec{x}.$$

We now define the tensors

$$\mathfrak{T}_{ij;k} = \frac{1}{2} \int_V \rho u_i u_j x_k\, d\vec{x}, \qquad \Pi_i = \int_V p x_i\, d\vec{x},$$

and

$$\mathfrak{W}_{ij;k} = -\frac{1}{2} \int_V \rho \mathfrak{B}_{ij} x_k\, d\vec{x}. \tag{73}$$

[Note that an index after the semicolon indicates that a moment with respect to the associated space co-ordinate is involved.] In terms of the tensors we have defined, equation (69) now becomes

$$\frac{d}{dt} \int_V \rho u_i x_j x_k\, d\vec{x} = 2(\mathfrak{T}_{ij;k} + \mathfrak{T}_{ik;j}) + \mathfrak{W}_{ij;k} + \mathfrak{W}_{ik;j}$$

$$+ \delta_{ij} \Pi_k + \delta_{ik} \Pi_j. \tag{74}$$

In particular, in a steady state

$$2(\mathfrak{T}_{ij;k} + \mathfrak{T}_{ik;j}) + \mathfrak{W}_{ij;k} + \mathfrak{W}_{ik;j} + \delta_{ij}\Pi_k + \delta_{ik}\Pi_j = 0. \tag{75}$$

Equation (75) represents a total of eighteen equations.

7. Tensor-potentials of higher rank

We have seen that the second-order virial equations require us to consider the superpotential

$$\chi(\vec{x}) = -G\int_V d\vec{x}'\rho(\vec{x}')\,|\vec{x}-\vec{x}'|, \tag{76}$$

and the potentials

$$\mathfrak{D}_i(\vec{x}) = G\int_V d\vec{x}'\rho(\vec{x}')\,\frac{x_i'}{|\vec{x}-\vec{x}'|}, \tag{77}$$

and

$$\mathfrak{B}_{ij}(\vec{x}) = G\int_V d\vec{x}'\rho(\vec{x}')\,\frac{(x_i - x_i')(x_j - x_j')}{|\vec{x}-\vec{x}'|^3} \tag{78}$$

which can be derived in terms of χ and which are related among themselves. Similarly, the third-order virial equations of §6 require us to consider the "super-superpotential"

$$\Phi(\vec{x}) = \frac{1}{3}G\int_V d\vec{x}'\rho(\vec{x}')\,|\vec{x}-\vec{x}'|^3, \tag{79}$$

and the potentials

$$\chi_i(\vec{x}) = -G\int_V d\vec{x}'\rho(\vec{x}')x_i'\,|\vec{x}-\vec{x}'|, \tag{80}$$

$$\mathfrak{D}_{ij}(\vec{x}) = G\int_V d\vec{x}'\rho(\vec{x}')\,\frac{x_i'x_j'}{|\vec{x}-\vec{x}'|}, \tag{81}$$

$$\mathfrak{D}_{ij;k}(\vec{x}) = G\int_V d\vec{x}'\rho(\vec{x}')x_k'\,\frac{(x_i - x_i')(x_j - x_j')}{|\vec{x}-\vec{x}'|^3} \tag{82}$$

and

$$\mathfrak{B}_{ijk\ell}(\vec{x}) = G\int_V d\vec{x}'\rho(\vec{x}')\frac{(x_i - x_i')(x_j - x_j')(x_k - x_k')(x_\ell - x_\ell')}{|\vec{x} - \vec{x}'|^5}, \quad (83)$$

which can all be derived from Φ. And there are also various relations among the potentials (80)-(83). Thus, since $\mathfrak{D}_{ij;k}$ is the tensor potential "$\mathfrak{B}_{ij}$" induced by the "fictitious" density ρx_k, the relations (56) and (58) give

$$\mathfrak{D}_{ij;k} = \frac{\partial \mathfrak{D}_{ik}}{\partial x_j} - x_i\frac{\partial \mathfrak{D}_k}{\partial x_j} = \delta_{ij}\mathfrak{D}_k + \frac{\partial^2 \chi_k}{\partial x_i \partial x_j} . \quad (84)$$

Also, it is apparent that

$$\nabla^2\Phi = -4\chi \quad (85)$$

and

$$\chi_i = \frac{\partial \Phi}{\partial x_i} + x_i\chi; \quad (86)$$

these are the present analogues of equations (55) and (59).

Virial equations of the fourth and higher orders can be similarly written down; and they will require us to consider potentials of correspondingly higher ranks and orders. There is no formal difficulty in listing the relevant equations and identities; but it is unlikely that they will be of much practical use in view of the rapidly increasing complexity in the bookkeeping.

References

S. Chandrasekhar, Hydrodynamic and Hydromagnetic Stability (Oxford, Clarendon Press) 1961.

S. Chandrasekhar, "The Virial Theorem in Hydromagnetics", J. Math. Anal. Applic. 1, 240-252, 1960.

S. Chandrasekhar, "On the Point of Bifurcation Along the Sequence of the Jacobi Ellipsoids", Ap. J., 136, 1048-1068, 1962.

Norman R. Lebovitz, "The Virial Tensor and its Application to Self-gravitating Fluids", Ap. J., 134, 500-536, 1961.

S. Chandrasekhar and Norman R. Lebovitz, "On Super-potentials in the Theory of Newtonian Gravitation", Ap. J., 135, 238-247, 1962.

S. Chandrasekhar and Norman R. Lebovitz, "On the Superpotentials in the Theory of Newtonian Gravitation. II. Tensors of Higher Rank", Ap. J., 136, 1032-1036, 1962.

15

A THEOREM ON ROTATING POLYTROPES

In the theory of polytropic gas spheres, the formula

$$\mathfrak{W} = -\frac{3}{5-n}\frac{GM^2}{R}, \tag{1}$$

due to Emden, giving the potential energy of a configuration of mass M and radius R, is well known. There is a simple generalization of this formula for rotating polytropes (cf. Chandrasekhar 1933) which seems to have escaped notice. The formula in question can be derived most simply by combining the equation of the virial (appropriate for rotating configurations) with an elementary integral which the equations of hydrostatic equilibrium admits when the relation between the pressure and the density is polytropic.

In a frame of reference rotating with an angular velocity $\mathbf{\Omega}$, the equations of motion can be written in the form

$$\rho\frac{du_i}{dt} = -\frac{\partial p}{\partial x_i} + \rho\frac{\partial}{\partial x_i}\left(\mathfrak{B} + \tfrac{1}{2}|\mathbf{\Omega}\times \mathbf{r}|^2\right) + 2\epsilon_{ilm}u_l\Omega_m, \tag{2}$$

where $\mathfrak{B}$ denotes the gravitational potential and the rest of the symbols have their standard meanings. By multiplying equation (2) by x_k and integrating over the volume V of the configuration, we readily obtain (cf. Chandrasekhar 1960; Lebovitz 1961) the relation

$$\frac{d}{dt}\int_V \rho u_i x_k d\mathbf{x} = 2\mathfrak{T}_{ik} + \mathfrak{W}_{ik} + \delta_{ik}\int_V p\, d\mathbf{x} + \Omega^2 I_{ik} - \Omega_i\Omega_j I_{jk} + 2\epsilon_{ilm}\int_V \rho x_k u_l \Omega_m d\mathbf{x}, \tag{3}$$

where

$$\mathfrak{T}_{ik} = \tfrac{1}{2}\int_V \rho u_i u_k d\mathbf{x}, \qquad I_{ik} = \int_V \rho x_i x_k d\mathbf{x}, \tag{4}$$

and

$$\mathfrak{W}_{ik} = -\tfrac{1}{2}G\int_V\int_V \rho(\mathbf{x})\rho(\mathbf{x}')\frac{(x_i - x_i')(x_k - x_k')}{|\mathbf{x}-\mathbf{x}'|^3}d\mathbf{x}\,d\mathbf{x}'$$
$$= \int_V \rho x_k \frac{\partial \mathfrak{B}}{\partial x_i} d\mathbf{x} \tag{5}$$

is the symmetric potential-energy tensor (Chandrasekhar 1960).

In hydrostatic equilibrium $u_i = 0$, and equation (3) gives

$$\mathfrak{W}_{ik} + \delta_{ik}\int_V p\, d\mathbf{x} + \Omega^2 I_{ik} - \Omega_i\Omega_j I_{jk} = 0; \tag{6}$$

in particular, for $i \neq k$, we have

$$\mathfrak{W}_{ik} + \Omega^2 I_{ik} - \Omega_i\Omega_j I_{jk} = 0 \qquad (i\neq k), \tag{7}$$

while, on contracting with respect to i and k, we obtain

$$\mathfrak{W} + 3\int_V p\, d\mathbf{x} + \Omega^2 I_{ik} - \Omega_i\Omega_j I_{ji} = 0, \tag{8}$$

where $\mathfrak{W}$ now represents the gravitational potential energy.

If the axis of rotation be chosen as the z-axis, equation (8) takes the simpler form

$$\mathfrak{W} + 3\int_V p\,dx + \Omega^2 I_\varpi = 0\,, \tag{9}$$

where

$$I_\varpi = \int_V \rho\,(x^2 + y^2)\,dx\,. \tag{10}$$

Now, for a polytropic gas,

$$p = \text{Constant}\ \rho^{1+1/n}, \tag{11}$$

and the equations of hydrostatic equilibrium can be rewritten as

$$(n+1)\,\text{grad}(p/\rho) = \text{grad}(\mathfrak{B} + \tfrac{1}{2}\,|\,\boldsymbol{\Omega}\times \boldsymbol{r}\,|^2)\,; \tag{12}$$

and this equation integrates to give

$$(n+1)\,p = \rho\,(\mathfrak{B} + \tfrac{1}{2}\,|\,\boldsymbol{\Omega}\times \boldsymbol{r}\,|^2 - \mathfrak{B}_0)\,, \tag{13}$$

where $\mathfrak{B}_0$ is the gravitational potential at the pole of the configuration. On integrating equation (13) over the volume of the configuration, we obtain

$$(n+1)\int_V p\,dx = \int_V \rho\,\mathfrak{B}\,dx + \tfrac{1}{2}\Omega^2 I_\varpi - \mathfrak{B}_0 M\,, \tag{14}$$

where M denotes the mass of the configuration. But

$$\mathfrak{W} = -\tfrac{1}{2}\int_V \rho\,\mathfrak{B}\,dx\,. \tag{15}$$

Hence

$$(n+1)\int_V p\,dx = -2\mathfrak{W} + \tfrac{1}{2}\Omega^2 I_\varpi - \mathfrak{B}_0 M\,. \tag{16}$$

Inserting this last relation in equation (9), we obtain

$$\mathfrak{W} = \frac{1}{5-n}\,[-3\mathfrak{B}_0 M + (n+\tfrac{5}{2})\,\Omega^2 I_\varpi]\,; \tag{17}$$

this is the required generalization of Emden's formula.

If one assumes that the rotating polytrope is a tri-axial body (with the direction of $\boldsymbol{\Omega}$ being one of the axes), then one can derive simple expressions for all the components of the potential-energy tensor. Thus, with the co-ordinate axes chosen along the principal axes of the potential-energy tensor, the virial theorem as expressed by equations (6) and (16) gives

$$\begin{aligned}\mathfrak{W}_{zz} &= \mathfrak{W}_{xx} + \Omega^2 I_{xx}\\ &= \mathfrak{W}_{yy} + \Omega^2 I_{yy} = -\int_V p\,dx\\ &= \frac{1}{n+1}\,\{2\,(\mathfrak{W}_{xx} + \mathfrak{W}_{yy} + \mathfrak{W}_{zz}) - \tfrac{1}{2}\Omega^2 (I_{xx} + I_{yy}) + \mathfrak{B}_0 M\}\end{aligned} \tag{18}$$

These equations can be solved to give:

$$\mathfrak{W}_{xx} = \frac{1}{5-n} \{ -\mathfrak{V}_0 M + (n - \tfrac{5}{2}) \Omega^2 I_{xx} + \tfrac{5}{2} \Omega^2 I_{yy} \} ,$$

$$\mathfrak{W}_{yy} = \frac{1}{5-n} \{ -\mathfrak{V}_0 M + \tfrac{5}{2} \Omega^2 I_{xx} + (n - \tfrac{5}{2}) \Omega^2 I_{yy} \} , \tag{19}$$

$$\mathfrak{W}_{zz} = \frac{1}{5-n} \{ -\mathfrak{V}_0 M + \tfrac{5}{2} \Omega^2 (I_{xx} + I_{yy}) \} .$$

If the rotating polytrope has axial symmetry about **Ω**, then

$$\mathfrak{W}_{xx} = \mathfrak{W}_{yy} \quad \text{and} \quad I_{xx} = I_{yy} ; \tag{20}$$

but axial symmetry about **Ω** is not always unconditional. Genuine tri-axial configurations can exist; and the condition for their existence is simply

$$(\mathfrak{W}_{xx} - \mathfrak{W}_{yy}) = -\Omega^2 (I_{xx} - I_{yy}) , \tag{21}$$

or, equivalently,

$$I_{yy}(\mathfrak{W}_{xx} - \mathfrak{W}_{zz}) = I_{xx}(\mathfrak{W}_{yy} - \mathfrak{W}_{zz}) . \tag{22}$$

It can be readily verified that for ellipsoidal configurations of constant density, equations (21) and (22) characterize the classical Jacobi ellipsoids [cf. Lebovitz (1960), eqs. (59) and (60)].

The approach to the study of rotating gaseous masses via the virial equations (3) and (8) is a fruitful one; and as Lebovitz' investigation has demonstrated, it is also a powerful one. Further developments along the lines of this communication will be published in the near future.

S. CHANDRASEKHAR

UNIVERSITY OF CHICAGO

REFERENCES

Chandrasekhar, S. 1933, *M.N.R.A.S.*, **93**, 390.
———. 1960, *J. Math. Anal. Appl.*, **1**, 240.
Lebovitz, N. 1961, *Ap. J.*, **134**, 500.

16

ON SUPER-POTENTIALS IN THE THEORY OF NEWTONIAN GRAVITATION

S. CHANDRASEKHAR AND NORMAN R. LEBOVITZ
University of Chicago and Massachusetts Institute of Technology
Received September 1, 1961

ABSTRACT

The character of the gravitational equilibrium of bodies in rotation and with prevalent magnetic fields depends on the tensor potential,

$$\mathfrak{B}_{ij} = G \int_V \rho(x') \frac{(x_i - x_i')(x_j - x_j')}{|x - x'|^3} dx',$$

and the associated tensors,

$$\mathfrak{W}_{ij} = -\frac{1}{2} \int_V \rho \mathfrak{B}_{ij} dx \qquad \text{and} \qquad \mathfrak{W}_{pq;ij} = \int_V \rho x_p \frac{\partial \mathfrak{B}_{ij}}{\partial x_q} dx .$$

This paper is devoted to a consideration of these fundamental tensors. It is shown, in particular, that the tensor potential can be expressed in the form

$$\mathfrak{B}_{ij} = \mathfrak{B} \delta_{ij} + \frac{\partial^2 \chi}{\partial x_i \partial x_j},$$

where $\mathfrak{B}$ is the gravitational potential as usually defined and χ is a *super-potential* determined by the equation

$$\nabla^2 \chi = -2 \mathfrak{B} .$$

I. INTRODUCTION

The development of the virial theorem in its general tensor form (Chandrasekhar 1960, 1961*a*) requires the consideration of the symmetric tensor potential

$$\mathfrak{B}_{ij}(x) = G \int_V \rho(x') \frac{(x_i - x_i')(x_j - x_j')}{|x - x'|^3} dx', \tag{1}$$

where, for the sake of brevity, we have written $dx' = dx_1\, dx_2\, dx_3$ and abridged three integral signs into one. In equation (1) the integration is effected over the entire volume V occupied by the fluid, and G is the constant of gravitation. The gravitational potential,

$$\mathfrak{B}(x) = G \int_V \frac{\rho(x')}{|x - x'|} dx', \tag{2}$$

as usually defined, is the contracted form of $\mathfrak{B}_{ij}$; thus

$$\mathfrak{B} = \mathfrak{B}_{ii} . \tag{3}$$

Associated with the tensor potential $\mathfrak{B}_{ij}$ is the potential energy tensor,

$$\mathfrak{W}_{ij} = -\tfrac{1}{2} G \int_V \int_V \rho(x) \rho(x') \frac{(x_i - x_i')(x_j - x_j')}{|x - x'|^3} dx dx' ; \tag{4}$$

and this tensor represents a similar generalization of the usual definition of the gravitational potential energy:

$$\mathfrak{W} = \mathfrak{W}_{ii} = -\tfrac{1}{2}G\int_V\int_V \frac{\rho(\boldsymbol{x})\,\rho(\boldsymbol{x}')}{|\boldsymbol{x}-\boldsymbol{x}'|}\,d\boldsymbol{x}\,d\boldsymbol{x}' . \tag{5}$$

A further quantity which occurs when treating small oscillations about equilibrium is the super-matrix (cf. Chandraskehar 1960 and 1961*a*, § 118)

$$\mathfrak{W}_{pq;ij} = \int_V \rho(\boldsymbol{x})\,x_p \frac{\partial \mathfrak{B}_{ij}}{\partial x_q}\,d\boldsymbol{x} . \tag{6}$$

It appears that a knowledge of the tensor potential $\mathfrak{B}_{ij}$ and the associated tensors $\mathfrak{W}_{ij}$ and $\mathfrak{W}_{pq;ij}$ is essential to an understanding of the character of the gravitational equilibrium of bodies in rotation and with magnetic field. Thus many features of the equilibrium of rotating masses which appear unexpected and obscure can be readily interpreted and understood in terms of the tensors $\mathfrak{W}_{ij}$ and $\mathfrak{W}_{pq;ij}$. For these reasons, and also for their own interest, this paper will be devoted to a consideration of these fundamental tensors in gravitational potential theory.

II. THE TENSOR POTENTIAL

First, we observe that, in terms of the vector (cf. Lebovitz 1961, eqs. [48] and [49]),

$$\mathfrak{D}_i = G\int_V \rho(\boldsymbol{x}')\frac{x_i'}{|\boldsymbol{x}-\boldsymbol{x}'|}\,d\boldsymbol{x}' , \tag{7}$$

we can express $\mathfrak{B}_{ij}$ in the manner

$$\mathfrak{B}_{ij} = -x_i \frac{\partial \mathfrak{B}}{\partial x_j} + \frac{\partial \mathfrak{D}_i}{\partial x_j} . \tag{8}$$

Since $\mathfrak{B}_{ij}$ is symmetric in its indices,

$$-x_i \frac{\partial \mathfrak{B}}{\partial x_j} + \frac{\partial \mathfrak{D}_i}{\partial x_j} = -x_j \frac{\partial \mathfrak{B}}{\partial x_i} + \frac{\partial \mathfrak{D}_j}{\partial x_i} , \tag{9}$$

or

$$\frac{\partial \mathfrak{D}_i}{\partial x_j} - \frac{\partial \mathfrak{D}_j}{\partial x_i} = x_i \frac{\partial \mathfrak{B}}{\partial x_j} - x_j \frac{\partial \mathfrak{B}}{\partial x_i} . \tag{10}$$

Alternatively, we can write

$$\text{curl}\,\mathfrak{D} = \text{grad}\,\mathfrak{B} \times \boldsymbol{x} = \text{curl}\,(\mathfrak{B}\boldsymbol{x}) . \tag{11}$$

Consequently, $\mathfrak{D}$ can differ from $\mathfrak{B}\boldsymbol{x}$ only by the gradient of a scalar function; and, denoting this scalar function by χ, we can write

$$\mathfrak{D}_i = \mathfrak{B}x_i + \frac{\partial \chi}{\partial x_i} . \tag{12}$$

The equation governing χ can be obtained by substituting the foregoing expression for $\mathfrak{D}$ in the contracted version of equation (8), namely,

$$\mathfrak{B} = -x_i \frac{\partial \mathfrak{B}}{\partial x_i} + \frac{\partial \mathfrak{D}_i}{\partial x_i} ; \tag{13}$$

we find

$$\mathfrak{B} = -x_i \frac{\partial \mathfrak{B}}{\partial x_i} + \left(x_i \frac{\partial \mathfrak{B}}{\partial x_i} + 3\mathfrak{B} + \nabla^2 \chi \right). \tag{14}$$

Hence

$$\nabla^2 \chi = -2\mathfrak{B} . \tag{15}$$

Taking the Laplacian of equation (15) and making use of Poisson's equation

$$\nabla^2 \mathfrak{B} = -4\pi G \rho \tag{16}$$

we obtain

$$\nabla^4 \chi = 8\pi G \rho . \tag{17}$$

It would thus be appropriate to call χ the *super-potential* of the gravitational field. In its term the tensor potential $\mathfrak{B}_{ij}$ is given by

$$\mathfrak{B}_{ij} = \mathfrak{B}\,\delta_{ij} + \frac{\partial^2 \chi}{\partial x_i \partial x_j} . \tag{18}$$

There is an alternative way of exhibiting the relationship between $\mathfrak{B}$ and χ that is instructive. Define

$$\chi = -G \int_V \rho(\boldsymbol{x}') \, | \boldsymbol{x} - \boldsymbol{x}' | \, d\boldsymbol{x}' . \tag{19}$$

By differentiating this equation with respect to x_i and making use of equations (2) and (7) (which *define* $\mathfrak{B}$ and $\mathfrak{D}_i$), we obtain

$$\frac{\partial \chi}{\partial x_i} = -G \int_V \rho(\boldsymbol{x}') \frac{x_i - x_i'}{| \boldsymbol{x} - \boldsymbol{x}' |} d\boldsymbol{x}' = -x_i \mathfrak{B} + \mathfrak{D}_i , \tag{20}$$

in agreement with equation (12). And, by a further differentiation, we find

$$\begin{aligned} \frac{\partial^2 \chi}{\partial x_j \partial x_i} &= -G \int_V \rho(\boldsymbol{x}') \left[\frac{\delta_{ij}}{| \boldsymbol{x} - \boldsymbol{x}' |} - \frac{(x_i - x_i')(x_j - x_j)}{| \boldsymbol{x} - \boldsymbol{x}' |^3} \right] d\boldsymbol{x}' \\ &= -\mathfrak{B}\,\delta_{ij} + \mathfrak{B}_{ij} , \end{aligned} \tag{21}$$

in agreement with equation (18).

The very existence of equation (19) shows that the entire gravitational problem is reduced, in principle at least, to the determination of the single function χ: $\mathfrak{B}$ is determined from χ by means of equation (18) (or [21]). And, moreover, equations (2) and (19) show precisely in what sense χ is to be regarded as a super-potential.

III. THE POTENTIAL ENERGY TENSOR

It follows from the definitions of the respective quantities that (cf. eqs. [1] and [4])

$$\mathfrak{W}_{ij} = -\frac{1}{2} \int_V \rho \mathfrak{B}_{ij} d\boldsymbol{x} . \tag{22}$$

Also, it can be readily shown that (cf. Chandrasekhar 1960, eq. [10])

$$\mathfrak{W}_{ij} = \int_V \rho x_i \frac{\partial \mathfrak{B}}{\partial x_j} d\boldsymbol{x} = \int_V \rho x_j \frac{\partial \mathfrak{B}}{\partial x_i} d\boldsymbol{x} . \tag{23}$$

The contraction of equations (22) and (23) leads to known results in potential theory; thus

$$\mathfrak{W} = -\frac{1}{2}\int_V \rho \mathfrak{V} d\boldsymbol{x} = \int_V \rho x_i \frac{\partial \mathfrak{V}}{\partial x_i} d\boldsymbol{x} . \tag{24}$$

IV. THE SUPER-MATRIX $\mathfrak{W}_{pq;ij}$

As defined, the tensor

$$\mathfrak{W}_{pq;ij} = \int_V \rho x_p \frac{\partial \mathfrak{V}_{ij}}{\partial x_q} d\boldsymbol{x} \tag{25}$$

is clearly symmetric in its second pair of indices; and its contraction with respect to this pair gives

$$\mathfrak{W}_{pq;ii} = \int_V \rho x_p \frac{\partial \mathfrak{V}}{\partial x_q} d\boldsymbol{x} = \mathfrak{W}_{pq} , \tag{26}$$

which *is* symmetric in p and q. However, as we shall presently see, the uncontracted tensor is not, generally, symmetric in its first pair of indices.

By an integration by parts, we obtain from equation (25) the formula

$$\mathfrak{W}_{pq;ij} = -\int_V x_p \frac{\partial \rho}{\partial x_q} \mathfrak{V}_{ij} d\boldsymbol{x} - \delta_{pq}\int_V \rho \mathfrak{V}_{ij} d\boldsymbol{x} , \tag{27}$$

on the assumption that the density vanishes on the boundary of V (or that it vanishes at infinity with sufficient rapidity). Making use of equation (22), we can rewrite equation (27) in the form

$$\mathfrak{W}_{pq;ij} = -\int_V x_p \frac{\partial \rho}{\partial x_q} \mathfrak{V}_{ij} d\boldsymbol{x} + 2\,\delta_{pq}\mathfrak{W}_{ij} . \tag{28}$$

Now substituting for $\mathfrak{V}_{ij}$ in equation (25) its explicit expression given in equation (1), we obtain

$$\begin{aligned}
\mathfrak{W}_{pq;ij} &= G\int_V d\boldsymbol{x}\, \rho(\boldsymbol{x})\, x_p \frac{\partial}{\partial x_q}\int_V d\boldsymbol{x}'\rho(\boldsymbol{x}')\frac{(x_i - x_i')(x_j - x_j')}{|\boldsymbol{x} - \boldsymbol{x}'|^3} \\
&= \delta_{qi}G\int_V\int_V d\boldsymbol{x}\, d\boldsymbol{x}'\rho(\boldsymbol{x})\rho(\boldsymbol{x}')\frac{x_p(x_j - x_j')}{|\boldsymbol{x} - \boldsymbol{x}'|^3} \\
&\qquad + \delta_{qj}G\int_V\int_V d\boldsymbol{x}\, d\boldsymbol{x}'\rho(\boldsymbol{x})\rho(\boldsymbol{x}')\frac{x_p(x_i - x_i')}{|\boldsymbol{x} - \boldsymbol{x}'|^3} \\
&\qquad - 3G\int_V\int_V d\boldsymbol{x}\, d\boldsymbol{x}'\rho(\boldsymbol{x})\rho(\boldsymbol{x}')\frac{x_p(x_q - x_q')(x_i - x_i')(x_j - x_j')}{|\boldsymbol{x} - \boldsymbol{x}'|^5} .
\end{aligned} \tag{29}$$

By averaging equation (29) and the equation resulting from it by interchanging the primed and the unprimed variables of integration, we obtain

$$\begin{aligned}
\mathfrak{W}_{pq;ij} &= -\mathfrak{W}_{pi}\delta_{qj} - \mathfrak{W}_{pj}\delta_{qi} \\
&\quad - \tfrac{3}{2}G\int_V\int_V \rho(\boldsymbol{x})\rho(\boldsymbol{x}')\frac{(x_p - x_p')(x_q - x_q')(x_i - x_i')(x_j - x_j')}{|\boldsymbol{x} - \boldsymbol{x}'|^5} d\boldsymbol{x}\, d\boldsymbol{x}' ,
\end{aligned} \tag{30}$$

where it should be noted that the last term on the right-hand side is completely symmetric in all four indices.

Several elementary identities can be deduced from equation (30). Thus

$$\mathfrak{W}_{pp;ii} = \mathfrak{W}_{ii;pp} \quad \text{(summation convention suspended)} \tag{31}$$

and

$$\mathfrak{W}_{ij;ij} - \mathfrak{W}_{ji;ij} = \mathfrak{W}_{jj} - \mathfrak{W}_{ii} \quad \text{(summation convention suspended)} . \tag{32}$$

Next, by contracting equation (30) with respect to two selected indices out of the four, we obtain (in turn)

$$\mathfrak{W}_{pq;ii} = \mathfrak{W}_{pq} , \qquad \mathfrak{W}_{pp;ij} = \mathfrak{W}_{ij} , \qquad \mathfrak{W}_{pq;iq} = -\mathfrak{W}_{pi} ,$$

and
$$\mathfrak{W}_{pq;pj} = 2\mathfrak{W}_{qj} - \mathfrak{W}\delta_{qj} . \tag{33}$$

An identity of a different sort follows from inserting in the definition of $\mathfrak{W}_{pq;ij}$ the representation of $\mathfrak{V}_{ij}$ in terms of the super-potential χ. Thus, when $i \neq j$ and with the summation convention suspended, we have

$$\mathfrak{W}_{ij;ij} = \int_V \rho \, x_i \, \frac{\partial \mathfrak{V}_{ij}}{\partial x_j} \, d\mathbf{x}$$

$$= \int_V \rho \, x_i \, \frac{\partial}{\partial x_j} \left(\frac{\partial^2 \chi}{\partial x_i \partial x_j} \right) d\mathbf{x}$$

$$= \int_V \rho \, x_i \, \frac{\partial}{\partial x_i} \left(\frac{\partial^2 \chi}{\partial x_j^2} \right) d\mathbf{x} \tag{34}$$

$$= \int_V \rho \, x_i \, \frac{\partial}{\partial x_i} \left(\mathfrak{V}_{jj} - \mathfrak{V} \right) d\mathbf{x} .$$

Thus

$$\mathfrak{W}_{ij; ij} = \mathfrak{W}_{ii;jj} - \mathfrak{W}_{ii} \quad \text{(summation convention suspended)} . \tag{35}$$

Relation (32) is now seen to be a consequence of relations (31) and (35).

V. THE FUNDAMENTAL TENSORS FOR SYSTEMS WITH TRIPLANAR SYMMETRY

We shall say that a configuration has *triplanar symmetry* if its density distribution, with respect to a suitably chosen system of Cartesian co-ordinates, is an even function of the co-ordinates, i.e., if a co-ordinate system exists such that

$$\rho(-x_1, x_2, x_3) \equiv \rho(x_1, -x_2, x_3) \equiv \rho(x_1, x_2, -x_3) \equiv \rho(x_1, x_2, x_3) . \tag{36}$$

When the density distribution has this symmetry, the moment of inertia tensor, I_{ij}, is clearly diagonal:

$$I_{ij} = \int_V \rho \, x_i x_j \, d\mathbf{x} = 0 \qquad\qquad \text{if } i \neq j . \tag{37}$$

From equations (2) and (19) defining $\mathfrak{V}$ and χ, it follows that the potential and the super-potential have the same symmetry as ρ with respect to reflection. Therefore, if ρ is an even function of the co-ordinates, then so are $\mathfrak{V}$ and χ. From equation (18) we may now conclude that under these circumstances

$$\mathfrak{V}_{ij} \text{ is odd in } x_i \text{ and } x_j \text{ if } i \neq j , \text{ and} \tag{38}$$

is even in all three co-ordinates if $i = j$.

Considering, now, the potential energy tensor, we infer from the symmetry properties of $\mathfrak{V}_{ij}$ and the formula

$$\mathfrak{W}_{ij} = -\frac{1}{2} \int_V \rho \mathfrak{V}_{ij} d\mathbf{x} \tag{39}$$

that

$$\mathfrak{W}_{ij} = 0 \qquad \text{if } i \neq j\,. \tag{40}$$

The potential energy tensor and the moment of inertia tensor can therefore be brought to the diagonal form simultaneously if the object has triplanar symmetry.

Turning our attention next to the super-matrix,

$$\mathfrak{W}_{pq;ij} = \int_V \rho x_p \frac{\partial \mathfrak{B}_{ij}}{\partial x_q}\, d\mathbf{x}\,, \tag{41}$$

we observe that when $i = j$, the integrand is odd in two of the co-ordinates, x_p and x_q, if $p \neq q$, while it is even in all three co-ordinates if $p = q$. Therefore,

$$\mathfrak{W}_{pq;ij} = 0 \qquad \text{if} \qquad p \neq q \qquad \text{and} \qquad i = j\,. \tag{42}$$

On the other hand, if $i \neq j$, then the integrand is odd in x_i and x_j if $p = q$; the integral will therefore vanish under these circumstances. Similarly, if $p \neq q$ and one of them is not equal to *either* i *or* j, then the integrand will again be odd in two of the three co-ordinates and the integral will again vanish. The only circumstance under which $\mathfrak{W}_{pq;ij}$ will not vanish identically is when the integrand is even in all three co-ordinates; and, when $i \neq j$, this can happen only in two cases, namely, when the pair of indices (p, q) coincides with the pair (i, j) or (j, i). Thus

$$\begin{aligned} \mathfrak{W}_{pq;ij} \neq 0 \text{ when } i \neq j \text{ only when } p = i \text{ and } q = j\,, \\ \text{or } p = j \text{ and } q = i\,. \end{aligned} \tag{43}$$

VI. THE FUNDAMENTAL TENSORS FOR SYSTEMS WHICH ARE IN ADDITION AXIALLY SYMMETRIC ABOUT x_3

We shall now suppose that the systems considered in Section V are, in addition, axially symmetric about x_3. Then, in cylindrical polar co-ordinates, $\varpi (= \sqrt{[x_1^2 + x_2^2]})$, $z\ (= x_3)$, and ϕ (= the azimuthal angle), the density distribution is of the form

$$\rho \equiv \rho(\varpi, z) \qquad \text{and} \qquad \rho(\varpi, z) \equiv \rho(\varpi, -z)\,. \tag{44}$$

From equations (2) and (19) it follows that $\mathfrak{B}$ and χ have the same symmetries as ρ. However, not all components of the tensor potential, $\mathfrak{B}_{ij}$, share these symmetries. Indeed, as we shall see presently, with the exception of $\mathfrak{B}_{33}$, all of them explicitly depend on x_1 and/or x_2. The nature of these dependences can be readily established. Thus, considering $\mathfrak{B}_{11}$, for example, we have

$$\mathfrak{B}_{11} = \mathfrak{B} + \frac{\partial^2 \chi}{\partial x_1^2} = \mathfrak{B} + \frac{\partial}{\partial x_1}\left(\frac{x_1}{\varpi}\frac{\partial \chi}{\partial \varpi}\right) \tag{45}$$

or

$$\mathfrak{B}_{11} = \mathfrak{B} + \frac{1}{\varpi}\frac{\partial \chi}{\partial \varpi} + \frac{x_1^2}{\varpi}\frac{\partial}{\partial \varpi}\left(\frac{1}{\varpi}\frac{\partial \chi}{\partial \varpi}\right). \tag{46}$$

Similarly,

$$\mathfrak{B}_{22} = \mathfrak{B} + \frac{1}{\varpi}\frac{\partial \chi}{\partial \varpi} + \frac{x_2^2}{\varpi}\frac{\partial}{\partial \varpi}\left(\frac{1}{\varpi}\frac{\partial \chi}{\partial \varpi}\right), \tag{47}$$

$$\mathfrak{B}_{12} = \frac{x_1 x_2}{\varpi}\frac{\partial}{\partial \varpi}\left(\frac{1}{\varpi}\frac{\partial \chi}{\partial \varpi}\right). \tag{48}$$

$$\mathfrak{B}_{33} = \mathfrak{B} + \frac{\partial^2 \chi}{\partial z^2}, \tag{49}$$

$$\mathfrak{B}_{13} = \frac{x_1}{\varpi} \frac{\partial^2 \chi}{\partial \varpi \partial z}, \quad \text{and} \quad \mathfrak{B}_{32} = \frac{x_2}{\varpi} \frac{\partial^2 \chi}{\partial \varpi \partial z}. \tag{50}$$

Considering the diagonal elements of $\mathfrak{W}_{ij}$ (which are the only non-vanishing ones in view of the triplanar symmetry), we have

$$\mathfrak{W}_{11} = \int_V \rho x_1 \frac{\partial \mathfrak{B}}{\partial x_1} dx = \iiint_V \rho x_1^2 \frac{\partial \mathfrak{B}}{\partial \varpi} d\varpi\, dz\, d\phi$$
$$= \int_0^\infty \int_{-\infty}^{+\infty} \int_0^{2\pi} \rho \varpi^2 \cos^2 \phi \frac{\partial \mathfrak{B}}{\partial \varpi} d\varpi\, dz\, d\phi\,; \tag{51}$$ [1]

or, effecting the integration over ϕ, we have

$$\mathfrak{W}_{11} = \pi \int_0^\infty \int_{-\infty}^{+\infty} \rho \varpi^2 \frac{\partial \mathfrak{B}}{\partial \varpi} d\varpi\, dz\,. \tag{52}$$

Clearly, we shall obtain this same expression for $\mathfrak{W}_{22}$. Thus, as one might have expected,

$$\mathfrak{W}_{11} = \mathfrak{W}_{22}\,. \tag{53}$$ [2]

However,

$$\mathfrak{W}_{33} = 2\pi \int_0^\infty \int_{-\infty}^{+\infty} \rho \varpi z \frac{\partial \mathfrak{B}}{\partial z} d\varpi\, dz \tag{54}$$

will, in general, be different from $\mathfrak{W}_{11}$ (or $\mathfrak{W}_{22}$).

From the equality of $\mathfrak{W}_{11}$ and $\mathfrak{W}_{22}$ it follows from equation (32) that

$$\mathfrak{W}_{12;12} = \mathfrak{W}_{21;12}\,. \tag{55}$$

Considering, now, the non-vanishing elements of $\mathfrak{W}_{pq;ij}$ systematically, we first observe that the equations

$$\mathfrak{W}_{11;11} = \int_V \rho x_1 \frac{\partial}{\partial x_1} \left(\mathfrak{B} + \frac{\partial^2 \chi}{\partial x_1^2} \right) dx \tag{56}$$

and

$$\mathfrak{W}_{22;22} = \int_V \rho x_2 \frac{\partial}{\partial x_2} \left(\mathfrak{B} + \frac{\partial^2 \chi}{\partial x_2^2} \right) dx, \tag{57}$$

after integrations over ϕ, lead to identical expressions. Therefore,

$$\mathfrak{W}_{11;11} = \mathfrak{W}_{22;22}\,. \tag{58}$$

Similarly, we can readily verify that

$$\mathfrak{W}_{22;33} = \mathfrak{W}_{11;33}\,, \qquad \mathfrak{W}_{13;13} = \mathfrak{W}_{23;23}\,, \qquad \mathfrak{W}_{31;13} = \mathfrak{W}_{32;23}\,. \tag{59}$$

Also, we have the relations (cf. eqs. [33] and [35])

$$\mathfrak{W}_{13;13} = \mathfrak{W}_{33;11} - \mathfrak{W}_{11} = \mathfrak{W}_{11;33} - \mathfrak{W}_{11} = -(\mathfrak{W}_{11;22} + \mathfrak{W}_{11;11}) \tag{60}$$

and

$$\mathfrak{W}_{31;13} = \mathfrak{W}_{11;33} - \mathfrak{W}_{33} = -(\mathfrak{W}_{22;33} + \mathfrak{W}_{33;33})\,. \tag{61}$$

[1] We are writing the limits (0, ∞) and (−∞, +∞) for ϖ and z only formally; we are not implying that the system necessarily extends to infinity.

[2] It is clear that, under these same circumstances, $I_{11} = I_{22}$.

A less obvious identity (in case of axial symmetry) is the following:

$$\mathfrak{W}_{12;12} = \tfrac{1}{2}(\mathfrak{W}_{11;11} - \mathfrak{W}_{11;22}). \tag{62}$$

This can be established as follows. Considering, first, $\mathfrak{W}_{12;12}$, we have

$$\begin{aligned}\mathfrak{W}_{12;12} &= \int_V \rho x_1 \frac{\partial \mathfrak{V}_{12}}{\partial x_2}\, d\mathbf{x}\\ &= -\int_V x_1 \frac{\partial \rho}{\partial x_2} \mathfrak{V}_{12} d\mathbf{x}\\ &= -\int_V \frac{x_1^2 x_2^2}{\varpi^2} \frac{\partial \rho}{\partial \varpi} \frac{\partial}{\partial \varpi}\left(\frac{1}{\varpi}\frac{\partial \chi}{\partial \varpi}\right) d\mathbf{x}\\ &= -2\pi \int_0^\infty \int_{-\infty}^{+\infty} \langle \cos^2\phi \sin^2\phi \rangle \varpi^3 \frac{\partial \rho}{\partial \varpi} \frac{\partial}{\partial \varpi}\left(\frac{1}{\varpi}\frac{\partial \chi}{\partial \varpi}\right) d\varpi\, dz\\ &= -\tfrac{1}{4}\pi \int_0^\infty \int_{-\infty}^{+\infty} \varpi^3 \frac{\partial \rho}{\partial \varpi} \frac{\partial}{\partial \varpi}\left(\frac{1}{\varpi}\frac{\partial \chi}{\partial \varpi}\right) d\varpi\, dz.\end{aligned} \tag{63}$$

Similarly,

$$\begin{aligned}\mathfrak{W}_{11;11} - \mathfrak{W}_{11;22} &= \int_V \rho x_1 \frac{\partial}{\partial x_1}\left(\frac{\partial^2 \chi}{\partial x_1^2} - \frac{\partial^2 \chi}{\partial x_2^2}\right) d\mathbf{x}\\ &= -\int_V x_1 \frac{\partial \rho}{\partial x_1}\left(\frac{\partial^2 \chi}{\partial x_1^2} - \frac{\partial^2 \chi}{\partial x_2^2}\right) d\mathbf{x} - \int_V \rho \left(\frac{\partial^2 \chi}{\partial x_1^2} - \frac{\partial^2 \chi}{\partial x_2^2}\right) d\mathbf{x}\\ &= -\int_V \frac{x_1^2}{\varpi} \frac{\partial \rho}{\partial \varpi}\left[\frac{\partial}{\partial x_1}\left(\frac{x_1}{\varpi}\frac{\partial \chi}{\partial \varpi}\right) - \frac{\partial}{\partial x_2}\left(\frac{x_2}{\varpi}\frac{\partial \chi}{\partial \varpi}\right)\right] d\mathbf{x}\\ &= -\int_V \frac{1}{\varpi^2}\frac{\partial \rho}{\partial \varpi}\frac{\partial}{\partial \varpi}\left(\frac{1}{\varpi}\frac{\partial \chi}{\partial \varpi}\right)(x_1^4 - x_1^2 x_2^2)\, d\mathbf{x}\\ &= -2\pi \int_0^\infty \int_{-\infty}^{+\infty} \varpi^3 \frac{\partial \rho}{\partial \varpi}\frac{\partial}{\partial \varpi}\left(\frac{1}{\varpi}\frac{\partial \chi}{\partial \varpi}\right)\\ &\qquad \times (\langle \cos^4\phi \rangle - \langle \cos^2\phi \sin^2\phi \rangle)\, d\varpi\, dz\\ &= -\tfrac{1}{2}\pi \int_0^\infty \int_{-\infty}^{+\infty} \varpi^3 \frac{\partial \rho}{\partial \varpi}\frac{\partial}{\partial \varpi}\left(\frac{1}{\varpi}\frac{\partial \chi}{\partial \varpi}\right) d\varpi\, dz.\end{aligned} \tag{64}$$

From a comparison of the end results of equations (63) and (64), we obtain relation (62).

Summarizing the results of the preceding analysis, we can express the non-vanishing elements of $\mathfrak{W}_{pq;ij}$ in terms of four of them as follows:

$$\begin{aligned}&\mathfrak{W}_{11;11} = \mathfrak{W}_{22;22} = A \text{ (say)},\\ &\mathfrak{W}_{11;22} = \mathfrak{W}_{22;11} = B \text{ (say)},\end{aligned} \tag{65}$$

$$\mathfrak{W}_{11;33} = \mathfrak{W}_{22;33} = \mathfrak{W}_{33;11} = \mathfrak{W}_{33;22} = C \text{ (say)}, \tag{66}$$

$$\mathfrak{W}_{33;33} = D \text{ (say)}, \tag{67}$$

$$\mathfrak{W}_{12;12} = \mathfrak{W}_{21;12} = \tfrac{1}{2}(\mathfrak{W}_{11;11} - \mathfrak{W}_{22;11}) = \tfrac{1}{2}(A - B), \tag{68}$$

$$\begin{aligned}&\mathfrak{W}_{13;13} = \mathfrak{W}_{23;23} = -\mathfrak{W}_{11} + \mathfrak{W}_{11;33} = C - \mathfrak{W}_{11},\\ &\mathfrak{W}_{31;13} = \mathfrak{W}_{32;23} = \mathfrak{W}_{11;33} - \mathfrak{W}_{33} = C - \mathfrak{W}_{33}.\end{aligned} \tag{69}$$

Among the four constants A, B, C, and E there are three relations. These follow from the general relations (cf. eqs. [33])

$$\mathfrak{W}_{11;qq} = \mathfrak{W}_{11;11} + \mathfrak{W}_{11;22} + \mathfrak{W}_{11;33} = \mathfrak{W}_{11} ,$$

$$\mathfrak{W}_{1q;1q} = \mathfrak{W}_{11;11} + \mathfrak{W}_{12;12} + \mathfrak{W}_{13;13} = -\mathfrak{W}_{11} , \tag{70}$$

and

$$\mathfrak{W}_{qq;33} = \mathfrak{W}_{11;33} + \mathfrak{W}_{22;33} + \mathfrak{W}_{33;33} = \mathfrak{W}_{33} .$$

Expressing the various elements in the foregoing relations in accordance with equations (65)–(69), we find

$$A + B + C = \mathfrak{W}_{11} , \tag{71}$$

$$2C + D = \mathfrak{W}_{33} , \tag{72}$$

and

$$3A - B + 2C = 0 . \tag{73}$$

In case the configuration has spherical symmetry, the number of independent elements are further drastically reduced; for in this case we must clearly have

$$A = D \quad \text{and} \quad B = C ; \tag{74}$$

and, moreover,

$$\mathfrak{W}_{11} = \mathfrak{W}_{33} = \tfrac{1}{3}\mathfrak{W} . \tag{75}$$

From equations (71)–(73) it now follows that

$$A = D = -\tfrac{1}{15}\mathfrak{W} \quad \text{and} \quad B = C = \tfrac{1}{5}\mathfrak{W} \tag{76}$$

(in case of spherical symmetry).

VII. ON THE OCCURRENCE OF A POINT OF BIFURCATION IN THE SEQUENCE OF ROTATING EQUILIBRIUM CONFIGURATIONS

It is well known that in the equilibrium sequence of rotating incompressible fluid masses a point of bifurcation occurs at which the Jacobi ellipsoids branch off from the Maclaurin spheroids (cf. Jeans 1929). The origin of such a point of bifurcation has remained obscure (see, however, Lebovitz 1961, § IV, and Chandrasekhar 1961*b*); at least, the question whether a similar point of bifurcation can occur among rotating equilibrium configurations of compressible fluid masses has never been satisfactorily answered. However, we shall show that, in terms of the tensor $\mathfrak{W}_{pq;ij}$, we can give a general criterion for the occurrence of a point of bifurcation.

On the assumption that the rotating configuration has triplanar symmetry with respect to a co-ordinate system in which one of the axes (say x_3) coincides with the direction of $\mathbf{\Omega}$, it readily follows from the virial theorem that (cf. Chandrasekhar 1961*a*, eq. [18])

$$\mathfrak{W}_{11} + \Omega^2 I_{11} = \mathfrak{W}_{22} + \Omega^2 I_{22} = \mathfrak{W}_{33} = -\int_V p\,d\mathbf{x} . \tag{77}$$

Making use of the general relations (cf. eqs. [31] and [35])

$$\mathfrak{W}_{11} = \mathfrak{W}_{11;22} - \mathfrak{W}_{12;12} \quad \text{and} \quad \mathfrak{W}_{22} = \mathfrak{W}_{11;22} - \mathfrak{W}_{21;12} , \tag{78}$$

we can reduce the first pair of equalities in equation (77) to

$$-\mathfrak{W}_{12;12} + \Omega^2 I_{11} = -\mathfrak{W}_{21;12} + \Omega^2 I_{22} = \mathfrak{W}_{33} - \mathfrak{W}_{11;22} . \tag{79}$$

There are two obvious ways in which the equalities in (79) can be satisfied: *either* by requiring that

$$\mathfrak{W}_{12;12} = \mathfrak{W}_{21;12} \quad \text{and} \quad I_{11} = I_{22} \tag{80}$$

(i.e., by requiring that the object have axial symmetry, satisfying the first equality in [79] *identically* for all Ω^2, and determining Ω^2 by the second equality) *or* by requiring that

$$\Omega^2 I_{11} = \mathfrak{W}_{12;12}\,, \quad \Omega^2 I_{22} = \mathfrak{W}_{21;12}\,, \quad \text{and} \quad \mathfrak{W}_{33} = \mathfrak{W}_{11;22}\,. \tag{81}^*$$

The equalities in (79) can be satisfied in this latter manner only when Ω^2 exceeds a certain critical value: for, when $\Omega^2 \rightarrow 0$, $\mathfrak{W}_{12;12}$ tends to a finite positive value, namely, $-2\mathfrak{W}/15$ (cf. eqs. [68] and [76]), and it will not be possible to satisfy the first two equalities in equations (81). Therefore, for $\Omega^2 \rightarrow 0$, the configurations must be axisymmetric about the direction of $\mathbf{\Omega}$. However, as we proceed along the sequence of axially symmetric configurations and Ω^2 increases, a point *may* be reached where

$$\Omega^2 I_{11} = \Omega^2 I_{22} = \mathfrak{W}_{12;12} = \mathfrak{W}_{21;12}\,. \tag{82}$$

At this point it will become possible to satisfy conditions (81) for the *first time;* and, for values of Ω^2 larger than that required to satisfy equations (82), the possibility is open for satisfying the necessary conditions with unequal values of I_{11} and I_{22}; this is what must happen at a point of bifurcation; and this is what happens at the point where the Jacobi ellipsoids branch off from the Maclaurin spheroids.

Since $\mathfrak{W}_{33} = \mathfrak{W}_{11;22}$ at the point of bifurcation, this means that, in the notation of Section VI, at this point,

$$B = \mathfrak{W}_{33}\,. \tag{83}$$

This condition combined with equations (71)–(73) leads to the specific results

$$A = 3\mathfrak{W}_{33} - 2\mathfrak{W}_{11}\,, \quad B = \mathfrak{W}_{33}\,, \quad C = 3\mathfrak{W}_{11} - 4\mathfrak{W}_{33}\,, \quad \text{and} \quad D = -6\mathfrak{W}_{11} + 9\mathfrak{W}_{33}\,. \tag{84}$$

The research reported in this paper has in part been supported by the Office of Naval Research under Contract Nonr-2121(20) with the Enrico Fermi Institute for Nuclear Studies, University of Chicago. The work of the second author was supported in part by the United States Air Force under contract No. AF-49(638)-42, monitored by the Air Force Office of Scientific Research of the Air Research and Development Command.

REFERENCES

Chandrasekhar, S. 1960, *J. Math. Anal. and Appl.*, **1**, 240.

———. 1961*a*, *Hydrodynamic and Hydromagnetic Stability* (Oxford: Clarendon Press), chap. xiii, §§ 117 and 118.

———. 1961*b*, *Ap. J.*, **134**, 662.

Jeans, J. H. 1929, *Astronomy and Cosmogony* (Cambridge: Cambridge University Press), chaps. viii and ix.

Lebovitz, N. R. 1961, *Ap. J.*, **134**, 500.

*That this criterion is not sufficiently general—a constant C can be added to each righthand member—has been emphasized by J. Friedman and B. Schutz (*Astrophysical Journal Letters* 199, p. L157, 1975) and by C. Hunter (*Astrophysical Journal* 213, p. 497, 1977).

ON SUPERPOTENTIALS IN THE THEORY OF NEWTONIAN GRAVITATION. II. TENSORS OF HIGHER RANK

S. Chandrasekhar and Norman R. Lebovitz
University of Chicago and Massachusetts Institute of Technology
Received June 16, 1962

ABSTRACT

In addition to the tensors considered in the earlier paper, the following tensors are defined and studied:

$$\mathfrak{D}_{ij}(x) = G\int_V \rho(x') \frac{x'_i x'_j}{|x - x'|} dx',$$

$$\mathfrak{D}_{ij;k}(x) = G\int_V \rho(x') \frac{(x_i - x'_i)(x_j - x'_j)x'_k}{|x - x'|^3} dx',$$

and

$$\mathfrak{B}_{ijkl}(x) = G\int_V \rho(x') \frac{(x_i - x'_i)(x_j - x'_j)(x_k - x'_k)(x_l - x'_l)}{|x - x'|^5} dx'.$$

These tensors are useful in problems (such as the stability of the Jacobi ellipsoids) in which it is necessary to examine the effects of perturbations belonging to the third harmonic.

I. INTRODUCTION

In earlier papers (Chandrasekhar 1960, 1961; Lebovitz 1961; Chandrasekhar and Lebovitz 1962*a*, *b*) we have shown the utility of defining the quantities

$$\mathfrak{D}_i(x) = G\int_V \rho(x') \frac{x'_i}{|x - x'|} dx' \tag{1}$$

and

$$\mathfrak{B}_{ij}(x) = G\int_V \rho(x') \frac{(x_i - x'_i)(x_j - x'_j)}{|x - x'|^3} dx' \tag{2}$$

where G is the constant of gravitation, $dx' = dx'_1\, dx'_2 dx'_3$, and the integration is effected over the whole volume V occupied by the fluid. The vector $\mathfrak{D}_i$ and the tensor potential $\mathfrak{B}_{ij}$ are related between themselves and with the Newtonian potential,

$$\mathfrak{B}(x) = G\int_V \frac{\rho(x')}{|x - x'|} dx', \tag{3}$$

and the superpotential

$$\chi(x) = -G\int_V \rho(x')|x - x'| dx', \tag{4}$$

by the relations

$$\mathfrak{B}_{ij} = -x_i \frac{\partial \mathfrak{B}}{\partial x_j} + \frac{\partial \mathfrak{D}_i}{\partial x_j} \tag{5}$$

$$= \mathfrak{B}\delta_{ij} + \frac{\partial^2 \chi}{\partial x_i \partial x_j}, \tag{6}$$

and

$$\mathfrak{D}_i = x_i \mathfrak{B} + \frac{\partial \chi}{\partial x_i}. \tag{7}$$

Further, $\mathfrak{D}_i$ and χ are governed by the "Poisson" equations,

$$\nabla^2 \mathfrak{D}_i = -4\pi G \rho x_i \tag{8}$$

and

$$\nabla^2 \chi = -2\mathfrak{B} . \tag{9}$$

In the further developments of the theory (cf. Chandrasekhar 1962) we have found that it is necessary to introduce the following tensors of still higher rank:

$$\mathfrak{D}_{ij}(x) = G\int_V \rho(x') \frac{x_i' x_j'}{|x - x'|} dx' , \tag{10}$$

$$\mathfrak{D}_{ij;k}(x) = G\int_V \rho(x') \frac{(x_i - x_i')(x_j - x_j') x_k'}{|x - x'|^3} dx' , \tag{11}$$

and

$$\mathfrak{B}_{ijkl}(x) = G\int_V \rho(x') \frac{(x_i - x_i')(x_j - x_j')(x_k - x_k')(x_l - x_l')}{|x - x'|^5} dx' . \tag{12}$$

On contraction, the tensors $\mathfrak{D}_{ij;k}$ and $\mathfrak{B}_{ijkl}$ reduce to the corresponding ones of lower order:

$$\mathfrak{D}_{ii;k} = \mathfrak{D}_k \quad \text{and} \quad \mathfrak{B}_{iikl} = \mathfrak{B}_{kl} . \tag{13}$$

II. THE TENSORS $\mathfrak{D}_{ij}$ AND $\mathfrak{D}_{ij;k}$

Clearly, $\mathfrak{D}_{ij}$ is the Newtonian potential induced by the "density distribution" $\rho x_i x_j$; accordingly, it can be deduced from Poisson's equation

$$\nabla^2 \mathfrak{D}_{ij} = -4\pi G \rho x_i x_j . \tag{14}$$

The tensor $\mathfrak{D}_{ij;k}$ is the proper tensorial generalization of $\mathfrak{D}_k$ even as $\mathfrak{B}_{ij}$ is the proper tensorial generalization of $\mathfrak{B}$: for $\mathfrak{D}_{ij;k}$ is the "$\mathfrak{B}_{ij}$" induced by ρx_k even as $\mathfrak{D}_k$ is the "$\mathfrak{B}$" induced by ρx_k. With this last interpretation of $\mathfrak{D}_{ij;k}$, we can express it directly in terms of $\mathfrak{D}_k$ and $\mathfrak{D}_{ik}$; for, according to equation (5), the tensor potential $\mathfrak{B}_{ij}$ is derivable from the gravitational potentials $\mathfrak{B}$ (induced by the basic density distribution ρ) and $\mathfrak{D}_i$ (induced by the basic density distribution *times* x_i); and, since the basic density distribution for $\mathfrak{D}_{ij;k}$ (considered as a tensor potential in its first two indices) is ρx_k, it is clear that $\mathfrak{D}_k$ and $\mathfrak{D}_{ik}$ play for it the roles of "$\mathfrak{B}$" and "$\mathfrak{D}_i$" for $\mathfrak{B}_{ij}$. Hence

$$\mathfrak{D}_{ij;k} = -x_i \frac{\partial \mathfrak{D}_k}{\partial x_j} + \frac{\partial \mathfrak{D}_{ik}}{\partial x_j} , \tag{15}$$

a relation which can also be derived directly from the definitions (1), (10), and (11) of the respective quantities. In analogy with equation (6), we can also infer the existence of a superpotential, χ_k, in terms of which $\mathfrak{D}_{ij;k}$ must be expressible in the form

$$\mathfrak{D}_{ij;k} = \delta_{ij}\mathfrak{D}_k + \frac{\partial^2 \chi_k}{\partial x_i \partial x_j} , \tag{16}$$

where (cf. eq. [4])

$$\chi_k = -G\int_V \rho(x') x_k' |x - x'| dx' . \tag{17}$$

III. THE COMPLETELY SYMMETRIC TENSOR $\mathfrak{B}_{ijkl}$

From the definitions of the respective quantities, we can readily derive the relation

$$-x_k \frac{\partial \mathfrak{B}_{ij}}{\partial x_l} + \frac{\partial \mathfrak{D}_{ij;k}}{\partial x_l} = 3\mathfrak{B}_{ijkl} - \delta_{il}\mathfrak{B}_{jk} - \delta_{jl}\mathfrak{B}_{ik} . \tag{18}$$

Interchanging the indices k and l and remembering that $\mathfrak{V}_{ijkl}$ is symmetric in its indices, we obtain

$$-x_l \frac{\partial \mathfrak{V}_{ij}}{\partial x_k} + \frac{\partial \mathfrak{D}_{ij;l}}{\partial x_k} = 3\mathfrak{V}_{ijkl} - \delta_{ik}\mathfrak{V}_{jl} - \delta_{jk}\mathfrak{V}_{il}. \tag{19}$$

On the subtraction of one of equations (18) and (19) from the other, we obtain

$$\frac{\partial \mathfrak{D}_{ij;k}}{\partial x_l} - \frac{\partial \mathfrak{D}_{ij;l}}{\partial x_k} + \delta_{il}\mathfrak{V}_{jk} + \delta_{jl}\mathfrak{V}_{ik} - \delta_{ik}\mathfrak{V}_{jl} - \delta_{jk}\mathfrak{V}_{il} = x_k \frac{\partial \mathfrak{V}_{ij}}{\partial x_l} - x_l \frac{\partial \mathfrak{V}_{ij}}{\partial x_k}. \tag{20}$$

Substituting for the tensor potentials, on the left-hand side of equation (20) in terms of $\mathfrak{V}$ and χ (in accordance with eq. [6]), we find, after some further reductions, that

$$\frac{\partial}{\partial x_l}\left[\mathfrak{D}_{ij;k} + \left(x_i \frac{\partial}{\partial x_j} + x_j \frac{\partial}{\partial x_i}\right)\frac{\partial \chi}{\partial x_k}\right] - \frac{\partial}{\partial x_k}\left[\mathfrak{D}_{ij;l} + \left(x_i \frac{\partial}{\partial x_j} + x_j \frac{\partial}{\partial x_i}\right)\frac{\partial \chi}{\partial x_l}\right]$$
$$= x_k \frac{\partial \mathfrak{V}_{ij}}{\partial x_l} - x_l \frac{\partial \mathfrak{V}_{ij}}{\partial x_k}. \tag{21}$$

The left-hand side of this equation is manifestly the curl of the expression in the square brackets, considered as a vector with respect to the index after the semicolon; and the right-hand side is similarly the curl of $\mathfrak{V}_{ij}x_k$ with respect to the same index (k). Accordingly, we may write

$$\mathfrak{D}_{ij;k} + \left(x_i \frac{\partial}{\partial x_j} + x_j \frac{\partial}{\partial x_i}\right)\frac{\partial \chi}{\partial x_k} = x_k\mathfrak{V}_{ij} + \frac{\partial \chi_{ij}}{\partial x_k}, \tag{22}$$

where χ_{ij} is some symmetric tensor. Rewriting equation (22) in the manner

$$\mathfrak{D}_{ij;k} = \frac{\partial}{\partial x_k}\left[\chi_{ij} - \left(x_i \frac{\partial}{\partial x_j} + x_j \frac{\partial}{\partial x_i}\right)\chi\right] + \delta_{ik}\frac{\partial \chi}{\partial x_j} + \delta_{jk}\frac{\partial \chi}{\partial x_i} + x_k\mathfrak{V}_{ij}, \tag{23}$$

differentiating it with respect to x_l, and making further use of equation (6) relating the tensor potential and χ, we find

$$\frac{\partial \mathfrak{D}_{ij;k}}{\partial x_l} = \frac{\partial^2}{\partial x_k \partial x_l}\left[\chi_{ij} - \left(x_i \frac{\partial}{\partial x_j} + x_j \frac{\partial}{\partial x_i}\right)\chi\right] + \delta_{lk}\mathfrak{V}_{ij} + \delta_{ik}(\mathfrak{V}_{jl} - \mathfrak{V}\delta_{jl})$$
$$+ \delta_{jk}(\mathfrak{V}_{il} - \mathfrak{V}\delta_{il}) + x_k \frac{\partial \mathfrak{V}_{ij}}{\partial x_l}. \tag{24}$$

Inserting this last result in equation (18), we find

$$3\mathfrak{V}_{ijkl} = \frac{\partial^2}{\partial x_k \partial x_l}\left[\chi_{ij} - \left(x_i \frac{\partial}{\partial x_j} + x_j \frac{\partial}{\partial x_i}\right)\chi\right] + \delta_{ik}(\mathfrak{V}_{jl} - \mathfrak{V}\delta_{jl})$$
$$+ \delta_{jk}(\mathfrak{V}_{il} - \mathfrak{V}\delta_{il}) + \delta_{lk}\mathfrak{V}_{ij} + \delta_{il}\mathfrak{V}_{jk} + \delta_{jl}\mathfrak{V}_{ik}, \tag{25}$$

and, adding to this equation the identity

$$0 = -\delta_{ij}\frac{\partial^2 \chi}{\partial x_k \partial x_l} + \delta_{ij}(\mathfrak{V}_{kl} - \delta_{kl}\mathfrak{V}), \tag{26}$$

we obtain

$$3\mathfrak{V}_{ijkl} = \frac{\partial^2}{\partial x_k \partial x_l}\left[\chi_{ij} - \left(x_i \frac{\partial}{\partial x_j} + x_j \frac{\partial}{\partial x_i}\right)\chi - \delta_{ij}\chi\right] + \mathfrak{V}_{ij}\delta_{kl} + \mathfrak{V}_{jk}\delta_{li} + \mathfrak{V}_{kl}\delta_{ij}$$
$$+ \mathfrak{V}_{li}\delta_{jk} + \mathfrak{V}_{ik}\delta_{jl} + \mathfrak{V}_{jl}\delta_{ik} - (\delta_{ij}\delta_{kl} + \delta_{ik}\delta_{jl} + \delta_{jk}\delta_{li})\mathfrak{V}. \tag{27}$$

From the symmetry in all four indices of $\mathfrak{V}_{ijkl}$ and of the terms on the right-hand side in the tensor potential and in $\mathfrak{V}$, we conclude that

$$\chi_{ij} - \left(x_i \frac{\partial}{\partial x_j} + x_j \frac{\partial}{\partial x_i} \right) \chi - \delta_{ij} \chi = \frac{\partial^2 \Phi}{\partial x_i \partial x_j}, \tag{28}$$

where Φ is some scalar function. We can therefore write

$$\begin{aligned} 3\mathfrak{V}_{ijkl} = \frac{\partial^4 \Phi}{\partial x_i \partial x_j \partial x_k \partial x_l} - (\delta_{ij}\delta_{kl} + \delta_{ik}\delta_{jl} + \delta_{jk}\delta_{li})\mathfrak{V} + \mathfrak{V}_{ij}\delta_{kl} + \mathfrak{V}_{jk}\delta_{li} \\ + \mathfrak{V}_{kl}\delta_{ij} + \mathfrak{V}_{li}\delta_{jk} + \mathfrak{V}_{ik}\delta_{jl} + \mathfrak{V}_{jl}\delta_{ik} . \end{aligned} \tag{29}$$

It remains to find an equation which will determine Φ. The required equation can be obtained by contracting equation (29) with respect to k and l (say). We obtain

$$3\mathfrak{V}_{ij} = \frac{\partial^2}{\partial x_i \partial x_{\cdot}} \nabla^2 \Phi - 5\delta_{ij}\mathfrak{V} + \delta_{ij}\mathfrak{V} + 3\mathfrak{V}_{ij} + 4\mathfrak{V}_{ij}, \tag{30}$$

or

$$0 = \frac{\partial^2}{\partial x_i \partial x_j} \nabla^2 \Phi + 4(\mathfrak{V}_{ij} - \delta_{ij}\mathfrak{V}) = \frac{\partial^2}{\partial x_i \partial x_j} (\nabla^2 \Phi + 4\chi) . \tag{31}$$

From equation (31) we may conclude, without loss of generality, that

$$\nabla^2 \Phi = -4\chi . \tag{32}$$

Thus Φ may be considered as the superpotential for χ.

With χ_{ij} given by equation (28), equation (23) for $\mathfrak{D}_{ij;k}$ becomes

$$\mathfrak{D}_{ij;k} = \frac{\partial^3 \Phi}{\partial x_i \partial x_j \partial x_k} + \delta_{ij} \frac{\partial \chi}{\partial x_k} + \delta_{jk} \frac{\partial \chi}{\partial x_i} + \delta_{ik} \frac{\partial \chi}{\partial x_j} + x_k \mathfrak{V}_{ij}, \tag{33}$$

or, expressing $\mathfrak{V}_{ij}$ in terms of $\mathfrak{V}$ and χ, we have

$$\mathfrak{D}_{ij;k} = \frac{\partial^2}{\partial x_i \partial x_j} \left(\frac{\partial \Phi}{\partial x_k} + x_k \chi \right) + \delta_{ij} \left(\frac{\partial \chi}{\partial x_k} + x_k \mathfrak{V} \right). \tag{34}$$

An equivalent expression for $\mathfrak{D}_{ij;k}$ is (cf. eq. [7])

$$\mathfrak{D}_{ij;k} = \delta_{ij}\mathfrak{D}_k + \frac{\partial^2}{\partial x_i \partial x_j} \left(\frac{\partial \Phi}{\partial x_k} + x_k \chi \right). \tag{35}$$

A comparison with equation (16) now shows that

$$\chi_k = \frac{\partial \Phi}{\partial x_k} + x_k \chi . \tag{36}$$

From equations (4), (17), and (37) we find

$$\frac{\partial \Phi}{\partial x_k} = G \int_V \rho(\boldsymbol{x}') (x_k - x'_k) \, | \boldsymbol{x} - \boldsymbol{x}' | \, d\boldsymbol{x}', \tag{37}$$

or

$$\Phi = \tfrac{1}{3} G \int_V \rho(\boldsymbol{x}') \, | \boldsymbol{x} - \boldsymbol{x}' |^3 d'\boldsymbol{x} . \tag{38}$$

IV. CONCLUDING REMARKS

It is evident that the tensors described in this and in the earlier paper (Chandrasekhar and Lebovitz 1962*a*) are only the first few in an entire hierarchy of tensors that one may define:

$\mathfrak{D}_{ijk\ldots}$ = the completely symmetric tensor which is the Newtonian potential induced by the "density" $\rho x_i x_j x_k \ldots$,

$\mathfrak{D}_{ij;kl\ldots}$ = the tensor potential "$\mathfrak{B}_{ij}$" induced by $\rho x_k x_l \ldots$,

$\mathfrak{D}_{ijkl;mn\ldots}$ = the tensor potential "$\mathfrak{B}_{ijkl}$" induced by $\rho x_m x_n \ldots$,

$\mathfrak{B}_{ijklmn\ldots}$ = the completely symmetric tensor of *even* order, which, on contraction with respect to any two of its indices, yields the corresponding tensor of two ranks lower; and the Newtonian potential is the completely contracted scalar of this tensor.

And associated with these tensors, we shall have a sequence of superpotentials of *odd* orders:

$$\chi^{(2n-1)} = \frac{(-1)^n G}{2n-1} \int_V \rho(\boldsymbol{x}') \, |\boldsymbol{x} - \boldsymbol{x}'|^{2n-1} d\boldsymbol{x}' \qquad (n = 1, 2, \ldots) . \quad (39)$$

It is clear, however, that these tensors of higher rank are not likely to be of much practical use as their "bookkeeping" will become excessively complicated. Indeed, even the tensors described in this paper (in contrast to those described in the earlier paper) are useful in only very special problems (cf. Chandrasekhar 1962).

The work of the first author was supported in part by the Office of Naval Research under contract Nonr-2121(24) with the University of Chicago. The work of the second author was supported in part by the United States Air Force under contract AF 49(638)-42 monitored by the Air Force Office of Scientific Research of the Air Research and Development Command.

REFERENCES

Chandrasekhar, S. 1960, *J. Math. Ann. and Appl.*, **1**, 240.
———. 1961, *Hydrodynamic and Hydromagnetic Stability* (Oxford: Clarendon Press), chap. xiii, §§ 117 and 118.
———. 1962, *Ap. J.*, **136**, 1048.
Lebovitz, N. R. 1961, *Ap. J.*, **134**, 500.
Chandrasekhar, S., and Lebovitz, N. R. 1962*a*, *Ap. J.*, **135**, 238.
———. 1962*b*, *ibid.*, p. 248.

THE POTENTIALS AND THE SUPERPOTENTIALS OF HOMOGENEOUS ELLIPSOIDS

S. CHANDRASEKHAR AND NORMAN R. LEBOVITZ
University of Chicago and Massachusetts Institute of Technology
Received June 16, 1962

ABSTRACT

Explicit expressions are found for the various potentials and superpotentials of homogeneous ellipsoids.

I. INTRODUCTION

The development of the virial equations in tensor forms has led to a considerable simplification of the theory of the equilibrium and the stability of rotating incompressible masses (see Lebovitz 1961; Chandrasekhar 1962; and a forthcoming paper on the stability of the Jacobi ellipsoids by Chandrasekhar and Lebovitz). Since the alternative theory based on the virial equations presupposes a knowledge of the various potentials and superpotentials (for their definitions see Chandrasekhar and Lebovitz 1962*a*, *c*; these two papers will be referred to hereafter as "Paper I" and "Paper II") of homogeneous ellipsoids, it will be convenient to have all the necessary formulae and relations assembled in one place in a common systematic notation. The present paper serves that purpose.

It is important to state here at the outset that in this paper the summation convention is not adopted: *summation, whenever required, will be indicated; summation over a repeated index is not to be understood unless it is explicitly indicated.*

II. SOME DEFINITIONS AND SOME RELATIONS

Consider an ellipsoid of semiaxes a_1, a_2, and a_3; and let

$$a_1 \geq a_2 > a_3 \; . \tag{1}$$

In the first instance, it will be supposed that the a's are all different. The case when $a_1 = a_2$ (and the object considered is an oblate spheroid) is degenerate and will be treated separately. (We shall not be interested in the case when all the a's are equal.)

We define the integrals

$$I = \int_0^\infty \frac{du}{\Delta} \tag{2}$$

and

$$A_{ijk\ldots} = \int_0^\infty \frac{du}{\Delta(a_i^2+u)(a_j^2+u)(a_k^2+u)\ldots}, \tag{3}$$

where

$$\Delta^2 = (a_1^2+u)(a_2^2+u)(a_3^2+u) \; . \tag{4}$$

As defined, $A_{ijk\ldots}$ is completely symmetric in its indices. We shall mostly be interested in the symbols $A_{ijk\ldots}$ with four or fewer indices.

a) *The One-Index Symbols, A_i*

By definition,

$$A_i = \int_0^\infty \frac{du}{\Delta(a_i^2+u)} \; . \tag{5}$$

We verify that

$$\frac{\partial I}{\partial a_i^2} = -\int_0^\infty \frac{1}{\Delta^2}\frac{\partial \Delta}{\partial a_i^2}\,du = -\frac{1}{2}\int_0^\infty \frac{du}{\Delta(a_i^2+u)} = -\tfrac{1}{2}A_i. \tag{6}$$

From equation (6) and the further relation

$$\frac{1}{\Delta}\frac{d\Delta}{du} = \frac{1}{2}\sum_{i=1}^{3}\frac{1}{a_i^2+u}, \tag{7}$$

we deduce that

$$\sum_{i=1}^{3}\frac{\partial I}{\partial a_i^2} = -\int_0^\infty \frac{1}{\Delta^2}\frac{d\Delta}{du}\,du = \frac{1}{\Delta}\Big|_0^\infty = -\frac{1}{a_1a_2a_3}. \tag{8}$$

Hence

$$\sum_{i=1}^{3} A_i = \frac{2}{a_1a_2a_3}, \tag{9}$$

a well-known result (cf. Routh 1892, p. 101). Also, since I is a homogeneous function of degree $-\frac{1}{2}$ in a_i^2, we have, by Euler's theorem,

$$\sum_{i=1}^{3} a_i^2\frac{\partial I}{\partial a_i^2} = -\tfrac{1}{2}I, \tag{10}$$

or, by equation (6),

$$\sum_{i=1}^{3} a_i^2 A_i = I. \tag{11}$$

The integrals defining the A_i's can be expressed in terms of the incomplete elliptic integrals

$$E(\theta,\phi) = \int_0^\phi (1-\sin^2\theta\sin^2\varphi)^{1/2}d\varphi \tag{12}$$

and

$$F(\theta,\phi) = \int_0^\phi (1-\sin^2\theta\sin^2\varphi)^{-1/2}d\varphi, \tag{13}$$

with the definitions

$$\sin\theta = \sqrt{\frac{a_1^2-a_2^2}{a_1^2-a_3^2}} \quad\text{and}\quad \cos\phi = \frac{a_3}{a_1}. \tag{14}$$

We have (cf. Kellogg 1929)

$$A_1 = \frac{2}{a_1^3\sin^3\phi}\,\frac{1}{\sin^2\theta}[F(\theta,\phi)-E(\theta,\phi)], \tag{15}$$

$$A_2 = \frac{2}{a_1^3\sin^3\phi}\,\frac{1}{\sin^2\theta\cos^2\theta}\left[E(\theta,\phi)-F(\theta,\phi)\cos^2\theta-\frac{a_3}{a_2}\sin^2\theta\sin\phi\right], \tag{16}$$

and

$$A_3 = \frac{2}{a_1^3\sin^3\phi}\,\frac{1}{\cos^2\theta}\left[\frac{a_2}{a_3}\sin\phi-E(\theta,\phi)\right]. \tag{17}$$

The foregoing expressions apply when $a_1 > a_2 > a_3$. When $a_1 = a_2 > a_3$, the integrals defining the A_i's become elementary, and we have

$$A_1 = A_2 = \frac{1}{a_1^3 e^3}[\sin^{-1} e - e\sqrt{(1-e^2)}] \tag{18}$$

and

$$A_3 = \frac{2}{a_1^3 e^3}\left[\frac{e}{\sqrt{(1-e^2)}} - \sin^{-1} e\right], \tag{19}$$

where

$$e = \left(1 - \frac{a_3^2}{a_1^2}\right)^{1/2} \tag{20}$$

is the eccentricity of the meridional sections.

b) The Two-Index Symbols, A_{ij}

By definition,

$$A_{ij} = \int_0^\infty \frac{du}{\Delta(a_i^2+u)(a_j^2+u)}. \tag{21}$$

It can readily be verified that when $i \neq j$, the two-index symbols are very simply related to the one-index symbols. Thus

$$A_{ij} = -\frac{A_i - A_j}{a_i^2 - a_j^2} \qquad (i \neq j). \tag{22}$$

An equivalent form of this relation is

$$A_i - a_j^2 A_{ji} = A_j - a_i^2 A_{ij}. \tag{23}$$

Moreover, by integrating over the range of u the equation

$$\frac{d}{du}\frac{1}{\Delta(a_i^2+u)} = -\frac{1}{2\Delta(a_i^2+u)}\left(\sum_{j=1}^{3}\frac{1}{a_j^2+u} + \frac{2}{a_i^2+u}\right), \tag{24}$$

we obtain the further relation

$$-\frac{1}{2}\sum_{j=1}^{3} A_{ij} - A_{ii} = \frac{1}{\Delta(a_i^2+u)}\Bigg|_0^\infty = -\frac{1}{a_1 a_2 a_3 a_i^2}, \tag{25}$$

or, alternatively,

$$3A_{ii} + \sum_{l\neq i} A_{il} = \frac{2}{a_1 a_2 a_3 a_i^2}. \tag{26}$$

Relations (22) and (26), together, determine all the two-index symbols in terms of the one-index symbols.

Making use of relation (9), we can write, instead of equation (26),

$$3A_{ii}a_i^2 + \sum_{l\neq i} a_i^2 A_{il} = \sum_{i=1}^{3} A_i. \tag{27}$$

Rewriting equation (27) in the form

$$3A_{ii}a_i^2+\sum_{l\neq i}(a_i^2A_{il}+A_i-A_l)=3A_i \tag{28}$$

and making use of the result expressed by equation (23), we obtain the relation

$$3A_{ii}a_i^2+\sum_{l\neq i}A_{il}a_l^2=3A_i\,. \tag{29}$$

When $a_1=a_2$, the number of distinct two-index symbols are only three; they are

$$A_{11}=A_{12}=A_{22}\,,\qquad A_{23}=A_{13}\,,\qquad \text{and}\qquad A_{33}\,. \tag{30}$$

All of these can be expressed in terms of A_1 and A_3. Thus

$$A_{13}=A_{23}=-\frac{A_1-A_3}{a_1^2-a_3^2}=-\frac{A_1-A_3}{a_1^2e^2}\,, \tag{31}$$

where e is the eccentricity defined in equation (20). And equation (26) for $i=1$ and 3 provides further relations from which A_{11} and A_{33} can be deduced. Thus

$$4A_{11}+A_{13}=\frac{2}{a_1^4a_3}=\frac{2}{a_1^5(1-e^2)^{1/2}}\,, \tag{32}$$

and

$$3A_{33}+2A_{13}=\frac{2}{a_1^2a_3^3}=\frac{2}{a_1^5(1-e^2)^{3/2}}\,, \tag{33}$$

where the equalities enumerated in (30) have been used. We therefore have

$$A_{11}=\frac{1}{2a_1^5(1-e^2)^{1/2}}+\frac{A_1-A_3}{4a_1^2e^2} \tag{34}$$

and

$$A_{33}=\frac{2}{3a_1^5(1-e^2)^{3/2}}+\frac{2}{3}\,\frac{A_1-A_3}{a_1^2e^2}\,. \tag{35}$$

The expressions for the two-index symbols given by equations (31), (34), and (35) are in agreement with those given in an earlier paper in a different notation (Lebovitz 1961, eqs. [130]–[133]).

c) The Three-Index Symbols, A_{ijk}

Again, from the definition

$$A_{ijk}=\int_0^\infty\frac{du}{\Delta(a_i^2+u)(a_j^2+u)(a_k^2+u)}\,, \tag{36}$$

it follows that

$$\begin{aligned} A_{ijk}&=-\frac{A_{ij}-A_{ik}}{a_j^2-a_k^2} \qquad (j\neq k)\\ &=-\frac{A_{ji}-A_{jk}}{a_i^2-a_k^2} \qquad (i\neq k)\,. \end{aligned} \tag{37}$$

Therefore, so long as all the indices are not the same, the three-index symbols can be simply deduced from the two-index symbols. When all three indices are the same, we can make use of the relation

$$\sum_{l=1}^{3} A_{ijl} + 2A_{iij} + 2A_{ijj} = \frac{2}{a_1 a_2 a_3 a_i^2 a_j^2}, \tag{38}$$

which follows from integrating, over the range of u, the equation

$$-2\frac{d}{du}\frac{1}{\Delta(a_i^2+u)(a_j^2+u)} = \sum_{l=1}^{3}\frac{1}{\Delta(a_i^2+u)(a_j^2+u)(a_l^2+u)} + \frac{2}{\Delta(a_i^2+u)^2(a_j^2+u)} + \frac{2}{\Delta(a_j^2+u)^2(a_i^2+u)}. \tag{39}$$

A particular case ($i = j$) of equation (38) is

$$5A_{iii} + \sum_{l\neq i} A_{iil} = \frac{2}{a_1 a_2 a_3 a_i^4}. \tag{40}$$

In combination with equation (26), equation (40) gives

$$5A_{iii}a_i^2 + \sum_{l\neq i} A_{iil}a_i^2 = 3A_{ii} + \sum_{l\neq i} A_{il}. \tag{41}$$

But, according to equation (37),

$$a_i^2 A_{iil} = A_{iil}a_l^2 - A_{ii} + A_{il}. \tag{42}$$

Hence

$$5A_{iii}a_i^2 + \sum_{l\neq i} A_{iil}a_l^2 = 5A_{ii}, \tag{43}$$

which is analogous to relation (29) among the two-index symbols.

We shall not stop to specialize the foregoing equations for the case $a_1 = a_2$: they can be readily written down if they are needed.

d) The Four-Index Symbols, A_{ijkl}

We have

$$A_{ijkl} = -\frac{A_{ijk} - A_{ijl}}{a_k^2 - a_l^2} \quad (k \neq l) \tag{44}$$

and further relations obtained by selecting other pairs of indices besides k and l. Equation (44) enables us to determine, in terms of the three-index symbols, all the four-index symbols that have at least two indices different from one another. And when all four indices are the same, the relevant equation to use is

$$\sum_{l=1}^{3} A_{ijkl} + 2A_{iijk} + 2A_{ijjk} + 2A_{ijkk} = \frac{2}{a_1 a_2 a_3 a_i^2 a_j^2 a_k^2}, \tag{45}$$

or, more particularly,

$$7A_{iiii} + \sum_{l\neq i} A_{iiil} = \frac{2}{a_1 a_2 a_3 a_i^6}. \tag{46}$$

It is clear that the relations we have derived can be extended to symbols with any number of indices.

The various relations derived in this section cannot be new: they must occur, in one form or another, in Darwin's writings. But it is difficult to penetrate the different notations; it was simpler to derive the necessary relations in the forms in which we shall need them.

III. THE NEWTONIAN POTENTIAL, $\mathfrak{V}$

The well-known expression for the gravitational potential in the interior of an ellipsoid can be expressed in terms of the quantities defined in Section II. We have

$$\frac{\mathfrak{V}}{\pi G\rho a_1 a_2 a_3} = I - \sum_{i=1}^{3} A_i x_i^2 . \tag{47}$$

IV. THE TENSORS $\mathfrak{D}_i$, $\mathfrak{V}_{ij}$, $\mathfrak{W}_{ij}$, AND $\mathfrak{W}_{pq;ij}$

During the latter half of the nineteenth century, various investigations were carried out on the gravitational potential of heterogeneous ellipsoids; and Ferrers (1877), in particular, gave explicit formulae for the general case when the density inside a given ellipsoidal boundary is expressible as a sum of terms of the form $x_1^l x_2^m x_3^n$. An account of Ferrers' method (and much useful information besides) will be found in Routh's *A Treatise on Analytical Statics* (**2**, 120–128). Clearly, the results of Ferrers are exactly those needed for the specification, in the case of ellipsoids of the general tensor $\mathfrak{D}_{ijk\ldots}$ defined in Paper II (Sec. IV); and, as we have shown, all the remaining tensors can be deduced from $\mathfrak{D}_{ijk\ldots}$.

For the specification of the tensor potential $\mathfrak{V}_{ij}$ we require, in particular, the gravitational potential $\mathfrak{D}_i$ induced by the fictitious density distribution ρx_i; this is given by (see Routh 1892)

$$\frac{\mathfrak{D}_i}{\pi G\rho a_1 a_2 a_3} = a_i^2 x_i \int_0^\infty \frac{du}{\Delta(a_i^2+u)}\left(1 - \sum_{j=1}^{3} \frac{x_j^2}{a_j^2+u}\right). \tag{48}$$

In terms of the two-index symbols defined in Section II, we can write

$$\frac{\mathfrak{D}_i}{\pi G\rho a_1 a_2 a_3} = a_i^2 x_i \left(A_i - \sum_{j=1}^{3} A_{ij} x_j^2\right). \tag{49}$$

Equation (49) is in agreement with the expression given by Lebovitz (1961, eq. [67]) and derived by him by a method different from that of Ferrers.

Inserting the expressions for $\mathfrak{V}$ and $\mathfrak{D}_i$ given by equations (47) and (49) in the general relation (cf. Paper I, eq. [8]),

$$\mathfrak{V}_{ij} = -x_i \frac{\partial \mathfrak{V}}{\partial x_j} + \frac{\partial \mathfrak{D}_i}{\partial x_j}, \tag{50}$$

we obtain

$$\frac{\mathfrak{V}_{ij}}{\pi G\rho a_1 a_2 a_3} = 2x_i x_j (A_j - a_i^2 A_{ij}) + a_i^2 \delta_{ij}\left(A_i - \sum_{l=1}^{3} A_{il} x_l^2\right). \tag{51}$$

In view of relation (23), this tensor is symmetrical in its indices, as required. From equation (51) we have, in particular,

$$\frac{\mathfrak{B}_{ij}}{\pi G\rho a_1 a_2 a_3} = 2x_i x_j(A_j - a_i^2 A_{ij}) \qquad (i \neq j) \tag{52}$$

and

$$\frac{\mathfrak{B}_{ii}}{\pi G\rho a_1 a_2 a_3} = 2x_i^2(A_i - a_i^2 A_{ii}) + a_i^2\left(A_i - \sum_{l=1}^{3} A_{il}x_l^2\right). \tag{53}$$

Using the various relations among the two-index symbols listed in Section II, we can verify that, as required,

$$\mathfrak{B} = \sum_{i=1}^{3} \mathfrak{B}_{ii}. \tag{54}$$

a) The Potential-Energy Tensor $\mathfrak{W}_{ij}$

From the definition

$$\mathfrak{W}_{ij} = -\frac{1}{2}\int_V \rho\,\mathfrak{B}_{ij} d\mathbf{x} \tag{55}$$

and the solution for the non-diagonal components of $\mathfrak{B}_{ij}$ (eq. [52]), it follows that $\mathfrak{W}_{ij}$ is diagonal in the chosen representation. The diagonal components of $\mathfrak{W}_{ij}$ are themselves best determined with the aid of the formula (Paper I, eq. [23])

$$\mathfrak{W}_{ii} = \int_V \rho x_i \frac{\partial \mathfrak{B}}{\partial x_i}\, d\mathbf{x}\,. \tag{56}$$

From equations (47) and (56) we obtain

$$\frac{\mathfrak{W}_{ii}}{\pi G\rho a_1 a_2 a_3} = -2A_i\int_V \rho x_i^2 d\mathbf{x} = -2A_i I_{ii}, \tag{57}$$

where

$$I_{ii} = \tfrac{1}{5} M a_i^2 \tag{58}$$

defines the moment of inertia tensor (which is also diagonal in the representation). In equation (58),

$$M = \tfrac{4}{3}\pi\rho a_1 a_2 a_3 \tag{59}$$

is the mass of the ellipsoid.

b) The Elements of the Supermatrix, $\mathfrak{W}_{pq;ij}$

By definition,

$$\mathfrak{W}_{pq;ij} = \int_V \rho x_p \frac{\partial \mathfrak{B}_{ij}}{\partial x_q}\, d\mathbf{x}\,; \tag{60}$$

and the different types of elements we have to consider are

$$\mathfrak{W}_{11;11}\,, \qquad \mathfrak{W}_{22;11}\,, \qquad \mathfrak{W}_{12;12}\,, \qquad \text{and} \qquad \mathfrak{W}_{21;12}\,. \tag{61}$$

Considering these elements, in turn, we have

$$\frac{\mathfrak{W}_{11;11}}{\pi G\rho a_1a_2a_3}=\int_V \rho x_1\frac{\partial \mathfrak{V}_{11}}{\partial x_1}\,dx$$
$$=\int_V \rho x_1[4x_1(A_1-a_1^2A_{11})-2a_1^2A_{11}x_1]\,dx \tag{62}$$
$$=2(2A_1-3a_1^2A_{11})I_{11}\,;$$

$$\frac{\mathfrak{W}_{22;11}}{\pi G\rho a_1a_2a_3}=\int_V \rho x_2\frac{\partial \mathfrak{V}_{11}}{\partial x_2}\,dx$$
$$=\int_V \rho x_2(-2a_1^2A_{12}x_2)\,dx=-2a_1^2A_{12}I_{22}\,; \tag{63}$$

$$\frac{\mathfrak{W}_{12;12}}{\pi G\rho a_1a_2a_3}=\int_V \rho x_1\frac{\partial \mathfrak{V}_{12}}{\partial x_2}\,dx$$
$$=2(A_2-a_1^2A_{12})\int_V \rho x_1^2dx=2(A_2-a_1^2A_{12})I_{11}\,; \tag{64}$$

and

$$\frac{\mathfrak{W}_{21;12}}{\pi G\rho a_1a_2a_3}=\int_V \rho x_2\frac{\partial \mathfrak{V}_{12}}{\partial x_1}\,dx=2(A_2-a_1^2A_{12})I_{22}\,. \tag{65}$$

An important identity which follows from equations (64) and (65) is

$$\frac{\mathfrak{W}_{12;12}}{I_{11}}=\frac{\mathfrak{W}_{21;12}}{I_{22}}\,. \tag{66}$$

V. THE TENSORS $\mathfrak{D}_{ik}$ AND $\mathfrak{D}_{ij;k}$

As we have already remarked, $\mathfrak{D}_{ik}$ is the gravitational potential due to the fictitious density distribution ρx_ix_k; and the expression for it is given in Routh (*loc. cit.*, p. 123).

When $i \neq k$, the expression for $\mathfrak{D}_{ik}$ is particularly simple:

$$\frac{\mathfrak{D}_{ik}}{\pi G\rho a_1a_2a_3}=a_i^2a_k^2x_ix_k\int_0^\infty \frac{du}{\Delta(a_i^2+u)(a_k^2+u)}\left(1-\sum_{l=1}^{3}\frac{x_l^2}{a_l^2+u}\right)\quad (i\neq k). \tag{67}$$

In terms of the two- and three-index symbols, we can write

$$\frac{\mathfrak{D}_{ik}}{\pi G\rho a_1a_2a_3}=a_i^2a_k^2x_ix_k\left(A_{ik}-\sum_{l=1}^{3}A_{ikl}x_l^2\right)\quad (i\neq k). \tag{68}$$

When $i = k$, the corresponding result is more complicated; we have

$$\frac{\mathfrak{D}_{ii}}{\pi G\rho a_1a_2a_3}=\tfrac{1}{4}a_i^2\int_0^\infty \frac{du}{\Delta}\left(1-\frac{a_i^2}{a_i^2+u}\right)\left(1-\sum_{l=1}^{3}\frac{x_l^2}{a_l^2+u}\right)^2$$
$$+a_i^4x_i^2\int_0^\infty \frac{du}{\Delta(a_i^2+u)^2}\left(1-\sum_{l=1}^{3}\frac{x_l^2}{a_l^2+u}\right), \tag{69}$$

or,

$$\frac{\mathfrak{D}_{ii}}{\pi G\rho a_1 a_2 a_3} = \tfrac{1}{4} a_i^2\Big[(I - a_i^2 A_i) - 2\sum_{l=1}^{3} (A_l - a_i^2 A_{il}) x_l^2 + \sum_{l=1}^{3}\sum_{m=1}^{3} (A_{lm} - a_i^2 A_{ilm}) x_l^2 x_m^2 \Big] + a_i^4 x_i^2 \Big(A_{ii} - \sum_{l=1}^{3} A_{iil} x_l^2 \Big) . \tag{70}$$

In terms of the expressions for $\mathfrak{D}_k$ and $\mathfrak{D}_{ik}$, the tensor $\mathfrak{D}_{ij;k}$ can be deduced from the relation (Paper II, eq. [15])

$$\mathfrak{D}_{ij;k} = - x_i \frac{\partial \mathfrak{D}_k}{\partial x_j} + \frac{\partial \mathfrak{D}_{ik}}{\partial x_j} . \tag{71}$$

Thus, in the case $i \neq k$, we have

$$\frac{\mathfrak{D}_{ij;k}}{\pi G\rho a_1 a_2 a_3} = - a_k^2 x_i \Big[\delta_{jk} \Big(A_k - \sum_{l=1}^{3} A_{kl} x_l^2 \Big) - 2 A_{kj} x_j x_k \Big] + a_i^2 a_k^2 \Big[(x_k \delta_{ij} + x_i \delta_{kj}) \Big(A_{ik} - \sum_{l=1}^{3} A_{ikl} x_l^2 \Big) - 2 A_{ikj} x_i x_k x_j \Big] \quad (i \neq k) . \tag{72}$$

VI. SOME ILLUSTRATIONS

We shall illustrate the use of the formulae of the preceding sections by deriving the known properties of the Maclaurin spheroids with the aid of the general relations provided by the virial theorem.

Now the virial theorem applied to rotating axisymmetric configurations gives

$$\mathfrak{W}_{11} + \Omega^2 I_{11} = \mathfrak{W}_{33} . \tag{73}$$

An alternative form of this same equation is (cf. Paper I, eq. [79])

$$- \mathfrak{W}_{12;12} + \Omega^2 I_{11} = \mathfrak{W}_{33} - \mathfrak{W}_{11;22} . \tag{74}$$

Using the expression for $\mathfrak{W}_{ii}$ given in Section III (eq. [57]), we find from equation (73) that

$$\frac{\Omega^2}{\pi G\rho a_1^2 a_3} = \frac{2}{I_{11}} (A_1 I_{11} - A_3 I_{33}) = 2 \Big(A_1 - \frac{a_3^2}{a_1^2} A_3 \Big) , \tag{75}$$

or, alternatively,

$$\frac{\Omega^2}{2\pi G\rho} = a_1^3 (1 - e^2)^{1/2} [A_1 - (1 - e^2) A_3] . \tag{76}$$

On substituting for A_1 and A_3, in equation (76), their values given in equations (18) and (19), we recover the known formula

$$\frac{\Omega^2}{2\pi G\rho} = \frac{\sqrt{(1 - e^2)}}{e^3} (3 - 2 e^2) \sin^{-1} e - \frac{3(1 - e^2)}{e^2} . \tag{77}$$

We have shown (Paper I, eq. [81]; see also Chandrasekhar 1962, Sec. II) that, at the point of bifurcation where the Jacobi ellipsoids branch off,

$$\mathfrak{W}_{33} = \mathfrak{W}_{11;22} . \tag{78}$$

With $\mathfrak{W}_{33}$ and $\mathfrak{W}_{11;22}$ given by equations (57) and (63), the condition is

$$A_3 I_{33} = a_1^2 A_{11} I_{11} , \tag{79}$$

since in this case the indices 1 and 2 are equivalent. Equation (79) is clearly the same as

$$a_3^2 A_3 = a_1^4 A_{11} . \tag{80}$$

Now inserting for A_{11} from equation (34), we obtain

$$\frac{1}{2a_1^3\sqrt{(1-e^2)}} + \frac{1}{4e^2}(A_1 - A_3) = (1-e^2)A_3 . \tag{81}$$

On further simplification (in which we make use of the explicit expressions for A_1 and A_3) equation (81) leads to the known equation

$$\frac{\sin^{-1} e}{\sqrt{(1-e^2)}} = \frac{3e + 10e^3}{3 + 8e^2 - 8e^4} , \tag{82}$$

whose root is

$$e = 0.8126700 . \tag{83}$$

Now the Maclaurin spheroids become unstable when (Lebovitz 1961 or Chandrasekhar and Lebovitz 1962*b*, Table 1)

$$\Omega^2 I_{11} = 2\mathfrak{W}_{12;12} . \tag{84}$$

This condition, when inserted into equation (74) governing equilibrium, gives

$$\mathfrak{W}_{33} = \mathfrak{W}_{11;22} + \mathfrak{W}_{12;12} . \tag{85}$$

Equation (85) can be further transformed in the following manner (see Paper I, eqs. [26] and [62]):

$$\begin{aligned} \mathfrak{W}_{33} &= \mathfrak{W}_{11;22} + \tfrac{1}{2}(\mathfrak{W}_{11;11} - \mathfrak{W}_{11;22}) \\ &= \tfrac{1}{2}(\mathfrak{W}_{11;11} + \mathfrak{W}_{11;22}) = \tfrac{1}{2}(\mathfrak{W}_{11} - \mathfrak{W}_{11;33}) . \end{aligned} \tag{86}$$

With the values of $\mathfrak{W}_{11}$ and $\mathfrak{W}_{11;33}$ given in Section III, equation (86) reduces to

$$\begin{aligned} 2A_3 a_3^2 &= A_1 a_1^2 - a_1^2 a_3^2 A_{13} \\ &= A_1 a_1^2 + \frac{a_3^2}{e^2}(A_1 - A_3) , \end{aligned} \tag{87}$$

where we have made use of equation (31). On further simplification, equation (87) reduces to

$$A_1 = (1-e^2)(1+2e^2)A_3 . \tag{88}$$

This is the condition for the "ordinary instability" of the Maclaurin spheroid; and on substituting for A_1 and A_3 their values, we recover, once again, the known equation

$$\frac{\sin^{-1} e}{\sqrt{(1-e^2)}} = \frac{3e + 4e^3}{3 + 2e^2 - 4e^4} , \tag{89}$$

whose root is

$$e = 0.9528867 . \tag{90}$$

The work of the first author was supported in part by the Office of Naval Research under contract Nonr-2121(24) with the University of Chicago. The work of the second author was supported in part by the United States Air Force under contract AF 49(638)-42 monitored by the Air Force Office of Scientific Research of the Air Research and Development Command.

REFERENCES

Chandrasekhar, S. 1962, *Ap. J.*, **136**, 1048.
Chandrasekhar, S., and Lebovitz, N. R. 1962*a*, *Ap. J.*, **135**, 238 (referred to as "Paper I").
———. 1962*b*, *ibid.*, p. 248.
———. 1962*c*, *ibid.*, **136**, 1032 (referred to as "Paper II").
Ferrers, N. M. 1877, *Quart. J. of Pure and Applied Mathematics*, **14**, 1.
Kellogg, O. D. 1929, *Foundations of Potential Theory* (New York: Frederick Ungar Publishing Co.), p. 196.
Lebovitz, N. R. 1961, *Ap. J.*, **134**, 500.
Routh, E. J. 1892, *A Treatise on Analytical Statics* (Cambridge: Cambridge University Press, 1922), Vol. 2.

19

ON THE OSCILLATIONS AND THE STABILITY OF ROTATING GASEOUS MASSES

S. Chandrasekhar and Norman R. Lebovitz
University of Chicago and Massachusetts Institute of Technology
Received September 1, 1961

ABSTRACT

In this paper the oscillations and the stability of a rotating gaseous mass are considered on the basis of an appropriate tensor form of the virial theorem. On the assumption that the Lagrangian displacement $\boldsymbol{\xi}$ can be expressed in the form

$$\xi_j = X_{jr} x_r e^{\lambda t} \qquad (X_{jr} \text{ constants}),$$

a characteristic equation for λ (of order eighteen) is derived from the nine integral relations provided by the virial theorem. An examination of the roots of this characteristic equation enables the enumeration of the properties of all the natural modes of oscillation belonging essentially to harmonics not higher than the second. It is shown that there are three principal groups among these modes: a group of three modes, each of which exhibits a doublet character; a group of two modes, one of which becomes neutral at a point where the condition for the occurrence of a point of bifurcation is satisfied and both of which become over-stable at a higher angular velocity; and a group which represents the coupling of two modes, one of which is purely radial and the other of which is purely non-radial in the absence of rotation. In addition to these modes, there are two "trivial" modes, one of which is neutral and the other of which has a characteristic frequency equal to the angular velocity.

I. INTRODUCTION

The stability of rotating incompressible fluid masses has been the subject of many investigations; and the role of the point of bifurcation (at which the Jacobi ellipsoids branch off from the sequence of the Maclaurin spheroids) in determining stability has attracted much attention. In a recent paper (Lebovitz 1961; this paper will be referred to hereafter as "Paper I") the principal results of the classical investigations bearing on the point of bifurcation were clarified by an explicit evaluation of the frequencies of all the normal modes belonging to the second harmonic. The relative simplicity of the methods used in Paper I has encouraged us to attempt, by similar methods, the more general problem of the stability of rotating gaseous masses. The importance of this general problem for the "wider aspects of cosmogony" requires no emphasis, but it has been emphasized by Ledoux (1951; see also Ledoux and Walraven 1958) that the same problem is relevant for certain specific and practical questions raised by variable stars which exhibit the phenomena of multiple periods and beats. In this paper we shall survey the fundamental problems; in later papers we shall return to detailed examinations of specific questions.

II. THE VIRIAL THEOREM FOR SMALL PERTURBATIONS ABOUT EQUILIBRIUM OF ROTATING BODIES

In a frame of reference rotating with a constant angular velocity $\boldsymbol{\Omega}$, the virial theorem takes the form (Chandrasekhar 1960, 1961*a*, *b;* and Paper I)

$$\frac{d}{dt}\int_V \rho x_i u_j d\boldsymbol{x} = 2\mathfrak{T}_{ij} + \mathfrak{W}_{ij} + \delta_{ij}\int_V p\, d\boldsymbol{x} + \Omega^2 I_{ij} - I_{il}\Omega_l\Omega_j + 2\int_V \rho x_i \epsilon_{jlm} u_l \Omega_m d\boldsymbol{x}\,, \qquad (1)$$

where

$$\mathfrak{T}_{ij} = \frac{1}{2}\int_V \rho u_i u_j d\boldsymbol{x}\,, \qquad (2)$$

$$\mathfrak{W}_{ij} = -\frac{1}{2}\int_V \rho \mathfrak{B}_{ij} d\boldsymbol{x}\,, \qquad (3)$$

and

$$I_{ij} = \int_V \rho x_i x_j d\boldsymbol{x} \tag{4}$$

are the kinetic energy, the potential energy, and the moment of inertia tensors, respectively. In equation (3) $\mathfrak{V}_{ij}$ is the tensor potential discussed at length in the preceding paper (Chandrasekhar and Lebovitz 1962; this paper will be referred to hereafter as "Paper II"), and the remaining symbols have their usual meanings.

Before we write down the form which equation (1) takes for infinitesimal perturbations about equilibrium, we may note the relations that obtain in equilibrium. They are (since u_i is now zero)

$$\mathfrak{W}_{ij} + \Omega^2 I_{ij} - I_{il}\Omega_l\Omega_j = -\delta_{ij}\int_V p\, d\boldsymbol{x}\,. \tag{5}$$

If the x_3-axis is chosen in the direction of $\boldsymbol{\Omega}$,

$$\Omega_j = \Omega\,\delta_{j3}\,, \tag{6}$$

and equation (5) gives

$$\mathfrak{W}_{11} + \Omega^2 I_{11} = \mathfrak{W}_{22} + \Omega^2 I_{22} = \mathfrak{W}_{33} = -\int_V p\, d\boldsymbol{x}\,. \tag{7}$$

It has been shown in a different connection (Chandrasekhar 1961*b*) how we can deduce from equations (7) explicit expressions for the different components of the potential energy tensor in case the equation of state is polytropic.

Now suppose that the equilibrium is disturbed and that, as a result, each element of mass dm $(=\rho d\boldsymbol{x})$ suffers a Lagrangian displacement $\boldsymbol{\xi}(\boldsymbol{x}, t)$. Let δI_{ij}, $\delta\mathfrak{W}_{ij}$, and

$$\delta\int_V p\, d\boldsymbol{x}$$

be the first-order changes in the respective quantities caused by the perturbation; then, equation (1) gives

$$\frac{d^2}{dt^2}\int_V \rho x_i \xi_j d\boldsymbol{x} = \delta\mathfrak{W}_{ij} + \delta_{ij}\left(\delta\int_V p\, d\boldsymbol{x}\right) + \Omega^2 \delta I_{ij} - \delta I_{il}\Omega_l\Omega_j + 2\frac{d}{dt}\int_V \rho x_i \epsilon_{jlm}\xi_l \Omega_m d\boldsymbol{x}\,. \tag{8}$$

The required first-order changes in I_{ij}, etc., are given by (cf. Chandrasekhar 1961*a*, § 118, and Paper I, § III)

$$\delta I_{ij} = \int_V \rho(x_i\xi_j + \xi_i x_j)\, d\boldsymbol{x}\,, \tag{9}$$

$$\delta\mathfrak{W}_{ij} = -\int_V \rho\,\xi_l \frac{\partial\mathfrak{V}_{ij}}{\partial x_l}\, d\boldsymbol{x}\,, \tag{10}$$

and

$$\delta\int_V p\, d\boldsymbol{x} = \int_V \delta\left(\frac{p}{\rho}\right)\rho\, d\boldsymbol{x} = -(\gamma-1)\int_V p\,\operatorname{div}\boldsymbol{\xi}\, d\boldsymbol{x}\,, \tag{11}$$

where γ denotes the ratio of the specific heats. In deriving the last of the foregoing equations, the assumption has been made that the perturbation is accompanied by adiabatic changes, so that

$$\frac{\delta p}{p} = \gamma\frac{\delta\rho}{\rho} = -\gamma\,\operatorname{div}\boldsymbol{\xi}\,. \tag{12}$$

Inserting the results (9)–(11) in equation (8), we obtain the desired equation:

$$\frac{d^2}{dt^2}\int_V \rho x_i \xi_j dx = -\int_V \rho \xi_l \frac{\partial \mathfrak{B}_{ij}}{\partial x_l} dx - (\gamma - 1)\,\delta_{ij}\int_V p \operatorname{div} \boldsymbol{\xi} dx$$

$$+\Omega^2\int_V \rho(x_i\xi_j + \xi_i x_j)\,dx - \Omega_j\Omega_l\int_V \rho(x_l\xi_i + \xi_l x_i)\,dx + 2\frac{d}{dt}\int_V \rho x_i \epsilon_{jlm}\xi_l\Omega_m dx\,. \tag{13}$$ [1]

III. THE METHOD OF APPROXIMATION

Equation (13) is an *exact* integral relation which must be satisfied in all cases of infinitesimal perturbations; in particular, it must be satisfied by the proper solutions belonging to the natural modes of oscillation of the system. Since the solutions belonging to the natural modes will have a time dependence of the form

$$e^{\lambda t}\,, \tag{14}$$

it would appear that, by inserting for $\boldsymbol{\xi}$ in equation (13) a "trial function" with a space dependence which one might use, for example, in a variational treatment of the problem (i.e., if the problem should allow one), we should obtain an equation which would enable us to determine the frequencies of oscillation[2] with some precision. Clearly, this method has no "absolute" basis such as those which are derived from a strict minimal (or maximal) principle. And, moreover, when basing on equation (13), we cannot use "trial" functions with as many variational parameters as we like. In the present instance, since equation (13) provides nine equations, a "trial" function which we may wish to insert cannot involve more than nine constants. And the simplest "trial" function that suggests itself in the present connection is (cf. Chandrasekhar 1961*a*)

$$\xi_j = X_{jr}x_r e^{\lambda t}\,; \tag{15}$$

and the nine coefficients of this linear transformation play, in the present treatment, the role of the variational parameters in a proper variational treatment.

In assuming a "trial" function of the form (15), we have been guided by two considerations: first, it is known from the investigations of Ledoux (1945) that the purely radial oscillations of a non-rotating gas sphere treated on the basis of the usual contracted version of the virial theorem and the simple substitution,

$$\xi_j = \text{Constant}\ x_j e^{\lambda t}\,, \tag{16}$$

[1] In this equation it is customary (cf. Chandrasekhar 1961*a*) to transform the integral

$$\int_V p \operatorname{div} \boldsymbol{\xi} dx$$

by an integration by parts to

$$-\int_V \boldsymbol{\xi}\cdot \operatorname{grad} p\, dx\,;$$

then, after substitution for grad p from the equation of hydrostatic equilibrium, express the integral in terms of equilibrium quantities. However, in the present connection it is more convenient to leave the term as it is.

[2] For reasons which have been explained in Paper I, by the use of the virial theorem in the form of eq. (13), we are, in the case of a configuration of uniform density, limited to the modes of oscillation that belong to the second harmonic.

leads to a formula for λ^2 (eq. [44] below) which agrees exactly with what follows from a strict variational treatment based on the same trial function (cf. Ledoux and Pekeris 1941); and, second, the substitution (15), in fact, corresponds to the *exact* solution in case the density is uniform. For these reasons the results to be derived on the basis of the substitution (15) are not likely to be seriously in error: certainly, the qualitative features of the theory should be trustworthy.

Inserting, then, in equation (13) the form (15) for ξ, we obtain

$$\lambda^2 X_{jl} I_{li} = 2\lambda\Omega\epsilon_{jl3} X_{lm} I_{mi} + \Omega^2 (X_{jl} I_{li} + X_{il} I_{lj}) - \Omega^2 \delta_{j3} (X_{3l} I_{li} + X_{il} I_{l3}) - X_{lr} \mathfrak{W}_{rl;ij} + J X_{rr} \delta_{ij} , \tag{17}$$

where the choice of the co-ordinates appropriate for the direction of $\mathbf{\Omega}$ to coincide with the x_3-axis has been made. Moreover, in equation (17),

$$\mathfrak{W}_{rl;ij} = \int_V \rho x_r \frac{\partial \mathfrak{B}_{ij}}{\partial x_l} dx \tag{18}$$

is the super-matrix considered in Paper II, X_{rr} is the trace of $\boldsymbol{X}$, and

$$J = -(\gamma - 1) \int_V p\, dx . \tag{19}$$

If use is made of the last of the three equilibrium conditions given in equations (7), we can write

$$J = (\gamma - 1) \mathfrak{W}_{33} . \tag{20}$$

In our further considerations, we shall suppose that the equilibrium configuration has axial symmetry about the x_3-axis (i.e., the axis of rotation) and also that the (x_1, x_2)-plane is a plane of symmetry. Under these circumstances the tensors $\mathfrak{W}_{ij}$ and $\mathfrak{W}_{pq;ij}$ have the symmetries listed in Paper II (§ 6); and the moment-of-inertia tensor is, of course, also diagonal with $I_{11} = I_{22}$. With the simplifications introduced by these various symmetries, the nine equations which equation (17) represents take the following explicit forms:

$$\lambda^2 X_{11} I_{11} = +2\lambda\Omega X_{21} I_{11} + 2\Omega^2 X_{11} I_{11} + J X_{rr} - (X_{11} \mathfrak{W}_{11;11} + X_{22} \mathfrak{W}_{22;11} + X_{33} \mathfrak{W}_{33;11}) , \tag{21}$$

$$\lambda^2 X_{22} I_{11} = -2\lambda\Omega X_{12} I_{11} + 2\Omega^2 X_{22} I_{11} + J X_{rr} - (X_{11} \mathfrak{W}_{22;11} + X_{22} \mathfrak{W}_{11;11} + X_{33} \mathfrak{W}_{33;11}) , \tag{22}$$

$$\lambda^2 X_{33} I_{33} = J X_{rr} - (X_{11} + X_{22}) \mathfrak{W}_{33;11} - X_{33} \mathfrak{W}_{33;33} , \tag{23}$$

$$\lambda^2 X_{21} I_{11} = -2\lambda\Omega X_{11} I_{11} + (\Omega^2 I_{11} - \mathfrak{W}_{12;12})(X_{21} + X_{12}) , \tag{24}$$

$$\lambda^2 X_{12} I_{11} = +2\lambda\Omega X_{22} I_{11} + (\Omega^2 I_{11} - \mathfrak{W}_{12;12})(X_{12} + X_{21}) , \tag{25}$$

$$\lambda^2 X_{31} I_{11} = -X_{13} \mathfrak{W}_{31;13} - X_{31} \mathfrak{W}_{13;13} , \tag{26}$$

$$\lambda^2 X_{32} I_{11} = -X_{23} \mathfrak{W}_{31;13} - X_{32} \mathfrak{W}_{13;13} , \tag{27}$$

$$\lambda^2 X_{13} I_{33} = +2\lambda\Omega X_{23} I_{33} + (\Omega^2 I_{33} - \mathfrak{W}_{31;13}) X_{13} + (\Omega^2 I_{11} - \mathfrak{W}_{13;13}) X_{31} , \tag{28}$$

$$\lambda^2 X_{23} I_{33} = -2\lambda\Omega X_{13} I_{33} + (\Omega^2 I_{33} - \mathfrak{W}_{31;13}) X_{23} + (\Omega^2 I_{11} - \mathfrak{W}_{13;13}) X_{32} . \tag{29}$$

IV. THE MODES OF OSCILLATION OF A SPHERICAL MASS OF GAS IN THE ABSENCE OF ROTATION

Before we proceed to a general consideration of equations (21)–(29), it will be instructive to examine them in the special case when $\Omega = 0$ and the unperturbed configuration is spherical. Then

$$I_{11} = I_{22} = I_{33} = \tfrac{1}{3} I \,(\text{say}), \tag{30}$$

and

$$\mathfrak{W}_{11} = \mathfrak{W}_{22} = \mathfrak{W}_{33} = \tfrac{1}{3}\mathfrak{W}\,; \tag{31}$$

and there will be only three distinct components of the tensor $\mathfrak{W}_{pq;ij}$. The typical elements of $\mathfrak{W}_{pq;ij}$ are (cf. Paper II, eqs. [74]–[76])

$$\mathfrak{W}_{11;11} = -\tfrac{1}{15}\mathfrak{W}, \qquad \mathfrak{W}_{11;22} = \tfrac{1}{5}\mathfrak{W}, \qquad \text{and} \qquad \mathfrak{W}_{12;12} = -\tfrac{2}{15}\mathfrak{W}. \tag{32}$$

The equations governing the non-diagonal elements of $\boldsymbol{X}$ (namely, eqs. [24]–[29]) now become

$$\tfrac{1}{3}\lambda^2 I X_{21} = \tfrac{2}{15}\mathfrak{W}(X_{12} + X_{21}) \tag{33}$$

and

$$\tfrac{1}{3}\lambda^2 I X_{12} = \tfrac{2}{15}\mathfrak{W}(X_{21} + X_{12})\,; \tag{34}$$

and two similar pairs which are obtained by cyclically permuting the indices. From equations (33) and (34) it follows that

$$\lambda^2 = 0 \qquad (\text{if } X_{21} \neq X_{12}), \tag{35}$$

and

$$\lambda^2 = \tfrac{4}{5}\frac{\mathfrak{W}}{I} \qquad (\text{if } X_{21} = X_{12})\,; \tag{36}$$

and both these roots are repeated three times.

Turning next to equations (21)–(23) we have

$$\tfrac{1}{3}\lambda^2 I X_{11} = J X_{rr} + \tfrac{1}{15}\mathfrak{W} X_{11} - \tfrac{1}{5}\mathfrak{W}(X_{22} + X_{33}), \tag{37}$$

$$\tfrac{1}{3}\lambda^2 I X_{22} = J X_{rr} + \tfrac{1}{15}\mathfrak{W} X_{22} - \tfrac{1}{5}\mathfrak{W}(X_{33} + X_{11}), \tag{38}$$

and

$$\tfrac{1}{3}\lambda^2 X_{33} = J X_{rr} + \tfrac{1}{15}\mathfrak{W} X_{33} - \tfrac{1}{5}\mathfrak{W}(X_{11} + X_{22}), \tag{39}$$

where it may be recalled that now (cf. eqs. [20] and [31])

$$J = \tfrac{1}{3}(\gamma - 1)\mathfrak{W} \qquad \text{and} \qquad X_{rr} = X_{11} + X_{22} + X_{33}\,. \tag{40}$$

By subtracting equation (38) from equation (37), we have

$$\tfrac{1}{3}\lambda^2 I (X_{11} - X_{22}) = \tfrac{4}{15}\mathfrak{W}(X_{11} - X_{22})\,. \tag{41}$$

The same equation governs $X_{22} - X_{33}$ and $X_{33} - X_{11}$; but, of these three equations, only two are linearly independent. Therefore, the root

$$\lambda^2 = \tfrac{4}{5}\frac{\mathfrak{W}}{I} \qquad (X_{11} \neq X_{22}), \tag{42}$$

which follows from these equations, is of multiplicity 2.

Finally, adding all three equations (37)–(39), we obtain

$$\tfrac{1}{3}\lambda^2 I X_{rr} = \left(3J + \tfrac{1}{15}\mathfrak{W} - \tfrac{2}{5}\mathfrak{W}\right) X_{rr} = \left(\gamma - \tfrac{4}{3}\right)\mathfrak{W} X_{rr}\,; \tag{43}$$

and this equation leads to the root

$$\lambda^2 = (3\gamma - 4)\frac{\mathfrak{W}}{I}\,. \tag{44}$$

This last root represents a purely radial mode of oscillation; and the formula for λ^2 agrees with the one first derived by Ledoux (1945).

The foregoing discussion has shown that a spherical mass of gas in equilibrium under its own gravitation has two fundamental frequencies of oscillation corresponding to the two roots $\lambda^2 = \frac{4}{5}\mathfrak{W}/I$ and $\lambda^2 = (3\gamma - 4)\mathfrak{W}/I$; the former root is of multiplicity 5 and represents all the non-radial modes belonging to the second harmonic; and the latter root is non-degenerate and represents the fundamental of the purely radial modes. In addition, we have three neutral modes belonging to $\lambda^2 = 0$; the existence of three such neutral modes corresponds to the possibility of arbitrary rotations about three perpendicular axes. And, finally, it should be noted that when $\gamma = 1.6$ the two non-vanishing roots coincide and a case of *accidental degeneracy* arises.[3]

V. THE TRANSVERSE SHEAR MODES

Returning to equations (21)–(29), we observe that equations (26)-(29) involving X_{31}, X_{32}, X_{13}, and X_{23} are independent of the others; we have, therefore, four modes of oscillation in which all the elements of the transformation (15), except these four, vanish. The transformation appropriate to these modes is

$$\xi_1 = X_{13}x_3\,, \qquad \xi_2 = X_{23}x_3\,, \qquad \text{and} \qquad \xi_3 = X_{31}x_1 + X_{32}x_2\,, \tag{45}$$

where we have suppressed the time-dependent factor $e^{\lambda t}$. The predominant feature of these modes is the relative shearing of the northern and the southern hemispheres. For this reason, we shall call them the *transverse shear modes*, where the qualification "transverse" is with respect to the direction of $\mathbf{\Omega}$.

The four equations governing X_{31}, X_{32}, X_{13}, and X_{23} can be written in the matrix form

$$\begin{vmatrix} \lambda^2 I_{11} + \mathfrak{W}_{13;13} & 0 & \mathfrak{W}_{31;13} & 0 \\ 0 & \lambda^2 I_{11} + \mathfrak{W}_{13;13} & 0 & \mathfrak{W}_{31;13} \\ -\Omega^2 I_{11} + \mathfrak{W}_{13;13} & 0 & (\lambda^2 - \Omega^2) I_{33} + \mathfrak{W}_{31;13} & -2\lambda\Omega I_{33} \\ 0 & -\Omega^2 I_{11} + \mathfrak{W}_{13;13} & +2\lambda\Omega I_{33} & (\lambda^2 - \Omega^2) I_{33} + \mathfrak{W}_{31;13} \end{vmatrix} \begin{Vmatrix} X_{31} \\ X_{32} \\ X_{13} \\ X_{23} \end{Vmatrix} = 0\,. \tag{46}$$

The determinant of the matrix on the right-hand side of equation (46) must vanish; and the characteristic equation which follows is

$$\left\{ (\sigma^2 I_{11} - \mathfrak{W}_{13;13})\left[(\sigma^2 + \Omega^2) I_{33} - \mathfrak{W}_{31;13}\right] + (\Omega^2 I_{11} - \mathfrak{W}_{13;13})\mathfrak{W}_{31;13} \right\}^2 = 4\sigma^2\Omega^2 I_{33}^2 (\sigma^2 I_{11} - \mathfrak{W}_{13;13})^2\,, \tag{47}$$

[3] This occurrence of accidental degeneracy has been deduced on the basis of the present approximative treatment. An exact treatment should reveal the same phenomenon; but the value of γ at which it will occur will probably differ from $\gamma = 1.6$.

where we have written

$$\sigma^2 = -\lambda^2 ; \tag{48}$$

so that, for stable oscillations, σ should be real. Equation (47) factorizes to give

$$(\sigma^2 - L)(\sigma^2 + \Omega^2 - M) + (\Omega^2 - L)M = \pm 2\sigma\Omega(\sigma^2 - L), \tag{49}$$

where, for the sake of brevity, we have written

$$L = \frac{\mathfrak{W}_{13;13}}{I_{11}} \quad \text{and} \quad M = \frac{\mathfrak{W}_{31;13}}{I_{33}} . \tag{50}$$

The occurrence of two signs on the right-hand side of equation (49) means that these modes have a doublet character, with $+\Omega$ and $-\Omega$ playing equivalent roles (as in the normal Zeeman effect; cf. Pekeris, Alterman, and Jarosch 1961).

On further simplification, equation (49) reduces to the form

$$(\sigma \mp \Omega)[(\sigma^2 - L)(\sigma \mp \Omega) - M(\sigma \pm \Omega)] = 0 . \tag{51}$$

Therefore,

$$\sigma^2 = \Omega^2 \tag{52}$$

is an allowed characteristic root. The remaining roots are given by

$$(\sigma^2 - L)(\sigma \mp \Omega) - M(\sigma \pm \Omega) = 0 \tag{53}$$

and, expanding this equation, we have

$$\sigma^3 \mp \sigma^2\Omega - \sigma\mu \pm \Omega\nu = 0 , \tag{54}$$

where we have introduced the abbreviations

$$\mu = L + M \quad \text{and} \quad \nu = L - M . \tag{55}$$

With the substitution

$$\sigma = \zeta \pm \tfrac{1}{3}\Omega , \tag{56}$$

equation (54) becomes

$$\zeta^3 - (\tfrac{1}{3}\Omega^2 + \mu)\zeta \mp \tfrac{2}{27}\Omega\,[\Omega^2 + \tfrac{9}{2}(\mu - 3\nu)] = 0 . \tag{57}$$

The necessary and sufficient conditions for the reality of the roots of this cubic equation are

$$\tfrac{1}{3}\Omega^2 + \mu \geq 0 \tag{58}$$

and

$$4(\tfrac{1}{3}\Omega^2 + \mu)^3 \geq \tfrac{4}{27}\Omega^2\,[\Omega^2 + \tfrac{9}{2}(\mu - 3\nu)]^2 . \tag{59}$$

On expanding this last inequality and simplifying, we are left with

$$\Omega^4\nu + \Omega^2\,[\mu^2 - \tfrac{3}{4}(\mu - 3\nu)^2] + \mu^3 \geq 0 . \tag{60}$$

Sufficient conditions for the reality of the roots of equation (51) are, therefore,

$$\mu \geq 0 , \quad \nu \geq 0 , \quad \text{and} \quad 4\nu\mu^3 \geq [\mu^2 - \tfrac{3}{4}(\mu - 3\nu)^2]^2$$

or (61)

$$\frac{1}{2\nu}\{-[\mu^2 - \tfrac{3}{4}(\mu - 3\nu)^2] \pm [\![\{\mu^2 - \tfrac{3}{4}(\mu - 3\nu)^2\}^2 - 4\nu\mu^3]\!]^{1/2}\} < 0 .$$

It can be verified that when $\mu \geq 0$ and $\nu \geq 0$, the last inequalities in (61) will be satisfied if

$$\mu \geq \tfrac{9}{7}\nu . \tag{62}$$

In terms of L and M, sufficient conditions for the stability of these modes are

$$L \geq 0 , \quad M \geq 0 , \quad \text{and} \quad M \leq L \leq 8M . \tag{63}$$

According to the results of Paper II (§ 6), the conditions enumerated in (63) imply

$$\mathfrak{W}_{13;13} = C - \mathfrak{W}_{11} \geq 0 , \quad \mathfrak{W}_{31;13} = C - \mathfrak{W}_{33} \geq 0 , \tag{64}$$

and

$$\frac{C - \mathfrak{W}_{11}}{I_{11}} \geq \frac{C - \mathfrak{W}_{33}}{I_{33}} . \tag{65}$$

By making use of the equilibrium condition

$$\mathfrak{W}_{11} + \Omega^2 I_{11} = \mathfrak{W}_{33} , \tag{66}$$

the inequality (65) can be brought to the form

$$\Omega^2 I_{11}^2 + (I_{33} - I_{11})(C - \mathfrak{W}_{11}) \geq 0 . \tag{67}$$

It is evident that in the limit of zero angular velocity (when $\mathfrak{W}_{11} \to \frac{1}{3}\mathfrak{W}$, $\mathfrak{W}_{33} \to \frac{1}{3}\mathfrak{W}$, and $C \to \frac{1}{5}\mathfrak{W}$) conditions (64) and (67) are fulfilled. Therefore, these modes certainly start being stable.

Finally, we may note that when $\Omega \to 0$, the roots of equation (54) have the limiting behaviors:

$$\sigma_1 = +\sqrt{\mu} \pm \tfrac{1}{2}\Omega\left(1 - \frac{\nu}{\mu}\right), \quad \sigma_2 = -\sqrt{\mu} \pm \tfrac{1}{2}\Omega\left(1 - \frac{\nu}{\mu}\right), \quad \text{and } \sigma_3 = \pm\Omega\frac{\nu}{\mu} . \tag{68}$$

These formulae exhibit in a striking manner the "doublet" character of these modes.

VI. THE TOROIDAL MODES

Returning to the remaining equations (21)–(25), we obtain, on subtracting equation (22) from equation (21),

$$\lambda^2 I_{11}(X_{11} - X_{22}) = 2\lambda\Omega I_{11}(X_{21} + X_{12}) + 2\Omega^2 I_{11}(X_{11} - X_{22}) - (\mathfrak{W}_{11;11} - \mathfrak{W}_{22;11})(X_{11} - X_{22}) . \tag{69}$$

By making use of the relation (cf. Paper II, eq. [60])

$$\mathfrak{W}_{11;11} - \mathfrak{W}_{22;11} = 2\mathfrak{W}_{12;12} , \tag{70}$$

we can rewrite equation (69) in the form

$$[\lambda^2 I_{11} - 2(\Omega^2 I_{11} - \mathfrak{W}_{12;12})](X_{11} - X_{22}) - 2\lambda\Omega I_{11}(X_{21} + X_{12}) = 0 . \tag{71}$$

Next, by the addition of equations (24) and (25), we obtain

$$\lambda^2 I_{11}(X_{12} + X_{21}) = -2\lambda\Omega I_{11}(X_{11} - X_{22}) + 2(\Omega^2 I_{11} - \mathfrak{W}_{12;12})(X_{21} + X_{12}) , \tag{72}$$

or

$$[\lambda^2 I_{11} - 2(\Omega^2 I_{11} - \mathfrak{W}_{12;12})](X_{12} + X_{21}) + 2\lambda\Omega I_{11}(X_{11} - X_{22}) = 0 . \tag{73}$$

Equations (71) and (73) can be treated together, and they lead to the characteristic equation

$$[\lambda^2 I_{11} - 2(\Omega^2 I_{11} - \mathfrak{W}_{12;12})]^2 + 4\lambda^2\Omega^2 I_{11}^2 = 0 \,. \tag{74}$$

On simplification, equation (74) becomes

$$I_{11}^2\lambda^4 + 4I_{11}\mathfrak{W}_{12;12}\lambda^2 + 4(\Omega^2 I_{11} - \mathfrak{W}_{12;12})^2 = 0 \,. \tag{75}$$

The roots of this equation are

$$\lambda^2 = -2\frac{\mathfrak{W}_{12;12}}{I_{11}} \pm 2\Omega\left(2\frac{\mathfrak{W}_{12;12}}{I_{11}} - \Omega^2\right)^{1/2} . \tag{76}$$

From equation (76) it now follows that

$$\lambda^2 = 0 \text{ is a root when } \Omega^2 = \frac{\mathfrak{W}_{12;12}}{I_{11}} \,. \tag{77*}$$

But we have already seen in Paper II, § VII, that if the sequence of axially symmetric configurations should have a point of bifurcation at which objects with genuine triplanar symmetry branch off, then at such a point the equality, $\Omega^2 I_{11} = \mathfrak{W}_{12;12}$, must obtain. Therefore, in analogy with what happens along the sequence of the Maclaurin spheroids at the point of bifurcation where the Jacobi ellipsoids branch off (cf. Paper I, § VII [d]), we may now conclude that the occurrence of a neutral mode at $\Omega^2 = \mathfrak{W}_{12;12}/I_{11}$ means only that, at the point of bifurcation (should one occur), the associated neutral mode simply carries the axially symmetric configuration over into a neighboring equilibrium configuration of genuine triplanar symmetry.

In the limit $\Omega^2 = 0$, $\mathfrak{W}_{12;12}$ has a finite positive limit, so that in this limit the modes are definitely stable. Should $\mathfrak{W}_{12;12}$ continue to be positive, then, when

$$\Omega^2 I_{11} > 2\mathfrak{W}_{12;12} \,, \tag{78}$$

the configuration becomes unstable. Also, it can be verified that, should these unstable modes occur, the real part of the frequencies will be Ω, so that instability occurs as overstability.

VII. THE PULSATION MODES

Of the five equations (21)–(25), we have considered in Section VI two linear combinations of them. It remains to consider three other linear combinations; and we shall select them in the following manner.

We have already considered in Section VI the equation resulting from the addition of equations (24) and (25). Now, by the subtraction of one from the other, we obtain

$$\lambda^2(X_{21} - X_{12})I_{11} = -2\lambda\Omega I_{11}(X_{11} + X_{22}) \,. \tag{79}$$

Next, adding equations (21) and (22) and subtracting from the result equation (23) twice, we obtain

$$\begin{aligned}\lambda^2[(X_{11}+X_{22})I_{11} - 2X_{33}I_{33}] &= 2\lambda\Omega I_{11}(X_{21}-X_{12}) + 2\Omega^2 I_{11}(X_{11}+X_{22}) \\ &\quad - 2(\mathfrak{W}_{33;11} - \mathfrak{W}_{33;33})X_{33} - (\mathfrak{W}_{11;11} + \mathfrak{W}_{22;11} - 2\mathfrak{W}_{33;11})(X_{11}+X_{22}) \,.\end{aligned} \tag{80}$$

And we shall retain equation (23) as it is:

$$\lambda^2 I_{33}X_{33} = J[(X_{11}+X_{22}) + X_{33}] - \mathfrak{W}_{33;11}(X_{11}+X_{22}) - \mathfrak{W}_{33;33}X_{33} \,. \tag{81}$$

*See, however, the footnote to equation (81) of paper 16 in this volume.

First, we observe that, according to equation (79),

$$\lambda^2 = 0 \text{ is a root if } X_{21} \neq X_{12} , \qquad X_{11} + X_{22} = X_{33} = 0 . \tag{82}$$

This neutral mode is, in fact, the same one that occurred, under the same circumstances, in the absence of rotation (Sec. IV, eq. [35]); its continued existence in the presence of rotation is clearly to be expected on symmetry grounds.

If we ignore the root $\lambda^2 = 0$, we can eliminate $(X_{12} - X_{21})$ in equation (80) by making use of equation (79); and the equations we must now consider are

$$\begin{aligned}[\lambda^2 I_{11} + 2\Omega^2 I_{11} + (\mathfrak{W}_{11;11} + \mathfrak{W}_{22;11} - 2\mathfrak{W}_{33;11})](X_{11} + X_{22})& \\ - 2[\lambda^2 I_{33} - (\mathfrak{W}_{33;11} - \mathfrak{W}_{33;33})] X_{33} &= 0\end{aligned} \tag{83}$$

and

$$(\lambda^2 I_{33} - J + \mathfrak{W}_{33;33}) X_{33} + (\mathfrak{W}_{33;11} - J)(X_{11} + X_{22}) = 0 . \tag{84}$$

In the notation of Paper II, § VI,

$$\mathfrak{W}_{33;11} = C \quad \text{and} \quad \mathfrak{W}_{33;33} = D ; \tag{85}$$

and, making use of the various relations listed in the same section, as well as equation (66), we verify that

$$\begin{aligned}-2\Omega^2 I_{11} - (\mathfrak{W}_{11;11} + \mathfrak{W}_{22;11} - 2\mathfrak{W}_{11;33}) &= -2\Omega^2 I_{11} - (A + B - 2C) \\ &= -\Omega^2 I_{11} + C - D .\end{aligned} \tag{86}$$

Equations (83) and (84) can now be written in the matrix form

$$\begin{vmatrix} \lambda^2 I_{11} + \Omega^2 I_{11} - C + D & -2\lambda^2 I_{33} + 2(C - D) \\ -J + C & \lambda^2 I_{33} - J + D \end{vmatrix} \begin{vmatrix} X_{11} + X_{22} \\ X_{33} \end{vmatrix} = 0 . \tag{87}$$

The determinant of the matrix on the left-hand side of equation (87) must vanish, and we find that this leads to the characteristic equation

$$I_{11} I_{33} \lambda^4 - [\beta I_{11} + (\beta + \alpha) I_{33} - \Omega^2 I_{11} I_{33}] \lambda^2 - \Omega^2 I_{11} \beta + (\beta + 2\alpha)(\beta - \alpha) = 0 , \tag{88}$$

where we have introduced the abbreviations

$$\alpha = J - C \quad \text{and} \quad \beta = J - D . \tag{89}$$

Note that, with the foregoing definition of α and β (cf. eq. [20] and Paper II, eq. [70]),

$$2\alpha + \beta = 3J - (2C + D) = (3\gamma - 4)\mathfrak{W}_{33} \tag{90}$$

and

$$\beta - \alpha = C - D . \tag{91}$$

The roots of equation (88) can be readily written down; but greater interest attaches to the coupling between the radial and the non-radial modes of oscillation which equation (88) predicts. The nature of this coupling is best clarified by considering the limit $\Omega^2 \to 0$.

When $\Omega^2 = 0$,

$$I_{11} = I_{33} = \tfrac{1}{3} I , \qquad \mathfrak{W}_{33} = \tfrac{1}{3}\mathfrak{W} , \qquad \text{and } C - D = \tfrac{4}{15}\mathfrak{W} , \tag{92}$$

and equation (88) becomes

$$\tfrac{1}{9} I^2 \lambda^4 - \tfrac{1}{3}(2\beta^0 + \alpha^0) I \lambda^2 + (\beta^0 + 2\alpha^0)(\beta^0 - \alpha^0) = 0 , \tag{93}$$

where the superscript "0" distinguishes the value of the quantity in the spherically symmetric state. The roots of equation (93) are

$$\lambda^2 = 3\,\frac{\beta^0 + 2\alpha^0}{I} \qquad \text{and} \qquad \lambda^2 = 3\,\frac{\beta^0 - \alpha^0}{I}\,; \tag{94}$$

or, making use of equations (90), (91), and (92), we have

$$\lambda^2 = (3\gamma - 4)\frac{\mathfrak{W}}{I} = \lambda_r^2\,(\text{say})$$

and (95)

$$\lambda^2 = \frac{4}{5}\frac{\mathfrak{W}}{I} = \lambda_s^2\,(\text{say})\,;$$

these agree with the results for the radial and the non-radial modes derived in Section IV (eqs. [42] and [44]). We now see that *rotation couples these modes.* An important aspect of this coupling is that the non-radial mode, which in the absence of rotation is volume-conserving, is no longer so in the presence of rotation. This fact is apparent from the incompatibility of the equations (87) with the condition $X_{rr} = 0$ if $\Omega \neq 0$.

We shall now determine the extent of the coupling between the radial and the non-radial modes in the limit $\Omega^2 \to 0$. For this purpose we shall first rewrite equation (88) in the form

$$I_{11}I_{33}\lambda^4 - [(\beta - \alpha)I_{11} + (\beta + 2\alpha)I_{33} + \alpha(I_{11} - I_{33}) - \Omega^2 I_{11}I_{33}]\,\lambda^2 - \Omega^2 I_{11}\beta + (\beta + 2\alpha)(\beta - \alpha) = 0\,, \tag{96}$$

or

$$[\lambda^2 I_{11} - (\beta + 2\alpha)][\lambda^2 I_{33} - (\beta - \alpha)] = \lambda^2(\alpha\Delta I - \Omega^2 I_{11}I_{33}) + \Omega^2 I_{11}\beta\,, \tag{97}$$

where

$$\Delta I = I_{11} - I_{33} \tag{98}$$

is the change in the components of the moment of inertia caused by the rotation; ΔI is clearly of order Ω^2.

Now the terms on the right-hand side of equation (97) are all of order Ω^2; therefore, to this order, we can replace the various coefficients (such as I_{11}, α, etc.,) by their values in the absence of rotation, namely,

$$\alpha^0 = J^0 - C^0 = \tfrac{1}{3}(\gamma - 1)\mathfrak{W} - \tfrac{1}{5}\mathfrak{W} = \tfrac{1}{15}(5\gamma - 8)\mathfrak{W},$$

$$\beta^0 = J^0 - D^0 = \tfrac{1}{3}(\gamma - 1)\mathfrak{W} + \tfrac{1}{15}\mathfrak{W} = \tfrac{1}{15}(5\gamma - 4)\mathfrak{W}, \tag{99}$$

and

$$I_{11}^0 = I_{33}^0 = \tfrac{1}{3}I\,.$$

Substituting the foregoing values on the right-hand side of equation (97), we obtain

$$(\lambda^2 - \lambda_R^2)(\lambda^2 - \lambda_S^2) = \tfrac{3}{5}(5\gamma - 8)\frac{\mathfrak{W}}{I}\frac{\Delta I}{I}\,\lambda^2 - \tfrac{1}{5}\Omega^2\Big[5\lambda^2 - (5\gamma - 4)\frac{\mathfrak{W}}{I}\Big] + O(\Omega^4)\,, \tag{100}$$

where

$$\lambda_R^2 = \frac{\beta + 2\alpha}{I_{11}} \qquad \text{and} \qquad \lambda_S^2 = \frac{\beta - \alpha}{I_{33}}\,. \tag{101}$$

(Note that the values of λ_R^2 and λ_S^2 differ from their zero-order values, namely, λ_r^2 and λ_s^2 given by equations (95), by quantities of order Ω^2, i.e., by the same order as those retained in the present calculations.)

If the terms on the right-hand side of equation (100) are ignored (which we may not!) the roots of the equation are, of course, λ_R^2 and λ_S^2. Therefore, to the first order in Ω^2, the change in the root λ_R^2, for example, due to the presence of the terms on the right-hand side, is given by

$$\delta\lambda_R^2(\lambda_R^2-\lambda_S^2)=\tfrac{3}{5}(5\gamma-8)\frac{\mathfrak{W}}{I}\frac{\Delta I}{I}\lambda_R^2-\tfrac{1}{5}\Omega^2\left[5\lambda_R^2-(5\gamma-4)\frac{\mathfrak{W}}{I}\right]. \tag{102}$$

In this equation we can clearly replace λ_R^2 and λ_S^2 by their zero-order values λ_r^2 and λ_s^2; and we find

$$\delta\lambda_R^2=\frac{3(5\gamma-8)}{5(\lambda_r^2-\lambda_s^2)}\frac{\mathfrak{W}}{I}\left(\frac{\Delta I}{I}\lambda_r^2-\tfrac{2}{3}\Omega^2\right). \tag{103}$$

In the same way we find

$$\delta\lambda_S^2=-\frac{3(5\gamma-8)}{5(\lambda_r^2-\lambda_s^2)}\frac{\mathfrak{W}}{I}\left(\frac{\Delta I}{I}\lambda_s^2+\tfrac{1}{3}\Omega^2\right). \tag{104}$$

The foregoing formulae do not apply when $\gamma = 1.6$; for, in this case,

$$\lambda_r^2=\lambda_s^2=\tfrac{4}{5}\frac{\mathfrak{W}}{I}, \tag{105}$$

and the terms of order Ω^2 on the right-hand side of equation (100) vanish identically. Therefore, in this case, the required roots are given by equations (101) correctly to the first order in Ω^2. Because α, β, I_{11}, and I_{33} differ (on account of rotation) from their zero-order values, the two roots, λ_R^2 and λ_S^2, which are coincident in the absence of rotation, become separate. The accidental degeneracy which exists when $\gamma = 1.6$ is thus lifted by rotation.

VIII. SUMMARY OF RESULTS

The principal results of the preceding sections are summarized in Table 1, in which a comparison is further made with the corresponding results for incompressible fluids.

IX. CONCLUDING REMARKS

To some extent the theory presented in this paper is a formal one, since it presupposes a knowledge of the structure of the equilibrium configuration; and, except in the case of incompressible fluids, this knowledge is, in large measure, lacking. However, there is one case in which the formulae of this paper can be used to derive concrete results; this is the case of the rotationally distorted polytropes (Chandrasekhar 1933). In this theory of the distorted polytropes, the effect of rotation is treated by a perturbation method valid for small Ω^2. The first-order changes in the pressure, the density, and the gravitational potential have been evaluated and expressed in terms of two functions (ψ_0 and ψ_2 in the theory) which have been tabulated. For our present purposes, this information will have to be further completed by determining the super-potential χ to the same order as the other quantities; but this is a straightforward matter, and we shall return to it in another paper.

TABLE 1

THE CLASSIFICATION OF THE MODES

Modes	Parameters	Non-rotating Mass	Rotating Incompressible Mass	Rotating Compressible Mass
Transverse shear modes	X_{12}, X_{21}	$X_{12}\neq X_{21}$: $\lambda^2=0$	$\lambda^2=0$	$\lambda^2=0$
	X_{13}, X_{23}, X_{31}, X_{32}	$X_{13}\neq X_{31}$: $\lambda^2=0$ $X_{23}\neq X_{32}$: $\lambda^2=0$ $X_{13}=X_{31}$: $\lambda^2=\frac{4}{5}\mathfrak{W}/I$ $X_{23}=X_{32}$: $\lambda^2=\frac{4}{5}\mathfrak{W}/I$	$\lambda^2=-\Omega^2$ $\lambda^2=0$ Two stable modes	$\lambda^2=-\Omega^2$ Three coupled modes, each having a doublet character; probably stable
Toroidal modes	$X_{12}+X_{21}$ $X_{11}-X_{22}$	$X_{21}=X_{12}$: $\lambda^2=\frac{4}{5}\mathfrak{W}/I$ $X_{11}=-X_{22}$: $\lambda^2=\frac{4}{5}\mathfrak{W}/I$	Two modes, one of which becomes neutral at the point of bifurcation ($e=0.81$); both become unstable when $e=0.95$; real part of frequency beyond instability Ω	Two modes, one of which becomes neutral when $\Omega^2 I_{11}=\mathfrak{W}_{12;\,12}$; both become unstable when $\Omega^2 I_{11}=2\mathfrak{W}_{12;\,12}$; real part of frequency beyond instability Ω
Pulsation modes	$X_{11}+X_{22}$ X_{33}	$X_{22}=-X_{33}$, $X_{11}=0$: $\lambda^2=\frac{4}{5}\mathfrak{W}/I$ $X_{11}=X_{22}=X_{33}$: $\lambda^2=(3\gamma-4)\mathfrak{W}/I$* $X_{11}+X_{22}+X_{33}=0$; $\lambda^2=0$†	Stable mode . . . . $\lambda^2=0$	The radial and the non-radial modes are coupled

* Compressible case.
† Incompressible case.

The research reported in this paper has in part been supported by the Office of Naval Research under Contract Nonr-2121(20) with the Enrico Fermi Institute for Nuclear Studies, University of Chicago. The work of the second author was supported in part by the United States Air Force under contract No. AF-49(638)-42, monitored by the Air Force Office of Scientific Research of the Air Research and Development Command.

REFERENCES

Chandrasekhar, S. 1933, *M.N.*, **93**, 390.
———. 1960, *J. Math. Anal. and Appl.*, **1**, 240.
———. 1961*a*, *Hydrodynamic and Hydromagnetic Stability* (Oxford: Clarendon Press), chap. xiii, pp. 117 and 118.
———. 1961*b*, *Ap. J.*, **134**, 662.
Chandrasekhar, S., and Lebovitz, N. R. 1962, *Ap. J.*, **135**, 238.
Lebovitz, N. R. 1961, *Ap. J.*, **134**, 500.
Ledoux, P. 1945, *Ap. J.*, **102**, 143.
———. 1951, *ibid.*, **114**, 373.
Ledoux, P., and Pekeris, C. L. 1941, *Ap. J.*, **94**, 124.
Ledoux, P., and Walraven, T. 1958, *Hdb. d. Phys.*, **51**, 605.
Pekeris, C. L., Alterman, L. Z., and Jarosch, H. 1961, *Phys. Rev.*, **122**, 1692.

20

ON THE OSCILLATIONS AND THE STABILITY OF ROTATING GASEOUS MASSES. II. THE HOMOGENEOUS, COMPRESSIBLE MODEL

S. Chandrasekhar and Norman R. Lebovitz
University of Chicago and Massachusetts Institute of Technology
Received July 2, 1962

ABSTRACT

The pulsation frequencies of rotating, gaseous masses of uniform density, i.e., of the Maclaurin spheroids, are found as functions of the angular momentum M and the ratio of the specific heats γ. Numerical calculations for the pulsation frequencies and normal modes are given for $\gamma = 1.3, \frac{4}{3}, 1.4, 1.5, 1.6$, and $\frac{5}{3}$. One finds that the value of γ at which dynamical instability sets in is reduced from $\gamma = \frac{4}{3}$ by rotation. One also finds that, when $\gamma = 1.6$, the normal modes of oscillation one obtains in the limit $M \to 0$ are both very far from being radial.

I. INTRODUCTION

The theory of the radial pulsations of spherical, gaseous masses, in its simplest form, predicts the following approximate formula for the frequency σ of the fundamental mode of pulsation:

$$\sigma^2 = (3\gamma - 4)\frac{|\mathfrak{W}|}{I}, \tag{1}$$

where $\mathfrak{W}$ is the gravitational energy and I the moment of inertia of the spherical configuration of equilibrium about which the pulsations (assumed small in amplitude) take place; the ratio of the specific heats, γ, appears because of the assumption that the pulsations take place adiabatically. This formula is known to be a good approximation if the central condensation is not too high (Ledoux and Pekeris 1941); it becomes exact in the limit in which the density is uniform.[1]

The assumption that the equilibrium configuration is spherical implies that rotation is absent. We now wish to drop this assumption and to find the effect of rotation on the pulsation frequency. For this purpose, one may ask whether there is a generalization of equation (1), valid in the presence of rotation, that reduces to equation (1) in the limit of vanishing rotation. The answer is in the affirmative (Chandrasekhar and Lebovitz 1962*a*, eq. [88]; this paper will be referred to as "Paper I"), and this result provides the solution to the problem in principle.

In practice, however, it is necessary to know the distribution of mass throughout the equilibrium configuration to find the pulsation frequencies, for only then can one evaluate $\mathfrak{W}$, I, and the further coefficients that appear in the formulae of Paper I. Such information is largely unavailable for rotating configurations, but there are exceptions: (1) the mass distribution in slowly rotating polytropes has been worked out (Chandrasekhar 1933); and (2) if one *assumes* that the mass is distributed uniformly, the equilibrium configurations are the well-known Maclaurin spheroids. This paper will be devoted to finding the oscillation frequencies of the Maclaurin spheroids.[2] The oscillations of slowly rotating polytropes are treated in the following paper (Chandrasekhar and Lebovitz 1962*c*).

II. THE VIRIAL EQUATIONS

Let the equilibrium figure, a Maclaurin spheroid, be rotating about the x_3-axis, and let the semiaxes in the directions of the x_1-, x_2-, and x_3-co-ordinate axes be a_1, a_2 $(= a_1)$, and

[1] A gaseous configuration of uniform density will be called a "homogeneous, compressible model."

[2] The Maclaurin spheroids are not the only such models: for sufficiently large angular momentum the Jacobi ellipsoids provide another.

a_3. In order to find the oscillation frequencies, we shall use the tensor virial equations, as adapted to treat small perturbations of the equilibrium configuration (Paper I, eq. [13]); they may be written

$$\frac{d^2}{dt^2}\int_V \rho x_i \xi_j d\boldsymbol{x} = 2\Omega\epsilon_{jl3}\frac{d}{dt}\int_V \rho x_i \xi_l d\boldsymbol{x} + \Omega^2\int_V \rho(x_i\xi_j + x_j\xi_i)\,d\boldsymbol{x}$$

$$-\Omega^2\delta_{j3}\int_V \rho(x_3\xi_i + x_i\xi_3)\,d\boldsymbol{x} + \delta_{ij}(\gamma-1)\int_V \xi_l \frac{\partial p}{\partial x_l}\,d\boldsymbol{x} - \int_V \rho\,\xi_l \frac{\partial \mathfrak{B}_{ij}}{\partial x_l}\,d\boldsymbol{x}\,. \tag{2}$$

The results and equations of Paper I were obtained by replacing the Lagrangian displacement $\boldsymbol{\xi}$ appearing in equation (2) with the linear form

$$\xi_j = X_{jk}x_k e^{\lambda t}\,. \tag{3}$$

The justification for the substitution (3) lies, in part, in that the results so obtained become exact in the limit of uniform density and should therefore be good approximations if the central condensation is not too high.

If the substitution (3) is used in equation (2), the result is (Paper I, eq. [17])

$$\lambda^2 X_{jl}I_{li} = 2\lambda\Omega\epsilon_{jl3}X_{lm}I_{mi} + \Omega^2(X_{jl}I_{li} + X_{il}I_{lj})$$
$$-\Omega^2\delta_{j3}(X_{3l}I_{li} + X_{il}I_{l3}) + JX_{rr}\delta_{ij} - X_{lr}\mathfrak{W}_{rl;ij}\,, \tag{4}$$

where

$$J = -(\gamma-1)\int_V p\,d\boldsymbol{x} = (\gamma-1)\mathfrak{W}_{33}\,;$$

and the precise assertion is that the characteristic frequencies[3] determined by equation (4) become exact if the coefficients (I_{ij}, $\mathfrak{W}_{pq;ij}$, and J) are taken to be those appropriate to the homogeneous, compressible model. It is clearly important to prove this assertion for two reasons: (1) to establish the validity of the results of Paper I and (2) to emphasize that the results of the present paper, in which we use the equations of Paper I, are exact (i.e., involve no assumption as to the nature of $\boldsymbol{\xi}$, the Lagrangian displacement). We now turn to this proof.

III. THE EXACTNESS OF THE RESULTS FOR THE HOMOGENEOUS, COMPRESSIBLE MODEL

If a tensor N_{ij} is defined by the equation (cf. Lebovitz 1961, eq. [65]; this paper will be referred to as "Paper II")

$$N_{ij} = \int_V \rho x_i \xi_j d\boldsymbol{x}\,, \tag{5}$$

then all the terms of equation (2), with the exception of the last two, manifestly involve the tensor N_{ij} linearly. Further, if $\partial p/\partial x_l$ and $\partial\mathfrak{B}_{ij}/\partial x_l$ should be linear in the co-ordinates, the last two terms would also be linear combinations of the N_{ij}'s. But, for the homogeneous, compressible model, these quantities are, in fact, linear in the co-ordinates: the equilibrium pressure gradient is

$$\frac{\partial p}{\partial x_l} = \rho\frac{\partial}{\partial x_l}\left[\mathfrak{B} + \tfrac{1}{2}\,|\,\boldsymbol{\Omega}\times\boldsymbol{x}\,|^2\right], \tag{6}$$

where (cf. Chandrasekhar and Lebovitz 1962*b*, eq. [47]; this paper will be referred to as "Paper III")

$$\mathfrak{B} = \pi G\rho a_1^2 a_3[I - A_1(x_1^2 + x_2^2) - A_3 x_3^2]\,; \tag{7}$$ [4]

[3] The frequencies are given by $\sigma = i\lambda$, where $i = \sqrt{-1}$.

[4] In this and subsequent formulae, the constants A_i and A_{ij} appearing in the expressions for $\mathfrak{B}$ and $\mathfrak{B}_{ij}$ are those defined in Paper III; this represents a departure from the notation of Paper II.

and for $\mathfrak{B}_{ij}$ we have the formula (Paper III, eq. [51])

$$\mathfrak{B}_{ij} = \pi G\rho a_1^2 a_3\left[2x_i x_j(A_j - a_i^2 A_{ij}) + a_i^2\delta_{ij}\left(A_i - \sum_{l=1}^{3} A_{il}x_l^2\right)\right]. \tag{8}$$

In equation (8) and in the rest of this paper the summation convention is suspended.

If the foregoing formulae are used in equation (2), the result is

$$\begin{aligned}\frac{d^2}{dt^2}N_{ij} = {} & \sum_{l=1}^{3} 2\Omega\epsilon_{jl3}\frac{dN_{il}}{dt} + \Omega^2(N_{ij}+N_{ji}) - \Omega^2\delta_{j3}(N_{3i}+N_{i3}) \\ & - \delta_{ij}(\gamma-1)[(N_{11}+N_{22})(2\pi G\rho a_1^2 a_3 A_1 - \Omega^2) + 2\pi G\rho a_1^2 a_3 N_{33}] \\ & - \left[2(A_j - a_i^2 A_{ij})(N_{ij}+N_{ji}) - 2\delta_{ij}\sum_{l=1}^{3} a_i^2 A_{il}N_{ll}\right](\pi G\rho a_1^2 a_3).\end{aligned} \tag{9}$$

Equation (9), which is exact (i.e., no assumption concerning the nature of ξ has been made), represents a system of nine linear differential equations with constant coefficients in nine unknowns; its solutions are therefore of the form

$$N_{ij}(t) = N_{ij}(0)e^{\lambda t}. \tag{10}$$

The substitution (10) reduces equation (9) to a system whose characteristic equation determines the frequencies; these frequencies are unaltered by the linear transformation

$$N_{ij}(0) = \sum_{m=1}^{3} X_{jm}I_{mi} = X_{ji}I_{ii}. \tag{11}$$

This substitution puts all but the last two terms of equation (9) in manifest agreement with the corresponding terms of equation (4). We wish to show that the last two terms agree also.

The fourth term on the right-hand side of equation (9) becomes

$$-\delta_{ij}(\gamma-1)[(X_{11}+X_{22})(2\pi G\rho a_1^2 a_3 A_1 - \Omega^2)I_{11} + 2\pi G\rho a_1^2 a_3 A_3 I_{33}]; \tag{12}$$

and, since for the Maclaurin spheroid (Paper III, eq. [75])

$$(2\pi G\rho a_1^2 a_3 A_1 - \Omega^2)I_{11} = 2\pi G\rho a_1^2 a_3 A_3 I_{33}, \tag{13}$$

expression (12) becomes (Paper III, eq. [57] and Paper I, eq. [20])

$$-\delta_{ij}(\gamma-1)(\pi G\rho a_1^2 a_3)2A_3 I_{33}\sum_{l=1}^{3} X_{ll} = \delta_{ij}(\gamma-1)\mathfrak{W}_{33}\sum_{l=1}^{3} X_{ll} = \delta_{ij}J\sum_{l=1}^{3} X_{ll}. \tag{14}$$

This is the same as the fourth term on the right-hand side of equation (4).

To verify that the final terms of equations (4) and (9) agree, we note that, according to equation (60) of Paper III together with equation (8) above,

$$\begin{aligned}-\sum_{r=1}^{3}\sum_{l=1}^{3} X_{lr}\mathfrak{W}_{rl;\,ij} = {} & -\pi G\rho a_1^2 a_3\Big[2(A_j - a_i^2 A_{ij})(I_{ii}X_{ji} + I_{jj}X_{ji}) \\ & - 2a_i^2\delta_{ij}\sum_{l=1}^{3} A_{il}I_{ll}X_{il}\Big].\end{aligned} \tag{15}$$

This shows that the exact equation (9) is equivalent to the "approximate" equation (4).

IV. THE TRANSVERSE SHEAR MODES

The nine equations represented by equation (4) separate into smaller sets of equations. One such set consists of four equations for X_{13}, X_{31}, X_{23}, and X_{32} (Paper I, Sec. V), the other X_{ij}'s being set equal to zero to satisfy the remaining five equations. The corresponding normal modes represent a relative shearing of the northern and southern hemispheres and are therefore called "transverse shear modes." The equations governing them (Paper I, eq. [46]) can be shown, with the aid of the explicit formulae for the supermatrix elements (Paper III, eqs. [62]–[65]), to be the same as those already discussed in the incompressible case (Paper II, eqs. [141]–[144]). The oscillation frequencies have been tabulated (Paper II, Table 1) and have been found to correspond to stable oscillations.

V. THE TOROIDAL MODES

A second set, comprising two equations, yields the "toroidal modes," in which the motions are restricted to planes parallel to the equatorial plane. The equations, which involve only the combinations $(X_{11} - X_{22})$ and $(X_{12} + X_{21})$, yield characteristic roots (Paper I, eq. [76]) that can once again be shown to be identical with those found in the incompressible case (Paper II, eq. [189]). These are the modes that lead to neutral stability at the point of bifurcation where the Jacobi ellipsoids branch off from the Maclaurin spheroids (at an eccentricity $e = 0.8127$) and to instability when $e = 0.9529$. That these modes are unchanged in the homogeneous, compressible model means that the occurrence of neutral and unstable points along the sequence of Maclaurin spheroids is independent of the assumption that the fluid is intrinsically incompressible.

VI. THE PULSATION MODES

The transverse shear modes and the toroidal modes account for six of the nine modes and the associated characteristic values of the root λ^2 for six of the nine characteristic values that satisfy equation (4). Further, one of the equations represented by equation (4) is

$$\lambda^2(X_{21} - X_{12}) = -2\lambda\Omega(X_{11} + X_{22})\,, \tag{16}$$

which can be satisfied by taking $\lambda^2 = 0$, $X_{12} = -X_{21} \neq 0$, and setting the remaining X_{ij}'s equal to zero (Paper I, eqs. [79] and [82]).

Seven modes are now accounted for. The remaining two are called "pulsation modes" because they represent the generalization to the case when rotation is present of the radial pulsation. They satisfy the equations (Paper I, eq. [87])

$$[\lambda^2 I_{11} + (\Omega^2 I_{11} - \alpha - \beta)](X_{11} + X_{22}) - 2\alpha X_{33} = 0 \tag{17}$$

and

$$-\alpha(X_{11} + X_{22}) + (\lambda^2 I_{33} - \beta)X_{33} = 0\,, \tag{18}$$

where

$$\alpha = (\gamma - 1)\mathfrak{W}_{33} - \mathfrak{W}_{33;11} \quad \text{and} \quad \beta = (\gamma - 1)\mathfrak{W}_{33} - \mathfrak{W}_{33;33}\,. \tag{19}$$

Equations (17) and (18) are different from any of the equations in Paper II. Only in these equations does the effect of compressibility appear (in particular, through the coefficients α and β). They lead to (Paper I, eq. [88])

$$I_{11}I_{33}\lambda^4 - [\beta I_{11} + (\beta + \alpha - \Omega^2 I_{11})I_{33}]\lambda^2 - \Omega^2 I_{11}\beta + (\beta + 2\alpha)(\beta - \alpha) = 0\,. \tag{20}$$

Let the roots of equation (20) be λ_1^2 and λ_2^2. There will be associated with these roots certain normal modes, which can be specified by the ratio of $X_{11} + X_{22}$ to X_{33}. Let

$$R = \frac{X_{11} + X_{22}}{X_{33}} = 2\,\frac{X_{11}}{X_{33}} \qquad (\text{since } X_{11} = X_{22}) \tag{21}$$

be this ratio; then two numbers, R_1 and R_2, specify the normal modes that correspond to λ_1^2 and λ_2^2. There is a simple identity relating R_1 and R_2, which we shall now derive.

In view of equations (17) and (18) and the foregoing definitions, we may write

$$[\lambda_1^2 I_{11} + (\Omega^2 I_{11} - a - \beta)]R_1 = 2a \tag{22}$$

and

$$aR_2 = (\lambda_2^2 I_{33} - \beta) . \tag{23}$$

Multiplication of each side of equation (22) by R_2 leads to

$$[\lambda_1^2 I_{11} + (\Omega^2 I_{11} - a - \beta)]R_1 R_2 = 2(\lambda_2^2 I_{33} - \beta) , \tag{24}$$

where equation (23) has been used. On account of equation (20), the roots λ_1^2 and λ_2^2 must satisfy

$$\lambda_1^2 + \lambda_2^2 = \frac{\beta}{I_{33}} + \frac{\beta + a - \Omega^2 I_{11}}{I_{11}}$$

or

$$\lambda_1^2 I_{11} + (\Omega^2 I_{11} - a - \beta) = -\frac{I_{11}}{I_{33}}(\lambda_2^2 I_{33} - \beta) . \tag{25}$$

We infer from equations (24) and (25) that

$$R_1 R_2 = -2\frac{I_{33}}{I_{11}}, \tag{26}$$ [5]

which is the required identity.

a) The Case When Rotation Is Absent

If $\Omega^2 = 0$ and the equilibrium configuration is spherical, the roots λ_1^2 and λ_2^2 are (Paper I, eq. [95])

$$\lambda_1^2 = (3\gamma - 4)\frac{\mathfrak{W}}{I} \quad \text{and} \quad \lambda_2^2 = \frac{4}{5}\frac{\mathfrak{W}}{I} . \tag{27}$$

These roots are distinct if $\gamma \neq \frac{5}{8}$, and the normal modes can then be found. They are

$$R_1 = 2 \quad \text{and} \quad R_2 = -1 . \tag{28}$$

It is easy to see that, for a spherically symmetrical disturbance, $X_{11} = X_{22} = X_{33}$, and therefore $R = 2$. Hence the root λ_1^2 corresponds to a radial pulsation. The root λ_2^2 can be shown to belong to that second-order spherical harmonic which is symmetric about the x_3-axis. Such a mode is volume-conserving; this is reflected in $R_2 = -1$, which, according to the substitution (3), implies that div $\boldsymbol{\xi} = 0$.

If $\gamma = \frac{8}{5}$, then $\lambda_1^2 = \lambda_2^2$, and the normal modes are consequently unspecified.

If $\mathfrak{W}$ and I are evaluated for the homogeneous, compressible model, equation (27) becomes

$$\lambda_1^2 = -(3\gamma - 4)\tfrac{4}{3}\pi G\rho \quad \text{and} \quad \lambda_2^2 = -\tfrac{16}{15}\pi G\rho . \tag{29}$$

b) The Case When Rotation Is Present

With the exception of equation (29), the equations and remarks of this section have been general, applying to centrally condensed, as well as to uniform, mass distributions. We now want to restrict our attention to the homogeneous, compressible model. For this purpose it is convenient to make the followings definitions:

$$l^2 = (4\pi G\rho)^{-1}\lambda^2 , \quad f = -(4\pi G\rho I_{11})^{-1}(a + \beta - \Omega^2 I_{11}) ,$$
$$g = -(4\pi G\rho I_{33})^{-1}\beta , \quad \text{and} \quad k = -(4\pi G\rho I_{33})^{-1}a . \tag{30}$$

[5] In deriving this identity we have benefited by a discussion with Dr. Alar Toomre.

Equations (17) and (18) now become

$$(l^2 + f)(N_{11} + N_{22}) + 2kN_{33} = 0 \tag{31}$$

and

$$(1 - e^2)k(N_{11} + N_{22}) + (l^2 + g)N_{33} = 0\,, \tag{32}$$

where we have used equation (11), suppressing the argument, and have introduced the eccentricity e through the formula

$$\frac{I_{33}}{I_{11}} = \frac{a_3^2}{a_1^2} = 1 - e^2. \tag{33}$$

The characteristic equation now becomes

$$l^4 + (f + g)l^2 + [fg - 2(1 - e^2)k^2] = 0\,, \tag{34}$$

with roots

$$l^2 = -\tfrac{1}{2}(f + g) \pm \tfrac{1}{2}[(f - g)^2 + 8(1 - e^2)k^2]^{1/2}\,. \tag{35}$$

Let l_1^2 be that root which in the limit $\Omega^2 \rightarrow 0$ approaches $-(\gamma - \frac{4}{3})$, and l_2^2 the root that approaches $-\frac{4}{15}$ (cf. eq. [29]). The normal modes will be specified by the ratios r_1 and r_2, where $r = (N_{11} + N_{22})/N_{33}$. The identity (26) becomes

$$r_1 r_2 = -\frac{2}{1 - e^2}, \tag{36}$$

since

$$r = \frac{RI_{11}}{I_{33}} = \frac{R}{1 - e^2}.$$

The quantities f, g, and k introduced in equation (30) can be written in terms of A_i and A_{ij} by means of the formulae of Paper III. The results are

$$\begin{aligned} f &= \tfrac{1}{2}a_1^2 a_3\{A_1 + a_3^2 A_{13} + (1 - e^2)[2(\gamma - 1)A_3 - 2A_3]\}\,, \\ g &= \tfrac{1}{2}a_1^2 a_3[2A_3 - 3a_3^2 A_{33} + (\gamma - 1)A_3]\,, \end{aligned} \tag{37}$$

and

$$k = \tfrac{1}{2}a_1^2 a_3[-a_1^2 A_{13} + (\gamma - 1)A_3]\,.$$

Since the constants A_i and A_{ij} can be written in terms of the eccentricity e (Paper III, eqs. [18], [19], [34], and [35]), these equations give f, g, and k as functions of e.

VII. THE LIMIT OF VANISHING ANGULAR MOMENTUM

If the rotation is slow, so that the equilibrium configuration is only slightly oblate, the coefficients f, g, and k may be approximated by expansions in the eccentricity e. To terms of second order in e the results are

$$\begin{aligned} f &= [-\tfrac{2}{15} + \tfrac{2}{3}(\gamma - 1)] + [\tfrac{26}{105} - \tfrac{4}{5}(\gamma - 1)]e^2\,, \\ g &= [\tfrac{1}{15} + \tfrac{1}{3}(\gamma - 1)] + [\tfrac{2}{21} + \tfrac{2}{15}(\gamma - 1)]e^2\,, \end{aligned} \tag{38}$$

and

$$k = [-\tfrac{1}{5} + \tfrac{1}{3}(\gamma - 1)] + [-\tfrac{4}{35} + \tfrac{2}{15}(\gamma - 1)]e^2\,.$$

The expressions $f + g$ and $(f - g)^2 + 8(1 - e^2)k^2$ appearing in equation (34) become, to the same order,

$$f + g = [-\tfrac{1}{15} + (\gamma - 1)] + [\tfrac{12}{35} - \tfrac{4}{15}(\gamma - 1)]e^2 \tag{39}$$

and

$$\begin{aligned}(f - g)^2 + 8(1 - e^2)k^2 &= 9[-\tfrac{1}{5} + \tfrac{1}{3}(\gamma - 1)]^2 \\ &+ [-\tfrac{1}{5} + \tfrac{1}{3}(\gamma - 1)][\tfrac{8}{105} - \tfrac{8}{5}(\gamma - 1)]e^2 .\end{aligned} \tag{40}$$

Note that the right-hand side of equation (40) vanishes identically if $\gamma = \frac{8}{5}$; this case must therefore be treated separately.

a) The Case When $\gamma \neq \frac{8}{5}$

In this case formulae (39) and (40) are correct to second order in e, and one finds from equation (35) that the roots are

$$l_1^2 = -\tfrac{1}{3}(3\gamma - 4) + \tfrac{4}{45}(3\gamma - 5)e^2 \tag{41}$$

and

$$l_2^2 = -\tfrac{4}{15} - \tfrac{52}{315}e^2 . \tag{42}$$

It is clear from equation (41) that the effect of a small rotation is to stabilize the configuration with $\gamma = \frac{4}{3}$. In other words, the critical value of γ, γ_c, at which dynamic instability sets in, is reduced from the value $\frac{4}{3}$. The amount of the reduction for a given value of e is obtained by setting $l_1^2 = 0$ in equation (41). The result is

$$\gamma_c = \tfrac{4}{3} - \tfrac{4}{45}e^2 . \tag{43}$$

With the aid of equations (41) and (42) we can find the normal modes r_1 and r_2. They are

$$r_1 = 2 + \frac{4}{3(8 - 5\gamma)} e^2 \tag{44}$$

and

$$r_2 = -1 + \frac{15\gamma - 22}{3(8 - 5\gamma)} e^2 . \tag{45}$$

That r_2 is no longer precisely -1 means that this mode is no longer volume-preserving and hence is no longer a deformation involving only a second-order surface harmonic: rotation has "mixed" the modes.

If, in equations (41) and (42), the factor $4\pi G\rho$ is restored, $\Omega^2(= 8\pi G\rho e^2/15)$ is used instead of e^2, and $\sigma^2 = -\lambda^2$, these equations become

$$\sigma_1^2 = (3\gamma - 4)\tfrac{4}{3}\pi G\rho - \tfrac{2}{3}(3\gamma - 5)\Omega^2 \tag{46}$$

and

$$\sigma_2^2 = \tfrac{16}{15}\pi G\rho + \tfrac{26}{21}\Omega^2 . \tag{47}$$

Equation (46) agrees with a formula found by Ledoux (1945, eq. [54]). (For a somewhat more general comparison, see Chandrasekhar and Lebovitz 1962*c*, Sec. VIII, *c*).

b) The Case When $\gamma = \frac{8}{5}$

If $\gamma = \frac{8}{5}$, equation (40) is not adequate for finding the frequencies correctly to second order in e, and it is necessary to return to equations (38), which, for the present case, become

$$f = \tfrac{4}{15} + \tfrac{4}{525}e^2, \qquad g = \tfrac{4}{15} + \tfrac{92}{525}e^2, \qquad \text{and} \qquad k = -\tfrac{6}{175}e^2 . \tag{48}$$

The roots are easily found to be

$$l_1^2 = -\frac{4}{15} - \frac{48 - \sqrt{2584}}{525}\, e^2 \tag{49}$$

and

$$l_2^2 = -\frac{4}{15} - \frac{48 + \sqrt{2584}}{525}\, e^2 . \tag{50}$$

If these results are used together with equations (31) and (32), the normal modes, which were unspecified in the absence of rotation, may be found, now that rotation has "lifted the degeneracy" in the characteristic roots, but not to the same order in e: the normal modes are found in the limit as $e \to 0$. They are

$$r_1 = -\frac{36}{44 - \sqrt{2584}} = 5.2685 \quad \text{(approximately)} \tag{51}$$

and

$$r_2 = -\frac{36}{44 + \sqrt{2584}} = -0.3796 \quad \text{(approximately)} . \tag{52}$$

Since $r = 2$ for a spherically symmetric perturbation, it is clear that neither mode is even approximately spherically symmetric. This last fact is not peculiar to the homogeneous compressible case: it is found for distorted polytropes in general (see Table 4*B* in Chandrasekhar and Lebovitz 1962*a*).

VIII. THE LIMIT OF LARGE ANGULAR MOMENTUM

When the angular momentum becomes large, the Maclaurin spheroid becomes highly flattened, and $e \to 1$. It is convenient in this case to expand the coefficients f, g, and k in powers, not of e, but of $\eta = (\pi/2 - \sin^{-1} e)$, which becomes small in the limit under consideration. Expansions of f, g, and k must be taken to the second order in η. The results are

$$f = \tfrac{1}{4}\pi\eta - 2(2 - \gamma)\eta^2 ,$$
$$g = \gamma - \tfrac{1}{2}\pi(\gamma + 1)\eta + 2(\gamma + 2)\eta^2 , \tag{53}$$

and

$$k = (\gamma - 2) + \tfrac{1}{4}\pi(5 - 2\gamma)\eta - 2(3 - \gamma)\eta^2 .$$

The characteristic roots and normal modes are easily found to be

$$l_1^2 = -\gamma \quad \text{and} \quad l_2^2 = -\tfrac{1}{4}\pi\eta \tag{54}$$

and

$$r_1 = -\frac{2(2-\gamma)}{\gamma} \quad \text{and} \quad r_2 = \frac{\gamma}{2-\gamma}\frac{1}{\eta^2} . \tag{55}$$

These equations are helpful in interpreting the tables and graphs of Section IX.

IX. NUMERICAL RESULTS

When the angular momentum M is neither very small nor very large, the only satisfactory way of finding the behavior of the pulsation frequencies and normal modes is numerically.

Equation (43) shows that, at least for small values of e, the critical value of γ, γ_c, below which dynamical instability occurs, is reduced by rotation. Table 1 shows that this is a trend that is maintained for all values of e: γ_c is a monotonically decreasing function of e. In addition, Ω^2 (in the unit $4\pi G\rho$) and M (in the unit $[G\mathfrak{M}a_1^2a_3]^{1/2}$, where $\mathfrak{M}$ is the mass of the spheroid) are given in Table 1. In Figure 1, γ_c is plotted against M.

The frequencies of those modes that start, when $e = 0$, as radial pulsations are given in Table 2*A*, and their corresponding normal modes in terms of the ratio r_1 in Table 2*B*, for several values of γ. The frequencies are plotted against M in Figure 2. The frequency σ $(= il)$ is in the unit $(4\pi G\rho)^{1/2}$.

TABLE 1

THE CRITICAL VALUE OF γ, Ω^2, AND M, AS FUNCTIONS OF e

e	Ω^2	M	γ_c	e	Ω^2	M	γ_c
0	0	0	1.3333	0.8	0.1816	0.2934	1.2318
0.1	0.0027	0.0255	1.3324	0.9	.2203	0.4000	1.1535
0.2	.0107	0.0514	1.3297	0.95	.2213	0.5008	1.0578
0.3	.0243	0.0787	1.3249	0.99	.1552	0.7120	0.7828
0.4	.0436	0.1085	1.3176	0.995	.1219	0.7968	0.6551
0.5	.0690	0.1417	1.3071	0.999	.0627	0.9928	0.3879
0.6	.1007	0.1804	1.2920	0.9999	0.0214	1.2198	0.1495
0.7	0.1387	0.2283	1.2693	1.0000	0	∞	0

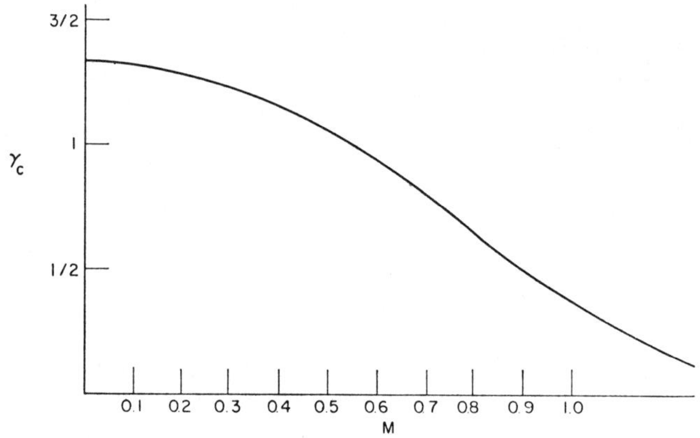

FIG. 1.—The critical value of γ, γ_c, as a function of the angular momentum M (the unit of M is $[G\mathfrak{M}a_1^2a_3]^{1/2}$, where $\mathfrak{M}$ is the mass of the spheroid).

In the same way, the frequencies of those modes that start, when $e = 0$, as second-order harmonics, are given in Table 3*A*, and their corresponding values of r_2 in Table 3*B*. The frequencies are also plotted against M in Figure 3.

Tables 2*B* and 3*B* for the normal modes are, in a sense, incomplete, for $N_{11} + N_{22}$ and N_{33} are not sufficient to specify the normal modes: $N_{12} - N_{21}$ is also needed (cf. Paper I, eq. [79]). It can, however, be determined from the information given in the tables through the equation

$$\lambda^2(N_{12} - N_{21}) = 2\lambda\Omega(N_{11} + N_{22}) . \tag{56}$$

TABLE 2*A*

THE FREQUENCIES OF THE MODE THAT STARTS AS A PURELY RADIAL PULSATION*

e	γ					
	1.3	$\frac{4}{3}$	1.4	1.5	1.6	$\frac{5}{3}$
0	i0.1825	0	0.2583	0.4083	0.5164	0.5773
0.1	i0.1800	0.0300	.2596	.4088	.5163	0.5773
0.2	i0.1715	.0600	.2636	.4104	.5161	0.5775
0.3	i0.1559	.0906	.2706	.4130	.5157	0.5783
0.4	i0.1296	.1225	.2804	.4167	.5147	0.5816
0.5	i0.0812	.1559	.2929	.4210	.5128	0.5923
0.6	0.0843	.1931	.3084	.4256	.5092	0.6155
0.7	0.1594	.2298	.3268	.4292	.5023	0.6540
0.8	0.2238	.2718	.3464	.4290	.4881	0.7143
0.9	0.2855	.3137	.3604	.4141	.4538	0.8198
0.95	0.3068	.3233	.3517	.3854	.4111	0.9184
0.99	0.2722	.2798	.2855	.2963	.3051	1.0941
0.995	0.2433	.2462	.2512	.2577	.2632	1.1457
0.999	0.1761	.1769	.1786	.1806	.1822	1.2219
0.9999	0.1034	0.1034	0.1039	0.1039	0.1044	1.2685
1.0000	0	0	0	0	0	1.2910

* The factors of $i(=\sqrt{-1})$ appearing in the first column indicate that the corresponding mode is unstable; the numerical values of these entries therefore represent the growth rate.

TABLE 2*B*

THE VALUES OF r $(=[N_{11}+N_{22}]/N_{33})$ FOR THE MODE THAT STARTS AS A PURELY RADIAL PULSATION

e	γ					
	1.3	$\frac{4}{3}$	1.4	1.5	1.6	$\frac{5}{3}$
0	2.0000	2.0000	2.0000	2.0000	....	+2.0000
0.1	2.0090	2.0101	2.0135	2.0270	5.30	+1.9589
0.2	2.0378	2.0425	2.0454	2.1117	5.48	+1.8201
0.3	2.0918	2.1031	2.1366	2.2664	5.75	+1.5538
0.4	2.1828	2.2050	2.2704	2.5172	6.21	+1.1111
0.5	2.3334*	2.3732	2.4899	2.9148	6.91	+0.5852
0.6	2.5911*	2.6820	2.8627	3.5634	8.02	+0.2051
0.7	3.0714	3.1963	3.5529	4.7142	9.92	−0.0090
0.8	4.1371	4.3845	5.0728	7.1142	1.372×10^{1}	−0.1384
0.9	7.7273	8.3751	1.0088×10^{1}	1.4515×10^{1}	2.500×10^{1}	−0.2356
0.95	1.5615×10^{1}	1.7074×10^{1}	2.0766×10^{1}	2.9465×10^{1}	4.669×10^{1}	−0.2866
0.99	8.4556×10^{1}	1.0880×10^{2}	1.1047×10^{2}	1.4943×10^{2}	2.1434×10^{2}	−0.3486
0.995	1.7358×10^{2}	1.8864×10^{2}	2.2454×10^{2}	2.9942×10^{2}	4.1995×10^{2}	−0.3634
0.999	9.0344×10^{2}	9.7438×10^{2}	1.1470×10^{3}	1.4994×10^{3}	2.0419×10^{3}	−0.3834
0.9999	9.1956×10^{3}	9.9442×10^{3}	1.1604×10^{4}	1.4999×10^{4}	2.0029×10^{4}	−0.3947
1.0000	∞	∞	∞	∞	∞	−0.4000

* Between $e = 0.5$ and $e = 0.6$ the "frequency" vanishes (see Table 2*A*). One can show that at this point, although the limiting ratio r is finite, $N_{11}+N_{22}=N_{33}=0$, and only $N_{12}-N_{21}\neq 0$ (cf. eq. [56j]). The interpretation of this is that the mode of neutral stability is one of pure rotation.

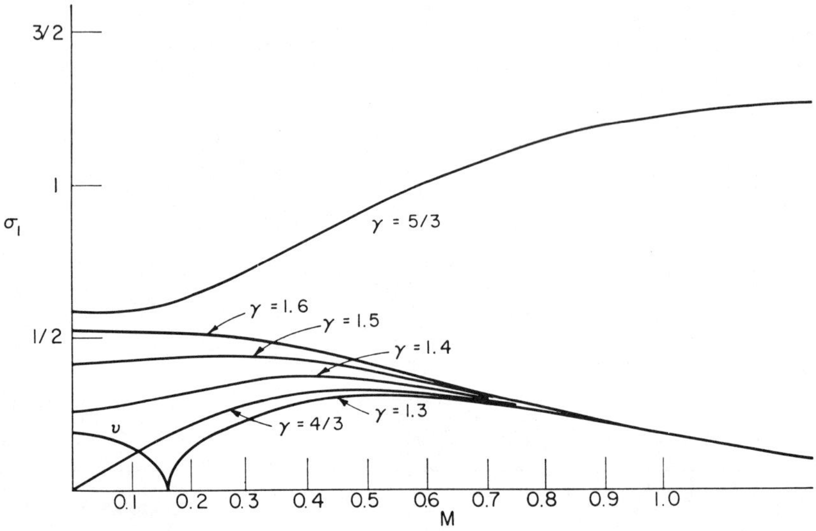

FIG. 2.—The frequencies of the modes that are radial pulsations when $M = 0$. When $\gamma = 1.3$, the radial pulsation starts out being unstable with growth rate ν. Stabilization occurs at $M = 0.165$; for $M > 0.165$, the curve labeled "$\gamma = 1.3$" gives the pulsation frequency.

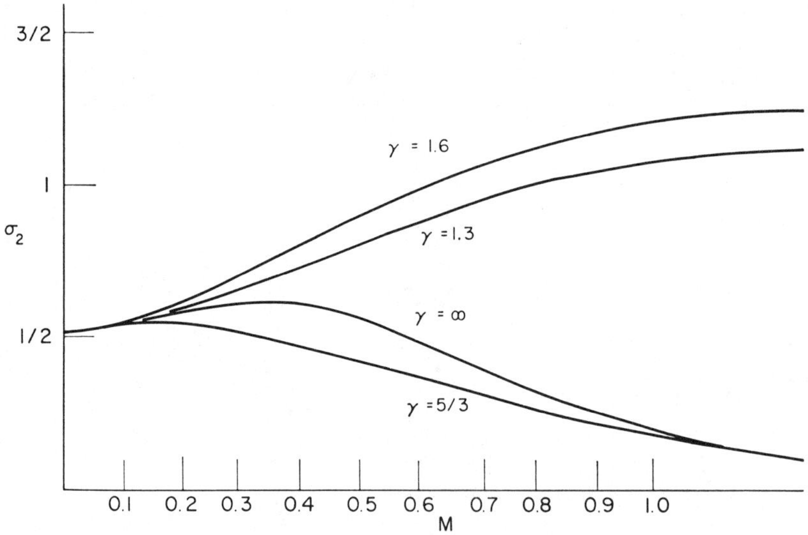

FIG. 3.—The frequencies of the modes that correspond to second-order harmonics when $M = 0$. The curves for $\gamma = \frac{4}{3}$, 1.4, and 1.5 lie between those for $\gamma = 1.3$ and $\gamma = 1.6$ and are similar in appearance; they are therefore not shown. The curve labeled "$\gamma = \infty$" is that for the incompressible case (cf. Paper II, Fig. 1, the curve labeled "σ_3").

TABLE 3A

THE FREQUENCIES OF THE MODE THAT STARTS AS A SECOND-ORDER HARMONIC DEFORMATION

e	γ					
	1.3	$\frac{4}{3}$	1.4	1.5	1.6	$\frac{5}{3}$
0	0.5164	0.5164	0.5164	0.5164	0.5164	0.5164
0.1	0.5180	0.5180	0.5180	0.5180	0.5183	.5180
0.2	0.5229	0.5229	0.5229	0.5229	0.5238	.5227
0.3	0.5312	0.5312	0.5312	0.5314	0.5336	.5303
0.4	0.5432	0.5433	0.5435	0.5441	0.5480	.5392
0.5	0.5597	0.5599	0.5604	0.5619	0.5683	.5453
0.6	0.5816	0.5821	0.5833	0.5867	0.5966	.5443
0.7	0.6111	0.6123	0.6151	0.6221	0.6365	.5358
0.8	0.6537	0.6563	0.6626	0.6760	0.6962	.5170
0.9	0.7279	0.7341	0.7479	0.7719	0.7996	.4744
0.95	0.8029	0.8125	0.8324	0.8637	0.8963	.4252
0.99	0.9537	0.9670	0.9933	1.0319	1.0696	.3102
0.995	1.0012	1.0151	1.0423	1.0820	1.1206	.2665
0.999	1.0734	1.0877	1.1158	1.1567	1.1963	.1833
0.9999	1.1182	1.1327	1.1611	1.2025	1.2424	0.1044
1.0000	1.1402	1.1547	1.1832	1.2247	1.2649	0

TABLE 3B

THE VALUES OF r ($= [N_{11} + N_{22}]/N_{33}$) FOR THE MODE THAT STARTS AS A SECOND-ORDER HARMONIC DEFORMATION

e	γ					
	1.3	$\frac{4}{3}$	1.4	1.5	1.6	$\frac{5}{3}$
0	−1.0000	−1.0000	−1.0000	−1.0000	....	−1.0000
0.1	−1.0056	−1.0050	−1.0033	−0.9966	−0.380	−1.0313
0.2	−1.0223	−1.0200	−1.0130	−0.9866	− .380	−1.1428
0.3	−1.0507	−1.0450	−1.0293	−0.9698	− .382	−1.4145
0.4	−1.0908	−1.0798	−1.0487	−0.9459	− .383	−2.1429
0.5	−1.1428	−1.1237	−1.0710	−0.9149	− .386	−4.5559
0.6	−1.2060	−1.1745	−1.0916	−0.8769	− .390	-1.5234×10^{1}*
0.7	−1.2768	−1.2269	−1.1038	−0.8319	− .395	$+4.3536\times10^{2}$*
0.8	−1.3428	−1.2671	−1.0951	−0.7809	− .405	$+4.0137\times10^{1}$
0.9	−1.3622	−1.2568	−1.0435	−0.7252	− .423	$+4.4685\times10^{1}$
0.95	−1.3137	−1.2014	−0.9878	−0.6962	− .439	$+7.1537\times10^{1}$
0.99	−1.1886	−1.0905	−0.9098	−0.6726	− .469	$+2.8829\times10^{2}$
0.995	−1.1551	−1.0629	−0.8930	−0.6696	− .477	$+5.5181\times10^{2}$
0.999	−1.1245	−1.0266	−0.8723	−0.6673	− .489	$+2.6098\times10^{3}$
0.9999	−1.0875	−1.0084	−0.8618	−0.6667	− .497	$+2.5338\times10^{4}$
1.0000	−1.0769	−1.0000	−0.8571	−0.6667	−0.500	$+\infty$

* Between $e = 0.6$ and $e = 0.7$, the quantity $1/r$ changes continuously from negative to positive values.

X. CONCLUDING REMARKS

Perhaps the only unexpected result of this work is that, when $\gamma = 1.6$, none of the normal modes becomes spherically symmetric as $M \rightarrow 0$. Even this might have been foreseen because the occurrence of the degeneracy was known (Paper I, Sec. IV), so there was no reason to believe that the normal modes when $\gamma = 1.6$ bear any resemblance to those when $\gamma \neq 1.6$.

This result is not, of course, restricted to the homogeneous model (cf. Chandrasekhar and Lebovitz 1962*a*, Sec. VIII). It is expressed in a more general way in another paper in this series (Chandrasekhar and Lebovitz 1962*d*), where it is applied to the question of double periods in the light- and velocity-curves of the β Canis Majoris stars.

The research reported in this paper has in part been supported by the Office of Naval Research under contract Nonr-2121(24) with the University of Chicago. The work of the second author was supported in part by the United States Air Force under contract No. AF-49(638)-42, monitored by the Air Force Office of Scientific Research of the Air Research and Development Command.

REFERENCES

Chandrasekhar, S. 1933, *M.N.*, **93**, 390.
Chandrasekhar, S., and Lebovitz, N. R. 1962*a*, *Ap. J.*, **135**, 248 (this paper will be referred to as "Paper I").
———. 1962*b*, *ibid.*, **136**, 1037 (this paper will be referred to as "Paper III").
———. 1962*c*, *ibid.*, p. 1082.
———. 1962*d*, *ibid.*, p. 1105.
Lebovitz, N. R. 1961, *Ap. J.*, **134**, 500 (this paper will be referred to as "Paper II").
Ledoux, P. 1945, *Ap. J.*, **102**, 143.
Ledoux, P., and Pekeris, C. 1941, *Ap. J.*, **94**, 124.

21

ON THE OSCILLATIONS AND THE STABILITY OF ROTATING GASEOUS MASSES. III. THE DISTORTED POLYTROPES

S. CHANDRASEKHAR AND NORMAN R. LEBOVITZ
University of Chicago and Massachusetts Institute of Technology
Received July 2, 1962

ABSTRACT

The theory of the oscillations of rotating gaseous masses, developed in an earlier paper of this series, is here applied to determine the effect of a small rotation (Ω) on the fundamental modes of oscillation of a polytrope. The basis for this application is provided by the theory of rotationally distorted polytropes; this theory is reviewed and amplified further to include a discussion of the superpotential. The various tensors, in terms of which the characteristic frequencies of oscillation are expressed, are evaluated for distorted polytropes, appropriately, to the first order in Ω^2. The final results on the effect of rotation on the characteristic frequencies are presented in the form of tables.

I. INTRODUCTION

We have recently developed a general theory of the oscillations of rotating gaseous masses (Chandrasekhar and Lebovitz 1962*b;* this paper will be referred to hereafter as "Paper III"). In this paper, we shall apply that theory to determine the effect of a small rotation on the fundamental modes of oscillation of a polytrope. Since the general theory presupposes a knowledge of the structure of the equilibrium configuration, it is clear that in this instance we must first determine the effect of rotation on the equilibrium of the polytrope itself; but this was determined in an earlier paper (Chandrasekhar 1933; this paper will be referred to hereafter as "Paper I"); the application of the general theory to the problem we have in mind is, therefore, possible.

The plan of this paper is as follows. In Section II, the theory of rotationally distorted polytropes is reviewed, and the basic equations are re-derived in a manner which is free of the objections that have sometimes been raised (cf. Jardetzky 1958, p. 165) against the analysis as given in Paper I. In Section III, the equation governing the superpotential is solved in the required approximation. In Sections IV, V, and VI, expressions are obtained for the various tensors defined in the general theory. In Section VII, numerical values for the relevant tensor components are listed for some values of n. In Section VIII, the formulae and the results of the earlier sections are combined to give explicit expressions for the different fundamental modes of oscillation of the distorted polytropes. Supplementary matters are considered in the three appendices. In Appendix I, the integrals in terms of which the various tensors are expressed in Sections IV, V, and VI are reduced to a minimum number; they are, then, numerically evaluated, and a table of values of the required integrals is provided. Appendix II is devoted to a brief consideration of the special case $n = 1$, for which all the relevant equations and integrals can be solved explicitly. And, finally, in Appendix III, the basic functions which describe the structure of the distorted polytropes are tabulated for $n = 1$, 1.5, 2, 3, and 3.5; a table of related constants is also included.

II. THE EQUILIBRIUM OF DISTORTED POLYTROPES

The fundamental assumption is, of course, that the pressure (p) and the density (ρ) are related in the manner

$$p = K \rho^{1+1/n} , \tag{1}$$

where K is a constant. For a configuration with this pressure-density relation, rotating uniformly with an angular velocity Ω, the equations of hydrostatic equilibrium integrate to give (cf. Paper I, eq. [32]; also Chandrasekhar 1961, eq. [13])

$$(n+1)p = \rho[\mathfrak{B} + \tfrac{1}{2}\Omega^2 r^2(1-\mu^2) - \mathfrak{B}_0], \tag{2}$$

where $\mathfrak{B}$ denotes the gravitational potential and $\mathfrak{B}_0$ its value at the pole of the configuration; also $\mu = \cos\vartheta$, where ϑ is the inclination of the radius vector to the axis of rotation.

With the substitutions

$$\rho = \lambda\Theta^n, \qquad p = K\lambda^{1+1/n}\Theta^{n+1}, \tag{3}$$

and

$$r = \alpha\,\xi = \left[\frac{(n+1)K}{4\pi G}\,\lambda^{-1+1/n}\right]^{1/2}\xi, \tag{4}$$

Poisson's equation governing $\mathfrak{B}$ (in view of eq. [2]) becomes (cf. Paper I, eq. [11])

$$\nabla^2\Theta = -\Theta^n + v, \tag{5}$$

where ∇^2 is the Laplacian operator appropriate for functions depending only on ξ and μ, and

$$v = \frac{\Omega^2}{2\pi G\lambda}. \tag{6}$$

We shall let λ denote the central density, so that

$$\Theta = 1 \quad \text{at} \quad \xi = 0. \tag{7}$$

In the subsequent analysis, the following quantities will be measured in the units which are specified (unless otherwise stated):

	Unit
Length	α (see eq. [4]),
Density	λ,
Pressure	$K\lambda^{1+1/n}$,
Mass	$4\pi\alpha^3\lambda$,
Moment of inertia	$4\pi\alpha^5\lambda$,
Gravitational potential	$(n+1)K\lambda^{1/n}$,
Superpotential	$(n+1)K\lambda^{1/n}\alpha^2$,
Potential energy	$4\pi(n+1)K\lambda^{1+1/n}\alpha^3$,
Frequency	$(4\pi G\lambda)^{1/2}$.

(8)

In the units specified, equation (2) takes the form

$$\mathfrak{B} = \Theta - \tfrac{1}{6}v\,\xi^2[1-P_2(\mu)] + \mathfrak{B}_0; \tag{9}$$

and the equation governing the superpotential becomes

$$\nabla^2\chi = -2\{\Theta - \tfrac{1}{6}v\,\xi^2[1-P_2(\mu)] + \mathfrak{B}_0\}. \tag{10}$$

Also, it may be noted that, in the chosen units, $\Omega^2 = \frac{1}{2}v$.

a) The Solutions for Θ and $\mathfrak{B}$ to the First Order in v

We shall suppose that the rotation is small, so that v may be treated as a small perturbation parameter. On this assumption, the solution of equation (5) will be carried out consistently to the first order in v.

We assume, then, for Θ, a solution of the form

$$\Theta = \theta(\xi) + v\left[\psi_0(\xi) + \sum_{l=1}^{\infty} A_l \psi_l(\xi) P_l(\mu)\right], \tag{11}$$

where the A_l's are constants unspecified for the present. Inserting this form for Θ in equation (5) and neglecting all terms of orders higher than the first in v, we obtain, for the radial functions θ, ψ_0, and ψ_l, the equations

$$\mathfrak{D}_0 \theta = -\theta^n , \tag{12}$$

$$\mathfrak{D}_0 \psi_0 = -n\theta^{n-1}\psi_0 + 1 , \quad \text{and} \quad \mathfrak{D}_l \psi_l = -n\theta^{n-1}\psi_l , \tag{13}$$

where

$$\mathfrak{D}_l = \frac{d^2}{d\xi^2} + \frac{2}{\xi}\frac{d}{d\xi} - \frac{l(l+1)}{\xi^2} . \tag{14}$$

Equation (12) is, of course, the Lane-Emden equation of index n.

The requirement that none of the radial functions ψ_l has a singularity at the origin evidently restricts them to vanish here; accordingly, condition (7) at $\xi = 0$ now restricts our choice of the solution of equation (12) to the Lane-Emden function (cf. Chandrasekhar 1939, chap. iv).

With Θ given by equation (11), we may write the solution for $\mathfrak{V}$ in the form

$$\mathfrak{V}^{(\text{int})} = \theta + c_0 + v\left\{\psi_0(\xi) + \sum_{l=1}^{\infty} A_l \psi_l(\xi) P_l(\mu) - \tfrac{1}{6}\xi^2[1 - P_2(\mu)] + c_{1;0}\right\}, \tag{15}$$

where the zero- and the first-order contributions to $\mathfrak{V}_0$ have been separated:

$$\mathfrak{V}_0 = c_0 + v\, c_{1;0} . \tag{16}$$

In equation (15) the superscript to $\mathfrak{V}$ distinguishes that this is the potential appropriate to the "interior" of the body, in contrast to the potential appropriate to the "exterior" of the body which satisfies Laplace's equation

$$\nabla^2 \mathfrak{V}^{(\text{ext})} = 0 . \tag{17}$$

The solution of equation (17) which should be associated with solution (15) is, clearly,

$$\mathfrak{V}^{(\text{ext})} = \frac{\kappa_0}{\xi} + v \sum_{l=0}^{\infty} \frac{\kappa_{1;l}}{\xi^{l+1}} P_l(\mu) . \tag{18}$$

It remains to determine the various constants which occur in the solutions for $\mathfrak{V}^{(\text{int})}$ and $\mathfrak{V}^{(\text{ext})}$ by imposing on them the boundary conditions which require the continuity of the gravitational potential and its gradient on the boundary of the configuration. For the application of the appropriate boundary conditions of solutions (15) and (18), it is clearly necessary that we first specify the *boundary;* let this be the surface

$$\Xi(\mu) = \xi_1 + v \sum_{l=0}^{\infty} q_l P_l(\mu) , \tag{19}$$

where ξ_1 is the first zero of the Lane-Emden function. The constants q_l introduced in the definition of $\Xi(\mu)$ become determinate by the additional requirement that Θ vanish on this surface; for Θ given by equation (11), this requirement is

$$\Theta(\Xi)=\theta(\xi_1)+v\left[\theta'(\xi_1)\sum_{l=0}^{\infty}q_lP_l(\mu)+\psi_0(\xi_1)+\sum_{l=1}^{\infty}A_l\psi_l(\xi_1)P_l(\mu)\right]=0\,. \tag{20}$$

Hence

$$q_0\theta'(\xi_1)=-\psi_0(\xi_1) \quad \text{and} \quad q_l\theta'(\xi_1)=-A_l\psi_l(\xi_1) \quad (l=1,2,\ldots)\,. \tag{21}$$

Now, evaluating $\mathfrak{V}^{(\text{int})}$ and $\mathfrak{V}^{(\text{ext})}$ and their derivatives on $\Xi(\mu)$, we find

$$\mathfrak{V}^{(\text{int})}(\Xi)=c_0+v\left\{c_{1;0}-\tfrac{1}{6}\xi^2[1-P_2(\mu)]\right\}, \tag{22}$$

$$\mathfrak{V}^{(\text{ext})}(\Xi)=\frac{\kappa_0}{\xi_1}+v\sum_{l=0}^{\infty}\left(\frac{\kappa_{1;l}}{\xi_1^{l+1}}-\frac{\kappa_0 q_l}{\xi_1^2}\right)P_l(\mu), \tag{23}$$

$$\left[\frac{\partial\mathfrak{V}^{(\text{int})}}{\partial\xi}\right]_{\Xi(\mu)}=\theta'(\xi_1)+v\left\{-\frac{2}{\xi_1}\theta'(\xi_1)\sum_{l=0}^{\infty}q_lP_l(\mu)+\psi_0'(\xi_1)+\sum_{l=1}^{\infty}A_l\psi_l'(\xi_1)P_l(\mu)-\tfrac{1}{3}\xi_1[1-P_2(\mu)]\right\}, \tag{24}$$

and

$$\left[\frac{\partial\mathfrak{V}^{(\text{ext})}}{\partial\xi}\right]_{\Xi(\mu)}=-\frac{\kappa_0}{\xi_1^2}+v\left[\frac{2\kappa_0}{\xi_1^3}\sum_{l=0}^{\infty}q_lP_l(\mu)-\sum_{l=0}^{\infty}\frac{(l+1)\kappa_{1;l}}{\xi_1^{l+2}}P_l(\mu)\right]. \tag{25}$$

Equating the respective expressions and making use of the relations (21), we find

$$c_0=-\xi_1\theta'(\xi_1), \qquad c_{1;0}=\tfrac{1}{2}\xi_1^2-\psi_0(\xi_1)-\xi_1\psi_0'(\xi_1); \tag{26}$$

$$\kappa_0=-\xi_1^2\theta'(\xi_1), \qquad \kappa_{1;0}=\tfrac{1}{3}\xi_1^3-\xi_1^2\psi_0'(\xi_1); \tag{27}$$

$$A_2=-\frac{5}{6}\frac{\xi_1^2}{\xi_1\psi_2'(\xi_1)+3\psi_2(\xi_1)}, \qquad A_l=0 \qquad (l\neq 2); \tag{28}$$

$$\kappa_{1;2}=\tfrac{1}{6}\xi_1^5\frac{\xi_1\psi_2'(\xi_1)-2\psi_2(\xi_1)}{\xi_1\psi_2'(\xi_1)+3\psi_2(\xi_1)}, \quad \text{and} \quad \kappa_{1;l}=0 \quad (l\neq 0,2)\,. \tag{29}$$

This completes the formal solution of the problem; and we may now write

$$\Theta=\theta(\xi)+v[\psi_0(\xi)+A_2\psi_2(\xi)P_2(\mu)], \tag{30}$$

$$\Xi(\mu)=\xi_1-\frac{v}{\theta'(\xi_1)}[\psi_0(\xi_1)+A_2\psi_2(\xi_1)P_2(\mu)], \tag{31}$$

$$\mathfrak{V}^{(\text{int})}=\theta+c_0+v\{\psi_0(\xi)+A_2\psi_2(\xi)P_2(\mu)-\tfrac{1}{6}\xi^2[1-P_2(\mu)]+c_{1\cdot 0}\}, \tag{32}$$

and

$$\mathfrak{B}^{(\text{ext})} = \frac{\kappa_0}{\xi} + v\left[\frac{\kappa_{1;0}}{\xi} + \frac{\kappa_{1;2}}{\xi^3} P_2(\mu)\right], \tag{33}$$

where the constants appearing in these equations have the values given in equations (26)–(29).

It will be observed that the solutions (30) and (31) for Θ and $\Xi(\mu)$ are in agreement with those given in Paper I (eqs. [36] and [38]).[1]

The solutions for ψ_0 and ψ_2 for $n = 1, 1.5, 2, 3$, and 3.5 are tabulated in Appendix III, together with other relevant information.

III. THE SOLUTION FOR THE SUPERPOTENTIAL, χ

Distinguishing the internal and the external parts of the superpotential by χ and ϕ, respectively, we have

$$\nabla^2\chi = -2(\theta + c_0) - 2v\{\psi_0(\xi) - \tfrac{1}{6}\xi^2 + c_{1;0} + [A_2\psi_2(\xi) + \tfrac{1}{6}\xi^2]P_2(\mu)\} \tag{34}$$

and

$$\nabla^2\phi = -\frac{2\kappa_0}{\xi} - 2v\left[\frac{\kappa_{1;0}}{\xi} + \frac{\kappa_{1;2}}{\xi^3} P_2(\mu)\right]. \tag{35}$$

From the continuity of $\mathfrak{B}$ on the boundary, it follows from equations (34) and (35) that

$$\nabla^2(\chi - \phi) = 0 \qquad \text{on } \Xi(\mu). \tag{36}$$

The required solutions for χ and ϕ are evidently of the forms

$$\chi = \chi_0(\xi) + v[\chi_{1;0}(\xi) + \chi_{1;2}(\xi)P_2(\mu)] \tag{37}$$

and

$$\phi = \phi_0(\xi) + v[\phi_{1;0}(\xi) + \phi_{1;2}(\xi)P_2(\mu)]; \tag{38}$$

and the radial functions in these solutions must be governed by the equations

$$\mathfrak{D}_0\chi_0 = -2(\theta + c_0), \tag{39}$$

$$\mathfrak{D}_0\chi_{1;0} = -2[\psi_0(\xi) - \tfrac{1}{6}\xi^2 + c_{1;0}], \tag{40}$$

$$\mathfrak{D}_2\chi_{1;2} = -2[A_2\psi_2(\xi) + \tfrac{1}{6}\xi^2], \tag{41}$$

$$\mathfrak{D}_0\phi_0 = -\frac{2\kappa_0}{\xi}, \qquad \mathfrak{D}_0\phi_{1;0} = -\frac{2\kappa_{1;0}}{\xi}, \qquad \text{and} \qquad \mathfrak{D}_2\phi_{1;2} = -\frac{2\kappa_{1;2}}{\xi^3}. \tag{42}$$

When the solutions have the forms (37) and (38), it follows from equation (36) (or, more directly from eqs. [39]–[42]) that

$$\mathfrak{D}_0(\chi_0 - \phi_0) = 0, \quad \mathfrak{D}_0(\chi_{1;0} - \phi_{1;0}) = 0, \quad \text{and} \quad \mathfrak{D}_2(\chi_{1;2} - \phi_{1;2}) = 0 \quad \text{for } \xi = \xi_1. \tag{43}$$

We must now seek solutions of equations (39)–(42) which are consistent with the requirement that the superpotential and its gradient are continuous on the boundary $\Xi(\mu)$. When this requirement is imposed on solutions (37) and (38), the conditions which follow from the terms of zero order in v are

$$\chi_0(\xi_1) = \phi_0(\xi_1) \qquad \text{and} \qquad \chi_0'(\xi_1) = \phi_0'(\xi_1); \tag{44}$$

[1] In Paper I, insufficient attention was given to a proper formulation of the boundary conditions; and this has been the source of the objections raised against the solutions given in that paper; however, these objections do not apply to the present derivation.

from these equalities and the first of the equations (43) we can clearly conclude that

$$\chi_0''(\xi_1) = \phi_0''(\xi_1). \tag{45}$$

When use is made of this last relation in the conditions which follow from the terms of the first order in v, when χ and ϕ and their derivatives are equated on $\Xi(\mu)$, we find that

$$\chi_{1;0}(\xi_1) = \phi_{1;0}(\xi_1), \qquad \chi_{1;0}'(\xi_1) = \phi_{1;0}'(\xi_1), \tag{46}$$

$$\chi_{1;2}(\xi_1) = \phi_{1;2}(\xi_1), \quad \text{and} \quad \chi_{1;2}'(\xi_1) = \phi_{1;2}'(\xi_1). \tag{47}$$

In other words, *the radial functions and their derivatives must be continuous on* $\xi = \xi_1$.

Now the solutions of equations (39)–(41) can be written in the forms

$$\chi_0 + \gamma_0 , \qquad \chi_{1;0} + \gamma_{1;0} , \qquad \text{and} \qquad \chi_{1;2} + \gamma_{1;2}\xi^2 , \tag{48}$$

where the χ's are particular integrals of the respective equations and the γ's are constants. Similarly, the solutions of equations (42) can be written in the forms

$$\phi_0 = -\kappa_0\xi + \frac{\beta_0}{\xi}, \qquad \phi_{1;0} = -\kappa_{1;0}\xi + \frac{\beta_{1;0}}{\xi}, \qquad \text{and} \qquad \phi_{1;2} = \frac{\kappa_{1;2}}{3\xi} + \frac{\beta_{1;2}}{\xi^3}, \tag{49}$$

where the β's are constants.

The constants which appear in the solutions (48) and (49) are to be determined by applying the boundary conditions (44), (46), and (47). We shall not stop to derive the explicit forms of these conditions, since we shall have no occasion to use them in this paper: the functions $\chi_{1;0}$ and $\chi_{1;2}$, rather surprisingly, do not appear in the final expressions for the different characteristic frequencies; and χ_0 appears only as χ_0'.

IV. THE MOMENT OF INERTIA TENSOR, I_{ij}

The formulae for the characteristic frequencies of the different modes of oscillation given in Paper III depend on the components of the moment of inertia and the potential-energy tensors and on the elements of the supermatrix (defined in Chandrasekhar and Lebovitz 1962*b;* this paper will be referred to hereafter as "Paper II"). We shall now obtain expressions for these various quantities.

We consider, first, the moment of inertia tensor, I_{ij}. This tensor is, of course, diagonal in the chosen representation and for the case under consideration; and in view of the axisymmetry, the two distinct components of this tensor are (in their natural units)

$$I_{11} = \tfrac{1}{2}\int_V \rho r^4(1-\mu^2)\,dr\,d\mu\,d\varphi \tag{50}$$

and

$$I_{33} = \int_V \rho r^4\mu^2 dr\,d\mu\,d\varphi . \tag{51}$$

In the units specified in (8), the foregoing expressions, under the circumstances of the present problem, become

$$I_{11} = \tfrac{1}{2}\int_0^1 d\mu(1-\mu^2)\int_0^{\Xi(\mu)} d\xi\,\xi^4\Theta^n \tag{52}$$

and

$$I_{33} = \int_0^1 d\mu\,\mu^2\int_0^{\Xi(\mu)} d\xi\,\xi^4\Theta^n . \tag{53}$$

With the solution for Θ given by equation (30), we can (neglecting terms of orders higher than the first in v) write[2]

$$I_{11}=\tfrac{1}{2}\int_0^{\xi_1}\int_0^1\{\theta^n+nv\,\theta^{n-1}[\psi_0+A_2\psi_2P_2(\mu)]\}\,\xi^4(1-\mu^2)\,d\mu\,d\xi \tag{54}$$

and

$$I_{33}=\int_0^{\xi_1}\int_0^1\{\theta^n+nv\,\theta^{n-1}[\psi_0+A_2\psi_2P_2(\mu)]\}\,\xi^4\mu^2\,d\mu\,d\xi\,. \tag{55}$$

The integrations over μ can be effected, and we are left with

$$I_{11}=\tfrac{1}{3}\int_0^{\xi_1}\theta^n\xi^4d\xi+\tfrac{1}{3}v\int_0^{\xi_1}n\,\theta^{n-1}\psi_0\xi^4d\xi-\tfrac{1}{15}vA_2\int_0^{\xi_1}n\,\theta^{n-1}\psi_2\xi^4d\xi \tag{56}$$

and

$$I_{33}=\tfrac{1}{3}\int_0^{\xi_1}\theta^n\xi^4d\xi+\tfrac{1}{3}v\int_0^{\xi_1}n\,\theta^{n-1}\psi_0\xi^4d\xi+\tfrac{2}{15}vA_2\int_0^{\xi_1}n\,\theta^{n-1}\psi_2\xi^4d\xi\,. \tag{57}$$

The trace of the tensor I_{ij} (the "moment of inertia" I as usually defined) is given by

$$I=\int_0^{\xi_1}\theta^n\xi^4d\xi+v\int_0^{\xi_1}n\,\theta^{n-1}\psi_0\xi^4d\xi\,. \tag{58}$$

Also, we may note that

$$\frac{\Delta I}{I}=\frac{I_{11}-I_{33}}{I}=-\tfrac{1}{5}vA_2\frac{\int_0^{\xi_1}n\,\theta^{n-1}\psi_2\xi^4d\xi}{\int_0^{\xi_1}\theta^n\xi^4d\xi}\,. \tag{59}$$

The different integrals which appear in the equations for I_{11} and I_{33} are further reduced in Appendix I; and numerical values are given in Section VII (Table 1).

V. THE POTENTIAL-ENERGY TENSOR, $\mathfrak{W}_{ij}$

The required components of the potential-energy tensor can readily be inferred from the general formulae (Chandrasekhar 1961, eqs. [19])

$$\mathfrak{W}_{11}=\frac{1}{5-n}(-\mathfrak{V}_0M+n\Omega^2I_{11}) \tag{60}$$

and

$$\mathfrak{W}_{33}=\frac{1}{5-n}(-\mathfrak{V}_0M+5\Omega^2I_{11})\,, \tag{61}$$

where M denotes the mass of the configuration and the different quantities are measured in their natural units. Now, expressing the different quantities in the units specified in (8), we have

$$\mathfrak{W}_{11}=\frac{1}{5-n}(-\mathfrak{V}_0M+\tfrac{1}{2}nvI_{11}) \tag{62}$$

and

$$\mathfrak{W}_{33}=\frac{1}{5-n}(-\mathfrak{V}_0M+\tfrac{5}{2}vI_{11})\,. \tag{63}$$

[2] It can readily be verified that the replacement of the limit $\Xi(\mu)$ by ξ_1 and inverting the order of the integrations introduces errors of orders higher than the first in v; the procedure is, therefore, justified when one is working only to the first order in v.

According to equations (26) and (32),

$$\mathfrak{B}_0 = -\xi_1\theta'(\xi_1) + v\left[\tfrac{1}{2}\xi_1^2 - \psi_0(\xi_1) - \xi_1\psi_0'(\xi_1)\right]. \tag{64}$$

The corresponding expression for M is (cf. Paper I, eq. [42])

$$M = -\xi_1^2\theta'(\xi_1) + v\left[\tfrac{1}{3}\xi_1^3 - \xi_1^2\psi_0'(\xi_1)\right]. \tag{65}$$

Inserting the foregoing expressions in equations (62) and (63), we obtain

$$\mathfrak{W}_{11} = -\frac{1}{5-n}\{\xi_1^3[\theta'(\xi_1)]^2 - v\xi_1^2\theta'(\xi_1)[\tfrac{5}{6}\xi_1^2 - \psi_0(\xi_1) - 2\xi_1\psi_0'(\xi_1)] - \tfrac{1}{6}nvI_0\} \tag{66}$$

and

$$\mathfrak{W}_{33} = -\frac{1}{5-n}\{\xi_1^3[\theta'(\xi_1)]^2 - v\xi_1^2\theta'(\xi_1)\tfrac{5}{6}[\xi_1^2 - \psi_0(\xi_1) - 2\xi_1\psi_0'(\xi_1)] - \tfrac{5}{6}vI_0\}, \tag{67}$$

where, consistently with the order of the approximation in which we are working, we have used for I_{11} its zero-order value (cf. eq. [58]),

$$I_{11} = \tfrac{1}{3}I_0 = \tfrac{1}{3}\int_0^{\xi_1}\theta^n\xi^4 d\xi. \tag{68}$$

A table of the values of $\mathfrak{W}_{11}$ and $\mathfrak{W}_{33}$ for different values of n is provided in Section VII.

VI. THE ELEMENTS OF THE SUPERMATRIX, $\mathfrak{W}_{pq;ij}$

For axisymmetric configurations, all the distinct elements of the supermatrix can be expressed in terms of one of them (see Paper II, eqs. [65]–[73]). We choose $\mathfrak{W}_{12;12}$ for an explicit evaluation; and we shall deduce the others from it.

By equation (63) of Paper II,

$$\mathfrak{W}_{12;12} = \tfrac{1}{16}\int_{-1}^{+1}\int_0^{\Xi(\mu)}\Theta^n\frac{1}{\varpi}\frac{\partial}{\partial\varpi}\left[\varpi^4\frac{1}{\varpi}\frac{\partial}{\partial\varpi}\left(\frac{1}{\varpi}\frac{\partial\chi}{\partial\varpi}\right)\right]\xi^2 d\xi d\mu, \tag{69}$$

where ϖ denotes the (horizontal) distance from the axis of rotation and the different quantities are expressed in the units specified in (8). We shall first reduce the foregoing expression for $\mathfrak{W}_{12;12}$, as far as we can, in the framework of the exact equations.

In the reduction of the integral representing $\mathfrak{W}_{12;12}$ it is convenient to introduce the following operators (cf. Chandrasekhar 1950):

$$D_\xi = \frac{1}{\xi}\frac{\partial}{\partial\xi} - \frac{\mu}{\xi^2}\frac{\partial}{\partial\mu} = \frac{1}{\varpi}\frac{\partial}{\partial\varpi} \quad\text{and}\quad D_\mu = \frac{1}{\xi}\frac{\partial}{\partial\mu}. \tag{70}$$

As defined, the operators D_ξ and D_μ commute, and we shall write

$$D_{\xi\mu} = D_{\mu\xi} = D_\xi D_\mu = D_\mu D_\xi. \tag{71}$$

Also, D_ξ commutes with $\xi\mu$:

$$D_\xi\xi\mu = \xi\mu D_\xi. \tag{72}$$

The Laplacian operator acting on functions of ξ and μ is equivalent to

$$\nabla^2 = \xi^2 D_{\xi\xi} + 2\xi\mu D_{\xi\mu} + D_{\mu\mu} + 3D_\xi. \tag{73}$$

An important property of D_ξ is that it is skew-symmetric with respect to functions which vanish on the boundary: *If* f *and* g *are any two functions of* ξ *and* μ, *then*

$$\iint_V f(D_\xi g)\xi^2 d\xi d\mu = -\iint_V g(D_\xi f)\xi^2 d\xi d\mu, \tag{74}$$

provided f *or* g *vanishes on the boundary of the volume* V *over which the integration is effected.*

The (exact) equation governing the superpotential can now be written in the form

$$\xi^2 D_{\xi\xi}\chi = -\Phi \,, \tag{75}$$

where

$$\Phi = 2\mathfrak{V} + 2\,\xi\mu D_{\xi\mu}\chi + D_{\mu\mu}\chi + 3D_{\xi}\chi \,, \tag{76}$$

and (it may be recalled that)

$$\mathfrak{V} = \Theta - \tfrac{1}{4}v\,\xi^2(1-\mu^2) + \mathfrak{V}_0 \,. \tag{77}$$

With the foregoing definitions, the expression for $\mathfrak{W}_{12;12}$ takes the form

$$\mathfrak{W}_{12;12} = \tfrac{1}{16}\int\!\!\int \Theta^n D_\xi[\,\xi^4(1-\mu^2)^2 D_{\xi\xi}\chi]\,\xi^2 d\xi d\mu \,, \tag{78}$$

where (as it entails no ambiguity) the limits of the integrations have been suppressed. Now, making use of equation (75), we can write

$$\mathfrak{W}_{12;12} = -\tfrac{1}{16}\int\!\!\int \Theta^n D_\xi[\,\xi^2(1-\mu^2)^2\Phi]\,\xi^2 d\xi d\mu \,. \tag{79}$$

With the various properties of the operators D_ξ and D_μ which have been listed, the further reduction of the integral for $\mathfrak{W}_{12;12}$ can be carried out entirely by algebraic processes. Thus a direct evaluation (during the course of which a further use of eq. [75] is made) leads to the following result:

$$\begin{aligned} D_\xi[\,\xi^2(1-\mu^2)^2\Phi] &= 2\,\xi^2(1-\mu^2)^2 D_\xi\mathfrak{V} + \tfrac{8}{7}[P_2(\mu) - P_4(\mu)]\Phi \\ &\quad + (1-\mu^2)^2(\,\xi^2 D_{\xi\mu\mu}\chi - 2\,\xi\mu D_\mu\Phi)\,. \end{aligned} \tag{80}$$

Moreover, by equation (77),

$$D_\xi\mathfrak{V} = D_\xi\Theta - \tfrac{1}{2}v \,. \tag{81}$$

The contribution to $\mathfrak{W}_{12;12}$ by the term in $D_\xi\mathfrak{V}$ in equation (80) is, therefore,

$$-\frac{1}{8(n+1)}\int\!\!\int (D_\xi\Theta^{n+1})\,\xi^4(1-\mu^2)^2 d\xi d\mu + \frac{v}{16}\int\!\!\int \Theta^n\xi^4(1-\mu^2)^2 d\xi d\mu \,. \tag{82}$$

The first of the two integrals in (82) can be reduced by an "integration by parts" by making use of the property expressed in equation (74). We find

$$\frac{1}{4(n+1)}\int\!\!\int \Theta^{n+1}\xi^2(1-\mu^4)\,d\xi d\mu + \frac{v}{16}\int\!\!\int \Theta^n\xi^4(1-\mu^2)^2 d\xi d\mu \,. \tag{83}$$

Together with the result of this last reduction, equations (79) and (80) give

$$\begin{aligned} \mathfrak{W}_{12;12} &= \frac{1}{4(n+1)}\int\!\!\int \Theta^{n+1}\xi^2(1-\mu^4)\,d\xi d\mu + \frac{v}{16}\int\!\!\int \Theta^n\xi^4(1-\mu^2)^2 d\xi d\mu \\ &\quad - \tfrac{1}{16}\int\!\!\int \Theta^n(\,\xi^2 D_{\xi\mu\mu}\chi - 2\,\xi\mu D_\mu\Phi)\,\xi^2(1-\mu^2)^2 d\xi d\mu \\ &\quad - \tfrac{1}{14}\int\!\!\int \Theta^n\Phi[P_2(\mu) - P_4(\mu)]\,\xi^2 d\xi d\mu = \mathfrak{I}_1 + \mathfrak{I}_2 + \mathfrak{I}_3\ (\text{say})\,. \end{aligned} \tag{84}$$

Equation (84) is exact; we shall now reduce it further by making use of the solutions found in Sections II and III and retaining terms only up to the first order in v. Thus, considering $\mathfrak{J}_1$, we have

$$\mathfrak{J}_1 = \frac{1}{4(n+1)}\int_0^{\xi_1}\int_{-1}^{+1}\{\theta^{n+1}+(n+1)v\theta^n[\psi_0+A_2P_2(\mu)]\}\xi^2(1-\mu^4)d\xi d\mu$$
$$+\frac{v}{16}\int_0^{\xi_1}\int_{-1}^{+1}\theta^n\xi^4(1-\mu^2)^2 d\xi d\mu. \tag{85}$$

After carrying out the integrations over μ, we are left with

$$\mathfrak{J}_1 = \frac{2}{5(n+1)}\int_0^{\xi_1}\theta^{n+1}\xi^2 d\xi + v\left(\tfrac{1}{15}\int_0^{\xi_1}\theta^n\xi^4 d\xi + \tfrac{2}{5}\int_0^{\xi_1}\theta^n\psi_0\xi^2 d\xi\right.$$
$$\left. -\tfrac{2}{35}A_2\int_0^{\xi_1}\theta^n\psi_2\xi^2 d\xi\right). \tag{86}$$

The zero-order term in equation (86) is a well-known integral in the theory of polytropes (cf. Milne 1929); and it has the value

$$\int_0^{\xi_1}\theta^{n+1}\xi^2 d\xi = \frac{n+1}{5-n}\,\xi_1^3[\theta'(\xi_1)]^2. \tag{87}$$

The further reduction of $\mathfrak{J}_2$ and $\mathfrak{J}_3$ is somewhat more complicated, since it involves, in addition, the superpotential. Writing the solution for χ in the form

$$\chi = \chi_0(\xi) + v[\chi_{1;0}(\xi) + \tfrac{1}{2}(3\mu^2-1)\chi_{1;2}(\xi)], \tag{88}$$

we first find by elementary calculations that

$$\Phi = 2(\theta + c_0) + v[2\psi_0 - \tfrac{1}{2}\xi^2 + 2c_{1\cdot 0} + A_2(3\mu^2-1)\psi_2 + \tfrac{1}{2}\xi^2\mu^2]$$
$$+3\left(\frac{\chi_0'}{\xi} + v\frac{\chi_{1;0}'}{\xi}\right) + \tfrac{3}{2}v(7\mu^2-1)\left(\frac{\chi_{1;2}'}{\xi} - 2\frac{\chi_{1;2}}{\xi^2}\right) \tag{89}$$

and

$$\xi^2 D_{\xi\mu\mu}\chi - 2\xi\mu D_\mu\Phi = v\left[3(1-14\mu^2)\left(\frac{\chi_{1;2}'}{\xi} - 2\frac{\chi_{1;2}}{\xi^2}\right) - 12A_2\mu^2\psi_2 - 2\xi^2\mu^2\right]. \tag{90}$$

Using these expressions in the integrands defining $\mathfrak{J}_2$ and $\mathfrak{J}_3$ and carrying out the integrations over μ, we find

$$\mathfrak{J}_2 = v\left[\tfrac{2}{105}\int_0^{\xi_1}\theta^n\xi^4 d\xi + \tfrac{4}{35}A_2\int_0^{\xi_1}\theta^n\psi_2\xi^2 d\xi - \tfrac{1}{5}\int_0^{\xi_1}\theta^{n-1}(3\theta + n\theta'\xi)\chi_{1;2}d\xi\right] \tag{91}$$

and

$$\mathfrak{J}_3 = v\left[-\tfrac{1}{105}\int_0^{\xi_1}\theta^n\xi^4 d\xi - \tfrac{2}{35}A_2\int_0^{\xi_1}\theta^n\psi_2\xi^2 d\xi\right.$$
$$\left.-\frac{n}{35}A_2\int_0^{\xi_1}\left(2\theta + 2c_0 + 3\frac{\chi_0'}{\xi}\right)\theta^{n-1}\psi_2\xi^2 d\xi + \tfrac{1}{5}\int_0^{\xi_1}\theta^{n-1}(3\theta + n\theta'\xi)\chi_{1;2}d\xi\right]. \tag{92}$$

Now, combining equations (85), (87), (91), and (92), we obtain

$$\mathfrak{W}_{12;12}=\frac{2}{5(5-n)}\,\xi_1^3[\theta'(\xi_1)]^2+v\left[\frac{8}{105}\int_0^{\xi_1}\theta^n\xi^4 d\xi+\frac{2}{5}\int_0^{\xi_1}\theta^n\psi_0\xi^2 d\xi-\frac{A_2}{35}\int_0^{\xi_1}\left(2\theta+2c_0+3\frac{\chi_0'}{\xi}\right)n\theta^{n-1}\psi_2\xi^2 d\xi\right]. \qquad (93)$$

We observe the remarkable fact that in this last expression for $\mathfrak{W}_{12;12}$ the terms in $\chi_{1;0}$ and $\chi_{1;2}$ have canceled and that the superpotential appears only in zero order.

Once $\mathfrak{W}_{12;12}$ has been evaluated in accordance with equation (93), the remaining elements of the supermatrix can be obtained with the aid of equations (65)–(69) and (71)–(73) of Paper II.

TABLE 1

THE MOMENT OF INERTIA AND THE POTENTIAL-ENERGY TENSORS AND THE ELEMENTS OF THE SUPERMATRIX

Quantity	$n=1$	$n=1.5$	$n=2$	$n=3$	$n=3.5$
I_{11}	$+4.05224+25.1865v$	$+3.70658+30.1773v$	$+3.53701+38.9479v$	$+3.61719+87.2093v$	$+3.91516+165.761v$
I_{33}	$+4.05224-1.32587v$	$+3.70658-0.91730v$	$+3.53701+0.43266v$	$+3.61719+12.0841v$	$+3.91516+36.5364v$
$I_{11}-I_{33}$	$+26.5124v$	$+31.0946v$	$+38.5153v$	$+75.1252v$	$+129.224v$
$\Delta I/I$	$+2.18089v$	$+2.79634v$	$+3.62974v$	$+6.92299v$	$+11.0021v$
$\mathfrak{W}_{11}$	$-0.785398-3.59692v$	$-0.576010-3.19798v$	$-0.445157-2.99604v$	$-0.295300-3.01334v$	$-0.249880-3.26243v$
$\mathfrak{W}_{33}$	$-0.785398-1.57080v$	$-0.576010-1.34470v$	$-0.445157-1.22753v$	$-0.295300-1.20474v$	$-0.249880-1.30485v$
$\mathfrak{W}_{12;12}$	$+0.314159+1.26510v$	$+0.230404+1.12031v$	$+0.178063+1.04684v$	$+0.118120+1.05030v$	$+0.0999518+1.13731v$
$\mathfrak{W}_{11;11}$	$+0.157080+0.198381v$	$+0.115202+0.162941v$	$+0.0890314+0.14448v$	$+0.0590600+0.137566v$	$+0.0499759+0.149501v$
$\mathfrak{W}_{11;22}$	$-0.471239-2.33182v$	$-0.345606-2.07768v$	$-0.267094-1.94919v$	$-0.177180-1.96303v$	$-0.149928-2.12512v$
$\mathfrak{W}_{11;33}$	$-0.471239-1.46348v$	$-0.345606-1.28325v$	$-0.267094-1.19131v$	$-0.177180-1.18787v$	$-0.149928-1.28681v$
$\mathfrak{W}_{33;33}$	$+0.157080+1.35616v$	$+0.115202+1.22180v$	$+0.0890314+1.15510v$	$+0.0590600+1.17099v$	$+0.0499759+1.26877v$
$\mathfrak{W}_{13;13}$	$+0.314159+2.13344v$	$+0.230404+1.91474v$	$+0.178063+1.80472v$	$+0.118120+1.82547v$	$+0.0999518+1.97562v$
$\mathfrak{W}_{31;13}$	$+0.314159+0.107317v$	$+0.230404+0.061448v$	$+0.178063+0.03622v$	$+0.118120+0.016875v$	$+0.0999518+0.018041v$

VII. NUMERICAL RESULTS

The different integrals which appear in the formulae for I_{11} (or I_{33}) and $\mathfrak{W}_{12;12}$ given in Sections IV and VI are further reduced in Appendix I, where it is shown how all of them can be expressed in terms of the following four:

$$\int_0^{\xi_1}\xi^2\psi_0 d\xi,\qquad \int_0^{\xi_1}\psi_2 d\xi,\qquad \int_0^{\xi_1}\theta\psi_2 d\xi,\qquad\text{and}\qquad \int_0^{\xi_1}n\theta^{n-1}\psi_2\chi_0'\xi d\xi. \qquad (94)$$

These integrals were evaluated numerically for different values of n with the aid of the functions tabulated in Appendix III; and their values, together with those of the others, are listed in Appendix I (Table 5). With the information provided by this table, it is a simple matter to evaluate the components of the different tensors which are needed for the application of the formulae of Paper III. The results of these calculations are summarized in Table 1.

VIII. THE EFFECT OF A SMALL ROTATION ON THE MODES OF OSCILLATION OF A POLYTROPE

With the data provided in Table 1, we can apply the general formulae of Paper III to determine the effect of a small rotation on the different characteristic frequencies (σ) of oscillation of a polytrope. But first we may recall that the theory developed in Paper III is an approximate one and that it is based on a "trial function" for the Lagrangian displacement $\boldsymbol{\xi}$ of the form

$$\xi_i = X_{ij}x_j \, , \tag{95}$$

where the X_{ij}'s are constants. With this assumed form of the Lagrangian displacement, the tensor virial equation (of the second order) leads to a set of nine linear homogeneous equations for the X_{ij}'s. The characteristic equation for σ which follows is of order 18; and the resulting nine modes of oscillation group themselves naturally into three classes. The three classes are distinguished by the particular set of constants, among the X_{ij}'s, which do not vanish. The three classes are as follows:

the transverse shear modes for which the non-vanishing X_{ij}'s are

X_{13}, X_{23}, X_{31}, and X_{32}; and

$$\xi_1 = X_{13}x_3, \ \xi_2 = X_{23}x_3, \text{ and } \xi_3 = X_{31}x_1 + X_{32}x_2 \, ; \tag{96}$$

the toroidal modes for which the non-vanishing X_{ij}'s are

X_{11}, X_{22}, X_{12}, and X_{21}, and, moreover, $X_{11} = -X_{22}$ and $X_{12} = X_{21}$; and

$$\xi_1 = X_{11}x_1 + X_{12}x_2, \ \xi_2 = X_{12}x_1 - X_{11}x_2, \text{ and } \xi_3 = 0 \, ; \tag{97}$$

and

the pulsation modes for which the non-vanishing X_{ij}'s are

X_{11}, X_{22}, X_{33}, X_{12}, and X_{21}, and, moreover, $X_{11} = X_{22}$ and $X_{12} = -X_{21}$; and

$$\xi_1 = X_{11}x_1 + X_{12}x_2, \ \xi_2 = -X_{12}x_1 + X_{11}x_2, \text{ and } \xi_3 = X_{33}x_3 \, . \tag{98}$$

We shall now consider these three classes of oscillations separately.

a) The Transverse Shear Modes

For small rotations, the characteristic frequencies of these modes are given by (Paper III, eqs. [50], [55], and [68])

$$\sigma_+ = +\sqrt{\mu} \pm \tfrac{1}{2}\Omega\left(1 - \frac{\nu}{\mu}\right), \quad \sigma_- = -\sqrt{\mu} \pm \tfrac{1}{2}\Omega\left(1 - \frac{\nu}{\mu}\right), \quad \text{and} \quad \sigma_0 = \pm\Omega\frac{\nu}{\mu}, \tag{99}$$

where

$$\mu = \frac{\mathfrak{W}_{13;13}}{I_{11}} + \frac{\mathfrak{W}_{31;13}}{I_{33}} \quad \text{and} \quad \nu = \frac{\mathfrak{W}_{13;13}}{I_{11}} - \frac{\mathfrak{W}_{31;13}}{I_{33}} \, . \tag{100}$$

For a configuration of uniform density, $\nu = 0$ by an identity which obtains for ellipsoids (Chandrasekhar and Lebovitz 1962*d*, eq. [66]), so that in this case

$$\sigma_+ = +\sqrt{\mu} \pm \tfrac{1}{2}\Omega\,, \qquad \sigma_- = -\sqrt{\mu} \pm \tfrac{1}{2}\Omega\,, \qquad \text{and} \qquad \sigma_0 = 0\,. \tag{101}$$

However, in general, $\nu \neq 0$ and $\sigma_0 \neq 0$.

The frequencies σ_+ and σ_0 calculated with the aid of Table 1 are listed in Table 2 for different values of n.

TABLE 2

THE CHARACTERISTIC FREQUENCIES OF THE TRANSVERSE SHEAR MODES

[The Frequencies Are Given in the Unit $(4\pi G\lambda)^{1/2}$]

n	σ_+	σ_0
1.0	$0.393770\,(1+0.311068v) \pm 0.353553\,(1+0.0466573v)\sqrt{v}$	$\pm 0.0329916\sqrt{v}$
1.5	$.352593\,(1+0.170742v) \pm .353553\,(1+0.17269v)\sqrt{v}$	$\pm 0.12211\sqrt{v}$
2.0	$.317310\,(1-0.19879v) \pm .353553\,(1+0.47864v)\sqrt{v}$	$\pm 0.33846\sqrt{v}$
3.0	$.255559\,(1-2.96330v) \pm .353553\,(1+2.72873v)\sqrt{v}$	$\pm 1.92951\sqrt{v}$
3.5	$0.225962\,(1-7.93102v) \pm 0.353553\,(1+6.71050v)\sqrt{v}$	$\pm 4.74504\sqrt{v}$

TABLE 3

THE CHARACTERISTIC FREQUENCIES OF THE TOROIDAL MODES

(σ^2 Is Listed in the Unit $4\pi G\lambda$)

n	σ^2
1.0	$0.155055-0.339341v \pm 0.556874\,(1-2.70660v)\sqrt{v}$
1.5	$.124322-0.407674v \pm .498642\,(1-3.65050v)\sqrt{v}$
2.0	$.100685-0.516767v \pm .448744\,(1-5.04923v)\sqrt{v}$
3.0	$.0653104-0.993887v \pm .361415\,(1-11.4368v)\sqrt{v}$
3.5	$0.0510589-1.58077v \pm 0.319559\,(1-20.3761v)\sqrt{v}$

b) The Toroidal Modes

The characteristic frequencies of these modes are given by

$$\sigma^2 = 2\,\frac{\mathfrak{W}_{12;12}}{I_{11}} \pm 2\Omega\left(2\,\frac{\mathfrak{W}_{12;12}}{I_{11}} - \Omega^2\right)^{1/2}. \tag{102}$$

The application of this formula to distorted polytropes gives the results summarized in Table 3.

c) The Pulsation Modes

In some ways, these modes are the most interesting among the ones considered: they represent a coupling of what, for a strictly spherical configuration, are a purely radial and a purely non-radial mode of oscillation. This is apparent from the features of these modes included in the description (98) and the more detailed information given in Table 1 of Paper III. Thus, for a purely radial oscillation of a sphere,

$$X_{11} = X_{33}\ (=X_{22}) \qquad \text{(radial or } R\text{-mode)}\,, \tag{103}$$

while, for the associated non-radial mode, we should have (as required by volume conservation)

$$X_{11} = -\tfrac{1}{2}X_{33}\ (=X_{22}) \qquad \text{(non-radial or } S\text{-mode)}\,. \tag{104}$$

And as we have shown in a different connection (Chandrasekhar and Lebovitz 1962*e*, eq. [26]) the property

$$\left(\frac{X_{11}}{X_{33}}\right)_R \left(\frac{X_{11}}{X_{33}}\right)_S = -\tfrac{1}{2} \tag{105}$$

is identically preserved in the limit $\Omega^2 \to 0$.

And, again, these "pulsation modes" disclose the critical role of $\gamma = 1.6$: when γ has this value, a case of accidental degeneracy arises, and, in the limit $\Omega^2 = 0$, the "pure states" (in the terminology of the quantum theory) are mixtures of the radial and the non-radial "states" (though still preserving relation [105]).

Considering, first, the case $\gamma \neq 1.6$, the characteristic frequencies (to the first order in Ω^2) are given by (Paper III, eqs. [85], [90], [91], [101], [103], and [104])[3]

$$\sigma_R^2 = -(3\gamma - 4)\frac{\mathfrak{W}_{33}}{I_{11}}\left(1 + \frac{\Delta I}{I}\right) + \tfrac{2}{3}\Omega^2 \tag{106}$$

and

$$\sigma_S^2 = \frac{\mathfrak{W}_{33;33} - \mathfrak{W}_{33;11}}{I_{33}}\left(1 - \frac{\Delta I}{I}\right) + \tfrac{1}{3}\Omega^2\,. \tag{107}$$

Equation (106) is equivalent to a formula first derived by Ledoux (1945). To show this, first observe that (to the first order in the departure of the configuration from spherical symmetry),

$$\begin{aligned}\frac{\mathfrak{W}}{I} = \frac{\mathfrak{W}_{33} - \tfrac{2}{3}(\mathfrak{W}_{33} - \mathfrak{W}_{11})}{I_{11} - \tfrac{1}{3}(I_{11} - I_{33})} &= \frac{\mathfrak{W}_{33}}{I_{11}}\left(1 - \frac{2}{3}\,\frac{\mathfrak{W}_{33} - \mathfrak{W}_{11}}{\mathfrak{W}_{33}} + \frac{1}{3}\,\frac{I_{11} - I_{33}}{I_{11}}\right) \\ &= \frac{\mathfrak{W}_{33}}{I_{11}}\left(1 + \frac{\Delta I}{I}\right) - \frac{2}{3}\,\frac{\mathfrak{W}_{33} - \mathfrak{W}_{11}}{I_{11}}\,.\end{aligned} \tag{108}$$

Making use of the general relation (cf. Paper III, eq. [7]),

$$\Omega^2 I_{11} = \mathfrak{W}_{33} - \mathfrak{W}_{11}\,, \tag{109}$$

we can rewrite equation (108) in the form

$$\frac{\mathfrak{W}_{33}}{I_{11}}\left(1 + \frac{\Delta I}{I}\right) = \frac{\mathfrak{W}}{I} + \tfrac{2}{3}\Omega^2\,. \tag{110}$$

Inserting this last result in equation (106), we obtain

$$\sigma_R^2 = -(3\gamma - 4)\frac{\mathfrak{W}}{I} + \tfrac{2}{3}(5 - 3\gamma)\Omega^2\,. \tag{111}$$

This is Ledoux's result (cf. Cowling and Newing 1949, eq. [19]). The present derivation of it underlines the fact (which Cowling and Newing have emphasized) that $\mathfrak{W}$ and I, being the potential energy and the moment of inertia of the distorted configuration, include, already in themselves, terms of order Ω^2 and that the term in Ω^2 which explicitly occurs in the equation is not the only one that determines the effect of rotation.

[3] We failed to notice in Paper III that the coefficients of the terms in parentheses on the right-hand sides of eqs. (103) and (104) (of that paper) are exactly unity.

The explicit forms of these relations, for distorted polytropes of different n's, are given in Table 4*A*.

When $\gamma = 1.6$, the degeneracy which exists in the absence of rotation is lifted by the presence of rotation.[4] And we may expect, from the perturbation theory familiar in the quantum theory, that the splitting of the mode will now be determined by a "secular equation" of order 2. This is, indeed, the case; by expressing σ^2 in the form

$$\sigma^2 = \frac{4}{5}\frac{|\mathfrak{W}_0|}{I_0} + \frac{\Omega^2}{2\pi G\lambda}\Delta\sigma^2 \tag{112}$$

(where $\mathfrak{W}_0$ and I_0 are the zero-order values of $\mathfrak{W}$ and I) we find that equation (97) of Paper III gives

$$\left[\Delta\sigma^2 + \frac{12}{5}\frac{\mathfrak{W}_0}{I_0}\left(\frac{\Delta\mathfrak{W}_{33}}{\mathfrak{W}_0} - \frac{\Delta I_{11}}{I_0}\right)\right]\left[\Delta\sigma^2 + \frac{4}{5}\frac{\mathfrak{W}_0}{I_0}\left(15\,\frac{\Delta\mathfrak{W}_{33;33} - \Delta\mathfrak{W}_{33;11}}{4\mathfrak{W}_0} - 3\,\frac{\Delta I_{33}}{I_0}\right)\right]$$

$$-\frac{4}{5}\frac{\mathfrak{W}_0}{I_0}\left[\frac{9}{I_0^2}(0.6\Delta\mathfrak{W}_{33} - \Delta\mathfrak{W}_{33;11})(\Delta I_{11} - \Delta I_{33}) + \frac{15}{8\mathfrak{W}_0}(0.6\Delta\mathfrak{W}_{33} - \Delta\mathfrak{W}_{33;33}) - \frac{3}{2}\frac{\Delta I_{33}}{I_0}\right] - \tfrac{1}{2}\Delta\sigma^2 = 0\,, \tag{113}$$

where $v\Delta$ prefixed to a quantity is the contribution to it by the term proportional to v (thus: $\mathfrak{W}_{33} = \frac{1}{3}\mathfrak{W}_0 + v\Delta\mathfrak{W}_{33}$). The determination of the splitting of σ^2 by equation (113) enables the specification, at the same time, of the values of $(X_{11}/X_{33})_R$ and $(X_{11}/X_{33})_S$ appropriate for the "pure states" in the limit $\Omega^2 = 0$. The values of these ratios, along with those of σ^2, are given in Table 4*B*.

TABLE 4*A*

THE CHARACTERISTIC FREQUENCIES σ_R AND σ_S OF THE PULSATION MODES FOR $\gamma \neq 1.6$

(σ^2 Is Given in the Unit $4\pi G\lambda$)

n	σ_R^2	σ_S^2
1.0	$(3\gamma-4)\,(0.193818 - 0.394338v) + \frac{1}{3}v$	$0.155055 + 0.575067v$
1.5	$(3\gamma-4)\,(\ .155402 - 0.467870v) + \frac{1}{3}v$	$.124322 + \ .525627v$
2.0	$(3\gamma-4)\,(\ .125857 - 0.581993v) + \frac{1}{3}v$	$.100685 + \ .452278v$
3.0	$(3\gamma-4)\,(\ .0816380 - 1.07003v) + \frac{1}{3}v$	$.0653104 + \ .148463v$
3.5	$(3\gamma-4)\,(0.0638237 - 1.66671v) + \frac{1}{3}v$	$0.0510589 - 0.218831v$

IX. CONCLUDING REMARKS

The solution of the problem of the oscillation of distorted polytropes, summarized in Section VIII and in Tables 2, 3, 4*A*, and 4*B*, is based on a comparison of a rotating and a non-rotating configuration of the same central density λ; this is apparent from the fact that the solution for Θ is expressed as the sum of the Lane-Emden function and a "correction" term of order v (see eq. [30]). This fact has the consequence that, other things being equal, the comparison is made between two configurations of different masses, the rotating configuration having a slightly larger mass (see eq. [65]). If we should now wish to make the comparison between configurations of equal mass, then, in a framework of polytropic equilibrium in which the constant K in the pressure-density

[4] The statements, relative to this case, in Paper III (following eq. [104]) are erroneous; the error was, however, corrected later (*Ap. J.*, **135**, 659, **1962**).

relation (1) is considered as invariable, the equality in the masses can be achieved only by an adjustment in the central density, which will secure

$$M(\lambda+\delta\lambda;v)=M(\lambda;0)\,. \tag{114}$$

From this requirement, we find from the mass relation (65) that (cf. Paper I, eq. [61])

$$\left(\frac{\delta\lambda}{\lambda}\right)_M=v\,\frac{2n}{3-n}\,\frac{\frac{1}{3}\xi_1-\psi_0'(\xi_1)}{\theta'(\xi_1)}\qquad(n\neq 3)\,. \tag{115}$$

(Values of $\delta\lambda/\lambda$ for different n's are included in Table 7 in Appendix III.)

So long as $n\neq 3$, the comparison sought can be accomplished (via the condition [114]) by simply increasing the values of the characteristic frequencies, as given in Section VIII in the unit $(4\pi G\lambda)^{1/2}$, by the factor $(1+\frac{1}{2}\delta\lambda/\lambda)$. However, the non-uniqueness in seeking a comparison between two configurations of equal mass, by an adjustment solely in the

TABLE 4*B*

THE SPLITTING OF THE PULSATION MODE IN CASE $\gamma=1.6$

(The Table Lists σ^2 in the Unit $4\pi G\lambda$)

n	σ^2	X_{11}/X_{33}
1.0	$0.155055\begin{cases}+0.665838v\\-0.0729082v\end{cases}$	−0.186094 +2.68681
1.5	$.124322\begin{cases}+0.624122v\\-0.139457v\end{cases}$	−0.179138 +2.79114
2.0	$.100685\begin{cases}+0.565546v\\-0.245529v\end{cases}$	−0.16742 +2.98652
3.0	$.0653104\begin{cases}+0.337868v\\-0.712094v\end{cases}$	−0.126351 +3.95723
3.5	$0.0510589\begin{cases}+0.072298v\\-1.29116v\end{cases}$	−0.09614 +5.20071

central density through equation (114), is made manifest by equation (115), which precludes such a comparison for the case $n=3$. Indeed, when a "real" star is set into rotation, we should not expect, for example, that the central temperature remains unchanged. An adjustment solely by a change in the central temperature, in the present context of polytropic equilibrium, means an adjustment by a change in K *at constant* λ. In this latter event, the results of Section VIII already provide the comparison sought. It is, however, clear that, in practice, the actual adjustment which a star will make when it is set into rotation will not be solely by a change in the central density or solely by a change in the central temperature; it will be by a combination of both. The two manners of adjustment we have considered—at constant K and at constant λ—are only the two extreme possibilities. A host of intermediate possibilities can be imagined. The resolution of the resulting ambiguity is, clearly, outside the scope of the present paper.

In conclusion, we wish to record our indebtedness to Miss Donna D. Elbert for her very essential assistance in the preparation of this paper: she not only carried out all the necessary integrations of the basic functions (tabulated in Appendix III); she also carried out all the auxiliary calculations, including the numerical evaluation of the many integrals that were needed.

The research reported in this paper has in part been supported by the Office of Naval Research under contract Nonr-2121(24) with the University of Chicago. The work of the second author was supported in part by the United States Air Force under contract No. AF-49(638)-42, monitored by the Air Force Office of Scientific Research of the Air Research and Development Command.

APPENDIX I

The Reduction of Certain Integrals and a Table of Their Values

The expressions for I_{11}, I_{33}, and $\mathfrak{W}_{12;12}$ given in Sections IV and VI involve the following integrals:

$$\text{I. } \int_0^{\xi_1} \theta^n \xi^4 d\xi\,; \qquad \text{II. } \int_0^{\xi_1} \theta^n \psi_0 \xi^2 d\xi\,; \qquad \text{III. } \int_0^{\xi_1} \theta^n \psi_2 \xi^2 d\xi\,;$$

$$\text{IV. } \int_0^{\xi_1} n\theta^{n-1}\psi_2 \xi^2 d\xi\,; \qquad \text{V. } \int_0^{\xi_1} n\theta^{n-1}\psi_0 \xi^4 d\xi\,; \qquad \text{VI. } \int_0^{\xi_1} n\theta^{n-1}\psi_2 \xi^4 d\xi\,. \tag{AI, 1}$$

[There is a further integral involving χ_0' which we shall retain in the same form.] We shall now reduce the foregoing integrals to simpler forms:

I.
$$\begin{aligned}\int_0^{\xi_1} \theta^n \xi^4 d\xi &= -\int_0^{\xi_1} \xi^2 \frac{d}{d\xi}\left(\xi^2 \frac{d\theta}{d\xi}\right) d\xi \\ &= -\xi_1^4 \theta'(\xi_1) + 2\int_0^{\xi_1} \xi^3 \frac{d\theta}{d\xi} d\xi \\ &= -\xi_1^4 \theta'(\xi_1) - 6\int_0^{\xi_1} \xi^2 \theta d\xi\,.\end{aligned} \tag{AI, 2}$$

II.
$$\begin{aligned}\int_0^{\xi_1} \theta^n \psi_0 \xi^2 d\xi &= -\int_0^{\xi_1} \psi_0 \frac{d}{d\xi}\left(\xi^2 \frac{d\theta}{d\xi}\right) d\xi \\ &= -\xi_1^2 \psi_0(\xi_1)\,\theta'(\xi_1) + \int_0^{\xi_1} \xi^2 \psi_0' \theta' d\xi\,.\end{aligned} \tag{AI, 3}$$

Alternatively, making use of the equation satisfied by ψ_0 (eq. [13]), we have

$$\begin{aligned}n\int_0^{\xi_1} \theta^n \psi_0 \xi^2 d\xi &= -\int_0^{\xi_1} \theta\left[\frac{d}{d\xi}\left(\xi^2 \frac{d\psi_0}{d\xi}\right) - \xi^2\right] d\xi \\ &= \int_0^{\xi_1} \xi^2 \psi_0' \theta' d\xi + \int_0^{\xi_1} \xi^2 \theta d\xi\,.\end{aligned} \tag{AI, 4}$$

Combining equations (AI, 3) and (AI, 4), we get

$$(n-1)\int_0^{\xi_1} \theta^n \psi_0 \xi^2 d\xi = \xi_1^2 \psi_0(\xi_1)\,\theta'(\xi_1) + \int_0^{\xi_1} \xi^2 \theta d\xi\,. \tag{AI, 5}$$

III.
$$\begin{aligned}\int_0^{\xi_1} \theta^n \psi_2 \xi^2 d\xi &= -\int_0^{\xi_1} \psi_2 \frac{d}{d\xi}\left(\xi^2 \frac{d\theta}{d\xi}\right) d\xi \\ &= -\xi_1^2 \psi_2(\xi_1)\,\theta'(\xi_1) + \int_0^{1} \xi^2 \theta' \psi_2' d\xi\,.\end{aligned} \tag{AI, 6}$$

Alternatively, making use of the equation satisfied by ψ_2 (eq. [13]), we have

$$\begin{aligned}n\int_0^{\xi_1} \theta^n \psi_2 \xi^2 d\xi &= -\int_0^{\xi_1} \theta\left[\frac{d}{d\xi}\left(\xi^2 \frac{d\psi_2}{d\xi}\right) - 6\psi_2\right] d\xi \\ &= \int_0^{\xi_1} \xi^2 \psi_2' \theta' d\xi + 6\int_0^{\xi_1} \theta \psi_2 d\xi\,.\end{aligned} \tag{AI, 7}$$

Combining equations (AI, 6) and (AI, 7), we get

$$(n-1)\int_0^{\xi_1}\theta^n\psi_2\xi^2 d\xi = \xi_1^2\psi_2(\xi_1)\,\theta'(\xi_1) + 6\int_0^{\xi_1}\theta\psi_2 d\xi . \qquad \text{(AI, 8)}$$

IV.

$$\int_0^{\xi_1} n\theta^{n-1}\psi_2\xi^2 d\xi = -\int_0^{\xi_1}\left[\frac{d}{d\xi}\left(\xi^2\frac{d\psi_2}{d\xi}\right) - 6\psi_2\right]d\xi$$

$$= -\xi_1^2\psi_2'(\xi_1) + 6\int_0^{\xi_1}\psi_2 d\xi ; \qquad \text{(AI, 9)}$$

V.

$$\int_0^{\xi_1} n\theta^{n-1}\psi_0\xi^4 d\xi = -\int_0^{\xi_1}\left[\xi^2\frac{d}{d\xi}\left(\xi^2\frac{d\psi_0}{d\xi}\right) - \xi^4\right]d\xi$$

$$= \tfrac{1}{5}\xi_1^5 - \xi_1^4\psi_0'(\xi_1) + 2\int_0^{\xi_1}\xi^3\frac{d\psi_0}{d\xi}d\xi \qquad \text{(AI, 10)}$$

$$= \tfrac{1}{5}\xi_1^5 - \xi_1^4\psi_0'(\xi_1) + 2\,\xi_1^3\psi_0(\xi_1) - 6\int_0^{\xi_1}\xi^2\psi_0 d\xi .$$

VI.

$$\int_0^{\xi_1} n\theta^{n-1}\psi_2\xi^4 d\xi = -\int_0^{\xi_1}\left[\xi^2\frac{d}{d\xi}\left(\xi^2\frac{d\psi_2}{d\xi}\right) - 6\,\xi^2\psi_2\right]d\xi$$

$$= -\xi_1^4\psi_2'(\xi_1) + 2\int_0^{\xi_1}\xi^3\frac{d\psi_2}{d\xi}d\xi + 6\int_0^{\xi_1}\xi^2\psi_2 d\xi \qquad \text{(AI, 11)}$$

$$= -\xi_1^4\psi_2'(\xi_1) + 2\,\xi_1^3\psi_2(\xi_1) .$$

Thus the six integrals listed can be expressed in terms of the following four:

$$\int_0^{\xi_1}\xi^2\theta d\xi , \quad \int_0^{\xi_1}\theta\psi_2 d\xi , \quad \int_0^{\xi_1}\psi_2 d\xi , \quad \text{and} \quad \int_0^{\xi_1}\xi^2\psi_0 d\xi . \qquad \text{(AI, 12)}$$

However, there is a further relation which expresses the first of these integrals in terms of the values of ψ_0 and its derivative at ξ_1. Thus, letting I_1, I_2, and I_3 denote the integrals

$$I_1 = \int_0^{\xi_1}\xi^2\theta d\xi , \quad I_2 = \int_0^{\xi_1}\theta^n\psi_0\xi^2 d\xi , \quad \text{and} \quad I_3 = \int_0^{\xi_1}\xi^2\theta'\psi_0' d\xi , \qquad \text{(AI, 13)}$$

we have already established among them the two relations (see eqs. [AI, 3] and [AI, 4])

$$I_2 - I_3 = -\xi_1^2\psi_0(\xi_1)\,\theta'(\xi_1) \qquad \text{(AI, 14)}$$

and

$$I_1 - nI_2 + I_3 = 0 . \qquad \text{(AI, 15)}$$

A third relation can be obtained as follows:

$$I_3 = \int_0^{\xi_1}\xi^2\theta'\psi_0' d\xi = \tfrac{1}{3}\xi_1^3\theta'(\xi_1)\psi_0'(\xi_1) - \tfrac{1}{3}\int_0^{\xi_1}\xi^3(\theta''\psi_0' + \theta'\psi_0'')\,d\xi$$

$$= \tfrac{1}{3}\xi_1^3\theta'(\xi_1)\psi_0'(\xi_1) + \tfrac{1}{3}\int_0^{\xi_1}\xi^3\left[\psi_0'\left(\theta^n + \frac{2}{\xi}\theta'\right) + \theta'\left(n\theta^{n-1}\psi_0 + \frac{2}{\xi}\psi_0' - 1\right)\right]d\xi$$

$$= \tfrac{1}{3}\xi_1^3\theta'(\xi_1)\psi_0'(\xi_1) + \tfrac{1}{3}\int_0^{\xi_1}[\,4\,\xi^2\psi_0'\theta' + \xi^3(\theta^n\psi_0)' - \xi^3\theta'\,]\,d\xi \qquad \text{(AI, 16)}$$

$$= \tfrac{1}{3}\xi_1^3\theta'(\xi_1)\psi_0'(\xi_1) + \tfrac{4}{3}I_3 - I_2 + I_1 .$$

Hence

$$I_1 - I_2 + \tfrac{1}{3} I_3 = -\tfrac{1}{3}\xi_1^3 \theta'(\xi_1)\psi_0'(\xi_1). \tag{AI, 17}$$

Equations (AI, 14), (AI, 15), and (AI, 17) can be solved for I_1, I_2, and I_3 separately. Thus we find

$$(3n-5) I_1 = \xi_1^2 \theta'(\xi_1)[(3-n)\psi_0(\xi_1) - (n-1)\xi_1 \psi_0'(\xi_1)] \tag{AI, 18}$$

and

$$(3n-5) I_2 = \xi_1^2 \theta'(\xi_1)[2\psi_0(\xi_1) - \xi_1 \psi_0'(\xi_1)]. \tag{AI, 19}$$

From equations (AI, 18) and (AI, 19) it is evident that

$$2\psi_0(\xi_1) = \xi_1 \psi_0'(\xi_1) \qquad \text{for } n = \tfrac{5}{3}; \tag{AI, 20}$$

and note also that $n = \frac{5}{3}$ corresponds to $\gamma = 1.6$!

The integrals listed in (AI, 12) were evaluated numerically with the aid of the functions tabulated in Appendix III. A comparison of the value of I_1 obtained by numerical integrations with the values given by equation (AI, 18) provided a useful check on the calculations.

In Table 5 we list the values of the four basic integrals, as well as those listed in (AI, 1) and

TABLE 5

A TABLE OF INTEGRALS

Integrals	$n=1.0$	$n=1.5$	$n=2.0$	$n=3.0$	$n=3.5$
$\int \xi^2 \theta d\xi$	3.14159	4.1854	5.8454	14.191	26.694
$\int \psi_2 d\xi$	....	8.4649	11.484	30.852	70.005
$\int \theta \psi_2 d\xi$	....	2.7212	3.2081	5.3351	8.1883
$\int \psi_0 \xi^2 d\xi$	7.19383	13.8132	32.037	358.52	2018.1
$\int \theta^n \xi^4 d\xi$	12.1567	11.1197	10.6110	10.8516	11.745
$\int \theta^n \psi_0 \xi^2 d\xi$	1.57080	1.3448	1.2276	1.2047	1.3053
$\int \theta^n \psi_2 \xi^2 d\xi$	8.34647	6.6682	5.6428	4.6254	4.4776
$n\int \theta^{n-1} \psi_2 \xi^2 d\xi$	36.2133	35.283	35.599	40.478	46.657
$n\int \theta^{n-1} \psi_0 \xi^4 d\xi$	49.0472	59.437	78.329	186.50	368.06
$n\int \theta^{n-1} \psi_2 \xi^4 d\xi$	241.764	260.03	299.80	519.32	847.15
$-n\int \theta^{n-1} \psi_2 \chi_0' \xi d\xi$	35.8652	29.308	25.258	21.315	20.893

also

$$n\int_0^{\xi_1} \theta^{n-1} \psi_2 \chi_0' \xi d\xi. \tag{AI, 21}$$

APPENDIX II

THE SOLUTION FOR THE CASE $n = 1$

In the case $n = 1$, the solutions for θ, ψ_0, and ψ_2 can be explicitly written down. We have

$$\theta = \frac{\sin \xi}{\xi} = \left(\frac{\pi}{2\xi}\right)^{1/2} J_{1/2}(\xi), \tag{AII, 1}$$

$$\psi_0 = 1 - \frac{\sin \xi}{\xi} = 1 - \theta(\xi), \tag{AII, 2}$$

and

$$\psi_2 = 15 \left(\frac{\pi}{2\xi}\right)^{1/2} J_{5/2}(\xi), \tag{AII, 3}$$

where the $J_{l+1/2}$'s are the usual Bessel functions. The factor 15 in the solution for ψ_2 is to provide the behavior

$$\psi_2 = \xi^2 + O(\xi^4) \qquad \text{as } \xi \to 0 \,, \tag{AII, 4}$$

in common with the solutions for the other values of n.

For the solutions given by equations (AII, 1)–(AII, 3),

$$\begin{aligned} &\xi_1 = \pi ; \qquad \psi_0(\xi_1) = 1 ; \qquad \psi_2(\xi_1) = \frac{45}{\pi^2} ; \\ &\theta'(\xi_1) = -\frac{1}{\pi} ; \qquad \psi_0'(\xi_1) = \frac{1}{\pi} ; \qquad \text{and} \qquad \psi_2'(\xi_1) = \frac{15}{\pi}\left(1 - \frac{9}{\pi^2}\right). \end{aligned} \tag{AII, 5}$$

From equations (26) and (28) we now find

$$c_0 = 1 \qquad \text{and} \qquad A_2 = -\frac{\pi^2}{18} . \tag{AII, 6}$$

The equation satisfied by the superpotential in zero order is

$$\mathfrak{D}_0 \chi_0 = -2\left[1 + \left(\frac{\pi}{2\xi}\right)^{1/2} J_{1/2}(\xi)\right] ; \tag{AII, 7}$$

and we find from this equation that

$$\chi_0' = -2\left[\tfrac{1}{3}\xi + \left(\frac{\pi}{2\xi}\right)^{1/2} J_{3/2}(\xi)\right] . \tag{AII, 8}$$

With the relevant solutions explicitly known, the different integrals which are needed can also be evaluated. We find

$$\begin{aligned} &\int_0^\pi \xi^2 \theta \, d\xi = \pi ; \qquad \int_0^\pi \xi^2 \psi_0 \, d\xi = \pi(\tfrac{1}{3}\pi^2 - 1) , \qquad \int_0^\pi \theta \psi_0 \xi^2 d\xi = \tfrac{1}{2}\pi ; \\ &\int_0^\pi \xi^2 \psi_2 \, d\xi = 15[3\,\mathrm{Si}(\pi) - \pi] , \qquad \int_0^\pi \xi^2 \theta \psi_2 \, d\xi = \tfrac{15}{2}[3\,\mathrm{Si}(2\pi) - \pi] , \\ &\int_0^\pi \xi \psi_2 \chi_0' \xi \, d\xi = -10[3\,\mathrm{Si}(\pi) - \pi] - 15\left[\mathrm{Si}(2\pi) - \frac{2}{\pi}\right] , \end{aligned} \tag{AII, 9}$$

where

$$\mathrm{Si}(x) = \int_0^x \frac{\sin \xi}{\xi} \, d\xi . \tag{AII, 10}$$

APPENDIX III

Tables of the Functions ψ_0, ψ_2, and χ_0'

The structure of the distorted polytropes is described, apart from θ, by the functions ψ_0 and ψ_2. These functions were tabulated for some values of n in Paper I; but their accuracy was limited and proved insufficient for the purposes of this paper. Also, the derivative, χ_0', of the superpotential is needed for the evaluation of the supermatrix elements.

We are grateful to Miss Donna D. Elbert for undertaking the task of integrating the relevant differential equations for $n = 1.5$, 2, 3, and 3.5. The results of her integrations[5] are given in Table 6.

The solutions for the case $n = 1$ are explicitly known (see Appendix II); but for completeness they are also included in Table 6.

[5] The equations for $\chi_{1;0}$ and $\chi_{1;2}$ were also integrated; but these functions are not included in Table 6, as they are not strictly needed in this paper. They are, however, available if a need for them should arise.

TABLE 6

THE RADIAL FUNCTIONS ψ_0 AND ψ_2 DESCRIBING THE DISTORTED POLYTROPE AND THE DERIVATIVE, χ_0', OF THE SUPERPOTENTIAL IN ZERO ORDER ($n = 1.0$)

ξ	ψ_0	ψ_2	χ_0'
	$n = 1.0$		
0	0	0	0
0.1	0.001666	0.009993	−0.133267
0.2	0.006653	0.039886	−0.266134
0.3	0.014933	0.089423	−0.398206
0.4	0.026454	0.158180	−0.529091
0.5	0.041149	0.245567	−0.658407
0.6	0.058929	0.350835	−0.785784
0.7	0.079689	0.473082	−0.910863
0.8	0.103305	0.611258	−1.033304
0.9	0.129637	0.764177	−1.152785
1.0	0.158529	0.930526	−1.269004
1.1	0.189812	1.108873	−1.381683
1.2	0.223301	1.297683	−1.490569
1.3	0.258801	1.495328	−1.595436
1.4	0.296107	1.700104	−1.696084
1.5	0.335003	1.910239	−1.792346
1.6	0.375266	2.123914	−1.884083
1.7	0.416668	2.339274	−1.971188
1.8	0.458974	2.554444	−2.053587
1.9	0.501947	2.767548	−2.131237
2.0	0.545351	2.976719	−2.204129
2.1	0.588948	3.180119	−2.272284
2.2	0.632502	3.375949	−2.335757
2.3	0.675780	3.562472	−2.394634
2.4	0.718557	3.738017	−2.449030
2.5	0.760611	3.901001	−2.499093
2.6	0.801730	4.049938	−2.544994
2.7	0.841711	4.183451	−2.586934
2.8	0.880361	4.300286	−2.625139
2.9	0.917500	4.399318	−2.659856
3.0	0.952960	4.479562	−2.691355
3.1	0.986587	4.540184	−2.719924
ξ_1	1.000000	4.559453	−2.731015

ξ	ψ_0	ψ_2	χ_0'
	$n = 1.5$		
0	0	0	0
0.1	0.001665	0.009989	−0.116121
0.2	0.006647	0.039829	−0.231843
0.3	0.014899	0.089138	−0.346771
0.4	0.026351	0.157289	−0.460520
0.5	0.040900	0.243426	−0.572715
0.6	0.058423	0.346477	−0.682999
0.7	0.078773	0.465183	−0.791034
0.8	0.101785	0.598122	−0.896506
0.9	0.127279	0.743736	−0.999125
1.0	0.155061	0.900368	−1.098631
1.1	0.184932	1.066288	−1.194794
1.2	0.216690	1.239734	−1.287413
1.3	0.250132	1.418936	−1.376322
1.4	0.285058	1.602153	−1.461386
1.5	0.321281	1.787699	−1.542505
1.6	0.358619	1.973973	−1.619609
1.7	0.396909	2.159479	−1.692660
1.8	0.436003	2.342850	−1.761653
1.9	0.475775	2.522862	−1.826607
2.0	0.516119	2.698450	−1.887572
2.1	0.556953	2.868718	−1.944621
2.2	0.598220	3.032946	−1.997849
2.3	0.639890	3.190594	−2.047373
2.4	0.681958	3.341303	−2.093326
2.5	0.724450	3.484898	−2.135857
2.6	0.767420	3.621380	−2.175127
2.7	0.810950	3.750931	−2.211307
2.8	0.855158	3.873909	−2.244576
2.9	0.900191	3.990844	−2.275117
3.0	0.946237	4.102446	−2.303118
3.1	0.993521	4.209605	−2.328765
3.2	1.042318	4.313414	−2.352245
3.3	1.092963	4.415190	−2.373741
3.4	1.145866	4.516536	−2.393428
3.5	1.201552	4.619456	−2.411479
3.6	1.260744	4.726644	−2.4281
ξ_1	1.294397	4.787439	−2.436408

ξ	ψ_0	ψ_2	χ_0'
	$n = 2.0$		
0	0	0	0
0.1	0.001665	0.009986	−0.103527
0.2	0.006640	0.039772	−0.206654
0.3	0.014867	0.088855	−0.308991
0.4	0.026249	0.156412	−0.410154
0.5	0.040657	0.241334	−0.509779
0.6	0.057935	0.342260	−0.607521
0.7	0.077902	0.457627	−0.703060
0.8	0.100361	0.585717	−0.796108
0.9	0.125105	0.724711	−0.886404
1.0	0.151922	0.872741	−0.973724
1.1	0.180604	1.027942	−1.057877
1.2	0.210951	1.188498	−1.138710
1.3	0.242778	1.352682	−1.216100
1.4	0.275916	1.518896	−1.289965
1.5	0.310219	1.685689	−1.360250
1.6	0.345564	1.851788	−1.426935
1.7	0.381854	2.016100	−1.490027
1.8	0.419020	2.177730	−1.549560
1.9	0.457018	2.335972	−1.605592
2.0	0.495833	2.490306	−1.658200
2.1	0.535473	2.640394	−1.707480
2.2	0.575972	2.786064	−1.753544
2.3	0.617386	2.927297	−1.796516
2.4	0.659793	3.064211	−1.836527
2.5	0.703290	3.197045	−1.873719
2.6	0.747992	3.326145	−1.908236
2.7	0.794027	3.451946	−1.940226
2.8	0.841541	3.574959	−1.969840
2.9	0.890692	3.695759	−1.997224
3.0	0.941648	3.814969	−2.022526
3.1	0.994589	3.933255	−2.045890
3.2	1.049706	4.051314	−2.067453
3.3	1.107197	4.169867	−2.087352
3.4	1.167272	4.289655	−2.105713
3.5	1.230148	4.411434	−2.122659
3.6	1.296051	4.535970	−2.138307
3.7	1.365218	4.664042	−2.152765
3.8	1.437894	4.796436	−2.166136
3.9	1.514336	4.933949	−2.178514
4.0	1.594813	5.077389	−2.189986
4.1	1.679603	5.227577	−2.200636
4.2	1.769001	5.385350	−2.210536
4.3	1.863316	5.551564	−2.21975
ξ_1	1.915281	5.643158	−2.22437

ξ	ψ_0	ψ_2	χ_0'
	$n = 3.0$		
0	0	0	0
0.1	0.001664	0.009979	−0.086109
0.2	.006627	0.039660	−0.171820
0.3	.014802	0.088297	−0.256744
0.4	.026050	0.154695	−0.340507
0.5	.040190	0.237286	−0.422761
0.6	.057011	0.334217	−0.503186
0.7	.076282	0.443448	−0.581497
0.8	.097766	0.562860	−0.657448
0.9	.121234	0.690341	−0.730833
1.0	.146471	0.823874	−0.801485
1.1	.173286	0.961602	−0.869279
1.2	.201521	1.101873	−0.934128
1.3	.231048	1.243275	−0.995981
1.4	.261775	1.384644	−1.054818
1.5	.293645	1.525065	−1.110651
1.6	.326634	1.663861	−1.163514
1.7	.360745	1.800572	−1.213465
1.8	.396009	1.934930	−1.260579
1.9	.432481	2.066837	−1.304945
2.0	.470232	2.196328	−1.346663
2.1	.509350	2.323556	−1.385842
2.2	.549934	2.448756	−1.422597
2.3	.592091	2.572234	−1.457043
2.4	.635936	2.694340	−1.489302
2.5	0.681586	2.815456	−1.519490

TABLE 6—*Continued*

ξ	ψ_0	ψ_2	χ_0'
2.6	0.729160	2.935980	−1.547725
2.7	0.778779	3.056319	−1.574123
2.8	0.830563	3.176878	−1.598794
2.9	0.884629	3.298054	−1.621847
3.0	0.941092	3.420232	−1.643384
3.1	1.000064	3.543783	−1.663505
3.2	1.061656	3.669058	−1.682302
3.3	1.125972	3.796391	−1.699864
3.4	1.193115	3.926094	−1.716276
3.5	1.263181	4.058462	−1.731616
3.6	1.336266	4.193768	−1.745959
3.7	1.412459	4.332269	−1.759373
3.8	1.491847	4.474203	−1.771925
3.9	1.574512	4.619788	−1.783674
4.0	1.660534	4.769230	−1.794678
4.1	1.749987	4.922717	−1.804989
4.2	1.842944	5.080424	−1.814656
4.3	1.939471	5.242510	−1.823725
4.4	2.039634	5.409125	−1.832239
4.5	2.143494	5.580405	−1.840237
4.6	2.251109	5.756475	−1.847755
4.7	2.362532	5.937450	−1.854827
4.8	2.477814	6.123434	−1.861485
4.9	2.597004	6.314523	−1.867757
5.0	2.720146	6.510805	−1.873672
5.1	2.847279	6.712357	−1.879252
5.2	2.978443	6.919249	−1.884523
5.3	3.113670	7.131544	−1.889504
5.4	3.252992	7.349298	−1.894216
5.5	3.396437	7.572558	−1.898676
5.6	3.544027	7.801364	−1.902903
5.7	3.695784	8.035750	−1.906910
5.8	3.851724	8.275744	−1.910714
5.9	4.011860	8.521365	−1.914326
6.0	4.176202	8.772626	−1.917760
6.1	4.344756	9.029536	−1.921027
6.2	4.517524	9.292093	−1.924138
6.3	4.694504	9.560290	−1.927101
6.4	4.875690	9.834116	−1.929927
6.5	5.061073	10.113550	−1.932623
6.6	5.250638	10.398564	−1.935198
6.7	5.444368	10.689127	−1.937658
6.8	5.642240	10.985196	−1.940011
ξ_1	5.837801	11.277139	−1.942192

$n = 3.5$

ξ	ψ_0	ψ_2	χ_0'
0	0	0	0
0.1	0.001664	0.009975	−0.079817
0.2	.006621	0.039604	−0.159237
0.3	.014770	0.088021	−0.237872
0.4	.025952	0.153855	−0.315352
0.5	.039965	0.235328	−0.391338
0.6	.056572	0.330378	−0.465520
0.7	.075527	0.436787	−0.537630
0.8	.096582	0.552305	−0.607442
0.9	.119505	0.674760	−0.674772
1.0	.144093	0.802148	−0.739478
1.1	.170173	0.932694	−0.801461
1.2	.197611	1.064892	−0.860660
1.3	.226314	1.197518	−0.917048
1.4	.256226	1.329625	−0.970629
1.5	.287326	1.460531	−1.021434
1.6	.319626	1.589786	−1.069515
1.7	.353163	1.717144	−1.114942
1.8	.387999	1.842529	−1.157800
1.9	.424210	1.966001	−1.198184
2.0	.461890	2.087725	−1.236198
2.1	.501138	2.207946	−1.271950
2.2	.542064	2.326963	−1.305549
2.3	0.584777	2.445113	−1.337107
2.4	0.629391	2.562753	−1.366736
2.5	0.676018	2.680247	−1.394544
2.6	0.724767	2.797959	−1.420637
2.7	0.775748	2.916244	−1.445118
2.8	0.829065	3.035444	−1.468085
2.9	0.884818	3.155885	−1.489635
3.0	0.943104	3.277875	−1.509854
3.1	1.004015	3.401700	−1.528829
3.2	1.067640	3.527629	−1.546642
3.3	1.134061	3.655909	−1.563365
3.4	1.203357	3.786769	−1.579073
3.5	1.275604	3.920420	−1.593830
3.6	1.350872	4.057056	−1.607701
3.7	1.429229	4.196852	−1.620744
3.8	1.510736	4.339972	−1.633012
3.9	1.595455	4.486564	−1.644559
4.0	1.683440	4.636763	−1.655431
4.1	1.774746	4.790692	−1.665673
4.2	1.869421	4.948466	−1.675327
4.3	1.967513	5.110186	−1.684430
4.4	2.069066	5.275948	−1.693021
4.5	2.174121	5.445837	−1.701130
4.6	2.282717	5.619931	−1.708790
4.7	2.394891	5.798302	−1.716030
4.8	2.510676	5.981016	−1.722877
4.9	2.630106	6.168132	−1.729356
5.0	2.753208	6.359704	−1.735490
5.1	2.880012	6.555782	−1.741301
5.2	3.010543	6.756412	−1.746809
5.3	3.144825	6.961634	−1.752034
5.4	3.282880	7.171486	−1.756992
5.5	3.424730	7.386001	−1.761701
5.6	3.570393	7.605211	−1.766175
5.7	3.719885	7.829144	−1.770429
5.8	3.873224	8.057823	−1.774476
5.9	4.030422	8.291272	−1.778329
6.0	4.191494	8.529510	−1.781998
6.1	4.356450	8.772555	−1.785495
6.2	4.525301	9.020423	−1.788830
6.3	4.698055	9.273127	−1.792011
6.4	4.874720	9.530678	−1.795049
6.5	5.055302	9.793087	−1.797950
6.6	5.239806	10.060362	−1.800724
6.7	5.428237	10.332509	−1.803376
6.8	5.620597	10.609533	−1.805914
6.9	5.816889	10.891437	−1.808344
7.0	6.017112	11.178224	−1.810672
7.1	6.221268	11.469894	−1.812903
7.2	6.429355	11.766447	−1.815042
7.3	6.641371	12.067882	−1.817095
7.4	6.857314	12.374194	−1.819066
7.5	7.077179	12.685381	−1.820958
7.6	7.300962	13.001438	−1.822777
7.7	7.528659	13.322358	−1.824525
7.8	7.760262	13.648135	−1.826207
7.9	7.995766	13.978762	−1.827826
8.0	8.235163	14.314229	−1.829384
8.1	8.478444	14.654526	−1.830885
8.2	8.725602	14.999644	−1.832332
8.3	8.976628	15.349572	−1.833726
8.4	9.231511	15.704299	−1.835071
8.5	9.490241	16.063812	−1.836369
8.6	9.752809	16.428099	−1.837622
8.7	10.019202	16.797147	−1.838832
8.8	10.289410	17.170943	−1.840001
8.9	10.563420	17.549473	−1.841130
9.0	10.841222	17.932722	−1.842223
9.1	11.122803	18.320677	−1.843279
9.2	11.408150	18.713322	−1.844301
9.3	11.697251	19.110643	−1.845291
9.4	11.990094	19.512626	−1.846249
9.5	12.286666	19.919256	−1.847177
ξ_1	12.39376	20.06598	−1.847502

The only remark that needs to be made about the accompanying tabulation is that at $\xi = 0$, the functions ψ_2 have the behavior

$$\psi_2 = \xi^2 + O(\xi^4) \qquad (\xi \to 0).$$

The values of the different functions and their derivatives are needed in many calculations. For convenience, they are listed separately in Table 7; this table also includes the values of the constants enumerated in equations (26)–(29).

TABLE 7

A TABLE OF CONSTANTS*

Constant	$n=1.0$	$n=1.5$	$n=2.0$	$n=3.0$	$n=3.5$
ξ_1	3.141593	3.653754	4.352875	6.896849	9.535805
$c_0=-\xi_1\theta_1'$	1.000000	0.742813	0.553897	0.292632	0.198259
$\psi_0(\xi_1)$	1.000000	1.29440	1.91528	5.83780	12.3938
$\psi_0'(\xi_1)$	0.318310	0.64074	0.99698	2.0391	2.9976
$\psi_0''(\xi_1)$	....	0.64927	0.54192	0.40869	0.37128
$\psi_2(\xi_1)$	4.559453	4.78744	5.64316	11.27714	20.0660
$\psi_2'(\xi_1)$	0.420691	1.16154	1.7578	3.0407	4.1061
$\psi_2''(\xi_1)$	....	1.51586	0.97935	0.54072	0.46283
$-\chi_0'(\xi_1)$	2.73101	2.43641	2.2244	1.9422	1.8475
$-\chi_0''(\xi_1)$	0.261382	0.151979	0.085771	0.022052	0.0090303
$c_{1;0}$	2.934802	3.0395	3.2188	3.8822	4.4870
$-A_2$	0.548311	0.597912	0.642354	0.72330	0.762698
κ	3.141593	2.71406	2.41105	2.01824	1.89056
$\kappa_{1;0}$	7.193833	7.7053	8.6018	12.3610	16.454
$-\kappa_{1;2}$	26.51241	31.095	38.515	75.125	129.2
Δ_2	1.51982	1.28651	1.14788	1.02889	1.00983
$-(\delta\lambda/\lambda)_M$	2.28987	5.67808	14.2707	....	−121.847

* The constants $c_{1;0}$, A_2, κ, $\kappa_{1;0}$, and $\kappa_{1;2}$ are defined in eqs. (26)–(29); the additional constants Δ_2 and $(\delta\lambda/\lambda)_M$ are

$$\Delta_2 = \frac{5\psi_2(\xi_1)}{3\psi_2(\xi_1)+\xi_1\psi_2'(\xi_1)} \quad \text{and} \quad \left(\frac{\delta\lambda}{\lambda}\right)_M = \frac{2n}{3-n}\,\frac{\frac{1}{3}\xi_1 - \psi_0'(\xi_1)}{\theta'(\xi_1)}.$$

REFERENCES

Chandrasekhar, S. 1933, *M.N.*, **93**, 390 (this paper will be referred to as "Paper I").
———. 1939, *An Introduction to the Study of Stellar Structure* (Chicago: University of Chicago Press; Dover reprint, 1957).
———. 1950, *Phil. Trans. R. Soc.*, **242**, 557.
———. 1961, *Ap. J.*, **134**, 662.
Chandrasekhar, S., and Lebovitz, N. R. 1962*a*, *Ap. J.*, **135**, 238 (this paper will be referred to as "Paper II").
———. 1962*b*, *ibid.*, p. 248 (this paper will be referred to as "Paper III").
———. 1962*c*, *ibid.*, p. 305.
———. 1962*d*, *ibid.*, **136**, 1037.
———. 1962*e*, *ibid.*, p. 1069.
Cowling, T. G., and Newing, R. A. 1949, *Ap. J.*, **109**, 149.
Jardetzky, W. S. 1958, *Theories of Figures of Celestial Bodies* (New York: Interscience Publishers).
Ledoux, P. 1945, *Ap. J.*, **102**, 143.
Milne, E. A. 1929, *M.N.*, **89**, 739.

22

ON THE OCCURRENCE OF MULTIPLE FREQUENCIES AND BEATS IN THE β CANIS MAJORIS STARS

S. CHANDRASEKHAR AND NORMAN R. LEBOVITZ
University of Chicago and Massachusetts Institute of Technology
Received July 7, 1962

ABSTRACT

An explanation is suggested for the occurrence of two nearly equal frequencies and associated beats in the light- and in the velocity-variations of the β Canis Majoris stars. It is shown that if the ratio of the specific heats γ is 1.6 and the star is rotating, any disturbance will excite two normal modes with nearly equal frequencies.

I. INTRODUCTION

In the two preceding papers (Chandrasekhar and Lebovitz 1962*c*, *d;* these papers will be referred to hereafter as "Paper I" and "Paper II," respectively) we have determined the effect of rotation on the different modes of oscillation of a homogeneous compressible mass and of a distorted polytrope. The detailed results of these two papers confirm, what had in fact been disclosed by the general theory (Chandrasekhar and Lebovitz 1962*a*), that a rotation couples two modes of oscillation which are, in the absence of rotation, purely radial and purely non-radial and, further, that a ratio of the specific heats $\gamma = 1.6$ plays a critical role in this phenomenon. The explanation which we wish to suggest for the occurrence of multiple periods and beats in the β Canis Majoris stars is on this theoretical basis; it clarifies and extends an earlier suggestion of ours (Chandrasekhar and Lebovitz 1962*b*). But first we shall briefly describe the nature of the coupling which is predicted and the origin of the critical role of $\gamma = 1.6$.

II. THE CASE WHEN ROTATION IS ABSENT

For the sake of simplicity we shall describe the phenomena in the framework of the approximate theory in which the Lagrangian displacement ξ is assumed to be a linear function of the co-ordinates. The arguments can be made more exact for the homogeneous model considered in Paper I, and they can be made more general for the others; but these refinements are not necessary for our present purposes.

The two modes of oscillation which are coupled by rotation are, in the absence of rotation, characterized by the displacement

$$\xi_1 = X_R x_1 , \qquad \xi_2 = X_R x_2 , \qquad \xi_3 = X_R x_3 \qquad (R\text{-mode}), \quad (1)$$

and

$$\xi_1 = X_S x_1 , \qquad \xi_2 = X_S x_2 , \qquad \xi_3 = -2 X_S x_3 \qquad (S\text{-mode}), \quad (2)$$

where X_R and X_S are two arbitrary constants. The corresponding characteristic frequencies are (in the same approximation)

$$\sigma_R^2 = (3\gamma - 4)\frac{|\mathfrak{W}|}{I} \qquad \text{and} \qquad \sigma_S^2 = \frac{4}{5}\frac{|\mathfrak{W}|}{I}, \quad (3)$$

where $\mathfrak{W}$ is the gravitational potential energy and I is the moment of inertia of the configuration.

The "radial" character of the displacement belonging to the R-mode and the solenoidal character of the displacement belonging to the S-mode are apparent from equations (1) and (2).

We notice that

$$\sigma_R^2 = \sigma_S^2 \qquad \text{if} \qquad \gamma = 1.6 . \quad (4)$$

Therefore, when $\gamma = 1.6$, the two characteristic frequencies coincide; and we have a case of degeneracy. Accordingly, in this case any linear combination of the proper solutions represented by equations (1) and (2) can be considered as belonging to the common characteristic frequency. This is the origin of the critical role of $\gamma = 1.6$.

III. THE CASE WHEN ROTATION IS PRESENT

Now when the configuration is set in rotation, the R- and S-modes (as we have designated them) get coupled, and the corresponding Lagrangian displacements are (cf. Paper II, eq. [98])

$$\xi_1 = X_R x_1 + Y_R x_2 , \qquad \xi_2 = X_R x_2 - Y_R x_1 , \qquad \xi_3 = Z_R x_3 , \tag{5}$$

and

$$\xi_1 = X_S x_1 + Y_S x_2 , \qquad \xi_2 = X_S x_2 - Y_S x_1 , \qquad \xi_3 = Z_S x_3 , \tag{6}$$

where

$$\frac{X_R}{Z_R} \neq 1 , \qquad \frac{X_S}{Z_S} \neq -\tfrac{1}{2} \qquad (\Omega \neq 0), \tag{7}$$

$$Y_R = -2i\frac{\Omega}{\sigma_R} X_R , \qquad Y_S = -2i\frac{\Omega}{\sigma_S} X_S , \tag{8}$$[1]

and σ_R and σ_S are the corresponding characteristic frequencies. A general property of solutions (5) and (6) is (cf. Paper I, eq. [26])

$$\left(\frac{X_R}{Z_R}\right)\left(\frac{X_S}{Z_S}\right) = -\tfrac{1}{2} + O(\Omega^2) \qquad (\Omega \to 0) . \tag{9}$$

So long as $\gamma \neq 1.6$, the proper solutions for the two modes given by equations (5) and (6) reduce to those given by equations (1) and (2) when $\Omega \to 0$. But this does not happen when $\gamma = 1.6$: in this case, the proper solutions belonging to the two *distinct* frequencies of oscillation (when Ω is different from zero), in the limit $\Omega = 0$, become determinate linear combinations of the solutions represented by equations (1) and (2). Thus, for the homogeneous compressible model (for which the foregoing arguments can be made exact; see Paper I, eqs. [50] and [51]),

$$\frac{X_R}{Z_R} = 2.63 , \quad \frac{X_S}{Z_S} = -0.185 , \quad \text{and} \quad Y_R = Y_S = 0 \qquad (\Omega = 0) ; \tag{10}$$

and the corresponding limits are not very different for the distorted polytropes (see Paper II, Table 4*B*). In other words, neither of the proper solutions belonging to the two distinct modes which obtain when $\Omega \neq 0$ become even approximately radial when $\Omega \to 0$: indeed, they are very far from being so.

IV. THE SUGGESTED EXPLANATION

The theoretical results summarized in Sections II and III suggest an explanation for the phenomenon of multiple periods and beats which is observed in the light- and in the velocity-variations of the β Canis Majoris stars (cf. Ledoux and Walraven 1958, Sec. 25). Earlier attempts (cf. Ledoux 1951) to explain the same phenomenon postulated the excitation of non-radial modes (*besides* the radial modes) under the influence of rotation. But one is generally reluctant to accept suggestions which appeal *directly* to the excitation of non-radial modes (besides the radial modes) on the grounds that such modes should be highly damped relative to the radial modes and, further, that their excitation would be "difficult" in view of the possible source of such excitation being in the deep interior. These arguments do not apply to the present interpretation, which is based on

[1] The occurrence of the imaginary in these relations signifies only that the terms in the Y's, in eqs. (5) and (6), oscillate 90° out of phase relative to the terms in the X's and the Z's.

the assumption that, for the stars in question, $\gamma = 1.6$; for, on this assumption, the degeneracy of the unperturbed state ($\Omega = 0$) makes indeterminate the specification of the proper solution belonging to the fundamental mode of oscillation; and the removal of this degeneracy, by the slightest amount of rotation, leads to two normal modes *neither* of which is even remotely radial. The excitation of *both* modes by any disturbance (spherically symmetric or otherwise) is not only to be expected: it is natural under the circumstances. We can "picture" what will happen as follows.

Consider a slowly rotating gaseous mass and suppose that it experiences a disturbance which has approximate spherical symmetry. The disturbance will set the mass into oscillations; and if $\gamma \neq 1.6$, only those normal modes which have the character of radial pulsations (approximately) will be excited (in agreement with one's normal expectations). However, when $\gamma = 1.6$, what will happen is quite different; for, in this case, the analysis of the original disturbance into normal modes must include the lowest modes, and neither of them, as we have seen, has a radial character. The excitation of the lowest modes with slightly different frequencies would appear to be inescapable, and, with their excitation, beats must ensue.

While we have supposed that $\gamma = 1.6$, an *exact* coincidence is not necessary for the foregoing explanation to be valid: it should clearly suffice if

$$|\gamma - 1.6| = O(e^2); \tag{11}$$

for then the difference in the unperturbed values of σ_R^2 and σ_S^2 would be "masked" by the displacements that each will undergo on account of rotation. It should be remembered in this connection that, in a star, γ (allowing for the effects of radiation pressure and ionization) is not, strictly, a constant: it will be variable, and for our present purposes it should suffice if γ takes the value 1.6 somewhere in the interior, so that condition (11) will be satisfied over a substantial part of the star.

It remains to verify that the assumption of $\gamma = 1.6$ is a reasonable one for the β Canis Majoris stars. Now it is known that, for stars of solar mass, γ is very nearly $\frac{5}{3}$ throughout the interior; and that, for stars of larger mass, γ will be less, on account of the greater importance of radiation pressure. It can be estimated[2] on the basis of our current ideas on the physical conditions in the interiors of stars that a value of $\gamma = 1.6$ will be attained among stars with masses 7–13$\odot$; this is eminently reasonable for the β Canis Majoris stars, which have spectral types and luminosity classes in the range B1 III–B2 IV.

Finally, it should perhaps be stated that the particular value $\gamma = 1.6$ has been deduced on the basis of a theory which is only approximate for configurations which are not strictly homogeneous. An exact theory may lead to a somewhat different value for γ, at which the kind of accidental degeneracy we have considered occurs. And the present interpretation will hold whenever such a degeneracy occurs.

The research reported in this paper has in part been supported by the Office of Naval Research under contract Nonr-2121(24) with the University of Chicago. The work of the second author was supported in part by the United States Air Force under contract No. AF-49(638)-42, monitored by the Air Force Office of Scientific Research of the Air Research and Development Command.

REFERENCES

Chandrasekhar, S., and Lebovitz, N. R. 1962*a*, *Ap. J.*, **135**, 248.
———. 1962*b*, *ibid.*, p. 305.
———. 1962*c*, *ibid.* **136**, 1069.
———. 1962*d*, *ibid.*, p. 1082.
Ledoux, P. 1951, *Ap. J.*, **114**, 373.
Ledoux, P., and Walraven, Th. 1958, *Hdb. d. Phys.*, ed. S. Flügge (Berlin: Springer Verlag), **51**, 398.

[2] We are indebted to Professor Nelson Limber for making this estimate for us.

NON-RADIAL OSCILLATIONS AND CONVECTIVE INSTABILITY OF GASEOUS MASSES

S. CHANDRASEKHAR AND NORMAN R. LEBOVITZ
University of Chicago and Massachusetts Institute of Technology
Received February 18, 1963

ABSTRACT

Modes of non-radial oscillation of gaseous masses belonging to spherical harmonics of orders $l = 1$ and 3 are considered on the basis of the first- and the third-order virial equations. For an assumed Lagrangian displacement $\boldsymbol{\xi}$ of the form

$$\xi_i = (L_{i;jk}x_jx_k + L_i)e^{i\sigma t}$$

(where $L_{i;jk}$ and L_i represent a total of twenty-one unspecified constants and σ is the characteristic frequency to be determined), the theory predicts the occurrence of modes of oscillation of two different types: modes (belonging to $l = 3$) which are analogous to the Kelvin modes of an incompressible sphere and modes (belonging to $l = 1$) which are analogous to those discovered by Pekeris for a homogeneous compressible sphere and which exhibit its convective instability. For the latter modes, the virial equations lead to a characteristic equation for σ^2 of degree 2 whose coefficients are integrals over the variables of the unperturbed configuration, including its superpotential. The theory is applied to the polytropic gas spheres, and it is shown that they are convectively unstable (for the modes belonging to $l = 1$) if the ratio of the specific heats γ is less than a certain critical value. The critical values of γ predicted by the (approximate) theory differ from $1 + 1/n$ (where n is the polytropic index) by less than 1 per cent over the range of n (≤ 3.5) considered; the extent of this agreement is a measure of the accuracy of the method based on the virial equations and the assumed form of the Lagrangian displacement.

I. INTRODUCTION

As Cowling (1942) and Ledoux (cf. Ledoux and Walraven 1958) have emphasized, a theory of the non-radial oscillations of gaseous masses is relevant in two different connections: in discovering modes of oscillation which may lead to "explosive instability" (particularly under external influences such as tidal action) and in establishing criteria for convective stability on a sound theoretical basis. It is with the latter aspect of the theory of non-radial oscillations that we shall be concerned in this paper.

The criterion for convective stability as commonly applied is that of Karl Schwarzschild, which requires the prevailing temperature gradient, at every point, not to exceed the adiabatic gradient. The manner in which one generally establishes this criterion is to examine how a small isolated element will react to a fluctuation which makes its physical state slightly different from that of its immediate environment. While the arguments as usually presented are reasonable, they are, really, no substitute for a proper treatment of the stability of the system in terms of normal modes and initial conditions: a system reacts to a perturbation *in toto* and is indivisible. From this strict point of view, the qualitative arguments which are used to validate the Schwarzschild criterion are no more than a suggestion that instability via modes of sufficiently high order (i.e., modes belonging to spherical harmonics of high orders l and radial functions with many nodes) will arise if the criterion is violated in any small isolated region. On this account, Cowling and Ledoux have attempted to establish the Schwarzschild criterion by exploring methods of approximation which may be suitable for treating these modes of non-radial oscillation which have many nodes both in the transverse and in the radial directions. The principal approximation underlying these treatments is the neglect of the change in the gravitational potential caused by the perturbation. While this approximation may be a good one[1] for treating modes of oscillations of high orders, it is hardly one which can

[1] See, however, the remarks in the last Sec. VII.

be applied to modes belonging to $l = 1$; and these are the modes that are most relevant for the discernment of convective instability in many important cases. Thus, Pekeris' (1939) first exact treatment of the non-radial adiabatic oscillations revealed the instability of a homogeneous compressible sphere for modes belonging to all l's including $l = 1$. The instability in this instance clearly derives from the fact that *a uniform density is superadiabatic everywhere for any finite ratio of the specific heats* γ. That, under these circumstances, instability should arise already for $l = 1$ is entirely to be expected: it is, for example, the mode by which convective instability first manifests itself in a *viscous* fluid sphere heated within (Chandrasekhar 1961). What is true of the homogeneous sphere must be equally true of the polytropic gas spheres in the sense that when they become convectively unstable, they must also do so for modes of oscillation belonging to $l = 1$. This is apparent when it is noted that a polytropic distribution in which the pressure and the density are related by

$$p = \text{constant } \rho^{1+1/n} \tag{1}$$

is superadiabatic everywhere if

$$\gamma < 1 + \frac{1}{n}. \tag{2}$$

Consequently, if $1 + 1/n$ should exceed γ, the Schwarzschild criterion will be violated simultaneously throughout the entire mass, and the instability can assert itself with the largest permissible pattern of circulation, i.e., by a mode of oscillation belonging to $l = 1$. We may conclude, then, that *a polytrope of index* n *will be convectively unstable for* $\gamma < 1 + 1/\mathrm{n}$ *and that the instability will be manifested already by a mode of oscillation belonging to* $\mathrm{l} = 1$. But this fundamental result has never been properly established.

In this paper we shall develop a method based on the third-order virial equations (see Chandrasekhar 1962; this paper will be referred to hereafter as "Paper I") which will enable us to treat modes of oscillation belonging to $l = 1$ and 3, taking full account of the variations in the gravitational potential during the oscillations. The linearized form of the virial equations permits exact and explicit solutions of problems associated with homogeneous masses. Thus, the problem of the oscillations and the stability of the Jacobi ellipsoids has recently been solved with their aid (Chandrasekhar and Lebovitz 1963*a;* see also the further paper 1963*b* on the Maclaurin spheroids; these papers will be referred to hereafter as "Papers II" and "III," respectively). And we shall see in Section III below that the Pekeris instability of the homogeneous compressible sphere for $l = 1$ can be derived equally with their aid. However, configurations which are not homogeneous cannot be treated exactly with the virial equations only; but they do provide a basis for an approximative treatment. And since the answer may be considered "known" in the case of the polytropes, the application of the method to them will provide a useful test of the precision of the method.

II. THE VIRIAL EQUATIONS OF THE THIRD ORDER FOR THE TREATMENT OF THE OSCILLATIONS OF A SPHERICAL DISTRIBUTION OF MASS

For departures from equilibrium described by a Lagrangian displacement of the form

$$\boldsymbol{\xi}(\boldsymbol{x})e^{i\sigma t}, \tag{3}$$

where σ denotes the characteristic frequency to be determined, the virial equations of the third order give (cf. Paper I, eq. [25])

$$-\sigma^2 V_{i;jk} = \delta\mathfrak{W}_{ij;k} + \delta\mathfrak{W}_{ik;j} + \delta_{ij}\delta\Pi_k + \delta_{ik}\delta\Pi_j, \tag{4}$$

where

$$V_{i;jk} = \int_V \rho\xi_i x_j x_k d\boldsymbol{x} \tag{5}$$

defines the third-order virial and $\delta\mathfrak{W}_{ij;k}$ and $\delta\Pi_k$ are the first variations (due to the displacement $\boldsymbol{\xi}$) of the quantities

$$\mathfrak{W}_{ij;k} = -\tfrac{1}{2}\int_V \rho\mathfrak{B}_{ij}x_k d\boldsymbol{x} \quad \text{and} \quad \Pi_k = \int_V p x_k d\boldsymbol{x}\,, \tag{6}$$

which have been defined in Paper I.

We have (Paper I, eq. [72])

$$-2\,\delta\mathfrak{W}_{ij;k} = \int_V \rho\mathfrak{B}_{ij}\xi_k d\boldsymbol{x} + \int_V \rho\xi_l \frac{\partial\mathfrak{B}_{ij}}{\partial x_l} x_k d\boldsymbol{x} + \int_V \rho\xi_l \frac{\partial\mathfrak{D}_{ij;k}}{\partial x_l} d\boldsymbol{x}\,, \tag{7}$$

where $\mathfrak{B}_{ij}$ is the tensor potential and $\mathfrak{D}_{ij;k}$ is the same tensor potential for the fictitious density distribution ρx_k.

If the oscillations are assumed to take place adiabatically with a ratio of the specific heats γ, then

$$\delta\left(\frac{p}{\rho}\right) = -(\gamma-1)\frac{p}{\rho}\operatorname{div}\boldsymbol{\xi}\,, \tag{8}$$

and

$$\delta\Pi_k = \delta\int_V \frac{p}{\rho} x_k \rho d\boldsymbol{x} = -(\gamma-1)\int_V p x_k \operatorname{div}\boldsymbol{\xi}\, d\boldsymbol{x} + \int_V p\xi_k d\boldsymbol{x}\,. \tag{9}$$

An alternative form of $\delta\Pi_k$, which we obtain after an integration by parts, is

$$\delta\Pi_k = (\gamma-1)\int_V x_k \boldsymbol{\xi}\cdot\operatorname{grad} p\, d\boldsymbol{x} + \gamma\int_V p\xi_k d\boldsymbol{x}\,. \tag{10}$$

In treating the eighteen equations represented by equation (4), we shall find it convenient to introduce the symmetrized virial,

$$V_{ijk} = V_{i;jk} + V_{j;ki} + V_{k;ij}\,, \tag{11}$$

and combine and group the equations into different non-combining sets. Such a grouping was accomplished in Paper II for the more general system of equations governing a uniformly rotating configuration with no special symmetry (except that implied by the rotation itself). In the present instance when there is no rotation and the unperturbed configuration is spherically symmetric, the desired reduction is, of course, much simpler. However, in writing the different equations, we shall refer to the equations in Papers II and III and specialize them appropriately; but they can be derived quite readily, *ab initio*, from equation (4).

(*a*) *First*, we have *six* equations of the form (cf. Paper II, eqs. [62], [76], [77], [96], and [97])

$$\sigma^2(V_{122} - \tfrac{1}{3}V_{111}) = \delta S_{122}\,, \tag{12}$$

where

$$\delta S_{122} = -4\delta\mathfrak{W}_{12;2} - 2\delta\mathfrak{W}_{22;1} + 2\delta\mathfrak{W}_{11;1}\,. \tag{13}$$

The other five equations of this set can be written down by selecting other pairs of indices $(i, j, i \neq j)$ besides (1, 2) to which equations (12) and (13) belong.

(*b*) *Second*, we have a set of *three* equations of the form (Paper II, eqs. [30] and [34])

$$\sigma^2(V_{1;22} - V_{1;33}) = -2\delta\mathfrak{W}_{12;2} + 2\delta\mathfrak{W}_{13;3}\,. \tag{14}$$

(*c*) *Third*, we have the *three* equations (Paper II, eqs. [38], [42], [43], and [93])

$$\sigma^2 V_{123} = \delta S_{123}\,, \tag{15}$$

$$\sigma^2(V_{1;23} - V_{2;31}) = 0\,, \quad \text{and} \quad \sigma^2(V_{2;13} - V_{3;12}) = 0\,, \tag{16}$$

where

$$\delta S_{123} = -2\delta\mathfrak{W}_{12;3} - 2\delta\mathfrak{W}_{23;1} - 2\delta\mathfrak{W}_{31;2} = -6\delta\mathfrak{W}_{12;3}\,. \tag{17}$$

In writing equations (16) and (17) we have made use of the fact (which we shall verify later; see eq. [51] below) that, under circumstances of spherical symmetry,

$$\delta\mathfrak{W}_{12;3} = \delta\mathfrak{W}_{23;1} = \delta\mathfrak{W}_{31;2}\,. \tag{18}$$

(*d*) *Fourth*, we have *three pairs* of equations of the form (Paper I, eqs. [30], [34], [48], [51], and [87])

$$\tfrac{1}{3}\sigma^2 V_{111} + 2\delta\mathfrak{W}_{11;1} = -\,2\delta\Pi_1 \tag{19}$$

and

$$\sigma^2(V_{1;22} + V_{1;33}) = -\,2\delta\mathfrak{W}_{12;2} - 2\delta\mathfrak{W}_{13;3}\,. \tag{20}$$

It will be noticed that, among the four groups of equations, only those in the fourth group (*d*) involve the $\delta\Pi_k$'s.

III. THE MODES OF OSCILLATION OF A COMPRESSIBLE HOMOGENEOUS SPHERE BELONGING TO THE HARMONICS OF ORDERS ONE AND THREE

For a homogeneous sphere, the equations of the preceding section can be solved exactly and explicitly: for, in this case, the various $\delta\mathfrak{W}_{ij;k}$'s, δS_{ijk}'s and $\delta\Pi_k$'s can all be expressed linearly in terms of the virials themselves with simple numerical coefficients. The required coefficients for homogeneous ellipsoids and spheroids have been tabulated in Paper II (Tables 1 and 2) and Paper III (Tables 1, 2, and 3). For a sphere, the coefficients are much simplified, since the symbols $A_{ijk\ldots}$ and $B_{ijk\ldots}$, in terms of which they are expressed, depend only on the *number* of the indices and not on what they are. Thus, suppressing a common factor $1/a^3$ and measuring length in units of the radius (a) of the sphere, we have

$$A_i = A_1 = \tfrac{2}{3}\,,\qquad A_{ij} = A_{11} = \tfrac{2}{5}\,,\qquad A_{ijk} = A_{111} = \tfrac{2}{7} \tag{21}$$

and

$$B_i = B_1 = \tfrac{4}{3}\,,\qquad B_{ij} = B_{11} = \tfrac{4}{15}\,,\qquad B_{ijk} = B_{111} = \tfrac{4}{35}\,. \tag{22}$$

The subsidiary symbols $A_{ij;k}$ and $B_{ij;k}$ (Paper II, eq. [123]) have the values

$$A_{ij;k} = A_{11} + A_{111} = \tfrac{24}{35} \quad\text{and}\quad B_{ij;k} = B_{11} + B_{111} = \tfrac{8}{21}\,. \tag{23}$$

With the coefficients listed in Paper III (Tables 1 and 3), we now find

$$\delta S_{122} = 6(B_{11} + B_{111})(V_{122} - \tfrac{1}{3}V_{111}) = \tfrac{16}{7}(V_{122} - \tfrac{1}{3}V_{111}) \tag{24}$$

and

$$\delta S_{123} = 6(B_{11} + B_{111}) = \tfrac{16}{7}V_{123}\,, \tag{25}$$

where a common factor $\pi G\rho$ has been suppressed.

In view of equations (24) and (25), the six equations of group (*a*) and equation (15) of group (*c*) all lead to the same root,

$$\sigma^2 = \tfrac{16}{7}\pi G\rho\,, \tag{26}$$

where the factor $\pi G\rho$ has been restored. The root (26) is of multiplicity 7.

The characteristic root given by equation (26) coincides with the Kelvin frequency for $l = 3$; its multiplicity 7 is in agreement with this identification.

In the equations of the remaining groups (*b*) and (*d*) we must set

$$V_{122} = V_{133} = \tfrac{1}{3}V_{111}\,,\qquad \text{etc.,} \tag{27}$$

to exclude the root (26) and not to be inconsistent with the equations of group (*a*).

In view of the equalities (27), we now have (cf. Paper III, Table 2)

$$-2\delta\mathfrak{W}_{12;2} = -2\delta\mathfrak{W}_{13;3} = \tfrac{1}{3}(2B_{11} + 7B_{111})V_{111} = \tfrac{4}{9}V_{111}\,. \tag{28}$$

The equations of group (*b*) and the two equations (16) of group (*c*) lead to the root

$$\sigma^2 = 0\,, \tag{29}$$

with multiplicity 5.

Finally, considering the equations of the group (*d*), we first note (cf. Paper II, Table 1)

$$\begin{aligned} -2\delta\mathfrak{W}_{11;1} &= [2(B_{11} + 2B_{111}) - A_{11;1}]V_{111} + (2B_{111} - A_{11;1})(V_{122} + V_{133}) \\ &= [2(\tfrac{4}{15} + \tfrac{8}{35}) - \tfrac{24}{35} + \tfrac{2}{3}(\tfrac{8}{35} - \tfrac{24}{35})]V_{111} = 0\,. \end{aligned} \tag{30}$$

The equations to be considered are, therefore,

$$\tfrac{1}{3}\sigma^2 V_{111} = -2\delta\Pi_1 \tag{31}$$

and

$$\sigma^2(V_{1;22} + V_{1;33}) = \tfrac{8}{9}V_{111}\,, \tag{32}$$ [2]

where we have made use of equation (28).

It remains to determine $\delta\Pi_1$. Now the pressure distribution in a homogeneous sphere is given by

$$p = \tfrac{2}{3}\rho(1 - r^2)\,, \tag{33}$$

where the same factor $\pi G\rho$ has been suppressed and r is measured in the unit a. For this pressure distribution, equation (10) gives

$$\delta\Pi_1 = -\tfrac{4}{3}(\gamma - 1)\int_V \rho\xi_j x_j x_1 dx - \tfrac{2}{3}\gamma\int_V \rho\xi_1 x_j x_j dx + \tfrac{2}{3}\int_V \rho\xi_1 dx\,. \tag{34}$$

The last term vanishes by the condition requiring the stationariness of the center of mass (cf. Paper II, Sec. II); and the remaining terms can be combined to give

$$\begin{aligned} \delta\Pi_1 &= -\tfrac{4}{3}(\gamma - 1)V_{j;j1} - \tfrac{2}{3}\gamma V_{1;jj} \\ &= -\tfrac{2}{3}(\gamma - 1)V_{1jj} - \tfrac{2}{3}(\tfrac{1}{3}V_{111} + V_{1;22} + V_{1;33}) \\ &= -\tfrac{2}{9}(5\gamma - 4)V_{111} - \tfrac{2}{3}(V_{1;22} + V_{1;33})\,. \end{aligned} \tag{35}$$

Inserting this value of $\delta\Pi_1$ in equation (31), we obtain

$$\tfrac{1}{3}\sigma^2 V_{111} = \tfrac{4}{9}(5\gamma - 4)V_{111} + \tfrac{4}{3}(V_{1;22} + V_{1;33})\,. \tag{36}$$

Equations (32) and (36) now lead to the characteristic equation

$$\sigma^4 - \tfrac{4}{3}(5\gamma - 4)\sigma^2 - \tfrac{32}{9} = 0\,. \tag{37}$$

The roots of this equation are

$$\sigma^2 = \tfrac{2}{3}(5\gamma - 4) \pm \tfrac{2}{3}[(5\gamma - 4)^2 + 8]^{1/2}\,; \tag{38}$$

and each of these roots is of multiplicity 3. The roots given by equation (38) agree with those found by Pekeris for $l = 1$.

[2] Actually, we can set $V_{1;22} = V_{1;33}$ in this and in the subsequent equations for consistency with the equations of group (*b*).

One of the two roots given by equation (38) is clearly negative; and this implies instability. We have thus derived the convective instability of the homogeneous sphere for modes of oscillation belonging to $l = 1$.

IV. THE SUPERPOTENTIALS AND THE TENSOR POTENTIALS OF A SPHERICAL DISTRIBUTION OF MASS

The quantities $\delta\mathfrak{W}_{ij;k}$ which occur in the virial equations of Section II are, as we have seen, expressible as integrals over the tensor potentials $\mathfrak{B}_{ij}$ and $\mathfrak{D}_{ij;k}$. In this section we shall assemble the necessary formulae.

The superpotentials χ and Φ are, in case of spherical symmetry, governed by the differential equations

$$\chi'' + \frac{2}{r}\chi' = -2\mathfrak{B} \tag{39}$$

and

$$\Phi'' + \frac{2}{r}\Phi' = -4\chi\,, \tag{40}$$

where the primes denote differentiations with respect to r. Once χ and Φ have been determined as solutions of these differential equations (satisfying the appropriate boundary conditions), the required tensor potentials follow from the equations (Chandrasekhar and Lebovitz 1962*b*, eqs. [6] and [33])

$$\mathfrak{B}_{ij} = \mathfrak{B}\,\delta_{ij} + \frac{\partial^2\chi}{\partial x_i \partial x_j} \tag{41}$$

and

$$\mathfrak{D}_{ij;k} = \frac{\partial^3\Phi}{\partial x_i \partial x_j \partial x_k} + \frac{\chi'}{r}(x_i\delta_{jk} + x_j\delta_{ki} + x_k\delta_{ij}) + x_k\mathfrak{B}_{ij}\,. \tag{42}$$

Inserting the foregoing expressions in equation (7), we obtain

$$\begin{aligned} -2\,\delta\mathfrak{W}_{ij;k} = {} & 2\int_V \rho\mathfrak{B}_{ij}\xi_k d\boldsymbol{x} + 2\int_V \rho\xi_l \frac{\partial\mathfrak{B}_{ij}}{\partial x_l} x_k d\boldsymbol{x} + \int_V \rho\xi_l \frac{\partial^4\Phi}{\partial x_l \partial x_i \partial x_j \partial x_k} d\boldsymbol{x} \\ & - \int_V \rho\xi_l x_l (x_i\delta_{jk} + x_j\delta_{ki} + x_k\delta_{ij})\left(2\frac{\mathfrak{B}}{r^2} + 3\frac{\chi'}{r^3}\right) d\boldsymbol{x} \\ & + (\delta_{li}\delta_{jk} + \delta_{lj}\delta_{ki} + \delta_{lk}\delta_{ij})\int_V \rho\xi_l \frac{\chi'}{r} d\boldsymbol{x}\,. \end{aligned} \tag{43}$$

The derivatives of χ and Φ which occur in equations (41)–(43) can all be expressed, in view of the spherical symmetry of these functions, in terms of the derivatives with respect to r; and in the resulting expressions all derivatives of χ and Φ of orders second and higher can be reduced to the first by successive use of the defining equations (39) and (40). We thus find

$$\frac{\partial^2\chi}{\partial x_i \partial x_j} = \frac{\chi'}{r}\delta_{ij} + \frac{B}{r^2} x_i x_j\,, \tag{44}$$

$$\frac{\partial^3\chi}{\partial x_i \partial x_j \partial x_k} = \frac{B}{r^2}(x_i\delta_{jk} + x_j\delta_{ki} + x_k\delta_{ij}) + \frac{F}{r^2} x_i x_j x_k\,, \tag{45}$$

$$\frac{\partial^3\Phi}{\partial x_i \partial x_j \partial x_k} = C(x_i\delta_{jk} + x_j\delta_{ki} + x_k\delta_{ij}) + \frac{D}{r^2} x_i x_j x_k\,, \tag{46}$$

$$\frac{\partial^4\Phi}{\partial x_i\partial x_j\partial x_k\partial x_l}=C(\delta_{li}\delta_{jk}+\delta_{lj}\delta_{ki}+\delta_{lk}\delta_{ij})+\frac{E}{r^4}x_ix_jx_kx_l$$
$$+\frac{D}{r^2}(x_lx_i\delta_{jk}+x_lx_j\delta_{ki}+x_lx_k\delta_{ij}+x_jx_k\delta_{il}+x_kx_i\delta_{jl}+x_ix_j\delta_{kl}), \tag{47}$$

where we have introduced the abbreviations

$$A=\mathfrak{V}+\frac{\chi'}{r},\qquad D=-\left(5C+4\frac{\chi'}{r}\right),$$
$$B=-\left(2\mathfrak{V}+3\frac{\chi'}{r}\right),\qquad E=8\mathfrak{V}+35C+40\frac{\chi'}{r}, \tag{48}$$
$$C=-\frac{1}{r^2}\left(4\chi+3\frac{\Phi'}{r}\right),\qquad F=-\frac{1}{r^2}(5B+2\mathfrak{V}'r).$$

In terms of these abbreviations, we can now write

$$\mathfrak{V}_{ij}=A\,\delta_{ij}+\frac{B}{r^2}x_ix_j, \tag{49}$$

and

$$\frac{\partial\mathfrak{V}_{ij}}{\partial x_l}=\frac{\mathfrak{V}'}{r}x_l\delta_{ij}+\frac{B}{r^2}(x_i\delta_{jl}+x_j\delta_{li}+x_l\delta_{ij})+\frac{F}{r^2}x_ix_jx_l. \tag{50}$$

Returning to equation (43) and inserting for the various quantities in accordance with equations (47), (49), and (50), we obtain, after some further regrouping of the terms,

$$-2\delta\mathfrak{W}_{ij;k}=2\delta_{ij}\int_V\rho A\,\xi_k d\boldsymbol{x}+2\delta_{ij}\int_V\rho\left(\frac{\mathfrak{V}'}{r}+\frac{B}{r^2}\right)\xi_lx_lx_kd\boldsymbol{x}$$
$$+\int_V\rho\left(\frac{2B}{r^2}+\frac{D}{r^2}\right)(\xi_ix_jx_k+\xi_jx_kx_i+\xi_kx_ix_j)\,d\boldsymbol{x}$$
$$+\int_V\rho\left(\frac{B}{r^2}+\frac{D}{r^2}\right)\xi_lx_l(x_i\delta_{jk}+x_j\delta_{ki}+x_k\delta_{ij})\,d\boldsymbol{x} \tag{51}$$
$$+\int_V\rho\left(C+\frac{\chi'}{r}\right)(\xi_k\delta_{ij}+\xi_i\delta_{jk}+\xi_j\delta_{ik})\,d\boldsymbol{x}$$
$$+\int_V\rho\left(2\frac{F}{r^2}+\frac{E}{r^4}\right)\xi_lx_lx_ix_jx_kd\boldsymbol{x}.$$

V. THE VIRIAL EQUATIONS FOR AN ASSUMED LAGRANGIAN DISPLACEMENT OF THE FORM $\xi_i = L_{i;jk}x_jx_k + L_i$

In using the virial equations of Section II in conjunction with the formulae of Section IV, we shall suppose that it will suffice to consider a Lagrangian displacement of the form

$$\xi_i = L_{i;jk}x_jx_k + L_i\,, \tag{52}$$

where $L_{i;jk}$ and L_i represent a total of twenty-one unspecified constants. The form (52) for $\boldsymbol{\xi}$ is, in the present context of the third-order virial equations, the proper generalization of the form

$$\xi_i = L_{i;j}x_j \tag{53}$$

assumed in an earlier paper (Chandrasekhar and Lebovitz 1962*a*) in the context of the second-order virial equations.[3] And the justification for assuming the form (52) is that it conforms to the *exact* solution for a homogeneous sphere and that we may, therefore, expect it to be adequate as a first "trial function" for configurations with not excessive degrees of central concentrations. We shall verify that this expectation is borne out by applications to the polytropes.

As has been pointed out in Paper II (Sec. II), the virial equations of the third order must be considered together with the equations of the first order which require

$$V_i = \int_V \rho \xi_i dx = 0 \tag{54}$$

as a condition for the stationariness of the center of mass. For the form of the Lagrangian displacement assumed, condition (54) gives

$$\tfrac{1}{3}L_{i;jj}\int_V \rho r^2 dx + L_i \int_V \rho dx = 0\,, \tag{55}$$

or

$$L_i = -\frac{\int_V \rho r^2 dx}{3\int_V \rho dx} L_{i;jj}\,. \tag{56}$$

Relations (56) reduce the number of unspecified constants in (52) to eighteen: the same as the number of the virial equations of the third order.

a) The Kelvin Modes

We shall first consider equations (12) and (15) in groups (*a*) and (*c*) (in Sec. II). We find, by direct calculation,

$$V_{122} - \tfrac{1}{3}V_{111} = \tfrac{2}{15}(L_{122} - \tfrac{1}{3}L_{111})\int_V \rho r^4 dx \tag{57}$$

and

$$V_{123} = \tfrac{2}{15} L_{123}\int_V \rho r^4 dx\,, \tag{58}$$

where L_{ijk} without the semicolon (like the symmetrized virial V_{ijk}) is defined in the manner

$$L_{ijk} = L_{i;jk} + L_{j;ki} + L_{k;ij}\,. \tag{59}$$

[3] It should be noted that greater generality is not achieved by an assumption of the form

$$\xi_i = L_{i;jk}x_j x_k + L_{i;j}x_j + L_i\,,$$

since the terms in $L_{i;j}$ will vanish identically in all the equations provided by the third-order virial equations. On the other hand, if one went to the virial equations of the fourth order, the form for the Lagrangian displacement one would assume is

$$\xi_i = L_{i;jkl}x_j x_k x_l + L_{i;j}x_j\,;$$

and the second- and the fourth-order virial equations will together provide a number of equations equal to the number of the constants $L_{i;jkl}$ and $L_{i;j}$ even as the first- and the third-order virial equations provide a number of equations equal to the number of the constants $L_{i;jk}$ and L_i. And, in general, the consideration of the virial equations of increasing (even or odd) orders corresponds to solving the exact problem with "trial functions" for ξ_i which are polynomials in the co-ordinates of increasing (odd or even) orders.

The reduction of the expressions for the required $\delta\mathfrak{W}_{ij;k}$'s and δS_{ijk}'s, in accordance with equation (51) for the assumed form of $\boldsymbol{\xi}$, to integrals over functions only of $\boldsymbol{r}$ is long but straightforward. We shall omit all the details of the reduction and quote only the final results.

We find

$$\delta S_{122} = -\tfrac{8}{35}(L_{122} - \tfrac{1}{3}L_{111}) \int_V \rho \mathfrak{V}' r^3 d\boldsymbol{x} \tag{60}$$

and

$$\delta S_{123} = -\tfrac{8}{35} L_{123} \int_V \rho \mathfrak{V}' r^3 d\boldsymbol{x}\,. \tag{61}$$

From equations (57), (58), (60), and (61) it is apparent that the equations of group (*a*) and equation (15) all lead to the same root:

$$\sigma^2 = -\tfrac{12}{7} \frac{\int_V \rho \mathfrak{V}' r^3 d\boldsymbol{x}}{\int_V \rho r^4 d\boldsymbol{x}}\,. \tag{62}$$

This root is of multiplicity 7 and belongs to $l = 3$; it represents the analogue of the corresponding Kelvin mode for the homogeneous sphere.

The integral in the numerator of the expression for σ^2 can be transformed in this manner:

$$\int_V \rho \frac{d\mathfrak{V}}{dr} r^3 d\boldsymbol{x} = 4\pi \int \frac{dp}{dr} r^5 dr = -20\pi \int p r^4 dr = -5 \int_V p r^2 d\boldsymbol{x}\,. \tag{63}$$

We may accordingly write

$$\sigma^2 = \tfrac{60}{7} \frac{\int_V p r^2 d\boldsymbol{x}}{\int_V \rho r^4 d\boldsymbol{x}}\,. \tag{64}$$

For polytropes, we find, on expressing $\boldsymbol{r}$, ρ, and p in terms of the usual Emden variables (cf. Chandrasekhar and Lebovitz 1962*c*, eq. [8]), that

$$\frac{\sigma^2}{4\pi G \rho_c} = \frac{60}{7(n+1)} \frac{\int_0^{\xi_1} \theta^{n+1} \xi^4 d\xi}{\int_0^{\xi_1} \theta^n \xi^6 d\xi}\,. \tag{65}$$

The integrals occurring in this expression for σ^2 are among those reduced and tabulated in the appendix (Table 1). And using the results given in the appendix, we obtain the values listed in the accompanying tabulation (eq. [66]).

n	$\sigma^2/(4\pi G\rho_c)$	n	$\sigma^2/(4\pi G\rho_c)$
0	0.57143	2.0	0.16609
1.0	0.29866	3.0	0.084302
1.5	0.22328	3.5	0.054834

(66)

b) The Characteristic Equation for the Modes of Oscillation Belonging to l = 1

We turn now to a consideration of equations (19) and (20). In considering them in the framework of the assumption (52), we must set

$$L_{122} = L_{133} = \tfrac{1}{3}L_{111}\,, \text{ etc.}\,, \tag{67}$$

to be consistent with the equations of group (*a*) and exclude the root (62).

We find, by direct calculation, that

$$V_{1;11} = \tfrac{1}{5}(L_{1;11} + \tfrac{1}{3}M_1)\int_V \rho r^4 dx + \tfrac{1}{3}L_1\int_V \rho r^2 dx \tag{68}$$

and

$$V_{1;22} + V_{1;33} = \tfrac{2}{15}(L_{1;11} + 2M_1)\int_V \rho r^4 dx + \tfrac{2}{3}L_1\int_V \rho r^2 dx\,, \tag{69}$$

where we have introduced the abbreviation

$$M_1 = L_{1;22} + L_{1;33}\,. \tag{70}$$ [4]

From equation (51) we similarly find (after some lengthy reductions) that, for the assumed form for $\boldsymbol{\xi}$ (with the additional restrictions [67]),

$$\begin{aligned}- 2\,\delta\mathfrak{W}_{12;2} - 2\,\delta\mathfrak{W}_{13;3} = &-\tfrac{4}{15}\int_V \rho(L_{1;11}r^2 + L_1)(\mathfrak{V} + 2\mathfrak{V}' r)\,dx \\ &+ \tfrac{2}{15}(2L_{1;11} - M_1)\int_V \rho\left(4\mathfrak{V} + 5\,\frac{\chi'}{r}\right) r^2 dx\end{aligned} \tag{71}$$

and

$$\begin{aligned}- 2\,\delta\mathfrak{W}_{11;1} = &\tfrac{2}{15}\int_V \rho(L_{1;11}r^2 + L_1)(2\mathfrak{V} - \mathfrak{V}' r)\,dx \\ &+ \tfrac{1}{15}(2L_{1;11} - M_1)\int_V \rho\left(2\mathfrak{V} + 5\,\frac{\chi'}{r}\right) r^2 dx\,.\end{aligned} \tag{72}$$

The coefficients of L_1 in equations (71) and (72) can be simplified by making use of the relations

$$\mathfrak{W} = -\tfrac{1}{2}\int_V \rho\mathfrak{V}\,dx = -3\int_V p\,dx \tag{73}$$

and

$$\int_V \rho\,\frac{d\mathfrak{V}}{dr}\,r\,dx = 4\pi\int_V \frac{dp}{dr}\,r^3 dr = -12\pi\int_V p r^2 dr = \mathfrak{W}. \tag{74}$$

Thus,

$$\int_V \rho(\mathfrak{V} + 2\mathfrak{V}' r)\,dx = 0\,, \qquad \text{and} \qquad \tfrac{2}{15}\int_V \rho(2\mathfrak{V} - \mathfrak{V}' r)\,dx = 2\int_V p\,dx\,; \tag{75}$$

and we can write

$$\begin{aligned}- 2\,\delta\mathfrak{W}_{12;2} - 2\,\delta\mathfrak{W}_{13;3} = &\tfrac{4}{15}L_{1;11}\int_V \rho\left(3\mathfrak{V} - 2\mathfrak{V}' r + 5\,\frac{\chi'}{r}\right) r^2 dx \\ &- \tfrac{2}{15}M_1\int_V \rho\left(4\mathfrak{V} + 5\,\frac{\chi'}{r}\right) r^2 dx\end{aligned} \tag{76}$$

[4] In this equation we can set $L_{1;22} = L_{1;33}$ to be consistent with eq. (14) and exclude the zero root.

and

$$- 2\delta \mathfrak{W}_{11;1} = \tfrac{2}{15} L_{1;11} \int_V \rho \left(4\mathfrak{V} - \mathfrak{V}' r + 5\frac{\chi'}{r}\right) r^2 dx$$
$$- \tfrac{1}{15} M_1 \int_V \rho \left(2\mathfrak{V} + 5\frac{\chi'}{r}\right) r^2 dx + 2L_1 \int_V p\, dx \,. \tag{77}$$

It remains to determine $\delta \Pi_1$ in accordance with equation (9). For the assumed form of ξ (with the additional restrictions [67]), we find

$$\delta \Pi_1 = - \tfrac{2}{3}(\gamma - 1) L_{j;j1} \int_V p\, r^2 dx + \tfrac{1}{3} L_{1;jj} \int_V p\, r^2 dx + L_1 \int_V p\, dx$$
$$= [- \tfrac{1}{3}(4\gamma - 5) L_{1;11} + \tfrac{1}{3}\gamma M_1] \int_V p\, r^2 dx + L_1 \int_V p\, dx \,. \tag{78}$$

Now substituting from equations (68), (69), (76), (77), and (78) in equations (19) and (20), we obtain

$$- \sigma^2 \left[\tfrac{1}{5} (L_{1;11} + \tfrac{1}{3} M_1) \int_V \rho r^4 dx + \tfrac{1}{3} L_1 \int_V \rho r^2 dx \right]$$
$$+ \tfrac{2}{15} L_{1;11} \int_V \rho \left(4\mathfrak{V} - \mathfrak{V}' r + 5\frac{\chi'}{r}\right) r^2 dx - \tfrac{1}{15} M_1 \int_V \rho \left(2\mathfrak{V} + 5\frac{\chi'}{r}\right) r^2 dx \tag{79}$$
$$= \tfrac{2}{3} [- (4\gamma - 5) L_{1:11} + \gamma M_1] \int_V p\, r^2 dx$$

and

$$- \sigma^2 \left[\tfrac{2}{15} (L_{1;11} + 2M_1) \int_V \rho r^4 dx + \tfrac{2}{3} L_1 \int_V \rho r^2 dx \right]$$
$$+ \tfrac{4}{15} L_{1;11} \int_V \rho \left(3\mathfrak{V} - 2\mathfrak{V}' r + 5\frac{\chi'}{r}\right) r^2 dx - \tfrac{2}{15} M_1 \int_V \rho \left(4\mathfrak{V} + 5\frac{\chi'}{r}\right) r^2 dx = 0 \,. \tag{80}$$

We notice that in these final equations the "super-superpotential" Φ does not appear: the terms in it canceled at an earlier stage.

Equations (79) and (80) must be supplemented by the further equation (cf. eq. [56])

$$L_1 = - \frac{\int_V \rho r^2 dx}{3 \int_V \rho\, dx} (L_{1;11} + M_1) \,. \tag{81}$$

With the aid of this last equation, L_1 can be eliminated from equations (79) and (80). After the elimination of L_1, we shall be left with a system of two linear homogeneous equations for $L_{1;11}$ and M_1; and the requirement that the determinant of the system vanish will lead to the desired characteristic equation of σ^2. It is to be particularly noted that the coefficients of the characteristic equation for σ^2 are integrals which involve, in addition to the usual quantities, the superpotential χ explicitly.

VI. THE CONVECTIVE INSTABILITY OF THE POLYTROPES FOR $\gamma < 1 + 1/n$

The characteristic equation for σ^2 which follows from equations (79)–(81) has been obtained in an explicit form for a few polytropes. Before we write down the characteristic equations which were obtained, we may note that when r, ρ, p, $\mathfrak{V}$, and χ are expressed in terms of the Emden variables ξ and θ, equations (79)–(81) take the forms

$$-\sigma^2\left[(3L_{1;11}+M_1)\int_0^{\xi_1}\theta^n\xi^6 d\xi+5L_1\int_0^{\xi_1}\theta^n\xi^4 d\xi\right]$$

$$+2L_{1;11}\int_0^{\xi_1}\theta^n\left[4(\theta+c_0)-\theta'\xi+5\frac{\chi'}{\xi}\right]\xi^4 d\xi-M_1\int_0^{\xi_1}\theta^n\left[2(\theta+c_0)+5\frac{\chi'}{\xi}\right]\xi^4 d\xi \tag{82}$$

$$=\frac{10}{n+1}[-(4\gamma-5)L_{1;11}+\gamma M_1]\int_0^{\xi_1}\theta^{n+1}\xi^4 d\xi,$$

$$-\sigma^2\left[(L_{1;11}+2M_1)\int_0^{\xi_1}\theta^n\xi^6 d\xi+5L_1\int_0^{\xi_1}\theta^n\xi^4 d\xi\right] \tag{83}$$

$$+2L_{1;11}\int_0^{\xi_1}\theta^n\left[3(\theta+c_0)-2\theta'\xi+5\frac{\chi'}{\xi}\right]\xi^4 d\xi$$

$$-M_1\int_0^{\xi_1}\theta^n\left[4(\theta+c_0)+5\frac{\chi'}{\xi}\right]\xi^4 d\xi=0,$$

and

$$L_1=-\frac{\int_0^{\xi_1}\theta^n\xi^4 d\xi}{3\xi_1^2|\theta'(\xi_1)|}(L_{1;11}+M_1), \tag{84}$$

where a prime denotes differentiation with respect to ξ, ξ_1 is the first zero of the Lane-Emden function, and σ^2 is measured in the unit $4\pi G\rho_c$; also in the chosen units (cf. Chandrasekhar and Lebovitz 1962*c*)

$$\mathfrak{V}=\theta+c_0, \qquad \text{where} \qquad c_0=-\xi_1\left(\frac{d\theta}{d\xi}\right)_{\xi_1}, \tag{85}$$

and

$$\chi''+\frac{2}{\xi}\chi'=-2(\theta+c_0). \tag{86}$$

The various integrals occurring in equations (82)–(84) are reduced and tabulated in the appendix (Table 1). And we finally obtain

$$\begin{aligned}
&\sigma^4-0.76513\,(\gamma-0.91079)\,\sigma^2+0.029036\,(\gamma-1.9955)=0 \quad (n=1.0)\,,\\
&\sigma^4-0.55037\,(\gamma-0.94660)\,\sigma^2+0.018768\,(\gamma-1.6609)=0 \quad (n=1.5)\,,\\
&\sigma^4-0.39698\,(\gamma-0.97624)\,\sigma^2+0.011071\,(\gamma-1.4931)=0 \quad (n=2.0)\,,\\
&\sigma^4-0.19192\,(\gamma-1.02493)\,\sigma^2+0.0029138\,(\gamma-1.3258)=0 \quad (n=3.0)\,,\\
&\sigma^4-0.12241\,(\gamma-1.04520)\,\sigma^2+0.0012184\,(\gamma-1.2784)=0 \quad (n=3.5)\,.
\end{aligned} \tag{87}$$

From these equations it is apparent that the polytropes are unstable when

$$\begin{aligned}
&\gamma<1.995 \text{ for } n=1.0, \text{ whereas } 1+1/n=2.000\,,\\
&\gamma<1.661 \text{ for } n=1.5, \text{ whereas } 1+1/n=1.667\,,\\
&\gamma<1.493 \text{ for } n=2.0, \text{ whereas } 1+1/n=1.500\,,\\
&\gamma<1.326 \text{ for } n=3.0, \text{ whereas } 1+1/n=1.333\,,\\
&\gamma<1.278 \text{ for } n=3.5, \text{ whereas } 1+1/n=1.286\,.
\end{aligned} \tag{88}$$

The accuracy with which the critical values of γ are predicted is remarkable: even for $n = 3.5$ (when $\rho_c = 153\bar{\rho}$), the error does not exceed 0.7 per cent; and this is indeed gratifying.

VII. CONCLUDING REMARKS

As we have remarked in the introductory section, the fact that the convective instability of polytropes asserts itself already for a mode of oscillation belonging to $l = 1$ should not cause any surprise: it is very much to be expected. However, in cases where, as in Cowling's point-source model for a star, the effective polytropic index decreases from a relatively high value in the exterior to a subcritical value in the interior, the circumstances are, of course, different. But even in these cases it would seem that the most relevant modes are again those which belong to spherical harmonics of low orders (including $l = 1$) and radial functions having one or more nodes. If this presumption is valid, then the modes of oscillation belonging to really high values of l are not as basic to the theory of convective instability as one might be inclined to suppose.

A further aspect of the analysis of the preceding sections that is noteworthy is the following. Apart from the fact that the virial equations of the third order appear to provide an adequate basis for treating non-radial oscillations belonging to $l = 1$ and 3, an important feature of the equations is the role of the superpotential χ. By virtue of its relationship to the tensor potential $\mathfrak{B}_{ij}$, the superpotential χ expresses, even more than the Newtonian potential $\mathfrak{B}$ does, the co-operative aspects of the gravitational interaction between the different parts of a body. The dependence of the characteristic frequencies of non-radial oscillations on the superpotential χ may be considered, then, as an indication of their dependence in some significant way on a delicate balance in the co-operative aspects of the gravitational interactions. It is indeed clear from the point of view of the virial equations that the higher modes of non-radial oscillations will depend on the superpotentials of still higher orders such as Φ, etc. It would not, therefore, seem that, when dealing with compressible masses, the usual assumption (derived from a knowledge of the reactions of incompressible masses) that the variations in the gravitational potential can be neglected in the treatment of *all* manner of modes of oscillation (belonging to a given l) is necessarily a valid one. An examination of the modes of oscillation belonging to $l = 4$ from the point of view of the virial equations of the fourth order may contribute to a clarification of these issues; and this examination is now under consideration.

We are greatly indebted to Miss Donna D. Elbert for her assistance with all the numerical work in connection with obtaining the characteristic equations (87).

The research reported in this paper has in part been supported by the Office of Naval Research under contract Nonr-2121(24) with the University of Chicago. The work of the second author was supported in part by the United States Air Force under contract No. AF-49(638)-42, monitored by the Air Force Office of Scientific Research of the Air Research and Development Command.

APPENDIX

The Reduction of Certain Integrals in the Theory of Polytropes

Equations (82)–(84) involve the following integrals:

$$\text{I.}\ \int_0^{\xi_1} \theta^n \xi^4 d\xi\,; \qquad \text{II.}\ \int_0^{\xi_1} \theta^n \xi^6 d\xi\,; \qquad \text{III.}\ \int_0^{\xi_1} \theta^{n+1} \xi^4 d\xi\,;$$

$$\text{IV.}\ \int_0^{\xi_1} \theta^n \theta' \xi^5 d\xi\,; \qquad \text{and} \qquad \text{V.}\ \int_0^{\xi_1} \theta^n \chi' \xi^3 d\xi\,. \tag{A.1}$$

We shall reduce these integrals to simpler forms:

I. $$\int_0^{\xi_1} \theta^n \xi^4 d\xi = -\xi_1^4 \theta_1' - 6\int_0^{\xi_1} \theta \xi^2 d\xi\,, \tag{A.2}$$

where a subscript 1 to a quantity indicates that its value at ξ_1 (the first zero of θ) is meant. The reduction (A. 2) was effected in an earlier paper (Chandrasekhar and Lebovitz 1962*c*, Appendix I). In an exactly similar way we find

II. $$\int_0^{\xi_1} \theta^n \xi^6 d\xi = -\xi_1^6 \theta_1' + 4\int_0^{\xi_1} \theta' \xi^5 d\xi\,; \tag{A.3}$$

III. $$\int_0^{\xi_1} \theta^{n+1} \xi^4 d\xi = \frac{n+1}{n+11}\left[\xi_1^5 (\theta_1')^2 - 3\int_0^{\xi_1} \theta^2 \xi^2 d\xi\right]. \tag{A.4}$$

Equation (A.4) is a special case of the following general formula due to Milne (1929, eq. [53]):

$$\begin{aligned}[n(s-3)+3s-1]\int_0^{\xi_1} \theta^{n+1}\xi^s d\xi &= (n+1)\Big[\xi_1^{s+1}(\theta_1')^2 \\ &\quad -\tfrac{1}{2}(s-3)(s-2)(s-1)\int_0^{\xi_1} \xi^{s-2}\theta^2 d\xi\Big] \quad (s>1)\,;\end{aligned} \tag{A.5}$$

IV. $$\int_0^{\xi_1} \theta^n \theta' \xi^5 d\xi = -\frac{5}{n+1}\int_0^{\xi_1} \theta^{n+1}\xi^4 d\xi\,. \tag{A.6}$$

This is equivalent to the general relation (63) established in the text.

V. $$\begin{aligned}\int_0^{\xi_1} \theta^n \frac{d\chi}{d\xi}\xi^3 d\xi &= -\int_0^{\xi_1} \xi \frac{d\chi}{d\xi}\frac{d}{d\xi}\left(\xi^2 \frac{d\theta}{d\xi}\right) d\xi \\ &= -\xi_1^3 \chi_1' \theta_1' + \int_0^{\xi_1} \xi^3 \frac{d\theta}{d\xi}\left(\frac{d^2\chi}{d\xi^2} + \frac{1}{\xi}\frac{d\chi}{d\xi}\right) d\xi \\ &= -\xi_1^3 \chi_1' \theta_1' - \int_0^{\xi_1} \xi^3 \frac{d\theta}{d\xi}\left[2(\theta + c_0) + \frac{1}{\xi}\frac{d\chi}{d\xi}\right] d\xi \\ &= -\xi_1^3 \chi_1' \theta_1' - \int_0^{\xi_1} \xi^3 \frac{d}{d\xi}(\theta + c_0)^2 d\xi - \int_0^{\xi_1} \frac{d\theta}{d\xi}\left(\xi^2 \frac{d\chi}{d\xi}\right) d\xi \\ &= -\xi_1^3 \chi_1' \theta_1' - \xi_1^3 c_0^2 + 3\int_0^{\xi_1} \xi^2 (\theta + c_0)^2 d\xi - 2\int_0^{\xi_1} \theta(\theta + c_0)\xi^2 d\xi \\ &= -\xi_1^3 \chi_1' \theta_1' + 4c_0\int_0^{\xi_1} \theta\xi^2 d\xi + \int_0^{\xi_1} \theta^2 \xi^2 d\xi\,,\end{aligned} \tag{A.7}$$

where, in the reductions, use has been made (at two different times) of the equation (cf. eq [86])

$$\chi'' + \frac{2}{\xi}\chi' = -2(\theta + c_0) \qquad (c_0 = -\xi_1 \theta_1') \tag{A.8}$$

governing χ.

Thus the five integrals listed in (A. 1) can be expressed in terms of the following three:

$$\int_0^{\xi_1} \theta \xi^2 d\xi\,, \quad \int_0^{\xi_1} \theta' \xi^5 d\xi\,, \quad \text{and} \quad \int_0^{\xi_1} \theta^2 \xi^2 d\xi\,. \tag{A.9}$$

Of these three integrals, the first has already been evaluated in a different connection (Chandrasekhar and Lebovitz 1962*c*, Appendix I and Table 5). The remaining two integrals have now been evaluated; and their values, along with the others are listed in Table 1.

TABLE 1

A TABLE OF INTEGRALS

	$n=1.0$*	$n=1.5$	$n=2.0$	$n=3.0$	$n=3.5$
$\int \theta \xi^2 d\xi$	3.14159	4.18545	5.84540	14.1915	26.6943
$-\int \theta' \xi^5 d\xi$	60.7836	104.6806	198.4622	1114.165	3868.37
$\int \theta^2 \xi^2 d\xi$	1.57080	1.91883	2.41105	4.32705	6.45430
$\int \theta^n \xi^4 d\xi$	12.15672	11.1197	10.6110	10.8516	11.7455
$\int \theta^n \xi^6 d\xi$	62.8853	64.9770	71.7372	109.7484	158.689
$\int \theta^{n+1} \xi^4 d\xi$	4.38231	4.23148	4.17018	4.31761	4.56828
$-\int \theta^n \chi' \xi^3 d\xi$	12.8169	9.8058	7.9827	6.0957	5.6829

* The values of the integrals down this column are π, $5\pi^3 - 30\pi$, $\pi/2$, $\pi^3 - 6\pi$, $\pi^5 - 20\pi^3 + 120\pi$, $(\pi^3 - \frac{3}{2}\pi)/6$, and $\frac{2}{3}\pi^3 - 2.5\pi$, respectively.

REFERENCES

Chandrasekhar, S. 1961, *Hydrodynamic and Hydromagnetic Stability* (Oxford: Clarendon Press).
———. 1962, *Ap. J.*, **136**, 1048.
Chandrasekhar, S., and Lebovitz, N. R. 1962*a*, *Ap. J.*, **135**, 248.
———. 1962*b*, *ibid.*, **136**, 1032.
———. 1962*c*, *ibid.*, p. 1082.
———. 1963*a*, *ibid.*, **137**, 1142.
———. 1963*b*, *ibid.*, p. 1162.
Cowling, T. G. 1942, *M.N.*, **101**, 367.
Ledoux, P., and Walraven, Th. 1958, *Hdb. d. Phys.*, ed. S. Flügge (Berlin: Springer-Verlag), p. 509.
Milne, E. A. 1929, *M.N.*, **89**, 739.
Pekeris, C. L. 1939, *Ap. J.*, **88**, 189.

A GENERAL VARIATIONAL PRINCIPLE GOVERNING THE RADIAL AND THE NON-RADIAL OSCILLATIONS OF GASEOUS MASSES

S. CHANDRASEKHAR
University of Chicago
Received September 30, 1963

ABSTRACT

In this paper a general variational principle, applicable to radial as well as non-radial oscillations of gaseous masses, is formulated and proved. And it is, further, shown that when the normal modes are analyzed in vector spherical harmonics, the variational principle requires that the square of the characteristic frequency of oscillation, σ^2, belonging to a particular spherical harmonic, is stationary with respect to simultaneous variations of two independent radial functions. A consequence of this result is that σ^2 (belonging to a particular harmonic) emerges as a characteristic root of a 2×2 matrix.

Two simple illustrations of the variational principle are given.

I. INTRODUCTION

A general variational principle applicable to radial as well as non-radial oscillations of gaseous masses has been formulated recently (Chandrasekhar 1963). In this paper the principle will be cast in a form in which it can be applied, directly, to determining the characteristic frequencies of the normal modes of oscillation belonging to the different spherical harmonics. For the sake of completeness, the principle will be rederived with, however, some amplifications.

II. THE VARIATIONAL PRINCIPLE

Consider a spherically symmetric configuration in equilibrium under its own gravitation and governed by the equations

$$\frac{dp}{dr}=\rho\frac{d\mathfrak{B}}{dr}=-\frac{GM(r)}{r^2}\rho\,, \tag{1}$$

and

$$\frac{1}{r^2}\frac{d}{dr}\left(r^2\frac{d\mathfrak{B}}{dr}\right)=-4\pi G\rho\,, \tag{2}$$

where ρ is the density, p is the pressure, $\mathfrak{B}$ is the gravitational potential, $M(r)$ is the mass interior to r, and G is the constant of gravitation.

The boundary condition on the equilibrium configuration requires that

$$p=0 \qquad \text{on} \qquad r=R\,, \tag{3}$$

where R denotes the radius of the configuration. We shall suppose, in addition, that

$$\rho=0 \qquad \text{on} \qquad r=R\,; \tag{4}$$

from equation (1) it then follows that also

$$\frac{dp}{dr}=0 \qquad \text{on} \qquad r=R\,. \tag{5}$$

Let the equilibrium configuration considered be subject to a small perturbation; let $\delta\rho$, δp, and $\delta\mathfrak{B}$ be the resulting Eulerian changes in the respective variables; and, finally, let u_i denote the velocity of the ensuing motions. The linearized form of the equation of motion which governs the perturbation is

$$\rho \frac{\partial u_i}{\partial t} = - \frac{\partial}{\partial x_i} \delta p + \delta \rho \frac{\partial \mathfrak{B}}{\partial x_i} + \rho \frac{\partial}{\partial x_i} \delta \mathfrak{B} ; \tag{6}$$

or using the equilibrium value $\partial p / \rho \partial x_i$ for $\partial \mathfrak{B} / \partial x_i$ we have

$$\rho \frac{\partial u_i}{\partial t} = - \frac{\partial}{\partial x_i} \delta p + \frac{\delta \rho}{\rho} \frac{\partial p}{\partial x_i} + \rho \frac{\partial}{\partial x_i} \delta \mathfrak{B} . \tag{7}$$

Let the perturbed state be described by a Lagrangian displacement of the form

$$\xi (x) e^{i \sigma t} , \tag{8}$$

where σ is a characteristic frequency to be determined. Equation (7) then becomes

$$\sigma^2 \rho \xi_i = \frac{\partial}{\partial x_i} \delta p - \frac{\delta \rho}{\rho} \frac{\partial p}{\partial x_i} - \rho \frac{\partial}{\partial x_i} \delta \mathfrak{B} , \tag{9}$$

where a common factor $e^{i\sigma t}$ in δp, $\delta \rho$, and $\delta \mathfrak{B}$ has been suppressed.

The Eulerian change in the density ensuing a Lagrangian displacement is given (quite generally) by

$$\delta \rho = - \xi_k \frac{\partial \rho}{\partial x_k} - \rho \operatorname{div} \xi = - \operatorname{div} \rho \xi . \tag{10}$$

Equation (10) is no more than the expression of the conservation of mass. In terms of $\delta \rho$, the change in the gravitational potential is given by Poisson's integral formula,

$$\delta \mathfrak{B} (x) = G \int_V \frac{\delta \rho (x')}{| x - x' |} d x' = - G \int_V \frac{[\operatorname{div} (\rho \xi)]_{x'}}{| x - x' |} d x' . \tag{11}$$

To express the Eulerian change in the pressure, similarly, in terms of the Lagrangian displacement, some additional assumption concerning the nature of the oscillations is necessary. We shall assume that the oscillations take place adiabatically in accordance with the laws appropriate to a gas with a ratio of the specific heats, γ. Then,

$$\delta p = - \xi_k \frac{\partial p}{\partial x_k} - \gamma p \operatorname{div} \xi . \tag{12}$$

The boundary conditions with respect to which equations (9)–(12) must be solved are that none of the physical variables has a singularity at the origin (or anywhere else); and, also, that

$$\delta p = 0 \quad \text{on} \quad r = R . \tag{13}$$

The particular form of the boundary condition (13) depends, explicitly, on the assumption (that we have made) that ρ vanishes on the boundary: for what is required, strictly, is that the Lagrangian change in the pressure vanish; that the Eulerian change vanishes follows from the fact that, when the density vanishes on the boundary, the gradient of the pressure (in addition to the pressure) vanishes on the boundary.

In the usual treatments of the problem (cf. Pekeris 1938; also Ledoux and Walraven 1958, p. 513) the boundary conditions on $\delta \mathfrak{B}$ (determined in those treatments with the aid of Poisson's and Laplace's equations) are also specified. We do not need to specify these conditions since, by expressing $\delta \mathfrak{B}$ in terms of $\delta \rho$ by Poisson's integral formula (11), the requisite conditions have been satisfied already.

Returning to equation (9), let $\sigma^{(\lambda)}$ denote a particular characteristic frequency; and let the proper solutions belonging to it be distinguished by the same superscript. Consider equation (9) belonging to $\sigma^{(\lambda)}$ and after multiplication by $\xi_i^{(\mu)}$, belonging to a different characteristic value $\sigma^{(\mu)}$, integrate over the volume V occupied by the fluid to obtain

$$[\sigma^{(\lambda)}]^2 \int_V \rho \boldsymbol{\xi}^{(\lambda)} \cdot \boldsymbol{\xi}^{(\mu)} d\boldsymbol{x} = \int_V \xi_i^{(\mu)} \frac{\partial}{\partial x_i} \delta p^{(\lambda)} d\boldsymbol{x} - \int_V \frac{\delta \rho^{(\lambda)}}{\rho} \xi_i^{(\mu)} \frac{\partial p}{\partial x_i} d\boldsymbol{x} - \int_V \rho \xi_i^{(\mu)} \frac{\partial}{\partial x_i} \delta \mathfrak{B}^{(\lambda)} d\boldsymbol{x} . \tag{14}$$

Integrating by parts the first and the last of the three integrals on the right-hand side of equation (14), and remembering that (under the assumed conditions) both $\delta p^{(\lambda)}$ and ρ vanish on the boundary of V, we obtain

$$[\sigma^{(\lambda)}]^2 \int_V \rho \boldsymbol{\xi}^{(\lambda)} \cdot \boldsymbol{\xi}^{(\mu)} d\boldsymbol{x} = -\int_V \delta p^{(\lambda)} \operatorname{div} \boldsymbol{\xi}^{(\mu)} d\boldsymbol{x} - \int_V \frac{\delta \rho^{(\lambda)}}{\rho} \xi_i^{(\mu)} \frac{\partial p}{\partial x_i} d\boldsymbol{x} + \int_V \delta \mathfrak{B}^{(\lambda)} \operatorname{div} \rho \boldsymbol{\xi}^{(\mu)} d\boldsymbol{x} . \tag{15}$$

Now substituting for $\delta\rho^{(\lambda)}$ and $\delta p^{(\lambda)}$ in accordance with equations (10) and (12), we have

$$[\sigma^{(\lambda)}]^2 \int_V \rho \boldsymbol{\xi}^{(\lambda)} \cdot \boldsymbol{\xi}^{(\mu)} d\boldsymbol{x} = \int_V \left[\xi_k^{(\lambda)} \frac{\partial p}{\partial x_k} + \gamma p \operatorname{div} \boldsymbol{\xi}^{(\lambda)} \right] \operatorname{div} \boldsymbol{\xi}^{(\mu)} d\boldsymbol{x} + \int_V \left[\frac{\xi_k^{(\lambda)}}{\rho} \frac{\partial \rho}{\partial x_k} + \operatorname{div} \boldsymbol{\xi}^{(\lambda)} \right] \xi_i^{(\mu)} \frac{\partial p}{\partial x_i} d\boldsymbol{x} - \int_V \delta \rho^{(\mu)} \delta \mathfrak{B}^{(\lambda)} d\boldsymbol{x} . \tag{16}$$

Regrouping the terms on the right-hand side of equation (16) and expressing $\delta\mathfrak{B}^{(\lambda)}$ in terms of $\delta\rho^{(\lambda)}$ by Poisson's integral formula, we obtain

$$[\sigma^{(\lambda)}]^2 \int_V \rho \boldsymbol{\xi}^{(\lambda)} \cdot \boldsymbol{\xi}^{(\mu)} d\boldsymbol{x} = \gamma \int_V p \operatorname{div} \boldsymbol{\xi}^{(\lambda)} \operatorname{div} \boldsymbol{\xi}^{(\mu)} d\boldsymbol{x} + \int_V \frac{1}{r} \frac{dp}{dr} \{ [\boldsymbol{x} \cdot \boldsymbol{\xi}^{(\lambda)}] \operatorname{div} \boldsymbol{\xi}^{(\mu)} + [\boldsymbol{x} \cdot \boldsymbol{\xi}^{(\mu)}] \operatorname{div} \boldsymbol{\xi}^{(\lambda)} \} d\boldsymbol{x} + \int_V \frac{[\boldsymbol{x} \cdot \boldsymbol{\xi}^{(\lambda)}][\boldsymbol{x} \cdot \boldsymbol{\xi}^{(\mu)}]}{r^2 \rho} \frac{d\rho}{dr} \frac{dp}{dr} d\boldsymbol{x} - G \int_V \int_V \frac{\delta \rho^{(\lambda)}(\boldsymbol{x}') \, d\rho^{(\mu)}(\boldsymbol{x})}{|\boldsymbol{x} - \boldsymbol{x}'|} d\boldsymbol{x} d\boldsymbol{x}' , \tag{17}$$

where the fact that p and ρ are functions of r only has been incorporated.

The right-hand side of equation (17) is manifestly symmetric in λ and μ; accordingly,

$$\int_V \rho \boldsymbol{\xi}^{(\lambda)} \cdot \boldsymbol{\xi}^{(\mu)} d\boldsymbol{x} = 0 \qquad (\lambda \neq \mu) . \tag{18}$$

The underlying characteristic value problem is, therefore, self-adjoint; and the equation obtained by setting $\lambda = \mu$ in equation (17) and suppressing the distinguishing superscripts, namely,

$$\sigma^2 \int_V \rho |\xi|^2 dx = \int_V \left[\gamma p (\operatorname{div} \xi)^2 + \frac{2}{r} \frac{dp}{dr} (x \cdot \xi) \operatorname{div} \xi \right] dx$$
$$+ \int_V \frac{(x \cdot \xi)^2}{r^2 \rho} \frac{d\rho}{dr} \frac{dp}{dr} dx - G \int_V \int_V \frac{(\operatorname{div} \rho \xi)_x (\operatorname{div} \rho \xi)_{x'}}{|x - x'|} dx dx' \tag{19}$$

provides a variational base for determining the characteristic frequencies.

While the fact that equation (19) provides a variational base for determining σ^2 is clear on general grounds, a direct proof that this is the case is of some interest; and such a proof is given below.

Proof of the Variational Principle

Rewriting equation (19) in the form

$$\sigma^2 = \frac{I_2}{I_1}, \tag{20}$$

where

$$I_1 = \int_V \rho |\xi|^2 dx \tag{21}$$

and

$$I_2 = \int_V \left[\gamma p (\operatorname{div} \xi)^2 + \frac{2}{r} \frac{dp}{dr} (x \cdot \xi) \operatorname{div} \xi \right] dx$$
$$+ \int_V \frac{(x \cdot \xi)^2}{r^2 \rho} \frac{d\rho}{dr} \frac{dp}{dr} dx - G \int_V \int_V \frac{(\operatorname{div} \rho \xi)_x (\operatorname{div} \rho \xi)_{x'}}{|x - x'|} dx dx', \tag{22}$$

we shall show that σ^2 given by these equations has a stationary property when I_1 and I_2 are evaluated in terms of the true proper solutions.

To prove the stationary property, let σ^2 be evaluated in accordance with equations (20)–(22) in terms of a displacement ξ which is arbitrary except for the requirement (apart from boundedness and continuity) that it satisfy the boundary condition (13) (with δp given by eq. [12]). And let $\delta\sigma^2$ be the change in σ^2 (evaluated in accordance with the same equations) when ξ is subject to a small variation $\delta\xi$ which is, again, restricted to be compatible only with the boundary conditions. According to equation (20)

$$\delta \sigma^2 = \frac{1}{I_1} (\delta I_2 - \sigma^2 \delta I_1), \tag{23}$$

where δI_1 and δI_2 are the changes in I_1 and I_2 consequent to the variation $\delta\xi$ in ξ. We have

$$\delta I_1 = 2 \int_V \rho \xi_i \delta \xi_i dx \tag{24}$$

and (cf. eq. [10])

$$\delta I_2 = \int_V \left(2 \gamma p \operatorname{div} \xi \operatorname{div} \delta \xi + 2 \delta \xi_i \frac{\partial p}{\partial x_i} \operatorname{div} \xi + 2 \xi_k \frac{\partial p}{\partial x_k} \operatorname{div} \delta \xi \right.$$
$$\left. + \frac{1}{\rho} \delta \xi_i \frac{\partial \rho}{\partial x_i} \xi_k \frac{\partial p}{\partial x_k} + \frac{\xi_k}{\rho} \frac{\partial \rho}{\partial x_k} \delta \xi_i \frac{\partial p}{\partial x_i} \right) dx \tag{25}$$
$$+ 2G \int_V dx \operatorname{div}(\rho \delta \xi) \int_V dx' \frac{\delta \rho(x')}{|x - x'|}.$$

Regrouping the terms, we can write

$$\delta I_2 = \int_V \left[2 \left(\gamma p \operatorname{div} \xi + \xi_k \frac{\partial p}{\partial x_k} \right) \operatorname{div} \delta \xi + 2 \delta \xi_i \frac{\partial p}{\rho \partial x_i} \left(\xi_k \frac{\partial \rho}{\partial x_k} + \rho \operatorname{div} \xi \right) \right.$$

$$\left. + \frac{\xi_k \delta \xi_i}{\rho} \frac{\partial(\rho, p)}{\partial(x_i, x_k)} \right] dx + 2G \int_V dx \operatorname{div}(\rho \delta \xi) \int_V dx' \frac{\delta \rho(x')}{|x - x'|}. \tag{26}$$

Since p and ρ are functions only of r, the Jacobian $\partial(\rho, p)/\partial(x_i, x_k)$ clearly vanishes; and making further use of equations (10)–(12) (all of which are valid as *definitions*), we can rewrite the expression for δI_2 in the form

$$\delta I_2 = -2 \int_V \delta p \operatorname{div} \delta \xi dx - 2 \int_V \frac{\delta \rho}{\rho} \delta \xi_i \frac{\partial p}{\partial x_i} dx + 2 \int_V \delta \mathfrak{B} \operatorname{div} \rho \delta \xi dx. \tag{27}$$

Now integrating by parts the first and the last integrals on the right-hand side of equation (27), we observe that the integrated parts vanish in both cases by virtue of the boundary conditions on δp and ρ; and we are left with

$$\delta I_2 = 2 \int_V \delta \xi_i \left(\frac{\partial}{\partial x_i} \delta p - \frac{\delta \rho}{\rho} \frac{\partial p}{\partial x_i} - \rho \frac{\partial}{\partial x_i} \delta \mathfrak{B} \right) dx. \tag{28}$$

Equations (23), (24), and (28) now give

$$\delta \sigma^2 = \frac{2}{I_1} \int_V \delta \xi_i \left(\frac{\partial}{\partial x_i} \delta p - \frac{\delta \rho}{\rho} \frac{\partial p}{\partial x_i} - \rho \frac{\partial}{\partial x_i} \delta \mathfrak{B} - \sigma^2 \rho \xi_i \right) dx. \tag{29}$$

From equation (29) it follows that $\delta\sigma^2 = 0$ if

$$\sigma^2 \rho \xi_i = \frac{\partial}{\partial x_i} \delta p - \frac{\delta \rho}{\rho} \frac{\partial p}{\partial x_i} - \rho \frac{\partial}{\partial x_i} \delta \mathfrak{B}, \tag{30}$$

that is, if ξ is a proper solution belonging to σ^2; and, conversely, if $\delta\sigma^2 = 0$ for an arbitrary $\delta\xi$, restricted only by the boundary conditions of the problem, then equation (30) must hold and the displacement ξ in terms of which σ^2 was initially evaluated must have been a proper solution belonging to it. This completes the proof of the variational principle.

III. THE FORM OF THE VARIATIONAL PRINCIPLE FOR THE NORMAL MODES BELONGING TO THE DIFFERENT SPHERICAL HARMONICS

It is known that, when the normal modes of the Lagrangian displacement are analyzed in vector spherical harmonics (for a brief account of the basic properties of these harmonics see Chandrasekhar 1961), the solution involves two functions which determine the radial dependence of the components, ξ_r, ξ_ϑ, and ξ_φ, of the displacement in spherical coordinates. In conformity with what is required, we shall find it convenient to express ξ_r, ξ_ϑ, and ξ_φ in the forms

$$\xi_r = \frac{\psi(r)}{r^2} Y_l^m(\vartheta, \varphi),$$

$$\xi_\vartheta = \frac{1}{l(l+1)r} \frac{d\chi(r)}{dr} \frac{\partial Y_l^m(\vartheta, \varphi)}{\partial \vartheta}, \tag{31}$$

and

$$\xi_\varphi = \frac{1}{l(l+1)r \sin\vartheta} \frac{d\chi(r)}{dr} \frac{\partial Y_l^m(\vartheta, \varphi)}{\partial \varphi},$$

where $\psi(r)$ and $\chi(r)$ are the two radial functions and $Y_l^m(\vartheta, \varphi)$ is a spherical harmonic.

For $\boldsymbol{\xi}$ given by equations (31)

$$\boldsymbol{x}\cdot\boldsymbol{\xi} = r\xi_r = \frac{\psi}{r} Y_l^m, \tag{32}$$

$$\operatorname{div}\boldsymbol{\xi} = \frac{1}{r^2}\frac{d}{dr}(\psi-\chi)Y_l^m, \tag{33}$$[1]

and

$$\delta\rho(\boldsymbol{x}) = -\operatorname{div}\rho\boldsymbol{\xi} = -\delta\rho(r)Y_l^m, \tag{34}$$

where

$$\delta\rho(r) = \frac{1}{r^2}\left[\frac{d}{dr}(\rho\psi) - \rho\frac{d\chi}{dr}\right] = \frac{1}{r^2}\left[\rho\frac{d}{dr}(\psi-\chi) + \psi\frac{d\rho}{dr}\right]. \tag{35}$$

Also, when averaged over a unit sphere,

$$\int_0^{2\pi}\int_0^{\pi} |\boldsymbol{\xi}|^2 \sin\vartheta\, d\vartheta\, d\varphi = \frac{N_l^m}{r^2}\left[\frac{\psi^2}{r^2} + \frac{1}{l(l+1)}\left(\frac{d\chi}{dr}\right)^2\right], \tag{36}$$

where

$$N_l^m = \frac{4\pi}{2l+1}\frac{(l+|m|)!}{(l-|m|)!}. \tag{37}$$

Making use of the foregoing results, we find that, for $\boldsymbol{\xi}$ of the chosen form, equation (19) becomes

$$\begin{aligned}\sigma^2\int_0^R \rho\left[\frac{\psi^2}{r^2} + \frac{1}{l(l+1)}\left(\frac{d\chi}{dr}\right)^2\right]dr &= \gamma\int_0^R p\left[\frac{d}{dr}(\psi-\chi)\right]^2\frac{dr}{r^2}\\ &+\int_0^R\left[2\frac{dp}{dr}\psi\frac{d}{dr}(\psi-\chi) + \frac{\psi^2}{\rho}\frac{dp}{dr}\frac{d\rho}{dr}\right]\frac{dr}{r^2} - \frac{G}{N_l^m}\int_V\int_V\frac{\delta\rho(\boldsymbol{x})\,\delta\rho(\boldsymbol{x}')}{|\boldsymbol{x}-\boldsymbol{x}'|}d\boldsymbol{x}\,d\boldsymbol{x}'.\end{aligned} \tag{38}$$

Since $\delta\rho(x)$ is now expressed as a product of a radial function and a spherical harmonic (cf. eq. [34]), the last term on the right-hand side of equation (38) can be reduced, in the usual manner, by expanding $|\boldsymbol{x}-\boldsymbol{x}'|^{-1}$ in spherical harmonics and carrying out the integrations over the angles. We thus find

$$\begin{aligned}\delta^2 W &= -G\int_V\int_V\frac{\delta\rho(\boldsymbol{x})\,\delta\rho(\boldsymbol{x}')}{|\boldsymbol{x}-\boldsymbol{x}'|}d\boldsymbol{x}\,d\boldsymbol{x}'\\ &= -\frac{4\pi G N_l^m}{2l+1}\int_0^R dr\, r^2\delta\rho(r)\left[\frac{1}{r^{l+1}}\int_0^r ds\, s^{l+2}\delta\rho(s) + r^l\int_r^R\frac{ds}{s^{l-1}}\delta\rho(s)\right].\end{aligned} \tag{39}$$

The two double integrals over r and s which occur in this expression for $\delta^2 W$ are equal: this becomes apparent when the order of the integrations in one of them is inverted. We can, therefore, write

$$\begin{aligned}\delta^2 W &= -\frac{8\pi G N_l^m}{2l+1}\int_0^R dr\, r^{l+2}\delta\rho(r)\int_r^R\frac{ds}{s^{l-1}}\delta\rho(s)\\ &= +\frac{4\pi G N_l^m}{2l+1}\int_0^R dr\, r^{2l+1}\frac{d}{dr}\left[\int_r^R\frac{ds}{s^{l-1}}\delta\rho(s)\right]^2\\ &= -4\pi G N_l^m\int_0^R dr\, r^{2l}\left[\int_r^R\frac{ds}{s^{l-1}}\delta\rho(s)\right]^2.\end{aligned} \tag{40}$$

[1] Note that the condition for the velocity field to be solenoidal is $\psi = \chi$.

With this reduction of the last term of equation (38), we can write

$$\sigma^2\int_0^R \rho\left[\frac{\psi^2}{r^2}+\frac{1}{l(l+1)}\left(\frac{d\chi}{dr}\right)^2\right]dr=\gamma\int_0^R p\left[\frac{d}{dr}(\psi-\chi)\right]^2\frac{dr}{r^2}$$
$$+\int_0^R\left[\frac{1}{\rho}\frac{dp}{dr}\frac{d}{dr}(\rho\psi^2)-2\frac{dp}{dr}\psi\frac{d\chi}{dr}\right]\frac{dr}{r^2}-4\pi G\int_0^R dr\,r^{2l}\left[\int_r^R\frac{ds}{s^{l-1}}\,\delta\rho(s)\right]^2, \tag{41}$$

where it may be recalled that

$$\delta\rho(r)=\frac{1}{r^2}\left[\frac{d}{dr}(\rho\psi)-\rho\frac{d\chi}{dr}\right]. \tag{42}$$

The variational principle requires that σ^2 determined by equation (41) be stationary with respect to *simultaneous* variations in ψ and χ. In other words, the variations to be effected are not only with respect to the chosen *forms* of ψ and χ but also with respect to their *relative amplitudes*. The latter variation, with respect to the amplitudes, can be effected independently of the forms. Thus, writing $A\psi$ and $B\chi$ in place of ψ and χ in equation (41), we find that the requirement that σ^2 be stationary with respect to variations in A and B (for fixed ψ and χ) leads to the characteristic equation

$$\begin{Vmatrix}
\begin{array}{l}-\sigma^2\int_0^R\rho\psi^2\frac{dr}{r^2}+\int_0^R\left[\gamma p\left(\frac{d\psi}{dr}\right)^2+\frac{1}{\rho}\frac{dp}{dr}\frac{d}{dr}(\rho\psi^2)\right]\frac{dr}{r^2}\\ \quad -4\pi G\int_0^R dr\,r^{2l}\left[\int_r\frac{ds}{s^{l+1}}\frac{d}{ds}(\rho\psi)\right]^2\end{array} &
\begin{array}{l}-\int_0^R\left(\gamma p\frac{d\psi}{dr}\frac{d\chi}{dr}+\frac{dp}{dr}\psi\frac{d\chi}{dr}\right)\frac{dr}{r^2}\\ \quad +4\pi G\int_0^R dr\,r^{2l}\left[\left(\int_r^R\frac{ds}{s^{l+1}}\frac{d}{ds}\rho\psi\right)\left(\int_r^R\frac{ds}{s^{l+1}}\rho\frac{d\chi}{ds}\right)\right]\end{array}\\
\begin{array}{l}-\int_0^R\left(\gamma p\frac{d\psi}{dr}\frac{d\chi}{dr}+\frac{dp}{dr}\psi\frac{d\chi}{dr}\right)\frac{dr}{r^2}\\ \quad +4\pi G\int_0^R dr\,r^{2l}\left[\left(\int_r^R\frac{ds}{s^{l+1}}\frac{d}{ds}\rho\psi\right)\left(\int_r^R\frac{ds}{s^{l+1}}\rho\frac{d\chi}{ds}\right)\right]\end{array} &
\begin{array}{l}-\frac{\sigma^2}{l(l+1)}\int_0^R\rho\left(\frac{d\chi}{dr}\right)^2dr+\gamma\int_0^R p\left(\frac{d\chi}{dr}\right)^2\frac{dr}{r^2}\\ \quad -4\pi G\int_0^R dr\,r^{2l}\left[\int_r^R\frac{ds}{s^{l+1}}\rho\frac{d\chi}{ds}\right]^2\end{array}
\end{Vmatrix}=0\,. \tag{43}$$

The fact that the variational principle leads to a characteristic equation which is quadratic in σ^2 corresponds to the circumstance (first noted by Cowling 1942) that the characteristic frequencies belonging to a particular spherical harmonic, apparently fall into two distinct spectra with normal modes of widely different attributes. Thus, it appears that one may distinguish between a set of modes which are principally radial—the p-modes of Cowling—and a set of modes which are principally transversal— the g-modes of Cowling. From the present point of view, the extreme aspects of these two sets of modes will be described by the following equations which are obtained from equation (43) by setting $\chi = 0$ and $\psi = 0$, respectively:

$$\sigma^2 \int_0^R \rho\psi^2 \frac{dr}{r^2} = \int_0^R \left[\gamma p \left(\frac{d\psi}{dr}\right)^2 + \frac{1}{\rho}\frac{dp}{dr}\frac{d}{dr}(\rho\psi^2) \right] \frac{dr}{r^2} - 4\pi G \int_0^R dr\, r^{2l} \left[\int_r^R \frac{ds}{s^{l+1}} \frac{d}{ds}(\rho\psi) \right]^2 , \tag{44}$$

and

$$\frac{\sigma^2}{l(l+1)} \int_0^R \rho \left(\frac{d\chi}{dr}\right)^2 dr = \gamma \int_0^R p \left(\frac{d\chi}{dr}\right)^2 \frac{dr}{r^2} - 4\pi G \int_0^R dr\, r^{2l} \left[\int_r^R \frac{ds}{s^{l+1}} \rho \frac{d\chi}{ds} \right]^2 . \tag{45}$$

The approximation considered by Cowling and others, of neglecting the variation of the gravitational potential during the oscillations, is equivalent to suppressing the last term ($\delta^2 W$) in equation (38); and when this term is suppressed, equation (43) becomes

$$\begin{Vmatrix} \displaystyle\int_0^R \left[\gamma p \left(\frac{d\psi}{dr}\right)^2 + \frac{1}{\rho}\frac{dp}{dr}\frac{d}{dr}(\rho\psi^2) \right] \frac{dr}{r^2} - \sigma^2 \int_0^R \rho\psi^2 \frac{dr}{r^2} & \displaystyle -\int_0^R \left(\gamma p \frac{d\psi}{dr}\frac{d\chi}{dr} + \frac{dp}{dr}\psi\frac{d\chi}{dr} \right) \frac{dr}{r^2} \\ \displaystyle -\int_0^R \left(\gamma p \frac{d\psi}{dr}\frac{d\chi}{dr} + \frac{dp}{dr}\psi\frac{d\chi}{dr} \right) \frac{dr}{r^2} & \displaystyle \gamma \int_0^R p \left(\frac{d\chi}{dr}\right)^2 \frac{dr}{r^2} - \frac{\sigma^2}{l(l+1)} \int_0^R \rho \left(\frac{d\chi}{dr}\right)^2 dr \end{Vmatrix} = 0 . \tag{46}$$

IV. TWO ILLUSTRATIONS

Detailed applications of equation (43) to the problem of determining the characteristic frequencies of non-radial oscillations of polytropes will be considered in a later paper. In this paper we shall limit ourselves to illustrating its applications by deriving two known results.

a) The Variational Principle for Purely Radial Oscillations

The case of purely radial oscillations is obtained by setting $l = 0$ and $\chi = 0$. Equation (43) then becomes

$$\sigma^2 \int_0^R \rho\psi^2 \frac{dr}{r^2} = \int_0^R \left[\gamma p \left(\frac{d\psi}{dr}\right)^2 + \frac{1}{\rho}\frac{dp}{dr}\frac{d}{dr}(\rho\psi^2) \right] \frac{dr}{r^2} - 4\pi G \int_0^R dr \left[\int_r^R \frac{ds}{s} \frac{d}{ds}(\rho\psi) \right]^2 . \tag{47}$$

The last term on the right-hand side of equation (47) can be reduced in the following manner:

$$
\begin{aligned}
4\pi G\int_0^R dr\left[\int_r^R \frac{ds}{s}\frac{d}{ds}(\rho\psi)\right]^2 &= 8\pi G\int_0^R dr\,\frac{d}{dr}(\rho\psi)\int_r^R \frac{ds}{s}\frac{d}{ds}(\rho\psi)\\
&= 8\pi G\int_0^R \frac{ds}{s}\frac{d}{ds}(\rho\psi)\int_0^s dr\,\frac{d}{dr}(\rho\psi)\\
&= 4\pi G\int_0^R \frac{dr}{r}\frac{d}{dr}(\rho\psi)^2 = 4\pi G\int_0^R \rho^2\psi^2\frac{dr}{r^2}\\
&= -\int_0^R \rho\frac{\psi^2}{r^4}\frac{d}{dr}\left(r^2\frac{d\mathfrak{V}}{dr}\right)dr = \int_0^R r^2\frac{d\mathfrak{V}}{dr}\frac{d}{dr}\left(\rho\frac{\psi^2}{r^4}\right)dr\\
&= \int_0^R \frac{r^2}{\rho}\frac{dp}{dr}\frac{d}{dr}\left(\rho\frac{\psi^2}{r^4}\right)dr\,.
\end{aligned}
\tag{48}
$$

Inserting the result of this reduction in equation (47) and simplifying, we are left with

$$
\sigma^2\int_0^R \rho\psi^2\frac{dr}{r^2} = \int_0^R\left[\gamma p\left(\frac{d\psi}{dr}\right)^2 + \frac{4}{r}\frac{dp}{dr}\psi^2\right]\frac{dr}{r^2}\,. \tag{49}
$$

It can be readily verified that this last equation is equivalent to the one which was first derived by Ledoux and Pekeris (cf. Ledoux and Walraven 1958, p. 465) for treating radial oscillations.

b) The "Kelvin Modes"

As a second illustration, consider the following particular case which leads to a simple class of solenoidal velocity fields (cf. n. 1 above):

$$
\psi = \chi = r^{l+1}\,. \tag{50}
$$

For this choice of the radial functions

$$
\delta\rho = r^{l-1}\frac{d\rho}{dr}\,. \tag{51}
$$

With the substitutions (50) and (51), equation (41) becomes

$$
\begin{aligned}
\frac{2l+1}{l}\sigma^2\int_0^R \rho r^{2l}dr &= \int_0^R \frac{r^{2l}}{\rho}\frac{dp}{dr}\frac{d\rho}{dr}dr - 4\pi G\int_0^R \rho^2 r^{2l}dr\\
&= -G\int_0^R \frac{d\rho}{dr}M(r)\,r^{2l-2}dr - 4\pi G\int_0^R \rho^2 r^{2l}dr\,.
\end{aligned}
\tag{52}
$$

On further simplification, we obtain

$$
\sigma^2 = \frac{2l(l-1)}{2l+1}G\frac{\displaystyle\int_0^R \rho r^{2l-3}M(r)\,dr}{\displaystyle\int_0^R \rho r^{2l}dr} = \frac{2l(l-1)(2l-1)}{2l+1}\frac{\displaystyle\int_0^R p r^{2l-2}dr}{\displaystyle\int_0^R \rho r^{2l}dr}\,. \tag{53}
$$

The analogy with the Kelvin modes of non-radial oscillations of an incompressible sphere is manifest.

Equation (53) is a generalization of the results known for the cases $l = 2$ and $l = 3$ (Chandrasekhar and Lebovitz 1962, eq. [3], and 1963, eq. [64]).

The research reported in this paper has in part been supported by the Office of Naval Research under contract Nonr-2121(24) with the University of Chicago.

APPENDIX

THE FORM OF THE VARIATIONAL PRINCIPLE FOR CYLINDRICAL SYSTEMS

The variational principle formulated in the paper, in the context of spherical systems, can be adapted readily to cylindrical systems in which, in the equilibrium state, the physical variables are functions only of the distance ϖ from the axis. The appropriately modified form of equation (19) is clearly

$$\sigma^2\int_V \rho\,|\,\xi\,|^2 dx = \int_V\left[\gamma p(\operatorname{div}\xi)^2 + 2\frac{dp}{d\varpi}\xi_\varpi \operatorname{div}\xi\right]dx + \int_V \frac{\xi_\varpi^2}{\rho}\frac{d\rho}{d\varpi}\frac{dp}{d\varpi}dx - \int_V \delta\rho(x)\,\delta\mathfrak{V}(x)\,dx\,, \tag{A.1}$$

where the integrations over V are now to be interpreted as averages per unit length of the cylinder; also, in equation (A.1), ξ_ϖ is the ϖ-component of the Lagrangian displacement $\xi = (\xi_\varpi, \xi_\varphi, \xi_z)$ in a cylindrical system of coordinates (ϖ, φ, z).

Without loss of generality, we may assume that, under the circumstances considered, a normal mode of oscillation has a (φ, z)-dependence given by $\cos kz \cos m\varphi$ where $0 \leq |k| < \infty$ and m is an integer (positive or zero). The Lagrangian displacement belonging to such a normal mode can be expressed in the form

$$\xi_\varpi = \frac{\psi(\varpi)}{\varpi}\cos kz\cos m\varphi\,, \qquad \xi_\varphi = \frac{m}{\varpi}\chi(\varpi)\cos kz \sin m\varphi\,, \qquad \text{and } \xi_z = k\chi(\varpi)\sin kz\cos m\varphi\,, \tag{A.2}$$

where ψ and χ are two radial functions. For ξ of this chosen form

$$\operatorname{div}\xi = \left[\frac{1}{\varpi}\frac{d\psi}{d\varpi} + \left(\frac{m^2}{\varpi^2} + k^2\right)\chi\right]\cos kz\cos m\varphi \tag{A.3}$$

and

$$\delta\rho(x) = -\operatorname{div}\rho\,\xi = -\,\delta\rho(\varpi)\cos kz\cos m\varphi\,, \tag{A.4}$$

where

$$\delta\rho(\varpi) = \frac{1}{\varpi}\frac{d}{d\varpi}(\rho\psi) + \left(\frac{m^2}{\varpi^2} + k^2\right)\rho\chi\,. \tag{A.5}$$

For $\delta\rho(x)$ given by equation (A.4), $\delta\mathfrak{V}$, determined as the solution of Poisson's equation, is given by

$$\delta\mathfrak{V} = -4\pi G\left[I_m(k\varpi)\int_\varpi^R \varpi'\delta\rho(\varpi')K_m(k\varpi')\,d\varpi' + K_m(k\varpi)\int_0^\varpi \varpi'\delta\rho(\varpi')I_m(k\varpi')\,d\varpi'\right]\cos kz\cos m\varphi\,, \tag{A.6}$$

where I_m and K_m are the Bessel functions of order m for a purely imaginary argument and R is the radius of the cylinder.

With the foregoing expressions for the various quantities which occur in equation (A.1), we find that the present analogue of equation (41) is

$$\sigma^2\int_0^R \rho\left[\frac{\psi^2}{\varpi^2}+\left(\frac{m^2}{\varpi^2}+k^2\right)\chi^2\right]\varpi d\varpi$$

$$=\gamma\int_0^R p\left[\frac{1}{\varpi}\frac{d\psi}{d\varpi}+\left(\frac{m^2}{\varpi^2}+k^2\right)\chi\right]^2\varpi d\varpi$$

$$+2\int_0^R \frac{dp}{d\varpi}\psi\left[\frac{1}{\varpi}\frac{d\psi}{d\varpi}+\left(\frac{m^2}{\varpi^2}+k^2\right)\chi\right]d\varpi+\int_0^R \frac{\psi^2}{\rho}\frac{d\rho}{d\varpi}\frac{dp}{d\varpi}\frac{d\varpi}{\varpi} \tag{A.7}$$

$$-8\pi G\int_0^R d\varpi\varpi\,\delta\rho(\varpi)K_m(k\varpi)\int_0^{\varpi} d\varpi'\varpi'\,\delta\rho(\varpi')I_m(k\varpi').$$

As in Section III of the paper, we may, by writing $A\psi$ and $B\chi$ in place of ψ and χ in equation (A.7), derive a characteristic equation of order 2 for σ^2 from the requirement that σ^2 be stationary with respect to small variations in A and B; but we shall not write out the resulting equation explicitly.

REFERENCES

Chandrasekhar, S. 1961, *Hydrodynamic and Hydromagnetic Stability* (Oxford: Clarendon Press), Appendix iii, pp. 622–25.
———. 1963, *Ap. J.*, **138**, 896.
Chandrasekhar, S., and Lebovitz, N. R. 1962, *Ap. J.*, **136**, 1105.
———. 1963, *ibid.*, **138**, 185.
Cowling, T. G. 1942, *M.N.*, **101**, 367.
Ledoux, P., and Walraven, Th. 1958, *Hdb. d. Phys.*, ed. S. Flügge (Berlin: Springer-Verlag), pp. 509–26.
Pekeris, C. L. 1938, *Ap. J.*, **88**, 189.

25

NON-RADIAL OSCILLATIONS OF GASEOUS MASSES

S. Chandrasekhar and Norman R. Lebovitz
University of Chicago
Received June 22, 1964; revised in proof October 19, 1964

ABSTRACT

In this paper the fundamental frequencies of non-radial oscillations of polytropic gas spheres, belonging to spherical harmonics of orders $l = 1$ and 2, are evaluated, in a "second approximation," by a variational method. Also, the value of the ratio of the specific heats γ is determined for which an accidental degeneracy occurs between the fundamental modes of radial oscillation and non-radial oscillation belonging to $l = 2$; it is found that this value of γ varies from 1.6 for a homogeneous compressible sphere to 1.5719 for a polytrope of index $n = 3$.

I. INTRODUCTION

In a recent paper (Chandrasekhar and Lebovitz 1963; this paper[1] will be referred to hereafter as "Paper I") the non-radial oscillations of gaseous masses belonging to the spherical harmonics of orders $l = 1$ (in a "second approximation") and $l = 3$ (in a "first approximation") were considered on the basis of the virial equations of orders 1 and 3 and trial functions of suitable forms for the Lagrangian displacement. The modes of oscillation belonging to $l = 2$ (in a first approximation) had been considered earlier (Chandrasekhar and Lebovitz 1962*a*, *c*); since their evaluation in a second approximation by the method of Paper I would have required the use of the virial equations of order 4, as well, it was not attempted. However, soon afterward, a general variational principle governing the non-radial oscillations of gaseous masses and belonging to the spherical harmonics of the different orders was established (Chandrasekhar 1964; this paper will be referred to hereafter as "Paper II"). In this paper the variational principle will be used to evaluate the modes belonging to $l = 2$ in a second approximation with the principal object of determining, more precisely than hitherto, the value of the ratio of the specific heats γ for which an accidental degeneracy occurs between the fundamental modes of radial oscillation and non-radial oscillation belonging to $l = 2$. The facts that such a degeneracy must occur and, further, that it may lie at the base of the beat phenomenon exhibited by the β Canis Majoris stars have been pointed out (Chandrasekhar and Lebovitz 1962*a*, *b*, and *d*); and the value $\gamma = 1.6$ (independently of the structure of the configuration) for the occurrence of such a degeneracy was deduced by equating the first approximations to the characteristic frequencies given by the two formulae (Ledoux 1945; Chandrasekhar and Lebovitz 1962*a*)

$$\sigma_R^2 = (3\gamma - 4)\frac{|\mathfrak{W}|}{I} \quad \text{and} \quad \sigma_S^2 = \tfrac{4}{5}\frac{|\mathfrak{W}|}{I}, \tag{1}$$

where $\mathfrak{W}$ denotes the gravitational potential energy and I the moment of inertia of the configuration. By restricting ourselves to polytropic gas spheres and evaluating σ_R^2 and σ_S^2 in a second approximation, we shall determine the dependence of γ (for which $\sigma_R^2 = \sigma_S^2$) on the density distribution in the configuration.

II. THE CHARACTERISTIC EQUATION IN THE SECOND APPROXIMATION

As shown in Paper II (Sec. III) the non-radial oscillations of gaseous masses, belonging to spherical harmonics of a particular order l, are determined by two radial functions

[1] The following errata for this paper may be noted here: In eq. (66) the entry opposite 3.5 should be 0.054251 (in place of 0.054834); the last of the equations in (87) should read $\sigma^4 - 0.12097\,(\gamma - 1.04641)\sigma^2 + 0.0011820\,(\gamma - 1.2784) = 0$; and finally in Table 1 on p. 199 the entry opposite $\int\theta^n\xi^6 d\xi$ in the last column should read 160.393 (in place of 158.689).

ψ and χ. And the variational principle, which expresses σ^2 as a functional of ψ and χ, requires that σ^2 be stationary with respect to arbitrary, infinitesimal variations of ψ and χ compatible with the boundary conditions.

Now it can be deduced from Pekeris' (1939) exact treatment of the non-radial oscillations of a homogeneous compressible sphere that the proper solutions for ψ and χ, in that case, are of the forms

$$\psi = Ar^{l+1} + Br^{l+3} \qquad \text{and} \qquad \chi = Ar^{l+1} + Cr^{l+3} . \tag{2}$$

We shall accordingly assume for ψ and χ these forms and treat the constants A, B, and C, as variational parameters.

It will be observed that we have made the coefficients of r^{l+1} equal in the assumed forms of ψ and χ. This equality in the coefficients follows, as it has been pointed out to us by Professor Paul H. Roberts, from the requirement that div $\boldsymbol{\xi}$, as given by equation (33) of Paper II, should behave like r^l at the origin[2]—a fact which becomes apparent when we examine the behavior at the origin of the expression for $\delta\rho$ given by equation (35) of Paper II.

By substituting for ψ and χ the chosen forms in Paper II, equation (41), we find that the characteristic equation for σ^2 becomes

$$|\boldsymbol{M}| = 0 , \tag{3}$$

where $\boldsymbol{M}$ is a symmetric matrix whose elements are

$$\begin{aligned}
M_{11} &= \sigma^2 \frac{2l+1}{l} \rho_{2l} - 2(2l-1)(l+1)p_{2l-2} - V_{2l+2} \\
&\qquad + 4\pi G \int_0^R r^{2l}[F_{l+1} - (l+1)f]^2 dr , \\
M_{12} &= \sigma^2 \rho_{2l+2} - (2l+1)(l+1)p_{2l} - V_{2l+4} \\
&\qquad + 4\pi G \int_0^R r^{2l} F_{l+3}[F_{l+1} - (l+1)f] dr , \\
M_{13} &= \sigma^2 \frac{l+3}{l} \rho_{2l+2} - (2l+1)(l+3)p_{2l} \\
&\qquad - 4\pi G(l+3) \int_0^R r^{2l} g[F_{l+1} - (l+1)f] dr , \\
M_{22} &= \sigma^2 \rho_{2l+4} - \gamma(l+3)^2 p_{2l+2} - V_{2l+6} + 4\pi G \int_0^R r^{2l} F^2_{l+3} dr , \\
M_{23} &= (l+3)[(l+3)\gamma - (2l+3)]p_{2l+2} - 4\pi G(l+3) \int_0^R r^{2l} F_{l+3} g \, dr , \\
M_{33} &= \sigma^2 \frac{(l+3)^2}{l(l+1)} \rho_{2l+4} - \gamma(l+3)^2 p_{2l+2} + 4\pi G(l+3)^2 \int_0^R r^{2l} g^2 dr ,
\end{aligned} \tag{4}$$

[2] This circumstance, that $\psi - \chi$ must behave like r^{l+3} as $r \to 0$, makes the reduction of the variational principle given by eq. (41) of Paper II to the form of a determinant of a 2×2 matrix (by varying ψ and χ independently of each other) strictly incorrect. However, it appears that the determinantal form retains a certain usefulness in providing an adequate approximation (see n. 4 on p. 1522) and clarifying the origin of the two physically distinct classes of modes.

and the following abbreviations have been used:

$$\rho_m=\int_0^R \rho r^m dr\,, \qquad p_m=\int_0^R p r^m dr\,, \tag{5}$$

$$V_m=\int_0^R \frac{1}{\rho}\frac{dp}{dr}\frac{d}{dr}(\rho r^m)\frac{dr}{r^2}=\int_0^R \frac{d\mathfrak{B}}{dr}\frac{d}{dr}(\rho r^m)\frac{dr}{r^2} \tag{6}$$

$$f(r)=\int_r^R \frac{ds}{s}\rho(s)\,, \qquad g(r)=\int_r^R ds\, s\rho(s)\,, \tag{7}$$

and

$$F_{l+1}(r)=\int_r^R \frac{ds}{s^{l+1}}\frac{d}{ds}(\rho s^{l+1})=-\rho+(l+1)f(r)\,,$$

$$F_{l+3}(r)=\int_r^R \frac{ds}{s^{l+1}}\frac{d}{ds}(\rho s^{l+3})=-\rho r^2+(l+1)g(r)\,. \tag{8}$$

The integrals which are explicitly written out in equations (4) can be reduced by integrations by parts and expressed in terms of the moments,

$$f_m=\int_0^R f\rho r^m dr\,, \qquad g_m=\int_0^R g\rho r^m dr\,. \tag{9}$$

We find

$$\int_0^R r^{2l}f^2 dr=\frac{2}{2l+1}f_{2l}\,,$$

$$\int_0^R r^{2l}fg\,dr=\frac{1}{2l+1}(f_{2l+2}+g_{2l})\,.$$

$$\int_0^R r^{2l}g^2 dr=\frac{2}{2l+1}g_{2l+2}\,,$$

$$\int_0^R r^{2l}F_{l+1}f\,dr=\frac{1}{2l+1}f_{2l}\,,$$

$$\int_0^R r^{2l}F_{l+1}g\,dr=\frac{1}{2l+1}[(l+1)f_{2l+2}-lg_{2l}]\,,$$

$$\int_0^R r^{2l}F_{l+3}f\,dr=\frac{1}{2l+1}[(l+1)g_{2l}-lf_{2l+2}]\,, \tag{10}$$

$$\int_0^R r^{2l}F_{l+3}g\,dr=\frac{1}{2l+1}g_{2l+2}\,,$$

$$\int_0^R r^{2l}F^2_{l+1}dr=\int_0^R r^{2l}\rho^2 dr-\frac{2l(l+1)}{2l+1}f_{2l}\,,$$

$$\int_0^R r^{2l}F_{l+1}F_{l+3}dr=\int_0^R r^{2l+2}\rho^2 dr-\frac{l(l+1)}{2l+1}(f_{2l+2}+g_{2l})\,,$$

$$\int_0^R r^{2l}F^2_{l+3}dr=\int_0^R r^{2l+4}\rho^2 dr-\frac{2l(l+1)}{2l+1}g_{2l+2}\,.$$

The integral defining V_m (eq. [6]) can also be reduced by integrations by parts and by making use of Poisson's equation and the equation governing equilibrium. Thus,

$$\begin{aligned} V_m &= \int_0^R \frac{1}{r^4}\left(r^2 \frac{d\mathfrak{V}}{dr}\right)\frac{d}{dr}(\rho r^m)\,dr \\ &= -\int_0^R \rho r^m\left[\frac{1}{r^4}\frac{d}{dr}\left(r^2\frac{d\mathfrak{V}}{dr}\right)-\frac{4}{r^3}\frac{d\mathfrak{V}}{dr}\right]dr \\ &= \int_0^R \rho r^m\left(\frac{4\pi G\rho}{r^2}+\frac{4}{r^3}\frac{1}{\rho}\frac{dp}{dr}\right)dr \\ &= 4\pi G\int_0^R \rho^2 r^{m-2}dr - 4(m-3)p_{m-4} \qquad (m \geq 4). \end{aligned} \tag{11}$$

Using equations (10) and (11), we find that the expressions for the matrix elements of $\boldsymbol{M}$ become

$$\begin{aligned} M_{11} &= \sigma^2\frac{2l+1}{l}\rho_{2l} - 2(2l-1)(l-1)p_{2l-2}, \\ M_{12} &= \sigma^2\rho_{2l+2} - (2l+1)(l-3)p_{2l} - 4\pi G(l+1)g_{2l}, \\ M_{13} &= \sigma^2\frac{l+3}{l}\rho_{2l+2} - (2l+1)(l+3)p_{2l} + 4\pi G(l+3)g_{2l}, \\ M_{22} &= \sigma^2\rho_{2l+4} - [\gamma(l+3)^2 - 4(2l+3)]p_{2l+2} - 8\pi G\frac{l(l+1)}{2l+1}g_{2l+2}, \\ M_{23} &= (l+3)[(l+3)\gamma - (2l+3)]p_{2l+2} - 4\pi G\frac{l+3}{2l+1}g_{2l+2}, \\ M_{33} &= \sigma^2\frac{(l+3)^2}{l(l+1)}\rho_{2l+4} - \gamma(l+3)^2 p_{2l+2} + 8\pi G\frac{(l+3)^2}{2l+1}g_{2l+2}. \end{aligned} \tag{12}$$

The approximation considered by Cowling (1942) and others of neglecting the variations in the gravitational potential during the oscillations is equivalent, in the present context, to ignoring the integrals over F_{l+1}, F_{l+3}, f, and g in the expressions for the matrix elements of $\boldsymbol{M}$ given in equations (4). Ignoring, then, these terms and substituting for the remaining terms in V_m in accordance with equation (11), we obtain the simpler expressions

$$\begin{aligned} M_{11}^{(a)} &= \sigma^2\frac{2l+1}{l}\rho_{2l} - 2(2l-1)(l-1)p_{2l-2} - 4\pi G\int_0^R \rho^2 r^{2l}dr, \\ M_{12}^{(a)} &= \sigma^2\rho_{2l+2} - (2l+1)(l-3)p_{2l} - 4\pi G\int_0^R \rho^2 r^{2l+2}dr, \\ M_{13}^{(a)} &= \sigma^2\frac{l+3}{l}\rho_{2l+2} - (2l+1)(l+3)p_{2l}, \\ M_{22}^{(a)} &= \sigma^2\rho_{2l+4} - [\gamma(l+3)^2 - 4(2l+3)]p_{2l+2} - 4\pi G\int_0^R \rho^2 r^{2l+4}dr, \\ M_{23}^{(a)} &= (l+3)[(l+3)\gamma - (2l+3)]p_{2l+2}\,. \\ M_{33}^{(a)} &= \sigma^2\frac{(l+3)^2}{l(l+1)}\rho_{2l+4} - \gamma(l+3)^2 p_{2l+2}, \end{aligned} \tag{13}$$

The Characteristic Equation for the Radial Mode of Oscillation in the Second Approximation

The variational expression for the characteristic frequencies of radial oscillation is given in Paper II (eq. [49]). With the trial function

$$\psi = ar^3 + br^5 \tag{14}$$

with the two variational parameters a and b, we obtain the characteristic equation

$$\begin{Vmatrix} \sigma^2 \rho_4 - 3(3\gamma - 4)p_2 & \sigma^2 \rho_6 - 5(3\gamma - 4)p_4 \\ \sigma^2 \rho_6 - 5(3\gamma - 4)p_4 & \sigma^2 \rho_8 - (25\gamma - 28)p_6 \end{Vmatrix} = 0 \,, \tag{15}$$

where ρ_m and p_m continue to have the same meanings as in equations (4). It is known (cf. Ledoux and Walraven 1958) that this "second approximation" gives the frequency of the fundamental mode of radial oscillation to well within 5 per cent in most cases of physical interest.

III. APPLICATION TO POLYTROPES

We shall use the characteristic equations derived in the preceding section to evaluate the frequencies of the fundamental modes of the radial and the non-radial oscillations of the polytropes.

When r, ρ, and p are expressed in terms of the usual Emden variables ξ and θ (cf. Chandrasekhar and Lebovitz 1962*c*, eq. [8]) and σ^2 is measured in the unit $4\pi G\rho_c/(n+1)$ (where n is the polytropic index), it can be readily verified that the elements of the secular matrix continue to be given by equations (12) if we replace $4\pi G$, wherever it occurs, by $n + 1$ and define ρ_m, p_m, and g_m in terms of the corresponding dimensionless variables as follows:

$$\rho_m = \int_0^{\xi_1} \theta^n \xi^m d\xi \,, \qquad p_m = \int_0^{\xi_1} \theta^{n+1} \xi^m d\xi \,,$$

and

$$g_m = \int_0^{\xi_1} g\,\theta^n \xi^m d\xi = \int_0^{\xi_1} d\xi\, \xi^m \theta^n \int_{\xi}^{\xi_1} d\eta\, \eta\, \theta^n(\eta)\,, \tag{16}$$

where ξ_1 is the first zero of θ.

In the case under consideration, g_m can be expressed directly in terms of ρ_m and p_m; thus,

$$g(\xi) = \int_{\xi}^{\xi_1} \theta^n(\eta)\, \eta\, d\eta = -\int_{\xi}^{\xi_1} \left(\eta \frac{d^2\theta}{d\eta^2} + 2\frac{d\theta}{d\eta}\right) d\eta = \xi \frac{d\theta}{d\xi} + \theta + \xi_1 |\theta_1'| \,, \tag{17}$$

where θ_1' is the value of $d\theta/d\xi$ at ξ_1, and

$$g_m = \xi_1 |\theta_1'| \rho_m + p_m + \int_0^{\xi_1} \theta^n \frac{d\theta}{d\xi} \xi^{m+1} d\xi = \xi_1 |\theta_1'| \rho_m + \frac{n-m}{n+1} p_m \,. \tag{18}$$

Also, it may be noted that

$$\rho_2 = \xi_1^2 |\theta_1'| \qquad \text{and} \qquad p_2 = \frac{n+1}{5-n} \xi_1^3 |\theta_1'|^2 \,. \tag{19}$$

In Table 1, we list the values of the various integrals which are needed for the determination (in the second approximation) of the lowest modes belonging to $l = 0$, 1, and 2.

a) The Convective Instability of the Polytropes for $\gamma < 1 + 1/n$ by Modes Belonging to $l = 1$

We have already shown in Paper I, by an application of the virial equations, that the manifestation of the convective instability of the polytropes for $\gamma < 1 + 1/n$, by modes belonging to $l = 1$, can be explicitly demonstrated: the critical value of γ for marginal stability predicted by the (approximate) theory differs from $1 + 1/n$ by less than 0.7 per cent for $n \leq 3.5$. It is evident that the present characteristic equation $|\boldsymbol{M}| = 0$ for $l = 1$ must predict the *same* critical values of γ (for the different polytropes) as were derived in Paper I: for the trial function assumed for the Lagrangian displacement in both the treatments are the same; the characteristic equations which follow cannot, therefore, be different.[3] And the fact that the present characteristic equation $|\boldsymbol{M}| = 0$ for $l = 1$ leads to the same critical values for γ as were derived in Paper I can, indeed, be verified directly.[4]

TABLE 1

A TABLE OF INTEGRALS

INTEGRAL	n				
	1.0	1.5	2.0	3.0	3.5
ρ_2	3.14159	2.71406	2.41105	2.01824	1.89056
ρ_4	12.1567	11.1197	10.6110	10.8516	11.7454
ρ_6	62.8853	64.9770	71.7372	109.748	160.39
ρ_8	379.112	460.821	617.802	1625.30	3585.89
p_0	1.41815	1.34001	1.27421	1.16855	1.12515
p_2	1.57080	1.44002	1.33547	1.18120	1.12446
p_4	4.38231	4.23148	4.17017	4.31761	4.56828
p_6	17.4550	18.7887	21.0149	30.4548	40.9726
g_2	2.35619	1.72803	1.33547	0.885900	0.749639
g_4	5.58325	4.02839	3.09731	2.09611	1.82105
g_6	19.2477	14.4461	11.7151	9.2748	9.0368

b) The "Kelvin Modes" Belonging to l = 2*; and the Modes Exhibiting Convective Instability*

Using the same matrix equation $|\boldsymbol{M}| = 0$ for $l = 2$, we have determined, with the aid of the integrals listed in Table 1, the lowest characteristic values for polytropes with the indices 1, 1.5, 2, 3, and 3.5. The calculated characteristic values together with the constants of the corresponding characteristic vector $(1, B/A, C/A)$ are listed in Table 2.

We notice the remarkably slight dependence of σ^2 on γ for $n \leq 2$. We also observe that for moderate central condensations ($n \leq 1.5$) the approximation $\psi \simeq \chi \simeq r^{l+1}$

[3] The fact that the two methods, the variational and the virial, must lead to the same characteristic equation (albeit by different routes) can be seen as follows: in the variational method, the characteristic equation for σ^2 appropriate to a trial function of the form

$$\xi_i = L_{i;jk}x_jx_k + L_i$$

(assumed in Paper I) follows from inserting this expression for $\boldsymbol{\xi}$ in a certain Hermitian form $(\boldsymbol{\xi}, H\boldsymbol{\xi})$, where $H\boldsymbol{\xi}$ gives the time-dependence of $\boldsymbol{\xi}$ for small departures from equilibrium in accordance with the linearized equations of motion, and making the result stationary for small variations of the constants $L_{i;jk}$ and L_i; whereas, in the virial method, we start with the three zero-order and the eighteen second-order x_jx_k-moments of the exact non-linear equations of motion, linearize these moment equations for small departures from equilibrium, and *then* make the same assumption for the Lagrangian displacement. The two sets of homogeneous equations which follow for the constants $L_{i;jk}$ and L_i, clearly, cannot differ (cf. Clement [1964] where the analysis exhibiting this equivalence is set out in full in another context).

[4] We find, for example, that the equation $|\boldsymbol{M}| = 0$ for $l = 1$ leads to the values $\gamma = 1.996$, 1.662, 1.494, 1.327, and 1.279 for the marginal stability of the polytropes $n = 1$, 1.5, 2, 3, and 3.5, respectively; and these values should be compared with 1.995, 1.661, 1.493, 1.326, and 1.278 derived in Paper I (eq. [88]).

provides a satisfactory representation.[5] As we have seen in Paper II (Sec. IV*b*) the *assumption* $\psi = \chi = r^{l+1}$ leads to an expression for σ^2 which is *exactly* analogous to Kelvin's formula for the non-radial modes of an incompressible sphere. On this account, it would seem proper to describe the modes obtained in this section as the "Kelvin modes."

TABLE 2

THE SQUARES OF THE CHARACTERISTIC FREQUENCIES AND RELATED CONSTANTS BELONGING TO THE FUNDAMENTAL MODES OF RADIAL AND NON-RADIAL ($l = 2$) OSCILLATIONS

(σ^2 Is Listed in the Unit $4\pi G\rho_c/[n + 1]$)

n	γ	RADIAL		NON-RADIAL ($l=2$)		
		σ^2	a_3/a_1	σ^2	B/A	C/A
1.0	1.55	0.25002	0.01594	0.30305	0.04938	+0.02537
	1.60	.30726	.01928	.30332	.04604	+ .02287
	1.65	.36437	.02252	.30357	.04293	+ .02055
	5/3	.38338	.02357	.30365	.04194	+ .01982
1.5	1.55	.24711	.02615	.29302	.07114	+ .02754
	1.60	.30287	.03206	.29390	.06429	+ .02284
	1.65	.35825	.03790	.29475	.05810	+ .01859
	5/3	.37663	.03983	.29503	.05616	+ .01727
2.0	1.55	.23357	.03925	.26848	.09219	+ .02462
	1.60	.28473	.04910	.27062	.08041	+ .01753
	1.65	.33509	.05922	.27271	.07020	+ .01143
	5/3	.35171	.06265	.27339	.06709	+ .00959
3.0	1.55	.17088	.09509	.18184	.14326	+ .01203
	1.60	.20195	.13147	.18845	.11701	+ .00195
	1.65	.23125	.17596	.19503	.09691	− .00538
	5/3	.24069	.19299	.19721	.09123	− .00738
3.5	1.55	.11740	.20435	.12214	.18433	+ .00533
	1.60	.13442	.34067	.12990	.14854	− .00457
	1.65	.15028	.59891	.13780	.12296	− .01097
	5/3	0.15538	0.74444	0.14044	0.11597	−0.01259

To test how good the approximation is of ignoring the variations in the gravitational potential during the oscillation, we have evaluated σ^2 from the simplified equation $|M^{(\alpha)}| = 0$ (cf. eq. [13]) for the case $n = 3$, $l = 2$, and $\gamma = 1.6$ and $\frac{5}{3}$. We find

$$\sigma^2 = \begin{cases} 0.2264 & (\gamma = 1.6), \\ 0.2398 & (\gamma = \frac{5}{3}); \end{cases} \tag{20}$$

and these values should be compared with $\sigma^2 = 0.1884$ and 0.1972 listed in Table 2. It would not appear from this comparison that the approximation is a very good one under the circumstances considered.

In addition to the Kelvin modes we have considered, there exist also modes which exhibit the convective instability of the polytropes for $\gamma < 1 + 1/n$. Indeed, we find

[5] If we had varied ψ and χ independently of each other, then for the same assumed form of the trial functions we should have found $\sigma^2 = 0.1800, 0.1868, 0.1935$, and 0.1957 for the case $n = 3$ and $\gamma = 1.55$, 1.60, 1.65, and $\frac{5}{3}$, respectively; and these values should be contrasted with the values $\sigma^2 = 0.1818, 0.1885$, 0.1950, and 0.1972 listed in Table 2 derived after satisfying the proper boundary condition, $\psi - \chi \rightarrow$ constant r^{l+3} as $r \rightarrow 0$.

from the same secular equation $|\boldsymbol{M}| = 0$ and $l = 2$ that a neutral mode occurs when (cf. Paper I, eq. [88]; also n. 3 on p. 1522)

$$\gamma = \begin{cases} 1.9946 & \text{for} \quad n = 1.0\,, \\ 1.6588 & \text{for} \quad n = 1.5\,, \\ 1.4899 & \text{for} \quad n = 2.0\,, \\ 1.3188 & \text{for} \quad n = 3.0\,, \\ 1.2686 & \text{for} \quad n = 3.5. \end{cases} \tag{21}$$

The departures of these values from $1 + 1/n$ is a measure of the accuracy attained by the present manner of application of the variational method.

c) The Radial Modes

The characteristic equation (15) governing the radial modes has also been solved for the same values of γ and n for which the Kelvin modes for $l = 2$ have been determined. The results of the calculations are included in Table 2.

TABLE 3

THE VALUES OF γ FOR WHICH THE ACCIDENTAL DEGENERACY OCCURS AND THE CONSTANTS OF THE CORRESPONDING MODES

n	γ	RADIAL		NON-RADIAL ($l=2$)		
		σ^2	a_3/a_1	σ^2	B/A	C/A
0.......	1.6000	0.26667	0	0.26667	0	0
1.0.....	1.5965	.30330	0.01906	.30330	0.04626	+0.02304
1.5.....	1.5918	.29376	.03109	.29376	.06537	+ .02358
2.0.....	1.5855	.27001	.04622	.27001	.08364	+ .01947
3.0.....	1.5719	.18473	.1101	.18473	.13240	+ .00790
3.5.....	1.5745	0.12592	0.2619	0.12591	0.16526	−0.00005

d) The Value of γ for Which the Accidental Degenercy Occurs

By interpolating among the values of σ^2 listed in Table 2, the values of γ, for which σ^2 for the radial mode and the non-radial mode belonging to $l = 2$ are equal, were determined. The results of this interpolation are given in Table 3.

We observe that, in contrast to what follows from the solutions in the first approximation, the value of γ for which the accidental degeneracy occurs depends on the density distribution in the configuration. However, the dependence is not very pronounced.

For the polytrope $n = 3$, the accidental degeneracy occurs for $\gamma = 1.572$. While this value of γ differs from 1.6 by only 0.028, the stellar requirements for the coincidence of the two frequencies are appreciably different: on the assumption that differences in γ arise from different admixtures of a monatomic gas and radiation, the change in the effective ratio of the specific heats from 1.6 to 1.572 means that $1 - \beta =$ (radiation pressure/total pressure) changes from 0.0532 to 0.0868; and this change in $(1 - \beta)$ implies, on the standard model, a change in $(M\mu^2/\odot)$ from 4.62 to 6.36. It would therefore appear that the suggested interpretation of the beat phenomenon exhibited by the β Canis Majoris stars will not be found inadequate on this account: its validity will have to be tested on other grounds.

We are greatly indebted to Miss Donna D. Elbert for her assistance with all the numerical work in connection with the preparation of this paper.

The research report in this paper has in part been supported by the Office of Naval Research under contract Nonr-2121(24) with the University of Chicago. The work of the second author was supported by the Air Force Office of Scientific Research under contract No. AFOSR-62-136 with the University of Chicago.

APPENDIX

AN ALTERNATIVE FORMULATION OF THE VARIATIONAL PRINCIPLE

In deriving the variational principle in Paper II, the assumption was made that the density vanishes on the boundary; and this assumption was explicitly used in the further reductions. Nevertheless, it was found, after suitable transformations of the basic equations, that the variational principle gives correctly the characteristic frequencies of the Kelvin modes of an incompressible sphere (cf. Paper II, Sec. IV*b*, eq. [53]). The fact that such "suitable transformations" are possible suggests that there is an alternative formulation of the variational principle which is valid without the assumption that the density vanishes on the surface. We shall show that such an alternative formulation exists which, moreover, permits the density to be discontinuous in the interior.

It is convenient for our present purposes to adopt the Lagrangian, instead of the Eulerian, formulation of the perturbation equations.

If $\Delta F(\boldsymbol{x})$ and $\delta F(\boldsymbol{x})$ denote, respectively, the Lagrangian and the Eulerian changes in a variable $F(\boldsymbol{x})$, then (cf. Lebovitz 1961, eq. [21])

$$\Delta F(\boldsymbol{x}) = \delta F(\boldsymbol{x}) + \boldsymbol{\xi}\cdot\text{grad}\, F(\boldsymbol{x}), \tag{A.1}$$

where $\boldsymbol{\xi}(\boldsymbol{x})$ is the Lagrangian displacement. Also, it may be verified that the Lagrangian change in the gradient of $F(\boldsymbol{x})$ is given by

$$\Delta\left(\frac{\partial F}{\partial x_i}\right) = \frac{\partial \Delta F}{\partial x_i} - \frac{\partial F}{\partial x_k}\frac{\partial \xi_k}{\partial x_i}. \tag{A.2}$$

Using the foregoing equations, we find that the Lagrangian form of the equations governing the perturbations are (cf. Paper I, eqs. [9]–[13])

$$\sigma^2 \rho\, \xi_i = \frac{\partial \Delta p}{\partial x_i} - \Delta\rho\, \frac{\partial \mathfrak{B}}{\partial x_i} - \rho\, \frac{\partial \Delta \mathfrak{B}}{\partial x_i}, \tag{A.3}$$

$$\Delta\rho = -\rho\, \text{div}\, \xi, \qquad \Delta p = -\gamma p\, \text{div}\, \xi, \tag{A.4}$$

and

$$\Delta \mathfrak{B} = \delta \mathfrak{B} + \boldsymbol{\xi}\cdot\text{grad}\, \mathfrak{B}, \tag{A.5}$$

where

$$\delta \mathfrak{B}(\boldsymbol{x}) = G\int_V \rho(\boldsymbol{x}')\, \xi_i(\boldsymbol{x}') \frac{\partial}{\partial x_i'} \frac{1}{|\boldsymbol{x}-\boldsymbol{x}'|}\, d\boldsymbol{x}'; \tag{A.6}$$

and the boundary condition is that Δp vanish on the boundary.

Multiplying equation (A.3) by ξ_i, contracting, and integrating over the volume occupied by the fluid, we obtain

$$\begin{aligned}\sigma^2\int_V \rho\, |\boldsymbol{\xi}|^2 d\boldsymbol{x} &= \int_V \gamma p (\text{div}\, \boldsymbol{\xi})^2 d\boldsymbol{x} + \int_V \xi_i \frac{\partial p}{\partial x_i}\, \text{div}\, \boldsymbol{\xi}\, d\boldsymbol{x} \\ &\quad - \int_V \rho\, \xi_i \frac{\partial}{\partial x_i}\left(\xi_k \frac{\partial \mathfrak{B}}{\partial x_k}\right) d\boldsymbol{x} - \int_V \rho\, \xi_i \frac{\partial \delta \mathfrak{B}}{\partial x_i}\, d\boldsymbol{x},\end{aligned} \tag{A.7}$$

where we have performed an integration by parts in the first term on the right-hand side of equation (A.7) and have used equations (A.4) and (A.5) as well as the equilibrium condition grad $p = \rho$ grad $\mathfrak{B}$.

It can be verified that equation (A.7) provides a variational base for determining the characteristic frequencies. The verification is sufficiently similar to that given in Paper II (Sec. II) that it is omitted here.

We shall now specialize equation (A.7) to the case when the equilibrium configuration is spherically symmetric. One can, without loss of generality, restrict consideration to a single spherical harmonic $Y_l^m(\vartheta, \varphi)$. The Lagrangian displacement may then be written (cf. Paper II, eq. [31]; see also eqs. [32] and [33])

$$\xi_r = \frac{\psi(r)}{r^2} Y_l^m(\vartheta, \varphi), \qquad \xi_\vartheta = \frac{1}{l(l+1)r} \frac{d\chi(r)}{dr} \frac{\partial Y_l^m(\vartheta, \varphi)}{\partial \vartheta},$$

and

$$\xi_\varphi = \frac{1}{l(l+1)r \sin\vartheta} \frac{d\chi(r)}{dr} \frac{\partial Y_l^m(\vartheta, \varphi)}{\partial \varphi}. \tag{A.8}$$

These expressions for the components of $\boldsymbol{\xi}$ must be substituted into equation (A.7) and the integrations over the angles performed. This reduction has already been performed for the left-hand side and the first two terms on the right-hand side of equation (A.7) (Paper II, eq. [38]) with the result

$$\sigma^2 \int_0^R \left[\frac{\psi^2}{r^2} + \frac{1}{l(l+1)} \left(\frac{d\chi}{dr} \right)^2 \right] dr = \int_0^R \gamma p \left[\frac{d}{dr}(\psi - \chi) \right]^2 \frac{dr}{r^2}$$
$$+ \int_0^R \frac{dp}{dr} \psi \frac{d}{dr}(\psi - \chi) \frac{dr}{r^2} - \int_V \rho \xi_i \frac{\partial}{\partial x_i} \left(\xi_k \frac{\partial \mathfrak{B}}{\partial x_k} \right) dx - \int_V \rho \xi_i \frac{\partial \delta \mathfrak{B}}{\partial x_i} dx, \tag{A.9}$$

where we have suppressed a factor $N_{lm} = 4\pi(l + |m|)!/(2l + 1)\,(l - |m|)!$ since it will ultimately appear in every term of equation (A.10) and hence may be canceled.

To facilitate evaluating the final two terms of equation (A.9), we state the following lemma: Let $\boldsymbol{\xi}$ be given by equation (A.8), $F(r)$ be arbitrary, and let S_1 denote the unit sphere and dS_1 the associated element of area. Then

$$\iint_{S_1} \boldsymbol{\xi} \cdot \text{grad}\,[F(r) Y_\lambda^\mu(\vartheta, \varphi)]\, dS_1 = \frac{N_{lm}}{r^2} \left(\psi \frac{dF}{dr} + F \frac{d\chi}{dr} \right) \delta_{l\lambda} \delta_{m\mu}. \tag{A.10}$$

The proof is omitted, since it is an easy consequence of a known result (cf. Chandrasekhar 1961, p. 625).

Turning now to the third term on the right-hand side of equation (A.10), we find

$$- \int_V \rho \boldsymbol{\xi} \cdot \text{grad} \left(\xi_r \frac{d\mathfrak{B}}{dr} \right) dx = - \int_0^R dr\, r^2 \rho(r) \iint_{S_1} dS_1 \boldsymbol{\xi} \cdot \text{grad} \left(\frac{\psi}{r^2} \frac{d\mathfrak{B}}{dr} Y_l^m \right)$$
$$= - N_{lm} \int_0^R \rho(r) \left[\psi \frac{d}{dr} \left(\frac{\psi}{r^4} r^2 \frac{d\mathfrak{B}}{dr} \right) + \frac{\psi}{r^2} \frac{d\mathfrak{B}}{dr} \frac{d\chi}{dr} \right] dr$$
$$= - N_{lm} \int_0^R \rho(r) \left(-4 \frac{\psi^2}{r^3} \frac{d\mathfrak{B}}{dr} + \frac{\psi}{r^2} \frac{d\psi}{dr} \frac{d\mathfrak{B}}{dr} - 4\pi G \rho \frac{\psi^2}{r^2} + \frac{\psi}{r^2} \frac{d\mathfrak{B}}{dr} \frac{d\chi}{dr} \right) dr$$
$$= - N_{lm} \int_0^R \left[-4 \frac{\psi^2}{r^3} \frac{dp}{dr} + \frac{\psi}{r^2} \frac{d}{dr}(\psi + \chi) \frac{dp}{dr} - 4\pi G \rho^2 \frac{\psi^2}{r^2} \right] dr. \tag{A.11}$$

To reduce the final term in equation (A.9) we first find $\delta\mathfrak{B}$ with the aid of equation (A.7) and the known expansion of $|\boldsymbol{x}-\boldsymbol{x}'|^{-1}$ in spherical harmonics,

$$\frac{1}{|\boldsymbol{x}-\boldsymbol{x}'|}=\sum_{l=0}^{\infty}f_l(r,r')\sum_{m=-l}^{l}\frac{(l-|m|)!}{(l+|m|)!}Y_l^m(\vartheta,\varphi)Y_l^m(\vartheta',\varphi'), \tag{A.12}$$

where

$$f_l(r,r')=\begin{cases} r'^l/r^{l+1} & \text{if} \quad r'<r\,, \\ r^l/r'^{l+1} & \text{if} \quad r'>r\,. \end{cases} \tag{A.13}$$

Applying the lemma (eq. [A.10]), we find

$$\begin{aligned}\delta\mathfrak{B}(\boldsymbol{x})&=G\int_V\rho(\boldsymbol{x}')\xi_i(\boldsymbol{x}')\frac{\partial}{\partial x_i'}\frac{1}{|\boldsymbol{x}-\boldsymbol{x}'|}d\boldsymbol{x}'\\ &=G\int_0^R dr'r'^2\rho(r')\iint_{S_1}dS_1\xi_i(\boldsymbol{x}')\frac{\partial}{\partial x_i'}\frac{1}{|\boldsymbol{x}-\boldsymbol{x}'|}\\ &=GN_{lm}\frac{(l-|m|)!}{(l+|m|)!}Y_l^m(\vartheta,\varphi)\int_0^R\rho(r')\bigg[\psi(r')\frac{\partial f_l(r,r')}{\partial r'}\\ &+\frac{d\chi(r')}{dr'}f_l(r,r')\bigg]dr'=\frac{4\pi G}{2l+1}Y_l^m(\vartheta,\varphi)\bigg[\frac{J_l(r)}{r^{l+1}}-r^lK_l(r)\bigg],\end{aligned} \tag{A.14}$$

where, in virtue of equation (A.13),

$$J_l(r)=\int_0^r\rho(s)s^l\bigg[l\frac{\psi(s)}{s}+\frac{d\chi(s)}{ds}\bigg]ds$$

and

$$K_l(r)=\int_r^R\frac{\rho(s)}{s^{l+1}}\bigg[(l+1)\frac{\psi(s)}{s}-\frac{d\chi(s)}{ds}\bigg]ds\,. \tag{A.15}$$

We cannot eliminate both J_l and K_l from the final formula, but we can eliminate one of them, say J_l. For this purpose we need the following formula, which is an easy consequence of equations (A.15):

$$\frac{d}{dr}\left(\frac{J_l}{r^{l+1}}-r^lK_l\right)=-\frac{l+1}{r^{l+2}}J_l-lr^{l-1}K_l+(2l+1)\frac{\rho\psi}{r^2}. \tag{A.16}$$

Using equations (A.14) and (A.16) and the lemma (A.10), we now find

$$\begin{aligned}\int\rho\xi\cdot\text{grad }\delta\mathfrak{B}dx&=\frac{4\pi G}{2l+1}\int_0^R dr\rho r^2\iint_{S_1}dS_1\xi\cdot\text{grad}\left[\left(\frac{J_l}{r^{l+1}}-r^lK_l\right)Y_l^m\right]\\ &=\frac{4\pi GN_{lm}}{2l+1}\int_0^R\rho\bigg[\psi\frac{d}{dr}\left(\frac{J_l}{r^{l+1}}-r^lK_l\right)+\frac{d\chi}{dr}\left(\frac{J_l}{r^{l+1}}-r^lK_l\right)\bigg]dr\\ &=4\pi GN_{lm}\int_0^R\rho^2\frac{\psi^2}{r^2}dr+\frac{4\pi GN_{lm}}{2l+1}\int_0^R\left(J_l\frac{dK_l}{dr}-K_l\frac{dJ_l}{dr}\right)dr\\ &=4\pi GN_{lm}\int_0^R\rho^2\frac{\psi^2}{r^2}dr-\frac{8\pi GN_{lm}}{2l+1}\int_0^R K_l\frac{dJ_l}{dr}dr\\ &=4\pi GN_{lm}\int_0^R\rho^2\frac{\psi^2}{r^2}dr-\frac{8\pi GN_{lm}}{2l+1}\int_0^R\rho K_lr^l\left(l\frac{\psi}{r}+\frac{d\chi}{dr}\right)dr\,.\end{aligned} \tag{A.17}$$

On account of equations (A.11) and (A.17), equation (A.9) becomes

$$\sigma^2\int_0^R \rho\left[\frac{\psi^2}{r^2}+\frac{1}{l(l+1)}\left(\frac{d\chi}{dr}\right)^2\right]dr = \int_0^R \gamma p\left[\frac{d}{dr}(\psi-\chi)\right]^2\frac{dr}{r^2}$$
$$+2\int_0^R \frac{dp}{dr}\frac{\psi}{r^2}\left(2\frac{\psi}{r}-\frac{d\chi}{dr}\right)dr+\frac{8\pi G}{2l+1}\int_0^R \rho K_l r^l\left(l\frac{\psi}{r}+\frac{d\chi}{dr}\right)dr, \tag{A.18}$$

which is the desired formula.

It may be useful to note that the formula (A.18) can be deduced also from equation (41) of Paper II, if one *continues* to assume that $\rho(R) = 0$. Straightforward manipulations involving integration by parts and explicit use of the condition $\rho(R) = 0$ alter two of the terms of that equation as follows:

$$\int_0^R \frac{1}{\rho}\frac{dp}{dr}\frac{d}{dr}(\rho\psi^2)\frac{dr}{r^2} = \int_0^R \frac{d\mathfrak{B}}{dr}\frac{d}{dr}(\rho\psi^2)\frac{dr}{r^2}$$
$$= 4\int_0^R \frac{dp}{dr}\frac{\psi^2}{r^3}dr + 4\pi G\int_0^R \rho^2\frac{\psi^2}{r^2}dr \tag{A.19}$$

and

$$\int_0^R r^{2l}\left[\int_r^R \frac{\delta\rho(s)}{s^{l-1}}ds\right]^2 dr = \int_0^R \rho^2\frac{\psi^2}{r^2}dr + \int_0^R r^{2l}K_l\left(K_l-2\frac{\rho\psi}{r^{l+1}}\right)dr$$
$$= \int_0^R \rho^2\frac{\psi^2}{r^2}dr - \frac{2}{2l+1}\int_0^R \rho K_l r^l\left(l\frac{\psi}{r}+\frac{d\chi}{dr}\right)dr, \tag{A.20}$$

where

$$\delta\rho(r) = \frac{1}{r^2}\left[\frac{d}{dr}(\rho\psi) - \rho\frac{d\chi}{dr}\right],$$

and we have made the observation that

$$\int_r^R \frac{\delta\rho(s)}{s^{l-1}}ds = -\frac{\rho\psi}{r^{l+1}} + K_l(r). \tag{A.21}$$

Using equations (A.19) and (A.20) in equation (41) of Paper II, we recover equation (A.18).

REFERENCES

Chandrasekhar, S. 1961, *Hydrodynamic and Hydromagnetic Stability* (Oxford: Clarendon Press), Appendix iii, pp. 622–25.
———. 1964, *Ap. J.*, **139**, 664.
Chandrasekhar, S., and Lebovitz, N. R. 1962*a*, *Ap. J.*, **135**, 248.
———. 1962*b*, *ibid.*, p. 305.
———. 1962*c*, *ibid.*, **136**, 1082.
———. 1962*d*, *ibid.*, p. 1105.
———. 1963, *ibid.*, **138**, 185.
Clement, M. 1964, *Ap. J.*, **141**, 1045.
Cowling, T. G. 1942, *M.N.*, **101**, 367.
Lebovitz, N. R. 1961, *Ap. J.*, **134**, 500.
Ledoux, P. 1945, *Ap. J.*, **102**, 143.
Ledoux, P., and Walraven, Th. 1958, *Hdb. d. Phys.*, ed. S. Flügge (Berlin: Springer-Verlag).
Pekeris, C. L. 1939, *Ap. J.*, **88**, 189.

THE PULSATIONS AND THE DYNAMICAL STABILITY OF GASEOUS MASSES IN UNIFORM ROTATION

S. CHANDRASEKHAR AND N. R. LEBOVITZ
University of Chicago
Received August 24, 1967

ABSTRACT

A variational principle, applicable to axisymmetric oscillations of uniformly rotating axisymmetric configurations, is established. On the assumption that the Lagrangian displacement (describing the oscillation) at any point is normal to the level surface (of constant total potential) through that point, it is shown how the variational expression, for the frequencies of oscillation, can be reduced to simple quadratures. The reduction is explicitly carried out for certain stratifications of special interest.

Some new results on the oscillations of slowly rotating configurations are included; and a number of related observations on their stability are also made.

I. INTRODUCTION

A slow uniform rotation affects the radial modes of adiabatic oscillation of an initial spherical distribution of mass in two essentially different ways (cf. Cowling and Newing 1949; see also Chandrasekhar and Lebovitz 1962*d*): *first*, the term in the centrifugal acceleration in the equation governing equilibrium modifies the initial distribution of density and pressure; and *second*, the term in the Coriolis acceleration, in the linearized equations governing small departures from equilibrium, further modifies the characteristic frequencies. In general both these effects are of order Ω^2 (the square of the angular velocity of rotation); and they are found to contribute terms of opposite signs to σ^2 (the square of the characteristic frequency of oscillation). In view particularly of this last circumstance, one cannot be certain whether, in a given situation, rotation will have a stabilizing or a destabilizing effect on the radial oscillations. There is, however, one important exception.

It is known that the condition for the dynamical stability of spherical masses (determined by the stability of the fundamental mode of radial oscillation) is that the ratio of the specific heats γ exceeds $\frac{4}{3}$, in case it is a constant.[1] Now if $\gamma - \frac{4}{3}$ is $O(\Omega^2)$ and $\Omega \to 0$, then it is an immediate consequence of a formula due to Ledoux (1945) that the effect of rotation on σ^2 can be written down at once, without having to determine its effect on the equilibrium distribution.

Now Ledoux's formula for σ^2, for a slowly rotating configuration, is

$$\sigma^2 = (3\gamma - 4)\frac{|\mathfrak{W}|}{I} + \tfrac{2}{3}(5 - 3\gamma)\Omega^2, \tag{1}$$

where $\mathfrak{W}$ is the gravitational potential energy and

$$I = \int_{\mathfrak{R}} \rho(x)\,|x|^2 dx \tag{2}$$

(where the integration is effected over the domain $\mathfrak{R}$ occupied by the fluid) is the moment of inertia about the center of mass. (Note that in eq. [1] both $\mathfrak{W}$ and I refer to the *rotating* configuration and therefore include, implicitly, terms of order Ω^2.)

[1] In case γ is not a constant, the condition is replaced by $\bar{\gamma} > \frac{4}{3}$, where $\bar{\gamma}$ is a pressure weighted average (cf. § IV*a*, however). In this introductory section, we shall restrict our remarks to the case γ = constant; the restriction will not be made in the later analytical developments.

While Ledoux originally derived his formula from an application of the scalar form of the virial theorem,[2] it was soon shown by Cowling and Newing (1949) that the formula equally follows from a variational expression for σ^2 with the *same* linear substitution for the Lagrangian displacement $\boldsymbol{\xi}$ (in terms of which σ^2 is expressed) that was made by Ledoux. And since in the limit $\gamma \to \frac{4}{3}$, the correct proper solution for $\boldsymbol{\xi}$ is indeed a linear function of the coordinates (cf. Rosseland 1949, p. 20, remarks following eq. [3.14]), it follows that *the formula*

$$\sigma^2 = (3\gamma - 4)\frac{|\mathfrak{W}_0|}{I_0} + \tfrac{2}{3}\Omega^2 + O(\Omega^4) \qquad [\gamma - \tfrac{4}{3} = O(\Omega^2)] \tag{3}$$

(where $\mathfrak{W}_0$ and I_0 now refer to the *non-rotating* configuration) *is an exact one.* The critical value of γ for marginal stability is therefore

$$\gamma_c(\text{Rot.}) = \tfrac{4}{3} - \frac{2\Omega^2 I_0}{9|\mathfrak{W}_0|} + O(\Omega^4). \tag{4}$$

The result (4) is of importance when effects besides rotation alter the value of γ_c from its "classical value" $\frac{4}{3}$. Thus, it is known that the post-Newtonian effects of general relativity have a destabilizing effect on radial pulsations and that (cf. Chandrasekhar 1964*b*)

$$\gamma_c(\text{G.R.}) = \tfrac{4}{3} + K\frac{R_S}{R} + O\left(\frac{R_S{}^2}{R}\right), \tag{5}$$

where K is a certain calculable constant, $R_S = 2GM/c^2$ is the Schwarzschild radius, and R is the radius for the configuration. Now if the effects of general relativity and of rotation are both present, and are both considered as first-order effects, then under their combined influence we must have

$$\gamma_c(\text{Rot.} + \text{G.R.}) = \tfrac{4}{3} + K\frac{R_S}{R} - \frac{2\Omega^2 I_0}{9|\mathfrak{W}_0|} + O\left(\Omega^4, \Omega^2\frac{R_S}{R}, \frac{R_S{}^2}{R^2}\right). \tag{6}$$

This last result is the essential content of some recent papers by Fowler (1966) and Durney and Roxburgh (1967).

While the stabilizing effect of rotation in the limit $\gamma \to \frac{4}{3}$ is an unambiguous result, it is not clear how small $\gamma - \frac{4}{3}$ must be to be of "order Ω^2" in a given practical situation. For the destabilizing effect of the distortion of the spherical distribution is the more dominating effect for centrally condensed configurations when γ is not too close to the value $\frac{4}{3}$ (see Table 1 in § IV*c* below). And it is pertinent to observe in this connection that the stabilizing contribution $\frac{2}{3}\Omega^2$ in equation (3) represents in turn a somewhat delicate balance between the stabilizing contribution $(+8\Omega^2/3)$ derived from the requirement of the conservation of angular momentum and the destabilizing contribution $(-2\Omega^2)$ derived from the distortion of the configuration (see § IV*b* below). It is natural to wonder whether this balance might not be upset when the distortion ceases to be of order Ω^2. The question can be stated differently. Consider the variation of γ_c along a sequence of configurations of increasing Ω^2. We know that $\gamma_c(\Omega^2)$, along such a sequence, has a negative slope at $\Omega^2 = 0$. What is the behavior of the curve $\gamma_c(\Omega^2)$ as Ω^2 increases? Does it always remain below[3] $\gamma = \frac{4}{3}$? The present paper is the first of two devoted to a consideration of these questions.

[2] For a somewhat more complete treatment, along the same lines but without any restriction on Ω^2, based on the tensor form of the virial theorem see Chandrasekhar and Lebovitz (1962*a*, § VII); and for the relation of *this* treatment to a variational formulation see Clement (1964, § III).

[3] It is known that for compressible Maclaurin spheroids this is the case (Chandrasekhar and Lebovitz 1962*c*, Fig. 1). But these configurations are, of course, atypical since their mass distribution is highly unrealistic.

The plan of this paper is as follows. In § II, we formulate and discuss the variational principle appropriate for axisymmetric oscillations of a uniformly rotating axisymmetric configuration. In § III we introduce a class of trial functions that appears specially convenient for the present problem. The formulae of § III are applied in § IV to some familiar problems; and it is shown how they lead to results in agreement with those obtained by other methods. In § V we introduce a coordinate system that enables the reduction of the various multiple integrals appearing in the variational expression to simple quadratures; and in §§ VI, VII, and VIII we carry out the necessary reductions for some special choices of equilibrium stratification that we intend to consider in our second paper.

II. THE VARIATIONAL PRINCIPLE

In this section we shall show that the characteristic value problem for axisymmetric perturbations of an axisymmetric uniformly rotating configuration can be put in the standard Rayleigh-Ritz form[4] (see also Lynden-Bell and Ostriker 1967) just as in the absence of rotation (cf. Chandrasekhar 1964*a*).

The equation governing equilibrium is

$$\text{grad } p = \rho \text{ grad } U \,, \tag{7}$$

where p is the pressure, ρ is the density, and

$$U = \mathfrak{B} + \tfrac{1}{2}\Omega^2(x_1^2 + x_2^2) \,. \tag{8}$$

In equation (8) $\mathfrak{B}$ denotes the gravitational potential and it has been assumed that the rotation is about the x_3-axis. Equation (7) implies that surfaces of constant p, constant ρ, and constant U all coincide. Accordingly, we may now write

$$p = p(U) \qquad \text{and} \qquad \rho = \rho(U) \,. \tag{9}$$

The linear equations governing small perturbations of such an equilibrium configuration are (cf. Clement 1965; Lebovitz 1967)

$$\rho\left(\frac{\partial^2 \boldsymbol{\xi}}{\partial t^2} + 2\boldsymbol{\Omega}\times\frac{\partial \boldsymbol{\xi}}{\partial t}\right) = \boldsymbol{L}[\,\boldsymbol{\xi}\,] \,, \tag{10}$$

where $\boldsymbol{\xi}$ is the Lagrangian displacement and the operator $\boldsymbol{L}$ is defined by

$$\boldsymbol{L}[\boldsymbol{\xi}] = -\text{grad } \Delta p + \Delta\rho \text{ grad } U + \rho \text{ grad } \Delta U \,, \tag{11}$$

where Δp, $\Delta\rho$, and ΔU are the Lagrangian changes in the respective variables resulting from the displacement $\boldsymbol{\xi}$. These changes are given by

$$\Delta\rho = -\rho \text{ div } \boldsymbol{\xi}, \qquad \Delta p = -\gamma p \text{ div } \boldsymbol{\xi} \,, \tag{12}$$

and

$$\Delta U = \delta\mathfrak{B} + \boldsymbol{\xi}\cdot\text{grad } U \,. \tag{13}$$

In equation (13), $\delta\mathfrak{B}$ represents the *Eulerian change*[5] in $\mathfrak{B}$ and is given by

$$\delta\mathfrak{B} = G\int_{\mathfrak{R}} \rho(\boldsymbol{x}')\,\xi_j(\boldsymbol{x}',t)\frac{\partial}{\partial x_j'}\frac{1}{|\boldsymbol{x}-\boldsymbol{x}'|}\,d\boldsymbol{x}' \,. \tag{14}$$

[4] One of us (N. R. L.) wishes to thank Dr. James Bardeen for a useful conversation on this point.

[5] Quite generally the operations Δ and δ leading to the Lagrangian and the Eulerian changes, respectively, are related by

$$\Delta = \delta + \boldsymbol{\xi}\cdot\text{grad} \,.$$

It can be proved in great generality (cf. Clement 1964; also Lynden-Bell and Ostriker 1967) that the operator $\boldsymbol{L}$ is *symmetric*, i.e., for any smooth functions $\boldsymbol{\xi}$ and $\boldsymbol{\eta}$,

$$\int_{\Re} \boldsymbol{\eta} \cdot \boldsymbol{L}[\boldsymbol{\xi}]\, d\boldsymbol{x} = \int_{\Re} \boldsymbol{\xi} \cdot \boldsymbol{L}[\boldsymbol{\eta}]\, d\boldsymbol{x}. \tag{15}$$

(It is this symmetry of $\boldsymbol{L}$ that insures the general variational formulation of the underlying characteristic value problem.)

We now suppose that the perturbation considered is also axisymmetric. In cylindrical polar coordinates ϖ, $z(=x_3)$, and φ, the assumption of axisymmetry means that the components ξ_ϖ, ξ_z, and ξ_φ are all independent of φ. Two immediate consequences of this assumption are (1) that ξ_φ *does not occur in* $\boldsymbol{L}[\boldsymbol{\xi}]$ *and* (2) *that the* φ*-component of* $\boldsymbol{L}$ *vanishes.* The latter consequence implies, according to equation (10), that

$$\frac{\partial \xi_\varphi}{\partial t} + 2\Omega \xi_\varpi = f(\varpi, z), \tag{16}$$

where we have integrated with respect to t, and $f(\varpi,z)$ is the "constant" of integration. As one can readily verify, equation (16) expresses the conservation of the z-component of the angular momentum per unit mass. In what follows we shall set $f(\varpi,z) = 0$. For the present, we shall justify this assumption by observing that, for a normal mode for which

$$\boldsymbol{\xi}(x,t) = \boldsymbol{\xi}(x)e^{i\sigma t}, \tag{17}$$

where σ denotes a characteristic frequency, $f(\varpi,z)$ necessarily vanishes as long as $\sigma \neq 0$ (for a further justification see below).

Considering next the ϖ-component of either side of equation (10), we have

$$\rho \left(\frac{\partial^2 \xi_\varpi}{\partial t^2} - 2\Omega \frac{\partial \xi_\varphi}{\partial t} \right) = L_\varpi[\boldsymbol{\xi}], \tag{18}$$

or eliminating ξ_φ with the aid of equation (16) (with $f = 0$), we have

$$\rho \frac{\partial^2 \xi_\varpi}{\partial t^2} = L_\varpi[\boldsymbol{\xi}] - 4\rho\Omega^2 \xi_\varpi. \tag{19}$$

The z-component of either side of equation (10) gives

$$\rho \frac{\partial^2 \xi_z}{\partial t^2} = L_z[\boldsymbol{\xi}]. \tag{20}$$

Since neither $L_\varpi[\boldsymbol{\xi}]$ nor $L_z[\boldsymbol{\xi}]$ involves ξ_φ, equations (19) and (20) contain no reference to ξ_φ: it has effectively been eliminated. For this reason it is convenient to *redefine* $\boldsymbol{\xi}$ *to be the two-component vector*

$$\boldsymbol{\xi} = \xi_\varpi \mathbf{1}_\varpi + \xi_z \mathbf{1}_z, \tag{21}$$

and this redefinition underlies the rest of this paper. As a consequence $\boldsymbol{\xi}$ no longer defines the complete three-component Lagrangian displacement. To obtain the latter, we must add to $\boldsymbol{\xi}$ the vector $\xi_\varphi \mathbf{1}_\varphi$, where ξ_φ is to be determined in terms of ξ_ϖ with the aid of equation (16) (with f set equal to zero).

With $\boldsymbol{\xi}$ redefined in the manner (21), we may formally combine equations (19) and (20) into the single equation

$$\rho \frac{\partial^2 \boldsymbol{\xi}}{\partial t^2} = \mathfrak{L}[\boldsymbol{\xi}], \tag{22}$$

where $\mathfrak{L}$ is now an operator acting on vectors in the $(\mathbf{1}_\varpi, \mathbf{1}_z)$-space and whose components are given by

$$\mathfrak{L}_\varpi[\boldsymbol{\xi}] = L_\varpi[\boldsymbol{\xi}] - 4\rho\Omega^2\xi_\varpi$$

and (23)[6]

$$\mathfrak{L}_z[\boldsymbol{\xi}] = L_z[\boldsymbol{\xi}] .$$

The symmetry of this operator $\mathfrak{L}$ follows trivially from that of $\boldsymbol{L}$.

With a dependence of $\boldsymbol{\xi}$ on time of the form (17), equation (22) leads to a characteristic value problem associated with the equation

$$-\sigma^2\rho\boldsymbol{\xi} = \mathfrak{L}[\boldsymbol{\xi}] . \tag{24}$$

From the symmetry of the operator $\mathfrak{L}$ it follows that the formula

$$\sigma^2 = -\frac{\int_{\mathfrak{R}} \boldsymbol{\xi}\cdot\mathfrak{L}[\boldsymbol{\xi}]\,d\boldsymbol{x}}{\int_{\mathfrak{R}} \rho|\boldsymbol{\xi}|^2 d\boldsymbol{x}} = -\frac{\int_{\mathfrak{R}} \{\boldsymbol{\xi}\cdot\boldsymbol{L}[\boldsymbol{\xi}] - 4\rho\Omega^2\xi_\varpi^2\}\,d\boldsymbol{x}}{\int_{\mathfrak{R}} \rho(\xi_\varpi^2 + \xi_z^2)\,d\boldsymbol{x}} \tag{25}$$

provides a variational basis for the determination of the least characteristic value of σ^2: *it is the minimum value that the ratio appearing in equation (25) can assume for any smooth function* $\boldsymbol{\xi}$. This minimum principle for σ^2 has the important consequence that if for some admissible choice of $\boldsymbol{\xi}$

$$\int_{\mathfrak{R}} \{\boldsymbol{\xi}\cdot\boldsymbol{L}[\boldsymbol{\xi}] - 4\rho\Omega^2\xi_\varpi^2\}\,d\boldsymbol{x} > 0 , \tag{26}$$

the equilibrium configuration is unstable; for the least characteristic value σ^2 is then necessarily negative and leads to an exponentially growing mode.

That the inequality (26) is a sufficient condition for instability has been deduced directly from equation (22) and independently of the theory of characteristic values by Laval, Mercier, and Pellat (1965). Their result is important, from the mathematical point of view, in our present context for two reasons: *first*, the theory in terms of which we have argued has not, to the authors' knowledge, been rigorously established in sufficient generality to apply to the problem being considered; and *second*, since a solution of equation (22) of unstable type leads to a solution of equation (10), also of unstable type, through the simple expedient of choosing an initial value for $\partial\xi_\varphi/\partial t$ so as to make f vanish in equation (16), we are justified in setting f equal to zero as long as we are interested only in establishing sufficient conditions for instability.

III. THE FORM OF THE VARIATIONAL EXPRESSION FOR σ^2 FOR A SPECIAL CHOICE OF $\boldsymbol{\xi}$

In using the variational expression for σ^2 given in equation (25), we must exercise judgment in the choice of suitable trial functions for $\boldsymbol{\xi}$: they must satisfy the requirements which experience and physical considerations suggest as necessary; at the same time they must make the evaluation of the various integrals as simple as possible.

Now the principal obstacle to the requirement of simplicity is the presence of the term in $\delta\mathfrak{B}$ in the operator $\boldsymbol{L}$; for $\delta\mathfrak{B}$ must be determined in terms of the chosen $\boldsymbol{\xi}$, either from equation (14) or, equivalently, from a solution of Poisson's equation,

$$\nabla^2\delta\mathfrak{B} = -4\pi G\delta\rho = 4\pi G \operatorname{div}(\rho\boldsymbol{\xi}) , \tag{27}$$

[6] Note that under the assumption of axisymmetry $\boldsymbol{L}$, like $\mathfrak{L}$, is also an operator in the $(\mathbf{1}_\varpi, \mathbf{1}_z)$-space.

together with appropriate conditions on the boundary $\partial\mathfrak{R}$ of $\mathfrak{R}$. The choice of $\boldsymbol{\xi}$ we shall presently make (eq. [34] below) is motivated, in part, by the desire to avoid this auxiliary calculation.

First, we observe that equation (27) implies, quite generally, that

$$\text{grad } \delta\mathfrak{V} = 4\pi G\rho\boldsymbol{\xi} + \text{curl } \boldsymbol{A} \tag{28}$$ [7]

for some vector $\boldsymbol{A}$. [Note that since the right-hand side of eq. (27) does not involve ξ_φ, and the left-hand side of eq. (28) has no φ-component, eq. (28) is in fact an equation in the $(\mathbf{1}_\varpi,\mathbf{1}_z)$-space.] The contribution to

$$\int_{\mathfrak{R}} \boldsymbol{\xi}\cdot\mathfrak{L}[\,\boldsymbol{\xi}\,]\,d\boldsymbol{x}$$

by the term involving $\delta\mathfrak{V}$ is, therefore,

$$\int_{\mathfrak{R}} \rho\,\boldsymbol{\xi}\cdot\text{grad }\delta\mathfrak{V}\,d\boldsymbol{x} = 4\pi G\int_{\mathfrak{R}}\rho^2\,|\,\boldsymbol{\xi}\,|^2 d\boldsymbol{x} + \int_{\mathfrak{R}}\rho\,\boldsymbol{\xi}\cdot\text{curl}\boldsymbol{A}\,d\boldsymbol{x}\,. \tag{29}$$

The integral involving $\boldsymbol{A}$ on the right-hand side of equation (29) can be made to vanish if we choose $\boldsymbol{\xi}$ so that

$$\rho\boldsymbol{\xi} = \text{grad }\phi \text{ and, on } \partial\mathfrak{R},\ \phi = \phi_0 = \text{constant}; \tag{30}$$

for, then,

$$\begin{aligned}\int_{\mathfrak{R}}\text{grad }\phi\cdot\text{curl}\boldsymbol{A}\,d\boldsymbol{x} &= \int_{\mathfrak{R}}\text{div}\,(\,\phi\text{ curl}\boldsymbol{A}\,)\,d\boldsymbol{x} = \int_{\partial\mathfrak{R}}\phi\text{ curl }\boldsymbol{A}\cdot d\boldsymbol{S}\\ &= \phi_0\int_{\partial\mathfrak{R}}\text{curl}\,\boldsymbol{A}\cdot d\boldsymbol{S} = 0\end{aligned} \tag{31}$$

(where $d\boldsymbol{S}$ is the vector element of area of $\partial\mathfrak{R}$). Hence for $\boldsymbol{\xi}$ of the chosen form

$$\begin{aligned}\int_{\mathfrak{R}}\rho\,\boldsymbol{\xi}\cdot\text{grad }\delta\,\mathfrak{V}\,d\boldsymbol{x} &= 4\pi G\int_{\mathfrak{R}}\rho^2\,|\,\boldsymbol{\xi}\,|^2 d\boldsymbol{x}\\ &= 4\pi G\int_{\mathfrak{R}}\rho^2\,(\,\xi_\varpi^2 + \xi_z^2\,)\,d\boldsymbol{x}\,;\end{aligned} \tag{32}$$

and this contribution to σ^2 (derived from the term in $\delta\mathfrak{V}$) is formally the same as in the absence of rotation.

The substitution (30) corresponds to the correct solution both when rotation is absent and also when allowance is made for it to $O(\Omega^2)$ (see § IV*b* below); it, therefore, appears as a reasonable choice for the general case even though it cannot correspond to the exact solution.

In this paper, we shall further specialize ϕ by requiring it to be a function of U only (which automatically insures the constancy of ϕ on $\partial\mathfrak{R}$); and since ρ is also a function of U (cf. eq. [9]), the assumption is equivalent to

$$\boldsymbol{\xi} = \chi(U)\text{ grad } U\,. \tag{33}$$

The present assumption (33) implies that, *at every point* $\boldsymbol{x}$, $\boldsymbol{\xi}(\boldsymbol{x})$ *is normal to the level surface through* $\boldsymbol{x}$, while the original assumption (30) required this normality only on the boundary $\partial\mathfrak{R}$ of the configuration. (But as we shall see in § IV*b* even this more specialized choice for $\boldsymbol{\xi}$ is in agreement with the exact solution for the case of slow rotation, and a fortiori for zero rotation.)

[7] In the absence of rotation $\boldsymbol{A} = 0$ (cf. Rosseland 1949).

a) The Reduction of the Variational Expression for σ^2 for the Chosen Form of the Trial Function

It is convenient to define the effective gravity g at any point; it is given by

$$g = |\text{grad } U| \,. \tag{34}$$

It follows from this definition that for ξ of the form (33)

$$\xi \cdot \text{grad } U = g^2 \chi \,, \tag{35}$$

and

$$\text{div } \xi = g^2 \chi' + \chi \nabla^2 U \,, \tag{36}$$

where a prime denotes differentiation with respect to the argument U. By making use of the known relation

$$\nabla^2 U = -4\pi G \rho + 2\Omega^2 \,, \tag{37}$$

we can rewrite the relation (36) in the form

$$\text{div } \xi = g^2 \chi' - (4\pi G \rho - 2\Omega^2)\chi \,. \tag{38}$$

We turn now to the reduction of the variational expression for σ^2 for the chosen form of the trial function. We have (cf. eqs. [11] and [23])

$$\int_{\mathfrak{R}} \xi \cdot \mathfrak{L}[\xi]\, d\boldsymbol{x} = -4\Omega^2 \int_{\mathfrak{R}} \rho\, \xi_{\varpi}^2 d\boldsymbol{x} - \int_{\mathfrak{R}} \xi \cdot \text{grad } \Delta p\, d\boldsymbol{x} + \int_{\mathfrak{R}} \xi \cdot (\Delta \rho \text{ grad } U + \rho \text{ grad } \Delta U)\, d\boldsymbol{x} \,. \tag{39}$$

With the expression for Δp given in equations (12), the second integral on the right-hand side of equation (39) becomes

$$-\int_{\mathfrak{R}} \xi \cdot \text{grad } \Delta p\, d\boldsymbol{x} = \int_{\mathfrak{R}} \Delta p \text{ div } \xi\, d\boldsymbol{x} = -\int_{\mathfrak{R}} \gamma p (\text{div } \xi)^2 d\boldsymbol{x} \,, \tag{40}$$

where we have integrated by parts and used the condition that Δp vanishes on $\partial\mathfrak{R}$. Similarly, by making use of equations (12), (13), and (32), the third integral on the right-hand side of equation (39) becomes

$$\int_{\mathfrak{R}} \xi \cdot (\Delta \rho \text{ grad } U + \rho \text{ grad } \Delta U)\, d\boldsymbol{x} = \int_{\mathfrak{R}} \rho [-(\text{div } \xi)(\xi \cdot \text{grad } U) + 4\pi G \rho |\xi|^2 + \xi \cdot \text{grad}(\xi \cdot \text{grad } U)]\, d\boldsymbol{x} \,. \tag{41}$$

Now combining equations (39), (40), and (41) and making further use of the relations (34), (35), and (38), we find

$$\sigma^2 \int_{\mathfrak{R}} \rho \chi^2 g^2 d\boldsymbol{x} = \int_{\mathfrak{R}} \gamma p [g^2 \chi' - (4\pi G \rho - 2\Omega^2)\chi]^2 d\boldsymbol{x} - \int_{\mathfrak{R}} \rho \chi^2 \left[2(4\pi G \rho - \Omega^2) g^2 + \text{grad } U \cdot \text{grad } g^2 - 4\Omega^2 \left(\frac{\partial U}{\partial \varpi} \right)^2 \right] d\boldsymbol{x} \,. \tag{42}$$

In this form, the variational expression, besides the known pressure and density distributions in the unperturbed configuration, involves only the single scalar function χ.

There are two alternative forms of equation (42) which we shall find useful. The first of these is obtained with the help of the definition

$$\begin{aligned} C(\boldsymbol{x}) &= \operatorname{div} \mathbf{1}_U = \operatorname{div}\left(\frac{1}{g}\operatorname{grad} U\right) \\ &= \frac{1}{g^2}(g\nabla^2 U - \operatorname{grad} U\cdot\operatorname{grad} g). \end{aligned} \tag{43}$$

Apart from a factor -2, C represents the mean curvature of the level surface passing through $\boldsymbol{x}$. With the aid of equation (37) and the definition (43), we find (cf. the integrand of the second integral on the right-hand side of eq. [42])

$$\begin{aligned} 2g^2(4\pi G\rho - \Omega^2) + \operatorname{grad} U\cdot\operatorname{grad} g^2 &= 2g^2(\Omega^2 - \nabla^2 U) + \operatorname{grad} U\cdot\operatorname{grad} g^2 \\ &= 2g^2\Omega^2 - 2g^3 C = 2\Omega^2\left[\left(\frac{\partial U}{\partial\varpi}\right)^2 + \left(\frac{\partial U}{\partial z}\right)^2\right] - 2g^3 C. \end{aligned} \tag{44}$$

Inserting this last relation in equation (42), we obtain

$$\begin{aligned} \sigma^2\int_{\mathfrak{R}}\rho\chi^2 g^2 d\boldsymbol{x} &= \int_{\mathfrak{R}}\gamma p[g^2\chi' - (4\pi G\rho - 2\Omega^2)\chi]^2 d\boldsymbol{x} \\ &\quad + 2\int_{\mathfrak{R}}\rho\chi^2 g^3 C d\boldsymbol{x} + 2\Omega^2\int_{\mathfrak{R}}\rho\chi^2\left[\left(\frac{\partial U}{\partial\varpi}\right)^2 - \left(\frac{\partial U}{\partial z}\right)^2\right]d\boldsymbol{x}. \end{aligned} \tag{45}$$

The second form of equation (42) which we shall need is obtained by noting that if ρ vanishes on $\partial\mathfrak{R}$,

$$\begin{aligned} \int_{\mathfrak{R}}\rho\chi^2 \operatorname{grad} U\cdot\operatorname{grad} g^2 d\boldsymbol{x} &= -\int_{\mathfrak{R}}[\rho\chi^2 g^2\nabla^2 U + g^2\operatorname{grad} U\cdot\operatorname{grad}(\rho\chi^2)]d\boldsymbol{x} \\ &= \int_{\mathfrak{R}}\left[\rho\chi^2 g^2(4\pi G\rho - 2\Omega^2) - g^4\frac{d}{dU}(\rho\chi^2)\right]d\boldsymbol{x} \end{aligned} \tag{46}$$

and hence

$$\begin{aligned} \sigma^2\int_{\mathfrak{R}}\rho\chi^2 g^2 d\boldsymbol{x} &= \int_{\mathfrak{R}}\gamma p[\chi' g^2 - (4\pi G\rho - 2\Omega^2)\chi]^2 d\boldsymbol{x} \\ &\quad + \int_{\mathfrak{R}} g^4\frac{d}{dU}(\rho\chi^2)d\boldsymbol{x} - 4\int_{\mathfrak{R}}\rho\chi^2\left[(3\pi G\rho - \Omega^2)g^2 - \Omega^2\left(\frac{\partial U}{\partial\varpi}\right)^2\right]d\boldsymbol{x}. \end{aligned} \tag{47}$$

IV. APPLICATIONS TO PREVIOUSLY CONSIDERED STRATIFICATIONS

In this section we shall apply the formulae of § III to the non-rotating and the slowly rotating cases, and also to spheroidally stratified configurations.

a) The Spherical Case: $\Omega = 0$

In this case the following formulae hold:

$$\begin{gathered} g = -\frac{d\mathfrak{B}}{dr}, \qquad \frac{dp}{dr} = -g\rho, \qquad \frac{1}{r^2}\frac{d}{dr}(r^2 g) = 4\pi G\rho, \\ \chi' = -\frac{1}{g}\frac{d\chi}{dr}, \qquad \text{and} \qquad C = -\frac{2}{r}. \end{gathered} \tag{48}$$

Also the radial Lagrangian displacement, ξ_r, is given by

$$\xi_r = -g\chi\,. \tag{49}$$

Inserting these relations in equation (45), letting

$$\psi = r^2\xi_r\,, \tag{50}$$

and simplifying, we readily obtain the formula

$$\sigma^2\int_0^R \rho\psi^2\frac{dr}{r^2} = \int_0^R\left[\gamma p\left(\frac{d\psi}{dr}\right)^2 + \frac{4}{r}\frac{dp}{dr}\psi^2\right]\frac{dr}{r^2}, \tag{51}$$

where R denotes the radius of the configuration. Equation (51) is one of the standard forms in which the variational principle for the radial oscillations of a spherical mass is generally expressed (cf. Chandrasekhar 1964*a*, eq. [49]; also Ledoux and Walraven 1958).

An alternative form of equation (51), which we obtain by integrating by parts the term in dp/dr in equation (51), is

$$\sigma^2\int_0^R \rho\psi^2\frac{dr}{r^2} = \int_0^R (\gamma - \tfrac{4}{3})p\left(\frac{d\psi}{dr}\right)^2\frac{dr}{r^2} + 12\int_0^R p\left(\psi - \tfrac{1}{3}r\frac{d\psi}{dr}\right)^2\frac{dr}{r^4}. \tag{52}$$

In view of the minimum character of the underlying variational principle (see remarks following eq. [25]), the least proper value of σ^2 must be less than *any* value given by equation (52) so long as the chosen ξ_r (i.e., ψ) is smooth. And with the particular choice

$$\psi = r^3 \quad \text{or} \quad \xi_r = r\,, \tag{53}$$

we obtain the inequality

$$\sigma^2 \leq \frac{9\int_0^R (\gamma - \frac{4}{3})pr^2dr}{\int_0^R \rho r^4 dr} = 9(\bar\gamma - \tfrac{4}{3})\frac{\int_0^R pr^2dr}{\int_0^R \rho r^4dr} = 3(\bar\gamma - \tfrac{4}{3})\frac{|\mathfrak{W}|}{I}, \tag{54}$$

where, in the latter two forms, $\bar\gamma$ denotes the pressure weighted average of γ. From the inequality (54) it follows that *a sufficient condition for instability is that* $\bar\gamma < \frac{4}{3}$. This result is of course well known: the basic formula, in precisely the form given in equation (54), occurs in one of Ledoux's early papers (1946, eq. [6]).[8]

A slightly sharper result (than the one we have stated) can be deduced from equation (52) in case γ is not constant. Writing

$$\psi = r^3 + \epsilon\phi(r) \tag{55}$$

(where ϵ will be assumed small), we find from equation (52) that

$$\sigma^2\int_0^R \rho\psi^2\frac{dr}{r^2} = 9(\bar\gamma - \tfrac{4}{3})\int_0^R pr^2dr + 6\epsilon\int_0^R(\gamma - \tfrac{4}{3})p\frac{d\phi}{dr}dr + O(\epsilon^2). \tag{56}$$

If, for the sake of simplicity, we suppose that $\bar\gamma = \frac{4}{3}$, we easily see that the right-hand side of equation (56) can be made negative by appropriate choices of $\phi(r)$ and of ϵ, the "appropriateness" depending on the precise nature of the variability of γ. In any event, it follows that *when γ is not a constant and $\bar\gamma = \frac{4}{3}$, the configuration is already dynamically unstable.*

[8] It should also be noted that Ledoux made use of the relation (54) (with the equality sign) to estimate the extent of the regions in which γ may take different constant values ($\gamma < \frac{4}{3}$ in some regions and $\gamma > \frac{4}{3}$ in other regions) before instability can set in.

b) The Case When Ω^2 Is Small and Effects of Order Ω^2, Only, Are Taken into Account

In this case, we write

$$f(r,\mu) = f_0(r) + \Omega^2 f_1(r,\mu) + O(\Omega^4) , \tag{57}$$

where μ is the cosine of the colatitude and f stands for any of the variables ρ, p, U, etc. The variables distinguished with the subscript "0" represent some appropriately chosen undistorted configuration (cf. Chandrasekhar 1933, or Chandrasekhar and Lebovitz 1962*d*, § II). In the present approximation, in which all terms of order Ω^4 and higher are neglected, it will suffice to extend the range of integration, in evaluating an integral, over the volume of the undistorted spherical configuration provided only the integrand vanishes over the boundary (for an explicit proof see, e.g., Clement 1965, Appendix I). For the integrals that appear in equation (45), the required condition is met if we restrict our consideration to equilibrium configurations for which the density ρ vanishes on the boundary; and this we shall do in the present subsection.

A conclusion of some interest may be drawn immediately from equation (45) if we suppose that the trial function χ may also be decomposed in the same manner (57). In that case the last term of equation (45) makes a contribution to σ^2 of the amount

$$2\Omega^2 \frac{\int_{\mathfrak{R}} \rho\chi^2[(\partial U/\partial \varpi)^2 - (\partial U/\partial z)^2]\,dx}{\int_{\mathfrak{R}} \rho\chi^2[(\partial U/\partial \varpi)^2 + (\partial U/\partial z)^2]\,dx} . \tag{58}$$

Consistently with neglecting terms of $O(\Omega^2)$, we must suppose that in the foregoing expression ρ, U, and χ are functions of r only. Accordingly, we may write

$$\frac{\partial U}{\partial \varpi} = \frac{dU}{dr}\sin\vartheta \quad \text{and} \quad \frac{\partial U}{\partial z} = \frac{dU}{dr}\cos\vartheta . \tag{59}$$

On evaluating the expression (58) under these conditions, we obtain the value $\frac{2}{3}\Omega^2$. This stabilizing contribution, already discussed in § I, is the *only* contribution of $O(\Omega^2)$ to σ^2 if $\gamma - \frac{4}{3}$ is small enough. It is therefore of interest to note that this stabilizing contribution is made up of two parts. Thus, tracing the effect of the conservation of angular momentum through the equations, we find that its contribution is (cf. eq. [42])

$$\frac{4\Omega^2 \int_{\mathfrak{R}} \rho\chi^2(\partial U/\partial \varpi)^2 dx}{\int_{\mathfrak{R}} \rho\chi^2[(\partial U/\partial \varpi)^2 + (\partial U/\partial z)^2]\,dx} = \tfrac{8}{3}\Omega^2 ; \tag{60}$$

and the remaining contribution $-2\Omega^2$ may be traced to the distortion of the equilibrium configuration and the perturbation of the gravitational potential. The particular coefficients $\frac{8}{3}$ and -2 that we find here are due to the replacement of ρ, U, and χ by functions of r only. If the configuration is strongly distorted, and these replacements become invalid, it may be that this stabilizing influence is strongly modified or even destroyed. Reference to this possibility has already been made in § I.

Returning to equation (57), we next observe that $f_1(r,\mu)$ will normally have the further decomposition

$$f_1(r,\mu) = f_{10}(r) + f_{12}(r)P_2(\mu) , \tag{61}$$

corresponding to the fact that, in the present approximation, the effect of the rotation is a P_2-deformation superposed on a uniform expansion (cf. Milne 1923). We shall adopt the same decomposition (61) for the trial function χ, as well.

We have already remarked that, under the assumption that ρ vanishes on the boundary, it is legitimate to extend all integrations only over the undistorted sphere. Under these circumstances, the coefficients ρ_{12}, p_{12}, χ_{12}, etc., do not contribute to any of the integrals. And since we shall prescribe χ_0 (for example, identify it with the proper solution belonging to σ_0^2, the characteristic value in the absence of rotation), only χ_{10} is left at our disposal in the selection of a trial function. In fact, it may appear that we do not have a choice even in the selection of χ_{10}, since we have also required χ to be a function of U only. We shall show that it is nevertheless legitimate to regard χ_{10} as at our disposal.

The condition, that χ be a function of U only, is

$$\frac{\partial \chi}{\partial r}\frac{\partial U}{\partial \mu}-\frac{\partial \chi}{\partial \mu}\frac{\partial U}{\partial r}=0\,. \tag{62}$$

On substituting for χ and U in accordance with equations (57) and (61), we obtain

$$\Omega^2\left(\frac{d\chi_0}{dr}+\Omega^2\frac{\partial \chi_1}{\partial r}\right)\frac{\partial U_1}{\partial \mu}-\Omega^2\left(\frac{dU_0}{dr}+\Omega^2\frac{\partial U_1}{\partial r}\right)\frac{\partial \chi_1}{\partial \mu}=0\,. \tag{63}$$

Neglecting terms of $O(\Omega^4)$, as we must, and substituting for χ_1 and U_1 in accordance with equation (61), we are left with

$$\frac{d\chi_0}{dr}U_{12}=\frac{dU_0}{dr}\chi_{12}\,. \tag{64}$$

Equation (64) uniquely specifies χ_{12} if χ_0 is prescribed; but χ_{10} is left entirely arbitrary.

We now return to equation (42) and simplify the right-hand side of this equation correctly to $O(\Omega^2)$ when the functions representing the various variables have the decompositions assumed.

To facilitate comparison with the non-rotating case, we shall write (cf. eqs. [49] and [50])

$$\chi_0=\frac{\psi_0(r)}{r^2 dU_0/dr}\,. \tag{65}$$

Since the coefficients of $P_2(\mu)$ in the expansion of the various quantities do not contribute to any of the integrals, it will suffice to include only the f_{10}-terms in writing out the expansions of the various quantities to $O(\Omega^2)$. Thus, we find

$$C=-\frac{2}{r}+\ldots\,, \tag{66}$$

and

$$\chi^2 g^2=\frac{\psi_0^2}{r^4}+2\Omega^2\frac{\psi_0 q}{r^4}+\ldots\,, \tag{67}$$

where

$$q=\frac{dU_{10}/dr}{dU_0/dr}\psi_0+r^2\frac{dU_0}{dr}\chi_{10}\,. \tag{68}$$

In equations (66) and (67) the dots stand for those terms of order Ω^2 that occur with the factor $P_2(\mu)$; and for the reasons we have stated, we have not written out these terms.

Turning now to the reduction of equation (45) and considering the first integral on the right-hand side, we find after a straightforward calculation,

$$\int_{\mathfrak{R}}\gamma p\left[g^2\frac{d\chi}{dU}-(4\pi G\rho-2\Omega^2)\chi\right]^2 dx=\int_{\mathfrak{R}}\gamma p\left(\frac{d\psi_0}{dr}\right)^2\frac{dx}{r^4}+2\Omega^2\int_{\mathfrak{R}}\gamma p_0\frac{d\psi_0}{dr}\frac{dq}{dr}\frac{dx}{r^4}\,. \tag{69}$$

In considering the second integral on the right-hand side of equation (45), we observe that

$$\rho g = |\text{grad}\, p| = \left[\left(\frac{\partial p}{\partial r}\right)^2 + \frac{1}{r^2}\left(\frac{\partial p}{\partial \mu}\right)^2\right]^{1/2} = -\frac{\partial p}{\partial r} + O(\Omega^4). \tag{70}$$

Consequently, by virtue of equations (66) and (67), we have

$$\begin{aligned} 2\int_{\mathfrak{R}} \rho g^3 C\chi^2 dx &= 4\int_{\mathfrak{R}} \frac{\partial p}{\partial r}(\psi_0^2 + 2\Omega^2\psi_0 q)\frac{dx}{r^5} \\ &= -4\int_{\mathfrak{R}} p\frac{d}{dr}\left(\frac{\psi_0^2}{r^3}\right)\frac{dx}{r^2} + 8\Omega^2\int_{\mathfrak{R}} q\psi_0 \frac{dp_0}{dr}\frac{dx}{r^5}, \end{aligned} \tag{71}$$

where we have performed an integration by parts. With the foregoing reductions, equation (45) becomes

$$\begin{aligned} \sigma^2\int_{\mathfrak{R}} \rho\chi^2 g^2 dx = \int_{\mathfrak{R}} p\left[\gamma\left(\frac{d\psi_0}{dr}\right)^2 - 4r^2\frac{d}{dr}\left(\frac{\psi_0^2}{r^3}\right)\right]\frac{dx}{r^4} + \tfrac{2}{3}\Omega^2\int_{\mathfrak{R}} \rho_0\psi_0^2\frac{dx}{r^4} \\ + 2\Omega^2\int_{\mathfrak{R}}\left(\gamma p_0\frac{d\psi_0}{dr}\frac{dq}{dr} + 4q\frac{\psi_0}{r}\frac{dp_0}{dr}\right)\frac{dx}{r^4}. \end{aligned} \tag{72}$$

We can recover Ledoux's formula from equation (72). (Note that the variables ρ and p still contain terms of order Ω^2.) Putting $\psi_0 = r^3$ we obtain

$$\sigma^2\int_{\mathfrak{R}} \rho\chi^2 g^2 dx = 9\int_{\mathfrak{R}}(\gamma - \tfrac{4}{3})p\,dx + \tfrac{2}{3}\Omega^2\int_{\mathfrak{R}}\rho_0 r^2 dx + 2\Omega^2\int_{\mathfrak{R}}(3\gamma - 4)p_0\frac{dq}{dr}\frac{dx}{r^2}, \tag{73}$$

after an integration by parts in the third integral on the right-hand side. Since χ_{10} is still at our disposal, we may choose it in such a way that q vanishes. By equation (68), we accomplish this with the choice

$$\chi_{10} = -r\frac{dU_{10}/dr}{(dU_0/dr)^2}. \tag{74}$$

With χ_{10} so chosen, equation (73) becomes (cf. also eq. [67])

$$\sigma^2\int_{\mathfrak{R}}\rho r^2 dx = 9\int_{\mathfrak{R}}(\gamma - \tfrac{4}{3})p\,dx + \tfrac{2}{3}\Omega^2\int_{\mathfrak{R}}\rho_0 r^2 dx. \tag{75}$$

And equation (75) reduces to Ledoux's formula (1) if we make use of the relation

$$3\int_{\mathfrak{R}} p\,dx = |\mathfrak{W}| - \tfrac{2}{3}I\Omega^2, \tag{76}$$

provided by the virial theorem in the case of slow rotation.

We now return to equation (72) to complete the reduction. Substituting for p and ρ their expansions according to equations (57) and (61) and writing $\sigma^2 = \sigma_0^2 + \Omega^2\sigma_2^2$, we obtain by equating the terms independent of Ω^2 and proportional to Ω^2,

$$\sigma_0^2\int_{\mathfrak{R}}\rho_0\psi_0^2\frac{dx}{r^4} = \int_{\mathfrak{R}} p_0\left[\gamma\left(\frac{d\psi_0}{dr}\right)^2 - 4r^2\frac{d}{dr}\left(\frac{\psi_0^2}{r^3}\right)\right]\frac{dx}{r^4}, \tag{77}$$

and

$$\begin{aligned} \sigma_2^2\int_{\mathfrak{R}}\rho_0\psi_0^2\frac{dx}{r^4} = 2\int_{\mathfrak{R}}\left[\gamma p_0\frac{d\psi_0}{dr}\frac{dq}{dr} + \left(\frac{4}{r}\frac{dp_0}{dr} - \sigma_0^2\rho_0\right)\psi_0 q\right]\frac{dx}{r^4} \\ + \int_{\mathfrak{R}} p_{10}\left[\gamma\left(\frac{d\psi_0}{dr}\right)^2 - 4r^2\frac{d}{dr}\left(\frac{\psi_0^2}{r^3}\right)\right]\frac{dx}{r^4} - \sigma_0^2\int_{\mathfrak{R}}\rho_{10}\psi_0^2\frac{dx}{r^4} + \tfrac{2}{3}\int_{\mathfrak{R}}\rho_0\psi_0^2\frac{dx}{r^4}. \end{aligned} \tag{78}$$

If we now assume that ψ_0 is the proper solution belonging to the characteristic value σ_0^2, then ψ_0 satisfies the Euler equation[9]

$$\frac{d}{dr}\left(\frac{\gamma p_0}{r^2}\frac{d\psi_0}{dr}\right)-\frac{4}{r^3}\frac{dp_0}{dr}\psi_0=-\sigma_0^2\frac{\rho_0\psi_0}{r^2}. \tag{79}$$

Equation (79) not only implies equation (77) but also implies (as one can readily verify) that the first term on the right-hand side of equation (78) vanishes for arbitrary q.[10] We therefore obtain

$$\sigma_2^2\int_0^R \rho_0\psi_0^2\frac{dr}{r^2}=\int_0^R p_{10}\left[\gamma\left(\frac{d\psi_0}{dr}\right)^2-4r^2\frac{d}{dr}\left(\frac{\psi_0^2}{r^3}\right)\right]\frac{dr}{r^2} -\sigma_0^2\int_0^R \rho_{10}\psi_0^2\frac{dr}{r^2}+\tfrac{2}{3}\int_0^R \rho_0\psi_0^2\frac{dr}{r^2}. \tag{80}$$

We notice the remarkable fact that this formula does not contain χ_{10}. Consequently to find the coefficient σ_2^2, one needs to know only the distortion of the equilibrium configuration (i.e., p_{10} and ρ_{10}) and the proper solution ψ_0 appropriate to the undistorted configuration. For the special case of polytropes, a formula equivalent to (80) was derived by Clement (1965, eq. [65]).

The present formulation of the variational principle with the trial function selected according to equation (33) is now seen to imply no loss of generality, since it leads to equation (80) which, in view of Clement's work, is exact.

c) The Evaluation of σ_2^2 for Distorted Polytropes

As we have already stated, the form which equation (80) takes for polytropes was derived by Clement. Using his formula, we have evaluated σ_2^2 for polytropic indices $n = 1.5$, 2.0, and 3.0 and for various values of γ. The results are listed in Table 1. We see that σ_2^2 becomes negative when γ is sufficiently different from $\frac{4}{3}$; and also that the value of γ at which σ_2^2 changes sign decreases with increasing n. We have referred to these facts in § I. However, it should be noted that there is some ambiguity in the comparison implied by the listings in Table 1, namely, that the characteristic frequencies of oscillation of a non-rotating and a rotating polytrope having the same parameters, n, ρ_c (central density), and K (the constant of proportionality in the pressure-density relation), are strictly comparable. It is by no means obvious that this is the case: the two configurations that are "compared" have, for example, different masses and different volumes (cf. Chandrasekhar and Lebovitz 1962*d*, § IX).

d) The Compressible Maclaurin Spheroid

The pulsation frequency σ^2 and the critical value γ_c of the ratio of specific heats for marginal stability have been worked out exactly for this model (Chandrasekhar and Lebovitz 1962*c*). This model therefore provides a useful test case for the application of equation (45).

The equilibrium configuration is described by the solution

$$p=\rho U \quad \text{and} \quad U=\kappa\left(1-\frac{\varpi^2}{a_1^2}-\frac{z^2}{a_3^2}\right), \tag{81}$$

where

$$\kappa=\pi G\rho a_1^2 a_3^3 A_3 \tag{82}$$

[9] Eddington's pulsation equation in this context.

[10] In fact the integral in question is proportional to the first variation of σ_0^2 as defined by eq. (77).

and the notation is that of Chandrasekhar and Lebovitz (1962*b*). With this solution, we readily find that

$$\operatorname{grad} U = -2\kappa\left(\frac{\varpi}{a_1^2}\,1_\varpi + \frac{z}{a_3^2}\,1_z\right), \qquad g = 2\kappa\left(\frac{\varpi^2}{a_1^4} + \frac{z^2}{a_3^4}\right)^{1/2},$$

and (83)

$$C = -\left(\frac{\varpi^2}{a_1^4} + \frac{z^2}{a_3^4}\right)^{-3/2}\left[\left(\frac{1}{a_1^2} + \frac{1}{a_3^2}\right)\frac{\varpi^2}{a_1^4} + \frac{2}{a_1^2}\frac{z^2}{a_3^4}\right].$$

And as a suitable trial function we shall choose

$$\chi = \text{constant} \tag{84}$$

since, in the absence of rotation, this choice leads to the exact solution. With the foregoing substitutions, we find from equation (45), after elementary integrations,

$$\sigma^2 = \frac{2\kappa}{a_1^2}\left[\left(2 + \frac{a_1^2}{a_3^2}\right)\gamma - 4\,\frac{2a_1^2 + a_3^2}{a_1^2 + 2a_3^2}\right] + 2\Omega^2\,\frac{2a_3^2 - a_1^2}{2a_3^2 + a_1^2}, \tag{85}$$

or, in terms of the eccentricity e of the meridional section, we have

$$\frac{\sigma^2}{\pi G\rho} = \left[2(3 - 2e^2)\gamma - 8\,\frac{(3 - e^2)(1 - e^2)}{3 - 2e^2}\right]A_3 + 2\,\frac{\Omega^2}{\pi G\rho}\,\frac{1 - 2e^2}{3 - 2e^2}. \tag{86}$$

For small values of e, equation (86) gives

$$\frac{\sigma^2}{\pi G\rho} = 4(\gamma - \tfrac{4}{3}) + \tfrac{16}{45}(5 - 3\gamma)e^2 + O(e^4); \tag{87}$$

this agrees with the exact formula (and with eq. [1]) through terms of $O(\Omega^2)$.

TABLE 1

VALUES OF σ_0^2 AND σ_2^2 FOR POLYTROPES*

γ	σ_0^2	σ_2^2	γ	σ_0^2	σ_2^2
	$n=1.5$			$n=3.0$	
1.35	0.007755	+0.30985	1.35	0.004010	+0.27697
1.375	.019334	+0.27430	1.375	.009755	+0.18085
1.40	.030850	+0.23837	1.40	.015182	+0.06946
1.45	.053710	+0.16546	1.4140	....	0
1.55	.098812	+0.01594	1.45	.025110	−0.20294
1.5605	....	0	1.50	.033894	−0.54011
1.60	.121094	−0.06051	1.55	.041680	−0.92946
1.6667	0.150560	−0.16394	1.60	.048642	−1.35145
			1.65	.054951	−1.78510
	$n=2.0$		1.6667	.056935	−1.93002
			1.70	0.060755	−2.21784
1.40	0.024752	+0.21248			
1.45	.042805	+0.11626			
1.5076	....	0			
1.55	.077773	−0.08913			
1.60	.094757	−0.19761			
1.6667	0.116956	−0.34747			

* The values of σ_0^2 listed are those derived by Roberts (1963) by a direct numerical integration of the pulsation equation; and his "exact" proper solutions were used in the evaluation of σ_2^2.

Values of σ^2 and γ_c, calculated with the aid of equation (86), are listed in Tables 2 and 3 and are further compared with the exact values given in Chandrasekhar and Lebovitz (1962*c*, Tables 1 and 2A). The agreement is satisfactory and encourages the hope that the chosen formulation of the variational principle will give equally satisfactory results in other cases as well.

TABLE 2

THE EXACT AND THE APPROXIMATE VALUES OF σ^2 FOR THE COMPRESSIBLE MACLAURIN MODEL

(σ^2 Is Measured in the Unit $\pi G\rho$)

e	σ^2 (exact)	σ^2 (approx.)	σ^2 (exact)	σ^2 (approx.)	σ^2 (exact)	σ^2 (approx.)
	$\gamma=1.3$		$\gamma=1.33\ldots$		$\gamma=1.4$	
0.0	−0.13322	−0.1332	0.00	0.00	0.26688	0.2669
.05	− .13237	− .1324	.00088	.0009	.26737	.2674
.10	− .12948	− .1294	.00350	.0036	.26945	.2695
.15	− .12447	− .1244	.00806	.0081	.27311	.2732
.20	− .11749	− .1174	.01440	.0145	.27817	.2783
.25	− .10839	− .1081	.02267	.0230	.28477	.2851
.30	− .09708	− .0965	.03295	.0336	.29294	.2938
.35	− .08341	− .0823	.04536	.0466	.30277	.3044
.40	− .06723	− .0651	.06004	.0625	.31436	.3175
.45	−0.04829	−0.0443	.07720	.0816	.32780	.3335
.50	....	....	0.09708	0.1049	.34324	.3531
0.60	....	....	....	....	0.38059	0.4089
	$\gamma=1.5$		$\gamma=1.6$		$\gamma=1.666\ldots$	
0.0	0.66684	0.6668	1.06668	1.0667	1.33310	1.3331
.05	.66711	.6671	1.06684	1.0669	1.33334	1.3334
.10	.66838	.6685	1.06654	1.0674	1.33330	1.3333
.15	.67066	.6707	1.06616	1.0683	1.33353	1.3334
.20	.67375	.6740	1.06561	1.0697	1.33402	1.3335
.25	.67771	.6784	1.06482	1.0717	1.33520	1.3339
.30	.68250	.6840	1.06369	1.0743	1.33775	1.3345
.35	.68810	.6911	1.06204	1.0778	1.34298	1.3356
.40	.69444	.7000	1.05972	1.0826	1.35314	1.3376
.45	.70142	.7113	1.05647	1.0890	1.37178	1.3409
.50	.70891	.7255	1.05197	1.0979	1.40306	1.3462
0.60	0.72434	0.7686	1.03721	1.1283	1.51536	1.3681

TABLE 3

COMPARISON OF THE EXACT AND THE APPROXIMATE VALUES OF γ_c

e	γ_c (exact)	γ_c (approx.)	e	γ_c (exact)	γ_c (approx.)
0.0	1.3333	1.3333	0.5	1.3071	1.3052
.1	1.3324	1.3324	.6	1.2920	1.2863
.2	1.3297	1.3297	.7	1.2693	1.2533
.3	1.3249	1.3247	.8	1.2318	1.1848
0.4	1.3176	1.3170	0.9	1.1535	1.0009

V. THE REDUCTION TO QUADRATURES OF THE INTEGRALS IN THE VARIATIONAL EXPRESSIONS FOR σ^2

In order to facilitate the evaluation of the integrals appearing in equation (42), and in its alternative forms (45) and (47), we shall introduce a system of coordinates s, ϑ, and φ (say) such that s is constant on the level surfaces, U = constant. Since the trial function $\chi(U)$, as well as $\rho(U)$ and $p(U)$, are then independent of ϑ and φ, the integrations with respect to these variables may be performed, independently of the choice of χ, for any given (or assumed) stratification. In this manner the triple integrals, which occur in the variational expression for σ^2, can be reduced to quadratures over s and put into a form analogous to that of equation (51).

The derivation of the basic formulae (eqs. [98] and [99] below) does not require that the surfaces in question be the level surfaces of a rotating mass; so we shall avoid this terminology. Let the surface $\mathfrak{S}$, inclosing the simply connected domain $\mathfrak{R}(s)$, be specified by a smooth function $S(\boldsymbol{x},s)$ as follows:

$$\mathfrak{S}(s) = \{\boldsymbol{x}: S(\boldsymbol{x},s) = 0\} ,$$

with (88)

$$\mathfrak{R}(s) = \{\boldsymbol{x}: S(\boldsymbol{x},s) \leq 0\} .$$

For definiteness we shall suppose that

$$\frac{\partial S}{\partial s} < 0 \qquad \text{in } \mathfrak{R} , \tag{89}$$

where $\mathfrak{R}$ is defined by equation (88) with s set equal to its maximum value s_1 (say), i.e.,

$$\mathfrak{R} = \mathfrak{R}(s_1) . \tag{90}$$

The condition (89) implies in a natural way that $\mathfrak{S}(t_1)$ is interior to $\mathfrak{S}(t_2)$ if $t_1 < t_2$. It further implies that there is a unique surface $\mathfrak{S}$ passing through a given point $\boldsymbol{x}$. Finally, we shall also require that there is a minimum value of s, s_0 (say), such that $S(\boldsymbol{x},s_0)$ is positive unless $x = 0$; this insures that s_0 corresponds to the origin.

Now let the position vector of a point lying on the surface $\mathfrak{S}(s)$ be given by

$$\boldsymbol{x} = \boldsymbol{X}(s,\vartheta,\varphi) , \tag{91}$$

where the surface parameters ϑ and φ have a domain $\mathfrak{D}(s)$ (say); as ϑ and φ range over $\mathfrak{D}(s)$, they should provide all points of $\mathfrak{S}(s)$ in a one-one fashion. With these definitions, the element of volume in $\mathfrak{R}$ is

$$|J|\,ds\,d\vartheta\,d\varphi , \tag{92}$$

where

$$J = \frac{\partial \boldsymbol{X}}{\partial s}\cdot\frac{\partial \boldsymbol{X}}{\partial \vartheta}\times\frac{\partial \boldsymbol{X}}{\partial \varphi} \tag{93}$$

is the Jacobian of the transformation.

If we substitute from equation (91) into the equation $S(\boldsymbol{x},s) = 0$ defining $\mathfrak{S}(s)$, we obtain an identity in s, ϑ, and φ; and differentiating this identity, we find

$$\frac{\partial \boldsymbol{X}}{\partial s}\cdot\text{grad } S = -\frac{\partial S}{\partial s}, \qquad \frac{\partial \boldsymbol{X}}{\partial \vartheta}\cdot\text{grad } S = 0 , \qquad \text{and} \qquad \frac{\partial \boldsymbol{X}}{\partial \varphi}\cdot\text{grad } S = 0 . \tag{94}$$

The last two of these equations imply that there exists a scalar function $\alpha(\boldsymbol{x})$ such that

$$\frac{\partial \boldsymbol{X}}{\partial \vartheta}\times\frac{\partial \boldsymbol{X}}{\partial \varphi} = \alpha(\boldsymbol{x})\,\text{grad } S . \tag{95}$$

The expression (93) for the Jacobian now becomes

$$J = \frac{\partial \boldsymbol{X}}{\partial s} \cdot \boldsymbol{\alpha} \operatorname{grad} S = -\alpha \frac{\partial S}{\partial s} . \tag{96}$$

Eliminating α in equation (96) with the aid, once again, of equation (95), we finally obtain

$$|J| = -\frac{|(\partial \boldsymbol{X}/\partial \vartheta) \times (\partial \boldsymbol{X}/\partial \varphi)|}{|\operatorname{grad} S|} \frac{\partial S}{\partial s}, \tag{97}$$

where the minus sign has been inserted to accord with our convention that $\partial S/\partial s$ is negative.

Returning to our principal problem, we are primarily interested in evaluating integrals whose integrands have the form $G(s)H(\boldsymbol{x})$, where H, in general, depends on ϑ and φ, as well as s (cf. eq. [42]). We now see that

$$\int_{\mathfrak{R}} G(s) H(\boldsymbol{x}) d\boldsymbol{x} = \int_{s_0}^{s_1} G(s) W(s) ds , \tag{98}$$

where

$$W(s) = \iint_{\mathfrak{D}(s)} |J| H(\boldsymbol{X}) d\vartheta d\varphi , \tag{99}$$

where, as the notation implies, the dependence of H on $\boldsymbol{x}$ must be replaced, by means of the transformation equation (91), by its dependence on s, ϑ, and φ.

In our present problem we may choose for φ the azimuthal angle. All the integrands are then independent of φ, and the formula (99) is accordingly simplified. We shall find in §§ VI–VIII that the evaluation of four integrals like the one on the right-hand side of equation (99) will reduce all the integrals appearing in equation (42) to single integrals over s.

VI. SPHEROIDAL STRATIFICATIONS

It is possible to show that, in a rigorous theory, a uniformly rotating configuration cannot be stratified on spheroids unless the density is everywhere the same (cf. Dive 1952). Since, however, on an approximate theory, slowly rotating configurations *are* stratified on spheroids, it is reasonable, when seeking approximate equilibrium configurations, to assume spheroidal stratification and try to find that set of spheroidal surfaces which is, in some sense, the best. This is the approach taken by Roberts (1962, 1963) and by Hurley and Roberts (1964*a*, *b*), who have constructed rapidly rotating polytropes selecting the "best" stratification on the basis of a variational principle. We shall adapt equation (47), appropriately, for a study of the oscillations of these models of Hurley and Roberts.

The function S of equations (88) may, in the present case, be written as

$$S(\varpi, z, s) = \tfrac{1}{2}\left(-s^2 + \varpi^2 + \frac{z^2}{1-e^2}\right), \tag{100}$$

where e is, in general, a function of s. The coordinate s is the equatorial radius of a given level surface and $e(s)$ is the eccentricity of its meridional section. From equation (100) it follows that

$$\frac{\partial S}{\partial s} = -s + \frac{e e'}{(1-e^2)^2} z^2 \qquad \left(e' = \frac{de}{ds}\right), \tag{101}$$

and

$$|\operatorname{grad} S| = \left[\varpi^2 + \frac{z^2}{(1-e^2)^2}\right]^{1/2} . \tag{102}$$

It is convenient to introduce the coordinates s, ϑ, and φ through the transformation formulae

$$x_1 = s \sin \vartheta \cos \varphi, \quad x_2 = s \sin \vartheta \sin \varphi, \quad \text{and} \quad x_3 = (1 - e^2)^{1/2} s \cos \vartheta\ ; \tag{103}$$

ϑ is then the "mean anomaly" and φ is the usual azimuthal angle.

We now calculate $|J|$ as given by equation (97). Since $e'(s) \geq 0$ in all cases of interest (cf. Hurley and Roberts 1964*b;* in the slowly rotating case, this inequality follows from Clairaut's equation) we may define, without ambiguity,

$$\beta(s) = \left(\frac{s e e'}{1 - e^2}\right)^{1/2} \qquad (e' \geq 0). \tag{104}$$

With this definition, equation (101) takes the alternative form

$$\frac{\partial S}{\partial s} = - s (1 - \beta^2 \cos^2 \vartheta). \tag{105}$$

Our earlier requirement (89) that $\partial S/\partial s \leq 0$ now implies that $\beta < 1$; we shall assume that this is the case. [It can be verified that the difficulty encountered by Hurley and Roberts (1964*a*) with the crossing of the level surfaces is related to a violation of this inequality.]

A straightforward calculation now gives

$$J = (1 - e^2)^{1/2} s^2 (1 - \beta^2 \cos^2 \vartheta) \sin \vartheta\ ; \tag{106}$$

and equation (99), for the stratification we are presently considering, becomes

$$W(s) = 2\pi (1 - e^2)^{1/2} s^2 \int_{-1}^{+1} (1 - \beta^2 \mu^2) H(s, \mu) d\mu, \tag{107}$$

where we have used $\mu = \cos \vartheta$ as the variable of integration.

Inspection of equation (47) now shows that to reduce the various integrals occurring in this equation to quadratures over s, it is necessary to evaluate four functions $W(s)$ with the choices

$$H_0 = 1, \qquad H_1 = \left(\frac{\partial U}{\partial \varpi}\right)^2, \qquad H_2 = g^2 = \left(\frac{\partial U}{\partial \varpi}\right)^2 + \left(\frac{\partial U}{\partial z}\right)^2, \qquad \text{and } H_3 = g^4. \tag{108}$$

We shall denote the corresponding functions by $W_i(s)$ ($i = 0, 1, 2$, and 3). It will appear that all these functions can be expressed in terms of elementary functions.

First, we note that since the potential U is a function of s only, we may write (to avoid ambiguity)

$$U = u(s)\ ; \tag{109}$$

and we have

$$\left(\frac{\partial U}{\partial \varpi}\right)^2 = \left(\frac{du}{ds}\right)^2 \left(\frac{\partial S}{\partial \varpi}\right)^2 \quad \text{and} \quad \left(\frac{\partial U}{\partial z}\right)^2 = \left(\frac{du}{ds}\right)^2 \left(\frac{\partial S}{\partial z}\right)^2. \tag{110}$$

On the other hand, the derivatives of $S(\varpi, z)$ may be obtained by implicit differentiation of equation (100); we find

$$\frac{\partial S}{\partial \varpi} = \frac{(1 - \mu^2)^{1/2}}{1 - \beta^2 \mu^2} \quad \text{and} \quad \frac{\partial S}{\partial z} = \frac{\mu}{(1 - e^2)^{1/2} (1 - \beta^2 \mu^2)}. \tag{111}$$

Writing out the functions $W_i(s)$, and simultaneously defining the new functions $V_i(s)$ (for later convenience), we have

$$4\pi V_0(s) \equiv W_0(s) = 4\pi(1-e^2)^{1/2}s^2\int_0^1(1-\beta^2\mu^2)\,d\mu = 4\pi(1-e^2)^{1/2}(1-\tfrac{1}{3}\beta^2)\,s^2\,, \tag{112}$$

$$4\pi\left(\frac{du}{ds}\right)^2 V_1(s) \equiv W_1(s) = 4\pi(1-e^2)^{1/2}s^2\left(\frac{du}{ds}\right)^2\int_0^1\frac{1-\mu^2}{1-\beta^2\mu^2}\,d\mu = 4\pi\,\frac{(1-e^2)^{1/2}}{\beta^2}\left[1-\frac{1-\beta^2}{2\beta}\log\left(\frac{1+\beta}{1-\beta}\right)\right]s^2\left(\frac{du}{ds}\right)^2, \tag{113}$$

and

$$4\pi\left(\frac{du}{ds}\right)^{2(i-1)}V_i(s) \equiv W_i(s) = 4\pi(1-e^2)^{1/2}s^2\left(\frac{du}{ds}\right)^{2(i-1)} \times\int_0^1\left(1+\frac{e^2\mu^2}{1-e^2}\right)^{i-1}\frac{d\mu}{(1-\beta^2\mu^2)^{2i-3}}\qquad(i=2,3)\,. \tag{114}$$

The integrals over μ appearing in the definitions of the functions $V_2(s)$ and $V_3(s)$ are also readily evaluated, and we find

$$\int_0^1\left(1+\frac{e^2\mu^2}{1-e^2}\right)\frac{d\mu}{1-\beta^2\mu^2} = \frac{1}{\beta^2}\left[\left(\beta^2+\frac{e^2}{1-e^2}\right)f(\beta)-\frac{e^2}{1-e^2}\right],$$

and

$$\int_0^1\left(1+\frac{e^2\mu^2}{1-e^2}\right)^2\frac{d\mu}{(1-\beta^2\mu^2)^3} = \frac{1}{8\beta^4}\left\{\left[3\beta^4-\frac{2e^2}{1-e^2}\beta^2+\frac{3e^4}{(1-e^2)^2}\right]f(\beta) + \frac{1}{(1-\beta^2)^2}\left[(5-3\beta^2)\beta^4+\frac{2e^2}{1-e^2}(1+\beta^2)\beta^2+\frac{e^4}{(1-e^2)^2}(5\beta^2-3)\right]\right\}, \tag{115}$$

where

$$f(\beta) = \frac{1}{2\beta}\log\left(\frac{1+\beta}{1-\beta}\right) = \sum_{k=0}^{\infty}\frac{\beta^{2k}}{2k+1}\,. \tag{116}$$

With the foregoing reductions equation (47) finally takes the form

$$\sigma^2\int_{s_0}^{s_1}\rho\chi^2\left(\frac{du}{ds}\right)^2V_2\,ds = \int_{s_0}^{s_1}\left(\frac{du}{ds}\right)^3\frac{d}{ds}(\rho\chi^2)V_3\,ds + \int_{s_0}^{s_1}\gamma p\left[\left(\frac{d\chi}{ds}\right)^2\left(\frac{du}{ds}\right)^2V_3-2(4\pi G\rho-2\Omega^2)\frac{d\chi}{ds}\frac{du}{ds}\chi V_2+(4\pi G\rho-2\Omega^2)^2\chi^2V_0\right]ds - 4\int_{s_0}^{s_1}\rho\chi^2\left(\frac{du}{ds}\right)^2[(3\pi G\rho-\Omega^2)V_2-\Omega^2V_1]\,ds\,. \tag{117}$$

This equation represents a generalization, to the case when rotation is present, of equation (51) valid in the absence of rotation. The Euler equation based on equation (117) will provide a generalization of Eddington's pulsation equation analogous to Roberts' generalization of the Lane-Emden equation (Roberts 1962, eq. [2.31]). However, we shall have no reason to write down this generalization.

VII. THE ROCHE STRATIFICATION

The spheroidal stratification considered in § VI is an extrapolation to larger rotations of the stratification that is known to be correct for slow rotation. In contrast, a stratification that will apply to all rotations, with sufficient approximation under most circumstances of practical interest, is that of Roche which one obtains when all the mass (M) is concentrated at the center. The level surfaces in this limit are given by (cf. Jeans 1929, p. 252)

$$U_{\text{Roche}} = \frac{GM}{r} + \tfrac{1}{2}\Omega^2 r^2 \sin^2 \vartheta . \tag{118}$$

While these level surfaces are strictly applicable only in the limit of infinite central condensation, they do provide a very good approximation to the true level surfaces of rotating masses of even moderate central condensations such as a polytrope of index $n = 3$. For example, the ellipticity ϵ of a polytrope $n = 3$ for slow rotation is given by (cf. Chandrasekhar 1933)

$$\epsilon = 0.7717 \frac{\Omega^2}{2\pi G \rho_m} \tag{119}$$

(where ρ_m denotes the mean density), while for the Roche surfaces

$$\epsilon = 0.75 \frac{\Omega^2}{2\pi G \rho_m} \qquad \left(\rho_m = \frac{M}{\frac{4}{3}\pi R^3}\right). \tag{120}$$

The discrepancy, for slow rotation, is thus less than 3 per cent even for the outermost boundary. The assumption that even for rapid rotation the Roche surfaces provide a very good approximation to the true level surfaces of rotating masses, with central condensations comparable to the polytrope $n = 3$, has been confirmed by Monaghan and Roxburgh (1965*a*, *b*) and by Ostriker and Mark.[11] Indeed Roxburgh, Griffith, and Sweet (1965) have made the approximation provided by the Roche surfaces the first step in a rapidly converging iteration scheme for obtaining the exact solutions under a wide variety of circumstances. It would therefore appear that, for a discussion of the oscillations and the stability of rapidly rotating masses, the reduction of equation (45) for the Roche stratification will provide a good starting point.

It is important to observe that the foregoing remarks on the good agreement of the level surfaces in rotating configurations with the *Roche surfaces* are not meant to suggest that the march of U in these configurations agrees well with the march of U in the *Roche model*. They are meant rather to point out that the level surfaces in rotating configurations (of even moderate central condensations) can be *geometrically* well approximated by the Roche surfaces so that in the same approximation we may consider U as a function of U_{Roche}, regarding the latter only as a labeling parameter. In other words, the assumption that we propose to make is

$$U = f(U_{\text{Roche}}) , \tag{121}$$

where f is some function (to be determined empirically from the integrations such as those of Ostriker and Mark).

Returning to Roche's equation (118), we first observe that the condition, that for equilibrium the effective gravity should be directed inward, requires that the equatorial radius of the outermost level surface does not exceed the critical value

$$\varpi_c = (GM/\Omega^2)^{1/3} . \tag{122}$$

[11] We are greatly indebted to Dr. J. Ostriker and Mr. J. Mark for providing us with the results of their exact numerical integrations for rapidly rotating polytropes.

By choosing ϖ_c as the unit of length and Ω as the unit of frequency, we can rewrite equation (118) in the form

$$\nu = \frac{1}{r} + \tfrac{1}{2} r^2 \sin^2 \vartheta \,, \tag{123}$$

if U_{Roche} is measured in the unit $(GM\Omega)^{2/3}$; and we suppose that U (also measured in the same unit) is some determinate function only of ν. Accordingly, the corresponding form of the function S of equations (88) is

$$S \equiv S(r, \mu, s) = \frac{1}{r} + \tfrac{1}{2} r^2 (1 - \mu^2) - \nu[U(s)]. \tag{124}$$

We now write the variational expression (42) for σ^2 in the form

$$\begin{aligned}\left(\frac{\sigma}{\Omega}\right)^2 \int_{\Re} \rho \chi^2 g^2 d\boldsymbol{x} &= \int_{\Re} \gamma p \left[\left(\frac{d\chi}{dU}\right)^2 g^4 - 4(2Q-1)\chi \frac{d\chi}{dU} g^2 + 4(2Q-1)^2 \chi^2\right] d\boldsymbol{x} \\ &\quad - \int_{\Re} \rho \chi^2 \left[2(4Q-1) g^2 + \operatorname{grad} U \cdot \operatorname{grad} g^2 - 4\left(\frac{\partial U}{\partial \varpi}\right)^2\right] d\boldsymbol{x} \,,\end{aligned} \tag{125}$$

where g is measured in the unit $(GM\Omega^4)^{1/3}$, and

$$Q = \pi G \rho / \Omega^2 \,; \tag{126}$$

also the unit in which p is measured is $(GM\Omega)^{2/3}$ times the unit of ρ.

Since we have supposed that U is some function of ν only,

$$\operatorname{grad} U = \frac{dU}{d\nu} \operatorname{grad} \nu \,, \qquad g^2 = \left(\frac{dU}{d\nu}\right)^2 |\operatorname{grad} \nu|^2 \,,$$

and

$$\operatorname{grad} U \cdot \operatorname{grad} g^2 = \left(\frac{dU}{d\nu}\right)^3 [\operatorname{grad} \nu \cdot \operatorname{grad}(|\operatorname{grad} \nu|^2)] + 2\left(\frac{dU}{d\nu}\right)^2 \frac{d^2U}{d\nu^2} |\operatorname{grad} \nu|^4 \,; \tag{127}$$

whereas from equation (123) we find

$$\begin{aligned}&\left(\frac{\partial \nu}{\partial \varpi}\right)^2 = \frac{(1 - r^3)^2}{r^4}(1 - \mu^2) \,, \qquad |\operatorname{grad} \nu|^2 = \frac{1}{r^4} - \left(\frac{2}{r} - r^2\right)(1 - \mu^2) \,, \\ &\text{and} \\ &\operatorname{grad} \nu \cdot \operatorname{grad}(|\operatorname{grad} \nu|^2) = \frac{4}{r^7} - 2\left(\frac{3}{r^4} + \frac{3\mu^2}{r} - r^2\right)(1 - \mu^2) \,.\end{aligned} \tag{128}$$

Further, taking $s = -\nu$ (the minus sign is needed because ν decreases outward, whereas s, by assumption, must increase outward), we find that the required Jacobian is

$$|J| = \frac{r^4(\nu, \mu)}{1 - r^3(\nu, \mu)(1 - \mu^2)} = \frac{r^4(\nu, \mu)}{3 - 2\nu r(\nu, \mu)} \,, \tag{129}$$ [12]

where the latter form is obtained by making use of equation (123).

With the definitions (cf. eq. [107])

[12] Eq. (129) follows directly from a formula given in § VIII (eq. [133]) which is applicable to entirely general axisymmetric level surfaces: we have only to identify S in that formula with the surface S defined by equation (123) and let $s = -\nu$.

$$V_0(\nu) = \nu^4 \int_0^1 |J| \, d\mu \,, \qquad V_1(\nu) = \int_0^1 |J| \, |\operatorname{grad} \nu|^2 d\mu \,,$$

$$V_2(\nu) = \frac{1}{\nu^4} \int_0^1 |J| \, |\operatorname{grad} \nu|^4 d\mu \,, \qquad V_4(\nu) = \int_0^1 |J| \left(\frac{\partial \nu}{\partial \varpi}\right)^2 d\mu \,, \tag{130}$$

$$V_3(\nu) = \frac{1}{\nu^4} \int_0^1 |J| \operatorname{grad} \nu \cdot \operatorname{grad}(|\operatorname{grad} \nu|^2) \, d\mu \,,$$

[13]

equation (24) may now be written in the form

$$\left(\frac{\sigma}{\Omega}\right)^2 \int \rho \chi^2 \left(\frac{dU}{d\nu}\right)^2 V_1 d\nu$$

$$= -\int \rho \chi^2 \left[2(4Q-1) \left(\frac{dU}{d\nu}\right)^2 V_1 + \nu^4 \left(\frac{dU}{d\nu} V_3 + 2 \frac{d^2 U}{d\nu^2} V_2\right) \left(\frac{dU}{d\nu}\right)^2 - 4V_4 \right] d\nu \tag{131}$$

$$+ \int \gamma p \left[\left(\frac{d\chi}{d\nu}\right)^2 \left(\frac{dU}{d\nu}\right)^2 \nu^4 V_2 - 4(2Q-1) \chi \frac{d\chi}{d\nu} \frac{dU}{d\nu} V_1 + 4(2Q-1)^2 \chi^2 \frac{V_0}{\nu^4} \right] d\nu \,,$$

where the integrations are over the appropriate range of ν. Table 4 gives V_0, V_1, V_2, V_3, and V_4 as functions of $1/\nu$.

In applying equation (129), we may substitute for $p(\nu)$ and $\rho(\nu)$ the distributions derived for configurations of at least moderate central condensations: as we have stated, the Roche stratification may be expected to provide a sufficiently good approximation for them.

VIII. GENERAL AXISYMMETRIC CONFIGURATIONS

A general axisymmetric configuration may be specified by giving the distance $R(s,\mu)$ from the center to a point on the level surface at colatitude $\vartheta = \cos^{-1} \mu$. The associated coordinate transformation (91) has the form

$$x_1 = R(s,\mu)(1-\mu^2)^{1/2} \cos \varphi \,, \quad x_2 = R(s,\mu)(1-\mu^2)^{1/2} \sin \varphi \,, \quad \text{and} \quad x_3 = R(s,\mu)\mu \,; \tag{132}$$

and a straightforward calculation gives (cf. eq. [97])

$$J = -R^2(s,\mu) \frac{\partial S/\partial s}{|\partial S/\partial r|} \,. \tag{133}$$

We may take

$$S(r,s,\mu) = r - R(s,\mu) \,, \tag{134}$$

as the equation defining the level surfaces; but to be in agreement with our convention (89) regarding s we must assume $\partial R/\partial s > 0$. With this definition of S equation (131) gives

$$|J| = R^2 \frac{\partial R}{\partial s} = -R^2 \frac{\partial R}{\partial U} \,, \tag{135}$$

where, as in the preceding section (see n. 12), we have taken $s = -U$. It also follows from equation (132) (with $s = -U$) that

$$\left(\frac{\partial U}{\partial \varpi}\right)^2 = \frac{(R + \mu \partial R/\partial \mu)^2}{R^2 (\partial R/\partial U)^2} (1-\mu^2) \quad \text{and} \quad g^2 = \frac{R^2 + (1-\mu^2)(\partial R/\partial \mu)^2}{R^2 (\partial R/\partial U)^2} \,. \tag{136}$$

[13] The factor ν^4 in the definition of $V_0(\nu)$ and the factor ν^{-4} in the definitions of $V_2(\nu)$ and $V_3(\nu)$ have been inserted in order that these functions may not vary over too wide a range of values.

TABLE 4

THE WEIGHT FUNCTIONS FOR THE ROCHE STRATIFICATION

ν^{-1}	V_0	V_1	V_2	V_3	V_4
0.0	1.000000	1.000000	1.000000	0	0.666667
.01	1.000000	0.999988	1.000008	0.040000	.666667
.02	1.000001	0.999990	0.999978	0.079999	.666662
.03	1.000048	0.999980	0.999912	0.119996	.666646
.04	1.000125	0.999956	0.999788	0.159986	.666616
.05	1.000248	0.999916	0.999584	0.199967	.666567
.06	1.000430	0.999855	0.999281	0.239931	.666494
.07	1.000685	0.999771	0.998857	0.279872	.666392
.08	1.001024	0.999658	0.998294	0.319781	.666257
.09	1.001460	0.999513	0.997572	0.359650	.666083
.10	1.002004	0.999332	0.996669	0.399466	.665866
.11	1.002670	0.999111	0.995568	0.439218	.665600
.12	1.003470	0.998846	0.994247	0.478893	.665282
.13	1.004417	0.998532	0.992688	0.518474	.664905
.14	1.005524	0.998165	0.990872	0.557947	.664465
.15	1.006805	0.997742	0.988778	0.597294	.663957
.16	1.008273	0.997258	0.986387	0.636495	.663375
.17	1.009943	0.996708	0.983682	0.675531	.662715
.18	1.011829	0.996089	0.980642	0.714380	.661972
.19	1.013948	0.995396	0.977250	0.753018	.661139
.20	1.016313	0.994623	0.973488	0.791422	.660211
.21	1.018944	0.993768	0.969337	0.829564	.659183
.22	1.021856	0.992824	0.964780	0.867419	.658049
.23	1.025068	0.991788	0.959801	0.904955	.656804
.24	1.028600	0.990654	0.954382	0.942143	.655440
.25	1.032472	0.989416	0.948507	0.978951	.653952
.26	1.036706	0.988071	0.942161	1.015343	.652333
.27	1.041327	0.986611	0.935329	1.051284	.650577
.28	1.046358	0.985032	0.927996	1.086737	.648676
.29	1.051828	0.983328	0.920148	1.121661	.646624
.30	1.057766	0.981492	0.911773	1.156015	.644412
.31	1.064204	0.979518	0.902858	1.189754	.642033
.32	1.071176	0.977399	0.893392	1.222832	.639478
.33	1.078721	0.975128	0.883365	1.255202	.636739
.34	1.086879	0.972697	0.872767	1.286810	.633806
.35	1.095698	0.970099	0.861590	1.317605	.630668
.36	1.105227	0.967325	0.849827	1.347529	.627316
.37	1.115522	0.964367	0.837472	1.376522	.623739
.38	1.126646	0.961214	0.824521	1.404521	.619924
.39	1.138669	0.957856	0.810971	1.431459	.615858
.40	1.151669	0.954282	0.796820	1.457267	.611527
.41	1.165733	0.950481	0.782069	1.481868	.606916
.42	1.180962	0.946439	0.766721	1.505185	.602010
.43	1.197470	0.942143	0.750779	1.527132	.596789
.44	1.215386	0.937577	0.734250	1.547620	.591234
.45	1.234860	0.932725	0.717144	1.566553	.585324
.46	1.256065	0.927569	0.699471	1.583827	.579036
.47	1.279202	0.922087	0.681245	1.599333	.572342
.48	1.304508	0.916258	0.662484	1.612951	.565213
.49	1.332263	0.910056	0.643209	1.624552	.557617
.50	1.362800	0.903452	0.623444	1.633996	.549517
.51	1.396521	0.896416	0.603216	1.641129	.540869
.52	1.433917	0.888910	0.582559	1.645782	.531625
.53	1.475589	0.880892	0.561510	1.647770	.521730
.54	1.522291	0.872313	0.540112	1.646884	.511118
.55	1.574979	0.863115	0.518414	1.642890	.499712
.56	1.634889	0.853231	0.496473	1.635520	.487419
.57	1.703651	0.842577	0.474353	1.624470	.474128
.58	1.783476	0.831052	0.452127	1.609379	.459702
.59	1.877447	0.818531	0.429880	1.589821	.443965
.60	1.990038	0.804849	0.407711	1.565273	.426694
.61	2.128060	0.789791	0.385734	1.535080	.407587
.62	2.302565	0.773057	0.364084	1.498384	.386227
.63	2.533135	0.754208	0.342927	1.453995	.361994
.64	2.859343	0.732543	0.322466	1.400126	.333888
.65	3.380898	0.706770	0.302972	1.333695	.300056
0.66	4.506312	0.673723	0.284843	1.247630	0.255881

It is now clear from equations (98) and (99) that to reduce equation (47) to simple quadratures over s, we need the following functions:

$$V_0(U)=\int_{-1}^{+1}|J|\,d\mu,\quad V_1(U)=\int_{-1}^{+1}g^2|J|\,d\mu,\quad V_2(U)=\int_{-1}^{+1}g^4|J|\,d\mu,$$

and

$$V_3(U)=\int_{-1}^{+1}\left(\frac{\partial U}{\partial \varpi}\right)^2|J|\,d\mu. \tag{137}$$

In terms of these functions equation (47) becomes

$$\sigma^2\int_{U_0}^{U_1}\rho\chi^2V_1dU=\int_{U_1}^{U_0}\frac{d}{dU}(\rho\chi^2)V_2dU-4\int_{U_0}^{U_1}\rho\chi^2[(3\pi G\rho-\Omega^2)V_1-\Omega^2V_3]dU$$
$$+\int_{U_0}^{U_1}\gamma p\left[\left(\frac{d\chi}{dU}\right)^2V_2-2(4\pi G\rho-2\Omega^2)\chi\frac{d\chi}{dU}V_1+(4\pi G\rho-2\Omega^2)^2\chi^2V_0\right]dU. \tag{138}$$

Equation (136) should be especially suitable for the numerical evaluation of σ^2 for the rotating polytropes of James (1964, 1967).

IX. CONCLUDING REMARKS

That slow rotation has a stabilizing effect, in lowering the critical values of γ (for marginal stability) below the value $\frac{4}{3}$, is an unambiguous theoretical result. When one seeks the origin of the stabilizing effect in the analytical treatment, one finds that it is directly traceable to the Coriolis term in the equations of motion which leads to the relation $\xi_\varphi=(2i\Omega/\sigma)\xi_\varpi$ (cf. eq. [16]) between the φ- and the ϖ-components of the Lagrangian displacement. But this is not the only effect of rotation: its effect on the equilibrium distribution of the pressure and the density, resulting particularly in a general expansion of the configuration, may have a destabilizing, and perhaps even a decisive, influence. The principal reason for adapting the general variational principle so as to be applicable to a wide variety of initial conditions is the hope that it will resolve these basic ambiguities. We shall return to these questions in a second paper which will include the results of the application of the different formulae developed in this paper.

On a somewhat different aspect of the problem, the present paper has been restricted to a consideration of the effects of uniform rotation. The extension of the analysis to allow for non-uniform rotation should be of considerable interest in view of the greater amount of angular momentum that can be stored in the configuration. Such an extension should be particularly simple when Ω is a function of ϖ only; for then not only are the considerations of § II almost unaltered (the only difference is that $4\Omega^2$ in eq. [19] is replaced by Rayleigh's discriminant $d[\varpi^4\Omega^2]/\varpi^3d\varpi$) but also the form (33) of the trial function would be applicable since equations (9) continue to hold. We hope to return to a consideration of these extensions in later papers.

We are grateful to Professor P. H. Roberts for providing us with the results of his exact numerical integrations of the pulsation equation which were used in the calculations summarized in Table 1; to Dr. M. Aizenman for evaluating the weight functions listed in Table 4; and to Miss Donna D. Elbert for evaluating the various matrix elements needed in the preparation of Table 1 and also for the results included in Tables 2 and 3.

The work of the first author was supported by the Office of Naval Research under contract Nonr-2121(24), and that of the second author by the Air Force Office of Scientific Research, through grant AF-AFOSR-712-67.

REFERENCES

Chandrasekhar, S. 1933, *M.N.R.A.S.*, **93**, 390.
———. 1964*a*, *Ap. J.*, **139**, 664.
———. 1964*b*, *ibid.*, **140**, 417.
Chandrasekhar, S., and Lebovitz, N. R. 1962*a*, *Ap. J.*, **135**, 248.
———. 1962*b*, *ibid.*, **136**, 1037.
———. 1962*c*, *ibid.*, p. 1069.
———. 1962*d*, *ibid.*, p. 1082.
Clement, M. 1964, *Ap. J.*, **140**, 1045.
———. 1965, *ibid.*, **141**, 210.
Cowling, T. G., and Newing, R. A. 1949, *Ap. J.*, **109**, 149.
Dive, P. 1952, *Bull. Sci. Math.*, **76**, 38.
Durney, B. R., and Roxburgh, I. W. 1967, *Proc. Roy. Soc. London, A*, **296**, 189.
Fowler, W. 1966, *Ap. J.*, **144**, 180.
Hurley, M., and Roberts, P. H. 1964*a*, *Ap. J.*, **140**, 583.
———. 1964*b*, *Ap. J. Suppl.*, **11**, 95.
James, R. A. 1964, *Ap. J.*, **140**, 552.
———. 1967 (private communication).
Jeans, J. H. 1929, *Astronomy and Cosmogony* (2d ed.; Cambridge: Cambridge University Press; reprinted 1964, Dover Publications, New York).
Laval, G., Mercier, C., and Pellat, R. 1965, *Nucl. Fusion*, **5**, 156.
Lebovitz, N. R. 1967, *Ap. J.*, **150**, 203.
Ledoux, P. 1945, *Ap. J.*, **102**, 143.
———. 1946, *ibid.*, **104**, 333.
Ledoux, P., and Walraven, Th. 1958, *Hdb. d. Phys.*, ed. S. Flügge (Berlin: Springer-Verlag).
Lynden-Bell, D., and Ostriker, J. 1967 ,*M.N.R.A.S.*, **136**, 293.
Milne, E. A. 1923, *M.N.R.A.S.*, **83**, 118.
Monaghan, J. J., and Roxburgh, I. W. 1965*a*, *M.N.R.A.S.*, **129**, 5.
———. 1965*b*, *ibid.*, **131**, 13.
Roberts, P. H. 1962, *Ap. J.*, **137**, 1129.
———. 1963, *ibid.*, **138**, 809.
Rosseland, S. 1949, *The Pulsation Theory of Variable Stars* (Oxford: Clarendon Press; reprinted 1964 Dover Publications, New York).
Roxburgh, I. W., Griffith, J. S., and Sweet, P. A. 1965, *Zs. f. Ap.*, **61**, 203.

THE ELLIPTICITY OF A SLOWLY ROTATING CONFIGURATION

S. Chandrasekhar and P. H. Roberts
University of Chicago
Received April 29, 1963

ABSTRACT

The second-order virial theorem is used to set upper and lower bounds for m/ϵ_R for a slowly rotating configuration, where m is the ratio of the centrifugal acceleration at the equator to the (average) gravitational acceleration on its surface and ϵ_R is the ellipticity of its slightly oblate figure of equilibrium. The bounds obtained are explicitly evaluated for the polytropes, for a model consisting of a core and a mantle of constant densities, and for a particular model for the earth.

I. INTRODUCTION

For a slowly rotating configuration, the ratio m, of the centrifugal acceleration at the equator to the average gravitational acceleration on its surface, and the ellipticity ϵ_R, of its slightly oblate figure of equilibrium, are in a relationship of cause and effect. It is, in fact, one of Newton's well-known theorems (cf. Todhunter 1873, Sec. 27) that, for a homogeneous configuration,

$$\frac{m}{\epsilon_R} = \tfrac{4}{5}. \tag{1}$$

Quite generally, m/ϵ_R is functionally dependent on the distribution of the density in the configuration; and the establishment of this dependence by Clairaut is one of the great achievements that followed in Newton's wake.

If one restricts one's self to configurations in which (in the non-rotating state) the density $\rho_0(r)$ at a point, does not exceed the mean density, $\bar{\rho}_0(r)$, interior to that point, then it is known that (cf. Chandrasekhar 1933, eq. [100])

$$1 \leq \tfrac{5}{4}\frac{m}{\epsilon_R} < 2.5\,. \tag{2}$$

And if the configuration is one in which the departures from homogeneity are not very great, then an approximate formula, derived by Darwin (1899) on the basis of a transformation of Clairaut's equation due to Radau, relates m/ϵ_R to the moment of inertia,

$$I = \int_V \rho_0(r)\, r^2 dx = 4\pi \int_0 \rho_0(r)\, r^4 dr\,, \tag{3}$$

of the configuration by

$$\frac{I}{MR^2} = 1 - \tfrac{2}{5}\left(\tfrac{5}{2}\frac{m}{\epsilon_R} - 1\right)^{1/2}, \tag{4}$$

where M is its mass and R is its (mean) radius. (In eq. [3], $\rho_0(r)$ is the zero-order spherically symmetric part of the density distribution; see eq. [10] below.)

In Table 1, the values of I/MR^2 for the polytropes are compared with those deduced with the aid of equation (4) and the known values of $5m/2\epsilon_R$. It would appear from this table that Darwin's formula (4) cannot be applied if the central density of the configuration exceeds, say, six times its mean density.

In this paper, we shall set certain upper and lower bounds to $5m/4\epsilon_R$ from an application to this problem of the second-order virial equations which have already proved their utility in other connections (see the various papers of Chandrasekhar, Lebovitz, and Roberts in the recent issues of the *Astrophysical Journal*).

II. AN INTEGRAL RELATION

Let the origin of the co-ordinate system be at the center of mass of the configuration; and let the z-axis coincide with the axis of rotation; and, finally, let Ω denote the uniform angular velocity of the rotation. We shall assume that the configuration has symmetry about the z-axis; this assumption is justified for slowly rotating configurations in which we are presently interested.

A fundamental relation provided by the virial theorem is

$$\Omega^2 I_{11} = \mathfrak{W}_{33} - \mathfrak{W}_{11}, \tag{5}$$

where

$$\mathfrak{W}_{ij} = \int_V \rho x_i \frac{\partial \mathfrak{B}}{\partial x_j} dx \quad \text{and} \quad I_{ij} = \int_V \rho x_i x_j dx \tag{6}$$

are the potential energy and the moment of inertia tensors. In equations (6), $\mathfrak{B}$ denotes the gravitational potential and ρ the density.

TABLE 1

APPLICATION OF DARWIN'S FORMULA TO POLYTROPES*

n	$\rho_c/\bar{\rho}$	$5m/2\epsilon_R$	I/MR^2	
			Exact	By Darwin's Formula
0.........	1.0	2.0000	0.60000	0.6000
1.0.......	3.29	3.2899	.39207	.3947
1.5.......	5.99	3.8865	.30690	.3204
2.0.......	11.40	4.3559	.23227	.2673
3.0.......	54.18	4.8596	.11304	.2142
3.5.......	152.9	4.9513	0.06832	0.2049

* The values of $5m/2\epsilon_R$ given in Chandrasekhar (1933, Table 3) have been revised in accordance with the results of the more precise integrations tabulated in Chandrasekhar and Lebovitz (1962, Table 7).

In the case of axisymmetry, $\mathfrak{W}_{33}$ and $\mathfrak{W}_{11}$ are the two distinct components of the potential energy tensor; and in spherical polar co-ordinates (r, ϑ, φ), the expressions for them become

$$\mathfrak{W}_{33} = \int_V \rho r \left[\mu^2 \frac{\partial \mathfrak{B}}{\partial r} + \frac{\mu(1-\mu^2)}{r} \frac{\partial \mathfrak{B}}{\partial \mu} \right] dx \tag{7}$$

and

$$\mathfrak{W}_{11} = \tfrac{1}{2} \int_V \rho r \left[(1-\mu^2) \frac{\partial \mathfrak{B}}{\partial r} - \frac{\mu(1-\mu^2)}{r} \frac{\partial \mathfrak{B}}{\partial \mu} \right] dx, \tag{8}$$

where $\mu = \cos \vartheta$. With the foregoing expressions for $\mathfrak{W}_{33}$ and $\mathfrak{W}_{11}$, equation (5) gives

$$\Omega^2 I_{11} = \int_V \rho r \left[P_2(\mu) \frac{\partial \mathfrak{B}}{\partial r} + \tfrac{3}{2} \frac{\mu(1-\mu^2)}{r} \frac{\partial \mathfrak{B}}{\partial \mu} \right] dx. \tag{9}$$

Equation (9) is exact. We shall now apply it to a slowly rotating configuration in which the departures from spherical symmetry, considered small, are governed by the Legendre function, $P_2(\mu)$. We shall write, then,

$$\rho = \rho_0(r) + \rho_2(r) P_2(\mu) \tag{10}$$

and

$$\mathfrak{B} = \mathfrak{B}_0(r) + \mathfrak{B}_2(r) P_2(\mu), \tag{11}$$

where

$$|\rho_2(r)| \ll |\rho_0(r)| \qquad \text{and} \qquad |\mathfrak{B}_2(r)| \ll |\mathfrak{B}_0(r)| \, . \tag{12}$$

Under these same circumstances, we may also suppose that the boundary of the configuration is given by

$$R(\mu) = R[1 - \tfrac{2}{3}\epsilon_R P_2(\mu)] \, , \tag{13}$$

where ϵ_R defines the ellipticity of the figure of equilibrium.

In the framework of the foregoing assumptions, we may clearly replace I_{11} in equation (9) by its zero-order value,

$$\tfrac{1}{3}I = \tfrac{1}{3}\int_V \rho_0(r)\, r^2 dx = \tfrac{4}{3}\pi \int_0^R \rho_0(r)\, r^4 dr \, . \tag{14}$$

Now substituting for ρ and $\mathfrak{B}$ in accordance with equations (10) and (11) in equation (9), we find (on ignoring all quantities of the second order and after performing the integrations over the angles)

$$\tfrac{1}{3}\Omega^2 I = \tfrac{4}{5}\pi \int_0^R \rho_0(r)\, r^3 \left(\frac{d\mathfrak{B}_2}{dr} + \frac{3}{r}\mathfrak{B}_2\right) dr + \tfrac{4}{5}\pi \int_0^R \rho_2(r)\, r^3 \frac{d\mathfrak{B}_0}{dr}\, dr \, . \tag{15}$$

Since the gravitational potential is derived in a linear fashion from the distribution of density, it is clear that $\mathfrak{B}_0$ and $\mathfrak{B}_2$ should be expressible in terms of ρ_0 and ρ_2, respectively. Thus, by expanding $|x - x'|^{-1}$ in spherical harmonics in the usual manner, we readily find, from Poisson's integral,

$$\mathfrak{B}_0(r) + \mathfrak{B}_2(r) P_2(\mu) = G \int_V \frac{\rho_0(r') + \rho_2(r') P_2(\mu')}{|x - x'|}\, dx' \, , \tag{16}$$

that

$$\mathfrak{B}_0(r) = 4\pi G \left[\frac{1}{r}\int_0^r \rho_0(s)\, s^2 ds + \int_r^R \rho_0(s)\, s\, ds\right] \tag{17}$$

and

$$\mathfrak{B}_2(r) = \tfrac{4}{5}\pi G \left[\frac{1}{r^3}\int_0^r \rho_2(s)\, s^4 ds + r^2 \int_r^R \frac{\rho_2(s)}{s}\, ds\right] . \tag{18}$$

Relations which follow from equations (17) and (18) are[1]

$$\frac{d\mathfrak{B}_0}{dr} = -\frac{4\pi G}{r^2}\int_0^r \rho_0(s)\, s^2 ds = -\frac{GM(r)}{r^2} \tag{19}$$

and

$$\frac{1}{r^3}\frac{d}{dr}(r^3 \mathfrak{B}_2) = \frac{d\mathfrak{B}_2}{dr} + \frac{3}{r}\mathfrak{B}_2 = 4\pi G r \int_r \frac{\rho_2(s)}{s}\, ds \, , \tag{20}$$

where, in equation (19), $M(r)$ is the mass interior to r in the zeroth approximation.

Inserting relations (19) and (20) in equation (15), we obtain

$$\Omega^2 I = \frac{48\pi^2 G}{5}\left[\int_0^R dr\, \rho_0(r)\, r^4 \int_r^R ds\, \frac{\rho_2(s)}{s} - \int_0^R dr\, \rho_2(r)\, r \int_0^r ds\, \rho_0(s)\, s^2\right] . \tag{21}$$

[1] These relations are readily seen to be the first integrals of the equations,

$$\frac{d^2\mathfrak{B}_0}{dr^2} + \frac{2}{r}\frac{d\mathfrak{B}_0}{dr} = -4\pi G \rho_0$$

and

$$\frac{d^2\mathfrak{B}_2}{dr^2} + \frac{2}{r}\frac{d\mathfrak{B}_2}{ar} - \frac{6}{r^2}\mathfrak{B}_2 = -4\pi G \rho_2 \, ,$$

governing $\mathfrak{B}_0$ and $\mathfrak{B}_2$.

Inverting the order of the integrations in the second term on the right-hand side, we can write

$$\Omega^2 I = \frac{48\pi^2 G}{5}\int_0^R d\,r\,\rho_0(\,r\,)\,r^2\Big[\,r^2\int_r^R d\,s\,\frac{\rho_2(\,s\,)}{s} - \int_r^R d\,s\,\rho_2(\,s\,)\,s\,\Big]\,, \tag{22}$$

or

$$\Omega^2 I = \frac{12\pi G}{5}\int_0^R d\,r\,\frac{dM(\,r\,)}{d\,r}\Big[\,r^2\int_r^R d\,s\,\frac{\rho_2(\,s\,)}{s} - \int_r^R d\,s\,\rho_2(\,s\,)\,s\,\Big]\,. \tag{23}$$

An integration by parts now leads to the result

$$\Omega^2 I = -\frac{24\pi G}{5}\int_0^R d\,rM(\,r\,)\,r\int_r^R d\,s\,\frac{\rho_2(\,s\,)}{s}\,. \tag{24}$$

In virtue of equation (20), we can also write

$$\Omega^2 I = -\tfrac{6}{5}\int_0^R \frac{M(\,r\,)}{r^3}\,\frac{d}{d\,r}\,(\,r^3\mathfrak{B}_2)\,d\,r\,; \tag{25}$$

and this is the desired integral relation.

III. AN INTEGRAL EQUATION GOVERNING THE ELLIPTICITY OF THE SURFACES OF EQUAL GEOPOTENTIAL

The definition of the geopotential is

$$\Psi = \mathfrak{B} + \tfrac{1}{2}\Omega^2 r^2(1-\mu^2)\,, \tag{26}$$

so that

$$\text{grad}\ p = \rho\ \text{grad}\ \Psi\,. \tag{27}$$

Let the surfaces of equal geopotential be

$$r[1 - \tfrac{2}{3}\epsilon(r)P_2(\mu)]\,; \tag{28}$$

$\epsilon(r)$ defines, then, the varying ellipticity of these surfaces. At the boundary of the configuration, $\epsilon(r)$ takes the value ϵ_R.

For the assumed form of $\mathfrak{B}$,

$$\Psi = \mathfrak{B}_0(r) + \mathfrak{B}_2(r)P_2(\mu) + \tfrac{1}{3}\Omega^2 r^2[1 - P_2(\mu)]\,. \tag{29}$$

The condition that Ψ so defined is constant over the surfaces (28) requires (in our present approximation) that

$$\mathfrak{B}_2(\,r\,) = \tfrac{2}{3}\,r\epsilon(\,r\,)\,\frac{d\mathfrak{B}_0}{d\,r} + \tfrac{1}{3}\Omega^2 r^2\,; \tag{30}$$

or (cf. eq. [19])

$$\mathfrak{B}_2(\,r\,) = -\tfrac{2}{3}G\epsilon(\,r\,)\,\frac{M(\,r\,)}{r} + \tfrac{1}{3}\Omega^2 r^2\,. \tag{31}$$

From this last equation it follows that

$$\frac{d}{d\,r}\,(\,r^3\mathfrak{B}_2) = -\tfrac{2}{3}G\frac{d}{d\,r}[\epsilon(\,r\,)M(\,r\,)\,r^2] + \tfrac{5}{3}\Omega^2 r^4\,. \tag{32}$$

Returning to equation (25), we may, in view of equation (32), write it in the form

$$\Omega^2 I = -2\Omega^2\int_0^R M(\,r\,)\,r\,d\,r + \tfrac{4}{5}G\int_0^R \frac{M(\,r\,)}{r^3}\,\frac{d}{d\,r}[\epsilon(\,r\,)M(\,r\,)\,r^2]\,d\,r\,. \tag{33}$$

On the other hand,

$$2\int_0^R M(r)\,r\,dr = MR^2 - 4\pi\int_0^R \rho_0(r)\,r^4 dr = MR^2 - I\,. \tag{34}$$

Equation (33), therefore, becomes

$$\Omega^2 MR^2 = \tfrac{4}{5}G\int_0^R \frac{M(r)}{r^3}\frac{d}{dr}[\epsilon(r)M(r)\,r^2]\,dr\,. \tag{35}$$

Now, by definition,

$$m = \frac{\Omega^2 R}{GM/R^2} = \frac{\Omega^2 R^3}{GM}\,. \tag{36}$$

In terms of m, we can rewrite equation (35) in the form

$$\tfrac{5}{4}m = \frac{R}{M^2}\int_0^R \frac{M(r)}{r^3}\frac{d}{dr}[\epsilon(r)M(r)\,r^2]\,dr\,; \tag{37}$$

and an integration by parts gives

$$\tfrac{5}{4}m = \epsilon_R + \frac{R}{M^2}\int_0^R \epsilon(r)M(r)\,r^2\frac{d}{dr}\left[-\frac{M(r)}{r^3}\right]dr\,. \tag{38}$$

Making use of the relation

$$-\frac{d}{dr}\frac{M(r)}{r^3} = -\tfrac{4}{3}\pi\frac{d\bar{\rho}_0(r)}{dr} = \frac{4\pi}{r}[\bar{\rho}_0(r) - \rho_0(r)]\,, \tag{39}$$

we can also write

$$\tfrac{5}{4}m = \epsilon_R + 4\pi\frac{R}{M^2}\int_0^R \epsilon(r)M(r)\,r\,[\bar{\rho}_0(r) - \rho_0(r)]\,dr\,. \tag{40}$$

The integrals which occur in equations (37), (38), and (40) are, therefore, all positive definite, so long as

$$\bar{\rho}_0(r) \geq \rho_0(r)\,. \tag{41}$$

IV. UPPER AND LOWER BOUNDS FOR m/ϵ_R

It is known (cf. Jeffreys 1959, p. 149) that, for configurations in which condition (41) is met, $\epsilon(r)$ is a monotonic increasing function of r and that

$$\epsilon_R \geq \epsilon(r) \geq \epsilon_R\left(\frac{r}{R}\right)^3. \tag{42}$$

Therefore, by inserting these bounds for $\epsilon(r)$ in equation (37), (38), or (40), we shall obtain corresponding bounds for m/ϵ_R. Thus, by replacing $\epsilon(r)$ by ϵ_R in equation (38), we obtain

$$\tfrac{5}{4}\frac{m}{\epsilon_R} \leq 1 - \frac{R}{M^2}\int_0^R M(r)\,r^2\frac{d}{dr}\frac{M(r)}{r^3}\,dr\,. \tag{43}$$

Now, by successive integrations by parts, we find

$$\begin{aligned}\int_0^R M(r)\,r^2\frac{d}{dr}\frac{M(r)}{r^3}\,dr &= \int_0^R \frac{M(r)\,dM(r)}{r} + 3\int_0^R M^2(r)\frac{d}{dr}\left(\frac{1}{r}\right)dr \\ &= 3\frac{M^2}{R} - 5\int_0^R \frac{M(r)\,dM(r)}{r} = 3\frac{M^2}{R} - 5\frac{|\mathfrak{W}|}{G}\,,\end{aligned} \tag{44}$$

where $\mathfrak{W}$ $(=\mathfrak{W}_{ii})$ is the gravitational potential energy of the configuration. We thus obtain the inequality,

$$\tfrac{5}{4}\frac{m}{\epsilon_R} \leq 5\,\frac{|\mathfrak{W}|}{GM^2/R} - 2\,, \tag{45}$$

which gives the desired upper bound.

Similarly, by replacing $\epsilon(r)$ by $\epsilon_R(r/R)^3$ in equation (37), we obtain

$$\tfrac{5}{4}\frac{m}{\epsilon_R} \geq \frac{1}{M^2R^2}\int_0^R \frac{M(r)}{r^3}\frac{d}{dr}[M(r)\,r^5]\,dr\,. \tag{46}$$

The integral on the right-hand side of this inequality is

$$\tfrac{5}{2}\int_0^R M^2(r)\,d(r^2) + \int_0^R r^2M(r)\,dM(r) = \tfrac{5}{2}\,M^2R^2 - 4\int_0^R r^2M(r)\,dM(r)\,. \tag{47}$$

We thus obtain the inequality,

$$\tfrac{5}{4}\frac{m}{\epsilon_R} \geq 2.5 - \frac{4}{M^2R^2}\int_0^R r^2M(r)\,dM(r)\,, \tag{48}$$

which gives the desired lower bound.

V. SOME ILLUSTRATIONS

We shall now consider some applications of the inequalities found in the preceding section.

a) The Polytropes

For the gravitational potential energy of a polytrope, we have Emden's well-known formula,

$$\mathfrak{W} = -\frac{3}{5-n}\frac{GM^2}{R}\,; \tag{49}$$

and the inequality (45) giving the upper bound of m/ϵ_R becomes, in this case,

$$\tfrac{5}{4}\frac{m}{\epsilon_R} \leq \frac{5+2n}{5-n}\,. \tag{50}$$

It should be noted that inequality (50) does not provide a meaningful bound for $n > \frac{5}{3}$: for we know that, in all cases, $5\,m/4\epsilon_R < 2.5$ (cf. eq. [2]); and only for $n > \frac{5}{3}$ is the right-hand side of (50) less than 2.5.

Turning to inequality (48) giving the lower bound of m/ϵ_R and expressing the various quantities in terms of Emden's variables and units (cf. Chandrasekhar and Lebovitz 1962, eq. [8]), we have

$$\tfrac{5}{4}\frac{m}{\epsilon_R} \geq 2.5 + \frac{4}{\xi_1^6(\theta_1')^2}\int_0^{\xi_1} \xi^6\theta^n\frac{d\theta}{d\xi}\,d\xi\,. \tag{51}$$

After an integration by parts, the inequality (51) becomes

$$\tfrac{5}{4}\frac{m}{\epsilon_R} \geq 2.5 - \frac{24}{(n+1)\,\xi_1^6(\theta_1')^2}\int_0^{\xi_1} \theta^{n+1}\xi^5 d\xi\,, \tag{52}$$

where it may be noted that, according to a reduction formula due to Milne (1929),

$$\int_0^{\xi_1} \theta^{n+1}\xi^5 d\xi = \frac{n+1}{2(n+7)}\left[\xi_1^6(\theta_1')^2 - 12\int_0^{\xi_1}\xi^3\theta^2 d\xi\right]. \tag{53}$$

In Table 2, the upper and the lower bounds of $5m/4\epsilon_R$ given by the inequalities (50) and (52) are listed, together with their known values and also the values deduced from Darwin's formula (4).

b) A Model Consisting of a Core and a Mantle of Different Densities

A model often used as an idealization of the density distribution which occurs inside the earth consists of a core of a certain constant density and a mantle of a different constant density. We shall consider the implications of the inequalities (45) and (48) for this model.

TABLE 2

UPPER AND LOWER BOUNDS OF $5m/4\epsilon_R$ FOR POLYTROPES

n	Lower Bound	Upper Bound	Exact Value	By Darwin's Formula
0	1.000	1.000	1.0000	1.000
1.0	1.456	1.750	1.6449	1.655
1.5	1.661	2.286	1.9432	2.001
2.0	1.850	....	2.1779	2.312
3.0	2.117	....	2.4298	....
3.5	2.293	....	2.4757	....
5.0	2.500	....	2.5000	....

Let the core occupy a fraction q of the radius R and let α be the ratio of the densities of the core and the mantle. The integrals which occur in the inequalities (45) and (48) are readily evaluated for this model, and we find

$$\frac{5}{4}\frac{m}{\epsilon_R} \leq 3\,\frac{\alpha^2 q^5 + (1-q^5) - 2.5(1-\alpha)q^3(1-q^2)}{(\alpha q^3 + 1 - q^3)^2} - 2 \tag{54}$$

and

$$\frac{5}{4}\frac{m}{\epsilon_R} \geq 2.5 - 1.5\,\frac{\alpha^2 q^8 + (1-q^8) - 1.6(1-\alpha)q^3(1-q^5)}{(\alpha q^3 + 1 - q^3)^2}\,. \tag{55}$$

For the earth

$$m = 0.003499 \qquad \text{and} \qquad \epsilon_R = 1/297\,; \tag{56}$$

and it is further estimated that

$$q = 0.545\,. \tag{57}$$

These values for m, ϵ_R, and q, when inserted in formulae (54) and (55), lead to two quadratic inequalities for α, which, when solved, give

$$4.68 \geq \alpha \geq 2.62\,. \tag{58}$$

The value of α generally used, when the present model for the earth is assumed, is 2.9. The lower bound for α determined by formula (54) is, thus, very close to its "true value."

c) A Particular Model for the Earth

As a last illustration, we shall consider a particular model for the earth which Bullard (1946) has used for a direct integration of Clairaut's equation. Using the values of $\rho(r)$ and $\bar{\rho}(r)$ tabulated in Bullard's paper, we find (by numerical integration) that

$$\frac{|\mathfrak{W}|}{GM^2/R} = 0.6669 \qquad \text{and} \qquad \frac{1}{M^2R^2}\int_0^R r^2 M(r)\,dM(r) = 0.3326\,; \tag{59}$$

and these values, when inserted into inequalities (45) and (48), give

$$1.170 \le \tfrac{5}{4}\frac{m}{\epsilon_R} \le 1.335. \tag{60}$$

The actual value of $5m/4\epsilon_R$ for the earth is 1.299; it thus differs from the upper bound given by formula (60) by less than 3 per cent.

The research reported in this paper (as it pertains to S. Chandrasekhar) has in part been supported by the Office of Naval Research under contract Nonr-2121(24) with the University of Chicago; and (as it pertains to P. H. Roberts) has been supported in part by a National Science Foundation Grant No. N.S.F.-GP-975.

APPENDIX

Clairaut's Equation

It will be noticed that nowhere in the text was there an occasion to define the ellipticity of the surfaces of constant density. However, if the equation of state is barotropic, then it follows from equation (27) that the surfaces of equal geopotential are also surfaces of equal pressure and density. The requirement (in these circumstances) that the density specified in equation (10) is constant over the same surfaces (28) the geopotential is constant, yields the additional relation,

$$\rho_2(r) = \tfrac{2}{3}\epsilon(r)\, r\frac{d\rho_0}{dr}. \tag{A, 1}$$

This relation when inserted in equation (20) gives

$$\frac{d}{dr}(r^3\mathfrak{B}_2) = \tfrac{8}{3}\pi G r^4 \int_r^R \epsilon(s)\frac{d\rho_0(s)}{ds}\, ds\,; \tag{A, 2}$$

and this equation in combination with equation (32) gives

$$4\pi G\int_r^R \epsilon(s)\frac{d\rho_0(s)}{ds}\, ds = -\frac{G}{r^4}\frac{d}{dr}[\epsilon(r)M(r)r^2] + \tfrac{5}{2}\Omega^2. \tag{A, 3}$$

Differentiating this equation with respect to r, we obtain

$$4\pi\epsilon(r)\frac{d\rho_0}{dr} = \frac{d}{dr}\left\{\frac{1}{r^4}\frac{d}{dr}[\epsilon(r)M(r)r^2]\right\}. \tag{A, 4}$$

Equation (A, 4), as can be verified by expansion, is the same as Clairaut's equation; its present derivation emphasizes (what is sometimes obscured) that its validity strictly depends on the equation of state being barotropic.

REFERENCES

Bullard, E. C. 1946, *Geophys. Suppl. M.N.*, **5**, 186.
Chandrasekhar, S. 1933, *M.N.*, **93**, 539.
Chandrasekhar, S., and Lebovitz, N. R. 1962, *Ap. J.*, **136**, 1082.
Darwin, G. H. 1899, *M.N.*, **60**, 92.
Jeffreys, H. 1959, *The Earth* (Cambridge: Cambridge University Press).
Milne, E. A. 1929, *M.N.*, **89**, 739.
Todhunter, I. 1873, *History of the Mathematical Theories of Attraction and the Figure of the Earth* (original ed.; London: Constable & Co., 1873; reprint ed.; New York: Dover Publications, Inc., 1962).

THE POINTS OF BIFURCATION ALONG THE MACLAURIN, THE JACOBI, AND THE JEANS SEQUENCES

S. Chandrasekhar
University of Chicago
Received January 23, 1963

ABSTRACT

The role which the second- and the third-order virial equations governing equilibrium can play in isolating points of neutral stability along equilibrium sequences is discussed and clarified. It is shown that a necessary condition for the occurrence of a neutral point is that a non-trivial Lagrangian displacement exists for which the first variations of *all* of the integral relations (five in the second order and fifteen in the third order), provided by the virial equations, vanish. By using this condition, it is possible, for example, to isolate the point of bifurcation along the Jacobian sequence without *any* prior specification of the nature of the sequence which follows bifurcation. As further illustrations of the method, the known points of neutral stability along the Maclaurin and the Jeans sequences are also derived.

I. INTRODUCTION

In an earlier paper (Chandrasekhar 1962; this paper[1] will be referred to hereafter as "Paper I") the occurrence of points of bifurcation along the Maclaurin and the Jacobian sequences was deduced from certain general integral relations provided by the second- and third-order virial equations. The point of bifurcation along the Jacobian sequence was deduced, for example, from the required invariance of an integral property (provided by the third-order virial equations) to an infinitesimal Lagrangian displacement which deforms the Jacobi ellipsoid into a pear-shaped configuration. However, the preoccupation with this matter of the exhibition and the isolation of the point of bifurcation along the Jacobian sequence was allowed to obscure (and to some extent confuse) the entirely general nature of the relations provided by the virial theorem and its extensions and the role they can be called upon to play in distinguishing equilibrium configurations which are neutrally stable along a given sequence. In this paper an attempt will be made to clarify this entire matter. As an instance of the generality achieved in the interpretation, it may be stated here that the point of bifurcation along the Jacobian sequence can be isolated by four (apparently) independent relations which do not presuppose any knowledge of the kind of equilibrium configurations which succeed the Jacobi ellipsoids beyond the point of bifurcation.

II. SOME REMARKS ON THE TERMINOLOGY

In view of the diverse (and sometimes conflicting) terminology which is current in the subject, it may be useful to clarify the different circumstances which must be distinguished and the particular terminology which will be adopted in the present discussion.

First, there is the concept of the *point of bifurcation* itself. There is no ambiguity in what can be meant by this term: a sequence of equilibrium configurations is given which can be arranged "linearly" with respect to some parameter; as we follow the sequence, we come to a point where a parting of the ways occurs; beyond such a point we can distinguish two equilibrium configurations where there was only one before the point was reached. A point of bifurcation defined in this manner, clearly presupposes that the *existence* of equilibrium configurations on both prongs of the fork has been established.

The occurrence of a point of bifurcation implies that a non-trivial infinitesimal Lagrangian displacement $\boldsymbol{\xi}$ exists such that the deformation, by this displacement, of the

[1] The subsequent papers by Chandrasekhar and Lebovitz (1963*a*, *b*, and *c*) will be referred to as "Papers II, III, and IV," respectively.

member of the sequence at the point of bifurcation will leave its equilibrium unaffected. The Lagrangian displacement in question is, in fact, the one which will deform the configuration from the "shape" it has in one of the branches to the shape it has in the branch after bifurcation. From the existence of a displacement $\boldsymbol{\xi}$ which leaves the equilibrium unaffected, we may draw two inferences: *first*, that at the point of bifurcation the equilibrium configuration must have a definite non-trivial neutral mode of oscillation; and *second*, that if J is any integral property (or, more generally, a functional) of the configuration which vanishes as a condition of equilibrium, then its first variation due to the displacement $\boldsymbol{\xi}$ must also vanish at the point of bifurcation. More explicitly, the content of the second statement is the following. The functionals J with which we shall be mostly concerned are of the general form

$$J=\int_V \rho(\boldsymbol{x})Q_1(\boldsymbol{x})\,d\boldsymbol{x}+\int_V\int_V \rho(\boldsymbol{x})\,\rho(\boldsymbol{x}')Q_2(\boldsymbol{x},\,\boldsymbol{x}')\,d\boldsymbol{x}d\boldsymbol{x}', \tag{1}$$

where $Q_1(\boldsymbol{x})$ and $Q_2(\boldsymbol{x},\,\boldsymbol{x}')$ are functions which are defined for all points $\boldsymbol{x}$ and pairs of points $(\boldsymbol{x},\,\boldsymbol{x}')$, respectively, and the integrations over $\boldsymbol{x}$ and $\boldsymbol{x}'$ are effected over the volume V occupied by the fluid. And by the first variation of such a functional due to the displacement $\boldsymbol{\xi}$ we mean

$$\begin{aligned}\delta J=&\int_V \rho(\boldsymbol{x})\,\xi_l(\boldsymbol{x})\frac{\partial Q_1}{\partial x_l}\,d\boldsymbol{x}\\&+\int_V\int_V \rho(\boldsymbol{x})\,\rho(\boldsymbol{x}')\Big[\,\xi_l(\boldsymbol{x})\frac{\partial}{\partial x_l}+\xi_l(\boldsymbol{x}')\frac{\partial}{\partial x_l'}\Big]Q_2(\boldsymbol{x},\,\boldsymbol{x}')\,d\boldsymbol{x}d\boldsymbol{x}'.\end{aligned} \tag{2}$$

The statement now is that, if

$$J=0 \text{ as a condition of equilibrium}\,, \tag{3}$$

then

$$\delta J=0 \text{ at the point of bifurcation for some } \boldsymbol{\xi}\,. \tag{4}$$

It seems to be generally assumed that, from the existence of a neutral mode of oscillation along an equilibrium sequence, we may, conversely, infer the occurrence of bifurcation. It is not clear to what extent this converse statement is true. In any event, it would appear that the concepts of a point of bifurcation and of a point of neutral stability are logically distinct. On this account, we shall define a *neutral point* along an equilibrium sequence as one at which the configuration has a definite non-trivial mode belonging to a zero characteristic frequency in a proper analysis of the normal modes. Clearly, a point of bifurcation is also a neutral point; but it would appear that a neutral point need not necessarily be a point of bifurcation.

Neutral points can be of two kinds. The distinction arises in this way. Let ϑ be a parameter which labels the members of a branch of a sequence; and let ϑ_0 be a neutral point along that branch. By definition, we can infer the existence of a characteristic frequency of oscillation σ such that

$$\sigma^2(\vartheta)=0 \quad \text{when} \quad \vartheta=\vartheta_0\,, \tag{5}$$

and σ^2 is real in a finite neighborhood of ϑ_0. The two cases to be distinguished are

$$\begin{aligned}&\text{case (i)}:\ \sigma^2(\vartheta)>0 \ \text{ for } \ \vartheta<\vartheta_0 \ \text{ and } \ \sigma^2(\vartheta)<0 \ \text{ for } \ \vartheta>\vartheta_0\,,\\&\quad\textit{or}\ \ \sigma^2(\vartheta)<0 \ \text{ for } \ \vartheta<\vartheta_0 \ \text{ and } \ \sigma^2(\vartheta)>0 \ \text{ for } \ \vartheta>\vartheta_0\,,\end{aligned} \tag{6}$$

and

$$\text{case (ii)}:\ \sigma^2(\vartheta)>0 \ \text{ for } \ \vartheta\neq\vartheta_0 \ \text{ in a finite neighborhood of } \vartheta_0\,. \tag{7}$$

The physical distinction between the two cases is that, while in case (i) the configurations along the branch considered are definitely unstable on one or the other side of the neutral point, in case (ii) the configurations are stable on both sides of the neutral point. We shall call a neutral point as of the *first kind* in case (i) and as of the *second kind* in case (ii). In the literature on the subject it is customary to distinguish the two cases by referring to a *point of ordinary instability* in case (i) and to a *point of secular instability* in case (ii). The reason for attributing instability in case (ii), as well, is the *belief* that we will, in fact, find instability if any dissipative mechanism is operative and the *presumption* that such mechanisms must be operative in any real physical system. What is implied by the terminology is that if we were to investigate the stability of the system by the standard methods of small oscillations, we will find in case (ii) that instability does, in fact, arise when the limit set by ϑ_0 is transgressed and, further, that the growth rate of the instability depends directly on the magnitude of the dissipative mechanisms which are operative; whereas in case (i), when instability arises, the growth rate of the instability will be comparable to the natural frequencies of oscillation of the system. While all this is implicitly and generally understood, the attributes of stability and instability are often ascribed without a detailed investigation of the stability of the system by the method of small oscillations.[2]

The principal objection to the current terminology is, however, its ambiguity and its inadequacy. For example, systems which are stable (in the absence of any dissipative mechanisms) on both sides of a neutral point often become unstable at a subsequent point ϑ_1 (say) by *overstability* (i.e., by oscillations of increasing amplitude). What generally happens in these cases is that the characteristic frequency which has the behavior (7) at ϑ_0 becomes coincident with another characteristic frequency (belonging to another mode) at ϑ_1, beyond which point the two frequencies become complex conjugates of one another. Under these circumstances we should say that the system becomes "ordinarily unstable" at ϑ_1 (since the instability occurs in the absence of any dissipative mechanisms); but such an "ordinary instability" has no bearing on the occurrence or otherwise of a point of bifurcation.

Returning to the question of isolating neutral points and points of bifurcation, we may recall that at these points we must necessarily have Lagrangian displacements which will leave the equilibrium unaffected. Consequently, if J is any functional of the configuration which vanishes as a condition of equilibrium, then *a necessary condition for the occurrence of a neutral point* (and *a fortiori* for a point of bifurcation) *is that a nontrivial Lagrangian displacement exists for which* $\delta J = 0$. It is necessary to emphasize, however, that the location of a neutral point by considering the vanishing of the first variations of such functionals will not enable us to discriminate the kind of neutral point that is located.

III. THE FIRST POINT OF BIFURCATION ALONG THE MACLAURIN SEQUENCE

The origin of the first point of bifurcation along the Maclaurin sequence has been considered in Paper I (Sec. II) on the basis of certain integral properties provided by the second-order virial theorem; and the arguments given on page 1050 (Paper I) leading to the condition for its occurrence are, of course, valid. But the arguments as given do not depend on the invariance of the same integral properties to first variations at the point of bifurcation. We shall now show how the point of bifurcation can also be deduced from these invariance requirements. The demonstration illustrates in the simplest context the essential elements of the present method.

[2] For example, it has always been stated that the Maclaurin spheroid *is* "secularly unstable" at the point of bifurcation where the Jacobi ellipsoids branch off. But only very recently has the problem of the oscillations of a *viscous* Maclaurin spheroid been treated successfully by Roberts and Stewartson (1963). And they do show that the viscous Maclaurin spheroid becomes unstable at the point of bifurcation.

After the elimination of Π $(= \int p d\mathbf{x})$, the second-order virial equations governing equilibrium are

$$\mathfrak{W}_{13} = \mathfrak{W}_{23} = 0\,, \quad I_{13} = I_{23} = 0\,, \quad \mathfrak{W}_{12} + \Omega^2 I_{12} = 0\,, \tag{8}$$

$$\mathfrak{W}_{11} - \mathfrak{W}_{22} + \Omega^2(I_{11} - I_{22}) = 0\,, \tag{9}$$

and

$$\mathfrak{W}_{11} + \mathfrak{W}_{22} - 2\mathfrak{W}_{33} + \Omega^2(I_{11} + I_{22}) = 0\,, \tag{10}$$

where

$$\mathfrak{W}_{ij} = -\tfrac{1}{2}\int_V \rho \mathfrak{B}_{ij} d\mathbf{x} \quad \text{and} \quad I_{ij} = \int_V \rho x_i x_j d\mathbf{x} \tag{11}$$

are the potential energy and the moment of inertia tensors. *At a neutral point (or a point of bifurcation) there must exist a non-trivial Lagrangian displacement* $\boldsymbol{\xi}$ *for which the first variations of all of the foregoing equations must vanish.*

The question might occur whether one is justified in treating Ω as a constant in carrying out the variations of equations (8)–(10). Actually the question is irrelevant, since $\boldsymbol{\xi}$, as we have seen, is the displacement belonging to a well-defined mode among the normal modes of oscillations of the system; and in the analysis of such modes, the various parameters (such as Ω) specifying the initial state of the system are certainly to be kept constant. For the same reason, the Lagrangian displacement can be restricted by the assumption (cf. Paper I, eqs. [104] and [AII, 2])

$$V_i = \langle \rho \xi_i \rangle = \int_V \rho \xi_i d\mathbf{x} = 0\,, \tag{12}$$

since this assumption merely corresponds to keeping the center of mass stationary in studying the oscillations of a system (cf. Paper II, Sec. II).

Returning to the first variations of equations (8)–(10), we first observe that

$$\delta I_{ij} = V_{ij}\,, \tag{13}$$

where

$$V_{ij} = \int_V \rho(x_i \xi_j + x_j \xi_i) d\mathbf{x} = V_{j;i} + V_{i;j} \tag{14}$$

is the symmetrized second-order virial. The corresponding first variations of the potential energy tensor of a homogeneous ellipsoid have been given in an earlier paper (Paper IV, eqs. [47] and [48]); they are

$$\delta\mathfrak{W}_{ij} = -2B_{ij}V_{ij} \qquad (i \neq j) \tag{15}$$

and

$$\delta\mathfrak{W}_{ii} = -(2B_{ii} - a_i^2 A_{ii})V_{ii} + a_i^2 \sum_{l \neq i} A_{il} V_{ll} \tag{16}$$

(no summation over repeated indices in eqs. [15] and [16]),

where the A_{ij}'s and the B_{ij}'s are the two index symbols which have been defined in earlier papers (Chandrasekhar and Lebovitz 1962*a;* also Paper II, eq. [114]). A common factor $\pi G \rho a_1 a_2 a_3$ has been suppressed in writing equations (15) and (16).

For the Maclaurin spheroids, $a_1 = a_2$; and this equality implies that the values of A_{ij} and B_{ij} are unaltered if the index 1 (wherever it may occur) is replaced by the index 2 (and conversely).

Now, making use of equations (13), (15), and (16), we readily find that the vanishing of the first variations of equations (8)–(10) requires

$$B_{13}V_{13} = 0\,, \quad B_{23}V_{23} = 0\,, \quad V_{13} = V_{23} = 0\,, \tag{17}$$

$$(\Omega^2 - 2B_{11})V_{12} = 0\,, \quad (\Omega^2 - 2B_{11})(V_{11} - V_{22}) = 0\,, \tag{18}$$

and

$$[-2(B_{11} - a_1^2A_{11} + a_3^2A_{13}) + \Omega^2](V_{11} + V_{22})$$
$$+ 2(2B_{33} - a_3^2A_{33} + a_1^2A_{13})V_{33} = 0\,. \tag{19}$$

If the additional assumption that the Lagrangian displacement is divergence-free is made, then the foregoing equations must be supplemented by the condition (Lebovitz 1961, eq. [83])

$$V_{11} + V_{22} = -\frac{a_1^2}{a_3^2}V_{33}\,. \tag{20}$$

From equations (17) and (18) it follows that

$$V_{13} = V_{23} = 0\,, \tag{21}$$

and

$$V_{12} = 0 \quad \text{and} \quad V_{11} = V_{22} \quad \text{if} \quad \Omega^2 - 2B_{11} \neq 0\,. \tag{22}$$

And, finally, equations (19) and (20) give

$$\Big[2(B_{11} - a_1^2A_{11} + a_3^2A_{13}) - \Omega^2$$
$$+ 2\frac{a_3^2}{a_1^2}(2B_{33} - a_3^2A_{33} + a_1^2A_{13})\Big](V_{11} + V_{22}) = 0\,. \tag{23}$$

An examination of the coefficient of $(V_{11} + V_{22})$ in equation (23) shows that it does not vanish along the Maclaurin sequence. Accordingly,

$$V_{11} + V_{22} = 0 \quad \text{and} \quad V_{33} = 0\,. \tag{24}$$

Hence, if $\Omega^2 \neq 2B_{11}$, all the six symmetrized virials must vanish; and this means that there is no non-trivial $\boldsymbol{\xi}$ belonging to the second harmonics which satisfies equations (8)–(10). Therefore, a *necessary* condition for the occurrence of a neutral point (where a neutral mode belonging to the second harmonics can exist) is

$$\frac{\Omega^2}{\pi G\rho} = 2a_1^2a_3B_{11} = 2a_1^2a_3(A_1 - a_1^2A_{11})\,; \tag{25}$$ [3]

and, moreover, the Lagrangian displacement must be such that

$$V_{13} = V_{23} = V_{33} = 0\,, \quad V_{11} = -V_{22}\,, \quad \text{and} \quad V_{12} \neq 0\,. \tag{26}$$

The condition is also a *sufficient* one, since there are, in fact, three linearly independent Lagrangian displacements which satisfy the requirements (26); thus

$$\xi_1 = \alpha x_1\,, \quad \xi_2 = -\alpha x_2\,, \quad \xi_3 = 0\,,$$

and

$$\xi_1 = \beta x_2\,, \quad \xi_2 = \gamma x_1\,, \quad \text{and} \quad \xi_3 = 0\,, \tag{27}$$

where α, β, and γ are arbitrary (non-zero) constants.

The condition (25) is exactly the one which determines the point along the Maclaurin sequence where the Jacobi ellipsoids branch off.[4] It should be noted that, in deducing

[3] The factor $\pi G\rho a_1a_2a_3$ which had been suppressed in writing equations (15) and (16) has been restored in this equation.

[4] The condition

$$\delta(\mathfrak{W}_{12} + \Omega^2 I_{12}) = (-2B_{12} + \Omega^2)V_{12} = 0$$

is satisfied identically all along the Jacobian sequence, since $\Omega^2 = 2B_{12}$ is the equation which determines Ω^2 along this sequence (cf. Paper I, eq. [AI, 7]). Consequently, the mode which is neutral at the point of bifurcation remains neutral along the entire Jacobian sequence.

that a neutral point occurs when condition (25) is met, we do not need to *demand* that the Lagrangian displacement be solenoidal. On this account, a neutral point will occur at the same place along the sequence of *compressible* homogeneous Maclaurin spheroids (as has, indeed, been found by Chandrasekhar and Lebovitz 1962*b*).

IV. THE POINT OF BIFURCATION ALONG THE JACOBIAN SEQUENCE

The point of bifurcation along the Jacobian sequence, where the pearshaped configurations first become possible as figures of equilibrium, was exhibited and isolated in Paper I by considering the vanishing of the first variation of the functional

$$J = \mathfrak{W}_{13;3} + \mathfrak{W}_{12;2} + \mathfrak{W}_{33;1} - \mathfrak{W}_{22;1}\,, \tag{28}$$

where

$$\mathfrak{W}_{ij;k} = -\tfrac{1}{2}\int_V \rho\,\mathfrak{B}_{ij} x_k d\,x \tag{29}$$

for the particular Lagrangian displacement (defined in Paper I, eqs. [10], [11], and [75]–[77]) which deforms an ellipsoidal into a pear. However, the statement in Paper I (in n. 3 on p. 1057) that (28) is the only functional "available for our present purposes" is an error.[5] It is clear now that the first variations of all of the fifteen equations in Paper I, equations (57)–(65), must, in fact, vanish at the point of bifurcation for the Lagrangian displacement considered.

The proper way of treating the fifteen third-order virial equations (obtained after the elimination of the Π_k's) will now be described.

First, we shall rewrite the fifteen equations in Paper I, equations (57)–(65), in four non-combining groups into which they fall (in what sense they are non-combining will be made clear presently):

A:
$$\mathfrak{W}_{13;2} = \mathfrak{W}_{23;1} = \Omega^2 I_{123} + \mathfrak{W}_{12;3} = 0\,; \tag{30}$$

B:
$$\begin{gathered}\Omega^2 I_{113} + \mathfrak{W}_{11;3} - \mathfrak{W}_{33;3} = \Omega^2 I_{223} + \mathfrak{W}_{22;3} - \mathfrak{W}_{33;3} = 0\,,\\ \mathfrak{W}_{13;1} = \mathfrak{W}_{23;2} = 0\,;\end{gathered} \tag{31}$$

C:
$$\begin{gathered}\Omega^2 I_{122} + 2\mathfrak{W}_{12;2} = \Omega^2 I_{133} + 2\mathfrak{W}_{13;3} = 0\,,\\ \Omega^2 I_{111} - 2(\mathfrak{W}_{22;1} - \mathfrak{W}_{12;2} - \mathfrak{W}_{11;1}) = 0\,,\\ \mathfrak{W}_{13;3} + \mathfrak{W}_{12;2} + \mathfrak{W}_{33;1} - \mathfrak{W}_{22;1} = 0\,;\end{gathered} \tag{32}$$

D:
$$\begin{gathered}\Omega^2 I_{112} + 2\mathfrak{W}_{12;1} = \Omega^2 I_{233} + 2\mathfrak{W}_{23;3} = 0\,,\\ \Omega^2 I_{222} - 2(\mathfrak{W}_{11;2} - \mathfrak{W}_{12;1} - \mathfrak{W}_{22;2}) = 0\,,\\ \mathfrak{W}_{32;3} + \mathfrak{W}_{12;1} + \mathfrak{W}_{33;2} - \mathfrak{W}_{11;2} = 0\,.\end{gathered} \tag{33}$$

And as we have stated, at a neutral point (or a point of bifurcation) the first variations of *all* of the foregoing equations must vanish.

[5] This has already been pointed out in the "Note added in proof" (Paper I, p. 1068). But the qualification in this note that "Ω^2 can be considered as invariable for deformations belonging to the third harmonic" is unnecessary: Ω^2 can be treated as a constant without any *proviso*.

It is found that the first variations of all the quantities which occur in equations (30)–(33) can be expressed as linear combinations of the symmetric third-order virials:[6]

$$V_{ijk} = \int_V \rho(\xi_i x_j x_k + \xi_j x_k x_i + \xi_k x_i x_j)\, d\mathbf{x}$$
$$= V_{i;jk} + V_{j;ki} + V_{k;ij} \,. \qquad (34)$$

Thus

$$\delta I_{ijk} = V_{ijk} \,. \qquad (35)$$

The expressions for the first variations of the $\mathfrak{W}_{ij;k}$'s have been derived in Paper II (Sec. V); and the coefficients of the virials in the expansions of the $\delta\mathfrak{W}_{ij;k}$'s have been tabulated for all combinations of indices (Paper II, Table 1). It will be seen that the expansion of a particular $\delta\mathfrak{W}_{ij;k}$ involves, at most, only three of the virials; and, more particularly, that the variations of the $\mathfrak{W}_{ij;k}$'s included in any of the groups A, B, C, and D involve only the same virials and are mutually exclusive. It is in this sense that the four groups are non-combining. Designating by δA, δB, δC, and δD the equations which are obtained by taking the first variations of the equations in the respective groups, we observe that the association of the groups and the virials is the following:

$$\delta A:\ V_{123}\,; \quad \delta B:\ V_{113}\,, \quad V_{223}\,, \quad V_{333}\,;$$
$$\delta C:\ V_{111}\,, \quad V_{122}\,, \quad V_{133}\,; \quad \delta D:\ V_{222}\,, \quad V_{233}\,, \quad V_{211}\,. \qquad (36)$$

We must now investigate whether a non-trivial Lagrangian displacement exists which will satisfy all the variational equations included in δA, δB, δC, and δD.

The discussion of the group δA is particularly simple: the equations belonging to this system are (cf. Paper II, eq. [119])

$$\delta A:\ (\Omega^2 - 2B_{12;3})V_{123} = -2B_{13;2}V_{123} = -2B_{23;1}V_{123} = 0\,, \qquad (37)$$

where

$$B_{ij;k} = B_{ij} + a_k^2 B_{ijk}\,; \qquad (38)$$

and these equations clearly demand that

$$V_{123} = 0\,. \qquad (39)$$

The discussion of the groups δB, δC, and δD is not so simple. Each of these groups provides a system of four linear homogeneous equations for the three virials which they involve. For the existence of a non-trivial solution of the variational equations, it is clearly necessary that at least one of the three 4×3 rectangular matrices, representing the linear systems of equations in the three groups, is of rank at most 2.

For the Jacobi ellipsoids as customarily defined ($a_1 > a_2 > a_3$ in a right-handed system of co-ordinates), it can be directly verified (by the method to be described presently in connection with the group δC) that the equations of the groups δB and δD do not allow any non-trivial solution. Accordingly (cf. eq. [36]),

$$V_{113} = V_{223} = V_{333} = V_{112} = V_{222} = V_{332} = 0\,. \qquad (40)$$

It remains to find out whether the group δC allows a non-trivial solution.

[6] A qualification is necessary: when $\delta\mathfrak{W}_{ij;k}$ is evaluated for an arbitrary Lagrangian displacement $\boldsymbol{\xi}$, we find (cf. Paper II, eqs. [120]–[122]) that it involves, in addition to the V_{ijk}'s, the first-order virials V_i also; but, as we have seen earlier, all of these can be set equal to zero (cf. eq. [12] and the remarks preceding it).

In discussing the equations of the group δC, it is convenient to choose the following functionals instead of those listed under C (eqs. [32]):

$$J_1 = -\Omega^2 I_{122} - 2\mathfrak{W}_{12;2}\,, \qquad J_2 = -\Omega^2 I_{133} - 2\mathfrak{W}_{13;3}\,,$$
$$J_3 = \Omega^2(I_{111} - 3I_{122}) + S_{122}\,, \quad J_4 = \Omega^2(I_{111} - I_{133}) + S_{133}\,, \tag{41}$$

where

$$S_{ijj} = -4\mathfrak{W}_{ij;j} - 2\mathfrak{W}_{jj;i} + 2\mathfrak{W}_{ii;i} \tag{42}$$
(no summation over repeated indices) .

The functionals J_1 and J_2 are, apart from sign, the same as the first two listed under C; and J_3 and J_4 are certain specific linear combinations.[7]

Equilibrium requires that all the J's vanish. Therefore, a necessary condition for the occurrence of a neutral point along the Jacobian sequence is that a Lagrangian displacement exists which will satisfy the requirements (39) and (40) and allow a non-trivial solution for the variational equations

$$\delta J_1 = -\Omega^2 V_{122} - 2\delta\mathfrak{W}_{12;2} = 0\,, \tag{43}$$

$$\delta J_2 = -\Omega^2 V_{133} - 2\delta\mathfrak{W}_{13;3} = 0\,, \tag{44}$$

$$\delta J_3 = \Omega^2(V_{111} - 3V_{122}) + \delta S_{122} = 0\,, \tag{45}$$

and

$$\delta J_4 = \Omega^2(V_{111} - V_{133}) + \delta S_{133} = 0\,. \tag{46}$$

The $\delta\mathfrak{W}_{ij;k}$'s and δS_{ijj}'s which occur in equations (43)–(46) are expressible as linear combinations of V_{111}, V_{122}, and V_{133} with coefficients which are listed in Paper II, Table 2. Accordingly, we may write

$$\delta J_i = \langle i|111\rangle V_{111} + \langle i|122\rangle V_{122} + \langle i|133\rangle V_{133} \qquad (i = 1, 2, 3, 4)\,, \tag{47}$$

where $\langle i|111\rangle$, etc., are certain matrix elements which are known.

If we should now require that the Lagrangian displacement be also solenoidal, then we should supplement equation (47) by the further condition (cf. Paper II, Sec. VII)

$$\frac{V_{111}}{a_1^2} + \frac{V_{122}}{a_2^2} + \frac{V_{133}}{a_3^2} = 0\,. \tag{48}$$

The existence of a non-trivial neutral mode belonging to the third harmonic requires, therefore, that, for some member of the Jacobian sequence, the 4×3 matrix representing equation (47) (or the 5×3 matrix if eq. [48] is also included) is at most of rank 2. Instead of examining the problem in this entirely general way from the outset, we shall find it instructive and useful to investigate Lagrangian displacements of progressively increasing generality.

a) The Lagrangian Displacement Which Deforms an Ellipsoid into a Pear

We may argue as in Paper I (Sec. III) that *if* a point of bifurcation occurs along the Jacobian sequence, then the new figure of equilibrium must be one which can be obtained by deforming an ellipsoid by the Lagrangian displacement

$$\xi_j = \text{Constant}\,\frac{\partial}{\partial x_j}\,x_1\left(\frac{x_1^2}{a_1^2+\lambda} + \frac{x_2^2}{a_2^2+\lambda} + \frac{x_3^2}{a_3^2+\lambda} - 1\right), \tag{49}$$

[7] The functional (28) considered in Paper I is $\frac{1}{4}(J_2 + J_3 - J_4 - 3J_1)$.

where λ is the numerically larger of the two roots of the equation

$$\frac{3}{a_1^2+\lambda}+\frac{1}{a_2^2+\lambda}+\frac{1}{a_3^2+\lambda}=0\,. \tag{50}$$

The components of this displacement can be written in the following forms (Paper I, eqs. [75] and [76]):

$$\xi_1 = (\alpha+\beta)x_1^2 - \alpha x_2^2 - \beta x_3^2 - 1\,,$$
$$\xi_2 = -2\alpha x_1 x_2\,, \qquad \text{and} \qquad \xi_3 = -2\beta x_1 x_3\,, \tag{51}$$

where

$$\alpha = -(a_2^2+\lambda)^{-1} \quad \text{and} \quad \beta = -(a_3^2+\lambda)^{-1}\,. \tag{52}$$

The Lagrangian displacement defined by the foregoing equations is clearly divergence-free; it satisfies the requirement (Paper I, eq. [104])

$$V_i = \int_V \rho\,\xi_i d\mathbf{x} = 0\,; \tag{53}$$

and, moreover, all the virials listed in equations (39) and (40) vanish. It would, in fact, appear that the displacement, as defined, is the only one (apart from a constant factor) which will satisfy all these requirements.

TABLE 1

VALUES OF δJ_i IN THE NEIGHBORHOOD OF THE POINT OF BIFURCATION FOR A LAGRANGIAN DISPLACEMENT WHICH DEFORMS AN ELLIPSOID INTO A PEAR

$\cos^{-1} a_3/a_1$	δJ_1	δJ_2	δJ_3	δJ_4
68°........	−0.0763	−0.0343	−1.297	−1.092
69°........	− .0277	− .0136	−0.534	−0.464
69°.8166...	+ .0001	+ .0000	+0.002	+0.001
70°........	+ .0051	+ .0027	+0.112	+0.100
71°........	+0.0262	+0.0149	+0.658	+0.599

For the Lagrangian displacement considered, it can be shown (by making use of Paper I, eqs. [103] and [104]) that

$$V_{111} = 6a_1^2[a_1^2(\alpha+\beta)-1]\,, \qquad V_{122} = -2a_2^2[(2a_1^2+a_2^2)\alpha+1]\,,$$
$$\text{and} \qquad V_{133} = -2a_3^2[(2a_1^2+a_3^2)\beta+1]\,. \tag{54}$$

Using these particular values for V_{111}, V_{122}, and V_{133}, we can readily evaluate the δJ_i's for different members of the Jacobian sequence. The results of the calculations are given in Table 1. It will be observed that the δJ_i's do seem to vanish simultaneously at one point along the Jacobian sequence; and the accuracy of Darwin's original determination of the point of bifurcation would appear to be fully confirmed.

b) A Lagrangian Displacement Which Is Divergence-free

If we did not know that it is a sequence of pear-shaped configurations which branches off from the Jacobian sequence, we might still wish to demand that the requisite Lagrangian displacement be divergence-free. In that case, the existence of a non-trivial solution will require that the 5×3 matrix representing equations (47) and (48) is (at most) of rank 2. All six determinants of the form

$$\Delta_{ij} = \begin{Vmatrix} \langle i|111\rangle & \langle i|122\rangle & \langle i|133\rangle \\ \langle j|111\rangle & \langle j|122\rangle & \langle j|133\rangle \\ \frac{1}{a_1^2} & \frac{1}{a_2^2} & \frac{1}{a_3^2} \end{Vmatrix} \quad (i \neq j) \qquad (55)$$

must, therefore, vanish simultaneously at the point of bifurcation. Table 2 shows that this does happen.

TABLE 2

VALUES OF Δ_{ij} IN THE NEIGHBORHOOD OF THE POINT OF BIFURCATION

$\cos^{-1} a_3/a_1$	69°	69°8166	70°
Δ_{12}	+0.000882	−0.000003	−0.000177
Δ_{23}	− .01365	+ .00004	+ .00309
Δ_{34}	+ .2994	− .0010	− .0742
Δ_{41}	+ .02148	− .00007	− .00473
Δ_{13}	+ .00687	− .00002	− .00152
Δ_{24}	−0.00426	+0.00001	+0.00097

TABLE 3

VALUES OF Δ_i IN THE NEIGHBORHOOD OF THE POINT OF BIFURCATION

$\cos^{-1} a_3/a_1$	Δ_1	Δ_2	Δ_3	Δ_4
69°	+0.00236	+0.00366	−0.000221	−0.000221
69°8166	− .00001	− .00001	+ .000001	+ .000001
70°	−0.00048	−0.00073	+0.000041	+0.000041

c) A Lagrangian Displacement Which Is Completely Unspecified

And, finally, if we do not wish to impose any restrictions whatsoever on the Lagrangian displacement, then the existence of a non-trivial solution will require that the 4×3 matrix representing equation (47) is (at most) of rank 2. All four determinants of order 3 which we can form from the 4×3 matrix must, therefore, vanish simultaneously at the point of bifurcation. Denoting by Δ_i the determinant of order 3 obtained by omitting the ith row in equation (47), we list in Table 3 its values in the neighborhood of the point of bifurcation. We observe that all the Δ_i's do vanish at the point of bifurcation, as required.

We can go even further. Suppose that we did not even wish to restrict the Lagrangian displacement by the requirement that the first-order virials V_i vanish. We should then have found an equation of the form

$$\delta J_i = \langle i|111\rangle V_{111} + \langle i|122\rangle V_{122} + \langle i|133\rangle V_{133} + \langle i|1\rangle\, V_1 \quad (i = 1, 2, 3, 4)\,, \qquad (56)[8]$$

[8] From Paper I, eqs. (114)–(117) it is apparent that the groups which we have designated δB, δC, and δD involve (besides the third-order virials enumerated in [36]) the first-order virials V_3, V_1, and V_2, respectively. And the consideration of the groups δB and δD would have led us (at an earlier stage) to infer the vanishing of V_2 and V_3 in addition to the requirements (39) and (40).

in place of equation (47). And the existence of a non-trivial solution will now simply require that the determinant of this system of equations vanish. From the results of Section IVa, b, and c, it is clear that we should now simply find that $V_1 = 0$. It appears, then, that we can isolate the point of bifurcation along the Jacobian sequence almost blindfolded!

It is not easy to see a priori why the different ways of looking at the problem are all consistent with one another; and, in particular, why the solenoidal character of the Lagrangian displacement and the vanishing of the first-order virials are *deducible* as necessary conditions for the occurrence of a neutral point. But a self-consistent way of looking at the problem is the following. Suppose it has been established (as it has been established by Darwin) that a pear-shaped configuration is a possible figure of equilibrium. Since the virial equations (30)–(33) are consequences of the same equations governing equilibrium, it is clearly *necessary* that the first variations of the virial equations vanish at the point of bifurcation for the particular Lagrangian displacement, considered in Section IVa, which deforms an ellipsoid into a pear; and this we simply verify.

V. THE SECOND AND THE THIRD NEUTRAL POINTS ALONG THE MACLAURIN SEQUENCE

We return once more to the Maclaurin sequence, but this time to isolate further neutral points, beyond the first, where neutral modes of oscillation belonging to the third harmonics may occur. For the exhibition and the isolation of these further neutral points, we may use the same functionals as those we considered in the discussion of the Jacobian sequence in Section IV.

When we consider the first variations of the equations listed under groups A and B (eqs. [30] and [31]), we find that these equations do not allow any non-trivial solution. Therefore, at a neutral point, we must necessarily have

$$V_{123} = V_{333} = V_{311} = V_{322} = 0 . \tag{57}$$

In view of the equality of a_1 and a_2, the variations of the equations in groups C and D will lead to identical equations governing the respective sets of virials; this is apparent from the fact that the coefficients in the expansions of the relevant quantities in terms of the two sets of virials are the same (see Paper III, Table 1). It will, therefore, suffice to consider equations (43)–(46) derived from group C.

Using the coefficients of the virials in the expansion of δS_{122} given in Paper III, Table 2, we find that, at a neutral point,

$$\delta J_3 = [\Omega^2 - 2(B_{11} + a_1^2 B_{111})](V_{111} - 3V_{122}) = 0 . \tag{58}$$

Hence a Lagrangian displacement for which

$$V_{111} \neq 3V_{122} \neq 0 \tag{59}$$

and all the remaining virials vanish will belong to a neutral mode, where

$$\Omega^2 = 2(B_{11} + a_1^2 B_{111}) . \tag{60}$$

This is the same neutral point as that isolated in Paper III (eq. [8]) from a direct investigation of the problem of small oscillations; and, as we have shown, the condition (60) is met along the Maclaurin sequence where

$$e = 0.89926 . \tag{61}$$

The Lagrangian displacement which belongs to the neutral mode of stability at $e =$ 0.89926 is given by

$$\xi_j = \text{Constant}\,\frac{\partial}{\partial x_j}\,x_1(x_1^2 - 3x_2^2)\,, \tag{62}$$

or

$$\xi_1 = \alpha(x_1^2 - x_2^2)\,, \quad \xi_2 = -2\alpha x_1 x_2\,, \quad \text{and} \quad \xi_3 = 0\,, \tag{63}$$

where α is a constant. It can be readily verified that the foregoing displacement satisfies all of the conditions (57) and (59).[9]

Turning to the consideration of the functionals J_1, J_2, and J_4, we should set in their variations

$$V_{111} = 3V_{122}\,, \tag{64}$$

TABLE 4

VALUES OF Δ_{ij} IN THE NEIGHBORHOOD OF THE NEUTRAL POINT

e	Δ_{24}	Δ_{12}	Δ_{41}
0.96.......	+0.015737	+0.0052456	+0.13906
.969373...	+ .000001	+ .0000003	+ .00001
0.98.......	−0.009929	−0.0033097	−0.17047

in order that we may not be inconsistent with equation (58). The coefficients of the virials V_{122} and V_{133} which then occur in the expansions of the relevant $\delta\mathfrak{W}_{ij;k}$'s and δS_{ijj}'s are those listed in Paper III, Table 2; and, using those coefficients, we find

$$\left|\begin{matrix}\delta J_1\\ \delta J_2\\ \delta J_4\end{matrix}\right| = \left|\begin{matrix} -\Omega^2 + 2(B_{11} + 3a_1^2 B_{111}) & a_1^2 B_{113}\\ 4a_3^2 B_{113} & 2B_{13} + 3a_2^2 B_{133} - \Omega^2\\ 3\Omega^2 + 4(2a_1^2 + a_3^2)B_{113} - 18a_1^2 B_{111} - 6B_{11} & -\Omega^2 + 3[B_{13} + B_{33} + (a_1^2 + 2a_3^2)B_{133} - a_1^2 B_{113}]\end{matrix}\right| \left|\begin{matrix}V_{122}\\ V_{133}\end{matrix}\right| = 0\,. \tag{65}$$

For the existence of a non-trivial solution for this system of equations, it is necessary that the rank of the 3×2 matrix on the right-hand side is 1. All three determinants of order 2 which can be formed from the 3×2 matrix must, therefore, vanish simultaneously for some member of the Maclaurin sequence; and this happens when (cf. Paper I, eq. [AII, 10])

$$e = 0.969373\,, \tag{66}$$

as is clear from Table 4, which gives the values of the determinant Δ_{ij} formed out of the ith and the jth rows of the matrix on the right-hand side of equation (65).

[9] The consideration of the functional $\Omega^2(V_{222} - 3V_{211}) + S_{112}$ (in group D) would have led to the same condition (60), but with the associated displacement

$$\xi_j = \text{Constant}\,\frac{\partial}{\partial x_j}\,x_2(x_2^2 - 3x_1^2)\,.$$

In deriving the conditions for the occurrence of neutral points based on equations (58) and (65), no demands were made on the Lagrangian displacement that it be solenoidal. If the solenoidal requirement were made, then equation (65) should be supplemented by the further condition

$$V_{122} = -\frac{a_1^2}{4a_3^2} V_{133}, \tag{67}$$

which is the form that the general condition

$$V_{111} + V_{122} = -\frac{a_1^2}{a_3^2} V_{133} \tag{68}$$

takes when relation (64) obtains.

When the equations requiring $\delta J_1 = \delta J_2 = 0$ are supplemented by the condition (67), they lead to the equations

$$\Omega^2 = 2(B_{11} + 3a_1^2 B_{111}) - 4a_3^2 B_{113} \tag{69}$$

and

$$\Omega^2 = 2B_{13} + 3a_3^2 B_{133} - a_1^2 B_{113}\,. \tag{70}$$

These two conditions are not independent, since it can be shown that the quantities on the right-hand sides of the equations are identically the same[10] and, moreover, determine the same value (66) for e; the value (66) was, in fact, determined with the aid of equations (69) and (70).

The vanishing of δJ_4, together with equation (67), leads to a further condition for the occurrence of a neutral point. In writing this last condition, it is more convenient to consider

$$\delta J_4 + 3\delta J_1 = \{-\Omega^2 + 3[B_{13} + B_{33} + (a_1^2 + 2a_3^2)B_{133}]\} V_{133} + 4(2a_1^2 + a_3^2)B_{113} V_{122} = 0\,; \tag{71}$$

or, making use of equation (67), we have

$$\Omega^2 = 3\,[B_{13} + B_{33} + (a_1^2 + 2a_3^2)B_{133}] - \frac{a_1^2}{a_3^2}(2a_1^2 + a_3^2)B_{113}\,. \tag{72}$$

It is found that this condition is also satisfied[11] at the same value of e; but the right-hand side of equation (72) is *not* identically the same as those of equations (69) and (70).

Again it is remarkable that the solenoidal character of the Lagrangian displacement should be deducible as a necessary condition for the occurrence of a neutral point.

[10] To prove the identity of the two sides, we first rewrite the equations in terms of the symbols A_{ijk}. We have

$$\Omega^2 = 2A_1 - a_1^2 A_{11} - 4a_3^2 A_{13} + 5a_1^2 a_3^2 A_{113}$$

and

$$\Omega^2 = 2A_1 - a_1^2 A_{11} + a_3^2 A_{13} + a_1^2 a_3^2 A_{113} - 3a_3^4 A_{133}\,;$$

and the equality of the right-hand sides now requires that

$$3a_3^2 A_{133} = -4a_1^2 A_{113} + 5A_{13}\,.$$

This last equality can be readily established by making use of the relations which express the symbols of a certain order in terms of those of the lower order. In the same way it can be shown that the condition obtained in Paper I, eq. (AII, 9) is identical with (69) and (70).

[11] Thus Ω^2 (in the unit $\pi G\rho$) given by the right-hand side of eq. (72), for e given by (66) is 0.41419, whereas the value found otherwise is 0.41413.

And, finally, we may note that the Lagrangian displacement which belongs to the third neutral point at $e = 0.96937$ is (cf. Paper I, eqs. [AII, 1]-[AII, 3])

$$\xi_j = \text{Constant}\, \frac{\partial}{\partial x_j} x_1 [x_1^2 + x_2^2 - 4x_3^2 - \tfrac{4}{5}(a_1^2 - a_3^2)] . \tag{73}$$

VI. THE NEUTRAL POINTS ALONG THE JEANS SEQUENCE

The equilibrium and stability of homogeneous masses under the action of a tidal potential of the form

$$\mathfrak{B}_T = -\tfrac{1}{2}\mu(x_1^2 + x_2^2 + x_3^2) + \tfrac{3}{2}\mu x_1^2 \tag{74}$$

was first studied by Jeans (1917) and more recently by Chandrasekhar and Lebovitz (Paper IV). In equation (74),

$$\mu = \frac{GM'}{R^3}, \tag{75}$$

where M' is the mass of the tidally distorting secondary and R is the distance between the centers of mass of the two objects.

It is known that prolate spheroidal forms are consistent with the equations of hydrostatic equilibrium so long as μ is less than a certain maximum value,

$$\mu_{\max} = 0.125536\ \pi G\rho , \tag{76}$$

which it attains when the eccentricity of the spheroid takes the value

$$e = 0.883026 . \tag{77}$$

In this section we shall isolate the neutral points along the Jeans sequence which belong to the second and the third harmonics.

a) The Neutral Point at $\mu_{\max}$

The second-order virial equations governing equilibrium are (Paper IV, eqs. [8] and [9])

$$\mathfrak{W}_{12} = \mathfrak{W}_{13} = 0 , \qquad \mathfrak{W}_{23} = \mu I_{23} , \tag{78}$$

$$\mathfrak{W}_{22} - \mathfrak{W}_{33} - \mu(I_{22} - I_{33}) = 0 , \tag{79}$$

and

$$\mathfrak{W}_{11} - \mathfrak{W}_{22} + \mu(2I_{11} + I_{22}) = 0 . \tag{80}$$

At a neutral point the first variations of all of these equations must vanish for a non-trivial Lagrangian displacement.

Combined with equation (15), the first variations of equations (78) give

$$B_{12}V_{12} = B_{13}V_{13} = 0 \quad \text{and} \quad (2B_{23} + \mu)V_{23} = 0 . \tag{81}$$

Since the B_{ij}'s are non-zero positive constants, it follows that

$$V_{12} = V_{13} = V_{23} = 0 \tag{82}$$

at a neutral point.

Considering next the first variation of equation (79) and making use of equation (16) (remembering that in the present context $a_2 = a_3$), we find

$$(2B_{22} + \mu)(V_{22} - V_{33}) = 0 . \tag{83}$$

Hence the Lagrangian displacement at a neutral point must satisfy the further condition

$$V_{22} = V_{33} \,. \tag{84}$$

Finally, the variation of equation (80), namely,

$$\delta\mathfrak{W}_{11} - \delta\mathfrak{W}_{22} + \mu(2V_{11} + V_{22}) = 0\,, \tag{85}$$

gives

$$-(2B_{11} - a_1^2A_{11} + a_2^2A_{21})V_{11} + 2(B_{22} - a_2^2A_{22} + a_1^2A_{12})V_{22} + \mu(2V_{11} + V_{22}) = 0\,. \tag{86}$$

By supplementing equation (86) by the further condition

$$V_{11} = -\frac{a_1^2}{a_2^2}(V_{22} + V_{33}) = -2\frac{a_1^2}{a_2^2}V_{22}\,, \tag{87}$$

which must be satisfied if the displacement is to be solenoidal, we obtain the necessary condition for the occurrence of a neutral point. The condition which we obtain in this manner is the same as that found in Paper IV (eq. [67]) by a direct solution of the problem of small oscillations.

From a comparison of equations (80) and (85) it is apparent that the neutral point must occur where μ attains its maximum value. While the present method cannot discriminate the *kind* of neutral point which occurs at $\mu_{\max}$, we know from the analysis of Paper IV that it is, in fact, of the first kind.

b) The Second Neutral Point along the Jeans Sequence

For the isolation of the second neutral point along the Jeans sequence, we must make use of the third-order virial equations which allows for the action of the tidal field (74). The required equation is readily obtained. We have (cf. Paper I, eq. [24])

$$\frac{d}{dt}\int_V \rho u_i x_j x_k d\boldsymbol{x} = 2(\mathfrak{T}_{ij;k} + \mathfrak{T}_{ik;j}) + \mathfrak{W}_{ij;k} + \mathfrak{W}_{ik;j} - \mu I_{ijk} + 3\mu\delta_{i1}I_{1jk} + \delta_{ij}\Pi_k + \delta_{ik}\Pi_j\,, \tag{88}$$

where the various symbols have their standard meanings.

When no relative motions are present and hydrostatic equilibrium prevails, equation (88) gives

$$\mathfrak{W}_{ij;k} + \mathfrak{W}_{ik;j} - \mu I_{ijk} + 3\mu\delta_{i1}I_{1jk} = -\delta_{ij}\Pi_k - \delta_{ik}\Pi_j\,. \tag{89}$$

Writing out explicitly the different components of equation (89), we obtain the following eighteen equations:

$$\mathfrak{W}_{11;1} + \mu I_{111} = -\Pi_1\,, \tag{90}$$

$$\mathfrak{W}_{22;1} + \mathfrak{W}_{21;2} - \mu I_{122} = -\Pi_1\,, \tag{91}$$

$$\mathfrak{W}_{33;1} + \mathfrak{W}_{31;3} - \mu I_{133} = -\Pi_1\,, \tag{92}$$

$$\mathfrak{W}_{12;2} + \mu I_{122} = \mathfrak{W}_{13;3} + \mu I_{133} = 0\,, \tag{93}$$

$$\mathfrak{W}_{23;1} + \mathfrak{W}_{21;3} = \mathfrak{W}_{13;2} + \mathfrak{W}_{23;1} = \mu I_{123}\,, \tag{94}$$

$$\mathfrak{W}_{12;3} + \mathfrak{W}_{13;2} = -2\mu I_{123}\,, \tag{95}$$

$$\mathfrak{W}_{11;2} + \mathfrak{W}_{12;1} + 2\mu I_{112} = -\Pi_2 , \tag{96}$$

$$\mathfrak{W}_{33;2} + \mathfrak{W}_{32;3} - \mu I_{233} = -\Pi_2 , \tag{97}$$

$$2\mathfrak{W}_{22;2} - \mu I_{222} = -2\Pi_2 , \tag{98}$$

$$2\mathfrak{W}_{12;1} = \mu I_{112} , \quad 2\mathfrak{W}_{23;3} = \mu I_{233} , \tag{99}$$

$$\mathfrak{W}_{11;3} + \mathfrak{W}_{13;1} + 2\mu I_{113} = -\Pi_3 , \tag{100}$$

$$\mathfrak{W}_{22;3} + \mathfrak{W}_{32;2} - \mu I_{223} = -\Pi_3 , \tag{101}$$

$$2\mathfrak{W}_{33;3} - \mu I_{333} = -2\Pi_3 , \tag{102}$$

$$2\mathfrak{W}_{13;1} = \mu I_{113} , \quad 2\mathfrak{W}_{23;2} = \mu I_{223} . \tag{103}$$

We next eliminate the Π_k's from equations (90)–(103). After their elimination we shall be left with fifteen equations; and, by suitably combining these remaining equations, we can arrange them in the following four "non-combining" groups:

$$A: \quad \mathfrak{W}_{12;3} = \mathfrak{W}_{13;2} = -\mu I_{123} , \quad \mathfrak{W}_{23;1} = 2\mu I_{123} ; \tag{104}$$

$$B: \quad 2\mathfrak{W}_{12;1} = \mu I_{112} , \quad 2\mathfrak{W}_{23;3} = \mu I_{233} , \quad S_{112} = \mu(I_{222} + 3I_{112}) , \quad S_{233} = \mu(I_{222} - 3I_{233}) ; \tag{105}$$

$$C: \quad 2\mathfrak{W}_{13;1} = \mu I_{113} , \quad 2\mathfrak{W}_{23;2} = \mu I_{223} , \quad S_{113} = \mu(I_{333} + 3I_{113}) , \quad S_{223} = \mu(I_{333} - 3I_{223}) ; \tag{106}$$

$$D: \quad \mathfrak{W}_{12;2} = -\mu I_{122} , \quad \mathfrak{W}_{13;3} = -\mu I_{133} , \quad S_{122} = S_{133} = -2\mu I_{111} . \tag{107}$$

In the foregoing equations, S_{ijj} has the same meaning as in equation (42).

At a neutral point a non-trivial Lagrangian displacement must exist such that the first variations of all of the equations in the groups A, B, C, and D vanish. When the variations are carried out, we find (as in the discussion of the Jacobi ellipsoids in Sec. IV) that the equations derived from the different groups involve different virials and are mutually exclusive. If δA, δB, δC, and δD denote the equations which are obtained by taking the first variations of the equations in the respective groups, then the association of the groups and the virials is the following:

$$\delta A: \ V_{123} ; \quad \delta B: \ V_{222} , \ V_{233} , \ V_{211} ; \quad \delta C: \ V_{333} , \ V_{311} , \ V_{322} ; \quad \delta D: \ V_{111} , \ V_{122} , \ V_{133} . \tag{108}$$

The coefficients of the virials in the expansions of the $\delta\mathfrak{W}_{ij;k}$'s and δS_{ijj}'s which occur in the varied forms of the equations (104)–(107) can be read off from Table 2 in Paper II. (There are some obvious simplifications arising from the present equality of a_2 and a_3.)

We readily find that the equations in the groups δA, δB, and δC do not allow any non-trivial solution. Therefore, at a neutral point, we must necessarily have

$$V_{123} = V_{222} = V_{233} = V_{211} = V_{333} = V_{311} = V_{322} = 0 . \tag{109}$$

The occurrence of a neutral point now depends on whether the remaining group δD allows a non-trivial solution. The equations to be considered can be written in the forms (cf. eq. [107])

$$-\delta\mathfrak{W}_{12;2} + \delta\mathfrak{W}_{13;3} - \mu(V_{122} - V_{133}) = 0 \,, \tag{110}$$

$$\delta S_{122} - \delta S_{133} = 0 \,, \tag{111}$$

$$-(\delta\mathfrak{W}_{12;2} + \delta\mathfrak{W}_{13;3}) - \mu(V_{122} + V_{133}) = 0 \,, \tag{112}$$

and

$$\tfrac{1}{2}(\delta S_{122} + \delta S_{133}) + 2\mu V_{111} = 0 \,. \tag{113}$$

Making use of the results of Paper II, Table 2, we find that equations (110) and (111) give

$$(B_{12} + a_2^2 B_{122} - \mu)(V_{122} - V_{133}) = 0 \tag{114}$$

and

$$[3(B_{22} + B_{12}) + (5a_2^2 + a_1^2)B_{122}](V_{122} - V_{133}) = 0 \,; \tag{115}$$

and these equations clearly require that

$$V_{122} = V_{133} \,. \tag{116}$$

Using this last result in equations (112) and (113), we find

$$a_2^2 B_{112} V_{111} + 2(B_{12} + 2a_2^2 B_{122} - \mu) V_{122} = 0 \tag{117}$$

and

$$[2\mu + (2a_1^2 + a_2^2)B_{112} - 5a_1^2 B_{111} - 2B_{11}]V_{111} + [3(B_{22} + B_{12}) + (5a_1^2 + 7a_2^2)B_{122} - 6a_1^2 B_{112}]V_{122} = 0 \,. \tag{118}$$

For the existence of a non-trivial solution, the determinant of equations (117) and (118) must vanish; and we find that this happens when

$$e = 0.94774 \quad \text{and} \quad \mu = 0.10913\, \pi G\rho \,. \tag{119}$$

These values agree with those derived by Jeans (1917) by a different argument.

In deriving the condition for the neutral point based on equations (117) and (118), no demands were made on the Lagrangian displacement that it be solenoidal. If the solenoidal requirement were made, then equations (117) and (118) should be supplemented by the further condition

$$V_{111} = -\frac{a_1^2}{a_2^2}(V_{122} + V_{133}) = -2\frac{a_1^2}{a_2^2} V_{122} \,. \tag{120}$$

With this additional condition, equations (117) and (118) give

$$\mu = B_{12} + 2a_2^2 B_{122} - a_1^2 B_{112} \tag{121}$$

and

$$[3(B_{22} + B_{12}) + (5a_1^2 + 7a_2^2)B_{122} - 6a_1^2 B_{112}] - 2\frac{a_1^2}{a_2^2}[2\mu + (2a_1^2 + a_2^2)B_{112} - 5a_1^2 B_{111} - 2B_{11}] = 0 \,. \tag{122}$$

We find that these two conditions are, in fact, satisfied at $e = 0.94774$. Thus the solenoidal character of the Lagrangian displacement is again deducible as a necessary condition for the occurrence of a neutral point.

And, finally, we may note that the Lagrangian displacement which belongs to this second neutral point at $e = 0.94774$ is

$$\xi_j = \text{Constant}\,\frac{\partial}{\partial x_j} x_1 [\tfrac{2}{3}x_1^2 - (x_2^2 + x_3^2) - \tfrac{2}{5}(a_1^2 - a_2^2)]\,. \qquad (123)$$

The research reported in this paper has in part been supported by the Office of Naval Research under contract Nonr-2121(24) with the University of Chicago.

REFERENCES

Chandrasekhar, S. 1962, *Ap. J.*, **136**, 1048 (referred to as "Paper I").
Chandrasekhar, S., and Lebovitz, N. R. 1962*a*, *Ap. J.*, **136**, 1037.
———. 1962*b*, *ibid.*, p. 1069.
———. 1963*a*, *ibid.*, **137**, 1142 (referred to as "Paper II").
———. 1963*b*, *ibid.*, p. 1162 (referred to as "Paper III").
———. 1963*c*, *ibid.*, p. 1172 (referred to as "Paper IV").
Jeans, J. H. 1917, *Mem. R.A.S.*, **62**, 1.
Lebovitz, N. R., 1961, *Ap. J.*, **134**, 500.
Roberts, P. H., and Stewartson, K. 1963, *Ap. J.*, **137**, 777.

CORRIGENDUM

The statement, "An examination of the coefficient of $(V_{11} + V_{22})$ in equation (23) shows that it does not vanish along the Maclaurin sequence," in my paper, "The Points of Bifurcation along the Maclaurin, the Jacobi, and the Jeans Sequences" (*Ap. J.*, **137**, 1189, 1963), is incorrect. The coefficient in question does in fact vanish at the point where Ω^2 attains its maximum value, $0.44933\pi G\rho$, along the Maclaurin sequence where $e = 0.92996$. The origin and the meaning of this result will be discussed in my paper, "The Equilibrium and the Stability of the Roche Ellipsoids," now in preparation.

S. CHANDRASEKHAR

THE EQUILIBRIUM AND THE STABILITY OF THE ROCHE ELLIPSOIDS

S. CHANDRASEKHAR
University of Chicago
Received July 15, 1963

ABSTRACT

Roche's problem is concerned with the equilibrium and the stability of rotating homogeneous masses which are, further, distorted by the constant tidal action of an attendant rigid spherical mass. This ancient problem is reconsidered in this paper with the principal object of determining the stability of the equilibrium configurations (the ellipsoids of Roche) by a direct evaluation of their characteristic frequencies of oscillation belonging to the second harmonics. The result of the evaluation is the demonstration that the Roche ellipsoid becomes unstable at a point subsequent to the Roche limit where the angular velocity of rotation, consistent with equilibrium, attains its maximum value. This result requires a revision of the current common view regarding the meaning that is to be attached to the Roche limit.

Among related matters which are considered are the following: the relationships that exist between the sequences of Roche and those of Maclaurin, Jacobi, and Jeans; the exhibition and the isolation of the second neutral point (belonging to the third harmonics) along the Roche sequences; and the effect of compressibility on the stability of the Roche ellipsoids. A result which emerges from these considerations is the universal instability of the Jacobi ellipsoids under the least tidal action.

The methods used in this paper are those derived from the virial theorem and its various extensions. The principal results are summarized in Section X and are exhibited in Figures 1, 2, and 3.

I. INTRODUCTION

Roche discovered in 1850 that no equilibrium configuration exists for an infinitesimal homogeneous satellite (of density ρ) rotating about a planet (of mass M') in a circular Keplerian orbit (of radius R'), if the angular velocity (Ω) of rotation exceeds the limit set by

$$\frac{\Omega^2}{\pi G\rho}=\frac{M'}{\pi\rho R'^3}\leq 0.090068\,, \qquad (1)^1$$

where G denotes the constant of gravitation. The lower limit to R' set by this inequality is called the *Roche limit*. It is generally believed that the non-existence of equilibrium configurations below the Roche limit implies some sort of instability for the satellite. Thus, Darwin (to whom we owe the term "Roche limit") describes Roche's result as follows. "Now Roche showed that instability will set in when the elongation of figure of the satellite has reached a certain degree. In other words, at a certain stage of proximity, the satellite cannot hold together by the force of its own gravitation, and it will be torn apart by the tide generating force" (Darwin 1911, p. 340). And Darwin's description expresses the still prevalent common view (e.g., Struve 1961, p. 18). Nevertheless, if Darwin's statement is taken to mean (what it apparently says) that the satellite is capable of a certain normal mode of oscillation with respect to which it becomes unstable as the Roche limit is approached (in the direction of increasing Ω^2), then, the statement (as we shall show in Sec. VIII below) is incorrect. However, other statements with other meanings have also been made concerning the nature of the "instability" which is supposed to set in at the Roche limit. Thus, Jeans (1919, pp. 52 and 53) has written: "We are dealing, it must be noted, with secular stability[2] only; the question means nothing except when dissipative forces are present . . . the instability is one of orbital motion only and

[1] Roche (1850) originally gave the value 0.092 for the constant on the right-hand side of this inequality. Darwin (1906) later gave the more precise value 0.09006. The value quoted, 0.090068, is that obtained in the present paper (see Table 2 in Sec. V below).

[2] For the meaning generally attached to this term see Chandrasekhar (1963*a*, p. 1187; this paper will be referred to hereinafter as "Paper I").

not one of the configurations of the masses." And, again, as we shall see, much of the implications of this statement are also incorrect.

The conflicting statements and views (which we have, in part, quoted) concerning the role of the Roche limit for the stability of the satellite arise, mostly, from the incompleteness of the available analytical information on the Roche ellipsoids, and the lack of any specific investigation on their normal modes of oscillation.[3] On these accounts, it has appeared worthwhile to reconsider this ancient problem and provide as complete and as explicit a solution as seemed necessary to clarify the basic issues.

II. ROCHE'S PROBLEM

Roche's problem is concerned with a particularly simple case in the equilibrium and the stability of a homogeneous body ("the primary") rotating about another ("the secondary") in a manner that their relative disposition remains the same.

Let the masses of the primary and the secondary be M and M', respectively.[4] Let the distance between their centers of mass be R'; and let the constant angular velocity of rotation about their common center of mass be Ω.

Choose a coordinate system in which the origin is at the center of mass of M, the x_1-axis is pointing to the center of mass of M', and the x_3-axis is parallel to the direction of Ω. In this coordinate system, the equation of the axis of rotation is

$$x_1 = \frac{M'}{M+M'} R' \qquad \text{and} \qquad x_2 = 0 . \tag{2}$$

In this frame of reference, rotating uniformly with the angular velocity Ω, the equation of motion governing the fluid elements of M is

$$\rho \frac{du_i}{dt} = -\frac{\partial p}{\partial x_i} + \rho \frac{\partial}{\partial x_i} \left\{ \mathfrak{B} + \mathfrak{B}_T + \tfrac{1}{2}\Omega^2 \left[\left(x_1 - \frac{M'R'}{M+M'} \right)^2 + x_2^2 \right] \right\} + 2\rho\Omega\epsilon_{il3}u_l , \tag{3}$$

where $\mathfrak{B}$ is the self-gravitational potential and $\mathfrak{B}_T$ is the tidal potential due to M'. Equation (3) can also be written as

$$\rho \frac{du_i}{dt} = -\frac{\partial p}{\partial x_i} + \rho \frac{\partial}{\partial x_i} \left[\mathfrak{B} + \mathfrak{B}_T + \tfrac{1}{2}\Omega^2 (x_1^2 + x_2^2) - \frac{M'R'}{M+M'} \Omega^2 x_1 \right] + 2\rho\Omega\epsilon_{il3}u_l . \tag{4}$$

In Roche's particular problem, the secondary is treated as a rigid sphere; then, over the primary, the tide-generating potential of $\mathfrak{B}_T$ can be expanded in the form

$$\mathfrak{B}_T = \frac{GM'}{R'} \left(1 + \frac{x_1}{R'} + \frac{x_1^2 - \frac{1}{2}x_2^2 - \frac{1}{2}x_3^2}{R'^2} + \ldots . \right) . \tag{5}$$

As is customary in the treatment of this problem, we shall retain in the expansion for $\mathfrak{B}_T$ only those terms which we have explicitly written out in equation (5). On this assumption concerning $\mathfrak{B}_T$, the equation of motion becomes

$$\begin{aligned} \rho \frac{du_i}{dt} = -\frac{\partial p}{\partial x_i} + \rho \frac{\partial}{\partial x_i} \Big[& \mathfrak{B} + \tfrac{1}{2}\Omega^2 (x_1^2 + x_2^2) + \mu (x_1^2 - \tfrac{1}{2}x_2^2 - \tfrac{1}{2}x_3^2) \\ & + \left(\frac{GM'}{R'^2} - \frac{M'R'}{M+M'} \Omega^2 \right) x_1 \Big] + 2\rho\Omega\epsilon_{il3}u_l , \end{aligned} \tag{6}$$

[3] Thus, in Figs. 7 and 15 in Jeans (1919, pp. 50 and 86) none of the lines representing the Roche sequences are based on any calculation; the same applies to Fig. 33 in Jeans (1929, p. 229). Milne's (1952, chap. ix; see particularly pp. 110–112) generous, but perceptive, account in his biography of Jeans brings out, by its accuracy, the inadequacy of the presently available information.

[4] By using the terms "primary" and "secondary" to describe the two bodies, we do not wish to imply that $M > M'$; indeed, in the important special case considered in Sec. I, $M \ll M'$.

where we have introduced the abbreviation

$$\mu = \frac{GM'}{R'^3}\,. \tag{7}$$

Now letting Ω^2 have its "Keplerian value"

$$\Omega^2 = \frac{G(M+M')}{R'^3} = \mu\left(1+\frac{M}{M'}\right), \tag{8}$$

we obtain the basic equation of this theory:

$$\rho\frac{du_i}{dt} = -\frac{\partial p}{\partial x_i} + \rho\frac{\partial}{\partial x_i}[\mathfrak{V} + \tfrac{1}{2}\Omega^2(x_1^2+x_2^2) + \mu(x_1^2 - \tfrac{1}{2}x_2^2 - \tfrac{1}{2}x_3^2)] + 2\rho\Omega\epsilon_{il3}u_l\,; \tag{9}$$

and *the problem of Roche is that of the equilibrium and the stability of homogeneous configurations governed by equation* (9).

a) The Second-Order Virial Theorem Appropriate to Roche's Problem

In treating Roche's problem, we shall use the methods, based on the virial theorem and its extensions, which have been developed recently (see the various papers by Chandrasekhar and by Chandrasekhar and Lebovitz in the *Astrophysical Journal* for the past three years).

By multiplying equation (9) by x_j and integrating over the volume V occupied by the fluid, we obtain in the usual manner the second-order tensor equation (cf. Chandrasekhar and Lebovitz 1963*b*, eq. [4]; this paper will be referred to hereafter as "Paper III")

$$\begin{aligned}\frac{d}{dt}\int_V \rho u_i x_j dx = 2\mathfrak{T}_{ij} + \mathfrak{W}_{ij} + (\Omega^2-\mu)I_{ij} - \Omega^2\delta_{i3}I_{3j} + 3\mu\delta_{i1}I_{1j}\\ + \delta_{ij}\Pi + 2\Omega\int_V \rho\epsilon_{il3}u_l x_j dx\,,\end{aligned} \tag{10}$$

where

$$\Pi = \int_V p\,dx\,, \tag{11}$$

and

$$\mathfrak{T}_{ij} = \tfrac{1}{2}\int_V \rho u_i u_j dx\,, \qquad \mathfrak{W}_{ij} = -\tfrac{1}{2}\int_V \rho\mathfrak{V}_{ij}dx\,, \qquad \text{an} \qquad I_{ij} = \int_V \rho x_i x_j dx \tag{12}$$

are the kinetic-energy, the potential-energy, and the moment of inertia tensors.

III. PROPERTIES OF THE EQUILIBRIUM ELLIPSOIDS

When no motions are present in the frame of reference considered and hydrostatic equilibrium prevails, equation (10) becomes

$$\mathfrak{W}_{ij} + (\Omega^2-\mu)I_{ij} - \Omega^2\delta_{i3}I_{3j} + 3\mu\delta_{i1}I_{1j} = -\Pi\delta_{ij}\,. \tag{13}$$

The diagonal elements of this relation give

$$\mathfrak{W}_{11} + (\Omega^2+2\mu)I_{11} = \mathfrak{W}_{22} + (\Omega^2-\mu)I_{22} = \mathfrak{W}_{33} - \mu I_{33} = -\Pi\,, \tag{14}$$

while the non-diagonal elements give:

$$\begin{aligned}&\mathfrak{W}_{12} + (\Omega^2+2\mu)I_{12} = \mathfrak{W}_{21} + (\Omega^2-\mu)I_{21} = 0\,,\\ &\mathfrak{W}_{23} + (\Omega^2-\mu)I_{23} = \mathfrak{W}_{32} - \mu I_{32} = 0\,,\\ &\mathfrak{W}_{31} - \mu I_{31} = \mathfrak{W}_{13} + (\Omega^2+2\mu)I_{13} = 0\,.\end{aligned} \tag{15}$$

In view of the symmetry of the tensors I_{ij} and $\mathfrak{W}_{ij}$, it follows from the foregoing equations that, so long as Ω^2 and μ are finite,

$$\mathfrak{W}_{ij} = 0 \qquad \text{and} \qquad I_{ij} = 0 \qquad (i \neq j)\,. \tag{16}$$

Therefore, *in the chosen coordinate system, the tensors I_{ij} and $\mathfrak{W}_{ij}$ are, necessarily, diagonal.*

Equations (13)–(16) are entirely general: they do not depend on any constitutive relations that may exist.

Now it can be shown quite readily that if the configuration is homogeneous (by assumption or by virtue of incompressibility), then an ellipsoidal figure is consistent with the equations of hydrostatic equilibrium as well as the condition which requires the pressure to be constant over the bounding surface. Once this has been established, the virial equations suffice (as we shall presently see) to determine the geometry and the properties of the equilibrium ellipsoids.

Letting

$$p = \frac{M}{M'} \qquad \text{so that} \qquad \Omega^2 = (1 + p)\,\mu\,, \tag{17}$$

we can rewrite equations (14) in the form

$$\mathfrak{W}_{11} + (3 + p)\mu I_{11} = \mathfrak{W}_{22} + p\mu I_{22} = \mathfrak{W}_{33} - \mu I_{33}\,, \tag{18}$$

or, alternatively,

$$\mu[(3 + p)I_{11} + I_{33}] = \mathfrak{W}_{33} - \mathfrak{W}_{11} \tag{19}$$

and

$$\mu(pI_{22} + I_{33}) = \mathfrak{W}_{33} - \mathfrak{W}_{22}\,. \tag{20}$$

The geometry of the ellipsoids (for an assigned p) is, therefore, determined by the equation

$$\frac{(3 + p)I_{11} + I_{33}}{pI_{22} + I_{33}} = \frac{\mathfrak{W}_{33} - \mathfrak{W}_{11}}{\mathfrak{W}_{33} - \mathfrak{W}_{22}}\,. \tag{21}$$

Expressions for the various tensors describing the properties of homogeneous ellipsoids have been given in an earlier paper (Chandrasekhar and Lebovitz 1962*a*); in particular, we have

$$\mathfrak{W}_{ii} = -2\pi G\rho(a_1a_2a_3)A_iI_{ii} \qquad \text{and} \qquad I_{ii} = \tfrac{1}{5}Ma_i^2\,, \tag{22}$$

(no summation over repeated indices)

where a_1, a_2, and a_3 are the semi-axes of the ellipsoid and the A_i's are the one-index symbols defined in the same paper.

Inserting the expressions for $\mathfrak{W}_{ii}$ and I_{ii} in equation (21), we obtain

$$\frac{(3 + p)a_1^2 + a_3^2}{pa_2^2 + a_3^2} = \frac{A_1a_1^2 - A_3a_3^2}{A_2a_2^2 - A_3a_3^2}\,. \tag{23}$$

Using the known expressions for the constants A_i (*loc. cit.*, eqs. [15]–[17]), we can reduce equation (23) to the following form which is convenient for numerical calculations:

$$\begin{aligned}[t] &[(p + 3)a_1^4(a_2^2 + a_3^2) + pa_2^4(a_3^2 + a_1^2) - a_3^4(a_1^2 + a_2^2) \\ &- 4(p+1)a_1^2a_2^2a_3^2]E(\theta, \phi) - (a_2^2 - a_3^2)[2(p+3)a_1^2a_2^2 + a_3^2(a_1^2 + a_2^2)]F(\theta, \phi) \\ &= \frac{a_2a_3}{a_1}(a_1^2 - a_2^2)(a_1^2 - a_3^2)^{1/2}[2(p + 3)a_1^2 - pa_2^2 + a_3^2]\,. \end{aligned} \tag{24}$$ [5]

[5] Notice that when $p \to \infty$, this equation tends to the one determining the geometry of the Jacobi ellipsoids (see Chandrasekhar 1962, Appendix I, eq. [AI, 5]).

In equation (24) $E(\theta, \phi)$ and $F(\theta, \phi)$ are the standard elliptic integrals of the two kinds with the arguments

$$\theta = \sin^{-1}\sqrt{\frac{a_1^2 - a_2^2}{a_1^2 - a_3^2}} \quad \text{and} \quad \phi = \cos^{-1}\frac{a_3}{a_1}. \tag{25}$$

For every pair of values $(a_2/a_1, a_3/a_1)$ determined consistently with equation (23) or (24), the associated values of Ω^2 and μ follow from the equations

$$\frac{\mu}{\pi G \rho} = 2(a_1 a_2 a_3)\frac{A_1 a_1^2 - A_3 a_3^2}{(3+p)a_1^2 + a_3^2} \quad \text{and} \quad \Omega^2 = (1+p)\mu. \tag{26}$$

For a few given values of p and different assigned values of ϕ (i.e., of a_3/a_1) equation (24) was solved for θ (i.e., for a_2/a_1) by a method of successive approximation; and the accuracy of the final solution was always tested against equation (23).

The solution of equations (23) and (24) has been carried out to determine adequately the Roche sequences for $p = 0, 1, 4, 20$, and 100. The results of the calculations are summarized in Table 1; in this table, in addition to the principal constants (θ, a_2/a_1, a_3/a_1, A_1, A_2, A_3, Ω^2, and μ), the semi-axes of the ellipsoid, in the unit $(a_1a_2a_3)^{1/3}$, are also listed.

IV. THE ARRANGEMENT OF THE SOLUTIONS

The solutions for the Roche sequences belonging to different values of p and the relationships which exist between these sequences and those of Maclaurin, Jacobi, and Jeans are most clearly exhibited in a plane in which each equilibrium ellipsoid is represented by a point whose coordinates are

$$\bar{a}_1 = \frac{a_1}{(a_1 a_2 a_3)^{1/3}} \quad \text{and} \quad \bar{a}_2 = \frac{a_2}{(a_1 a_2 a_3)^{1/3}}. \tag{27}$$

The utility of this plane for the exhibition of these relationships seems to have been recognized, first, by Jeans.

Before we describe the arrangement of the solutions in the $(\bar{a}_1, \bar{a}_2)$-plane, it is important to observe that when $p = -1$, and Ω^2 according to equation (17) is zero, we recover the pure tidally distorted spheroids of Jeans[6] (1917, 1919; see also Paper III); and, further, that when $p \to \infty$, and the term in μ in equation (9) becomes negligible, we similarly recover the pure rotationally distorted configurations of Maclaurin and Jacobi.

Now consider Figure 1 in which the solutions for all the sequences are exhibited. In this diagram, each equilibrium ellipsoid is represented by a point whose coordinates are $\bar{a}_1$ and $\bar{a}_2$. By the chosen normalization, the volume (and in view of the homogeneity, also the mass) of all the ellipsoids represented is unity. The undistorted sphere is, therefore, represented by the point $\bar{a}_1 = \bar{a}_2 = 1$; this is the point S in Figure 1.

The Jeans spheroids, being prolate, are represented by the pseudo-hyperbolic locus

$$\bar{a}_1 \bar{a}_2^2 = 1\,; \tag{28}$$

this is the curve ST.

The Maclaurin spheroids, being oblate, are represented by the straight line

$$\bar{a}_1 = \bar{a}_2 (\geq 1)\,; \tag{29}$$

[6] The fact, that the spheroids of Jeans are obtained when p is assigned the "unphysical" value -1, is the origin of the statement, that is sometimes made, that these spheroids are of "no physical interest." However, it will appear that for an understanding of "what happens" in the $(\bar{a}_1, \bar{a}_2)$-plane, it is essential that consideration is given to the Jeans spheroids.

TABLE 1

THE PROPERTIES OF THE ROCHE ELLIPSOIDS†

$$\phi = \cos^{-1} a_3/a_1;\ \theta = \sin^{-1}\sqrt{(a_1^2 - a_2^2)/(a_1^2 - a_3^2)};\ A_i^* = a_1a_2a_3A_i;\ \bar{a}_i = a_i/(a_1a_2a_3)^{1/3}$$

ϕ	θ	a_2/a_1	a_3/a_1	A_1^*	A_2^*	A_3^*	Ω^2	μ	$\bar{a}_1$	$\bar{a}_2$	$\bar{a}_3$
						$p=0$					
24°..	63°113	0.93188	0.91355	0.624238	0.679759	0.696004	0.022624	0.022624	1.0551	0.9832	0.9639
36...	66.948	.84112	.80902	.567104	.700085	.732811	.047871	.047871	1.1369	.9563	.9198
48...	72.143	.70687	.66913	.479968	.736033	.783999	.074799	.074799	1.2835	.9072	.8588
57...	76.685	.57787	.54464	.391944	.777221	.830835	.088267	.088267	1.4701	.8495	.8007
60...	78.263	.53013	.50000	.358101	.794144	.847756	.089946	.089946	1.5567	.8253	.7784
61...	78.797	.51373	.48481	.346293	.800216	.853492	.090068	.090068	1.5894	.8165	.7706
62...	79.327	.49714	.46947	.334272	.806436	.859292	.089977	.089977	1.6242	.8074	.7625
63...	79.846	.48040	.45399	.322096	.812728	.865177	.089689	.089689	1.6613	.7981	.7542
66...	81.416	.42898	.40674	.284086	.832949	.882963	.087201	.087201	1.7896	.7677	.7279
71...	83.941	.34052	.32557	.217086	.869951	.912963	.077474	.077474	2.0816	.7088	.6777
72...	84.424	.32254	.30902	.203288	.877769	.918942	.074648	.074648	2.1568	.6957	.6665
75...	85.810	.26827	.25882	.161540	.901812	.936647	.064426	.064426	2.4330	.6527	.6297
79...	87.465	0.19569	0.19081	0.106446	0.934365	0.959192	0.047111	0.047111	2.9919	0.5855	0.5709
						$p=1$					
12°..	51°71	0.98660	0.97815	0.657194	0.667954	0.674850	0.009293	0.004647	1.0119	0.9984	0.9898
24...	54.380	.94376	.91355	.627500	.672967	.699532	.036152	.018076	1.0507	.9916	.9598
36...	59.116	.86345	.80902	.573517	.685952	.740530	.076342	.038171	1.1270	.9731	.9118
48...	65.938	.73454	.66913	.488555	.715127	.796317	.118726	.059363	1.2671	.9308	.8479
54...	70.017	.64956	.58779	.432108	.739433	.828458	.134284	.067142	1.3784	.8954	.8102
59...	73.623	.56892	.51504	.377406	.766047	.856546	.140854	.070427	1.5056	.8566	.7754
60...	74 355	.55186	.50000	.365639	.772117	.862243	.141250	.070625	1.5360	.8477	.7680
61 ..	75.088	.53451	.48481	.353605	.778442	.867954	.141298	.070649	1.5685	.8384	.7604
66...	78.715	.44429	.40674	.289716	.813764	.896520	.135785	.067892	1.7688	.7859	.7194
69...	80.813	.38813	.35837	.248826	.837737	.913437	.127424	.063712	1.9300	.7491	.6917
71...	82.154	.35022	.32557	.220788	.854695	.924516	.119625	.059812	2.0622	.7222	.6714
72...	82.8038	.33119	.30902	.206604	.863429	.929965	.115054	.057527	2.1379	.7080	.6606
73...	83.437	.31213	.29237	.192351	.872298	.935348	.110044	.055022	2.2211	.6933	.6494
75...	84.648	.27405	.25882	.163768	.890351	.945879	.098753	.049376	2.4158	.6620	.6252
78...	86.304	.21726	.20791	.121325	.917769	.960904	.078934	.039467	2.8079	.6100	.5838
81...	87.728	0.16126	0.15643	0.080697	0.944630	0.974664	0.056499	0.028249	3.4097	0.5498	0.5334
						$p=4$					
12°..	38°67	0.99153	0.97815	0.658519	0.665285	0.676197	0.01452	0.002904	1.0103	1.0017	0.9882
24...	41.492	.96301	.91355	.632677	.662197	.705127	.05642	.011284	1.0436	1.0050	.9534
36...	46.922	.90315	.80902	.584433	.661930	.753636	.11911	.023822	1.1103	1.0027	.8982
48...	55.906	.78821	.66913	.504195	.677213	.818593	.18486	.036973	1.2377	0.9756	.8282
58...	66.148	.63120	.52992	.403460	.719650	.876890	.21593	.043187	1.4406	0.9093	.7634
60...	68.353	.59334	.50000	.379227	.732740	.888034	.21685	.043371	1.4994	0.8896	.7497
61...	69.456	.57380	.48481	.366628	.739884	.893488	.21648	.043296	1.5319	0.8790	.7427
70...	78.845	.38733	.34202	.241528	.820128	.938344	.18514	.037028	1.9616	0.7598	.6709
72...	80.682	.34526	.30902	.211856	.840934	.947210	.17110	.034220	2.1084	0.7280	.6515
74...	82.390	.30361	.27564	.182039	.862282	.955681	.15465	.030930	2.2862	0.6941	.6302
75...	83.194	.28300	.25882	.167150	.873109	.959736	.14555	.029110	2.3900	0.6764	.6186
80...	86.639	0.18299	0.17365	0.095083	0.926613	0.978290	0.09329	0.018658	3.1572	0.5777	0.5482
						$p=20$					
36°..	27°969	0.96125	0.80902	0.599366	0.629135	0.771500	0.16763	0.007983	1.0874	1.0453	0.8797
48...	38.306	.88758	.66913	.530024	.615046	.854929	.26374	.012559	1.1897	1.0559	.7960
58...	55.238	.71737	.52992	.428110	.651851	.920039	.30624	.014583	1.3804	0.9903	.7315
60...	59.136	.66886	.50000	.401536	.668988	.929475	.30559	.014552	1.4407	0.9636	.7203
66...	69.927	.51356	.40674	.313115	.735445	.951441	.28232	.013444	1.6854	0.8655	.6855
70...	75.865	.41187	.34202	.250124	.787015	.962860	.24980	.011895	1.9219	0.7916	.6573
72...	78.422	.36322	.30902	.218310	.813640	.968050	.22890	.010900	2.0731	0.7530	.6406
74...	80.711	.31631	.27564	.186679	.840313	.973008	.20522	.009772	2.2552	0.7133	.6216
75...	81.759	0.29352	0.25882	0.171026	0.853573	0.975400	0.19243	0.009163	2.3611	0.6930	0.6111
						$p=100$					
48°..	21°399	0.96254	0.66913	0.547202	0.574019	0.878779	0.30021	0.0029724	1.1580	1.1146	0.7748
52...	28.218	.92799	.61566	.516745	.567853	.915402	.33173	.0032845	1.2051	1.1184	.7420
56...	39.753	.84790	.55919	.473434	.583718	.942848	.34922	.0034576	1.2824	1.0874	.7171
58...	46.363	.78950	.52992	.446235	.602727	.951038	.35043	.0034696	1.3370	1.0556	.7085
60...	52.610	.72564	.50000	.416519	.626800	.956681	.34697	.0034353	1.4021	1.0174	.7010
70...	74.605	.42333	.34202	.253984	.772315	.973702	.27441	.0027170	1.9044	0.8062	.6513
75...	81.228	.29780	.25882	.172573	.845842	.981585	.20935	.0020728	2.3498	0.6998	.6082
83...	87.838	0.12749	0.12187	0.056598	0.949318	0.994088	0.08203	0.0008122	4.0075	0.5109	0.4884

† Ω^2 and μ are listed in the unit $\pi G\rho$.

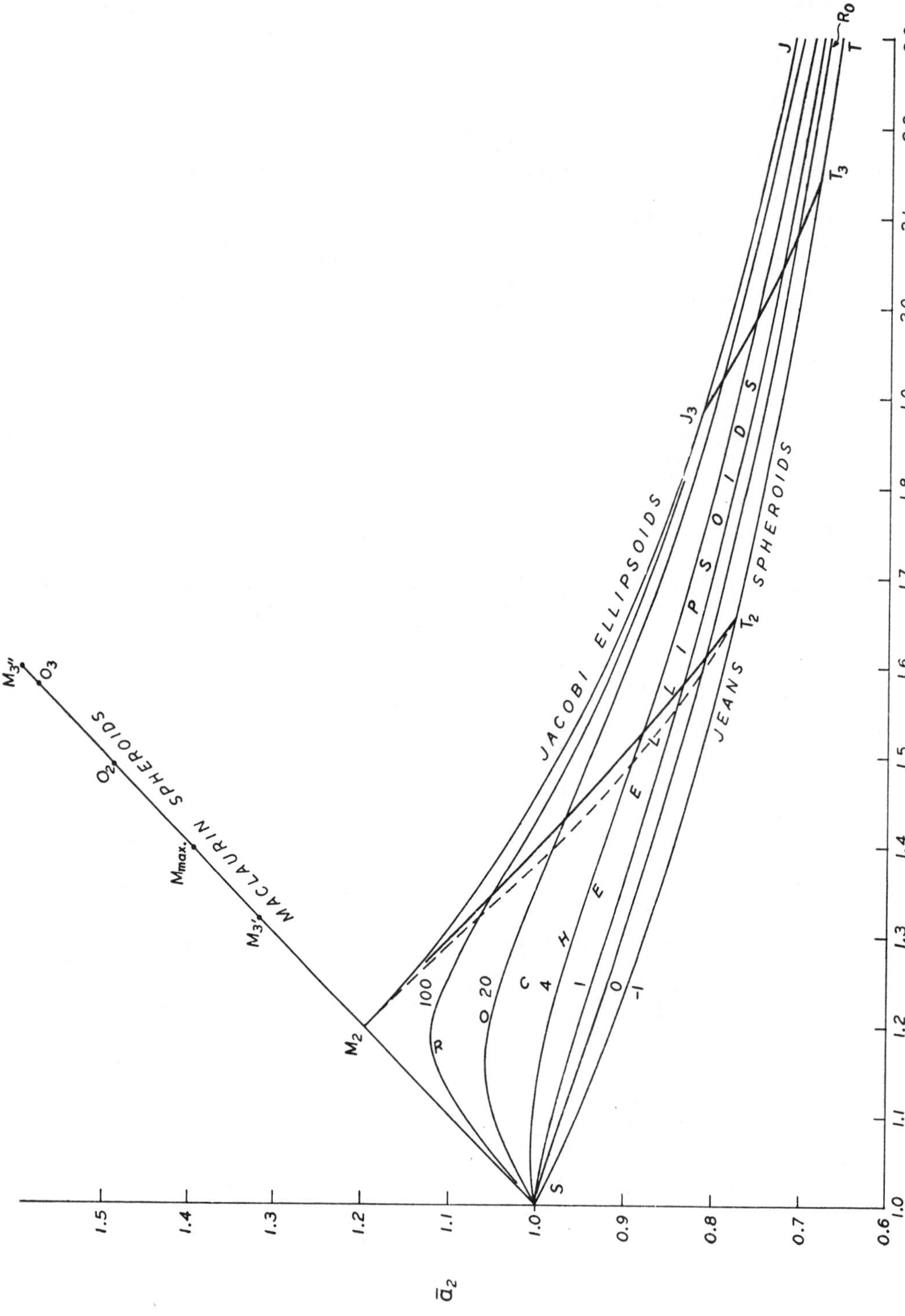

MACLAURIN SPHEROIDS
JACOBI ELLIPSOIDS
JEANS SPHEROIDS
M2
M3'
Mmax.
O2
O3
M3''
J3
T2
T3
J
T
Ro
S
100
20
4
1
0
-1
ā1
ā2
1.0
1.1
1.2
1.3
1.4
1.5
1.6
1.7
1.8
1.9
2.0
2.1
2.2
2.3
0.6
0.7
0.8
0.9

FIG. 1.—The relationships among the Maclaurin, the Jacobi, the Jeans, and the Roche sequences. The ordinates and the abscissa are the normalized values, $\bar{a}_1 = a_1/(a_1a_2a_3)^{1/3}$ and $\bar{a}_2 = a_2/(a_1a_2a_3)^{1/3}$, of two of the principal axes of the equilibrium ellipsoid (or spheroid). The undistorted sphere is represented by S, the Maclaurin sequence is SM_3'', the Jacobian sequence by M_2J, the Jeans sequence by ST, and the Roche sequences (labeled by the values of $p = M/M'$ to which they belong) are confined to the domain bounded by the combined Maclaurin-Jacobi sequence, SM_2J, and the Roche sequence, SR_0, for $p = 0$. The first point of bifurcation along the Maclaurin sequences occurs at M_2; at M_3' and M_3'' occur further neutral points belonging to the third harmonics. At the points O_2 and O_3 the Maclaurin spheroid becomes unstable by modes of overstable oscillation belonging to the second and the third harmonics; and at M_{max}, Ω^2 attains its maximum values along SM_3''. The Jacobi ellipsoids become unstable by a mode of oscillation belonging to the third harmonics at J_3 where the pear-shaped sequence branches off; also, along the entire sequence M_2J, the Jacobi ellipsoids are characterized by a neutral mode of oscillation belonging to the second harmonics. At the points T_2 and T_3 the Jeans spheroids become unstable by modes of oscillation belonging to the second and the third harmonics. The Roche limit where Ω^2 and μ attain their maxima along the different Roche sequences is represented by the dashed curve joining M_2 and T_2; and the locus of points where instability sets in by a mode of oscillation belonging to the second harmonics is shown by the heavy curve joining M_2 and T_2. The locus of the neutral point (belonging to the third harmonics) is shown by the curve joining T_3 and J_3. Note that the Jacobi ellipsoids are to be considered unstable in the limit $p \to \infty$.

this is the line $SM_{3''}$. The first point of bifurcation along the Maclaurin sequence occurs at M_2 where

$$\bar{a}_1 = \bar{a}_2 = 1.19723 \; ; \tag{30}$$

at this point $\Omega^2 = 0.37423\ \pi G\rho$.

The Jacobian sequence branches off from the Maclaurin sequence at M_2; it is represented by the locus M_2J. Since the Jacobi ellipsoids eventually become prolate, the locus M_2J becomes asymptotic to ST as $a_1 \to \infty$. Also, the Jacobi ellipsoids are known to become unstable at the point where the sequence of the pear-shaped configurations branches off; this occurs at $J_3(\bar{a}_1 = 1.8858,\ \bar{a}_2 = 0.81498)$.

The Roche sequences for the different p's are represented by a one-parameter family of loci in the domain bounded by SM_2, M_2J, and ST (strictly, we should rather say SM_2, M_2J, and SR_0 the Roche sequence for $p = 0$). All these loci start at the point S and all eventually become asymptotic to ST. It follows from our earlier remarks, and it is now apparent from Figure 1, that as $p \to \infty$, *the Roche sequence tends to the combined Maclaurin-Jacobi sequence represented by the broken curve* SM_2J.

V. THE ROCHE LIMIT

Since Ω^2 attains its maximum value along the Maclaurin sequence (at the point $M_{\max}$ in Fig. 1) subsequent to the point of bifurcation, and it decreases monotonically down the Jacobian sequence from the value $(=0.37423\ \pi G\rho)$ it has at the point of bifurcation (M_2), it is clear that along the combined Maclaurin-Jacobi sequence, SM_2J, Ω^2 has its maximum value at M_2. It is also known that along the Jeans sequence, μ attains its maximum value at (cf. Paper III, eq. [26])

$$e = 0.883026 \qquad \text{where} \qquad \mu = \mu_{\max} = 0.125536\ \pi G\rho \; ; \tag{31}$$

this point (denoted by T_2 in Fig. 1) occurs along ST where

$$\bar{a}_1 = 1.65584 \qquad \text{and} \qquad \bar{a}_2 = 0.777125 \, . \tag{32}$$

We should accordingly expect that, along each of the Roche sequences, Ω^2 and $\mu = \Omega^2/(1 + p)$ attain maxima, simultaneously, at some determinate point; that this is indeed the case is apparent from the results of Table 1 exhibited in Figure 2.

The place where Ω^2 and μ attain their maxima along a Roche sequence can be determined as follows.

Since the structure of the Roche ellipsoids is uniquely determined by equations (14), it is clear that, when Ω^2 and μ attain their maxima, not only these equations but also their first variations (with respect to a suitable infinitesimal solenoidal displacement which preserves the ellipsoidal shape) must be satisfied. Therefore, at the maximum, in addition to equations (14), the equations,

$$\delta\mathfrak{W}_{11} + (\Omega^2 + 2\mu)\delta I_{11} = \delta\mathfrak{W}_{22} + (\Omega^2 - \mu)\delta I_{22} = \delta\mathfrak{W}_{33} - \mu\delta I_{33} \, , \tag{33}$$

obtained by considering their first variations, must also be satisfied.[7] As two independent equations, equivalent to equations (33), we shall use

$$\delta\mathfrak{W}_{11} - \delta\mathfrak{W}_{22} + (\Omega^2 + 2\mu)\delta I_{11} - (\Omega^2 - \mu)\delta I_{22} = 0 \tag{34}$$

and

$$\delta\mathfrak{W}_{11} + \delta\mathfrak{W}_{22} - 2\delta\mathfrak{W}_{33} + (\Omega^2 + 2\mu)\delta I_{11} + (\Omega^2 - \mu)\delta I_{22} + 2\mu\delta I_{33} = 0 \, . \tag{35}$$

[7] Similar considerations can be applied equally to the location of the maximum of Ω^2 along the Maclaurin sequence. The relevant analysis is included between equations (17) and (23) in Paper I; and this analysis must be supplemented by the corrigendum (Chandrasekhar 1963*b*) relative to the remark which follows equation (23).

Now, the first variations of the moment of inertia and the potential-energy tensors can all be expressed in terms of the symmetrized second-order virials

$$V_{ij} = \int_V \rho (\xi_i x_j + \xi_j x_i) d\mathbf{x}, \tag{36}$$

where $\boldsymbol{\xi}$ denotes the displacement in question. Thus

$$\delta I_{ij} = V_{ij} \tag{37}$$

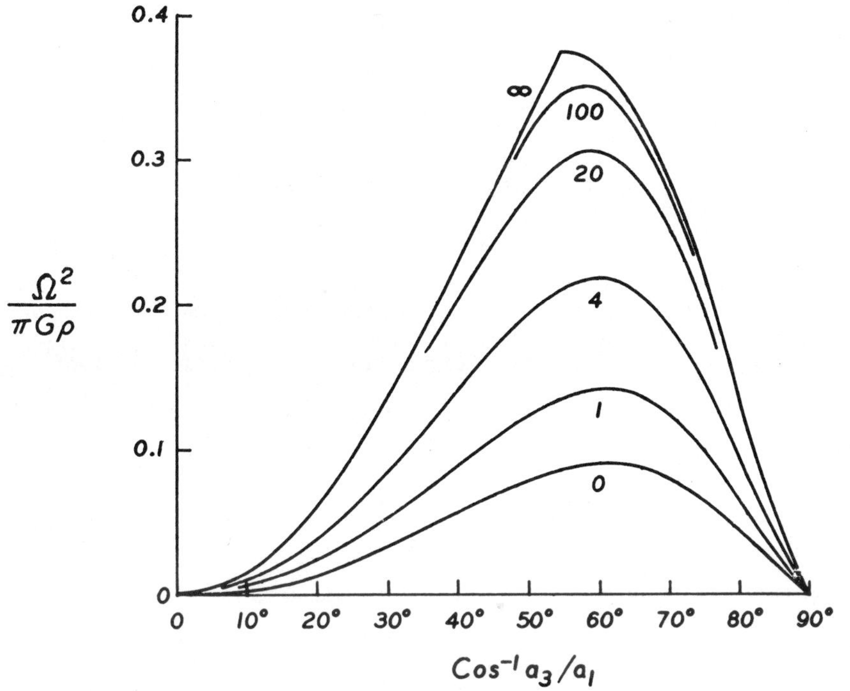

FIG. 2.—The variation of Ω^2 along the Roche sequences. The curves are labeled by the values of p to which they belong; and the curve labeled by ∞ belongs to the combined Maclaurin-Jacobi sequence. The maxima of the curves define the Roche limit.

and (cf. Paper III, eqs. [47] and [48])

$$\delta \mathfrak{W}_{ij} = -2B_{ij}V_{ij} \qquad (i \neq j) \tag{38}$$

and

$$\delta \mathfrak{W}_{ii} = -(2B_{ii} - a_i^2 A_{ii}) V_{ii} + a_i^2 \sum_{l \neq i} A_{il} V_{ll}, \tag{39}$$

where

$$B_{ij} = A_i - a_j^2 A_{ij} = A_j - a_i^2 A_{ij}, \tag{40}$$

(no summation over repeated indices in eqs. [38]–[40])

and A_i and A_{ij} are the one- and the two-index symbols defined in an earlier paper (Chandrasekhar and Lebovitz 1962*a*). (*Note that in writing eqs.* [38] *and* [39], *a common factor* $\pi G \rho a_1 a_2 a_3$ *has been suppressed.*)

From equations (38) and (39) we obtain, after some reductions in which use is made of equations (40),

$$\delta\mathfrak{W}_{11} - \delta\mathfrak{W}_{22} = -(3B_{11} - B_{12})V_{11} + (3B_{22} - B_{12})V_{22} + (B_{23} - B_{13})V_{33}\,, \tag{41}$$

and

$$\begin{aligned}\delta\mathfrak{W}_{11} + \delta\mathfrak{W}_{22} - 2\delta\mathfrak{W}_{33} = &-(3B_{11} + B_{12} - 2B_{13})V_{11} - (3B_{22} + B_{12} - 2B_{23})V_{22}\\ &+ (6B_{33} - B_{13} - B_{23})V_{33}\,.\end{aligned} \tag{42}$$

Now combining equations (34), (35), (37), (41), and (42), we obtain

$$\begin{aligned}[(\Omega^2 + 2\mu) - (3B_{11} - B_{12})]V_{11} - [(\Omega^2 - \mu) - (3B_{22} - B_{12})]V_{22}\\ + (B_{23} - B_{13})V_{33} = 0\end{aligned} \tag{43}$$

and

$$\begin{aligned}[(\Omega^2 + 2\mu) - (3B_{11} + B_{12} - 2B_{13})]V_{11} + [(\Omega^2 - \mu) - (3B_{22} + B_{12} - 2B_{23})]V_{22}\\ + [2\mu + (6B_{33} - B_{13} - B_{23})]V_{33} = 0\,.\end{aligned} \tag{44}$$

Since in writing the expressions for $\delta\mathfrak{W}_{ij}$ a common factor $\pi G\rho(a_1a_2a_3)$ was suppressed, it must be assumed that Ω^2 and μ in equations (43) and (44) (and in similar equations in the sequel) are divided out by the same factor.

To equations (43) and (44) we must adjoin the condition (cf. Lebovitz 1961, eq. [83])

$$\frac{V_{11}}{a_1^2} + \frac{V_{22}}{a_2^2} + \frac{V_{33}}{a_3^2} = 0 \tag{45}$$

which expresses the solenoidal requirement on $\boldsymbol{\xi}$.

The condition for the occurrence of a maximum for Ω^2 and μ is, then, the vanishing of the determinant of equations (43)–(45). We thus obtain

$$\begin{Vmatrix} \Omega^2 + 2\mu - (3B_{11} - B_{12}) & -(\Omega^2 - \mu) + (3B_{22} - B_{12}) & B_{23} - B_{13} \\ \Omega^2 + 2\mu - (3B_{11} + B_{12} - 2B_{13}) & +(\Omega^2 - \mu) - (3B_{22} + B_{12} - 2B_{23}) & 2\mu + 6B_{33} - B_{13} - B_{23} \\ \dfrac{1}{a_1^2} & \dfrac{1}{a_2^2} & \dfrac{1}{a_3^2} \end{Vmatrix} = 0\,. \tag{46}$$

By some elementary transformations, equation (46) can be brought to the somewhat simpler form

$$\begin{Vmatrix} \Omega^2 + 2\mu - (3B_{11} - B_{13}) & B_{23} - B_{12} & \mu + 3B_{33} - B_{13} \\ B_{13} - B_{12} & \Omega^2 - \mu - (3B_{22} - B_{23}) & \mu + 3B_{33} - B_{23} \\ \dfrac{1}{a_1^2} & \dfrac{1}{a_2^2} & \dfrac{1}{a_3^2} \end{Vmatrix} = 0\,. \tag{47}$$

By evaluating the determinant (47) along the different Roche sequences, the points listed in Table 2, where the maxima of Ω^2 are attained, were determined by interpolation. The maxima located in this manner agree exactly (as they should) with those determined by a direct interpolation among the values of Ω^2 given in Table 1.

In Figure 1 the locus of the points where the maxima of Ω^2 (and, or μ) are attained is shown by the dashed curve joining the points T_2 and M_2. This locus defines the Roche limit.

VI. THE SECOND-ORDER VIRIAL EQUATIONS GOVERNING SMALL OSCILLATIONS ABOUT EQUILIBRIUM

As stated in the introduction, it is generally believed that the Roche ellipsoids develop some sort of instability at the Roche limit where Ω^2 and μ attain their maxima. However, when we consider the following circumstances, it becomes clear that the matter requires a careful examination of the normal modes of oscillation of the Roche ellipsoids.

It is known (cf. Paper III, Sec. V) that the Jeans spheroid becomes unstable, by a mode of oscillation belonging to the second harmonics, when μ attains it maximum at T_2. And it is also known that the Maclaurin spheroid becomes neutral, with respect to the same mode of oscillation, at the point of bifurcation M_2; and that the Jacobi ellipsoid retains this neutrality along its entire sequence. The question which requires clarification

TABLE 2

THE ROCHE LIMIT AND THE CONSTANTS OF THE CRITICAL ELLIPSOID

(Ω^2 and μ Are Listed in the Unit $\pi G\rho$)

p	θ	$\Omega_{\max}^2$	$\mu_{\max}$	$\bar{a}_1$	$\bar{a}_2$	$\bar{a}_3$
− 1......	62°009	0	0.125536	1.6558	0.7771	0.7771
0......	61.156	0.090068	.090068	1.5947	0.8151	.7693
1......	60.638	.141322	.070661	1.5565	0.8418	.7632
4......	59.840	.216861	.043372	1.4944	0.8913	.7508
20......	58.646	.306396	.014590	1.3989	0.9821	.7279
100......	57.499	.350562	0.003471	1.3224	1.0642	.7106
∞.........	54.358	0.374230	0	1.1972	1.1972	0.6977

is: how does the mode, which for $p = -1$ is unstable beyond $\mu_{\max}$ at T_2, become neutral, beyond M_2, along the entire Jacobian part of the combined Maclaurin-Jacobi sequence, when $p \rightarrow \infty$? Clearly, this question cannot be fully answered without a detailed analysis of the normal modes of oscillation of the Roche ellipsoids belonging to the second harmonics. We shall now show how the required analysis can be carried out with the aid of the linearized form of the virial equation (10).

Suppose, then, that the equilibrium ellipsoid considered in Sections III–V is slightly perturbed; and, further, that the ensuing motions are described by a Lagrangian displacement of the form

$$\boldsymbol{\xi}(x)\, e^{\lambda t}\,, \tag{48}$$

where λ is a parameter whose characteristic values are to be determined. To the first order in $\boldsymbol{\xi}$, the virial equation (10) gives

$$\lambda^2 V_{i;j} - 2\lambda\Omega\epsilon_{il3}V_{l;j} = \delta\mathfrak{W}_{ij} + (\Omega^2 - \mu)\delta I_{ij} - \Omega^2\delta_{i3}\delta I_{3j} + 3\mu\delta_{i1}I_{1j} + \delta_{ij}\delta\Pi\,, \tag{49}$$

where

$$V_{i;j} = \int_V \rho\,\xi_i x_j dx \tag{50}$$

denotes the second-order (unsymmetrized) virial and $\delta\Pi$, $\delta\mathfrak{W}_{ij}$, and δI_{ij} are the first variations of Π, $\mathfrak{W}_{ij}$, and I_{ij} due to the deformation of the ellipsoid caused by the displacement $\boldsymbol{\xi}$.

We have already seen (cf. eqs. [37]–[39]) how the $\delta\mathfrak{W}_{ij}$'s and δI_{ij}'s can be expressed in terms of the symmetrized virials

$$V_{ij} = V_{i;j} + V_{j;i} . \tag{51}$$

In particular, replacing δI_{ij} by V_{ij} (in accordance with eq. [37]), we can rewrite equation (49) in the form

$$\lambda^2 V_{i;j} - 2\lambda\Omega\epsilon_{il3}V_{l;j} = \delta\mathfrak{W}_{ij} + (\Omega^2 - \mu)V_{ij} - \Omega^2\delta_{i3}V_{3j} + 3\mu\delta_{i1}V_{1j} + \delta_{ij}\delta\Pi . \tag{52}$$

Equation (52) represents a total of nine equations for the nine virials $V_{i;j}$. These nine equations fall into two non-combining groups of four and five equations, respectively, distinguished by their parity (i.e., oddness or evenness) with respect to the index 3. It is convenient to have these equations written out explicitly. The odd equations are

$$\lambda^2 V_{3;1} = \delta\mathfrak{W}_{31} - \mu V_{13} = -(2B_{13} + \mu)V_{13} , \tag{53}$$

$$\lambda^2 V_{3;2} = \delta\mathfrak{W}_{32} - \mu V_{23} = -(2B_{23} + \mu)V_{23} , \tag{54}$$

$$\lambda^2 V_{1;3} - 2\lambda\Omega V_{2;3} = \delta\mathfrak{W}_{13} + (\Omega^2 + 2\mu)V_{13} = -(2B_{13} - \Omega^2 - 2\mu)V_{13} , \tag{55}$$

$$\lambda^2 V_{2;3} + 2\lambda\Omega V_{1;3} = \delta\mathfrak{W}_{23} + (\Omega^2 - \mu)V_{23} = -(2B_{23} - \Omega^2 + \mu)V_{23} , \tag{56}$$

where we have substituted for the $\delta\mathfrak{W}_{ij}$'s in accordance with equation (38). And, similarly, the even equations are

$$\lambda^2 V_{3;3} = \delta\mathfrak{W}_{33} - \mu V_{33} + \delta\Pi , \tag{57}$$

$$\lambda^2 V_{1;1} - 2\lambda\Omega V_{2;1} = \delta\mathfrak{W}_{11} + (\Omega^2 + 2\mu)V_{11} + \delta\Pi , \tag{58}$$

$$\lambda^2 V_{2;2} + 2\lambda\Omega V_{1;2} = \delta\mathfrak{W}_{22} + (\Omega^2 - \mu)V_{22} + \delta\Pi , \tag{59}$$

$$\lambda^2 V_{1;2} - 2\lambda\Omega V_{2;2} = \delta\mathfrak{W}_{12} + (\Omega^2 + 2\mu)V_{12} = -(2B_{12} - \Omega^2 - 2\mu)V_{12} , \tag{60}$$

$$\lambda^2 V_{2;1} + 2\lambda\Omega V_{1;1} = \delta\mathfrak{W}_{21} + (\Omega^2 - \mu)V_{12} = -(2B_{12} - \Omega^2 + \mu)V_{12} . \tag{61}$$

VII. THE CHARACTERISTIC EQUATIONS

We shall now show how equations (53)–(61) can be used to determine the different characteristic frequencies of oscillation of the Roche ellipsoid belonging to the second harmonics.

a) The Characteristic Equation for the Odd Modes

Adding equations (53) and (55) and similarly equations (54) and (56), we obtain

$$(\lambda^2 + 4B_{13} - \Omega^2 - \mu)V_{13} - 2\lambda\Omega V_{23} + 2\lambda\Omega V_{3;2} = 0 \tag{62}$$

and

$$(\lambda^2 + 4B_{23} - \Omega^2 + 2\mu)V_{23} + 2\lambda\Omega V_{13} - 2\lambda\Omega V_{3;1} = 0 . \tag{63}$$

Eliminating $V_{3;1}$ and $V_{3;2}$ from the foregoing equations with the aid of equations (53) and (54), we have

$$\lambda(\lambda^2 + 4B_{13} - \Omega^2 - \mu)V_{13} - 2\Omega(\lambda^2 + 2B_{23} + \mu)V_{23} = 0 \tag{64}$$

and

$$\lambda(\lambda^2 + 4B_{23} - \Omega^2 + 2\mu)V_{23} + 2\Omega(\lambda^2 + 2B_{13} + \mu)V_{13} = 0 ; \tag{65}$$

and these two equations lead to the characteristic equation

$$\lambda^2(\lambda^2 + 4B_{13} - \Omega^2 - \mu)(\lambda^2 + 4B_{23} - \Omega^2 + 2\mu) + 4\Omega^2(\lambda^2 + 2B_{13} + \mu)(\lambda^2 + 2B_{23} + \mu) = 0 . \tag{66}$$

This is a cubic equation for λ^2; and it can be verified that, in the limits $\Omega^2 = 0$ and $\mu = 0$, the equation provides characteristic roots which are appropriate, respectively, for the Jeans spheroids and the Jacobi ellipsoids (or the Maclaurin spheroids if the indices 1 and 2 are not distinguished).

b) The Characteristic Equation for the Even Modes

Turning next to the even equations (57)–(61), we can combine them to give the following four equations in which $\delta\Pi$ no longer appears:

$$(\lambda^2 + 4B_{12} - 2\Omega^2 - \mu)V_{12} + \lambda\Omega(V_{11} - V_{22}) = 0 , \tag{67}$$

$$\lambda^2(V_{1;2} - V_{2;1}) = \lambda\Omega(V_{11} + V_{22}) + 3\mu V_{12} , \tag{68}$$

$$\tfrac{1}{2}\lambda^2(V_{11} - V_{22}) - 2\lambda\Omega V_{12} = \delta\mathfrak{W}_{11} - \delta\mathfrak{W}_{22} + (\Omega^2 + 2\mu)V_{11} - (\Omega^2 - \mu)V_{22} , \tag{69}$$

$$\tfrac{1}{2}\lambda^2(V_{11} + V_{22}) + 2\lambda\Omega(V_{1;2} - V_{2;1}) - \lambda^2 V_{33} = \delta\mathfrak{W}_{11} + \delta\mathfrak{W}_{22} - 2\delta\mathfrak{W}_{33} + (\Omega^2 + 2\mu)V_{11} + (\Omega^2 - \mu)V_{22} + 2\mu V_{33} . \tag{70}$$

Rearranging equation (69) and eliminating $(V_{1;2} - V_{2;1})$ from equation (70) with the aid of equation (68) (and rearranging), we obtain the pair of equations:

$$(\tfrac{1}{2}\lambda^2 - \Omega^2 - 2\mu)V_{11} - (\tfrac{1}{2}\lambda^2 - \Omega^2 + \mu)V_{22} - 2\lambda\Omega V_{12} = \delta\mathfrak{W}_{11} - \delta\mathfrak{W}_{22} \tag{71}$$

$$(\tfrac{1}{2}\lambda^2 + \Omega^2 - 2\mu)V_{11} + (\tfrac{1}{2}\lambda^2 + \Omega^2 + \mu)V_{22} - (\lambda^2 + 2\mu)V_{33} + \frac{6\Omega\mu}{\lambda}V_{12} = \delta\mathfrak{W}_{11} + \delta\mathfrak{W}_{22} - 2\delta\mathfrak{W}_{33} . \tag{72}$$

Now substituting for $\delta\mathfrak{W}_{11} - \delta\mathfrak{W}_{22}$ and $\delta\mathfrak{W}_{11} + \delta\mathfrak{W}_{22} - 2\delta\mathfrak{W}_{33}$ their expansions (41) and (42) in terms of the virials, and regrouping the terms, we find

$$(\tfrac{1}{2}\lambda^2 - \Omega^2 - 2\mu + 3B_{11} - B_{12})V_{11} - (\tfrac{1}{2}\lambda^2 - \Omega^2 + \mu + 3B_{22} - B_{12})V_{22} + (B_{13} - B_{23})V_{33} - 2\lambda\Omega V_{12} = 0 \tag{73}$$

and

$$(\tfrac{1}{2}\lambda^2 + \Omega^2 - 2\mu + 3B_{11} + B_{12} - 2B_{13})V_{11} + (\tfrac{1}{2}\lambda^2 + \Omega^2 + \mu + 3B_{22} + B_{12} - 2B_{23})V_{22} - (\lambda^2 + 2\mu + 6B_{33} - B_{13} - B_{23})V_{33} + \frac{6\Omega\mu}{\lambda}V_{12} = 0 . \tag{74}$$

Equations (67), (73), and (74) provide three relations among the four virials V_{11}, V_{12}, V_{22}, and V_{33}. A fourth relation is obtained by making use of the solenoidal character of the Lagrangian displacement. In the present context, the relation which expresses this requirement is (cf. Lebovitz 1961, eq. [83])

$$\frac{V_{11}}{a_1^2} + \frac{V_{22}}{a_2^2} + \frac{V_{33}}{a_3^2} = 0 . \tag{75}$$

The characteristic equation governing the even modes now follows from setting the determinant of equations (67) and (73)–(75) equal to zero. We thus obtain

$$\begin{Vmatrix} \frac{1}{2}\lambda^2-\Omega^2-2\mu+3B_{11}-B_{12} & -\frac{1}{2}\lambda^2+\Omega^2-\mu-3B_{22}+B_{12} & B_{13}-B_{23} & -2\lambda\Omega \\ \frac{1}{2}\lambda^2+\Omega^2-2\mu+3B_{11}+B_{12}-2B_{13} & +\frac{1}{2}\lambda^2+\Omega^2+\mu+3B_{22}+B_{12}-2B_{23} & -\lambda^2-2\mu-6B_{33}+B_{13}+B_{23} & 6\Omega\mu/\lambda \\ \lambda\Omega & -\lambda\Omega & 0 & \lambda^2+4B_{12}-2\Omega^2-\mu \\ \frac{1}{a_1^2} & \frac{1}{a_2^2} & \frac{1}{a_3^2} & 0 \end{Vmatrix} = 0\,. \tag{76}$$

After some elementary transformations, equation (76) can be brought to the following somewhat simpler form:

$$(\lambda^2+4B_{12}-2\Omega^2-\mu)\begin{Vmatrix} \frac{1}{2}\lambda^2-\Omega^2-2\mu+3B_{11}-B_{12} & -(B_{13}-B_{23}) & -3\mu+3(B_{11}-B_{22})+B_{13}-B_{23} \\ \Omega^2+B_{12}-B_{13} & \frac{1}{2}\lambda^2+\mu+3B_{33}-B_{23} & \Omega^2+3(B_{22}-B_{33})+B_{12}-B_{13} \\ \frac{1}{a_1^2} & -\frac{1}{a_3^2} & \frac{1}{a_1^2}+\frac{1}{a_2^2}+\frac{1}{a_3^2} \end{Vmatrix}$$
$$+\Omega^2\begin{Vmatrix} 2\lambda^2 & -(B_{13}-B_{23}) & -3\mu+3(B_{11}-B_{22})+B_{13}-B_{23} \\ -(\lambda^2+3\mu) & \frac{1}{2}\lambda^2+\mu+3B_{33}-B_{23} & \Omega^2+3(B_{22}-B_{33})+B_{12}-B_{13} \\ 0 & -\frac{1}{a_3^2} & \frac{1}{a_1^2}+\frac{1}{a_2^2}+\frac{1}{a_3^2} \end{Vmatrix} = 0\,. \tag{77}$$

And, again, it can be verified that, in the limits $\Omega^2 = 0$ and $\mu = 0$, equation (77) provides characteristic roots which are appropriate, respectively, for the Jeans spheroids and the Jacobi ellipsoids[8] (or the Maclaurin spheroids if the indices 1 and 2 are not distinguished).

[8] Since $\mu = 0$ and $\Omega^2 = 2B_{12}\pi G\rho a_1 a_2 a_3$ (after restoring the suppressed factor) for these ellipsoids, the occurrence of a neutral mode ($\lambda^2 = 0$) along the entire Jacobian sequence is manifest from equation (77).

VIII. THE CHARACTERISTIC FREQUENCIES OF OSCILLATION BELONGING TO THE SECOND HARMONICS: THE POINT AT WHICH INSTABILITY SETS IN ALONG THE ROCHE SEQUENCE

The characteristic equations (66) and (77) have been solved for the different Roche sequences for which the equilibrium properties have been tabulated in Section III; and the results are given in Table 3.

An examination of the roots listed in Table 3 shows that the Roche ellipsoids do become unstable by a mode of oscillation belonging to the second harmonics. Their instability arises, in fact, by the same mode by which the Jeans spheroid becomes unstable at $\mu_{\max}$ and for which the Jacobi ellipsoids are neutral along their entire sequence. Figure 3 exhibits the behavior of the corresponding characteristic root (σ_3^2) as we pass

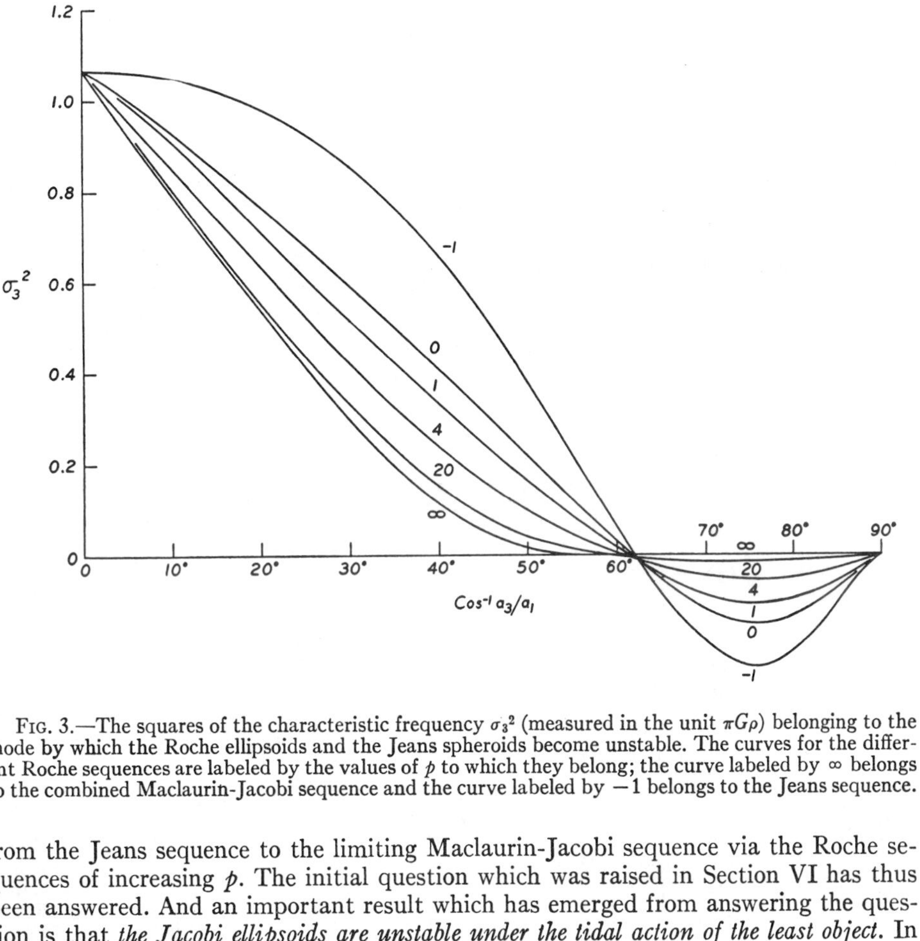

FIG. 3.—The squares of the characteristic frequency σ_3^2 (measured in the unit $\pi G\rho$) belonging to the mode by which the Roche ellipsoids and the Jeans spheroids become unstable. The curves for the different Roche sequences are labeled by the values of p to which they belong; the curve labeled by ∞ belongs to the combined Maclaurin-Jacobi sequence and the curve labeled by -1 belongs to the Jeans sequence.

from the Jeans sequence to the limiting Maclaurin-Jacobi sequence via the Roche sequences of increasing p. The initial question which was raised in Section VI has thus been answered. And an important result which has emerged from answering the question is that *the Jacobi ellipsoids are unstable under the tidal action of the least object.* In Section XII, we shall briefly consider the possible bearing of this result for cosmogony.

The points, beyond which the Roche ellipsoids are unstable, were determined by interpolation among the roots σ_3^2 listed in Table 3. Their positions, together with some additional information, are given in Table 4. In Figure 1 the locus of these points, at

TABLE 3

THE SQUARES OF THE CHARACTERISTIC FREQUENCIES BELONGING TO THE SECOND HARMONICS

(σ^2 Is Listed in the Unit $\pi G\rho$)

ϕ	Even Modes			Odd Modes		
	σ_1^2	σ_2^2	σ_3^2	σ_4^2	σ_5^2	σ_6^2
	$p=0$					
0°......	1.067	1.0667	+1.0667	1.067	1.0667	0
24......	1.323	1.1239	+0.7007	1.272	0.9370	0.0241
36......	1.429	1.1393	+0.4866	1.412	0.8718	.0551
48......	1.569	1.0445	+0.2574	1.577	0.7757	.0954
57......	1.697	0.8788	+0.0836	1.708	0.6640	.1234
60......	1.740	0.8070	+0.0285	1.752	0.6168	.1301
61......	1.754	0.7814	+0.0109	1.767	0.5997	.1318
62......	1.767	0.7552	−0.0063	1.781	0.5822	.1333
63......	1.783	0.7284	−0.0232	1.793	0.5642	.1344
66......	1.824	0.6442	−0.0695	1.833	0.5063	.1356
71......	1.886	0.4948	−0.1278	1.893	0.3987	.1284
72......	1.898	0.4640	−0.1357	1.903	0.3758	.1253
75......	1.928	0.3711	−0.1495	1.933	0.3049	.1125
79......	1.962	0.2482	−0.1416	1.963	0.2076	0.0865
	$p=1$					
0°......	1.067	1.0667	+1.0667	1.067	1.0667	0
12......	1.280	1.0333	+0.8740	1.181	0.9799	0.0094
24......	1.392	1.1398	+0.6294	1.313	0.9078	.0380
36......	1.494	1.1901	+0.4045	1.464	0.8440	.0854
48......	1.602	1.1410	+0.1952	1.625	0.7610	.1440
54......	1.673	1.0468	+0.1011	1.705	0.6994	.1713
59......	1.738	0.9309	+0.0303	1.770	0.6321	.1884
60......	1.752	0.9040	+0.0171	1.782	0.6166	.1909
61......	1.765	0.8760	+0.0044	1.795	0.6004	.1930
66......	1.830	0.7210	−0.0521	1.853	0.5088	.1965
72......	1.901	0.5127	−0.0976	1.914	0.3771	.1803
78......	1.955	0.3000	−0.1054	1.961	0.2298	.1354
81......	1.975	0.1999	−0.0905	1.978	0.1559	0.1017
	$p=4$					
0°......	1.067	1.0667	+1.0667	1.067	1.0667	0
12......	1.299	1.0886	+0.8070	1.206	0.9565	0.0146
24......	1.478	1.1560	+0.5453	1.368	0.8712	.0582
36......	1.583	1.2442	+0.3083	1.543	0.8050	.1275
48......	1.643	1.2650	+0.1241	1.708	0.7373	.2081
58......	1.736	1.0995	+0.0213	1.822	0.6467	.2576
60......	1.760	1.0430	+0.0059	1.841	0.6210	.2625
61......	1.772	1.0121	−0.0012	1.851	0.6069	.2641
72......	1.903	0.5882	−0.0532	1.936	0.3969	.2318
75......	1.932	0.4600	−0.0576	1.954	0.3241	.2042
80......	1.970	0.2560	−0.0520	1.977	0.1961	0.1388

TABLE 3—*Continued*

ϕ	Even Modes σ_1^2	σ_2^2	σ_3^2	Odd Modes σ_4^2	σ_5^2	σ_6^2
			$p=20$			
0°......	1.067	1.0667	+1.0667	1.067	1.0667	0
36.......	1.664	1.2959	+0.2204	1.637	0.7620	0.1714
48.......	1.656	1.4017	+0.0588	1.826	0.7038	.2738
58.......	1.712	1.2684	+0.0042	1.921	0.6452	.3242
60.......	1.739	1.2052	−0.0008	1.931	0.6275	.3253
66.......	1.827	0.9641	−0.0106	1.953	0.5528	.3063
72.......	1.903	0.6739	−0.0157	1.967	0.4403	.2533
75.......	1.933	0.5230	−0.0166	1.974	0.3699	0.2149
			$p=100$			
0°......	1.067	1.0667	+1.0667	1.067	1.0667	0
48.......	1.639	1.4624	+0.0332	1.886	0.6840	0.3027
52.......	1.605	1.4785	+0.0106	1.937	0.6666	.3349
56.......	1.631	1.4132	+0.0018	1.967	0.6500	.3532
58.......	1.665	1.3531	+0.0001	1.974	0.6401	.3549
60.......	1.709	1.2795	−0.0010	1.978	0.6274	.3519
70.......	1.876	0.8132	−0.0033	1.981	0.4996	.2804
75.......	1.933	0.5475	−0.0036	1.982	0.3892	.2146
83.......	1.986	0.1668	−0.0025	1.991	0.1592	0.0844

TABLE 4

The Point at Which Instability Sets in along the Roche Sequences

(Ω^2 and μ Are Listed in the Unit $\pi G\rho$)

p	θ	Ω^2	μ	$\tilde{a}_1$	$\tilde{a}_2$	$\tilde{a}_3$
− 1......	62°009	0	0.12554	1.656	0.7771	0.7771
0......	61.63	0.090034	.090034	1.611	0.8108	.7655
1......	61.35	.141229	.070614	1.581	0.8350	.7577
4......	60.83	.216581	.043316	1.526	0.8808	.7439
20......	59.66	.305937	.014568	1.430	0.9683	.7222
100......	58.18	.350303	0.003468	1.343	1.0523	.7078
∞.........	54.358	0.374230	0	1.197	1.1972	0.6977

which instability sets in along the Roche sequences, is the full-line curve joining T_2 and M_2. It is at once apparent that the Roche ellipsoids do not become unstable at the Roche limit but at a subsequent point.

That the two points, the point where Ω^2 and μ attain their maxima and the point where instability sets in, are distinct, follows from a comparison of equation (47), which determines the former point, and the equation

$$(4B_{12}-2\Omega^2-\mu)\begin{Vmatrix} -\Omega^2-2\mu+3B_{11}-B_{12} & -(B_{13}-B_{23}) & -3\mu+3(B_{11}-B_{22})+B_{13}+B_{23} \\ \Omega^2+B_{12}-B_{13} & \mu+3B_{33}-B_{23} & \Omega^2+3(B_{22}-B_{33})+B_{12}-B_{13} \\ \frac{1}{a_1^2} & -\frac{1}{a_3^2} & \frac{1}{a_1^2}+\frac{1}{a_2^2}+\frac{1}{a_3^2} \end{Vmatrix}$$

$$-3\mu\Omega^2\begin{Vmatrix} B_{13}-B_{23} & -3\mu+3(B_{11}-B_{22})+B_{13}-B_{23} \\ \frac{1}{a_3^2} & \frac{1}{a_1^2}+\frac{1}{a_2^2}+\frac{1}{a_3^2} \end{Vmatrix}=0, \tag{78}$$

which determines the latter point. That the two equations are necessarily distinct can be seen more directly from a comparison of equations (34) and (35) and the limiting forms of equations (70) and (71) for $\lambda = 0$. Thus while equation (71), in the limit $\lambda = 0$, is the same as equation (34), this is not the case with respect to equations (70) and (35). The difference in the latter case arises from the circumstance that, by virtue of equations (67) and (68), the additional term in equation (70), namely,

$$\begin{aligned} 2\lambda\Omega(V_{1;2}-V_{2;1}) &= 2\Omega^2(V_{11}+V_{22})+\frac{6\Omega\mu}{\lambda}V_{12} \\ &= 2\Omega^2(V_{11}+V_{22})-6\Omega^2\mu\frac{V_{11}-V_{22}}{\lambda^2+4B_{12}-2\Omega^2-\mu} \end{aligned} \tag{79}$$

does *not* tend to zero as $\lambda \to 0$; it tends, instead, to the finite limit:

$$\lim_{\lambda\to 0} 2\lambda\Omega(V_{1;2}-V_{2;1}) = 2\Omega^2\left(V_{11}+V_{22}-3\mu\frac{V_{11}-V_{22}}{4B_{12}-2\Omega^2-\mu}\right)\neq 0. \tag{80}$$ [9]

And this circumstance cannot have been foreseen by any simple consideration based on equilibrium only. The basic reason for the occurrence of the neutral point beyond the Roche limit is the presence of the Coriolis term in the equations of motion. The matter is further clarified by Lebovitz (1963) in the paper following this one.

And finally it should be noted that the fact that the neutral point does not coincide with the Roche limit is in agreement with a theorem due to Karl Schwarzschild (1898) that *along a Roche sequence there is no point of bifurcation where a second ellipsoidal sequence branches off.*

In Section XII we shall consider the bearing of the foregoing results for a proper interpretation of the Roche limit.

[9] However, the limit *is* zero for the Jeans sequence (for which $\Omega^2 = 0$); and this is in agreement with the fact that along this sequence the point where μ attains its maximum coincides with the point where instability sets in. And along the Maclaurin sequence (for which $\mu = 0$) the limit is also zero *if* $V_{11} + V_{22} = 0$, a condition which is necessary to *exclude* the point where Ω^2 attains its maximum (at M_{max}) and *determine* the point of bifurcation (at M^2) (cf. Chandrasekhar 1963*b* and n. 7 on p. 1190).

IX. THE NEUTRAL POINT ALONG THE ROCHE SEQUENCE BELONGING TO THE THIRD HARMONICS

It is known that along the Jacobian sequence a point of bifurcation occurs where the Jacobi ellipsoid becomes unstable by a mode of oscillation belonging to the third harmonics; this is the point J_3 in Figure 1. It is also known that along the Jeans sequence a second neutral point belonging also to the third harmonics, occurs; this is the point T_3 in Figure 1. The occurrence of these two neutral points, J_3 and T_3, suggests that a similar neutral point occurs along each Roche sequence. Such a neutral point, if one such exists, can be exhibited and isolated by the method which has been described in the contexts of the Jacobian and the Jeans sequences (Paper I, Secs. IV and VI). The method is based on the integral properties which follow, *as identities*, from the relevant third-order virial equation. Accordingly, we shall first derive the analogous properties appropriate for configurations which are both rotationally and tidally distorted.

a) *The Third-Order Virial Equations and the Integral Properties Governing Equilibrium*

By multiplying equation (9) by $x_j x_k$ and integrating over the volume V occupied by the fluid, we obtain in the usual manner the required third-order virial equation (cf. Chandrasekhar 1962, Sec. IV, and Paper I, eq. [88]):

$$\frac{d}{dt}\int_V \rho u_i x_j x_k d\boldsymbol{x} = 2(\mathfrak{T}_{ij;k} + \mathfrak{T}_{ik\cdot j}) + \mathfrak{W}_{ij;k} + \mathfrak{W}_{ik\cdot j}$$

$$+ (\Omega^2 - \mu) I_{ijk} - \Omega^2 \delta_{i3} I_{3jk} + 3\mu\delta_{i1} I_{1jk} + \delta_{ij}\Pi_k + \delta_{ik}\Pi_j \tag{81}$$

$$+ 2\Omega \int_V \rho \epsilon_{il3} u_l x_j x_k d\boldsymbol{x} \ ,$$

where

$$\mathfrak{T}_{ij;k} = \tfrac{1}{2}\int_V \rho u_i u_j x_k d\boldsymbol{x} \ , \qquad \mathfrak{W}_{ij\cdot k} = -\tfrac{1}{2}\int_V \rho \mathfrak{B}_{ij} x_k d\boldsymbol{x} \ ,$$

$$I_{ijk} = \int_V \rho x_i x_j x_k d\boldsymbol{x} \ , \qquad \text{and} \qquad \Pi_k = \int_V p x_k d\boldsymbol{x} \ . \tag{82}$$

When no relative motions are present in the frame of reference considered, and hydrostatic equilibrium prevails, equation (81) becomes

$$\mathfrak{W}_{ij;k} + \mathfrak{W}_{ik;j} + (\Omega^2 - \mu) I_{ijk} - \Omega^2 \delta_{i3} I_{3jk} + 3\mu\delta_{i1} I_{1jk} = -\delta_{ij}\Pi_k - \delta_{ik}\Pi_j \ . \tag{83}$$

Equation (83) represents a total of eighteen equations. These eighteen equations fall into four groups distinguished by their parity with respect to the indices 1, 2, and 3: a group of three equations which are odd in all three indices, and three groups of five equations each which are odd with respect to one of the three indices and even with respect to the remaining two. The four groups of equations are

$$\mathfrak{W}_{12;3} + \mathfrak{W}_{31;2} + (\Omega^2 + 2\mu) I_{123} = 0 \ ,$$

$$\mathfrak{W}_{23;1} + \mathfrak{W}_{12;3} + (\Omega^2 - \mu) I_{123} = 0 \ , \tag{84}$$

$$\mathfrak{W}_{31;2} + \mathfrak{W}_{23;1} - \mu I_{123} = 0 \ ;$$

$$
\begin{aligned}
&2\mathfrak{W}_{11;1} + (\Omega^2 + 2\mu)I_{111} = -2\Pi_1 , \\
&\mathfrak{W}_{22;1} + \mathfrak{W}_{21;2} + (\Omega^2 - \mu)I_{221} = -\Pi_1 , \\
&\mathfrak{W}_{33;1} + \mathfrak{W}_{31;3} - \mu I_{331} = -\Pi_1 , \\
&2\mathfrak{W}_{12;2} + (\Omega^2 + 2\mu)I_{122} = 2\mathfrak{W}_{13;3} + (\Omega^2 + 2\mu)I_{133} = 0 ;
\end{aligned} \tag{85}
$$

$$
\begin{aligned}
&2\mathfrak{W}_{22;2} + (\Omega^2 - \mu)I_{222} = -2\Pi_2 , \\
&\mathfrak{W}_{11;2} + \mathfrak{W}_{12;1} + (\Omega^2 + 2\mu)I_{112} = -\Pi_2 , \\
&\mathfrak{W}_{33;2} + \mathfrak{W}_{32;3} - \mu I_{332} = -\Pi_2 , \\
&2\mathfrak{W}_{21;1} + (\Omega^2 - \mu)I_{211} = 2\mathfrak{W}_{23;3} + (\Omega^2 - \mu)I_{233} = 0 ;
\end{aligned} \tag{86}
$$

and

$$
\begin{aligned}
&2\mathfrak{W}_{33;3} - \mu I_{333} = -2\Pi_3 , \\
&\mathfrak{W}_{22;3} + \mathfrak{W}_{23;2} + (\Omega^2 - \mu)I_{223} = -\Pi_3 , \\
&\mathfrak{W}_{11;3} + \mathfrak{W}_{13;1} + (\Omega^2 + 2\mu)I_{113} = -\Pi_3 , \\
&2\mathfrak{W}_{31;1} - \mu I_{311} = 2\mathfrak{W}_{32;2} - \mu I_{322} = 0 .
\end{aligned} \tag{87}
$$

After the elimination of the Π_k's from the foregoing equations we shall be left with a total of fifteen equations; and by suitably combining them, we obtain the following four groups of equations:

A:

$$\mathfrak{W}_{23;1} = -2\mathfrak{W}_{31;2} = 2\mu I_{123}, \quad \mathfrak{W}_{12;3} = -(\Omega^2 + \mu)I_{123} ; \tag{88}$$

B:

$$
\begin{aligned}
&2\mathfrak{W}_{12;2} + (\Omega^2 + 2\mu)I_{122} = 2\mathfrak{W}_{13;3} + (\Omega^2 + 2\mu)I_{133} = 0 , \\
&S_{122} + (\Omega^2 + 2\mu)I_{111} - 3\Omega^2 I_{122} = 0 , \\
&S_{133} + (\Omega^2 + 2\mu)I_{111} - \Omega^2 I_{133} = 0 ;
\end{aligned} \tag{89}
$$

C:

$$
\begin{aligned}
&2\mathfrak{W}_{21;1} + (\Omega^2 - \mu)I_{211} = 2\mathfrak{W}_{23;3} + (\Omega^2 - \mu)I_{233} = 0 , \\
&S_{211} + (\Omega^2 - \mu)I_{222} - 3(\Omega^2 + \mu)I_{211} = 0 , \\
&S_{233} + (\Omega^2 - \mu)I_{222} - (\Omega^2 - 3\mu)I_{233} = 0 ;
\end{aligned} \tag{90}
$$

D:

$$
\begin{aligned}
&2\mathfrak{W}_{31;1} - \mu I_{311} = 2\mathfrak{W}_{32;2} - \mu I_{322} = 0 , \\
&S_{322} - \mu I_{333} - (2\Omega^2 - 3\mu)I_{322} = 0 , \\
&S_{311} - \mu I_{333} - (2\Omega^2 + 3\mu)I_{311} = 0 ,
\end{aligned} \tag{91}
$$

where

$$S_{ijj} = -4\mathfrak{W}_{ij;j} - 2\mathfrak{W}_{jj;i} + 2\mathfrak{W}_{ii;i} . \tag{92}$$

(no summation over repeated indices)

b) The Neutral Point along the Roche Sequence Belonging to the Third Harmonics

Now a necessary condition for the occurrence of a neutral point (belonging to the third harmonics) is that a non-trivial Lagrangian displacement exists such that the first variations of all of the equations in the four groups A, B, C, and D vanish at that point.

It has been shown in an earlier paper (Chandrasekhar and Lebovitz 1963*a*, Sec. V;

this paper will be referred to hereafter as "Paper II") that the first variations of all the quantities which occur in equations (88)–(91) are expressible, linearly, in terms of the symmetrized third-order virials:

$$V_{ijk} = \int_V \rho\,(\xi_i x'_j x'_k + \xi_j x'_k x'_i + \xi_k x'_i x'_j)\,dx = \delta I_{ijk}. \tag{93}$$

Moreover, the equations derived (by first variation) from the different groups (A, B, C, and D) involve different virials and are mutually exclusive. Thus, if δA, δB, δC, and δD denote the equations which are obtained by taking the first variations of the equations in the respective groups, then the association of the groups and the virials is the following:

$$\begin{aligned} &\delta\text{A}: V_{123}\,; &&\delta\text{B}: V_{111},\ V_{122},\ V_{133}\,;\\ &\delta\text{C}: V_{222},\ V_{233},\ V_{211}\,; &&\delta\text{D}: V_{333},\ V_{311},\ V_{322}\,. \end{aligned} \tag{94}$$

The coefficients of the virials in the expansions of the $\delta\mathfrak{W}_{ij;k}$'s and δS_{ijj}'s which occur in the varied form of equations (88)–(91) are tabulated in Table 2, Paper II.

We readily verify that the equations in the groups δA, δC, and δD do not allow any non-trivial solutions. Therefore, at a neutral point, we must necessarily have

$$V_{123} = V_{222} = V_{233} = V_{211} = V_{333} = V_{311} = V_{322} = 0\,. \tag{95}$$

In this respect the situation is the same as in the cases of the Jacobi ellipsoids and the Jeans spheroids (cf. Paper I, eqs. [39], [40], and [109]).

The occurrence of a second neutral point now depends on whether the remaining group δB allows a non-trivial solution. The equations to be considered are

$$\delta J_1 = -2\delta\mathfrak{W}_{12;2} - (\Omega^2 + 2\mu)V_{122} = 0\,, \tag{96}$$

$$\delta J_2 = -2\delta\mathfrak{W}_{13;3} - (\Omega^2 + 2\mu)V_{133} = 0\,, \tag{97}$$

$$\delta J_3 = \delta S_{122} + (\Omega^2 + 2\mu)V_{111} - 3\Omega^2 V_{122} = 0\,, \tag{98}$$

and

$$\delta J_4 = \delta S_{133} + (\Omega^2 + 2\mu)V_{111} - \Omega^2 V_{133} = 0\,. \tag{99}$$

It will be observed that these equations represent simple generalizations of the ones considered in the contexts of the Jacobi ellipsoids and the Jeans spheroids (Paper I, eqs. [43]–[46] and [110]–[113]).

Since the $\delta\mathfrak{W}_{ij;k}$'s and δS_{ijj}'s which occur in equations (96)–(99) are expressible as linear combinations of V_{111}, V_{122}, and V_{133}, we may write

$$\delta J_i = \langle i|111\rangle V_{111} + \langle i|122\rangle V_{122} + \langle i|133\rangle V_{133} = 0 \quad (i = 1, 2, 3, 4)\,, \tag{100}$$

where $\langle i|111\rangle$, etc., are certain matrix elements which are known.

If we should now require that the Lagrangian displacement be also solenoidal, then we should supplement equation (100) by the condition (see Paper II, Sec. VII)

$$\frac{V_{111}}{a_1^2} + \frac{V_{122}}{a_2^2} + \frac{V_{133}}{a_3^2} = 0\,. \tag{101}$$

The existence of a non-trivial neutral mode belonging to the third harmonics requires, therefore, that for some member of the Roche sequence considered, the 5×3 matrix representing equations (100) and (101) is, at most, of rank 2.

Using the known expressions for the matrix elements $\langle i|111\rangle$, etc., we find that the 5×3 matrix whose rank we have to consider is given in equation (102).

$$\left|\begin{matrix} a_2^2 B_{112} & 2B_{12} + 3a_2^2 B_{122} - (\Omega^2 + 2\mu) & a_2^2 B_{123} \\ a_3^2 B_{113} & a_3^2 B_{123} & 2B_{13} + 3a_3^2 B_{133} - (\Omega^2 + 2\mu) \\ (a_2^2 + 2a_1^2) B_{112} - 5a_1^2 B_{111} - 2B_{11} + \Omega^2 + 2\mu & 3[B_{22} + B_{12} + (2a_2^2 + a_1^2) B_{122} - a_1^2 B_{112} - \Omega^2] & (a_2^2 + 2a_1^2) B_{123} - 3a_1^2 B_{113} \\ (a_3^2 + 2a_1^2) B_{113} - 5a_1^2 B_{111} - 2B_{11} + \Omega^2 + 2\mu & (a_3^2 + 2a_1^2) B_{123} - 3a_1^2 B_{112} & 3[B_{33} + B_{13} + (2a_3^2 + a_1^2) B_{133} - a_1^2 B_{113}] - \Omega^2 \\ \dfrac{1}{a_1^2} & \dfrac{1}{a_2^2} & \dfrac{1}{a_3^2} \end{matrix}\right| . \tag{102}$$

By evaluating (along a Roche sequence) the determinants of the six different sets of equations which we can form by considering equation (101) and two of the four equations (100), we find that all six determinants vanish, simultaneously, at a determinate point. This verifies that the rank of the matrix (102) does become two for a particular member of the sequence. A neutral point of the kind sought, therefore, exists. And the remarks made in Paper I (p. 1195), in the context of analogous considerations relative to the Jacobian sequence, apply, equally, to the Roche sequences.

It is a relatively simple matter to isolate the neutral point along a Roche sequence with the aid of the requirement on the rank of the matrix (102); and the results given in Table 5 were obtained with that aid. And in Figure 1, the corresponding locus of the neutral point is the curve joining T_3 and J_3.

X. SUMMARY OF THE PRINCIPAL RESULTS

Returning to Figure 1, we shall now recapitulate the principal results pertaining to the equilibrium and the stability of the configurations along the Maclaurin, the Jacobi, the Jeans, and the Roche sequences. All these configurations are ellipsoids and their structures are uniquely determined by the virial equations

$$\mathfrak{W}_{11} + (\Omega^2 + 2\mu)I_{11} = \mathfrak{W}_{22} + (\Omega^2 - \mu)I_{22} = \mathfrak{W}_{33} - \mu I_{33} ; \tag{103}$$

and their characteristic frequencies of oscillation, belonging to the second and the third harmonics, can be ascertained with the aid of the linearized forms of the second- and the third-order virial equations.

In the plane of Figure 1, an ellipsoid (or a spheroid) of equilibrium is represented by the normalized values,

$$\bar{a}_1 = \frac{a_1}{(a_1a_2a_3)^{1/3}} \quad \text{and} \quad \bar{a}_2 = \frac{a_2}{(a_1a_2a_3)^{1/3}}, \tag{104}$$

of two of its principal axes.

TABLE 5

THE NEUTRAL POINT ALONG THE ROCHE SEQUENCES BELONGING TO THE THIRD HARMONICS

(Ω^2 and μ Are Listed in the Unit $\pi G\rho$)

p	θ	Ω^2	μ	$\bar{a}_1$	$\bar{a}_2$	$\bar{a}_3$
− 1	71°395	0	0.1091	2.142	0.6833	0.6833
0	70°98	0.07754	.07754	2.080	.7091	.6780
1	70.74	.12073	.06036	2.044	.7258	.6741
4	70.40	.1825	.03649	1.989	.7535	.6671
20	70.02	.2496	0.01188	1.923	.7912	.6572
100	69.86	....	....	....	....	....
∞	69.817	0.2840	0	1.886	0.8150	0.6507

The undistorted sphere is represented by the point

$$S(\bar{a}_1 = \bar{a}_2 = \bar{a}_3 = 1) ; \tag{105}$$

and the Maclaurin spheroids by the line

$$SM_{3''}: \bar{a}_1 = \bar{a}_2(\geq 1) . \tag{106}$$

Along the Maclaurin sequence, the first point of bifurcation (where the Jacobian sequence branches off) occurs at

$$M_2(e = 0.81267, \bar{a}_1 = \bar{a}_2 = 1.1972, \bar{a}_3 = 0.69766) . \tag{107}$$

At this point, the Maclaurin spheroid becomes unstable if any dissipative mechanism is operative (cf. Roberts and Stewartson 1963). In the absence of such mechanisms, M_2 is a neutral point of the second kind (i.e., characterized by stability on either side of the point). In all events, the Maclaurin spheroid becomes unstable at

$$O_2(e = 0.95289, \bar{a}_1 = \bar{a}_2 = 1.4883, \bar{a}_3 = 0.45145) , \tag{108}$$

by a mode[10] of overstable oscillations belonging to the second harmonics. At

$$M_{\max}(e = 0.929955, \bar{a}_1 = \bar{a}_2 = 1.3959, \bar{a}_3 = 0.51322) , \tag{109}$$

[10] This is the same mode which is neutral at M_2.

intermediate between M_2 and O_2, Ω^2 attains its maximum value ($0.449332\ \pi G\rho$) along this sequence. At

$$M_{3'}(e = 0.89926,\ \bar{a}_1 = \bar{a}_2 = 1.3174,\ \bar{a}_3 = 0.57623) \tag{110}$$

and

$$M_{3''}(e = 0.96937,\ \bar{a}_1 = \bar{a}_2 = 1.5968,\ \bar{a}_3 = 0.39217) \tag{111}$$

occur two further neutral points of the second kind (belonging to the third harmonics). At each of these points, a (different) sequence of pear-shaped configurations (presumably) branches off. And finally at

$$O_3(e = 0.96696,\ \bar{a}_1 = \bar{a}_2 = 1.5771,\ \bar{a}_3 = 0.40205) \tag{112}$$

the mode, which is neutral at $M_{3'}$, becomes unstable by overstable oscillations.

The Jacobian sequence M_2J, which branches off from the Maclaurin sequence at M_2, becomes unstable at the point

$$J_3(\bar{a}_1 = 1.8858,\ \bar{a}_2 = 0.81498,\ \bar{a}_3 = 0.65066)\ , \tag{113}$$

where the sequence of the pear-shaped configurations branches off; the analogous point on the Maclaurin sequence is $M_{3''}$. (The point $M_{3'}$ has no analogue on the Jacobian sequence.) While the Jacobi ellipsoid becomes strictly unstable only at J_3, it is, nevertheless, characterized by a non-trivial neutral mode of oscillation along its entire sequence; the mode in question is, in fact, the one which becomes neutral at M_2.

It should also be noted that along the combined Maclaurin-Jacobi sequence, SM_2J, Ω^2 has its maximum ($=0.374230\ \pi G\rho$) at M_2.

The Jeans sequence of the pure tidally distorted prolate spheroids is represented by the "hyperbola"

$$ST\!: \bar{a}_1\bar{a}_2^2 = 1\ . \tag{114}$$

Along this sequence we have the two neutral points,

$$T_2(e = 0.88303,\ \bar{a}_1 = 1.6558,\ \bar{a}_2 = \bar{a}_3 = 0.77712) \tag{115}$$

and

$$T_3(e = 0.94774,\ \bar{a}_1 = 2.1417,\ \bar{a}_2 = \bar{a}_3 = 0.68331)\ , \tag{116}$$

where the Jeans spheroid becomes unstable by modes of oscillation belonging to the second and the third harmonics, respectively. Moreover, μ attains its maximum value ($0.125536\ \pi G\rho$) along this sequence at T_2.

The Roche sequences (belonging to the different p's) fill the domain bounded by SM_2, M_2J, and SR_0, the Roche sequence for $p = 0$. They are represented by continuous curves which start at S and eventually become asymptotic to ST. Along each Roche sequence, Ω^2 and, simultaneously, μ attain maxima. The locus of these maxima is the dotted curve joining M_2 and T_2. This locus defines the Roche limit. But the Roche limit does not limit the stability of the Roche ellipsoids. The Roche ellipsoid actually becomes unstable at a somewhat later point by a mode of oscillation belonging to the second harmonics; and the locus which limits the stability of the Roche ellipsoids is the other heavy curve joining M_2 and T_2. And finally, along each Roche sequence a neutral point occurs where instability, by a mode of oscillation belonging to the third harmonics, sets in; the locus of this neutral point is the curve joining T_3 and J_3.

In the limit $p \to \infty$, the Roche sequence tends to the combined Maclaurin-Jacobi sequence represented by the broken curve SM_2J. In this limit, we must regard the *entire* Jacobian part of the combined Maclaurin-Jacobi sequence as unstable. Stated less abstractly, the conclusion to be drawn is that the Jacobi ellipsoids are unstable under the least tidal action: they are unstable in the presence of a fly!

XI. THE EFFECT OF COMPRESSIBILITY ON THE STABILITY OF THE ROCHE ELLIPSOIDS

The analysis in Section VII can be readily extended to determine the effect of compressibility on the stability of the Roche ellipsoids. Specifically, the problem to be considered is that of the adiabatic oscillations of rotationally and tidally distorted homogeneous configurations. The assumption of *homogeneity* insures that in the equilibrium state the configurations will be indistinguishable from the incompressible Roche ellipsoids considered in Section III. But the present assumption, that the configurations are *gaseous*, has the consequence that the Lagrangian displacement describing the deformation can no longer be restricted to be solenoidal; instead, we must apply the laws appropriate to a gas subject to adiabatic changes. If the gas is assumed to have a ratio of specific heats γ, then the condition div $\xi = 0$ must be replaced by the condition

$$\frac{\delta p}{p} = \gamma \frac{\delta \rho}{\rho} = -\gamma \operatorname{div} \xi . \qquad (117)$$

These relations will enable us to express the first variation $\delta\Pi$ $(=\delta \int p dx)$ in terms of ξ; we have

$$\delta\Pi = (\gamma - 1) \int_V \xi \cdot \operatorname{grad} p dx , \qquad (118)$$

where p denotes the pressure in the equilibrium state.

For the case under consideration

$$\operatorname{grad} p = \rho \operatorname{grad} \left[I - \sum_{i=1}^{3} A_i x_i^2 + \tfrac{1}{2}\Omega^2 (x_1^2 + x_2^2) + \mu (x_1^2 - \tfrac{1}{2} x_2^2 - \tfrac{1}{2} x_3^2) \right] ; \qquad (119)$$

and equation (118) gives

$$\delta\Pi = -q[(A_1 - \mu - \tfrac{1}{2}\Omega^2)V_{11} + (A_2 + \tfrac{1}{2}\mu - \tfrac{1}{2}\Omega^2)V_{22} + (A_3 + \tfrac{1}{2}\mu)V_{33}] , \qquad (120)$$

where, for the sake of brevity, we have written

$$q = \gamma - 1 . \qquad (121)$$

In equation (120) a factor $\pi G \rho a_1 a_2 a_3$ (by which *every* quantity must be divided out) has been suppressed.

a) The Characteristic Equation for the Even Modes Which Are Affected by Compressibility

Turning to the problem of determining the characteristic frequencies of oscillation under the present more general circumstances, we first observe that the modes which we have designated as odd are unaffected: for, in deriving the corresponding characteristic equation (66), we made no use of the solenoidal condition (75). However, the modes which we have designated as even are affected; and the place where the analysis needs to be changed is clear: instead of supplementing equations (67), (73), and (74) by the solenoidal condition (as we did in Section VII*b*), we must now use one of the original equations (57)–(59) and assign to $\delta\Pi$ its present value (120). Thus, choosing equation (57) as the fourth equation, we have

$$\begin{aligned}(\tfrac{1}{2}\lambda^2 + \mu)V_{33} &- \delta\mathfrak{W}_{33} \\ &+ q[(A_1 - \mu - \tfrac{1}{2}\Omega^2)V_{11} + (A_2 + \tfrac{1}{2}\mu - \tfrac{1}{2}\Omega^2)V_{22} + (A_3 + \tfrac{1}{2}\mu)V_{33}] = 0 ;\end{aligned} \qquad (122)$$

or substituting for $\delta\mathfrak{W}_{33}$ in accordance with equation (39) and regrouping the terms, we have

$$\begin{aligned}&[q(A_1 - \mu - \tfrac{1}{2}\Omega^2) - a_3^2 A_{31}]V_{11} + [q(A_2 + \tfrac{1}{2}\mu - \tfrac{1}{2}\Omega^2) - a_3^2 A_{32}]V_{22} \\ &+ [\tfrac{1}{2}\lambda^2 + \mu + 2B_{33} - a_3^2 A_{33} + q(A_3 + \tfrac{1}{2}\mu)]V_{33} = 0 .\end{aligned} \qquad (123)$$

Equations (67), (73), (74), and (123) now lead to the characteristic equation[11] (cf. eq. [76])

[11] We may, once again, draw attention to the fact that *every* quantity in these and similar equations must be considered as having been divided out by $\pi G \rho a_1 a_2 a_3$.

$$\left\| \begin{matrix} \frac{1}{2}\lambda^2 - \Omega^2 - 2\mu + 3B_{11} - B_{12} & -\frac{1}{2}\lambda^2 + \Omega^2 - \mu - 3B_{22} + B_{12} & B_{13} - B_{23} & -2\lambda\Omega \\ \frac{1}{2}\lambda^2 + \Omega^2 - 2\mu + 3B_{11} + B_{12} - 2B_{13} & +\frac{1}{2}\lambda^2 + \Omega^2 + \mu + 3B_{22} + B_{12} - 2B_{23} & -\lambda^2 - 2\mu - 6B_{33} + B_{13} + B_{23} & 6\Omega\mu/\lambda \\ \lambda\Omega & -\lambda\Omega & 0 & \lambda^2 + 4B_{12} - 2\Omega^2 - \mu \\ q(A_1 - \mu - \frac{1}{2}\Omega^2) - a_3^2 A_{31} & q(A_2 + \frac{1}{2}\mu - \frac{1}{2}\Omega^2) - a_3^2 A_{32} & \frac{1}{2}\lambda^2 + \mu + 2B_{33} - a_3^2 A_{33} + q(A_3 + \frac{1}{2}\mu) & 0 \end{matrix} \right\| = 0 . \tag{124}$$

This secular equation can be simplified in the same manner as equation (76); and we find (cf. eq. [77])

$$(\lambda^2 + 4B_{12} - 2\Omega^2 - \mu) \left\| \begin{matrix} \frac{1}{2}\lambda^2 - \Omega^2 - 2\mu + 3B_{11} - B_{12} & -(B_{13} - B_{23}) & -3\mu + 3(B_{11} - B_{22}) + B_{13} - B_{23} \\ \Omega^2 + B_{12} - B_{13} & \frac{1}{2}\lambda^2 + \mu + 3B_{33} - B_{23} & \Omega^2 + 3(B_{22} - B_{33}) + B_{12} - B_{13} \\ q(A_1 - \mu - \frac{1}{2}\Omega^2) - a_3^2 A_{31} & -\frac{1}{2}\lambda^2 - \mu - 2B_{33} + a_3^2 A_{33} - q(A_3 + \frac{1}{2}\mu) & \frac{1}{2}\lambda^2 + \mu + 2A_3 - 2 + q(2 - \Omega^2) \end{matrix} \right\|$$
$$+ \Omega^2 \left\| \begin{matrix} 2\lambda^2 & -(B_{13} - B_{23}) & -3\mu + 3(B_{11} - B_{22}) + B_{13} - B_{23} \\ -(\lambda^2 + 3\mu) & \frac{1}{2}\lambda^2 + \mu + 3B_{33} - B_{23} & \Omega^2 + 3(B_{22} - B_{33}) + B_{12} - B_{13} \\ 0 & -\frac{1}{2}\lambda^2 - \mu - 2B_{33} + a_3^2 A_{33} - q(A_3 + \frac{1}{2}\mu) & \frac{1}{2}\lambda^2 + \mu + 2A_3 - 2 + q(2 - \Omega^2) \end{matrix} \right\| = 0 . \tag{125}$$

b) *The Effect of Compressibility on the Onset of Instability*

Greatest interest is attached to the effect of compressibility on the onset of instability along a Roche sequence; and to ascertain this, it is not necessary to solve equation (125) for all its roots, since by putting $\lambda^2 = 0$, we obtain, at once, the equation which governs marginal stability. It will suffice, then, to consider

$$(4B_{12} - 2\Omega^2 - \mu) \left\| \begin{matrix} -\Omega^2 - 2\mu + 3B_{11} - B_{12} & -(B_{13} - B_{23}) & -3\mu + 3(B_{11} - B_{22}) + B_{13} - B_{23} \\ \Omega^2 + B_{12} - B_{13} & \mu + 3B_{33} - B_{23} & \Omega^2 + 3(B_{22} - B_{33}) + B_{12} - B_{13} \\ q(A_1 - \mu - \frac{1}{2}\Omega^2) - a_3^2 A_{13} & -\mu - 2B_{33} + a_3^2 A_{33} - q(A_3 + \frac{1}{2}\mu) & \mu + 2A_3 - 2 + q(2 - \Omega^2) \end{matrix} \right\|$$
$$+ 3\mu\Omega^2 \left\| \begin{matrix} -(B_{13} - B_{23}) & -3\mu + 3(B_{11} - B_{22}) + B_{13} - B_{23} \\ -\mu - 2B_{33} + a_3^2 A_{33} - q(A_3 + \frac{1}{2}\mu) & \mu + 2A_3 - 2 + q(2 - \Omega^2) \end{matrix} \right\| = 0 . \tag{126}$$

Equation (126), considered as an equation for γ $(= 1 + q)$, will determine, for every equilibrium ellipsoid, a value for γ which we shall designate by γ_c. Since we must recover the results, which are valid for the incompressible case, by letting γ tend to infinity, it is clear that γ_c will be positive or negative according as the equilibrium ellipsoid considered is stable or unstable by the criterion which obtains in the incompressible case. It is also clear that γ_c will tend to infinity when the limit to stability, set by incompressibility, is approached.

It follows from the foregoing remarks that, in considering equation (126), we can restrict ourselves to ellipsoids which are stable according to the considerations of Sections VII and VIII. The deduced values of γ_c will then be positive; and the meaning to be attached to γ_c, in these cases, is that *the ellipsoid considered is stable if* $\gamma > \gamma_c$ *and unstable if* $\gamma < \gamma_c$.

We shall first consider the two special cases, $\Omega^2 = 0$ and $\mu = 0$.

By setting $\Omega^2 = 0$ in equation (126), and remembering that in this case the ellipsoid degenerates to a prolate spheroid and the indices 2 and 3 are indistinguishable, we find that the equation becomes

$$\left\| \begin{matrix} -2\mu + 3B_{11} - B_{12} & -3\mu + 3B_{11} - 4B_{22} + B_{12} \\ q(A_1 - \mu) - a_2^2 A_{12} & 2q + \mu - A_1 \end{matrix} \right\| = 0 . \tag{127}$$

In agreement with the results of Paper III (Sec. VI, Table 2 and Fig. 2) we find from equation (127) that

$$\gamma_c \rightarrow \tfrac{4}{3} \text{ as } \mu \rightarrow 0 \qquad \text{and} \qquad \gamma_c \rightarrow \infty \text{ as } \mu \rightarrow \mu_{\max} . \tag{128}$$

The equations determining the neutral modes along the Maclaurin and the Jacobian sequences must be considered separately. Setting $\mu = 0$ in equation (125) and remembering that $\Omega^2 = 2B_{12}$ along the Jacobian sequence, we observe that $\lambda^2 = 0$ is a characteristic root of the equation. Accordingly, *the neutral mode which obtains along the entire Jacobian sequence, in the incompressible case, is unaffected by compressibility.*[12] And along the Maclaurin sequence, the equation determining the neutral mode can be obtained from equation (126) by setting $\mu = 0$ and remembering that, in this case, the indices 1 and 2 are indistinguishable; we thus find

$$\left\| \begin{matrix} 3B_{33} - B_{23} & \Omega^2 + 4B_{11} - 3B_{33} - B_{13} \\ -2B_{33} + a_3^2 A_{33} - qA_3 & 2A_3 - 2 + q(2 - \Omega^2) \end{matrix} \right\| = 0 . \tag{129}$$

And this equation determines the same γ_c's as have been tabulated in an earlier paper (Chandrasekhar and Lebovitz 1962*b*, Table 1). We may note here in particular, that

$$\gamma_c = 1.22515 \text{ at the point of bifurcation, } M_2 . \tag{130}$$

The values of γ_c determined in accordance with equation (126), for the different Roche sequences, are given in Table 6; the table also includes the results for the Jeans sequence derived from equation (127).

[12] And the neutral mode *depending* on compressibility is determined by the equation

$$\left\| \begin{matrix} -\Omega^2 + 3B_{11} - B_{12} & -(B_{13} - B_{23}) & 3(B_{11} - B_{22}) + B_{13} - B_{23} \\ +\Omega^2 + B_{12} - B_{13} & 3B_{33} - B_{23} & \Omega^2 + 3(B_{22} - B_{33}) + B_{12} - B_{13} \\ q(A_1 - \tfrac{1}{2}\Omega^2) - a_3^2 A_{13} & -2B_{33} + a_3^2 A_{33} - qA_3 & 2A_3 - 2 + q(2 - \Omega^2) \end{matrix} \right\| = 0 .$$

For this equation we find:

$\cos^{-1} a_3/a_1$	γ_c	$\cos^{-1} a_3/a_1$	γ_c
54°.358	1.22515	65°	1.2194
55	1.22513	70	1.2118
56	1.2250	75	1.2000
60	1.2236	80	1.1820
63	1.2214	83	1.1664

From Figure 4, which exhibits the results, it is apparent that the principal effect of compressibility on the stability of the Roche ellipsoids is to bring closer toward the Maclaurin branch (SM_2 in Fig. 1) the locus of the points of marginal stability and reduce, still further, the domain of stability which prevails in the incompressible case. More precisely, the situation is the following: for any $\gamma < \frac{4}{3}$, the locus of the points of marginal stability, in the $(\bar{a}_1, \bar{a}_2)$-plane of Figure 1, is a curve which joins M_2 to a point on ST_2;

TABLE 6

THE VALUES OF γ_c WHICH LIMIT THE STABILITY OF THE HOMOGENEOUS COMPRESSIBLE ROCHE ELLIPSOIDS

$p = -1$

e	$\cos^{-1} a_3/a_1$	γ_c
0	0	1.3333
0.1	5°739	1.3334
.2	11.537	1.3337
.3	17.468	1.3353
.4	23.578	1.3403
.5	30	1.3530
.6	36.870	1.3837
.7	44.427	1.4646
.75	48.590	1.5594
.80	53.130	1.7768
.82	55.085	1.9628
.84	57.140	2.3230
.86	59.316	3.3098
.88	61.642	17.329
0.88303	62.009	∞

$p = 0$

$\cos^{-1} a_3/a_1$	γ_c
24°	1.3340
36	1.3541
48	1.4735
57	2.1339
60	4.0719
61	8.795
61.63	∞

$p = 1$

$\cos^{-1} a_3/a_1$	γ_c
12°	1.3319
24	1.3301
36	1.3403
48	1.4329
54	1.6599
59	2.8570
60	4.1806
61	13.023
61.35	∞

$p = 4$

$\cos^{-1} a_3/a_1$	γ_c
12°	1.3310
24	1.3246
36	1.3209
48	1.3707
58	2.1899
60	4.9399
60.83	∞

$p = 20$

$\cos^{-1} a_3/a_1$	γ_c
36°	1.3017
48	1.2952
58	2.1503
59.66	∞

$p = 100$

$\cos^{-1} a_3/a_1$	γ_c
48°	1.2634
52	1.2646
56	1.4692
58	5.9359
58.18	∞

for $\gamma = \frac{4}{3}$, the locus is an arc which joins M_2 and S; for $1.22515 < \gamma < \frac{4}{3}$, the locus similarly joins M_2 to a point on SM_2; and for $\gamma \leq 1.22515$ the entire domain is unstable.

And finally we may note explicitly that compressibility does not affect the instability of the Jacobi ellipsoids under the least tidal action.

XII. CONCLUDING REMARKS

The principal results bearing on the stability of the equilibrium configurations along the Maclaurin, the Jacobi, the Jeans, and the Roche sequences have been summarized in Section X. In this concluding section we shall consider only briefly two questions which occur.

First, since the Roche ellipsoids become unstable only after surpassing the Roche limit, what will be their behavior if they should so surpass the limit? And *second*, has the in-

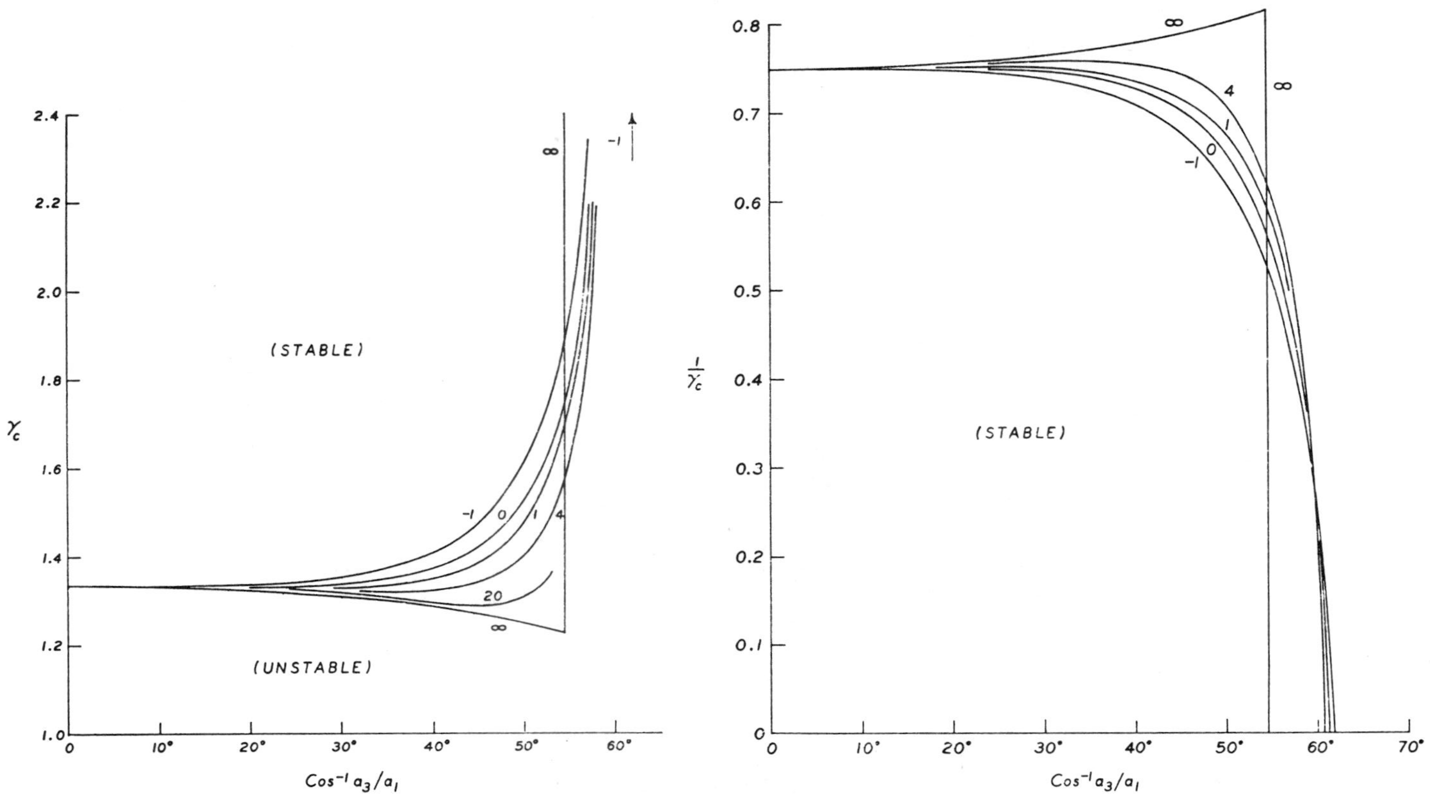

FIG. 4.—The critical values of γ_c and $1/\gamma_c$ at which marginal stability occurs for the homogeneous but compressible Roche ellipsoids. The curves are labeled by the values of p to which they belong.

stability of the Jacobi ellipsoid, under the least tidal action, any consequence for the "wider aspects of cosmogony?"

With respect to the first question, we may argue as Jeans (1919, pp. 118–128) has in the context of what may happen to one of his tidally distorted prolate spheroids if R' should decrease below the limit set by μ_{max}. On the assumption that the spheroid (secularly) retains its prolate form, Jeans derived an equation of motion for its eccentricity if μ varies in some prescribed manner. And he showed that if μ (during the course of its variation) should transgress μ_{max}, then, the character of the problem will change into "a truly dynamical one" and, in the (very) first instance, the spheroid will continue to evolve down the sequence toward higher eccentricities.[13] If similar considerations can be applied to the Roche ellipsoids and should the Roche limit be surpassed, then the character of the problem will change into "a truly dynamical one" and, in the (very) first instance, the ellipsoid will continue to evolve down the sequence toward higher elongations. Since the point at which instability sets in is only a very little beyond the Roche limit (cf. Fig. 1), the dynamical evolution has not far to proceed before it will be arrested by the instability of one of its normal modes. These considerations are qualitative; but they suggest that it might be useful to formulate them quantitatively in the manner of Jeans.

Turning to the second question, we observe that as $p \to \infty$,

$$\lambda^2 = O\left(\frac{1}{p}\right) \tag{131}$$

along the near Jacobian part of the Roche sequence. Consequently, the tidal action of a body, one hundredth as massive as the primary, will induce an instability with an *e*-folding time of the order of ten times the natural periods of oscillation. If we should now idealize the bar of a barred spiral galaxy as a uniform prolate spheroid (which is the limiting form of the Jacobi ellipsoid), then its natural period of oscillation will be of the order of 10^8 years (cf. Burbidge, Burbidge, and Prendergast 1960); and the tidal action of an external mass, which is 1 per cent of the mass of the bar, will induce an instability which will become manifest in 10^9 years. It would appear, then, that the present considerations may have some relevance for cosmogony.

It should be apparent that the magnitude of the numerical work involved in the preparation of this paper is a very large one; and further, that it is particularly true of the present investigation that one cannot, in Kelvin's well-known phrase, "obtain satisfaction from formulas without their numerical magnitude." For that "satisfaction," I am greatly indebted to Miss Donna Elbert: she carried out all the necessary numerical work. I am also indebted to Dr. Norman R. Lebovitz for his careful scrutiny of the present manuscript and for helpful discussions.

The research reported in this paper has in part been supported by the Office of Naval Research under contract Nonr-2121(24) with the University of Chicago.

REFERENCES

Burbidge, E. M., Burbidge, G. R., and Prendergast, K. H. 1960, *Ap. J.*, **132**, 654.
Chandrasekhar, S. 1962, *Ap. J.*, **136**, 1048.
———. 1963*a*, *ibid.*, **137**, 1185 (referred to as "Paper I").
———. 1963*b*, *ibid.*, **138**, 311.
Chandrasekhar, S., and Lebovitz, N. R. 1962*a*, *Ap. J.*, **136**, 1037.
———. 1962*b*, *ibid.*, p. 1069.
———. 1963*a*, *ibid.*, **137**, 1142 (referred to as "Paper II").
———. 1963*b*, *ibid.*, p. 1172 (referred to as "Paper III").

[13] Actually, Jeans's considerations, in the context of his particular problem, are, in a strict sense, unnecessary: for his spheroids (as we have seen) become unstable at μ_{max}, a fact of which Jeans does not seem to have been fully aware (cf. comments in Paper III, Sec. VII).

Darwin, G. H. 1906, *Phil. Trans. R. Soc. London, A*, **206**, 161; also *Scientific Papers*, Vol. 3 (Cambridge: Cambridge University Press, 1910) p. 436.
———. 1911, *The Tides* (3d ed.; London: John Murray), p. 340.
Jeans, J. H. 1917, *Mem. R.A.S.*, **62**, 1.
———. 1919, *Problems of Cosmogony and Stellar Dynamics* (Cambridge: Cambridge University Press), pp. 51–53, 85–86, and 118–128; and figs., pp. 50 and 86.
———. 1929, *Astronomy and Cosmogony* (Cambridge: Cambridge University Press), pp. 225–244; also fig. 33, p. 229.
Lebovitz, N. R. 1961, *Ap. J.*, **134**, 500.
———. 1963, *ibid.*, **138**, 1214.
Milne, E. A. 1952, *Sir James Jeans, a Biography* (Cambridge: Cambridge University Press), chap. ix, particularly pp. 110–112.
Roberts, P. H., and Stewartson, K. 1963, *Ap. J.*, **137**, 777.
Roche, M. E'd. 1850, *La Figure d'une masse fluide soumise à l'attraction d'un point é'loigné, Acad. des Sci. de Montepellier*, **1**, 243.
Schwarzschild, K. 1898, *Ann. d. Konig, Stern. München*, **3**, 233.
Struve, O. 1961, *The Universe* (Cambridge, Mass.: Massachusetts Institute of Technology Press), p. 18.

THE EQUILIBRIUM AND THE STABILITY OF THE RIEMANN ELLIPSOIDS. I

S. CHANDRASEKHAR
University of Chicago
Received April 23, 1965

ABSTRACT

Riemann's problem is concerned with the ellipsoidal figures of equilibrium of homogeneous masses rotating uniformly with an angular velocity Ω and with internal motions having a uniform vorticity ζ_Ω in the rotating frame. In this paper, the equilibrium and the stability of these *Riemann ellipsoids* are considered in the special case the axes of rotation and vorticity coincide with a principal axis of the ellipsoid. It is shown that for the case considered (1) the equilibrium figures can be delineated into sequences—*the Riemann sequences*—along which the ratio $f = \zeta_\Omega/\Omega$ is a constant; (2) an ellipsoid which is a figure of equilibrium for some given f is also a figure of equilibrium for $f\dagger = (a_1^2 + a_2^2)^2/a_1^2a_2^2f$, where a_1 and a_2 are the semi-axes of the ellipsoid in the equatorial plane; (3) the two states of internal motion, corresponding to f and $f\dagger$, lead to configurations which are *adjoint* in the sense of a theorem due to Dedekind; (4) the first member of a Riemann sequence is a Maclaurin spheroid which is stable in the absence of any dissipative mechanism; (5) from each point of the stable part of the Maclaurin sequence two Riemann sequences bifurcate; (6) there exist two self-adjoint sequences along which $f = f\dagger = \pm(a_1^2 + a_2^2)/a_1a_2$ and that limit the domain of the Riemann sequences in the $(a_2/a_1, a_3/a_1)$ = plane; and (7) the bifurcation of the Jacobian and the Dedekind sequences from, what is usually called, *the* point of bifurcation is a special case of a much more general phenomenon.

The stability of the Riemann ellipsoids with respect to modes of oscillation belonging to the second and the third harmonics is also investigated. With respect to modes of oscillation belonging to the second harmonics it is shown that (1) the Riemann ellipsoids allow a non-trivial neutral mode of oscillation; (2) the characteristic frequencies of oscillation of an ellipsoid and its adjoint are the same; (3) the Riemann ellipsoids with $f \geq -2$ are stable with respect to these modes; and (4) instability by one of these modes arises along the sequences for $f < -2$. With respect to modes of oscillation belonging to the third harmonics, it is shown that along all Riemann sequences instability first arises by a mode which deforms the ellipsoid into a pear-shaped configuration. The points at which instability sets in along the different Riemann sequences and the loci, which separate the regions of stability from the regions of instability in the domain of the Riemann ellipsoids considered, are also determined.

I. INTRODUCTION

In 1860 Dedekind proved the following remarkable theorem: *if a homogeneous ellipsoid with semi-axes* a_1, a_2, *and* a_3 *is in gravitational equilibrium, with a prevalent motion* $\mathbf{u}^{(0)}$ *whose components in an inertial frame are given by*

$$u^{(0)} = \begin{vmatrix} \alpha_{11} & \alpha_{12} & \alpha_{13} \\ \alpha_{21} & \alpha_{22} & \alpha_{23} \\ \alpha_{31} & \alpha_{32} & \alpha_{33} \end{vmatrix} \begin{vmatrix} x_1/a_1 \\ x_2/a_2 \\ x_3/a_3 \end{vmatrix} = A \begin{vmatrix} x_1/a_1 \\ x_2/a_2 \\ x_3/a_3 \end{vmatrix}, \tag{1}$$

then, the same ellipsoid will also be a figure of equilibrium if the prevalent motion is that derived from the transposed matrix $\mathbf{A}\dagger$, *i.e., for* $\mathbf{u}^{(0)}$ *given by*

$$u^{(0)} = \begin{vmatrix} \alpha_{11} & \alpha_{21} & \alpha_{31} \\ \alpha_{12} & \alpha_{22} & \alpha_{32} \\ \alpha_{13} & \alpha_{23} & \alpha_{33} \end{vmatrix} \begin{vmatrix} x_1/a_1 \\ x_2/a_2 \\ x_3/a_3 \end{vmatrix} = A\dagger \begin{vmatrix} x_1/a_1 \\ x_2/a_2 \\ x_3/a_3 \end{vmatrix}. \tag{2}$$

We shall call the configuration with the motion derived from $\boldsymbol{A}\dagger$ as the *adjoint* of the configuration with the motion derived from $\boldsymbol{A}$. Love (1888) seems to have been the first to have pointed out that *the Jacobian and the Dedekind sequences*[1] *are adjoint* in this sense. This fact can be verified quite readily as follows.

The motion of a Jacobi ellipsoid (rotating uniformly with an angular velocity Ω about the x_3-axis) can be represented in the manner

$$\boldsymbol{u}^{(0)} = \begin{vmatrix} 0 & -\Omega a_2 & 0 \\ \Omega a_1 & 0 & 0 \\ 0 & 0 & 0 \end{vmatrix} \begin{vmatrix} x_1/a_1 \\ x_2/a_2 \\ x_3/a_3 \end{vmatrix} . \tag{3}$$

By Dedekind's theorem, the motion in the adjoint configuration will be given by

$$\boldsymbol{u}^{(0)} = \begin{vmatrix} 0 & \Omega a_1 & 0 \\ -\Omega a_2 & 0 & 0 \\ 0 & 0 & 0 \end{vmatrix} \begin{vmatrix} x_1/a_1 \\ x_2/a_2 \\ x_3/a_3 \end{vmatrix} , \tag{4}$$

or, in terms of components,

$$u^{(0)}{}_1 = \frac{\Omega a_1}{a_2} x_2 , \qquad u^{(0)}{}_2 = -\frac{\Omega a_2}{a_1} x_1 , \qquad \text{and} \qquad u_3 = 0 ; \tag{5}$$

and this motion clearly satisfies the condition

$$\boldsymbol{u}^{(0)} \cdot \text{grad} \left(\frac{x_1^2}{a_1^2} + \frac{x_2^2}{a_2^2} + \frac{x_3^2}{a_3^2} - 1 \right) = 0 \tag{6}$$

required for the preservation of the ellipsoidal boundary. Also, the motion represented by (4) is one of uniform vorticity ζ, about the x_3-axis, given by

$$\zeta = -\Omega \left(\frac{a_2}{a_1} + \frac{a_1}{a_2} \right) = -\frac{a_1^2 + a_2^2}{a_1 a_2} \Omega ; \tag{7}$$

and this result is in agreement with the known relation between the Jacobian and the Dedekind sequences (cf. Paper I, eq. [18]).

Pursuing the ideas underlying Dedekind's theorem along somewhat different lines, Riemann (1860) investigated the type of motions, in accordance with Dedekind's initial *assumption* (1), that *can* lead to an ellipsoidal figure of equilibrium. And Riemann showed that *the most general type of motion compatible with an ellipsoidal figure of equilibrium consists of a superposition of uniform rotation and internal motions of uniform vorticity about axes that lie in a principal plane of the ellipsoid.*

In spite of the interest that would appear to be attached to Riemann's investigation of Dedekind's problem, it does not seem to have attracted any attention from those who, like Poincaré, Darwin, and Jeans, have concerned themselves with the ellipsoidal figures of equilibrium of homogeneous liquid masses. Indeed, apart from a brief account of an expository nature by Basset (1888), the subject seems to have been simply ignored. In restituting to its proper place this long-neglected work of Riemann, we shall consider

[1] The Dedekind sequence consists of ellipsoids that are stationary in an inertial frame and whose ellipsoidal figures are maintained by internal motions of uniform vorticity. The equilibrium and the stability of these *Dedekind ellipsoids* have recently been considered (Chandrasekhar 1965; this paper will be referred to hereinafter as "Paper I").

in this paper the equilibrium and the stability of the *Riemann ellipsoids* in the special case when the axes of rotation and of internal vorticity coincide with a principal axis of the ellipsoid. It will appear that the delineation of the resulting *Riemann sequences* not only discloses several features of general interest, it also clarifies many aspects of the "classical sequences" of Maclaurin and Jacobi that have lain obscured. In a later paper we shall consider the case when the axes of rotation and of internal vorticity do not coincide but lie in a principal plane of the ellipsoid.

II. THE SECOND-ORDER VIRIAL THEOREM AND THE EQUATIONS DETERMINING THE EQUILIBRIUM OF THE RIEMANN ELLIPSOIDS

The problem that is to be considered in this paper is that of a homogeneous mass, rotating uniformly with an angular velocity $\boldsymbol{\Omega}$, with internal motions having a uniform vorticity $\boldsymbol{\zeta}_\Omega$ in the direction of Ω and in the frame of reference rotating with the angular velocity $\boldsymbol{\Omega}$. In treating this problem, we shall use the methods, based on the virial theorem and its extensions, which have been developed recently in various papers published in this *Journal* during the past four years.

Let the directions of $\boldsymbol{\Omega}$ and $\boldsymbol{\zeta}_\Omega$ be along the x_3-axis; and let the figure of equilibrium be an ellipsoid with semi-axes a_1, a_2, and a_3. Without entailing any loss of generality, we may suppose that

$$a_1 > a_2 \,. \tag{8}$$

The components of the internal motion, having the assigned vorticity ζ_Ω, that will preserve the ellipsoidal boundary can be expressed in the form (cf. Paper I, eqs. [5] and [6])

$$u_1 = Q_1 x_2, \qquad u_2 = Q_2 x_1, \qquad \text{and } u_3 = 0 \,, \tag{9}$$

where

$$Q_1 = -\frac{a_1^2}{a_1^2 + a_2^2}\,\zeta_\Omega \qquad \text{and } Q_2 = +\frac{a_2^2}{a_1^2 + a_2^2}\,\zeta_\Omega \,. \tag{10}$$

Some elementary relations which follow from these definitions and which we shall find useful are

$$a_1^2 Q_2 = -a_2^2 Q_1, \qquad a_1^2 Q_2^2 = -a_2^2 Q_1 Q_2, \qquad \text{and } a_2^2 Q_1^2 = -a_1^2 Q_1 Q_2 \,, \tag{11}$$

where

$$Q_1 Q_2 = -\left(\frac{a_1 a_2}{a_1^2 + a_2^2}\,\zeta_\Omega\right)^2 . \tag{12}$$

The internal motion specified by equation (9) is with respect to the rotating frame of reference; the components of the same motion in the inertial frame follow from the equation

$$\boldsymbol{u}^{(0)} = \boldsymbol{u} + \boldsymbol{\Omega} \times \boldsymbol{x} \,, \tag{13}$$

or, explicitly,

$$u^{(0)}{}_1 = u_1 - \Omega x_2 = -\Omega\left(1 + \frac{a_1^2 f}{a_1^2 + a_2^2}\right) x_2 \,,$$

$$u^{(0)}{}_2 = u_2 + \Omega x_1 = +\Omega\left(1 + \frac{a_2^2 f}{a_1^2 + a_2^2}\right) x_1 \,, \tag{14}$$

and

$$u^{(0)}{}_3 = 0 \,,$$

where we have defined

$$f = \zeta_\Omega / \Omega \,. \tag{15}$$

We can rewrite the components of $u^{(0)}$, in conformity with Dedekind's assumption (1), in the forms

$$u^{(0)}{}_1 = U_1 \frac{x_2}{a_2}, \qquad u^{(0)}{}_2 = U_2 \frac{x_1}{a_1}, \qquad \text{and} \qquad u^{(0)}{}_3 = 0\,, \tag{16}$$

where

$$U_1 = -\,a_2\Omega\left(1 + \frac{a_1^2 f}{a_1^2 + a_2^2}\right) \qquad \text{and} \qquad U_2 = +\,a_1\Omega\left(1 + \frac{a_2^2 f}{a_1^2 + a_2^2}\right). \tag{17}$$

To obtain the conditions that the ellipsoid is also in gravitational equilibrium, we shall make use of the second-order virial theorem. According to this theorem

$$\frac{d}{dt}\int_V \rho u_i x_j d\boldsymbol{x} = 2\mathfrak{T}_{ij} + \mathfrak{W}_{ij} + \Omega^2(I_{ij} - \delta_{i3}I_{3j}) + 2\Omega\epsilon_{il3}\int_V \rho u_l x_j d\boldsymbol{x} + \delta_{ij}\Pi\,, \tag{18}$$

where $\mathfrak{T}_{ij}$, $\mathfrak{W}_{ij}$, I_{ij}, and Π have their usual meanings. Under conditions of equilibrium, equation (18) gives

$$\begin{aligned} -\Pi = \mathfrak{W}_{33} &= 2\mathfrak{T}_{11} + \Omega^2 I_{11} + \mathfrak{W}_{11} + 2\Omega\int_V \rho u_2 x_1 d\boldsymbol{x} \\ &= 2\mathfrak{T}_{22} + \Omega^2 I_{22} + \mathfrak{W}_{22} - 2\Omega\int_V \rho u_1 x_2 d\boldsymbol{x}\,. \end{aligned} \tag{19}$$

(We need not consider the non-diagonal components of eq. [18]: these equations are trivially satisfied.)

For the internal motion specified by equation (9),

$$\mathfrak{T}_{11} = \tfrac{1}{2}Q_1^2 I_{22}\,, \qquad \int_V \rho u_2 x_1 d\boldsymbol{x} = Q_2 I_{11}\,,$$

and

$$\mathfrak{T}_{22} = \tfrac{1}{2}Q_2^2 I_{11}\,, \qquad \int_V \rho u_1 x_2 d\boldsymbol{x} = Q_1 I_{22}\,; \tag{20}$$

and equation (19) gives

$$\begin{aligned} \mathfrak{W}_{33} &= Q_1^2 I_{22} + (\Omega^2 + 2Q_2\Omega)I_{11} + \mathfrak{W}_{11} \\ &= Q_2^2 I_{11} + (\Omega^2 - 2Q_1\Omega)I_{22} + \mathfrak{W}_{22}\,. \end{aligned} \tag{21}$$

Making use of the known expressions for I_{ij} and $\mathfrak{W}_{ij}$ (Chandrasekhar and Lebovitz 1962, eqs. [57] and [58]), we can rewrite the foregoing equations in the forms

$$\begin{aligned} -2A_3a_3^2 &= a_2^2Q_1^2 + a_1^2(\Omega^2 + 2Q_2\Omega) - 2A_1a_1^2 \\ &= a_1^2Q_2^2 + a_2^2(\Omega^2 - 2Q_1\Omega) - 2A_2a_2^2\,, \end{aligned} \tag{22}$$

where the index symbols A_i are so normalized that $\Sigma A_i = 2$ and Ω^2 and ζ_Ω^2 are measured in the unit $\pi G\rho$.

In view of equations (11), the relation expressed by the second equality in (22) becomes

$$-a_1^2Q_1Q_2 + a_1^2\Omega^2 - 2A_1a_1^2 = -a_2^2Q_1Q_2 + a_2^2\Omega^2 - 2A_2a_2^2\,, \tag{23}$$

or, alternatively,

$$\Omega^2 - Q_1Q_2 = 2\,\frac{A_1a_1^2 - A_2a_2^2}{a_1^2 - a_2^2} = 2B_{12}\,. \tag{24}$$

This last equation is a generalization of relations known to be valid for the Jacobi ($Q_1 = Q_2 = 0$) and the Dedekind ($\Omega = 0$) ellipsoids (cf. Paper I, eq. [46]).

Returning to equations (22), we have the pair of equations

$$\Omega^2 - Q_1Q_2 + 2Q_2\Omega = 2\left(A_1 - \frac{a_3^2}{a_1^2}A_3\right) = 2\,\frac{a_1^2 - a_3^2}{a_1^2}B_{13} \tag{25}$$

and

$$\Omega^2 - Q_1Q_2 - 2Q_1\Omega = 2\left(A_2 - \frac{a_3^2}{a_2^2}A_3\right) = 2\,\frac{a_2^2 - a_3^2}{a_2^2}B_{23}\,. \tag{26}$$

From equation (25) it follows that

$$a_1 > a_3\,; \tag{27}$$

for, by substituting for Q_1 and Q_2 their expressions in terms of ζ_Ω, we find

$$\Omega^2 - Q_1Q_2 + 2Q_2\Omega = \left(\Omega + \frac{a_2^2}{a_1^2 + a_2^2}\zeta_\Omega\right)^2 + \frac{a_2^2(a_1^2 - a_2^2)}{(a_1^2 + a_2^2)^2}\zeta_\Omega^2\,, \tag{28}$$

and this expression is positive definite since $a_1 > a_2$ (by definition!). Thus, *the choice of* a_1 *as the longer of the two axes in the* (x_1,x_2)*-plane makes it also the longest.*

Next, we obtain, by addition and subtraction, from equations (25) and (26)

$$(\Omega - Q_1)(\Omega + Q_2) = A_1 + A_2 - \frac{a_3^2(a_1^2 + a_2^2)}{a_1^2a_2^2}A_3$$

and (29)

$$(Q_1 + Q_2)\Omega = A_1 - A_2 + \frac{a_3^2(a_1^2 - a_2^2)}{a_1^2a_2^2}A_3\,.$$

Accordingly, for determining the geometry of the ellipsoid we may use the equation

$$\frac{(\Omega - Q_1)(\Omega + Q_2)}{\Omega(Q_1 + Q_2)} = \frac{A_1 + A_2 - a_3^2(a_1^2 + a_2^2)A_3/a_1^2a_2^2}{A_1 - A_2 + a_3^2(a_1^2 - a_2^2)A_3/a_1^2a_2^2}\,. \tag{30}$$

We shall now define a *Riemann sequence as one along which* $f = \zeta_\Omega/\Omega$ *is a constant.* According to this definition, *the Jacobian and the Dedekind sequences are Riemann sequences for* $f = 0$ *and* $f = \pm\infty$, *respectively*. Moreover, since, according to equation (13) the vorticity $\zeta^{(0)}$ in the inertial frame is related to the vorticity ζ_Ω in the rotating frame by

$$\zeta^{(0)} = \zeta_\Omega + 2\Omega = \Omega(2 + f)\,, \tag{31}$$

it follows that *the irrotational sequence,* $\zeta^{(0)} = 0$, *is a Riemann sequence for* $f = -2$.

In terms of f, we have the equations

$$Q_1 = -\frac{a_1^2\Omega}{a_1^2 + a_2^2}f\,,\quad Q_2 = +\frac{a_2^2\Omega}{a_1^2 + a_2^2}f\,,\quad\text{and}\quad Q_1Q_2 = -\frac{a_1^2a_2^2}{(a_1^2 + a_2^2)^2}\Omega^2f^2\,. \tag{32}$$

Making use of these relations, we can rewrite equations (24) and (30) in the forms

$$\Omega^2 = \frac{2B_{12}}{1 + a_1^2a_2^2f^2/(a_1^2 + a_2^2)^2} \tag{33}$$

and

$$\frac{[1 + a_1^2f/(a_1^2 + a_2^2)][1 + a_2^2f/(a_1^2 + a_2^2)]}{(a_1^2 - a_2^2)f/(a_1^2 + a_2^2)} = -\frac{A_1 + A_2 - a_3^2(a_1^2 + a_2^2)A_3/a_1^2a_2^2}{A_1 - A_2 + a_3^2(a_1^2 - a_2^2)A_3/a_1^2a_2^2}\,. \tag{34}$$

On simplifying this last equation, we find

$$a_1^2a_2^2\left[A_1-A_2+\frac{a_3^2(a_1^2-a_2^2)}{a_1^2a_2^2}A_3\right]f^2+2(a_1^2+a_2^2)(A_1a_1^2-A_2a_2^2)f$$
$$+(a_1^2+a_2^2)^2\left[A_1-A_2+\frac{a_3^2(a_1^2-a_2^2)}{a_1^2a_2^2}A_3\right]=0. \tag{35}$$

For a given f, equation (35) determines the ratios of the axes of the ellipsoids that are compatible with equilibrium; and the value of Ω^2 which is to be associated with a particular solution of equation (35), then, follows from equation (33).

III. THE ADJOINT RIEMANN ELLIPSOIDS AND DEDEKIND'S THEOREM

For reasons which will become apparent presently we shall let

$$x=\frac{a_1a_2}{a_1^2+a_2^2}f. \tag{36}$$

The relations given in equations (32) and (33) now take the forms

$$\Omega^2=\frac{2B_{12}}{1+x^2},\qquad -Q_1Q_2=2B_{12}\frac{x^2}{1+x^2},$$
$$Q_1\Omega=-2B_{12}\frac{a_1}{a_2}\frac{x}{1+x^2},\quad\text{and}\quad Q_2\Omega=+2B_{12}\frac{a_2}{a_1}\frac{x}{1+x^2}; \tag{37}$$

and equation (35) becomes

$$\left(A_1-A_2+\frac{a_1^2-a_2^2}{a_1^2a_2^2}a_3^2A_3\right)x^2+\frac{2}{a_1a_2}(A_1a_1^2-A_2a_2^2)x$$
$$+\left(A_1-A_2+\frac{a_1^2-a_2^2}{a_1^2a_2^2}a_3^2A_3\right)=0. \tag{38}$$

Since the constant term and the coefficient of x^2 are the same in equation (38), it is clear that *if* x *is a root, then so is* $1/x$. From equation (36) it now follows that *a Riemann ellipsoid belonging to a given* f *is equally a figure of equilibrium for*

$$f\dagger=\frac{(a_1^2+a_2^2)^2}{a_1^2a_2^2}\frac{1}{f}. \tag{39}$$

We shall call *a Riemann ellipsoid with internal motions corresponding to* f† *the adjoint of the one with internal motions corresponding to* f. We shall presently verify that this definition of adjointness is in accord with the earlier definition (§ I) based on Dedekind's theorem. But before verifying this fact, it is useful to notice the following features of the present definition.

i) The relationship between two adjoint configurations is a *reflexive* one.

ii) Since the Jacobian and the Dedekind sequences belong to $f=0$ and $f=\pm\infty$, respectively, it follows that the two sequences are adjoints of one another, also in the present sense.

iii) Since the constants appropriate to an adjoint configuration are obtained by simply replacing x by $1/x$ in the relevant formulae, it follows from equations (37) that *the values of* Ω^2 *and* $-Q_1Q_2$ *are simply interchanged when one passes from a configuration to its adjoint, while the values of* $Q_1\Omega$ *and* $Q_2\Omega$ *are unchanged.*

iv) A further relation between Ω and $\Omega\dagger$ (the angular velocity appropriate to the adjoint configuration) which follows from equations (37) is

$$(\Omega\dagger)^2 = 2B_{12}\frac{x^2}{1+x^2} = \Omega^2 x^2,$$

or (40)

$$\Omega\dagger = \pm\,\Omega x = \pm\frac{a_1 a_2}{a_1^2+a_2^2}\,\Omega f\,.$$

We now verify that the present definition of adjointness is in accord with Dedekind's theorem. For this verification, we should determine the components of the motion, in the adjoint configuration, in the inertial frame of reference; and these follow from equations (16) and (17) by simply replacing Ω and f by $\Omega\dagger$ and $f\dagger$. Thus

$$U_1\dagger = -\,a_2\Omega\dagger\left(1+\frac{a_1^2 f\dagger}{a_1^2+a_2^2}\right); \tag{41}$$

or making use of equations (39) and (40), we have

$$\begin{aligned} U_1\dagger &= \mp\frac{a_1 a_2^2}{a_1^2+a_2^2}\,\Omega f\left(1+\frac{a_1^2+a_2^2}{a_2^2}\,\frac{1}{f}\right) \\ &= \mp\,a_1\Omega\left(1+\frac{a_2^2 f}{a_1^2+a_2^2}\right) = \mp\,U_2\,. \end{aligned} \tag{42}$$

Similarly,

$$U_2\dagger = \mp U_1\,. \tag{43}$$

Hence, the matrix representing the motion in the adjoint configuration is the transposed of the matrix, representing the motion in the original configuration, or its negative. Since an alteration in the sign of Ω does not affect any of the conditions for equilibrium, it is clear that configurations which are adjoints of one another according to the present definition are also adjoints of one another in the sense of Dedekind's theorem.

We shall conclude this section by evaluating the angular momentum $\mathfrak{M}$ of a Riemann ellipsoid about the x_3-axis in the inertial frame. By definition

$$\mathfrak{M} = \int_V \rho\,[x_1 u^{(0)}{}_2 - x_2 u^{(0)}{}_1]\,d\boldsymbol{x}\,. \tag{44}$$

Inserting in this equation the expressions for $u^{(0)}{}_1$ and $u^{(0)}{}_2$ given in equations (14), we find

$$\mathfrak{M} = \tfrac{1}{5}M(a_1^2+a_2^2)\,\Omega\left[1+\frac{2a_1^2a_2^2}{(a_1^2+a_2^2)^2}\,f\right], \tag{45}$$

where M denotes the mass of the ellipsoid. The angular momentum, $\mathfrak{M}\dagger$, of the corresponding adjoint configuration is given by

$$\mathfrak{M}\dagger = \tfrac{1}{5}M(a_1^2+a_2^2)\,\Omega\dagger\left[1+\frac{2a_1^2a_2^2}{(a_1^2+a_2^2)^2}\,f\dagger\right], \tag{46}$$

or making use of equations (39) and (40), we have

$$\mathfrak{M}\dagger = \pm\tfrac{1}{5}Ma_1a_2\Omega f(1+2/f)\,. \tag{47}$$

In view of equation (31), we can also write

$$\mathfrak{M}\dagger = \pm\tfrac{1}{5}Ma_1a_2\zeta^{(0)}\,. \tag{48}$$

IV. THE STABLE MACLAURIN SPHEROIDS AS THE FIRST MEMBERS OF RIEMANN SEQUENCES

In this section we shall show how a Maclaurin spheroid can be considered as a limiting form of a Riemann ellipsoid. If a Maclaurin spheroid can be so considered, then the angular velocity Ω that would be ascribed to it will not necessarily be the same as the angular velocity Ω_{Mc} which would normally be attributed to it. On the other hand, since an object cannot be materially affected by simply viewing it from a different frame of reference, it is clear that, when viewed from a frame of reference rotating with an angular velocity Ω, a Maclaurin spheroid, rotating uniformly with an angular velocity Ω_{Mc} with respect to an inertial frame, will be described as having internal motions with a uniform vorticity ζ_Ω such that the net vorticity in the inertial frame is the same.

Now the vorticity associated with the uniform rotation Ω_{Mc} is

$$\zeta^{(0)} = 2\Omega_{Mc}\,; \tag{49}$$

and from equation (31) it follows that, when viewed from a frame of reference rotating with an angular velocity Ω, the Maclaurin spheroid will be described as having internal motions with the vorticity

$$\zeta_\Omega = \zeta^{(0)} - 2\Omega = 2(\Omega_{Mc} - \Omega)\,. \tag{50}$$

At the same time, from equation (12) it follows that when $a_1 = a_2$ (as is the case for a Maclaurin spheroid)

$$Q_1 Q_2 = -\tfrac{1}{4}\zeta_\Omega^2 = -(\Omega_{Mc} - \Omega)^2\,. \tag{51}$$

But in order that a Maclaurin spheroid may be considered as a Riemann ellipsoid, equation (24) must be satisfied. Hence, the angular velocity Ω which should be ascribed to a Maclaurin spheroid, when it is considered as a limiting form of a Riemann ellipsoid, is determined by the equation

$$\Omega^2 + (\Omega_{Mc} - \Omega)^2 = 2B_{11}\,, \tag{52}$$

where B_{11} is the index symbol determined by the eccentricity of the spheroid. Solving equation (52), we have

$$\Omega = \tfrac{1}{2}[\Omega_{Mc} \pm \surd(4B_{11} - \Omega_{Mc}^2)]\,. \tag{53}$$

The corresponding value of ζ_Ω is given by (cf. eq. [50])

$$\zeta_\Omega = \Omega_{Mc} \mp \surd(4B_{11} - \Omega_{Mc}^2)\,. \tag{54}$$

We now observe that *equation (54) determining ζ_Ω is the same as the characteristic equation that determines the natural frequencies of the toroidal modes of oscillation of a Maclaurin spheroid.*[2] *Consequently, only the stable members of the Maclaurin sequence can be considered as limiting forms of Riemann ellipsoids:* for the Maclaurin spheroid becomes unstable when $\Omega_{Mc}^2 > 4B_{11}$; and when this is the case, Ω and ζ_Ω given by equations (53) and (54) become complex and this is incompatible with their meanings as real quantities.

We may now state that a Maclaurin spheroid rotating uniformly with an angular velocity Ω_{Mc} may be considered as the first member of two Riemann sequences belonging to

$$f = 2\,\frac{\Omega_{Mc} - \surd(4B_{11} - \Omega_{Mc}^2)}{\Omega_{Mc} + \surd(4B_{11} - \Omega_{Mc}^2)} \quad \text{and} \quad f^\dagger = 2\,\frac{\Omega_{Mc} + \surd(4B_{11} - \Omega_{Mc}^2)}{\Omega_{Mc} - \surd(4B_{11} - \Omega_{Mc}^2)}\,. \tag{55}$$

There are several things to be noticed about this association of the Riemann sequences with the stable members of the Maclaurin sequence.

[2] Cf. Lebovitz (1961, eq. [189]). For a more direct comparison with equation (54), see Chandrasekhar (1964, p. 70, eq. [38]), where the oscillations of the Maclaurin spheroid are treated in the present notation.

i) Since $(a_1^2 + a_2^2)^2/a_1^2 a_2^2 = 4$ when $a_1 = a_2$, f and $f\dagger$, as defined by equations (55), are consistent with the general relation (39), which now requires that

$$f\dagger = 4/f\,. \tag{56}$$

ii) When

$$\Omega_{Mc}^2 = 2B_{11}, \quad f = 0, \quad \Omega = \Omega_{Mc}, \quad \text{and} \quad \zeta_\Omega = 0\,;$$
$$\text{and} \quad f\dagger = \infty, \quad \Omega\dagger = 0, \quad \text{and} \quad \zeta_\Omega\dagger = 2\Omega_{Mc}\,. \tag{57}$$

In other words, the Maclaurin spheroid at the usual point of bifurcation is the first member of both the Jacobian and the Dedekind sequences. This result is, of course, in accord with what we know about these sequences.

iii) When

$$\Omega_{Mc} = 0, \quad f = f\dagger = -2, \quad \Omega^2 = B_{11}, \quad \text{and } \zeta_\Omega^2 = 4B_{11}\,,$$

where

$$B_{11} = \tfrac{4}{15} \tag{58}$$

is the value appropriate for a sphere. *The first member of the irrotational sequence, $\zeta^{(0)} = 0$ and f = -2, is the sphere.* However, the sphere considered as the first member of the irrotational sequence is ascribed an angular velocity[3] $\Omega = \sqrt{(4\pi G\rho/15)}$ and internal motions with the vorticity, $\zeta_\Omega = \sqrt{(16\pi G\rho/15)}$, which exactly cancels the vorticity due to the rotation.

TABLE 1

THE PARAMETERS TO BE ASSOCIATED WITH THE MACLAURIN SPHEROIDS WHEN CONSIDERED AS THE FIRST MEMBERS OF RIEMANN SEQUENCES

e	f	Ω	$f\dagger$	$\Omega\dagger$	e	f	Ω	$f\dagger$	$\Omega\dagger$
0.	−2.00000	0.51640	− 2.0000	−0.51640	0.80. . .	−0.05014	0.61812	− 79.780	−0.01550
0.20.	−1.49613	.58136	− 2.6736	− .43490	.81. . .	−0.01083	.61316	−369.20	− .00332
.25.	−1.38358	.59502	− 2.8910	− .41163	.81267	0	.61174	± ∞	0
.30.	−1.27419	.60757	− 3.1392	− .38708	.82. . .	+0.03055	.60764	+130.93	+ .00928
.35.	−1.16686	.61891	− 3.4280	− .36109	.83. . .	+0.07434	.60148	+ 53.806	+ .02236
.40.	−1.06059	.62893	− 3.7715	− .33352	.84. . .	+0.12094	.59460	+ 33.074	+ .03596
.45.	−0.95434	.63748	− 4.1914	− .30419	.85. . .	+0.17086	.58691	+ 23.411	+ .05014
.50.	−0.84694	.64433	− 4.7229	− .27285	.86. . .	+0.22475	.57828	+ 17.798	+ .06498
.55.	−0.73698	.64918	− 5.4276	− .23921	.88. . .	+0.34824	.55749	+ 11.486	+ .09707
.60.	−0.62269	.65159	− 6.4237	− .20287	.90. . .	+0.50327	.53029	+ 7.9480	+ .13344
.65.	−0.50169	.65096	− 7.9731	− .16329	.92. . .	+0.71695	.49279	+ 5.5792	+ .17666
.70.	−0.37044	.64635	−10.798	− .11972	.94. . .	+1.08248	.43421	+ 3.6952	+ .23501
0.75.	−0.22331	0.63626	−17.912	−0.07104	0.95289	+2.00000	0.33175	+ 2.0000	+0.33175

iv) The Maclaurin spheroids with angular velocities in the range $0 \leq \Omega_{Mc}^2 < 2B_{11}$ are the first members of Riemann sequences belonging to values of f in the ranges $-2 \leq f < 0$ and $-2 \geq f > -\infty$; and the Maclaurin spheroids with angular velocities in the ranges $2B_{11} < \Omega_{Mc}^2 \leq 4B_{11}$ are the first members of Riemann sequences belonging to values of f in the ranges $0 < f \leq 2$ and $2 \leq f < +\infty$. Since every value of f has been accounted for in this enumeration of the first members, it is clear that no Riemann sequence has been left out. In other words, *there is no Riemann sequence which does not begin with a stable Maclaurin spheroid.*

The variation of f and $f\dagger$ along the Maclaurin sequence is exhibited in Figure 1; and the relevant numerical data are given in Table 1.

[3] Restoring the factor $\pi G\rho$, which had hitherto been suppressed.

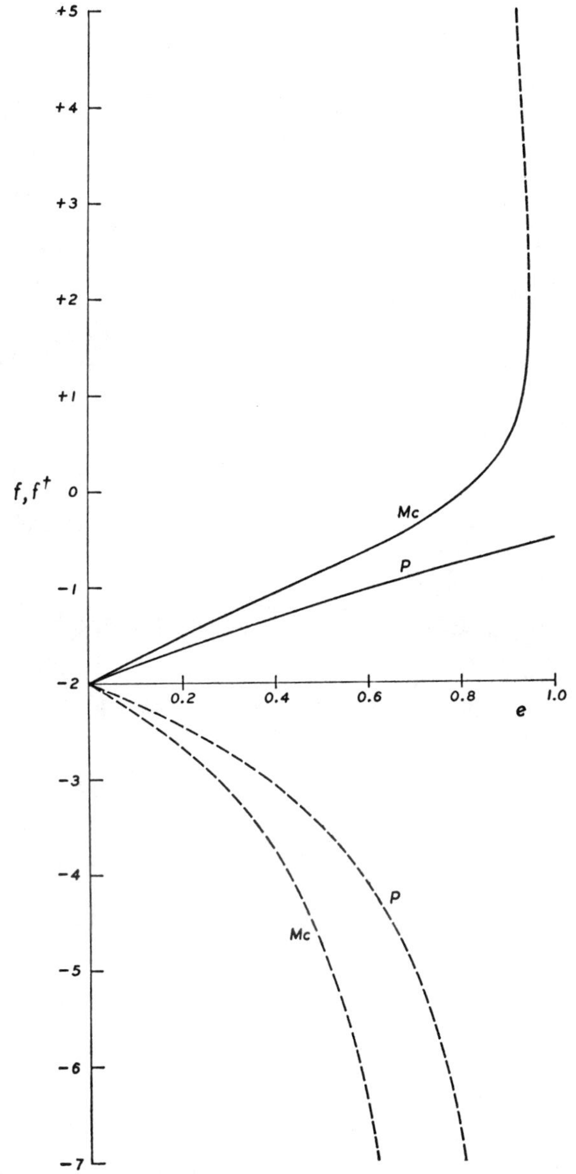

FIG. 1.—The variation of f (*full-line curves*) and $f\dagger$ (*dashed curves*) along the Maclaurin sequence (labeled "Mc") and the sequence of the prolate spheroids (labeled "P").

V. THE DELINEATION OF THE RIEMANN SEQUENCES

The Riemann sequences for different values of f are best delineated in an $(a_2/a_1, a_3/a_1)$-plane, where a_2 and a_3 are the two minor axes of the ellipsoid (see Fig. 2). In this plane the sphere is represented by the point $S = (1, 1)$. And the Maclaurin sequence is represented by points along the line $a_2/a_1 = 1$; on this line the point of bifurcation (where $\Omega_{\text{Mc}}^2 = 2B_{11}$) occurs at M_2, where $a_3/a_1 = 0.58272$ and $e = 0.81267$; and the point where $\Omega_{\text{Mc}}^2 = 4B_{11}$ and the Maclaurin spheroid becomes unstable by overstable oscillations occurs at O_2 where $a_3/a_1 = 0.30333$ and $e = 0.95289$. As we have seen in § IV, all Riemann sequences adjoin some member of the Maclaurin sequence represented by points on the segment O_2S of the line $a_2/a_1 = 1$. In particular, the Jacobian and the Dedekind sequences considered as Riemann sequences (belonging to $f = 0$ and $\pm\infty$, respectively) branch off at M_2 and join the origin becoming tangential to the line $a_2 = a_3$ at this point. We now proceed to a consideration of the Riemann sequences for some representative values of f.

a) The Irrotational Sequence, $\zeta^{(0)} = 0$ *and* $f = -2$

The Riemann sequence for $f = -2$ is of particular interest since its first member is a sphere; its generalization to compressible masses may be expected to have practical applications for astrophysics (see a forthcoming paper by Chandrasekhar and Clement where this matter is considered).

As we have already remarked, when $f = -2$ the vorticity of the internal motions exactly cancels the vorticity due to the rotation. It is on this account that the sequence is designated irrotational. It further follows from equation (48) that the adjoint to an irrotational ellipsoid has zero angular momentum.[4] *There exists then a sequence of zero angular momentum ellipsoids; and this sequence is the adjoint of the irrotational sequence.*

Turning to a more detailed consideration of the irrotational ellipsoids, we first observe that according to equations (32) we now have

$$Q_1 = \frac{2a_1^2\Omega}{a_1^2 + a_2^2} \quad \text{and} \quad Q_2 = -\frac{2a_2^2\Omega}{a_1^2 + a_2^2}. \tag{59}$$

These values of Q_1 and Q_2 when inserted into equations (25) and (26) give

$$\frac{a_1^2(a_1^2 - a_2^2)(a_1^2 + 3a_2^2)}{(a_1^2 + a_2^2)^2}\,\Omega^2 = 2(A_1a_1^2 - A_3a_3^2) = 2(a_1^2 - a_3^2)B_{13} \tag{60}$$

and

$$\frac{a_2^2(a_1^2 - a_2^2)(a_2^2 + 3a_1^2)}{(a_1^2 + a_2^2)^2}\,\Omega^2 = 2(A_3a_3^2 - A_2a_2^2) = 2(a_3^2 - a_2^2)B_{23}. \tag{61}$$

Since the left-hand side of equation (61) is necessarily positive (since $a_1 > a_2$ by definition) it follows that

$$a_3 > a_2 . \tag{62}$$

The irrotational ellipsoids are therefore prolate.[5] In particular, the axis in the direction of rotation is *not* the least axis.

The equation determining the geometry of the ellipsoid can be obtained by setting $f = -2$ in equation (35), or, more directly, by taking the ratio of equations (60) and (61). We find

$$a_1^2a_2^2[A_1(a_2^2 + 3a_1^2) + A_2(a_1^2 + 3a_2^2)] = A_3a_3^2(a_1^4 + 6a_1^2a_2^2 + a_2^4)\,; \tag{63}$$

[4] This fact was first noticed by Norman R. Lebovitz.

[5] We shall call an ellipsoid *prolate* if $a_1 > a_3 > a_2$ and *oblate* if $a_1 > a_2 > a_3$.

and the value of Ω^2 to be associated with a solution of equation (63) is given by (cf. eq. [33])

$$\Omega^2 = \frac{2(a_1^2 + a_2^2)^2}{a_1^4 + 6a_1^2a_2^2 + a_2^4} B_{12} . \tag{64}$$

Equations, essentially equivalent to the two foregoing, were derived by Greenhill (1879, 1880) in his investigations of this subject.

Equation (63) has been solved for a number of initially assigned values of the ratio a_2/a_1. The results of the calculations are summarized in Table 2. The irrotational sequence is delineated in Figure 2; and the variation of Ω^2 along the sequence is exhibited in Figure 3.

In many respects the most striking feature of the irrotational sequence is the extreme sensitivity of the figure to very small changes of Ω^2 as we progress along the sequence from its initial spherical form. Thus Ω^2 changes by less than 1 per cent from its

TABLE 2*

THE PROPERTIES OF THE IRROTATIONAL ELLIPSOIDS

ϕ	0°	30°	45°	55°	57°	59°	60°
θ	....	48°828	54°2405	59°822	61°190	62°656	63°4267
a_2/a_1	1.00000	0.86603	0.70711	0.57358	0.54464	0.51504	0.50000
a_3/a_1	1.00000	0.92647	0.81899	0.70608	0.67822	0.64829	0.63252
A_1	0.66667	0.60869	0.52665	0.44363	0.42361	0.40231	0.39117
A_2	0.66667	0.72345	0.79948	0.87005	0.88587	0.90211	0.91035
A_3	0.66667	0.66786	0.67387	0.68632	0.69052	0.69558	0.69848
Ω^2	0.26667	0.26715	0.26875	0.26882	0.26811	0.26686	0.26598
Q_1Q_2	−0.26667	−0.26170	−0.23889	−0.20029	−0.18921	−0.17688	−0.17023
Ω	0.51640	0.51687	0.51841	0.51848	0.51779	0.51659	0.51573
Q_1	0.51640	0.59070	0.69121	0.78026	0.79867	0.81657	0.82517
Q_2	−0.51640	−0.44303	−0.34561	−0.25670	−0.23691	−0.21661	−0.20629
$f\dagger$	−2.00000	−2.04167	−2.25000	−2.68430	−2.83391	−3.01755	−3.12500
$\Omega\dagger$	0.51640	0.51156	0.48876	0.44754	0.43499	0.42057	0.41259
$Q_1\dagger$	0.51640	0.59682	0.73314	0.90394	0.95071	1.00301	1.03147
$Q_2\dagger$	−0.51640	−0.44762	−0.36657	−0.29739	−0.28201	−0.26606	−0.25787
ϕ	70°	71°	72°	75°	78°	80°	85°
θ	72°618	73°6775	74°756	78°074	81°4035	83°535	87°991
a_2/a_1	0.34202	0.32557	0.30902	0.25882	0.20791	0.17365	0.08716
a_3/a_1	0.44248	0.42025	0.39752	0.32685	0.25417	0.20603	0.09389
A_1	0.26022	0.24522	0.22992	0.18264	0.13452	0.10313	0.03475
A_2	0.99102	0.99801	1.00458	1.02095	1.03063	1.03239	1.01973
A_3	0.74876	0.75677	0.76550	0.79641	0.83485	0.86449	0.94552
Ω^2	0.23768	0.23164	0.22475	0.19825	0.16228	0.13332	0.05285
Q_1Q_2	−0.08914	−0.08029	−0.07153	−0.04666	−0.02578	−0.01515	−0.00158
Ω	0.48753	0.48129	0.47407	0.44525	0.40283	0.36513	0.22988
Q_1	0.87294	0.87034	0.86550	0.83459	0.77228	0.70888	0.45630
Q_2	−0.10211	−0.09225	−0.08265	−0.05591	−0.03338	−0.02138	−0.00347
$f\dagger$	−5.33280	−5.77022	−6.28381	−8.49760	−12.58839	−17.59680	−66.82684
$\Omega\dagger$	0.29856	0.28335	0.26745	0.21601	0.16057	0.12310	0.03977
$Q_1\dagger$	1.42543	1.47832	1.53414	1.72030	1.93752	2.10269	2.63759
$Q_2\dagger$	−0.16674	−0.15669	−0.14650	−0.11524	−0.08375	−0.06340	−0.02004

* Cos $\phi = a_2/a_1$; sin $\theta = [(a_1^2 - a_3^2)/(a_1^2 - a_2^2)]^{1/2}$; A_1, A_2, and A_3 are so normalized that $\Sigma A_i = 2$; Ω, Q_1, and Q_2 are measured in the unit $\sqrt{(\pi G \rho)}$; and $f\dagger$, $\Omega\dagger$, etc., refer to the adjoint configurations having the same figure.

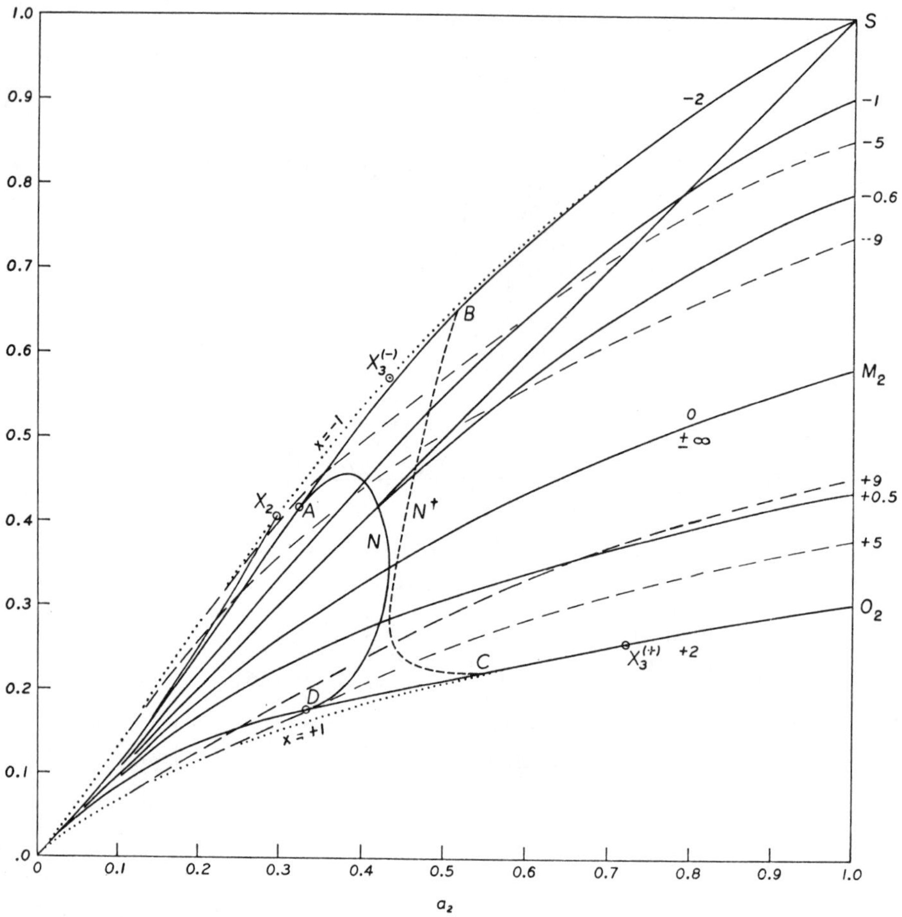

FIG. 2.—The delineation of the Riemann sequences in the (a_2, a_3)-plane (a_1 has been set equal to 1). The stable part of the Maclaurin sequence is represented by the segment O_2S of the line $a_2 = 1$. At O_2 the Maclaurin spheroid becomes unstable by overstable oscillations and at M_2 the Jacobian and the Dedekind sequences bifurcate (labeled by "0, $\pm\infty$").

The different Riemann sequences are labeled by the values of f to which they belong; these sequences are bounded by the two self-adjoint sequences (the dotted curves labeled $x = -1$ and $x = +1$) along which $f = f\dagger = \mp(a_1^2 + a_2^2)/a_1a_2$. The sequences belonging to f in the range $-2 \leq f \leq +2$ form a non-intersecting family of continuous curves which join points on the line O_2S to the origin. The sequences belonging to $f < -2$ and $f > +2$ are represented by curves which consist of two parts: a part which joins a point on the line SM_2 (or M_2O_2) to a point of the self-adjoint sequence for $x = -1$ (or $x = +1$) and a part which joins the point on the self-adjoint sequence to the origin. Along the self-adjoint sequence $x = -1$, instability by a mode of oscillation belonging to the second harmonics sets in at the point indicated by X_2 and the locus of points at which instability by this mode sets in, is the curve which joins X_2 to the origin. The curve labeled AND is the locus of neutral points, belonging to the third harmonics, along the Riemann sequences for $-2 \leq f \leq +2$; and the curve labeled $BN\dagger C$ is the corresponding locus for configurations adjoint to the Riemann ellipsoids represented in the domain included between the same sequences $f = -2$ and $f = +2$. The continuations of the loci AND and $BN\dagger C$ into the domains included between the sequences $x = -1$ and $f = -2$ (and, similarly, between the sequences $x = +1$ and $f = +2$) are represented by curves (not shown) joining the points A and B to $X_3^{(-)}$ on the sequence $x = -1$ (and, similarly, by curves joining the points D and C to the point $X_3^{(+)}$ on the sequence $x = +1$); $X_3^{(-)}$ and $X_3^{(+)}$ are the neutral points, belonging to the third harmonics, along the self-adjoint sequences $x = -1$ and $x = +1$, respectively.

initial value $4\pi G\rho/15$ for figures of equilibrium comprised in the range $1 \geq a_2/a_1 \geq 0.5$ and $1 \geq a_3/a_1 \geq 0.6325$.

b) Prolate Spheroids among the Riemann Ellipsoids

We have seen that the irrotational sequence is entirely prolate (i.e., $a_1 > a_3 > a_2$); but the Jacobian sequence, which is a Riemann sequence for $f = 0$, is entirely oblate (i.e., $a_1 > a_2 > a_3$). Moreover, since the Riemann sequences for $f \neq -2$, all have for their first members Maclaurin spheroids with $\Omega_{\text{Mc}}^2 > 0$, it is clear that these sequences begin as oblate objects. Indeed, it follows from remark (iv) in § IV (p. 898) that *all* Riemann ellipsoids belonging to positive values of f must be oblate since they are represented by points below the Jacobian sequence in the (a_2, a_3)-plane. However, Riemann sequences for at least some negative values of f *must* end as prolate objects. And the question arises as to which among the Riemann sequences belonging to negative values

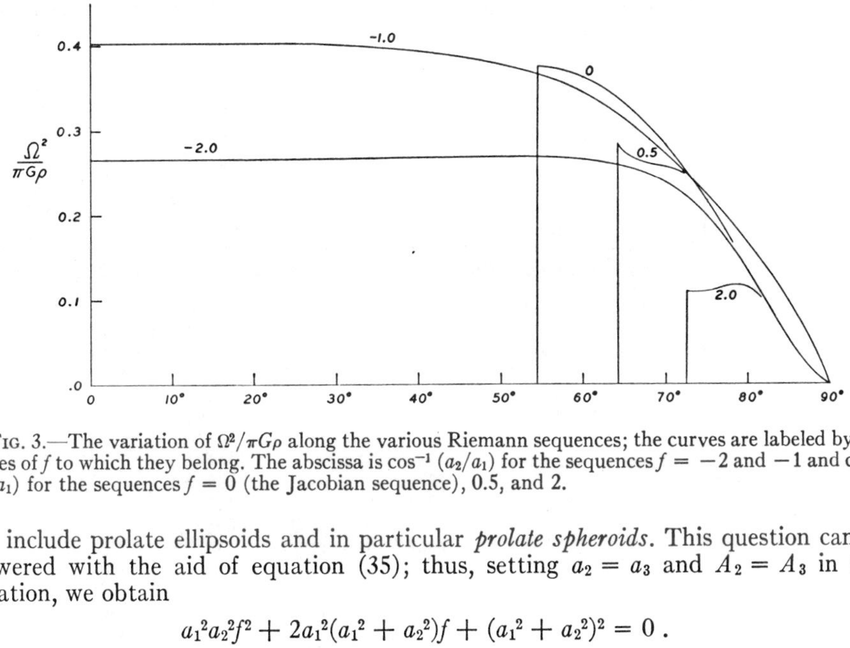

FIG. 3.—The variation of $\Omega^2/\pi G\rho$ along the various Riemann sequences; the curves are labeled by the values of f to which they belong. The abscissa is $\cos^{-1}(a_2/a_1)$ for the sequences $f = -2$ and -1 and $\cos^{-1}(a_3/a_1)$ for the sequences $f = 0$ (the Jacobian sequence), 0.5, and 2.

of f include prolate ellipsoids and in particular ***prolate spheroids***. This question can be answered with the aid of equation (35); thus, setting $a_2 = a_3$ and $A_2 = A_3$ in this equation, we obtain

$$a_1^2 a_2^2 f^2 + 2a_1^2(a_1^2 + a_2^2)f + (a_1^2 + a_2^2)^2 = 0\,. \tag{65}$$

Solving equation (65) for f, we have

$$f = -\left(1 + \frac{a_1^2}{a_2^2}\right)\left[1 \pm \left(1 - \frac{a_2^2}{a_1^2}\right)^{1/2}\right]. \tag{66}$$

In terms of the eccentricity

$$e = \surd(1 - a_2^2/a_1^2)\,, \tag{67}$$

the solution for f takes the simple form

$$f = -\frac{2 - e^2}{1 \pm e}\,. \tag{68}$$

The two solutions for f,

$$f = -\frac{2 - e^2}{1 + e} \quad \text{and} \quad f\dagger = -\frac{2 - e^2}{1 - e}\,, \tag{69}$$

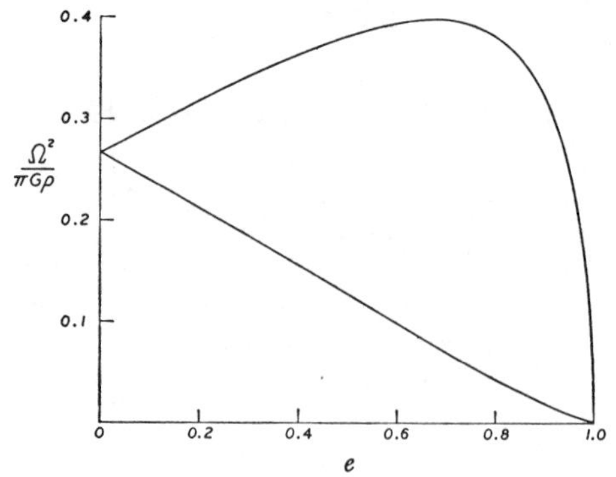

FIG. 4.—The variation of $\Omega^2/\pi G\rho$ along the sequence of prolate spheroids for the two adjoint configurations associated with each figure.

TABLE 3

THE PHYSICAL PARAMETERS OF THE PROLATE SPHEROIDS INCLUDED AMONG THE RIEMANN ELLIPSOIDS

e	f	Ω^2	Q_1	Q_2	$f^\dagger$	$(\Omega^\dagger)^2$	$Q_1^\dagger$	$Q_2^\dagger$
0.......	−2.00000	0.26667	0.51640	−0.51640	− 2.00000	0.26667	0.51640	−0.51640
0.05....	−1.90238	.28018	.50411	− .50285	− 2.10263	.25350	0.52999	− .52866
.10....	−1.80909	.29289	.49199	− .48707	− 2.21111	.23963	0.54391	− .53847
.15....	−1.71957	.30566	.48075	− .46994	− 2.32647	.22592	0.55919	− .54661
.20....	−1.63333	.31810	.47000	− .45120	− 2.45000	.21207	0.57563	− .55261
.25....	−1.55000	.33016	.45968	− .43095	− 2.58333	.19810	0.59344	− .55635
.30....	−1.46923	.34177	.44970	− .40923	− 2.72857	.18403	0.61284	− .55769
.35....	−1.39074	.35285	.44001	− .38611	− 2.88846	.16989	0.63412	− .55644
.40....	−1.31429	.36324	.43050	− .36162	− 3.06667	.15568	0.65760	− .55238
.45....	−1.23966	.37281	.42109	− .33582	− 3.26818	.14141	0.68372	− .54526
.50....	−1.16667	.38131	.41167	− .30875	− 3.50000	.12710	0.71303	− .53477
.55....	−1.09516	.38846	.40211	− .28047	− 3.77222	.11278	0.74628	− .52053
.60....	−1.02500	.39386	.39224	− .25103	− 4.10000	.09846	0.78448	− .50206
.65....	−0.95606	.39692	.38183	− .22051	− 4.50714	.08420	0.82904	− .47877
.70....	−0.88824	.39682	.37055	− .18898	− 5.03333	.07003	0.88209	− .44986
.75....	−0.82143	.39225	.35789	− .15658	− 5.75000	.05604	0.94688	− .41426
.80....	−0.75556	.38112	.34297	− .12347	− 6.80000	.04235	1.02892	− .37041
.82....	−0.72945	.37409	.33606	− .11009	− 7.37556	.03700	1.06860	− .35007
.84....	−0.70348	.36508	.32838	− .09668	− 8.09000	.03175	1.11360	− .32784
.86....	−0.67763	.35364	.31972	− .08325	− 9.00286	.02662	1.16536	− .30346
.88....	−0.65191	.33908	.30974	− .06988	−10.21333	.02164	1.22597	− .27658
.90....	−0.62632	.32045	.29794	− .05661	−11.90000	.01687	1.29868	− .24675
.92....	−0.60083	.29628	.28350	− .04354	−14.42000	.01234	1.38884	− .21333
.94....	−0.57546	.26412	.26491	− .03084	−18.60667	.00817	1.50636	− .17534
.96....	−0.55020	.21944	.23900	− .01874	−26.96000	.00448	1.67300	− .13116
0.98....	−0.52505	0.15146	0.19656	−0.00778	−51.98000	0.00153	1.95570	−0.07745

given by equation (68) define adjoint configurations; for, as can be readily verified, f and $f^\dagger$ as defined are consistent with the general relation (39) which now requires that

$$f = \frac{(2-e^2)^2}{1-e^2}\frac{1}{f^\dagger}. \tag{70}$$

The solution for f given by equation (68) allows all values of $f < -\frac{1}{2}$ (see Fig. 1). Therefore, along all Riemann sequences for $f < -\frac{1}{2}$, a prolate spheroid occurs; therefore, in the (a_2,a_3)-plane the curves representing these sequences must intersect the line $a_2 = a_3$. We conclude then that *the Riemann sequences for* f $< -\frac{1}{2}$ *start as oblate ellipsoids and end as prolate ellipsoids, while the Riemann sequences for* f $> -\frac{1}{2}$ *consist entirely of oblate ellipsoids.*

The values of Ω^2 that are to be associated with a prolate spheroid of eccentricity e follow from equation (33). We find

$$\Omega^2 = (1 \pm e)B_{12}(e)\,; \tag{71}$$

the variation of Ω^2 with e predicted by this equation is illustrated in Figure 4.

The occurrence of a sequence of prolate spheroids among the Riemann ellipsoids seems to have been first noticed by Greenhill (1879, 1880).

Finally, in Table 3 we list some of the physical parameters which characterize the sequence of prolate spheroids.

c) The Self-adjoint Sequences $x = \pm 1$

From equations (36)–(38) it is clear that any set of values a_1, a_2 ($<a_1$ by definition), and a_3 consistent with these equations provides a solution. The only restriction these equations impose is the requirement that equation (38) allows real roots for x; and the reality of x requires that

$$A_1a_1^2 - A_2a_2^2 \geq a_1a_2\left|A_1 - A_2 + \frac{a_1^2-a_2^2}{a_1^2a_2^2}\,a_3^2A_3\right|. \tag{72}$$

On simplification, this condition leads to the inequalities

$$\frac{a_1+a_2}{a_1a_2}\,a_3^2A_3 \leq a_2A_2 + a_1A_1 \qquad \text{(case i)}$$

and

$$\frac{a_1-a_2}{a_1a_2}\,a_3^2A_3 \geq a_2A_2 - a_1A_1 \qquad \text{(case ii)}. \tag{73}$$

When these inequalities degenerate into equalities, equation (38) allows two equal roots ($x = -1$ and $x = +1$, respectively), and we are led to two *self-adjoint sequences* along which

$$x = -1, f = f^\dagger = -(a_1^2 + a_2^2)/a_1a_2, \qquad \Omega^2 = -Q_1Q_2 = B_{12}\,,$$
$$Q_1\Omega = +a_1B_{12}/a_2\,, \qquad \text{and} \qquad Q_2\Omega = -a_2B_{12}/a_1\,;$$

and

$$x = +1, f = f^\dagger = +(a_1^2 + a_2^2)/a_1a_2\,, \qquad \Omega^2 = -Q_1Q_2 = B_{12}\,,$$
$$Q_1\Omega = -a_1B_{12}/a_2\,, \qquad \text{and} \qquad Q_2\Omega = +a_2B_{12}/a_1\,. \tag{74}$$

The properties of the Riemann ellipsoids along these two self-adjoint sequences are listed in Table 4. The sequences are further delineated in Figure 2. It is clear that these self-adjoint sequences limit the domain of occupancy of the Riemann sequences in the $(a_2/a_1,\ a_3/a_1)$-plane.

d) The Riemann Sequences for $f = -1.0, -0.6, +0.5$, *and* 2.0

The equations governing the equilibrium of the Riemann ellipsoids belonging to $f = -1.0, -0.6, +0.5$, and 2.0 have been solved numerically for a sufficient number of cases to determine their behavior along the respective sequences. The results of the calculations are summarized in Tables 5, 6, 7, and 8. The sequences are delineated in Figure 2; and the variation of Ω^2 along them is further exhibited in Figure 3.

And finally, in Table 9 we list the properties of a few ellipsoids which occur in the

TABLE 4

THE PROPERTIES OF THE RIEMANN ELLIPSOIDS ALONG THE SELF-ADJOINT SEQUENCES

a) The Sequence $x = -1$

a_2/a_1	0.68127	0.56284	0.51014	0.46101	0.37135	0.29066	0.21645	0.14641	0.077470
a_3/a_1	0.80	0.70	0.65	0.60	0.50	0.40	0.30	0.20	0.10
A_1	0.51198	0.43747	0.40070	0.36423	0.29223	0.22169	0.15342	0.08918	0.03320
A_2	0.81312	0.87786	0.90841	0.93764	0.99217	1.04057	1.08076	1.10810	1.10988
A_3	0.67490	0.68468	0.69089	0.69813	0.71561	0.73774	0.76582	0.80272	0.85692
$f=f^\dagger$	−2.1491	−2.3395	−2.4704	−2.6302	−3.0642	−3.7311	−4.8364	−6.9766	−12.986
$\Omega^2=-Q_1Q_2$	0.25115	0.23326	0.22209	0.20948	0.18027	0.14612	0.10783	0.06686	0.02670
$\Omega=\Omega^\dagger$	0.50115	0.48298	0.47127	0.45769	0.42458	0.38226	0.32838	0.25858	0.16340
$Q_1=Q_1^\dagger$	0.73561	0.85810	0.92381	0.99279	1.14335	1.31514	1.51711	1.76611	2.1092
$Q_2=Q_2^\dagger$	−0.34142	−0.27184	−0.24041	−0.21100	−0.15767	−0.11111	−0.07108	−0.03786	− 0.01266

b) The Sequence $x = +1$

a_2/a_1	0.75	0.70	0.60	0.55	0.50	0.40	0.30	0.20	0.10
a_3/a_1	0.26129	0.25172	0.23105	0.21986	0.20798	0.18181	0.15147	0.11513	0.068941
A_1	0.27694	0.26229	0.23091	0.21412	0.19653	0.15887	0.11787	0.074102	0.030198
A_2	0.40912	0.42560	0.46236	0.48299	0.50536	0.55641	0.61842	0.69642	0.80166
A_3	1.31395	1.31212	1.30672	1.30290	1.29811	1.28472	1.26370	1.22948	1.16814
$f=f^\dagger$	2.0833	2.1286	2.2667	2.3682	2.5000	2.9000	3.6333	5.2000	10.1000
$\Omega^2=-Q_1Q_2$	0.10699	0.10538	0.10072	0.097508	0.093583	0.083143	0.068368	0.048172	0.022406
$\Omega=\Omega^\dagger$	0.32709	0.32462	0.31737	0.31226	0.30591	0.28834	0.26147	0.21948	0.14969
$Q_1=Q_1^\dagger$	−0.43612	−0.46375	−0.52894	−0 56775	−0.61183	−0 72086	−0.87157	−1.0974	− 1.4969
$Q_2=Q_2^\dagger$	0.24532	0.22724	0.19042	0.17174	0.15296	0.11534	0.078442	0.043896	0.014969

TABLE 5*

THE PROPERTIES OF THE RIEMANN ELLIPSOIDS ALONG THE SEQUENCES FOR $f = -1.0$

ϕ	....	38°1727	60°	61°	62°	66°	67°	68°
θ	....	90°	74°627	75°026	75°460	77°492	78°063	78°656
a_2/a_1	1.00000	0.78615	0.50000	0.48481	0.46947	0.40674	0.39073	0.37461
a_3/a_1	0.90352	0.78615	0.55019	0.53489	0.51917	0.45231	0.43465	0.41664
A_1	0.63925	0.54342	0.36507	0.35374	0.34210	0.29259	0.27951	0.26616
A_2	0.63925	0.72829	0.86116	0.86822	0.87528	0.90340	0.91030	0.91709
A_3	0.72151	0.72829	0.77377	0.77804	0.78262	0.80401	0.81020	0.81674
Ω^2	0.40196	0.39527	0.34433	0.33907	0.33339	0.30578	0.29753	0.28867
Q_1Q_2	−0.10049	−0.09331	−0.05509	−0.05225	−0.04934	−0.03724	−0.03419	−0.03115
Ω	0.63400	0.62870	0.58679	0.58230	0.57740	0.55298	0.54546	0.53728
Q_1	0.31700	0.38856	0.46943	0.47148	0.47312	0.47448	0.47321	0.47116
Q_2	−0.31700	−0.24014	−0.11736	−0.11082	−0.10428	−0.07850	−0.07225	−0.06612
$f^\dagger$	−4.00000	−4.23607	−6.25000	−6.48963	−6.75754	−8.21012	−8.70271	−9.26638
$\Omega^\dagger$	0.31700	0.30547	0.23472	0.22858	0.22212	0.19299	0.18490	0.17650
$Q_1^\dagger$	0.63400	0.79972	1.17358	1.20108	1.22988	1.35954	1.39599	1.43424
$Q_2^\dagger$	−0.63400	−0.49426	−0.29340	−0.28230	−0.27107	−0.22492	−0.21313	−0.20127

* Cos $\phi = a_2/a_1$; sin $\theta = [(a_1^2 - a_3^2)/(a_1^2 - a_2^2)]^{1/2}$; and the remaining symbols have the same meanings as explained at the bottom of Table 2. The first and the second columns refer to configurations that are on the lines $a_2/a_1 = 1$ and $a_2 = a_3$, respectively.

domains bounded by the self-adjoint sequences $x = -1$ and $x = +1$ and the sequences $f = -2$ and $f = +2$, respectively.

e) The Riemann Sequences for $-2 > f > -\infty$ *and* $+2 < f < +\infty$

We have seen that Riemann sequences for f in the range $-2 \leq f \leq +2$ are represented in the $(a_2/a_1, a_3/a_1)$-plane by a continuous family of non-intersecting curves which join the origin to points along the line O_2S (see Fig. 2). The Riemann sequences for f outside this range are included in the domain bounded by the self-adjoint sequences for

TABLE 6*

THE PROPERTIES OF THE RIEMANN ELLIPSOIDS ALONG THE SEQUENCES FOR $f = -0.6$

ϕ	37°.5628	52°	67°.0221	72°
θ	0°	71°.085	90°	87°.400
a_2/a_1	1.00000	0.66655	0.39038	0.30902
a_3/a_1	0.79269	0.61566	0.39038	0.31201
A_1	0.60299	0.44796	0.26272	0.19916
A_2	0.60299	0.74086	0.86864	0.90508
A_3	0.79402	0.81118	0.86864	0.89576
Ω^2	0.42476	0.39711	0.29537	0.24232
Q_1Q_2	−0.03823	−0.03045	− 0.01220	− 0.00694
Ω	0.65173	0.63017	0.54347	0.49226
Q_1	0.19552	0.26179	0.28296	0.26961
Q_2	−0.19552	−0.11631	− 0.04312	− 0.02575
f†	−6.66667	−7.82511	−14.52393	−20.94605
Ω†	0.19552	0.17450	0.11046	0.08331
Q_1†	0.65173	0.94541	1.39218	1.59299
Q_2†	−0.65173	−0.42004	− 0.21216	− 0.15212

* In the first two columns $\cos\phi = a_3/a_1$ and $\sin\theta = [(a_1^2 - a_2^2)/(a_1^2 - a_3^2)]^{1/2}$; and in the second two columns $\cos\phi = a_2/a_1$ and $\sin\theta = [(a_1^2 - a_3^2)/(a_1^2 - a_2^2)]^{1/2}$; and the remaining symbols have the same meanings as explained at the bottom of Table 2. And the first and the third columns refer to configurations on the lines $a_2/a_1 = 1$ and $a_2 = a_3$, respectively.

TABLE 7*

THE PROPERTIES OF THE RIEMANN ELLIPSOIDS ALONG THE SEQUENCES FOR $f = 0.5$

ϕ	64°.1095	69°	72°	73°	74°	75°
θ	0°	52°.478	65°.214	68°.543	71°.498	74°.120
a_2/a_1	1.00000	0.67212	0.50444	0.45591	0.41113	0.36992
a_3/a_1	0.43665	0.35837	0.30902	0.29237	0.27564	0.25882
A_1	0.43545	0.32896	0.26099	0.23886	0.21726	0.19631
A_2	0.43545	0.55436	0.64248	0.67264	0.70249	0.73169
A_3	1.12910	1.11668	1.09652	1.08850	1.08026	1.07200
Ω^2	0.28182	0.25875	0.25141	0.24147	0.23000	0.21711
Q_1Q_2	−0.01761	−0.01387	− 0.01016	− 0.00860	− 0.00711	− 0.00575
Ω	0.53086	0.50868	0.50141	0.49139	0.47958	0.46595
Q_1	−0.13272	−0.17520	− 0.19985	− 0.20342	− 0.20512	− 0.20493
Q_2	0.13272	0.07914	0.05085	0.04228	0.03467	0.02804
f†	8.00000	9.33075	12.36863	14.03782	16.17040	18.88910
Ω†	0.13272	0.11775	0.10081	0.09274	0.08433	0.07581
Q_1†	−0.53086	−0.75682	− 0.99399	− 1.07783	− 1.16649	− 1.25960
Q_2†	0.53086	0.34189	0.25293	0.22403	0.19717	0.17237

* $\cos\phi = a_3/a_1$; $\sin\theta = [(a_1^2 - a_2^2)/(a_1^2 - a_3^2)]^{1/2}$; and the remaining symbols have the same meanings as explained at the bottom of Table 2.

$x = -1$ and $x = +1$; and they can be sketched in with the information already at our disposal by following the procedure outlined below.

The adjoint of an ellipsoid along a given sequence belongs to some other sequence; and the value of f to which the adjoint belongs is $f\dagger$ given by equation (39). Along the sequence for a given f, the value of $f\dagger$ varies; and this variation, for the sequences we have constructed, is shown in Figure 5. Figure 5 also includes the variation of $f(= f\dagger)$ along the self-adjoint sequences for $x = -1$ and $x = +1$. These variations together with those along the Maclaurin sequence and the sequence of the prolate spheroids (shown in Fig. 1) will enable us to delineate in Figure 2 the Riemann sequences for f outside the range $-2 \leq f \leq +2$. The sequences for $f = -5, -9, +9$, and $+5$ (determined graphically with the aid of Figs. 1 and 5) have been sketched in Figure 2.

We observe that the sequences for $f < -2$ and $f > +2$ are represented in the $(a_2/a_1, a_3/a_1)$-plane by curves which consist of two parts: a part which joins a point on the line

TABLE 8*

THE PROPERTIES OF THE RIEMANN ELLIPSOIDS ALONG THE SEQUENCES FOR $f = 2.0$

ϕ	72°3425	75°	76°	77°	79°	80°	83°
θ	0°	44°516	51°732	57°936	68°632	73°340	83°401
a_2/a_1	1.00000	0.73578	0.64781	0.56406	0.40538	0.33146	0.16692
a_3/a_1	0.30333	0.25882	0.24192	0.22495	0.19081	0.17365	0.12187
A_1	0.34132	0.27299	0.24681	0.22028	0.16591	0.13851	0.06863
A_2	0.34132	0.41391	0.44501	0.47975	0.56815	0.62574	0.81114
A_3	1.31737	1.31309	1.30818	1.29997	1.26594	1.23575	1.12024
Ω^2	0.11005	0.11159	0.11292	0.11453	0.11693	0.11554	0.08565
Q_1Q_2	−0.11005	−0.10171	−0.09405	−0.08389	−0.05670	−0.04122	− 0.00904
Ω	0.33175	0.33405	0.33604	0.33843	0.34196	0.33991	0.29267
Q_1	−0.33175	−0.43344	−0.47341	−0.51348	−0.58739	−0.61252	− 0.56946
Q_2	0.33175	0.23465	0.19867	0.16337	0.09653	0.06730	0.01587
$f\dagger$	2.00000	2.19426	2.40127	2.73061	4.12481	5.60587	18.95916
$\Omega\dagger$	0.33175	0.31892	0.30668	0.28963	0.23811	0.20303	0.09506
$Q_1\dagger$	−0.33175	−0.45400	−0.51873	−0.59999	−0.84355	−1.02547	− 1.75332
$Q_2\dagger$	0.33175	0.24579	0.21769	0.19089	0.13862	0.11267	0.04885

* Cos $\phi = a_3/a_1$; sin $\theta = [(a_1^2 - a_2^2)/(a_1^2 - a_3^2)]^{1/2}$; and the remaining symbols have the same meanings as explained at the bottom of Table 2.

TABLE 9

THE PROPERTIES OF THE RIEMANN ELLIPSOIDS WHICH OCCUR IN THE DOMAINS BOUNDED BY THE SEQUENCES $x = \mp 1$ AND $f = \mp 2$

	Ellipsoids in the Domain Bounded by the Sequences $x=-1$ and $f=-2$				Ellipsoids in the Domain Bounded by the Sequences $x=+1$ and $f=+2$			
a_2/a_1	0.30769	0.23077	0.16667	0.15385	0.3	0.2	0.2	0.1
a_3/a_1	0.4	0.3	0.2	0.2	0.15789	0.12500	0.11765	0.076923
A_1	0.23009	0.16041	0.098128	0.092535	0.12169	0.079039	0.075378	0.033011
A_2	1.00943	1.04656	1.04051	1.08239	0.63502	0.73231	0.70580	0.85354
A_3	0.76049	0.79303	0.86136	0.82508	1.24328	1.18865	1.21882	1.11345
f	−5.9801	−8.3140	−17.232	−11.8405	5.3534	9.6687	7.0137	25.359
Ω^2	0.077686	0.051209	0.016167	0.032899	0.044732	0.023252	0.034840	0.006770
Q_1Q_2	−0.21949	−0.16993	− 0.12624	− 0.10418	−0.097113	−0.080387	−0.063381	− 0.042676
Ω	0.27872	0.22629	0.12715	0.18138	0.21150	0.15249	0.18666	0.082279
Q_1	1.5226	1.7863	2.1318	2.0980	−1.03877	−1.4176	−1.2588	− 2.0658
Q_2	−0.14415	−0.095128	− 0.059217	− 0.049656	0.093489	0.056705	0.050351	0.020658
$f\dagger$	−2.1166	−2.5055	− 2.2068	− 3.7392	2.4659	2.7967	3.8553	4.0227
$\Omega\dagger$	0.46850	0.41222	0.35530	0.32277	0.31163	0.28353	0.25176	0.20658
$Q_1\dagger$	0.90585	0.98061	0.76289	1.1790	−0.70500	−0.76243	−0.93327	− 0.82279
$Q_2\dagger$	−0.085760	−0.052222	− 0.021191	− 0.02790	0.063450	0.030497	0.037331	0.008228

SM_2 (or M_2O_2) to a point on the self-adjoint sequence for $x = -1$ (or $x = +1$) and a part which joins the point on the self-adjoint sequence to the origin. It follows that the sequences for $f < -2$ intersect one another in the domain bounded by the sequences $f = -2$ and $x = -1$; similarly, the sequences for $f > +2$ intersect one another in the domain bounded by the sequences $f = +2$ and $x = +1$; and these facts are consistent with Dedekind's theorem which requires that each figure of equilibrium (except when it is self-adjoint) allows two distinct states of internal motions.

VI. THE SECOND-ORDER VIRIAL EQUATIONS GOVERNING SMALL OSCILLATIONS ABOUT EQUILIBRIUM

Suppose that an equilibrium ellipsoid determined consistently with respect to equations (33) and (35) is slightly perturbed. Let the ensuing motions be described in terms

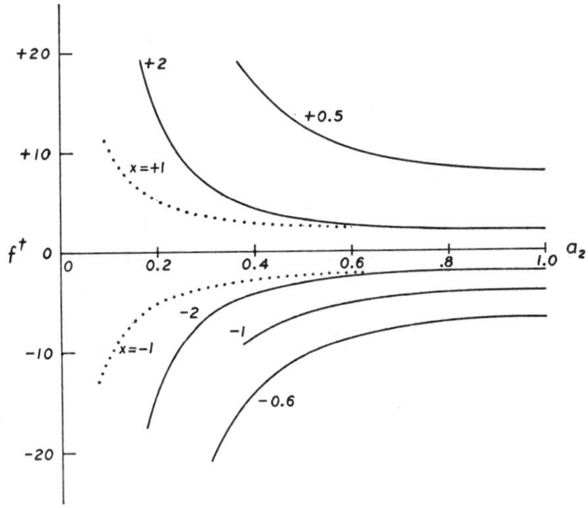

FIG. 5.—The variation of $f\dagger$ along the different Riemann sequences: the curves are labeled by the values of f to which they belong. The dotted curves represent the variation of $f = f\dagger$ along the two self-adjoint sequences for $x = +1$ and $x = -1$.

of a Lagrangian displacement $\boldsymbol{\xi}$. Then a fluid element originally at $\boldsymbol{x}$ with a velocity $\boldsymbol{u}$ will find itself at $\boldsymbol{x} + \boldsymbol{\xi}(\boldsymbol{x}, t)$ with a velocity $\boldsymbol{u} + \Delta\boldsymbol{u}$ where (to the first order in $\boldsymbol{\xi}$)

$$\Delta u_i = \frac{\partial \xi_i}{\partial t} + u_j \frac{\partial \xi_i}{\partial x_j}, \tag{75}$$

where u_j denotes the velocity in the equilibrium state. The virial equation (18) linearized to the first order in $\boldsymbol{\xi}$, gives

$$\frac{d}{dt}\int_V \rho(\Delta u_i x_j + u_i \xi_j)\, d\boldsymbol{x} - \int_V \rho(\Delta u_i u_j + \Delta u_j u_i)\, d\boldsymbol{x}$$
$$- 2\Omega\epsilon_{il3}\int_V \rho(\Delta u_l x_j + u_l \xi_j)\, d\boldsymbol{x} = \Omega^2 (V_{ij} - \delta_{i3} V_{3j}) + \delta\mathfrak{W}_{ij} + \delta_{ij}\delta\Pi\,, \tag{76}$$

where

$$V_{ij} = V_{i;j} + V_{j;i} = \int_V \rho \xi_i x_j\, d\boldsymbol{x} + \int_V \rho \xi_j x_i\, d\boldsymbol{x}\,. \tag{77}$$

Also, it follows from the definitions of the various quantities that (cf. Paper I, eq. [27])

$$\frac{dV_{i;j}}{dt} = \int_V \rho \Delta u_i x_j dx + \int_V \rho \xi_i u_j dx . \tag{78}$$

In the further reduction of equation (73) we shall assume that in the equilibrium configuration

$$u_j = Q_{jl} x_l , \tag{79}$$

where the Q_{jl}'s are certain constants. On this assumption equation (78) gives

$$\frac{dV_{i;j}}{dt} = \int_V \rho \Delta u_i x_j dx + Q_{jl} V_{i;l} . \tag{80}$$

A further application of this last equation gives

$$\int_V \rho \Delta u_i u_j dx = Q_{jl} \int_V \rho \Delta u_i x_l dx = Q_{jl} \frac{dV_{i;l}}{dt} - Q^2{}_{jl} V_{i;l} , \tag{81}$$

where Q^2 denotes the square of the matrix Q. Making use of the foregoing relations in equation (76) and simplifying, we obtain

$$\begin{aligned} \frac{d^2 V_{i;j}}{dt^2} - 2Q_{jl} \frac{dV_{i;l}}{dt} + Q^2{}_{jl} V_{i;l} + Q^2{}_{il} V_{j;l} &= \Omega^2 (V_{ij} - \delta_{i3} V_{3j}) \\ + 2\Omega \epsilon_{il3} \left(\frac{dV_{l;j}}{dt} - Q_{jk} V_{l;k} + Q_{lk} V_{j;k} \right) &+ \delta \mathfrak{W}_{ij} + \delta_{ij} \delta \Pi . \end{aligned} \tag{82}$$

We shall now suppose that the Lagrangian displacement is of the form

$$\xi(x,t) = e^{\lambda t} \xi(x) , \tag{83}$$

where λ is a parameter whose characteristic values are to be determined. Equation (82) then gives

$$\begin{aligned} \lambda^2 V_{i;j} - 2\lambda Q_{jl} V_{i;l} - 2\lambda \Omega \epsilon_{il3} V_{l;j} + Q_{jl}{}^2 V_{i;l} + Q_{il}{}^2 V_{j;l} \\ - 2\Omega \epsilon_{il3} (Q_{lk} V_{j;k} - Q_{jk} V_{l;k}) = \Omega^2 (V_{ij} - \delta_{i3} V_{3j}) + \delta \mathfrak{W}_{ij} + \delta_{ij} \delta \Pi . \end{aligned} \tag{84}$$

And this equation must be supplemented by the condition,

$$\frac{V_{11}}{a_1{}^2} + \frac{V_{22}}{a_2{}^2} + \frac{V_{33}}{a_3{}^2} = 0 , \tag{85}$$

required by the solenoidal character of ξ.

For the particular case of the Riemann ellipsoids, the matrices Q and Q^2 have the simple forms

$$Q = \begin{vmatrix} 0 & Q_1 & 0 \\ Q_2 & 0 & 0 \\ 0 & 0 & 0 \end{vmatrix} \quad \text{and} \quad Q^2 = \begin{vmatrix} Q_1 Q_2 & 0 & 0 \\ 0 & Q_1 Q_2 & 0 \\ 0 & 0 & 0 \end{vmatrix} . \tag{86}$$

Also, it is known that $\delta\mathfrak{W}_{ij}$ can be expressed in terms of the symmetrized virials V_{ij}. Thus (cf. Chandrasekhar and Lebovitz 1963*b*, eqs. [47] and [48])

$$\delta\mathfrak{W}_{ij} = -2B_{ij}V_{ij} \qquad (i \neq j) \tag{87}$$

and

$$\delta\mathfrak{W}_{ii} = -(2B_{ii} - a_i^2 A_{ii})V_{ii} + a_i^2 \sum_{l \neq i} A_{il}V_{ll} \tag{88}$$

(no summation over repeated indices in eqs. [87] and [88]).

(In writing eqs. [87] and [88] a common factor $\pi G\rho a_1 a_2 a_3$ has been suppressed. The suppression of this factor is equivalent to the normalization of the index symbols to give $\Sigma A_i = 2$ and the measurement of Ω^2 and ζ_Ω^2 in the unit $\pi G\rho$—conventions which we have consistently adopted.)

We shall now write down the explicit forms which the different components of equation (84) take in view of the special forms of the matrices $\boldsymbol{Q}$ and $\boldsymbol{Q}^2$. The five equations even in the index 3 are:

$$\tfrac{1}{2}\lambda^2 V_{33} = \delta\mathfrak{W}_{33} + \delta\Pi\,, \tag{89}$$

$$(\tfrac{1}{2}\lambda^2 + Q_1Q_2 - \Omega^2)V_{11} - 2\lambda Q_1 V_{1;2} - 2\lambda\Omega V_{2;1} - \Omega(Q_2V_{11} - Q_1V_{22}) = \delta\mathfrak{W}_{11} + \delta\Pi\,, \tag{90}$$

$$(\tfrac{1}{2}\lambda^2 + Q_1Q_2 - \Omega^2)V_{22} - 2\lambda Q_2 V_{2;1} + 2\lambda\Omega V_{1;2} + \Omega(Q_1V_{22} - Q_2V_{11}) = \delta\mathfrak{W}_{22} + \delta\Pi\,, \tag{91}$$

$$\lambda^2 V_{1;2} - \lambda Q_2 V_{11} + Q_1Q_2V_{12} - \lambda\Omega V_{22} = \delta\mathfrak{W}_{12} + \Omega^2 V_{12} = -(2B_{12} - \Omega^2)V_{12}\,, \tag{92}$$

$$\lambda^2 V_{2;1} - \lambda Q_1 V_{22} + Q_1Q_2V_{12} + \lambda\Omega V_{11} = \delta\mathfrak{W}_{12} + \Omega^2 V_{12} = -(2B_{12} - \Omega^2)V_{12}\,, \tag{93}$$

where in equations (92) and (93) we have substituted for $\delta\mathfrak{W}_{12}$ in accordance with equation (87). Similarly, the four equations odd in the index 3 are:

$$\begin{aligned}
\lambda^2 V_{1;3} - 2\lambda\Omega V_{2;3} + Q_1Q_2V_{3;1} - 2\Omega Q_2 V_{3;1} &= \delta\mathfrak{W}_{13} + \Omega^2 V_{13} = -(2B_{13} - \Omega^2)V_{13}\,,\\
\lambda^2 V_{2;3} + 2\lambda\Omega V_{1;3} + Q_1Q_2V_{3;2} + 2\Omega Q_1 V_{3;2} &= \delta\mathfrak{W}_{23} + \Omega^2 V_{23} = -(2B_{23} - \Omega^2)V_{23}\,,\\
\lambda^2 V_{3;1} - 2\lambda Q_1 V_{3;2} + Q_1Q_2V_{3;1} &= \delta\mathfrak{W}_{13} = -2B_{13}V_{13}\,,\\
\lambda^2 V_{3;2} - 2\lambda Q_2 V_{3;1} + Q_1Q_2V_{3;2} &= \delta\mathfrak{W}_{23} = -2B_{23}V_{23}\,.
\end{aligned} \tag{94}$$

VII. THE CHARACTERISTIC FREQUENCIES OF OSCILLATION OF THE RIEMANN ELLIPSOIDS BELONGING TO THE SECOND HARMONICS

Considering first equations (89)–(93) governing the even modes, we observe that, in view of the relation (24), equations (92) and (93) become

$$\lambda^2 V_{1;2} - \lambda Q_2 V_{11} - \lambda\Omega V_{22} = 0 \tag{95}$$

and

$$\lambda^2 V_{2;1} - \lambda Q_1 V_{22} + \lambda\Omega V_{11} = 0\,. \tag{96}$$

Excluding the possibility that λ may be zero—a possibility to which we shall return presently—we may conclude from equations (95) and (96) that

$$\lambda V_{1;2} = Q_2V_{11} + \Omega V_{22} \tag{97}$$

and

$$\lambda V_{2;1} = Q_1V_{22} - \Omega V_{11}\,. \tag{98}$$

Eliminating $V_{1;2}$ and $V_{2;1}$ from equations (90) and (91) with the aid of equations (97) and (98), we obtain

$$(\tfrac{1}{2}\lambda^2 + \Omega^2 - Q_1Q_2 - \Omega Q_2)V_{11} - 3\Omega Q_1 V_{22} = \delta\mathfrak{W}_{11} + \delta\Pi \tag{99}$$

and

$$(\tfrac{1}{2}\lambda^2 + \Omega^2 - Q_1Q_2 + \Omega Q_1)V_{22} + 3\Omega Q_2 V_{11} = \delta\mathfrak{W}_{22} + \delta\Pi\,. \tag{100}$$

Next eliminating $\delta\Pi$ from equations (99) and (100) with the aid of equation (89), we obtain the pair of equations

$$\begin{aligned}(\tfrac{1}{2}\lambda^2 + 2B_{12} - \Omega Q_2)V_{11} - 3\Omega Q_1 V_{22} - \tfrac{1}{2}\lambda^2 V_{33} &= \delta\mathfrak{W}_{11} - \delta\mathfrak{W}_{33} \\ &= -(3B_{11} - B_{13})V_{11} + (B_{23} - B_{12})V_{22} + (3B_{33} - B_{13})V_{33}\end{aligned} \tag{101}$$

and

$$\begin{aligned}(\tfrac{1}{2}\lambda^2 + 2B_{12} + \Omega Q_1)V_{22} + 3\Omega Q_2 V_{11} - \tfrac{1}{2}\lambda^2 V_{33} &= \delta\mathfrak{W}_{22} - \delta\mathfrak{W}_{33} \\ &= -(3B_{22} - B_{23})V_{22} + (B_{13} - B_{12})V_{11} + (3B_{33} - B_{23})V_{33}\,,\end{aligned} \tag{102}$$

where we have made use of the relation (24) and further substituted for $\delta\mathfrak{W}_{11} - \delta\mathfrak{W}_{33}$ and $\delta\mathfrak{W}_{22} - \delta\mathfrak{W}_{33}$ in accordance with equation (88).

Equations (101) and (102) must be supplemented by equation (85) which expresses the solenoidal condition on $\boldsymbol{\xi}$; and these three equations lead to the following characteristic equation for λ^2:

$$\begin{bmatrix} \tfrac{1}{2}\lambda^2+2B_{12}-\Omega Q_2+3B_{11}-B_{13} & B_{12}-B_{23}-3\Omega Q_1 & -\tfrac{1}{2}\lambda^2-3B_{33}+B_{13} \\ B_{12}-B_{13}+3\Omega Q_2 & \tfrac{1}{2}\lambda^2+2B_{12}+\Omega Q_1+3B_{22}-B_{23} & -\tfrac{1}{2}\lambda^2-3B_{33}+B_{23} \\ \dfrac{1}{a_1^2} & \dfrac{1}{a_2^2} & \dfrac{1}{a_3^2} \end{bmatrix} = 0\,. \tag{103}$$

A somewhat simpler form of equation (103) is

$$\begin{bmatrix} \tfrac{1}{2}\lambda^2+2B_{12}-\Omega Q_2+3B_{11}-B_{13} & B_{12}-B_{23}-3\Omega Q_1 & 3(B_{12}+B_{11}-B_{33})-B_{23}-\Omega(Q_2+3Q_1) \\ B_{12}-B_{13}+3\Omega Q_2 & \tfrac{1}{2}\lambda^2+2B_{12}+\Omega Q_1+3B_{22}-B_{23} & 3(B_{12}+B_{22}-B_{33})-B_{13}+\Omega(Q_1+3Q_2) \\ \dfrac{1}{a_1^2} & \dfrac{1}{a_2^2} & \dfrac{1}{a_1^2}+\dfrac{1}{a_2^2}+\dfrac{1}{a_3^2} \end{bmatrix} = 0\,. \tag{104}$$

Returning to equations (95) and (96) and the possibility of a non-trivial root $\lambda = 0$, we observe that we do indeed have such a root belonging to a proper solution associated with the possibility

$$V_{11} = V_{22} = V_{33} = 0 \qquad \text{and} \qquad V_{12} \neq 0\,. \tag{105}$$

In other words, *all Riemann ellipsoids allow a non-trivial neutral mode.* The Jacobian and the Dedekind ellipsoids share this property, as indeed, they must as special cases.

Turning next to equations (94) governing the odd modes of oscillation, we first rewrite them in the forms

$$(\lambda^2 + 2B_{13} - \Omega^2)V_{1;3} + (2B_{13} - \Omega^2 + Q_1Q_2 - 2\Omega Q_2)V_{3;1} - 2\lambda\Omega V_{2;3} = 0\,, \tag{106}$$

$$(\lambda^2 + 2B_{23} - \Omega^2)V_{2;3} + (2B_{23} - \Omega^2 + Q_1Q_2 + 2\Omega Q_1)V_{3;2} + 2\lambda\Omega V_{1;3} = 0\,, \quad (107)$$

$$(\lambda^2 + 2B_{13} + Q_1Q_2)V_{3;1} + 2B_{13}V_{1;3} - 2\lambda Q_1 V_{3;2} = 0\,, \quad (108)$$

$$(\lambda^2 + 2B_{23} + Q_1Q_2)V_{3;2} + 2B_{23}V_{2;3} - 2\lambda Q_2 V_{3;1} = 0\,. \quad (109)$$

On the other hand, according to equations (25) and (26)

$$2B_{13} - (\Omega^2 - Q_1Q_2 + 2\Omega Q_2) = 2\,\frac{a_3^2}{a_1^2}\,B_{13} \quad (110)$$

and

$$2B_{23} - (\Omega^2 - Q_1Q_2 - 2\Omega Q_1) = 2\,\frac{a_3^2}{a_2^2}\,B_{23}\,. \quad (111)$$

With this simplification of the coefficients of $V_{3;1}$ and $V_{3;2}$ in equation (106) and (107), equations (106)–(109) lead to the following characteristic equation:

$$\begin{bmatrix} \lambda^2 + 2B_{13} - \Omega^2 & 2a_3^2B_{13}/a_1^2 & -2\lambda\Omega & 0 \\ 2B_{13} & \lambda^2 + 2B_{13} + Q_1Q_2 & 0 & -2\lambda Q_1 \\ 2\lambda\Omega & 0 & \lambda^2 + 2B_{23} - \Omega^2 & 2a_3^2B_{23}/a_2^2 \\ 0 & -2\lambda Q_2 & 2B_{23} & \lambda^2 + 2B_{23} + Q_1Q_2 \end{bmatrix} = 0\,. \quad (112)$$

By a judicious manipulation of this determinant, it can be shown that equation (112) allows the roots

$$\lambda^2 = -\Omega^2 \quad \text{and} \quad \lambda^2 = Q_1Q_2\,; \quad (113)$$

and on factoring out $(\lambda^2 + \Omega^2)(\lambda^2 - Q_1Q_2)$, we find that the characteristic equation reduces to

$$\lambda^4 + (4B_{13} + 4B_{23} + 2B_{12})\lambda^2 + (4B_{13} - \Omega Q_1)(4B_{23} + \Omega Q_2) = 0\,. \quad (114)$$

The most striking feature of the characteristic equations (104) and (114) is that, apart from constants that depend only on the semi-axes of the ellipsoid, the equations involve Ω, Q_1, and Q_2 only in the combinations ΩQ_1 and ΩQ_2; and these combinations, as we have noted (see p. 895), have the same values for a configuration and its adjoint. Hence, the roots of equations (104) and (114) are unchanged as we pass from a configuration to its adjoint; and moreover, the roots $-\Omega^2$ and Q_1Q_2, which the characteristic equation (112) allows, are simply interchanged in the process. Thus, *all the characteristic frequencies of oscillation of a Riemann ellipsoid, belonging to the second harmonics, are the same for the two states of internal motions that can prevail in accordance with Dedekind's theorem.* This theorem generalizes the result established in Paper I for the special case of the Jacobi and the Dedekind ellipsoids.

An additional fact of some interest may be noted here. For an irrotational ellipsoid, it can be shown that one of the roots of equation (114) is again Q_1Q_2 so that in this case the four characteristic roots belonging to the odd modes of oscillation are

$$\lambda^2 = -\Omega^2, \qquad \lambda^2 = Q_1Q_2 \text{ (double root)}, \qquad \lambda^2 = -(4B_{13} + 4B_{23} + \Omega^2)\,. \quad (115)$$

In Table 10 we list the characteristic frequencies of oscillation, determined in accordance with equations (104) and (114), of the Riemann ellipsoids for which the equilibrium parameters have been determined; and in Figure 6 the variation of these frequencies along the sequences are illustrated for the two cases, $f = -2$ and $f = +2$.

TABLE 10*

THE SQUARES OF THE CHARACTERISTIC FREQUENCIES BELONGING TO THE SECOND HARMONICS

(σ^2 Is Listed in the Unit $\pi G\rho$)

ϕ	σ_1^2	σ_2^2	σ_3^2	σ_4^2
	$f=-2$			
0°	4.2667	1.0667	2.4000	0.2667
30°	4.2164	1.0667	2.4001	.2617
45°	3.9852	1.0662	2.3998	.2389
55°	3.5944	1.0614	2.3975	.2003
57°	3.4830	1.0584	2.3962	.1892
59°	3.3596	1.0537	2.3943	.1769
60°	3.2936	1.0505	2.3931	.1702
70°	2.5523	0.9227	2.3595	.0891
71°	2.4852	0.8904	2.3523	.0803
72°	2.4233	0.8524	2.3439	.0715
75°	2.2761	0.7026	2.3107	.0467
78°	2.1880	0.5119	2.2627	.0258
80°	2.1525	0.3787	2.2217	.0152
85°	2.0810	0.1074	2.0928	0.0016

ϕ	σ_1^2	σ_2^2	σ_3^2	σ_4^2
	$f=-1$			
25°.3753	3.6177	1.1959	2.3033	0.3879
60°	2.7215	1.1468	2.3050	.2820
61°	2.6701	1.1350	2.3007	.2776
62°	2.6186	1.1208	2.2959	.2728
66°	2.4226	1.0347	2.2723	.2508
67°	2.3785	1.0040	2.2653	.2444
68°	2.3375	0.9692	2.2576	0.2377
	$f=-0.6$			
37°.5628	2.8714	1.3419	2.2648	0.4383
52°	2.6173	1.3026	2.2486	0.4134

ϕ	σ_1^2	σ_2^2	σ_3^2	σ_4^2
	$f=+0.5$			
64°.1095	1.5131	0.6341	1.6737	0.8020
69°	1.5416	.6481	1.7127	.7493
72°	1.6152	.5896	1.7522	.6795
73°	1.6459	.5663	1.7711	.6454
74°	1.6790	.5374	1.7903	.6081
75°	1.7132	0.5036	1.8097	0.5683
	$f=+2$			
72°.3425	1.2612	0	1.0797	1.0797
75°	1.2550	0.0125	1.1776	0.9767
76°	1.2514	.0256	1.2189	0.9319
77°	1.2504	.0454	1.2658	0.8820
79°	1.2817	.1090	1.3902	0.7590
80°	1.3373	.1460	1.4736	0.6807
83°	1.6632	0.1532	1.7399	0.3990

e	σ_1^2	σ_2^2	σ_3^2	σ_4^2
	Sequence of the Prolate Spheroids			
0	4.2667	1.0667	2.4000	0.2667
0.20	4.1982	1.0800	2.3959	.2737
.30	4.1082	1.0969	2.3903	.2825
.40	3.9738	1.1211	2.3818	.2947
.50	3.7850	1.1527	2.3692	.3101
.55	3.6659	1.1710	2.3609	.3189
.60	3.5273	1.1906	2.3508	.3281
.65	3.3666	1.2106	2.3383	.3375
.70	3.1810	1.2292	2.3229	.3467
.80	2.7295	1.2405	2.2785	.3602
.85	2.4757	1.1998	2.2455	.3597
.90	2.2437	1.0624	2.1998	.3453
0.95	2.0939	0.7249	2.1321	0.2924

a_3/a_1	σ_1^2	σ_2^2	σ_3^2	σ_4^2
	Along the Self-adjoint Sequence $x=-1$			
0.80	3.9326	1.0644	2.4000	+0.2320
.70	3.5844	1.0543	2.4000	+ .1917
.65	3.3870	1.0430	2.3995	+ .1664
.60	3.1833	1.0250	2.3987	+ .1378
.50	2.7896	0.9555	2.3953	+ .0712
.40	2.4688	0.8178	2.3875	− .0059
.30	2.2671	0.6003	2.3719	− .0894
.20	2.1782	0.3412	2.3395	− .1675
0.10	2.1531	0.1127	2.2679	−0.2061

a_2/a_1	σ_1^2	σ_2^2	σ_3^2	σ_4^2
	Along the Self-adjoint Sequence $x=+1$			
0.60	1.2449	0.0297	1.2361	0.9077
.55	1.2411	.0389	1.2609	.8776
.50	1.2376	.0494	1.2874	.8446
.40	1.2359	.0728	1.3471	.7678
.30	1.2546	.0932	1.4195	.6716
.20	1.3274	.0950	1.5140	.5457
0.10	1.5206	0.0575	1.6556	0.3671

* The roots σ_1^2 and σ_2^2 belong to the even modes (eq. [104]) and the roots σ_3^2 and σ_4^2 belong to the odd modes (eq. [114]).

TABLE 10—*Continued*

a_3/a_1	a_2/a_1	σ_1^2	σ_2^2	σ_3^2	σ_4^2	a_2/a_1	a_3/a_1	σ_1^2	σ_2^2	σ_3^2	σ_4^2
	Ellipsoids Included between the Sequences $f=-2$ and $x=-1$						Ellipsoids Included between the Sequences $f=+2$ and $x=+1$				
0.4...	0.308	2.4364	0.8510	2.3514	+0.0609	0.3.....	0.158	1.3102	0.1230	1.4638	0.6580
.3...	.231	2.2518	.6225	2.3242	− .0093	.2.....	.125	1.4547	.1314	1.5973	.5112
.2...	.167	2.1547	.3603	2.2292	− .0047	.2.....	.118	1.3610	.1044	1.5359	.5372
0.2...	0.154	2.1746	0.3491	2.2976	−0.1038	0.1.....	0.077	1.6958	0.0746	1.7642	0.2990

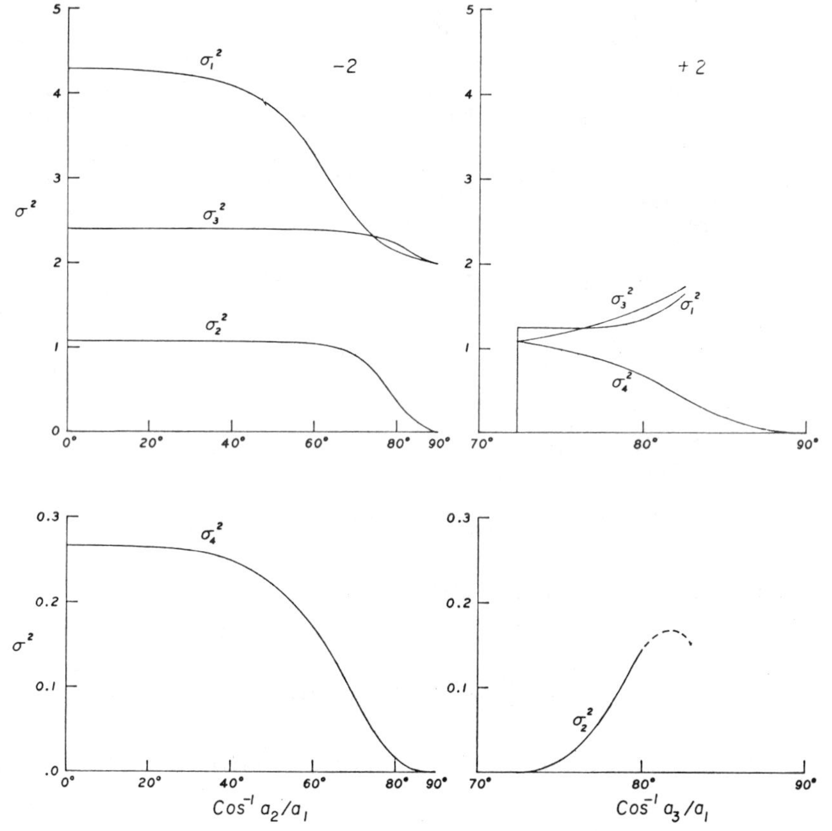

FIG. 6.—The squares of the characteristic frequencies belonging to the second harmonics for the Riemann sequences belonging to $f = -2$ (the curves on the left) and for $f = +2$ (the curves on the right). The roots denoted by σ_1^2 and σ_2^2 belong to the even modes (eq. [104]) and the roots denoted by σ_3^2 and σ_4^2 belong to the odd modes (eq. [114]).

It emerges from the foregoing calculations that all Riemann ellipsoids with $f \geq -2$ are stable with respect to oscillations belonging to the second harmonics. But instability by one of the odd modes sets in along the sequences for $f < -2$. Thus, along the self-adjoint sequence for $x = -1$, instability sets in where

$$a_2/a_1 = 0.29633\,, \qquad a_3/a_1 = 0.40733\,, \qquad f = f\dagger = -3.6710\,,$$

$$\Omega^2 = -Q_1Q_2 = 0.14878\,, \qquad Q_1 = Q_1\dagger = 1.3017\,, \qquad \text{and} \qquad Q_2 = Q_2\dagger = -0.1143\,. \tag{116}$$ [6]

This is the point denoted by X_2 in Figure 2; and the locus of points in $(a_2/a_1, a_3/a_1)$-plane along which instability sets in by this odd mode of oscillation is the curve which joins X_2 to the origin.

VIII. THE ISOLATION OF THE NEUTRAL POINTS BELONGING TO THE THIRD HARMONICS ALONG THE RIEMANN SEQUENCES

We have seen in § VII that the Riemann ellipsoids for $f \geq -2$ are stable with respect to all modes of oscillation belonging to the second harmonics. We may therefore expect, in analogy with the Jacobian and the Dedekind sequences, that instability along these Riemann sequences will be manifested, first, by a mode of oscillation belonging to the third harmonics. In this section we shall show how the various integral properties provided by the third-order virial theorem enable us to determine the locus of points, in the domain of the Riemann ellipsoids, that separate the regions of stability from the regions of instability.

Now the third-order virial theorem gives

$$\begin{aligned}\frac{d}{dt}\int_V \rho u_i x_j x_k d\mathbf{x} = {} & 2(\mathfrak{T}_{ij;k} + \mathfrak{T}_{ik;j}) + \Omega^2(I_{ijk} - \delta_{i3}I_{3jk}) + \mathfrak{W}_{ij;k} + \mathfrak{W}_{ik;j} \\ & + 2\Omega\epsilon_{il3}\int_V \rho u_l x_j x_k d\mathbf{x} + \Pi_k\delta_{ij} + \Pi_j\delta_{ik}\,,\end{aligned} \tag{117}$$

where

$$\mathfrak{T}_{ij;k} = \tfrac{1}{2}\int_V \rho u_i u_j x_k d\mathbf{x}\,, \qquad \mathfrak{W}_{ij;k} = -\tfrac{1}{2}\int_V \rho \mathfrak{B}_{ij} x_k d\mathbf{x}\,,$$

$$I_{ijk} = \int_V \rho x_i x_j x_k d\mathbf{x}\,, \qquad \text{and} \qquad \Pi_j = \int_V p x_j d\mathbf{x}\,. \tag{117'}$$

Under conditions of equilibrium, equation (117) gives

$$\begin{aligned}& 2(\mathfrak{T}_{ij;k} + \mathfrak{T}_{ik;j}) + \Omega^2(I_{ijk} - \delta_{i3}I_{3jk}) + \mathfrak{W}_{ij;k} + \mathfrak{W}_{ik;j} \\ & + 2\Omega\epsilon_{il3}\int_V \rho u_l x_j x_k d\mathbf{x} = -\Pi_k\delta_{ij} - \Pi_j\delta_{ik}\,.\end{aligned} \tag{118}$$

[6] It follows from eqs. (74) and (114) that instability along the sequence $x = -1$ sets in where

$$4B_{13} = \Omega Q_1 = a_1 B_{12}/a_2\,;$$

and, making use of eq. (25), which now takes the form

$$B_{13} = \frac{a_1(a_1 - a_2)}{a_1^2 - a_3^2}B_{12}\,,$$

we find that at the point where instability sets in,

$$a_1 - a_3 = 2a_2\,.$$

We observe that this relation is satisfied by the values quoted.

At a neutral point belonging to the third harmonics, the first variations of all the integral relations provided by equation (118) must vanish for a Lagrangian displacement which leads to a set of third-order virials

$$V_{i;jk} = \int_V \rho \xi_i x_j x_k dx \tag{119}$$

that are not all zero.

The first variation of equation (118) gives

$$\begin{aligned}2(\delta\mathfrak{T}_{ij;k}+\delta\mathfrak{T}_{ik;j}) + \Omega^2(V_{ijk}-\delta_{i3}V_{3jk}) + \delta\mathfrak{W}_{ij;k}+\delta\mathfrak{W}_{ik;j}\\ + 2\Omega\epsilon_{il3}\left(\int_V \rho\Delta u_l x_j x_k dx + \int_V \rho u_l \xi_j x_k dx + \int_V \rho u_l x_j \xi_k dx\right)\\ = -(\delta\Pi_k\delta_{ij}+\delta\Pi_j\delta_{ik}),\end{aligned} \tag{120}$$

where $\delta\mathfrak{T}_{ij;k}$, etc., are the first-order changes in the respective quantities induced by an appropriate Lagrangian displacement and V_{ijk} is the symmetrized virial

$$V_{ijk} = V_{i;jk} + V_{j;ki} + V_{k;ij} (= \delta I_{ijk}) . \tag{121}$$

Equation (120) must be supplemented by the three further conditions

$$\sum_{j=1}^{3} \frac{V_{ijj}}{a_j^2} = 0 \qquad (i = 1, 2, 3) \tag{122}$$

(summation only over the index indicated),

which express the solenoidal character of the displacement.

In an earlier paper (Chandrasekhar and Lebovitz 1963*a*, Table 1) it has been shown that first variations of $\delta\mathfrak{W}_{ij;k}$ can be expressed as linear combinations of the symmetrized virials V_{ijk}. And in Paper I it has been shown how, for a quasi-static deformation, $\delta\mathfrak{T}_{ij;k}$ can be expressed in terms of the virials $V_{i;jk}$ if the motion in the equilibrium configuration is linear in the coordinates and is of the form given by equation (76); thus (cf. Paper I, eq. [90])

$$2\delta\mathfrak{T}_{ij;k} = -Q_{jl}(Q_{ln}V_{i;kn} + Q_{kn}V_{i;ln}) - Q_{il}(Q_{ln}V_{j;kn} + Q_{kn}V_{j;ln}) + Q_{im}Q_{jn}V_{k;mn} . \tag{123}$$

This equation follows from the relation (Paper I, eq. [85])

$$\int_V \rho\Delta u_i x_l x_m dx = -(Q_{ln}V_{i;mn} + Q_{mn}V_{i;ln}) , \tag{124}$$

which obtains for any quasi-static deformation. By a further application of this last relation, the Coriolis term in equation (120) can also be expressed in terms of the virials; thus

$$2\Omega\epsilon_{il3}\delta\int_V \rho u_l x_j x_k dx = 2\Omega\epsilon_{il3}[Q_{ln}(V_{j;kn}+V_{k;jn}) - Q_{jn}V_{l;kn} - Q_{kn}V_{l;jn}] . \tag{125}$$

Equations (123) and (125), together with the known expansions of the $\delta\mathfrak{W}_{ij;k}$'s in terms of the virials, enable us to express the right-hand side of equation (120) as linear combinations of the virials. For our present purposes of isolating the neutral point, it will suffice to consider the five equations which are odd in the index 1 and even in the indices 2 and 3; these equations are

$$4\delta\mathfrak{T}_{11;1} + 2\delta\mathfrak{W}_{11;1} + \Omega^2 V_{111} + 4\Omega(Q_2V_{1;11} - Q_1V_{2;12}) = -2\delta\Pi_1 , \tag{126}$$

$$2\delta\mathfrak{T}_{12;2} + 2\delta\mathfrak{T}_{22;1} + \delta\mathfrak{W}_{12;2} + \delta\mathfrak{W}_{22;1} + \Omega^2 V_{122} + 2\Omega(Q_2V_{1;11} - Q_1V_{2;12}) = -\delta\Pi_1 , \tag{127}$$

$$2\delta\mathfrak{T}_{13;3} + 2\delta\mathfrak{T}_{33;1} + \delta\mathfrak{W}_{13;3} + \delta\mathfrak{W}_{33;1} = -\delta\Pi_1 , \tag{128}$$

$$4\delta\mathfrak{T}_{12;2} + 2\delta\mathfrak{W}_{12;2} + \Omega^2 V_{122} = 0 , \tag{129}$$

$$4\delta\mathfrak{T}_{13;3} + 2\delta\mathfrak{W}_{13;3} + \Omega^2 V_{133} + 4\Omega Q_2 V_{3;31} = 0 . \tag{130}$$

Eliminating $\delta\Pi_1$ appropriately from these equations, we obtain, in addition to equations (129) and (130) the pair of equations

$$2\delta R_{122} + \delta S_{122} + \Omega^2(V_{111} - 3V_{122}) = 0 \tag{131}$$

and

$$2\delta R_{133} + \delta S_{133} + \Omega^2(V_{111} - V_{133}) + 4\Omega Q_2 V_{1;11} - 4\Omega Q_1 V_{2;12} - 4\Omega Q_2 V_{3;13} = 0 , \tag{132}$$

where

$$\delta S_{ijj} = -4\delta\mathfrak{W}_{ij;j} - 2\delta\mathfrak{W}_{jj;i} + 2\delta\mathfrak{W}_{ii;i}$$

and

$$\delta R_{ijj} = -4\delta\mathfrak{T}_{ij;j} - 2\delta\mathfrak{T}_{jj;i} + 2\delta\mathfrak{T}_{ii;i} \tag{133}$$

(no summation over repeated indices).

And we may note here for reference the following explicit expressions for the particular quantities which occur in equations (129)–(132) (cf. Paper I, eqs. [95]–[98])

$$-4\delta\mathfrak{T}_{12;2} = 2Q_1Q_2(V_{2;21} + V_{1;22}) + 2Q_2^2 V_{1;11} , \tag{134}$$

$$-4\delta\mathfrak{T}_{13;3} = 2Q_1Q_2V_{3;31} , \tag{135}$$

$$+2\delta R_{122} = (2Q_2^2 - 4Q_1Q_2)V_{1;11} + (4Q_1Q_2 - 2Q_1^2)V_{1;22} + 12Q_1Q_2V_{2;12} , \tag{136}$$

$$+2\delta R_{133} = -4Q_1Q_2V_{1;11} - 2Q_1^2V_{1;22} + 4Q_1Q_2V_{3;31} . \tag{137}$$

Equations (129)–(132), together with the solenoidal condition,

$$\frac{V_{111}}{a_1^2} + \frac{V_{122}}{a_2^2} + \frac{V_{133}}{a_3^2} = 0 , \tag{138}$$

provide five homogeneous equations for the five virials $V_{1;11}$, $V_{1;22}$, $V_{1;33}$, $V_{2;12}$, and $V_{3:13}$. The condition for the occurrence of a neutral point is that the determinant of the system vanishes. The neutral points along the Jacobian and the Dedekind sequences have already been determined by making use of this condition (Chandrasekhar 1962, 1963, and Paper I). The neutral points along the other Riemann sequences were similarly determined; and by considering the adjoint configurations along the same sequences the places where the neutral points occur for these configurations were also determined. In the same way, the two neutral points along the sequence of the prolate spheroids, as well as the two self-adjoint sequences, were determined. The results of all these calculations are summarized in Table 11. And in Figure 2 the loci separating the regions of stability from the regions of instability, in the domain of these Riemann ellipsoids, are drawn.

IX. CONCLUDING REMARKS

In some respects, the most important result which has emerged from the present study is the fact that from every point of the stable part of the Maclaurin sequence two Riemann sequences branch off; and, further, that the branching of these sequences is, in no essential way, different from the branching of the Jacobian sequence. Indeed, from the present vantage point, the branching of the Jacobian sequence from the Maclaurin sequence does not appear as a unique or an isolated phenomenon: it appears as the manifestation of a cause which is operative at *every point* of the Maclaurin sequence. The restrictiveness of the common view which ascribes a special significance to the branching of the Jacobian sequence is nowhere more apparent than in its disregard for the simultaneous branching of the Dedekind sequence from the same "point of bifurcation." In view of the contrariness of these remarks, it may be useful to formulate the present interpretation in a manner in which its physical basis becomes transparent.

Consider a Maclaurin spheroid rotating uniformly with an angular velocity Ω_{Mc} in an

TABLE 11

THE NEUTRAL POINTS ALONG THE RIEMANN SEQUENCES BELONGING TO THE THIRD HARMONICS

	Neutral Points along the Riemann Sequences for					Along the Sequence of the Prolate Spheroids	
	$f=-2$	$f=-1$	$f=0$*	$f=0.5$	$f=2$	$f=-0.61563$	$f=-10.65076$
a_2/a_1	+0.32098	+0.40152	0.43216	+0.42253	+0.33197	+0.41815	+0.46386
a_3/a_1	+ .41397	+ .44657	.34503	+ .28004	+ .17378	+ .41815	+0.46386
Ω^2	+ .22981	+ .30315	0.28400	+ .23315	+ .11556	+ .31111	+0.02021
$-Q_1Q_2$	+ .07784	+ .03624	0	+ .00749	+ .04133	+ .01494	+0.33406
Q_1	+ .86922	+ .47415	0	− .20486	− .61240	+ .29228	+1.24601
Q_2	−0.08955	−0.07644	0	+0.03657	+0.06749	−0.05110	−0.26810

* The entries in this column refer to the Jacobian sequence.

	Neutral Points for the Adjoint Configurations along the Riemann Sequences for				
	$f=-2$	$f=-1$	$f=0$*	$f=0.5$	$f=2$
$f\dagger$	−3.02418	−6.53927	$\pm\infty$	+15.02537	+2.84745
a_2/a_1	+0.51407	+0.48185	+0.44133	+ 0.43336	+0.54226
a_3/a_1	+0.64729	+0.53188	+0.35041	+ 0.28412	+0.22038
$(\Omega\dagger)^2$	+0.17645	+0.05169	0	+ 0.00785	+0.08075
$-Q_1\dagger Q_2\dagger$	+0.26681	+0.33800	+0.28782	+ 0.23599	+0.11496
$Q_1\dagger$	+1.00481	+1.20657	−1.21560	− 1.12099	−0.62527
$Q_2\dagger$	−0.26553	−0.28014	+0.23677	+ 0.21052	+0.18386

* The entries in this column refer to the Dedekind sequence.

	Neutral Points along the Self-adjoint Sequences for	
	$x=-1$	$x=+1$
$f=f\dagger$	−2.7279	2.1067
a_2/a_1	0.43640	0.7224
a_3/a_1	0.57370	0.25606
$\Omega^2=(\Omega\dagger)^2=-Q_1Q_2$	0.20230	0.10615
$Q_1=Q_1\dagger$	1.03067	−0.45101
$Q_2=Q_2\dagger$	−0.19628	0.23536

inertial frame. When viewed from a frame of reference rotating with this same angular velocity Ω_{Mc}, the spheroid will appear as in hydrostatic equilibrium with no internal motions. However, when viewed from a frame of reference rotating with an angular velocity Ω different from Ω_{Mc}, the spheroid will appear as having internal motions with the vorticity

$$\zeta_\Omega = 2(\Omega_{Mc} - \Omega) . \tag{139}$$

Since the transverse sections of the spheroid are circular, the motion associated with ζ_Ω is purely rotational and is exactly the difference between Ω and Ω_{Mc}; the components of the motion, in the chosen frame, are, in fact, (cf. eq. [9])

$$u_1 = -(\Omega_{Mc} - \Omega)x_2 = Q_1 x_2 \text{ (say)}$$

and

$$u_2 = +(\Omega_{Mc} - \Omega)x_1 = Q_2 x_1 \text{ (say)} . \tag{140}$$

We now ask the question whether the Maclaurin spheroid, so described, can be deformed quasi-statistically into a triaxial ellipsoid without in any way affecting its equilibrium *as viewed* from the frame of reference rotating with the angular velocity Ω. We shall now show that such a deformation can be accomplished only if the angular velocity Ω of the frame is chosen properly.

Now an infinitesimal displacement $\boldsymbol{\xi}$ that will deform the spheroid into a triaxial ellipsoid without affecting its angular momentum is given by

$$\xi_1 = \alpha x_2, \qquad \xi_2 = \beta x_1, \qquad \text{and} \qquad \xi_3 = 0 , \tag{141}$$

where α and β are two infinitesimal constants. From the equations derived in § VI, the condition that the displacement (141) will have the properties requisite for a neutral mode of oscillation can be written down. But it should be noted, first, that, in deriving the linearized virial equations (89)–(94), no restrictive assumptions regarding the underlying equilibrium were made; the only assumption that was made was that the object, prior to its perturbation and in the chosen frame of reference, was in a steady state with internal motions compatible with the general form (9) and (79).

For the deformation represented by equation (141), the only non-vanishing virials are $V_{1;2}$ and $V_{2;1}$; therefore,

$$V_{i;j} \neq 0 \textit{ only for } i = 1, j = 2 \textit{ and } i = 2, j = 1 . \tag{142}$$

Now setting $\lambda = 0$ (as required for a neutral mode) in equations (89)–(94), we observe that these equations (and eq. [85] expressing the solenoidal character of $\boldsymbol{\xi}$) are consistent with the requirements expressed in (142) only if

$$\Omega^2 - Q_1 Q_2 = 2B_{11} . \tag{143}$$

The condition in this particular form follows from equations (92) and (93); the remaining equations are satisfied identically.

With the present definitions of Q_1 and Q_2 (see eqs. [140]), the condition (143) gives

$$\Omega^2 + (\Omega_{Mc} - \Omega)^2 = 2B_{11} . \tag{144}$$

But this last equation is the same as equation (52) derived in § IV for determining the particular Riemann sequences which branch off from a given point of the Maclaurin sequence.

The foregoing arguments, when carried out in the inertial frame ($\Omega = 0$) or in the frame in which the Maclaurin spheroid appears as in hydrostatic equilibrium ($\Omega = \Omega_{Mc}$), lead to the same condition

$$\Omega_{Mc}{}^2 = 2B_{11} ; \tag{145}$$

and this result is in agreement with the known fact that the Jacobian and the Dedekind sequences bifurcate from the Maclaurin sequence at the same point. The non-uniqueness of this phenomenon from the general point of view leading to equation (144) is apparent.

Finally, it should be stated that the investigation of the Riemann ellipsoids, in the case where the directions of $\mathbf{\Omega}$ and ζ_Ω do not coincide, leads to a point of view which is even more general than the one described in this section.[7]

I am greatly indebted to Miss Donna Elbert for her patience in carrying out the extensive numerical work which was required in the preparation of this paper. I am also grateful to Dr. M. Clement for his careful scrutiny of the analysis and to Dr. N. Lebovitz for many helpful discussions.

The research reported in this paper has in part been supported by the Office of Naval Research under contract Nonr-2121(24) with the University of Chicago.

REFERENCES

Basset, A. B. 1888, *A Treatise on Hydrodynamics* (Cambridge, Eng.: Deighton Bell & Co.; reprinted in 1961 by Dover Publications, New York), Vol. **2**.
Chandrasekhar, S. 1962, *Ap. J.*, **136**, 1048.
———. 1963, *ibid.*, **137**, 1185.
———. 1964, *Lectures in Theoretical Physics*, ed. W. E. Brittin and W. R. Chappell (Boulder: University of Colorado Press), p. 1.
———. 1965, *Ap. J.*, **141**, 1043.
Chandrasekhar, S., and Lebovitz, N. R. 1962, *Ap. J.*, **136**, 1037.
———. 1963*a*, *ibid.*, **137**, 1142.
———. 1963*b*, *ibid.*, p. 1172.
Dedekind, R. 1860, *J. f. Reine und Angew. Math.*, **58**, 217.
Greenhill, A. G. 1879, *Proc. Camb. Phil. Soc.*, **3**, 233.
———. 1880, *ibid.*, **4**, 4.
Lebovitz, N. R. 1961, *Ap. J.*, **134**, 500.
Love, A. E. 1888, *Phil. Mag.*, Ser. 5, **25**, 40.
Riemann, B. 1860, *Abh. d. Königl. Gessell. der Wis. zur Göttingen*, **9**, 3; also 1892, *Gesammelte Mathematische Werke* (Leipzig: Verlag Von B. G. Teubner), p. 182.

[7] The arguments of this section can be stated more generally. Consider the oscillations of a Maclaurin spheroid in a frame of reference rotating with an angular velocity Ω different from $\Omega_{\rm Mc}$. It can be shown that the frequencies of oscillation of the toroidal (σ_e) and the transverse sheer (σ_o) modes are then given by

$$\sigma_e = 2\Omega - \Omega_{\rm Mc} \pm \surd(4B_{11} - \Omega_{\rm Mc}^2) \tag{i}$$

and

$$2\sigma_o = 2\Omega - \Omega_{\rm Mc} \pm \surd(16B_{13} + \Omega_{\rm Mc}^2)\ . \tag{ii}$$

Accordingly, the even mode can be neutralized by choosing Ω so that

$$2\Omega - \Omega_{\rm Mc} = \surd(4B_{11} - \Omega_{\rm Mc}^2)\ ; \tag{iii}$$

and similarly the odd mode can be neutralized by choosing Ω so that

$$2\Omega - \Omega_{\rm Mc} = \surd(16B_{13} + \Omega_{\rm Mc}^2)\ . \tag{iv}$$

It can be readily verified that condition (iii) is the same as that given by eq. (144) and leads to the two manners of bifurcation described in the text. The choice provided by condition (iv) leads to two further manners of bifurcation corresponding to the existence of another class of Riemann ellipsoids in which the directions of $\mathbf{\Omega}$ and ζ_Ω do not coincide. We shall return to these matters in greater detail in a further paper on the Riemann ellipsoids; but it is important to notice meantime that *every point of the Maclaurin sequence is a point of bifurcation in four different ways.*

31

THE EQUILIBRIUM AND THE STABILITY OF THE RIEMANN ELLIPSOIDS. II

S. CHANDRASEKHAR
University of Chicago
Received February 21, 1966

ABSTRACT

In this paper we consider ellipsoidal figures of equilibrium (of semi-axes a_1, a_2, and a_3) of homogeneous masses rotating uniformly with an angular velocity $\mathbf{\Omega}$ and with internal motions having a uniform vorticity $\boldsymbol{\zeta}$ (in the rotating frame) in the case that the directions of $\mathbf{\Omega}$ and $\boldsymbol{\zeta}$ do not coincide. Riemann's theorem, that in this case $\mathbf{\Omega}$ and $\boldsymbol{\zeta}$ must lie in a principal plane of the ellipsoid, is shown to follow from a consideration of the non-diagonal components of the second-order tensor-virial theorem. The conditions for equilibrium are also derived; and the domains of occupancy of these Riemann ellipsoids in the $(a_2/a_1, a_3/a_1)$-plane (on the assumptions, which entail no loss of generality, that $\mathbf{\Omega}$ and $\boldsymbol{\zeta}$ have no components in the x_1-direction and that $a_2 \geq a_3$) are explicitly specified.

It is shown that the equilibrium ellipsoids are of three types: ellipsoids of type I which occupy the domain $2a_1 \geq (a_2 + a_3)$ and $a_2 \geq a_1 \geq a_3$; ellipsoids of type II for which $a_2 \geq 2a_1$ and a_3/a_1 (≤ 1) are limited by a locus along which $\int p dx = 0$; and ellipsoids of type III which occupy the domain limited by $2a_1 \leq (a_2 - a_3)$ and a locus along which $\Omega_2 = \zeta_2 = 0$ and $a_3 \geq a_1$. And quite generally, it is shown that an ellipsoid, represented by a point in the allowed domain of occupancy, is a figure of equilibrium for two different states of motion $(\mathbf{\Omega}, \boldsymbol{\zeta})$ and $(\mathbf{\Omega}^\dagger, \boldsymbol{\zeta}^\dagger)$; and that the two resulting configurations are adjoints of one another in the sense of Dedekind's theorem.

Ellipsoids of type I may be considered as branching off from the Maclaurin sequence with an odd mode of oscillation neutralized at the point of bifurcation by the choice of Ω_3 and ζ_3 (Ω_2 and ζ_2 being zero). And ellipsoids of type III may be similarly considered as branching off from the ellipsoids of type S (for which the directions of $\mathbf{\Omega}$ and $\boldsymbol{\zeta}$ coincide with the x_3-axis) along the curve where they are marginally unstable.

The stability of the Riemann ellipsoids with respect to oscillations belonging to the second harmonics is also investigated. It is first shown that the characteristic frequencies of oscillation of an ellipsoid and its adjoint are the same; and further that $|\mathbf{\Omega}|$ and $|\mathbf{\Omega}^\dagger|$ are allowed proper frequencies. The loci along which instability sets in, in the different domains of occupancy, are determined. Of particular interest are the facts that *all* ellipsoids of type II are unstable; that along the curve where the ellipsoids of type III branch off from ellipsoids of type S, the stability passes from the latter to the former; and that among the ellipsoids of type I there are some very highly flattened ones that are stable.

Several statements of Riemann concerning the stability of these ellipsoids are not substantiated by the present detailed investigation. The origin of Riemann's errors is clarified in the paper by Lebovitz following this one.

I. INTRODUCTION

Pursuing earlier investigations of Dirichlet and Dedekind, Riemann (1860; see also Hicks 1882 and Basset 1888) proved that *the most general type of motion (linear in the coordinates) compatible with an ellipsoidal figure of equilibrium of a homogeneous mass consists of a superposition of a uniform rotation* $\mathbf{\Omega}$ *and internal motions of a uniform vorticity* $\boldsymbol{\zeta}$ *(in the rotating frame) about axes that lie in a principal plane of the ellipsoid.* More precisely, according to Riemann's theorem there are three distinct circumstances (and *only* three) under which ellipsoidal figures of equilibrium can arise. These are: (*a*) the case of uniform rotation $\mathbf{\Omega}$ about the least axis of the ellipsoid; (*b*) the case when the directions of $\mathbf{\Omega}$ and $\boldsymbol{\zeta}$ coincide with a principal axis of the ellipsoid; and (*c*) the case when the directions of $\mathbf{\Omega}$ and $\boldsymbol{\zeta}$ do not coincide but lie in a principal plane of the ellipsoid. Case (*a*) leads to the classical sequences of Maclaurin and Jacobi; case (*b*) leads to the various *Riemann sequences* considered in an earlier paper (Chandrasekhar 1965*b;* this paper will be referred to hereafter as "Paper I"); and case (*c*) will be considered in this paper.

II. THE EQUATIONS DETERMINING THE EQUILIBRIUM ELLIPSOIDS: RIEMANN'S THEOREM

We shall consider quite generally the conditions under which a homogeneous ellipsoid, with semi-axes a_1, a_2, and a_3, can be a figure of equilibrium when subject to a uniform rotation $\boldsymbol{\Omega}$ and internal motions (linear in the coordinates) with a uniform vorticity $\boldsymbol{\zeta}$ in the rotating frame.

We shall suppose that the coordinate axes are along the principal axes of the ellipsoid and, further, that $\boldsymbol{\Omega}$ and $\boldsymbol{\zeta}$, in the chosen coordinate system, have the components Ω_1, Ω_2, and Ω_3 and ζ_1, ζ_2, and ζ_3. The condition that the internal motion associated with $\boldsymbol{\zeta}$ preserve the ellipsoidal boundary requires that it be expressible in the form

$$u_1 = -\frac{a_1^2}{a_1^2+a_2^2}\zeta_3 x_2 + \frac{a_1^2}{a_1^2+a_3^2}\zeta_2 x_3\,,$$

$$u_2 = -\frac{a_2^2}{a_2^2+a_3^2}\zeta_1 x_3 + \frac{a_2^2}{a_2^2+a_1^2}\zeta_3 x_1\,, \tag{1}$$

$$u_3 = -\frac{a_3^2}{a_3^2+a_1^2}\zeta_2 x_1 + \frac{a_3^2}{a_3^2+a_2^2}\zeta_1 x_2\,.$$

To obtain the conditions that the ellipsoid is also in gravitational equilibrium, we shall make use of the second-order virial theorem. According to this theorem

$$\frac{d}{dt}\int_V \rho u_i x_j d\boldsymbol{x} = 2\mathfrak{T}_{ij} + \Omega^2 I_{ij} - \Omega_i\Omega_k I_{kj} + \mathfrak{W}_{ij} + \delta_{ij}\Pi + 2\epsilon_{ilm}\Omega_m\int_V \rho u_l x_j d\boldsymbol{x}\,, \tag{2}$$

where the various symbols have their usual meanings. Under conditions of a stationary state, equation (2) gives

$$2\mathfrak{T}_{ij} + \Omega^2 I_{ij} - \Omega_i\Omega_k I_{kj} + \mathfrak{W}_{ij} + 2\epsilon_{ilm}\Omega_m\int_V \rho u_l x_j d\boldsymbol{x} = -\delta_{ij}\Pi\,. \tag{3}$$

Consider first the non-diagonal components of equation (3). The (2,3)- and the (3,2)-components of equation (3) give, for example,

$$2\mathfrak{T}_{23} - \Omega_2\Omega_3 I_{33} - 2\Omega_3\int_V \rho u_1 x_3 d\boldsymbol{x} = 0 \tag{4}$$

and

$$2\mathfrak{T}_{32} - \Omega_3\Omega_2 I_{22} + 2\Omega_2\int_V \rho u_1 x_2 d\boldsymbol{x} = 0\,, \tag{5}$$

since, in the chosen coordinate system, the tensors I_{ij} and $\mathfrak{W}_{ij}$ are diagonal and, moreover,

$$\int_V \rho u_i x_j d\boldsymbol{x} = 0 \qquad \text{if} \qquad i = j\,. \tag{6}$$

Adding and subtracting equations (4) and (5), we get

$$4\mathfrak{T}_{23} - \Omega_2\Omega_3(I_{22}+I_{33}) + 2\int_V \rho u_1(\Omega_2 x_2 - \Omega_3 x_3)\,d\boldsymbol{x} = 0$$

and

$$\Omega_2\Omega_3(I_{22}-I_{33}) - 2\int_V \rho u_1(\Omega_2 x_2 + \Omega_3 x_3)\,d\boldsymbol{x} = 0\,. \tag{7}$$

For the motions specified in equations (1)

$$2\mathfrak{T}_{23} = -\frac{a_2^2 a_3^2}{(a_1^2+a_2^2)(a_1^2+a_3^2)} \zeta_2\zeta_3 I_{11}, \tag{8}$$

$$\int_V \rho u_1 x_2 dx = -\frac{a_1^2}{a_1^2+a_2^2} \zeta_3 I_{22}, \quad \text{and} \quad \int_V \rho u_1 x_3 dx = +\frac{a_1^2}{a_1^2+a_3^2} \zeta_2 I_{33}. \tag{9}$$

Inserting the foregoing relations in equation (7) and substituting for I_{ij} its value in terms of the mass of the ellipsoid and its semi-axes, we find, after some rearrangements, the equations

$$a_2^2+a_3^2+\frac{2a_1^2a_3^2}{a_1^2+a_3^2}\frac{\zeta_2}{\Omega_2}+\frac{2a_1^2a_2^2}{a_1^2+a_2^2}\frac{\zeta_3}{\Omega_3}+\frac{2a_1^2a_2^2a_3^2}{(a_1^2+a_3^2)(a_1^2+a_2^2)}\frac{\zeta_2}{\Omega_2}\frac{\zeta_3}{\Omega_3}=0 \tag{10}$$

and

$$a_3^2+\frac{2a_1^2a_3^2}{a_1^2+a_3^2}\frac{\zeta_2}{\Omega_2}=a_2^2+\frac{2a_1^2a_2^2}{a_1^2+a_2^2}\frac{\zeta_3}{\Omega_3}, \tag{11}$$

where, in writing the equations in these forms, *we have supposed that* Ω_2 *and* Ω_3 *are different from zero.*

Now letting

$$\beta = -\frac{a_3^2}{a_1^2+a_3^2}\frac{\zeta_2}{\Omega_2} \quad \text{and} \quad \gamma = -\frac{a_2^2}{a_1^2+a_2^2}\frac{\zeta_3}{\Omega_3}, \tag{12}$$

we can rewrite equations (10) and (11) in the forms

$$\beta+\gamma-\beta\gamma=\frac{a_3^2+a_2^2}{2a_1^2} \tag{13}$$

and

$$\beta-\gamma=\frac{a_3^2-a_2^2}{2a_1^2}. \tag{14}$$

Equations (13) and (14) provide for β and γ the equations

$$\beta^2-\frac{4a_1^2+a_3^2-a_2^2}{2a_1^2}\beta+\frac{a_3^2}{a_1^2}=0 \tag{15}$$

and

$$\gamma^2-\frac{4a_1^2+a_2^2-a_3^2}{2a_1^2}\gamma+\frac{a_2^2}{a_1^2}=0. \tag{16}$$

The roots of these equations are

$$\beta=\frac{1}{4a_1^2}\{4a_1^2-a_2^2+a_3^2\pm\sqrt{[4a_1^2-(a_2+a_3)^2][4a_1^2-(a_2-a_3)^2]}\} \tag{17}$$

and

$$\gamma=\frac{1}{4a_1^2}\{4a_1^2+a_2^2-a_3^2\pm\sqrt{[4a_1^2-(a_2+a_3)^2][4a_1^2-(a_2-a_3)^2]}\}. \tag{18}$$

Thus, if Ω_2 and Ω_3 are assumed to be different from zero, the ratios ζ_2/Ω_2 and ζ_3/Ω_3 are determined by the foregoing equations. In particular, equation (15), expressed in terms of ζ_2/Ω_2, is

$$\left(\frac{\zeta_2}{\Omega_2}\right)^2+(4a_1^2+a_3^2-a_2^2)\frac{a_1^2+a_3^2}{2a_1^2a_3^2}\left(\frac{\zeta_2}{\Omega_2}\right)+\frac{(a_1^2+a_3^2)^2}{a_1^2a_3^2}=0. \tag{19}$$

On the other hand, if Ω_1 is also different from zero, then the (1,2)- and the (2,1)-components of equation (3) would have led to the equation

$$\left(\frac{\zeta_2}{\Omega_2}\right)^2+(4a_3^2+a_1^2-a_2^2)\frac{a_1^2+a_3^2}{2a_1^2a_3^2}\left(\frac{\zeta_2}{\Omega_2}\right)+\frac{(a_1^2+a_3^2)^2}{a_1^2a_3^2}=0\,. \tag{20}$$

Equations (19) and (20) are clearly incompatible unless $a_1 = a_3$; and the consideration of the equations governing ζ_3/Ω_3 would have similarly required that $a_1 = a_2$. It therefore follows that *non-trivial solutions are obtained only if no more than two of the three pairs of components* (ζ_1,Ω_1), (ζ_2,Ω_2), *and* (ζ_3,Ω_3) *are different from zero.* This is Riemann's theorem.

If we assume that (ζ_2,Ω_2) and (ζ_3,Ω_3) are different from zero, while (ζ_1,Ω_1) is zero, then, the (1,2)-, (2,1)-, (1,3)-, and (3,1)-components of equation (3) will be trivially satisfied and the only non-trivial relations are those that follow from the (2,3)- and the (3,2)-components; and these relations, as we have seen, determine the ratios ζ_2/Ω_2 and ζ_3/Ω_3. On the other hand, if two of the three components of $\boldsymbol{\Omega}$ and $\boldsymbol{\zeta}$, say (Ω_1,ζ_1) and (Ω_2,ζ_2), are assumed to vanish, then *all* the non-diagonal components of equation (3) will be trivially satisfied and the problem reduces to the one already considered in detail in Paper I.

In our further considerations, we shall suppose that Ω_2 and Ω_3 are different from zero while Ω_1 and ζ_1 are zero. Then, the internal motions specified in equation (1) become

$$\begin{aligned}
u_1&=-\frac{a_1^2}{a_1^2+a_2^2}\zeta_3x_2+\frac{a_1^2}{a_1^2+a_3^2}\zeta_2x_3=\Omega_3\gamma\frac{a_1^2}{a_2^2}x_2-\Omega_2\beta\frac{a_1^2}{a_3^2}x_3\,,\\
u_2&=+\frac{a_2^2}{a_2^2+a_1^2}\zeta_3x_1=-\Omega_3\gamma x_1\,,\\
u_3&=-\frac{a_3^2}{a_3^2+a_1^2}\zeta_2x_1=+\Omega_2\beta x_1\,;
\end{aligned} \tag{21}$$

and the ratios ζ_2/Ω_2 and ζ_3/Ω_3 are determined by the solutions for β and γ expressed in equations (17) and (18).

It remains to determine the values of Ω_2 and Ω_3 that are to be associated with the ellipsoid. The necessary additional relations follow from a consideration of the diagonal components of equation (3). And under the present circumstances ($\Omega_1 = \zeta_1 = 0$) equation (3) gives

$$2\mathfrak{T}_{11}+(\Omega_2^2+\Omega_3^2)I_{11}+\mathfrak{W}_{11}+2\int_V\rho x_1(\Omega_3u_2-\Omega_2u_3)\,d\boldsymbol{x}=-\Pi\,, \tag{22}$$

$$2\mathfrak{T}_{22}+\Omega_3^2I_{22}+\mathfrak{W}_{22}-2\Omega_3\int_V\rho u_1x_2\,d\boldsymbol{x}=-\Pi\,, \tag{23}$$

and

$$2\mathfrak{T}_{33}+\Omega_2^2I_{33}+\mathfrak{W}_{33}+2\Omega_2\int_V\rho u_1x_3\,d\boldsymbol{x}=-\Pi\,. \tag{24}$$

On evaluating the components of the kinetic-energy tensor and the moments of the velocities that occur in the foregoing equations, in accordance with equations (21), we find

$$\left[\Omega_3^2\left(1-2\gamma+\frac{a_1^2}{a_2^2}\gamma^2\right)+\Omega_2^2\left(1-2\beta+\frac{a_1^2}{a_3^2}\beta^2\right)\right]I_{11}+\mathfrak{W}_{11}=-\Pi\,, \tag{25}$$

$$\Omega_3^2\left(\gamma^2-2\gamma+\frac{a_2^2}{a_1^2}\right)I_{11}+\mathfrak{W}_{22}=-\Pi\,, \tag{26}$$

and

$$\Omega_2^2\left(\beta^2-2\beta+\frac{a_3^2}{a_1^2}\right)I_{11}+\mathfrak{W}_{33}=-\Pi\,, \tag{27}$$

where, by equations (15) and (16),

$$\beta^2-2\beta+\frac{a_3^2}{a_1^2}=\frac{a_3^2-a_2^2}{2a_1^2}\,\beta\,;\qquad \gamma^2-2\gamma+\frac{a_2^2}{a_1^2}=\frac{a_2^2-a_3^2}{2a_1^2}\,\gamma\,; \tag{28}$$

$$1-2\beta+\frac{a_1^2}{a_3^2}\,\beta^2=\frac{4a_1^2-a_2^2-3a_3^2}{2a_3^2}\,\beta\,; \tag{29}$$

and

$$1-2\gamma+\frac{a_1^2}{a_2^2}\,\gamma^2=\frac{4a_1^2-a_3^2-3a_2^2}{2a_2^2}\,\gamma\,. \tag{30}$$

Eliminating Π between equations (26) and (27) and making use of the relations (28), we find

$$\beta\Omega_2^2+\gamma\Omega_3^2=\frac{2a_1^2}{a_3^2-a_2^2}\,\frac{\mathfrak{W}_{22}-\mathfrak{W}_{33}}{I_{11}}\,. \tag{31}$$

The expressions for the components of the moment of inertia and the potential-energy tensors of homogeneous ellipsoids have been given in an earlier paper (Chandrasekhar and Lebovitz 1962, eqs. [57] and [58]); with their aid, equation (31) gives

$$\beta\Omega_2^2+\gamma\Omega_3^2=4\,\frac{A_3a_3^2-A_2a_2^2}{a_3^2-a_2^2}=4B_{23}\,, \tag{32}$$

where the index symbols A_i, A_{ij}, and B_{ij} are so normalized that $\Sigma A_i=2$ and Ω^2 and ζ^2 are measured in the unit $\pi G\rho$ (see eqs. [44] and [45] below).

Next, eliminating Π between equations (25) and (26), we have (on making use of eqs. [28])

$$\Omega_3^2\left(1-\frac{a_2^2}{a_1^2}\right)\left(1+\frac{a_1^2}{a_2^2}\,\gamma^2\right)+\Omega_2^2\left(1-2\beta+\frac{a_1^2}{a_3^2}\,\beta^2\right)=\frac{2}{a_1^2}(A_1a_1^2-A_2a_2^2)\,, \tag{33}$$

or, alternatively (cf. eq. [29]),

$$2B_{12}-\Omega_3^2\left(1+\frac{a_1^2}{a_2^2}\,\gamma^2\right)=\frac{a_1^2}{a_2^2-a_1^2}\,\frac{3a_3^2-4a_1^2+a_2^2}{2a_3^2}\,\Omega_2^2\beta\,. \tag{34}$$

Similarly, by eliminating Π between equations (25) and (27), we obtain

$$2B_{13}-\Omega_2^2\left(1+\frac{a_1^2}{a_3^2}\,\beta^2\right)=\frac{a_1^2}{a_3^2-a_1^2}\,\frac{3a_2^2-4a_1^2+a_3^2}{2a_2^2}\,\Omega_3^2\gamma\,. \tag{35}$$

Making use of the readily verified relation (cf. eq. [16])

$$\frac{1}{\gamma}\left(1+\frac{a_1^2}{a_2^2}\,\gamma^2\right)=\frac{4a_1^2+a_2^2-a_3^2}{2a_2^2}\,, \tag{36}$$

and eliminating Ω_3^2 between equations (32) and (34), we obtain

$$\Omega_2^2\beta=\frac{4a_3^2(a_2^2-a_1^2)}{a_2^2-a_3^2}\,\frac{(4a_1^2+a_2^2-a_3^2)B_{23}-a_2^2B_{12}}{4a_1^4-a_1^2(a_2^2+a_3^2)+a_2^2a_3^2}\,. \tag{37}$$

Similarly, by eliminating Ω_2^2 between equations (32) and (35), we obtain

$$\Omega_3^2\gamma = \frac{4a_2^2(a_3^2-a_1^2)}{a_3^2-a_2^2}\,\frac{(4a_1^2+a_3^2-a_2^2)B_{23}-a_3^2B_{13}}{4a_1^4-a_1^2(a_2^2+a_3^2)+a_2^2a_3^2}. \tag{38}$$

Equations (37) and (38) together with equations (17) and (18) determined the angular velocities and the vorticities that are to be associated with an ellipsoid with semi-axes a_1, a_2, and a_3.

To complete the solution, we must determine Π. First, we observe that, by making use of equations (28)–(30), we can rewrite equations (25)–(27) in the forms

$$(\Omega_2^2\beta+\Omega_3^2\gamma)-\tfrac{1}{2}(4a_1^2-a_2^2-a_3^2)\left(\frac{\Omega_2^2\beta}{a_3^2}+\frac{\Omega_3^2\gamma}{a_2^2}\right)+2A_1=\frac{5\Pi}{Ma_1^2}, \tag{39}$$

$$\frac{a_3^2-a_2^2}{2a_2^2}\,\Omega_3^2\gamma+2A_2=\frac{5\Pi}{Ma_2^2}, \tag{40}$$

and

$$\frac{a_2^2-a_3^2}{2a_3^2}\,\Omega_2^2\beta+2A_3=\frac{5\Pi}{Ma_3^2}, \tag{41}$$

where M denotes the mass of the ellipsoid. From equations (40) and (41) we obtain

$$\tfrac{1}{2}\left(\frac{\Omega_2^2\beta}{a_3^2}+\frac{\Omega_3^2\gamma}{a_2^2}\right)+2A_{23}=\frac{5\Pi}{Ma_2^2a_3^2}, \tag{42}$$

where we have made use of the relation $(A_2-A_3)/(a_3^2-a_2^2)=A_{23}$. Now combining equations (32), (39), and (42), we obtain

$$\frac{5\Pi}{2Ma_1^2a_2^2a_3^2}=\frac{2B_{23}+(4a_1^2-a_2^2-a_3^2)A_{23}+A_1}{4a_1^4-a_1^2(a_2^2+a_3^2)+a_2^2a_3^2}. \tag{43}$$

In the chosen normalization ($\Sigma A_i=2$) the index symbols have the values

$$A_i=a_1a_2a_3\int_0^\infty\frac{du}{(a_i^2+u)\Delta},\qquad A_{ij}=a_1a_2a_3\int_0^\infty\frac{du}{(a_i^2+u)(a_j^2+u)\Delta},$$

and

$$B_{ij}=a_1a_2a_3\int_0^\infty\frac{u\,du}{(a_i^2+u)(a_j^2+u)\Delta}=A_i-a_j^2A_{ij}=A_j-a_i^2A_{ij}, \tag{44}$$

where

$$\Delta^2=(a_1^2+u)(a_2^2+u)(a_3^2+u)\,. \tag{45}$$

Inserting for the index symbols that appear in equations (37), (38), and (43), in accordance with the foregoing definitions, we find

$$\Omega_2^2\beta=4a_1a_2a_3\frac{a_3^2(a_2^2-a_1^2)}{(a_2^2-a_3^2)D}$$
$$\times\int_0^\infty[(4a_1^2-a_3^2)u+a_1^2(4a_1^2+a_2^2-a_3^2)-a_2^2a_3^2]\frac{u\,du}{\Delta^3}, \tag{46}$$

$$\Omega_3^2\gamma=4a_1a_2a_3\frac{a_2^2(a_3^2-a_1^2)}{(a_3^2-a_2^2)D}$$
$$\times\int_0^\infty[(4a_1^2-a_2^2)u+a_1^2(4a_1^2+a_3^2-a_2^2)-a_2^2a_3^2]\frac{u\,du}{\Delta^3}, \tag{47}$$

and

$$\frac{5\Pi}{2a_1^3a_2^3a_3^3M}=\frac{1}{D}\int_0^\infty (3u^2+6a_1^2u+D)\frac{du}{\Delta^3}, \tag{48}$$

where

$$D = 4a_1^4 - a_1^2(a_2^2 + a_3^2) + a_2^2a_3^2 . \tag{49}$$

The equations in essentially these forms are due to Riemann.

III. THE DOMAIN OF OCCUPANCY IN THE $(a_2/a_1,\ a_3/a_1)$-PLANE

It is clear that any set of values (a_1, a_2, a_3) that is consistent with equations (17), (18), (37), (38), and (43) (or, equivalently, eqs. [17], [18], and [46]–[49]) and leads to realizable values for the various physical parameters provides an admissible solution. As we shall presently see in some detail, the physical requirements that β, γ, Ω_2, and Ω_3 are real and that $\Pi \geq 0$ limit the domain of occupancy of these ellipsoidal figures of equilibrium in the $(a_2/a_1, a_3/a_1)$-plane. In determining the nature of these limits, we shall follow Riemann's original discussion (see also Basset 1888). But we shall arrange the arguments somewhat differently; and shall, moreover, specify the domain of occupancy explicitly.

First, we observe that since all the equations are symmetric in the indices 2 and 3, the domain of occupancy in the $(a_2/a_1, a_3/a_1)$-plane must be symmetrically situated about the 45°-line, $a_2 = a_3$. Therefore, without loss of generality, we may restrict ourselves to the part of the plane

$$a_2 \geq a_3 . \tag{50}$$

Next, we observe that the reality of β and γ requires that

$$\textit{either } 2a_1 \geq (a_2 + a_3) \textit{ or } 2a_1 \leq |a_2 - a_3| = a_2 - a_3 ; \tag{51}$$

and these two cases must be considered separately.

a) *Case I:* $2a_1 \geq (a_2 + a_3)$ *and* $a_2 \geq a_3$

Under the restrictions of this case

$$4a_1^2 \pm (a_3^2 - a_2^2) \geq (a_2 + a_3)^2 \pm (a_3^2 - a_2^2) > 0 . \tag{52}$$

In view of this inequality, it follows from equations (15) and (16) that

$$\beta > 0 \qquad \text{and} \qquad \gamma > 0 . \tag{53}$$

The reality of Ω_2 and Ω_3 now requires that the quantities on the right-hand sides of equations (46) and (47) are positive. Clearly,

$$D \geq a_1^2(a_2 + a_3)^2 - a_1^2(a_2^2 + a_3^2) + a_2^2a_3^2 > 0 . \tag{54}$$

Also, making use of the inequality (52), we have

$$a_1^2[4a_1^2 \pm (a_2^2 - a_3^2)] - a_2^2a_3^2 \geq \tfrac{1}{4}(a_2 + a_3)^2[(a_2 + a_3)^2 \pm (a_2^2 - a_3^2)] - a_2^2a_3^2 . \tag{55}$$

The right-hand side of this inequality is

$$\tfrac{1}{2}a_2[(a_2 + a_3)^3 - 2a_2a_3^2] \qquad \text{or} \qquad \tfrac{1}{2}a_3[(a_3 + a_2)^3 - 2a_3a_2^2] ; \tag{56}$$

and in either case it is positive. Since $4a_1^2 - a_3^2$ and $4a_1^2 - a_2^2$ are also positive, the integrands on the right-hand sides of equations (46) and (47) are positive definite; the integrals are, therefore, positive. As D and β have already been shown to be positive, it follows that the reality of Ω_2 requires that

$$a_2 \geq a_1 \,. \tag{57}$$

Hence, we are, in this case, limited to the domain

$$2a_1 \geq (a_2 + a_3) \qquad \text{and} \qquad a_2 \geq a_1 \geq a_3 \,. \tag{58}$$

Under these circumstances the reality of Ω_3 is also assured. Moreover, since $D > 0$ the quantity on the right-hand side of equation (48) is manifestly positive definite and assures that $\Pi > 0$. All ellipsoids represented in the triangle $SMcR_{\rm I}$ in Figure 1, therefore, are allowed figures of equilibrium; we shall call them *Riemann ellipsoids of type I*.

b) The Case $2a_1 < (a_2 - a_3)$ and $a_2 \geq a_3$

In this case

$$4a_1^2 \leq a_2^2 - a_3^2 \,, \tag{59}$$

since $2a_1$ is necessarily less than $a_2 + a_3$. From equations (15) and (16) it now follows that

$$\beta < 0 \qquad \text{and} \qquad \gamma > 0 \,. \tag{60}$$

Also, under the circumstances of this case, the integrand appearing on the right-hand side of equation (47) defining $\Omega_3^2\gamma$ is clearly negative. The integral is accordingly negative; and since γ has been shown to be positive and $a_2 \geq a_3$ (by definition), the reality of Ω_3 requires

$$\frac{a_3^2 - a_1^2}{D} \geq 0 \,. \tag{61}$$

Hence

$$\textit{either } a_3 < a_1 \qquad \text{and} \qquad D < 0 \,,$$

$$\textit{or } a_3 > a_1 \qquad \text{and} \qquad D > 0 \,; \tag{62}$$

and these two cases must be considered separately.

c) Case II: $2a_1 \leq (a_2 - a_3)$ and $a_3 \leq a_1$

In this case we must require (cf. eq. [62])

$$D = 4a_1^4 - a_1^2(a_2^2 + a_3^2) + a_2^2a_3^2 \leq 0 \,. \tag{63}$$

This restriction on D implies that

$$\frac{a_3}{a_1} \leq \left(\frac{a_2^2 - 4a_1^2}{a_2^2 - a_1^2}\right)^{1/2} . \tag{64}$$

It can be verified that the further restriction

$$\frac{a_3}{a_1} \leq \frac{a_2}{a_1} - 2 \tag{65}$$

assures that inequality (64) is satisfied so long as (see Fig. 2)

$$2 \leq \frac{a_2}{a_1} \leq 1 + \sqrt{3} \,. \tag{66}$$

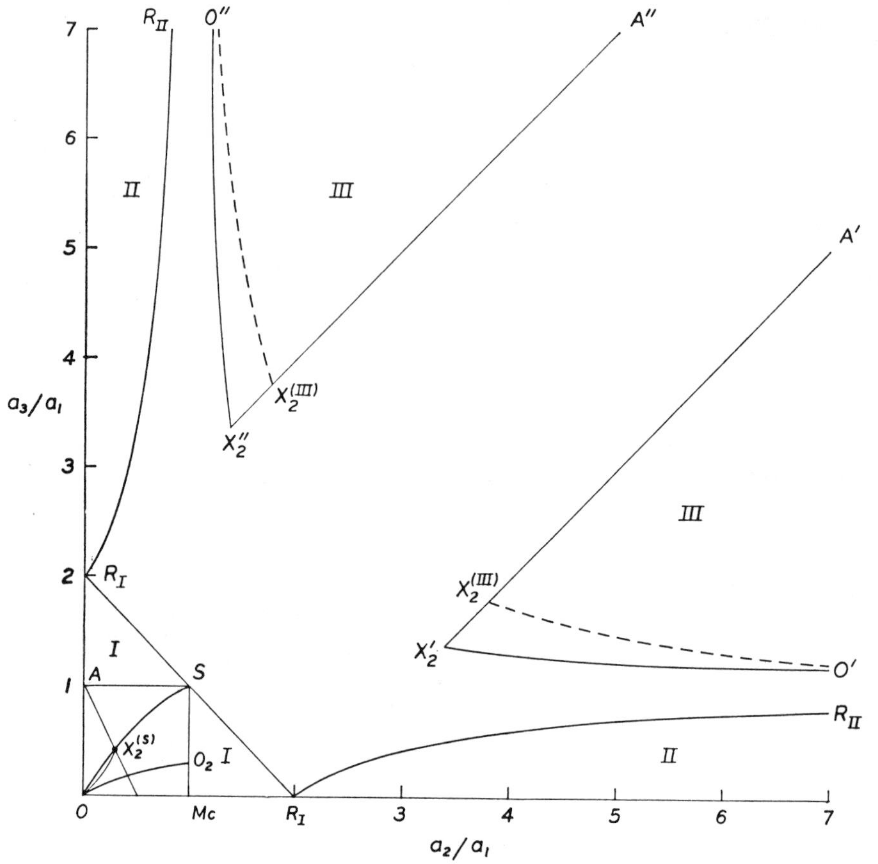

FIG. 1.—The domain of occupancy of the Riemann ellipsoids in the $(a_2/a_1, a_3/a_1)$-plane. The stable part of the Maclaurin sequence is represented by the segment O_2S on the line $a_2 = a_1$. At O_2 the Maclaurin spheroid becomes unstable by overstable oscillations.

The Riemann ellipsoids of type S (for which the directions of rotation and vorticity coincide with the a_3-axis) are included between the self-adjoint sequences represented by SO and O_2O. Along the arc $X_2^{(S)}O$ the Riemann ellipsoids of type S become unstable by an odd mode of oscillation belonging to the second harmonics.

The Riemann ellipsoids, in which the directions of rotation and vorticity do not coincide but lie in the (a_2,a_3)-plane, are of three types—I, II, and III—with the domains of occupancy shown. Type I ellipsoids adjoin the Maclaurin sequence and are bounded on one side (SR_I) by a self-adjoint sequence. Along the locus R_IR_{II}, which limits the domain of occupancy of the type II ellipsoids, the pressure is zero. And along the loci $X_2'O'$ and $X_2''O''$, limiting the domain of occupancy of the type III ellipsoids, the directions of Ω and ζ coincide with one of the principal axes (the a_3-axis in the case $a_2 > a_3$ and the a_2-axis in the case $a_2 < a_3$). The locus $X_2'O'$ (for the case $a_2 > a_3$) is transformed into $X_2^{(S)}O$ if the roles of a_1 and a_2 are interchanged; and simultaneously the domain of occupancy $A'X_2'O'$ similarly becomes transformed into the domain $AX_2^{(S)}O$. The dotted curve $X_2^{(III)}O'$ defines the locus of configurations, among the type III ellipsoids, that are marginally overstable by a mode of oscillation belonging to the second harmonics.

Hence, we must require (65) so long as $2 \le a_2/a_1 \le 1 + \sqrt{3}$ and (64) for $a_2/a_1 \ge 1 + \sqrt{3}$. With a_3/a_1 so restricted, the reality of Ω_3 is assured.

Turning next to equation (37) defining $\Omega_2^2\beta$ and rewriting it in the manner

$$\Omega_2^2\beta = 4a_1a_2a_3\frac{a_3^2(a_2^2-a_1^2)}{(a_2^2-a_3^2)D}\int_0^\infty\left(\frac{4a_1^2+a_2^2-a_3^2}{a_3^2+u}-\frac{a_2^2}{a_1^2+u}\right)\frac{udu}{(a_2^2+u)\Delta}, \tag{67}$$

we observe that the integrand is positive, since

$$\frac{4a_1^2-a_3^2}{a_3^2+u}+a_2^2\left(\frac{1}{a_3^2+u}-\frac{1}{a_1^2+u}\right)>0. \tag{68}$$

The integral on the right-hand side of equation (67) is therefore positive and since, further, $a_2 \ge a_1 \ge a_3$, $D < 0$, and $\beta < 0$, the reality of Ω_2 is assured. On the other hand,

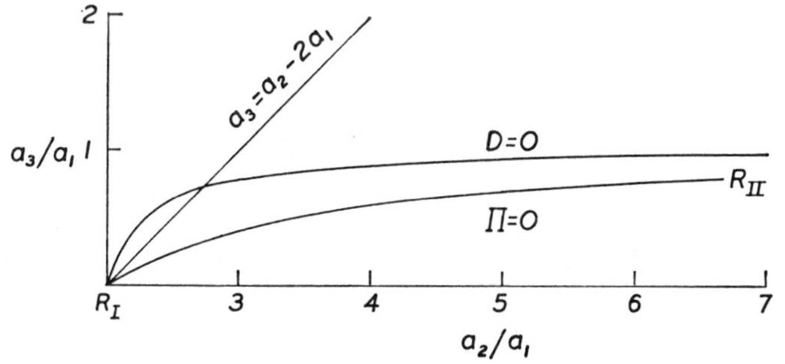

FIG. 2.—The loci $\Pi = 0$, $D = 0$, and $a_3 = a_2 - 2a_1$ which are used to determine the domain of occupancy of the type II ellipsoids.

TABLE 1

THE LOCUS ALONG WHICH $A_{23}(a_2^2 + a_3^2 - 2a_1^2) = 1$

a_2/a_1	a_3/a_1	a_2/a_1	a_3/a_1
2	0	5	0.70215
20/7	.37591	$6\frac{2}{3}$	0.79853
10/3	.49063	10	0.88676
4	0.59938	∞	1.00000

since $D < 0$, the positive definiteness of Π is not manifest from equation (48). Indeed the requirement (cf. the alternative expression [43] for Π),

$$2B_{23} + (4a_1^2 - a_2^2 - a_3^2)A_{23} + A_1 \le 0, \tag{69}$$

provides a limit on a_3/a_1, for an assigned a_2/a_1 (≥ 2), which, it will appear, is more stringent than either (64) or (65).

The condition (69) can be expressed more conveniently by making use of the relations

$$B_{23} = A_2 - a_3^2A_{23} = A_3 - a_2^2A_{23} \qquad \text{and} \qquad A_1 + A_2 + A_3 = 2\,; \tag{70}$$

thus,

$$A_{23}(a_2^2 + a_3^2 - 2a_1^2) \ge 1. \tag{71}$$

The locus in the $(a_2/a_1, a_3/a_1)$-plane along which the inequality (71) becomes an equality has been determined (see Table 1). And the curves, labeled R_IR_{II} in Figures

1 and 2, define this locus. It is apparent from Figure 2 that it is the requirement $\Pi > 0$ that limits the domain of occupancy in this case.

We shall call the ellipsoids, represented in the domain limited by the a_2-axis and the inequality (71), *Riemann ellipsoids of type II.*

d) Case III: $2a_1 \leq (a_2 - a_3)$ *and* $a_2 \geq a_3 > a_1$

In this case a_1 is the least axis; and

$$D = 3a_1^4 + (a_2^2 - a_1^2)(a_3^2 - a_1^2) > 0\,, \tag{72}$$

as required by (62). And since $D > 0$, Π is manifestly positive. The reality of Ω_3 has already been assured. It remains to insure the reality of Ω_2. Since β is negative, the reality of Ω_2 requires (cf. eq. [37])

$$(4a_1^2 + a_2^2 - a_3^2)B_{23} - a_2^2 B_{12} \leq 0\,. \tag{73}$$

TABLE 2

THE LOCUS DETERMINED BY EQUATION (74)

a_2/a_1	a_3/a_1
3.3746.	1.3746
4.1739.	1.3
5.8677.	1.2
∞	1.0

And this inequality requires that, for a given a_2 (≥ 3.3746, as we shall see presently), a_3 ($\leq a_2$) exceeds a certain lower limit. The limit is determined by the condition

$$(4a_1^2 + a_2^2 - a_3^2)B_{23} - a_2^2 B_{12} = 0\,. \tag{74}$$

A few pairs of values (a_2/a_1, a_3/a_1) along the locus defined by equation (74) are listed in Table 2; and the locus (labeled $X_2'O'$) is delineated in Figure 1.

We shall call the ellipsoids, limited by the locus (74) and the line $2a_1 = a_2 - a_3$, *Riemann ellipsoids of type III.*

It is to be particularly noted that, along the locus (74), $\Omega_2 = \zeta_2 = 0$ so that Ω_3 and ζ_3 are the only non-vanishing components of $\mathbf{\Omega}$ and $\boldsymbol{\zeta}$. Accordingly, the ellipsoids along this locus "belong" among the ellipsoids considered in Paper I; as to precisely "where" they belong, we shall return in § VI.

To distinguish the ellipsoids considered in Paper I from the ones which we have now designated as of types I, II, and III, we shall call them *Riemann ellipsoids of type S.*

IV. THE ADJOINT RIEMANN ELLIPSOIDS AND DEDEKIND'S THEOREM

For any pair of values (a_2/a_1, a_3/a_1), which represents a point in the permitted domain of occupancy of the Riemann ellipsoids of types I, II, and III, there are two states of motion, compatible with equilibrium, corresponding to the two roots of β and γ given by equations (17) and (18). It is clear on general grounds that the two physically distinct configurations which one obtains in this way must be adjoints of one another in the sense of Dedekind's theorem (cf. Paper I, § I). It is of interest to verify that this is the case.

Let β and $\beta^\dagger$ and γ and $\gamma^\dagger$ be the two roots of equations (15) and (16). Clearly,

$$\beta\beta^\dagger = a_3^2/a_1^2 \qquad \text{and} \qquad \gamma\gamma^\dagger = a_2^2/a_1^2 . \tag{75}$$

Also, since the right-hand sides of equations (37) and (38) depend only on the geometry of the ellipsoid, it is clear that

$$\Omega_2^2\beta = \Omega_2^{\dagger 2}\beta^\dagger \qquad \text{and} \qquad \Omega_3^2\gamma = \Omega_3^{\dagger 2}\gamma^\dagger . \tag{76}$$

From the foregoing relations it follows, for example, that

$$(\Omega_3\gamma a_1)\left(\Omega_3^\dagger\gamma^\dagger \frac{a_1^2}{a_2}\right) = (\Omega_3 a_1)(\Omega_3^\dagger a_2)$$

and

$$(\Omega_3\gamma a_1)(\Omega_3 a_1) = \left(\Omega_3^\dagger\gamma^\dagger \frac{a_1^2}{a_2}\right)(\Omega_3^\dagger a_2) \tag{77}$$

Hence

$$\Omega_3\gamma a_1 = \Omega_3^\dagger a_2 \qquad \text{and} \qquad \Omega_3 a_1 = \Omega_3^\dagger\gamma^\dagger \frac{a_1^2}{a_2} . \tag{78}$$

Equations (78) will continue to be valid if each quantity is replaced by its adjoint; thus

$$\Omega_3^\dagger\gamma^\dagger a_1 = \Omega_3 a_2 \qquad \text{and} \qquad \Omega_3^\dagger a_1 = \Omega_3\gamma \frac{a_1^2}{a_2} . \tag{79}$$

In exactly the same way,

$$\Omega_2\beta a_1 = \Omega_2^\dagger a_3 , \qquad \Omega_2 a_1 = \Omega_2^\dagger\beta^\dagger \frac{a_1^2}{a_3} ,$$

$$\Omega_2^\dagger\beta^\dagger a_1 = \Omega_2 a_3 , \qquad \text{and} \qquad \Omega_2^\dagger a_1 = \Omega_2\beta \frac{a_1^2}{a_3} . \tag{80}$$

Now the motion $(\boldsymbol{u})$ in the frame of reference rotating with the angular velocity $\boldsymbol{\Omega}$ is given by equations (21). The motion, $\boldsymbol{u}^{(0)}$, in the inertial frame follows from the equation

$$\boldsymbol{u}^{(0)} = \boldsymbol{u} + \boldsymbol{\Omega} \times \boldsymbol{x} . \tag{81}$$

Expressing $\boldsymbol{u}^{(0)}$ in the manner required in the enunciation of Dedekind's theorem, we have

$$\boldsymbol{u}^{(0)} = \begin{vmatrix} 0 & -\Omega_3(1-\gamma a_1^2/a_2^2)a_2 & +\Omega_2(1-\beta a_1^2/a_3^2)a_3 \\ +\Omega_3(1-\gamma)a_1 & 0 & 0 \\ -\Omega_2(1-\beta)a_1 & 0 & 0 \end{vmatrix} \begin{vmatrix} x_1/a_1 \\ x_2/a_2 \\ x_3/a_3 \end{vmatrix} . \tag{82}$$

And the motion $\boldsymbol{u}^{(0)\dagger}$ in the configuration derived from $\beta^\dagger$ and $\gamma^\dagger$ is, similarly, given by

$$\boldsymbol{u}^{(0)\dagger} = \begin{vmatrix} 0 & -\Omega_3^\dagger(1-\gamma^\dagger a_1^2/a_2^2)a_2 & +\Omega_2^\dagger(1-\beta^\dagger a_1^2/a_3^2)a_3 \\ +\Omega_3^\dagger(1-\gamma^\dagger)a_1 & 0 & 0 \\ -\Omega_2^\dagger(1-\beta^\dagger)a_1 & 0 & 0 \end{vmatrix} \begin{vmatrix} x_1/a_1 \\ x_2/a_2 \\ x_3/a_3 \end{vmatrix} . \tag{83}$$

And Dedekind's theorem requires that the two matrices expressing $\boldsymbol{u}^{(0)}$ and $\boldsymbol{u}^{(0)\dagger}$ are the transposed of one another. In other words, we must have

$$
\begin{aligned}
\Omega_3^\dagger(1-\gamma^\dagger)a_1 &= -\,\Omega_3(1-\gamma a_1^2/a_2^2)a_2\,,\\
\Omega_2^\dagger(1-\beta^\dagger)a_1 &= -\,\Omega_2(1-\beta a_1^2/a_3^2)a_3\,,\\
\Omega_3(1-\gamma)a_1 &= -\,\Omega_3^\dagger(1-\gamma^\dagger a_1^2/a_2^2)a_2\,,
\end{aligned}
\tag{84}
$$

and

$$\Omega_2(1-\beta)a_1 = -\,\Omega_2^\dagger(1-\beta^\dagger a_1^2/a_3^2)a_3\,;$$

and these relations are clearly valid in virtue of equations (78)–(80). *The configurations derived from the two roots for β and γ are, therefore, adjoints of one another in the sense of Dedekind's theorem.*

If $\beta = \beta^\dagger$ and $\gamma = \gamma^\dagger$, then the configuration is *self*-adjoint. Such self-adjoint configurations occur on the lines

$$2a_1 = a_2 + a_3 \qquad \text{and} \qquad 2a_1 = |a_2 - a_3|\,; \tag{85}$$

i.e., on one of the boundaries that limit the domains of occupancy of the ellipsoids of types I and III.

V. THE MACLAURIN SPHEROIDS AS LIMITING FORMS OF THE RIEMANN ELLIPSOIDS OF TYPE I

In this section we shall show how the Maclaurin spheroids may be considered as limiting forms of the Riemann ellipsoids of type I.

Let $a_2/a_1 \to 1$ while a_3 remains finite. From equations (17) and (18), we find that in this limit

$$\beta = \frac{1}{4a_1^2}[\,3a_1^2 + a_3^2 \pm \sqrt{(9a_1^2 - a_3^2)(a_1^2 - a_3^2)}\,],$$

and

$$\gamma = \frac{1}{4a_1^2}[\,5a_1^2 - a_3^2 \pm \sqrt{(9a_1^2 - a_3^2)(a_1^2 - a_3^2)}\,] \qquad (a_1 = a_2). \tag{86}$$

At the same time it follows from equations (37) and (38) that

$$\Omega_2^2\beta = 0 \qquad \text{and} \qquad \Omega_3^2\gamma = 4B_{13}\,. \tag{87}$$

From equation (12) we may now conclude that

$$\zeta_2 = 0 \qquad \text{and} \qquad \zeta_3 = -\,2\gamma\Omega_3\,. \tag{88}$$

Hence, on the line $a_2 = a_1$, the Riemann ellipsoids of type I become spheroids and are attributed the parameters,

$$\Omega_3^2 = 4B_{13}/\gamma\,, \qquad \zeta_3 = -2\gamma\Omega_3\,, \qquad \text{and} \qquad \zeta_2 = \Omega_2 = 0 \qquad (a_2 \to a_1)\,, \tag{89}$$

while ζ_2/Ω_2 tends to a finite value.

The relations (89) give

$$(\Omega_3 + \tfrac{1}{2}\zeta_3)^2 = \Omega_3^2(1-\gamma)^2 = 4B_{13}\,\frac{(1-\gamma)^2}{\gamma}\,. \tag{90}$$

Now, when $a_1 \rightarrow a_2$, the equation governing γ becomes

$$\gamma^2 - \left[2 + \tfrac{1}{2}\left(1 - \frac{a_3^2}{a_1^2}\right)\right]\gamma + 1 = 0\,, \tag{91}$$

or

$$(\gamma - 1)^2 = \tfrac{1}{2}\gamma\left(1 - \frac{a_3^2}{a_1^2}\right). \tag{92}$$

In view of this last relation, we can rewrite equation (90) in the form

$$(\Omega_3 + \tfrac{1}{2}\zeta_3)^2 = 2B_{13}\left(1 - \frac{a_3^2}{a_1^2}\right). \tag{93}$$

But it is known that the angular velocity of rotation Ω_{Mc}, associated with a Maclaurin spheroid of axes $a_1 = a_2$ and a_3, is given by

$$\Omega_{\text{Mc}}^2 = 2B_{13}\left(1 - \frac{a_3^2}{a_1^2}\right). \tag{94}$$

We have, therefore, the relation

$$\Omega_3 + \tfrac{1}{2}\zeta_3 = \Omega_{\text{Mc}}\,; \tag{95}$$

and this is exactly the relation which must be satisfied if what we are viewing is a Maclaurin spheroid, rotating uniformly with an angular velocity Ω_{Mc} (in an inertial frame), from a frame of reference rotating with an angular velocity Ω_3 different from Ω_{Mc}. The Riemann ellipsoids of type I, therefore, degenerate to Maclaurin spheroids when $a_2 \rightarrow a_1$ (from the right); but they are viewed from a frame of reference in which they are attributed internal motions with a vorticity ζ_3. We can arrive at this same conclusion, somewhat differently, by arguing along the lines of § IX of Paper I.

When a Maclaurin spheroid is viewed from a frame of reference rotating with an angular velocity Ω ($\neq \Omega_{\text{Mc}}$), it will be attributed internal motions having the components (cf. Paper I, eq. [140])

$$u_1 = -(\Omega_{\text{Mc}} - \Omega)x_2 = Q_1 x_2$$

and

$$u_2 = +(\Omega_{\text{Mc}} - \Omega)x_1 = Q_2 x_1\,. \tag{96}$$

(Note that $Q_1 = -Q_2$.) We now ask: can we deform the spheroid quasi-statically, without in any way affecting its equilibrium (*as viewed* from the chosen frame of reference), by a non-trivial (odd) Lagrangian displacement of the form

$$\xi_1 = \alpha_1 x_3\,, \qquad \xi_2 = \alpha_2 x_3\,, \qquad \text{and} \qquad \xi_3 = \alpha_3 x_1 + \alpha_4 x_2\,, \tag{97}$$

where $\alpha_1, \ldots, \alpha_4$ are constants? We shall show that such a quasi-static deformation is possible if Ω is chosen to be equal to Ω_3 given by equations (86) and (89).

We require that, in the chosen frame of reference, the displacement (97) have the properties requisite for a neutral mode of oscillation; and the conditions for this to be the case can be written down from the equations derived in Paper I, § VII.

For the deformation specified in (97), the only non-vanishing virials are those that are odd in the index 3:

$$V_{1;3}\,, \qquad V_{3;1}\,, \qquad V_{2;3}\,, \qquad \text{and} \qquad V_{3;2}\,. \tag{98}$$

Under these conditions, the equations governing the virials even in the index 3 (namely, Paper I, eqs. [101] and [102]) are trivially satisfied. Next, setting $\lambda = 0$ (as required for a neutral mode) in Paper I, equations (106)–(109), and remembering that in the case ($a_1 = a_2$) we are presently considering $B_{13} = B_{23}$, we find that the condition, that the virials listed in (97) do not vanish identically, is (cf. Paper I, eq. [112])

$$\begin{bmatrix} 2B_{13} - \Omega^2 & 2B_{13} - \Omega^2 + Q_1 Q_2 - 2\Omega Q_2 \\ 2B_{13} & 2B_{13} + Q_1 Q_2 \end{bmatrix} = 0 . \tag{99}$$

On simplification equation (99) becomes

$$\begin{bmatrix} \Omega & \Omega + 2Q_2 \\ 2B_{13} & 2B_{13} + Q_1 Q_2 \end{bmatrix} = 0 ; \tag{100}$$

and on expanding the determinant, we are left with

$$\Omega Q_1 - 4B_{13} = 0 . \tag{101}$$

And inserting the value of Q_1 ($= \Omega - \Omega_{\text{Mc}}$) in equation (101), we finally obtain

$$\Omega^2 - \Omega\Omega_{\text{Mc}} - 4B_{13} = 0 . \tag{102}$$

We observe that equation (102) for Ω is identical with the equation determining the characteristic frequencies of the odd modes of oscillation of the Maclaurin spheroid in the frame of reference rotating with the angular velocity Ω_{Mc} (Lebovitz 1961, eq. [169]; also Chandrasekhar 1964, p. 69, eq. [30]).

It remains to verify that the values of Ω which are given by equation (102) are the same as those that follow from equations (86) and (89). To verify this fact, set

$$\Omega^2 = 4B_{13}/z \tag{103}$$

in equation (102). On further substituting for Ω_{Mc} its value given by equation (94), we obtain

$$\frac{4B_{13}}{z} - \left[\frac{8B_{13}^2}{z}\left(1 - \frac{a_3^2}{a_1^2}\right)\right]^{1/2} - 4B_{13} = 0 . \tag{104}$$

On simplification, equation (104) becomes

$$z^2 - \left[2 + \tfrac{1}{2}\left(1 - \frac{a_3^2}{a_1^2}\right)\right] z + 1 = 0 ; \tag{105}$$

and this equation is identical with equation (91) for γ. Hence $z = \gamma$; and this completes the proof that Ω_3 determined by equation (102) agrees with Ω_3 appropriate for the Riemann ellipsoid when $a_2 \to a_1$.

An alternative, but equivalent, way of arriving at the relation (102) is to recall that the frequencies of the odd modes of oscillation of a Maclaurin spheroid, in a frame of reference rotating with an angular velocity Ω (different from Ω_{Mc}), is given by (cf. Paper I, n. 7, eq. [ii])

$$2\sigma_o = 2\Omega - \Omega_{\text{Mc}} \pm \sqrt{(16B_{13} + \Omega_{\text{Mc}}^2)} ; \tag{106}$$

accordingly, these modes can be "neutralized" by the choice

$$\Omega = \tfrac{1}{2}[\Omega_{\text{Mc}} \pm \sqrt{(16B_{13} + \Omega_{\text{Mc}}^2)}] ; \tag{107}$$

and these values of Ω are the same as those which follow from equation (102).

The parameters that are to be associated with the Maclaurin spheroids when considered as the first members of the Riemann ellipsoids of type I are listed in Table 3.

VI. THE ELLIPSOIDS OF TYPE III AS BRANCHING OFF FROM THE ELLIPSOIDS OF TYPE S ALONG A CURVE OF BIFURCATION

As we have already remarked in § III*d*, along the locus (74), which limits the domain of the Riemann ellipsoids of type III,

$$\Omega_2 = \zeta_2 = 0 . \tag{108}$$

Accordingly, for these ellipsoids the only non-vanishing components of $\mathbf{\Omega}$ and $\boldsymbol{\zeta}$ are Ω_3 and ζ_3 along the x_3-axis. These ellipsoids are, therefore, of the type S considered in Paper I. But to be in agreement with the convention adopted in that paper, namely,

TABLE 3

THE PARAMETERS TO BE ASSOCIATED WITH THE MACLAURIN SPHEROIDS WHEN CONSIDERED AS THE FIRST MEMBERS OF THE RIEMANN ELLIPSOIDS OF TYPE I

e	Ω_3	ζ_3	$\Omega_3\dagger$	$\zeta_3\dagger$
0	1.03280	−2.06559	1.03280	−2.06559
.20	0.96509	−2.22313	1.11156	−1.93018
.30	0.93498	−2.31093	1.15547	−1.86996
.40	0.90713	−2.40508	1.20254	−1.81426
.50	0.88124	−2.50544	1.25272	−1.76249
.60	0.85682	−2.61109	1.30555	−1.71365
.70	0.83274	−2.71875	1.35937	−1.66548
.80	0.80578	−2.81682	1.40841	−1.61156
.82	0.79932	−2.83247	1.41624	−1.59864
.84	0.79215	−2.84542	1.42271	−1.58430
.86	0.78398	−2.85449	1.42724	−1.56797
.88	0.77438	−2.85787	1.42894	−1.54875
.90	0.76261	−2.85266	1.42633	−1.52521
.92	0.74742	−2.83373	1.41687	−1.49483
.94	0.72636	−2.79117	1.39558	−1.45273
.96	0.69381	−2.70205	1.35102	−1.38762
0.98	0.63158	−2.49282	1.24641	−1.26316

that a_1 is the longest axis, we must interchange the roles of the indices 1 and 2 since, by our present convention, a_2 is the longest axis for the ellipsoids of type III. With the indices 1 and 2 interchanged, the loci limiting the domain of the ellipsoids become (see Fig. 1)

$$a_2 = 0 , \qquad 2a_2 + a_3 = a_1 , \tag{109}$$

and (cf. eq. [74])

$$(4a_2^2 + a_1^2 - a_3^2)B_{13} - a_1^2 B_{12} = 0 . \tag{110}$$

Now, it is clear on general grounds that, *along the locus (110), the Riemann ellipsoids must be characterized by a neutral mode of oscillation and, further, that stability must pass from the ellipsoids of type S to the ellipsoids of type III; in other words, that the locus (110) is a curve of bifurcation.*

As we have shown in Paper I, instability by an odd mode of oscillation sets in along the Riemann sequences for $f < -2$. And according to Paper I, equation (114), instability occurs when (cf. Paper I, eq. [37])

$$4B_{13} = \Omega Q_1 = -2B_{12}\frac{a_1}{a_2}\frac{1}{x+1/x}. \tag{111}$$

But by Paper I, equation (38),

$$\begin{aligned} x+\frac{1}{x} &= -\frac{2a_1a_2(A_1a_1^2 - A_2a_2^2)}{(a_1^2-a_2^2)a_3^2A_3 + a_1^2a_2^2(A_1 - A_2)} \\ &= -\frac{2a_1a_2B_{12}}{a_3^2A_3 - a_1^2a_2^2A_{12}}. \end{aligned} \tag{112}$$

TABLE 4

THE PROPERTIES OF A FEW RIEMANN ELLIPSOIDS OF TYPE I

a_2/a_1....	1.05263	1.25000	1.44065	1.66667	1.36444	1.69351	1.52303	1.78590
a_3/a_1....	0.41667	0.50000	0.49273	0.33333	0.09518	0.11813	0.05315	0.06233
Ω_2......	+0.14834	+0.39259	+0.57179	+0.71251	+0.05632	+0.15764	+0.03311	+0.08952
Ω_3......	+0.73257	+0.66536	+0.59896	+0.52815	+0.40707	+0.38504	+0.29600	+0.28558
ζ_2......	−1.41355	−2.19983	−2.24560	−2.37502	−6.68275	−6.27092	−9.85239	−9.19424
ζ_3......	−2.61578	−1.93895	−1.49425	−1.19714	−1.24612	−1.02536	−0.84580	−0.74657
$\Omega_2\dagger$.....	+0.50185	+0.87993	+0.89032	+0.71251	+0.63035	+0.73061	+0.52221	+0.57083
$\Omega_3\dagger$.....	+1.30617	+0.94583	+0.69996	+0.52815	+0.59414	+0.44893	+0.38805	+0.31825
$\zeta_2\dagger$......	−0.41783	−0.98148	−1.44219	−2.37502	−0.59714	−1.35309	−0.62474	−1.4418
$\zeta_3\dagger$......	−1.46707	−1.36398	−1.27866	−1.19714	−0.85376	−0.87944	−0.64518	−0.66992

Inserting from this last equation in equation (111), we obtain for the locus of marginal stability the equation

$$4a_2^2B_{13} - a_3^2A_3 + a_1^2a_2^2A_{12} = 0 \tag{113}$$

or, alternatively,

$$0 = 4a_2^2B_{13} - a_3^2A_3 + a_1^2(A_1 - B_{12}) = (4a_2^2 + a_1^2 - a_3^2)B_{13} - a_1^2B_{12}\,; \tag{114}$$

and this is the same as equation (110).

We have already verified in Paper I (n. 6 on page 916) that the point, on the self-adjoint sequence $x = -1$ (which limits on the side $a_3 > a_2$ the domain of the Riemann ellipsoids of type S), at which instability sets in satisfies the condition $2a_2 + a_3 = a_1$. This completes the demonstration that *the curve along which the Riemann ellipsoids of type S become marginally unstable is also the curve along which the Riemann ellipsoids of type III branch off*. And, finally, in § XI we shall show that along the curve of bifurcation stability passes from the ellipsoids of type S to the ellipsoids of type III.

VII. SOME NUMERICAL EXAMPLES

The properties of the Riemann ellipsoids of the three types, in their respective domains of occupancy, have been determined, with the aid of the formulae of § II, for a large number of cases. In Tables 4 and 5 we list them for a few typical cases. More extensive tables will be published elsewhere.

VIII. THE SECOND-ORDER VIRIAL EQUATIONS GOVERNING SMALL OSCILLATIONS ABOUT EQUILIBRIUM: THE CHARACTERISTIC EQUATION

Suppose that an equilibrium ellipsoid determined consistently with respect to the equations derived in § III is slightly perturbed. Let the ensuing motions be described in terms of a Lagrangian displacement of the form

$$\xi(x)e^{\lambda t}, \tag{115}$$

where λ is a parameter whose characteristic values are to be determined. Then proceeding exactly as in Paper I, § VI, we find that the linearized form of the virial equation (2) gives (cf. Paper I, eq. [84])

$$\begin{aligned}\lambda^2 V_{i;j} - 2\lambda Q_{jl} V_{i;l} - 2\lambda\epsilon_{ilm}\Omega_m V_{l;j} + Q_{jl}{}^2 V_{i;l} + Q_{il}{}^2 V_{j;l} \\ + 2\epsilon_{ilm}\Omega_m(Q_{jk}V_{l;k} - Q_{lk}V_{j;k}) - \Omega^2 V_{ij} + \Omega_i\Omega_k V_{kj} - \delta\mathfrak{W}_{ij} = \delta_{ij}\delta\Pi\,,\end{aligned} \tag{116}$$

TABLE 5

THE PROPERTIES OF A FEW RIEMANN ELLIPSOIDS OF TYPES II AND III

	Ellipsoids of Type II			Ellipsoids of Type III			
a_2/a_1	3.05590	4.31608	6.24270	5.00000	4.00000	3.75000	4.66667
a_3/a_1	0.10665	0.45115	0.21787	3.00000	2.00000	1.50000	1.40000
Ω_2	+0.70172	+1.13288	+0.83636	+0.44981	+0.40641	+0.36030	+0.45130
Ω_3	+0.16788	+0.17511	+0.06655	+0.43847	+0.43856	+0.32561	+0.20871
ζ_2	+0.26688	+0.18979	+0.05016	+1.49937	+1.01603	+0.36548	+0.17456
ζ_3	−0.86595	−1.69461	−1.32826	−2.28003	−1.86387	−1.81498	−2.10718
Ω_2†	+0.02814	+0.07114	+0.01044	+0.44981	+0.40641	+0.16868	+0.08256
Ω_3†	+0.25596	+0.37263	+0.20745	+0.43847	+0.43856	+0.45186	+0.43172
ζ_2†	+6.65457	+3.02219	+4.02055	+1.49937	+1.01603	+0.78066	+0.95417
ζ_3†	−0.56795	−0.79638	−0.42612	−2.28003	−1.86387	−1.30787	−1.01872

where the various symbols have their usual meanings and the assumption has been made that the internal motion in the equilibrium configuration is given by

$$u_j = Q_{jl}x_l\,, \tag{117}$$

where the Q_{jl}'s are certain constants. For the case of the Riemann ellipsoids, presently considered, the matrices $\boldsymbol{Q}$ and $\boldsymbol{Q}^2$ are of the forms (cf. eq. [21])

$$\boldsymbol{Q} = \begin{vmatrix} 0 & Q_{12} & Q_{13} \\ Q_{21} & 0 & 0 \\ Q_{31} & 0 & 0 \end{vmatrix} \tag{118}$$

and

$$\boldsymbol{Q}^2 = \begin{vmatrix} Q_{12}Q_{21} + Q_{13}Q_{31} & 0 & 0 \\ 0 & Q_{21}Q_{12} & Q_{21}Q_{13} \\ 0 & Q_{31}Q_{12} & Q_{31}Q_{13} \end{vmatrix}, \tag{119}$$

where

$$Q_{12} = +\Omega_3 \frac{a_1^2}{a_2^2}\gamma , \qquad Q_{21} = -\Omega_3\gamma ,$$
$$Q_{13} = -\Omega_2 \frac{a_1^2}{a_3^2}\beta , \quad \text{and} \quad Q_{31} = +\Omega_2\beta . \tag{120}$$

We shall now write down the explicit forms which the different components of equation (116) take in view of the special forms of the matrices $\boldsymbol{Q}$ and $\boldsymbol{Q}^2$. The three diagonal components of equation (116) are

$$\begin{aligned}\lambda^2 V_{1;1} &- 2\lambda(Q_{12}V_{1;2} + Q_{13}V_{1;3}) - 2\lambda(\Omega_3 V_{2;1} - \Omega_2 V_{3;1}) \\ &+ 2(Q_{12}Q_{21} + Q_{13}Q_{31})V_{1;1} + 2\Omega_3(Q_{12}V_{2;2} - Q_{21}V_{1;1} + Q_{13}V_{2;3}) \\ &- 2\Omega_2(Q_{13}V_{3;3} - Q_{31}V_{1;1} + Q_{12}V_{3;2}) - (\Omega_2^2 + \Omega_3^2)V_{11} - \delta\mathfrak{W}_{11} = \delta\Pi ,\end{aligned} \tag{121}$$

$$\begin{aligned}\lambda^2 V_{2;2} &- 2\lambda Q_{21}V_{2;1} + 2\lambda\Omega_3 V_{1;2} + 2(Q_{21}Q_{12}V_{2;2} + Q_{21}Q_{13}V_{2;3}) \\ &+ 2\Omega_3(Q_{12}V_{2;2} - Q_{21}V_{1;1} + Q_{13}V_{2;3}) - \Omega_3^2 V_{22} + \Omega_2\Omega_3 V_{32} - \delta\mathfrak{W}_{22} = \delta\Pi ,\end{aligned} \tag{122}$$

and

$$\begin{aligned}\lambda^2 V_{3;3} &- 2\lambda Q_{31}V_{3;1} - 2\lambda\Omega_2 V_{1;3} + 2(Q_{31}Q_{12}V_{3;2} + Q_{31}Q_{13}V_{3;3}) \\ &- 2\Omega_2(Q_{13}V_{3;3} - Q_{31}V_{1;1} + Q_{12}V_{3;2}) - \Omega_2^2 V_{33} + \Omega_2\Omega_3 V_{23} - \delta\mathfrak{W}_{33} = \delta\Pi .\end{aligned} \tag{123}$$

Eliminating $\delta\Pi$ from the foregoing equations, we obtain the pair of equations

$$\begin{aligned}&[\lambda^2 + 2(Q_{12}Q_{21} + Q_{13}Q_{31}) + 2\Omega_2 Q_{31} - 2(\Omega_2^2 + \Omega_3^2)]V_{1;1} - (\lambda^2 + 2Q_{12}Q_{21} - 2\Omega_3^2)V_{2;2} \\ &\quad - 2\Omega_2 Q_{13}V_{3;3} - 2\lambda(Q_{12} + \Omega_3)V_{1;2} - 2\lambda(\Omega_3 - Q_{21})V_{2;1} - 2\lambda Q_{13}V_{1;3} + 2\lambda\Omega_2 V_{3;1} \\ &\quad - (2Q_{21}Q_{13} + \Omega_2\Omega_3)V_{2;3} - (2\Omega_2 Q_{12} + \Omega_2\Omega_3)V_{3,2} - \delta\mathfrak{W}_{11} + \delta\mathfrak{W}_{22} = 0 ,\end{aligned} \tag{124}$$

and

$$\begin{aligned}&[\lambda^2 + 2(Q_{12}Q_{21} + Q_{13}Q_{31}) - 2\Omega_3 Q_{21} - 2(\Omega_2^2 + \Omega_3^2)]V_{1;1} - (\lambda^2 + 2Q_{13}Q_{31} - 2\Omega_2^2)V_{3;3} \\ &\quad + 2\Omega_3 Q_{12}V_{2;2} - 2\lambda(Q_{13} - \Omega_2)V_{1;3} + 2\lambda(\Omega_2 + Q_{31})V_{3;1} - 2\lambda Q_{12}V_{1,2} - 2\lambda\Omega_3 V_{2;1} \\ &\quad - (2Q_{31}Q_{12} + \Omega_2\Omega_3)V_{3;2} + (2\Omega_3 Q_{13} - \Omega_2\Omega_3)V_{2;3} - \delta\mathfrak{W}_{11} + \delta\mathfrak{W}_{33} = 0 .\end{aligned} \tag{125}$$

The remaining six non-diagonal components of equation (116) are

$$\begin{aligned}\lambda^2 V_{2;1} &+ 2\lambda\Omega_3 V_{1;1} - 2\lambda(Q_{12}V_{2,2} + Q_{13}V_{2;3}) + (Q_{12}Q_{21} + Q_{13}Q_{31})V_{2;1} + Q_{21}Q_{12}V_{1;2} \\ &+ Q_{21}Q_{13}V_{1;3} - \Omega_3^2 V_{21} + \Omega_2\Omega_3 V_{31} - \delta\mathfrak{W}_{12} = 0 ,\end{aligned} \tag{126}$$

$$\lambda^2 V_{1;2} - 2\lambda Q_{21} V_{1;1} - 2\lambda(\Omega_3 V_{2;2} - \Omega_2 V_{3;2}) + (Q_{12}Q_{21} + Q_{13}Q_{31}) V_{2;1} + Q_{12}Q_{21} V_{1;2}$$
$$+ Q_{21}Q_{13} V_{1;3} - 2\Omega_2(Q_{21} V_{3;1} - Q_{31} V_{2;1}) - (\Omega_2^2 + \Omega_3^2) V_{12} - \delta\mathfrak{W}_{12} = 0 \,, \tag{127}$$

$$\lambda^2 V_{3;1} - 2\lambda\Omega_2 V_{1;1} - 2\lambda(Q_{12} V_{3;2} + Q_{13} V_{3;3}) + (Q_{12}Q_{21} + Q_{13}Q_{31}) V_{3;1} + Q_{13}Q_{31} V_{1;3}$$
$$+ Q_{31}Q_{12} V_{1;2} - \Omega_2^2 V_{31} + \Omega_2\Omega_3 V_{21} - \delta\mathfrak{W}_{13} = 0 \,, \tag{128}$$

$$\lambda^2 V_{1;3} - 2\lambda Q_{31} V_{1;1} - 2\lambda(\Omega_3 V_{2;3} - \Omega_2 V_{3;3}) + (Q_{12}Q_{21} + Q_{13}Q_{31}) V_{3;1} + Q_{13}Q_{31} V_{1;3}$$
$$+ Q_{31}Q_{12} V_{1;2} - 2\Omega_3(Q_{21} V_{3;1} - Q_{31} V_{2;1}) - (\Omega_2^2 + \Omega_3^2) V_{13} - \delta\mathfrak{W}_{13} = 0 \,, \tag{129}$$

$$\lambda^2 V_{3;2} - 2\lambda Q_{21} V_{3;1} - 2\lambda\Omega_2 V_{1;2} + Q_{31}Q_{12} V_{2;2} + Q_{21}Q_{13} V_{3;3} + Q_{31}Q_{13} V_{2;3} + Q_{21}Q_{12} V_{3;2}$$
$$- 2\Omega_2(Q_{12} V_{2,2} - Q_{21} V_{1;1} + Q_{13} V_{2;3}) - \Omega_2^2 V_{32} + \Omega_2\Omega_3 V_{22} - \delta\mathfrak{W}_{32} = 0 \,, \tag{130}$$

and

$$\lambda^2 V_{2;3} - 2\lambda Q_{31} V_{2;1} + 2\lambda\Omega_3 V_{1;3} + Q_{31}Q_{12} V_{2,2} + Q_{21}Q_{13} V_{3;3} + Q_{31}Q_{13} V_{2;3} + Q_{21}Q_{12} V_{3;2}$$
$$+ 2\Omega_3(Q_{13} V_{3;3} - Q_{31} V_{1;1} + Q_{12} V_{3;2}) - \Omega_3^2 V_{32} + \Omega_2\Omega_3 V_{33} - \delta\mathfrak{W}_{23} = 0 \,. \tag{131}$$

The eight equations (124)–(131) must be supplemented by the condition

$$\frac{V_{1;1}}{a_1^2} + \frac{V_{2;2}}{a_2^2} + \frac{V_{3;3}}{a_3^2} = 0 \tag{132}$$

required by the solenoidal character of $\boldsymbol{\xi}$.

On inserting the values of Q_{12}, etc., in accordance with equations (120), we find that the system of equations (124)–(132) can be written in matrix notation in the form given on pages 862–863, where we have substituted for the $\delta\mathfrak{W}_{ij}$'s their known values (cf., for example, Paper I, eqs. [87] and [88]).

The required characteristic equation follows from setting the determinant of the matrix on the left-hand side of equation (133) equal to zero.

IX. THE EQUALITY OF THE CHARACTERISTIC FREQUENCIES OF AN ELLIPSOID AND ITS ADJOINT AND OTHER THEOREMS

Eliminating $V_{1;1}$ from the system of equations (133) and multiplying the different rows and columns of the resulting secular matrix by suitable factors, we find that the secular determinant can be brought to the form given on pages 864–865.

$\lambda^2 - 2(\boldsymbol{\Omega}^2 + \boldsymbol{\Omega}^{\dagger 2}) + 2\Omega_2^2\beta + 6B_{11} - 2B_{12}$	$-\lambda^2 + 2\Omega_3^{\dagger 2} + 2\Omega_3^2 - 6B_{22} + 2B_{12}$	$2\dfrac{a_1^2}{a_3^2}\Omega_2^2\beta + 2B_{13} - 2B_{23}$	$-2\lambda\Omega_3\left(1 + \dfrac{a_1^2}{a_2^2}\gamma\right)$	$-2\lambda\Omega_3(1+\gamma)$
$\lambda^2 - 2(\boldsymbol{\Omega}^2 + \boldsymbol{\Omega}^{\dagger 2}) + 2\Omega_3^2\gamma + 6B_{11} - 2B_{13}$	$2\dfrac{a_1^2}{a_2^2}\Omega_3^2\gamma + 2B_{12} - 2B_{23}$	$-\lambda^2 + 2\Omega_2^{\dagger 2} + 2\Omega_2^2 - 6B_{33} + 2B_{13}$	$-2\lambda\dfrac{a_1^2}{a_2^2}\Omega_3\gamma$	$-2\lambda\Omega_3$
$2\lambda\Omega_3$	$-2\lambda\Omega_3\dfrac{a_1^2}{a_2^2}\gamma$	0	$-(\Omega_3^2 + \Omega_3^{\dagger 2}) + 2B_{12}$	$+\lambda^2 - \boldsymbol{\Omega}^{\dagger 2} - \Omega_3^2 + 2B_{12}$
$2\lambda\Omega_3\gamma$	$-2\lambda\Omega_3$	0	$+\lambda^2 - \boldsymbol{\Omega}^2 - \Omega_3^{\dagger 2} + 2B_{12}$	$-(\boldsymbol{\Omega}^2 + \boldsymbol{\Omega}^{\dagger 2}) + 2\Omega_2^2\beta + 2B_{12}$
$-2\lambda\Omega_2$	0	$2\lambda\Omega_2\dfrac{a_1^2}{a_3^2}\beta$	$\Omega_2\Omega_3\left(1 + \dfrac{a_1^2}{a_2^2}\beta\gamma\right)$	$\Omega_2\Omega_3$
$-2\lambda\Omega_2\beta$	0	$2\lambda\Omega_2$	$\Omega_2\Omega_3\dfrac{a_1^2}{a_2^2}\beta\gamma$	$2\Omega_2\Omega_3\beta$
$-2\Omega_2\Omega_3\gamma$	$\Omega_2\Omega_3$	$\Omega_2\Omega_3\dfrac{a_1^2}{a_3^2}\beta\gamma$	$-2\lambda\Omega_2$	0
$-2\Omega_2\Omega_3\beta$	$\Omega_2\Omega_3\dfrac{a_1^2}{a_2^2}\beta\gamma$	$\Omega_2\Omega_3$	0	$-2\lambda\Omega_2\beta$
$\dfrac{1}{a_1^2}$	$\dfrac{1}{a_2^2}$	$\dfrac{1}{a_3^2}$	0	0

$$
\begin{array}{cccc||c|}
2\lambda\Omega_2 \frac{a_1^2}{a_3^2} \beta & 2\lambda\Omega_2 & -\Omega_2\Omega_3 \left(1 + 2 \frac{a_1^2}{a_3^2} \beta\gamma\right) & -\Omega_2\Omega_3 \left(1 + 2 \frac{a_1^2}{a_2^2} \gamma\right) & V_{1;1} \\
2\lambda\Omega_2 \left(1 + \frac{a_1^2}{a_3^2} \beta\right) & 2\lambda\Omega_2 (1 + \beta) & -\Omega_2\Omega_3 \left(1 + 2 \frac{a_1^2}{a_3^2} \beta\right) & -\Omega_2\Omega_3 \left(1 + 2 \frac{a_1^2}{a_2^2} \beta\gamma\right) & V_{2;2} \\
\Omega_2\Omega_3 \left(1 + \frac{a_1^2}{a_3^2} \beta\gamma\right) & \Omega_2\Omega_3 & 2\lambda\Omega_2 \frac{a_1^2}{a_3^2} \beta & 0 & V_{3;3} \\
\Omega_2\Omega_3 \frac{a_1^2}{a_3^2} \beta\gamma & 2\Omega_2\Omega_3\gamma & 0 & 2\lambda\Omega_2 & V_{1;2} \\
-\Omega_2^2 - \Omega_2^{\dagger 2} + 2B_{13} & \lambda^2 - \boldsymbol{\Omega}^{\dagger 2} - \Omega_2^2 + 2B_{13} & 0 & -2\lambda\Omega_3 \frac{a_1^2}{a_2^2} \gamma & V_{2;1} \\
\lambda^2 - \boldsymbol{\Omega}^2 - \Omega_2^{\dagger 2} + 2B_{13} & -(\boldsymbol{\Omega}^2 + \boldsymbol{\Omega}^{\dagger 2}) + 2\Omega_3^2\gamma + 2B_{13} & -2\lambda\Omega_3 & 0 & V_{1;3} \\
0 & 2\lambda\Omega_3\gamma & -\Omega_2^2 - \Omega_2^{\dagger 2} + 2\Omega_2^2 \frac{a_1^2}{a_3^2} \beta + 2B_{23} & \lambda^2 - \Omega_2^2 - \Omega_3^{\dagger 2} + 2B_{23} & V_{3;1} \\
2\lambda\Omega_3 & 0 & \lambda^2 - \Omega_3^2 - \Omega_2^{\dagger 2} + 2B_{23} & -\Omega_3^2 - \Omega_3^{\dagger 2} + 2\Omega_3^2 \frac{a_1^2}{a_2^2} \gamma + 2B_{23} & V_{2;3} \\
0 & 0 & 0 & 0 & V_{3;2}
\end{array}
= 0 \,. \quad (133)
$$

$$
\left[\begin{array}{cccc}
(1,1) & (1,2) & -2\lambda\left(\Omega_3+\frac{a_1}{a_2}\,\Omega_3^{\dagger}\right) & -2\lambda\left(\Omega_3+\frac{a_2}{a_1}\,\Omega_3^{\dagger}\right) \\
(2,1) & (2,2) & -2\lambda\Omega_3^{\dagger}\,\frac{a_1}{a_2} & -2\lambda\Omega_3 \\
-2\lambda\left(\Omega_3^{\dagger}+\frac{a_1}{a_2}\,\Omega_3\right) & -2\lambda\Omega_3\,\frac{a_1}{a_2} & -(\Omega_3^2+\Omega_3^{\dagger 2})+2B_{12} & \lambda^2-\mathbf{\Omega}^{\dagger 2}-\Omega_3^2+2B_{12} \\
-2\lambda\left(\Omega_3^{\dagger}+\frac{a_2}{a_1}\,\Omega_3\right) & -2\lambda\Omega_3^{\dagger} & \lambda^2-\mathbf{\Omega}^2-\Omega_3^{\dagger 2}+2B_{12} & -(\mathbf{\Omega}^2+\mathbf{\Omega}^{\dagger 2})+2\Omega_2^2\beta+2B_{12} \\
2\lambda\Omega_2\,\frac{a_1}{a_3} & 2\lambda\left(\Omega_2^{\dagger}+\frac{a_1}{a_3}\,\Omega_2\right) & 2\Omega_2\Omega_3^{\dagger}\,\frac{a_1}{a_3} & \Omega_2\Omega_3\,\frac{a_2}{a_3} \\
2\lambda\Omega_2^{\dagger} & 2\lambda\left(\Omega_2^{\dagger}+\frac{a_3}{a_1}\,\Omega_2\right) & \Omega_2^{\dagger}\Omega_3^{\dagger} & 2\Omega_2^{\dagger}\Omega_3\,\frac{a_2}{a_1} \\
2\Omega_2\Omega_3\,\frac{a_2}{a_3}+\Omega_2^{\dagger}\Omega_3^{\dagger} & 2\Omega_2^{\dagger}\Omega_3^{\dagger}+\frac{a_2}{a_3}\,\Omega_2\Omega_3 & -2\lambda\Omega_2\,\frac{a_1}{a_3} & 0 \\
2\Omega_2^{\dagger}\Omega_3^{\dagger}+\frac{a_3}{a_2}\,\Omega_2\Omega_3 & \Omega_2^{\dagger}\Omega_3^{\dagger}+2\,\frac{a_3}{a_2}\,\Omega_2\Omega_3 & 0 & -2\lambda\Omega_2^{\dagger}
\end{array}\right.
$$

$$
\left.\begin{matrix}
2\lambda\Omega_2^{\dagger}\dfrac{a_1}{a_3} & 2\lambda\Omega_2 & \Omega_2\Omega_3+2\dfrac{a_2}{a_3}\Omega_2^{\dagger}\Omega_3^{\dagger} & 2\Omega_2\Omega_3+\dfrac{a_3}{a_2}\Omega_2^{\dagger}\Omega_3^{\dagger} \\
2\lambda\left(\Omega_2+\dfrac{a_1}{a_3}\Omega_2^{\dagger}\right) & 2\lambda\left(\Omega_2+\dfrac{a_3}{a_1}\Omega_2^{\dagger}\right) & 2\Omega_2\Omega_3+\dfrac{a_2}{a_3}\Omega_2^{\dagger}\Omega_3^{\dagger} & \Omega_2\Omega_3+2\dfrac{a_3}{a_2}\Omega_2^{\dagger}\Omega_3^{\dagger} \\
2\Omega_2^{\dagger}\Omega_3\dfrac{a_1}{a_3} & \Omega_2\Omega_3 & -2\lambda\Omega_2^{\dagger}\dfrac{a_1}{a_3} & 0 \\
\Omega_2^{\dagger}\Omega_3^{\dagger}\dfrac{a_2}{a_3} & 2\Omega_2\Omega_3^{\dagger}\dfrac{a_2}{a_1} & 0 & -2\lambda\Omega_2 \\
\dfrac{a_2}{a_3}(-\Omega_2^2-\Omega_2^{\dagger 2}+2B_{13}) & \dfrac{a_2}{a_3}(\lambda^2-\mathbf{\Omega}^{\dagger 2}-\Omega_2^2+2B_{13}) & 0 & 2\lambda\Omega_3^{\dagger}\dfrac{a_1}{a_3} \\
\dfrac{a_2}{a_3}(\lambda^2-\mathbf{\Omega}^2-\Omega_2^{\dagger 2}+2B_{13}) & \dfrac{a_2}{a_3}(-\mathbf{\Omega}^2-\mathbf{\Omega}^{\dagger 2}+2\Omega_3^2\gamma+2B_{13}) & +2\lambda\Omega_3\dfrac{a_2}{a_3} & 0 \\
0 & 2\lambda\Omega_3^{\dagger}\dfrac{a_2}{a_3} & \dfrac{a_1}{a_3}\left(\Omega_2^2+\Omega_2^{\dagger 2}-2\Omega_2^2\dfrac{a_1^2}{a_3^2}\beta-2B_{23}\right) & \dfrac{a_1}{a_3}(-\lambda^2+\Omega_2^2+\Omega_3^{\dagger 2}-2B_{23}) \\
2\lambda\Omega_3\dfrac{a_1}{a_3} & 0 & \dfrac{a_1}{a_3}(-\lambda^2+\Omega_2^{\dagger 2}+\Omega_3^2-2B_{23}) & \dfrac{a_1}{a_3}\left(+\Omega_3^2+\Omega_3^{\dagger 2}-2\Omega_3^2\dfrac{a_1^2}{a_2^2}\gamma-2B_{23}\right)
\end{matrix}\right]
$$

$$= 0\,, \qquad (134)$$

where

$$
\begin{aligned}
(1,1) &= -\frac{a_1}{a_2}(\lambda^2-2\mathbf{\Omega}^2-2\mathbf{\Omega}^{\dagger 2}+2\Omega_2^2\beta+6B_{11}-2B_{12})-\frac{a_2}{a_1}(\lambda^2-2\Omega_3^2-2\Omega_3^{\dagger 2}+6B_{22}-2B_{12}),\\
(2,2) &= -\frac{a_1}{a_2}(\lambda^2-2\mathbf{\Omega}^2-2\mathbf{\Omega}^{\dagger 2}+2\Omega_3^2\gamma+6B_{11}-2B_{13})-\frac{a_3^2}{a_1a_2}(\lambda^2-2\Omega_2^2-2\Omega_2^{\dagger 2}+6B_{33}-2B_{13}),
\end{aligned}
\qquad (135)
$$

and

$$
(1,2) = (2,1) = -\frac{a_1}{a_2}\left\{\lambda^2-2\mathbf{\Omega}^2-2\mathbf{\Omega}^{\dagger 2}+6B_{11}-2\left[\left(\frac{a_2^2a_3^2}{a_1^2}-a_1^2\right)B_{123}+B_{23}\right]\right\}.
$$

In reducing the secular determinant to the form (134) use must be made of the relations (cf. eqs. [78] and [80])

$$\Omega_3{}^\dagger = \Omega_3 \frac{a_1}{a_2} \gamma \qquad \text{and} \qquad \Omega_2{}^\dagger = \Omega_2 \frac{a_1}{a_3} \beta \,. \tag{136}$$

From the form to which the secular equation has been reduced, it is manifest that if in each element of the secular matrix we replace the components of $\mathbf{\Omega}$ and $\mathbf{\Omega}^\dagger$, which occur, by their respective adjoints we obtain the transposed matrix. It, therefore, follows that *the characteristic frequencies of oscillation, belonging to these "second harmonics," of an ellipsoid and its adjoint are the same.* This theorem, first established in the contexts of the Dedekind and the Jacobian sequences (Chandrasekhar 1965*a*) and then generalized to the Riemann sequences (Paper I), is now seen to be an entirely general property.

We shall now prove that $|\mathbf{\Omega}|$ *and* $|\mathbf{\Omega}^\dagger|$ *are characteristic frequencies provided as roots of the characteristic equation.* We shall prove this theorem by showing that the following set of three equations, included as simple linear combinations in the system of equations (133), are linearly dependent if λ^2 is set equal to $-|\mathbf{\Omega}|^2 = -(\Omega_2{}^2 + \Omega_3{}^2)$:

$$\begin{aligned} \lambda^2(V_{1;2} - V_{2;1}) - 2\lambda(Q_{21}V_{1;1} - Q_{12}V_{2;2} - Q_{13}V_{2;3}) - 2\lambda(\Omega_3 V_{2;2} - \Omega_2 V_{3;2} + \Omega_3 V_{1;1}) \\ - 2\Omega_2(Q_{21}V_{3;1} - Q_{31}V_{2;1}) - \Omega_2{}^2 V_{12} - \Omega_2\Omega_3 V_{13} = 0 \,, \end{aligned} \tag{137}$$

$$\begin{aligned} \lambda^2(V_{1;3} - V_{3;1}) - 2\lambda(Q_{31}V_{1;1} - Q_{13}V_{3;3} - Q_{12}V_{3;2}) - 2\lambda(\Omega_3 V_{2;3} - \Omega_2 V_{3;3} - \Omega_2 V_{1;1}) \\ - 2\Omega_3(Q_{21}V_{3;1} - Q_{31}V_{2;1}) - \Omega_3{}^2 V_{13} - \Omega_2\Omega_3 V_{12} = 0 \,, \end{aligned} \tag{138}$$

and

$$\begin{aligned} \lambda^2(V_{2;3} - V_{3;2}) - 2\lambda(Q_{31}V_{2;1} - Q_{21}V_{3;1}) + 2\lambda(\Omega_3 V_{1;3} + \Omega_2 V_{1;2}) \\ + 2\Omega_3(Q_{13}V_{3;3} - Q_{31}V_{1;1} + Q_{12}V_{3;2}) + 2\Omega_2(Q_{12}V_{2;2} - Q_{21}V_{1;1} + Q_{13}V_{2;3}) \\ + (\Omega_2{}^2 - \Omega_3{}^2)V_{23} + \Omega_2\Omega_3(V_{33} - V_{22}) = 0 \,. \end{aligned} \tag{139}$$

These equations are obtained by subtracting equations (126), (128), and (130) from equations (127), (129), and (131), respectively. Eliminating $V_{1;1}$ from these equations with the aid of the divergence condition (132), we obtain

$$\begin{aligned} &(\lambda^2 - \Omega_2{}^2)V_{1;2} - (\lambda^2 + \Omega_2{}^2 - 2\Omega_2 Q_{31})V_{2;1} - \Omega_2\Omega_3 V_{1;3} \\ &- \Omega_2(2Q_{21} + \Omega_3)V_{3;1} + 2\lambda Q_{13}V_{2;3} + 2\lambda\Omega_2 V_{3;2} \\ &+ 2\lambda\left[Q_{12} - \Omega_3 + \frac{a_1{}^2}{a_2{}^2}(Q_{21} + \Omega_3)\right]V_{2;2} + 2\lambda \frac{a_1{}^2}{a_3{}^2}(Q_{21} + \Omega_3)V_{3;3} = 0 \,, \end{aligned} \tag{140}$$

$$\begin{aligned} &-\Omega_2\Omega_3 V_{1;2} + \Omega_3(2Q_{31} - \Omega_2)V_{2;1} + (\lambda^2 - \Omega_3{}^2)V_{1;3} \\ &- (\lambda^2 + \Omega_3{}^2 + 2\Omega_3 Q_{21})V_{3;1} - 2\lambda\Omega_3 V_{2;3} + 2\lambda Q_{12}V_{3;2} \\ &+ 2\lambda \frac{a_1{}^2}{a_2{}^2}(Q_{31} - \Omega_2)V_{2;2} + 2\lambda\left[Q_{13} + \Omega_2 + \frac{a_1{}^2}{a_3{}^2}(Q_{31} - \Omega_2)\right]V_{3;3} = 0 \,, \end{aligned} \tag{141}$$

and

$$\begin{aligned} &2\lambda\Omega_2 V_{1;2} - 2\lambda Q_{31}V_{2;1} + 2\lambda\Omega_3 V_{1;3} + 2\lambda Q_{21}V_{3;1} + (\lambda^2 + 2\Omega_2 Q_{13} + \Omega_2{}^2 - \Omega_3{}^2)V_{2;3} \\ &- (\lambda^2 - 2\Omega_3 Q_{12} - \Omega_2{}^2 + \Omega_3{}^2)V_{3;2} + 2\left[\Omega_2(Q_{12} - \Omega_3) + \frac{a_1{}^2}{a_2{}^2}(\Omega_3 Q_{31} + \Omega_2 Q_{21})\right]V_{2;2} \\ &+ 2\left[\Omega_3(Q_{13} + \Omega_2) + \frac{a_1{}^2}{a_3{}^2}(\Omega_3 Q_{31} + \Omega_2 Q_{21})\right]V_{3;3} = 0 \,. \end{aligned} \tag{142}$$

It can now be verified directly that all the three-rowed determinants of the 3×8 matrix representing the foregoing system of equations vanish identically if λ^2 is set equal to $-(\Omega_2^2 + \Omega_3^2)$. The equations are therefore linearly dependent in the case considered. The linear dependence, under the same circumstances, of the equations of the system (133) follows a fortiori. This establishment of the linear dependence proves that $|\boldsymbol{\Omega}|$ is a characteristic frequency; that $|\boldsymbol{\Omega}^\dagger|$ is also a characteristic frequency follows from the theorem proved earlier that for an ellipsoid and its adjoint the characteristic equation (134) gives the same frequencies.

Finally, we shall show that $\lambda^2 = 0$ is a non-trivial double root of the characteristic equation. More precisely, we shall show that for the proper solutions belonging to the zero root the only non-vanishing virials are $V_{1;2}$, $V_{2;1}$, $V_{1;3}$, and $V_{3;1}$.

By setting

$$\lambda = 0 \qquad \text{and} \qquad V_{1;1} = V_{2;2} = V_{3;3} = V_{2;3} = V_{3;2} = 0\,, \tag{143}$$

we satisfy trivially equations (124), (125), (130), (131), and (132). Considering (127), (129), (137), and (138) as the remaining four equations, we find that under the circumstances specified (namely, [143]) equations (127) and (129) give

$$\begin{aligned}(Q_{12}Q_{21} + Q_{13}Q_{31})V_{2;1} + Q_{21}Q_{12}V_{1;2} + Q_{21}Q_{13}V_{1;3}&\\ -2\Omega_2(Q_{21}V_{3;1} - Q_{31}V_{2;1}) - (\Omega_2^2 + \Omega_3^2 - 2B_{12})V_{12} &= 0\end{aligned} \tag{144}$$

and

$$\begin{aligned}(Q_{12}Q_{21} + Q_{13}Q_{31})V_{3;1} + Q_{31}Q_{13}V_{1;3} + Q_{31}Q_{12}V_{1;2}&\\ - 2\Omega_3(Q_{21}V_{3;1} - Q_{31}V_{2;1}) - (\Omega_2^2 + \Omega_3^2 - 2B_{13})V_{13} &= 0\,,\end{aligned} \tag{145}$$

while equations (137) and (138) provide the *single* equation

$$2(Q_{21}V_{3;1} - Q_{31}V_{2;1}) + \Omega_2 V_{12} + \Omega_3 V_{13} = 0\,. \tag{146}$$

Equations (144) and (145) can be rewritten in the forms

$$[Q_{12}Q_{21} - (\Omega_2^2 + \Omega_3^2) + 2B_{12}]V_{12} + Q_{21}Q_{13}V_{13} - (2\Omega_2 + Q_{13})(Q_{21}V_{3;1} - Q_{31}V_{2;1}) = 0 \tag{147}$$

and

$$[Q_{13}Q_{31} - (\Omega_2^2 + \Omega_3^2) + 2B_{13}]V_{13} + Q_{31}Q_{12}V_{12} - (2\Omega_3 - Q_{12})(Q_{21}V_{3;1} - Q_{31}V_{2;1}) = 0\,. \tag{148}$$

We observe that equations (146)–(148) are linear and homogeneous in the virials

$$V_{12},\ V_{13}\,, \qquad \text{and} \qquad Q_{21}V_{3;1} - Q_{31}V_{2;1}\,; \tag{149}$$

and if these virials do not vanish identically, it must be true that

$$\begin{vmatrix} 2B_{12} - \boldsymbol{\Omega}^2 - \dfrac{a_1^2}{a_2^2}\Omega_3^2\gamma^2 & \dfrac{a_1^2}{a_3^2}\Omega_2\Omega_3\beta\gamma & -\Omega_2\left(2 - \dfrac{a_1^2}{a_3^2}\beta\right) \\ \dfrac{a_1^2}{a_2^2}\Omega_2\Omega_3\beta\gamma & 2B_{13} - \boldsymbol{\Omega}^2 - \dfrac{a_1^2}{a_3^2}\Omega_2^2\beta^2 & -\Omega_3\left(2 - \dfrac{a_1^2}{a_2^2}\gamma\right) \\ \Omega_2 & \Omega_3 & 2 \end{vmatrix} = 0\,. \tag{150}$$

By making use of the relations (34) and (35), the requirement (150) can be reduced to the form

$$\begin{bmatrix} -\dfrac{a_1^2(3a_3^2-4a_1^2+a_2^2)\beta}{2a_3^2(a_1^2-a_2^2)}-1 & \dfrac{a_1^2}{a_3^2}\beta\gamma & -2+\dfrac{a_1^2}{a_3^2}\beta \\ \dfrac{a_1^2}{a_2^2}\beta\gamma & -\dfrac{a_1^2(3a_2^2-4a_1^2+a_3^2)\gamma}{2a_2^2(a_1^2-a_3^2)}-1 & -2+\dfrac{a_1^2}{a_2^2}\gamma \\ 1 & 1 & 2 \end{bmatrix}=0\,, \tag{151}$$

and by direct evaluation it can be verified that the determinant is in fact zero. Equations (124)–(132) can, therefore, be satisfied non-trivially for $\lambda^2 = 0$ with non-vanishing values for the virials listed in (149). Moreover, it is clear that there are two linearly independent solutions; and since the original characteristic equation (134) is even in λ^2 (of degree *eight*), $\lambda^2 = 0$ is a double root.

Excluding the roots $-|\boldsymbol{\Omega}|^2$, $-|\boldsymbol{\Omega}^\dagger|^2$, and zero (of multiplicity two) we have four roots of the characteristic equation yet to determine. These remaining roots have been determined numerically for a number of ellipsoids of types I, II, and III and are considered in § XI below.

X. THE ASYMPTOTIC PROPERTIES OF THE DISKLIKE RIEMANN ELLIPSOIDS ON THE a_2-AXIS

As $a_3 \to 0$, the ellipsoids of types I and II become disklike and their asymptotic properties are of interest.

It can be readily verified that, in the limit

$$\epsilon = a_3/a_1 \to 0\,, \tag{152}$$

the index symbols A_i (cf. eq. [44]) have the behavior

$$A_1 = \alpha_1\epsilon\,, \qquad A_2 = \alpha_2\epsilon\,, \qquad \text{and} \qquad A_3 = 2\,, \tag{153}$$

where α_1 and α_2 are certain constants expressible in terms of the *complete* elliptic integrals

$$E(\theta) = \int_0^{\pi/2} d\phi(1-\sin^2\theta\sin^2\phi)^{1/2}$$

and

$$F(\theta) = \int_0^{\pi/2} d\phi(1-\sin^2\theta\sin^2\phi)^{-1/2}, \tag{154}$$

with the argument

$$\theta = \sec^{-1}(a_2/a_1)\,. \tag{155}$$

Thus,

$$\alpha_1 = \frac{2}{\sin^2\theta}[E(\theta)-F(\theta)\cos^2\theta] \qquad \text{and} \qquad \alpha_2 = \frac{2}{\tan^2\theta}[F(\theta)-E(\theta)]. \tag{156}$$

The corresponding asymptotic forms of the two-index symbols are

$$B_{11} = \beta_{11}\epsilon\,, \qquad B_{22} = \beta_{22}\epsilon\,, \qquad B_{33} = \tfrac{4}{3}\,,$$

$$B_{12} = \beta_{12}\epsilon\,, \qquad B_{23} = \beta_{23}\epsilon\,, \qquad \text{and} \qquad B_{31} = \beta_{31}\epsilon\,, \tag{157}$$

where

$$\beta_{11} = \frac{(\alpha_1 - \alpha_2) a_2^2}{3(a_2^2 - a_1^2)}, \qquad \beta_{22} = \frac{(\alpha_1 - \alpha_2) a_1^2}{3(a_2^2 - a_1^2)},$$
$$\beta_{12} = \frac{\alpha_2 a_2^2 - \alpha_1 a_1^2}{a_2^2 - a_1^2}, \qquad \beta_{23} = \alpha_2, \qquad \text{and} \qquad \beta_{31} = \alpha_1. \tag{158}$$

Similarly, from equations (17), (18), and (75), we conclude that in this same limit,

$$\beta = \frac{4a_1^2 - a_2^2}{2a_1^2}, \quad \gamma = 2, \quad \beta^\dagger = \frac{2a_1^2}{4a_1^2 - a_2^2}\epsilon^2, \qquad \text{and} \qquad \gamma^\dagger = \frac{a_2^2}{2a_1^2}. \tag{159}$$

Inserting the foregoing asymptotic forms of the various constants in equations (37), (38), and (136), we find

$$\Omega_2 = \omega_2 \epsilon^{3/2}, \qquad \Omega_3 = \omega_3 \epsilon^{1/2}, \qquad \Omega_2^\dagger = \omega_2^\dagger \epsilon^{1/2}, \qquad \text{and} \qquad \Omega_3^\dagger = \omega_3^\dagger \epsilon^{1/2}, \tag{160}$$

where

$$\omega_2 = \frac{2a_1^2}{a_2(4a_1^2 - a_2^2)}[2(3\alpha_2 + \alpha_1)a_2^2 - 8\alpha_2 a_1^2]^{1/2}, \qquad \omega_3 = (2\alpha_2)^{1/2},$$
$$\omega_2^\dagger = \omega_2 \beta = \frac{1}{a_2}[2(3\alpha_2 + \alpha_1)a_2^2 - 8\alpha_2 a_1^2]^{1/2}, \qquad \text{and} \qquad \omega_3^\dagger = \frac{a_1}{a_2}(8\alpha_2)^{1/2}. \tag{161}$$

The corresponding asymptotic forms of the components of the vorticity are

$$\zeta_2 = z_2 \epsilon^{-1/2}, \qquad \zeta_3 = z_3 \epsilon^{1/2}, \qquad \zeta_2^\dagger = z_2^\dagger \epsilon^{1/2}, \qquad \text{and} \qquad \zeta_3^\dagger = z_3^\dagger \epsilon^{1/2}, \tag{162}$$

where

$$z_2 = -\frac{4a_1^2 - a_2^2}{2a_1^2}\omega_2, \qquad z_3 = -2\frac{a_1^2 + a_2^2}{a_2^2}\omega_3,$$
$$z_2^\dagger = -\frac{2a_1^2}{4a_1^2 - a_2^2}\omega_2^\dagger, \qquad z_3^\dagger = -\frac{a_1^2 + a_2^2}{2a_1^2}\omega_3^\dagger. \tag{163}$$

The properties of the disklike ellipsoids of type I, determined with the aid of the foregoing formulae, are included in Table 10, § XI below.

a) The Asymptotic Form of the Characteristic Equation

Turning next to the stability of the disklike objects, we find that in the limit considered

$$\lambda = x\epsilon^{1/2}, \tag{164}$$

where the constant of proportionality x is determined by the appropriate limiting form of the characteristic equation (134).

On inserting the asymptotic forms of the various constants in the different elements of the secular matrix (134), we find that, while all the elements in the first four columns occur with a factor ϵ, the remaining four columns tend to finite limits. After the removal of the factor ϵ^4 the secular determinant takes a finite form that is, moreover, manifestly the product of the two determinants

$$\begin{bmatrix} x^2 - 3\omega_3^2 & -2x\omega_3^\dagger \\ +2x\omega_3^\dagger & x^2 + \omega_3^2 - (\omega_3^\dagger)^2 \end{bmatrix} = 0 \tag{165}$$

and

$$
\begin{bmatrix}
(1,1) & (1,2) & -2x\left(\omega_3+\frac{a_1}{a_2}\omega_3^\dagger\right) & -2x\left(\omega_3+\frac{a_2}{a_1}\omega_3^\dagger\right) & 2x\omega_2^\dagger & -2\omega_2^\dagger\omega_3^\dagger\frac{a_2}{a_1} \\
(2,1) & (2,2) & -2x\omega_3^\dagger\frac{a_1}{a_2} & -2x\omega_3 & 2x\omega_2^\dagger & -\omega_2^\dagger\omega_3^\dagger\frac{a_2}{a_1} \\
-2x\left(\omega_3^\dagger+\frac{a_1}{a_2}\omega_3\right) & -2x\omega_3\frac{a_1}{a_2} & -\omega_3^2-(\omega_3^\dagger)^2+2\beta_{12} & x^2-\omega_3^2-(\omega_2^\dagger)^2-(\omega_3^\dagger)^2+2\beta_{12} & 2\omega_2^\dagger\omega_3 & 2x\omega_2^\dagger \\
-2x\left(\omega_3^\dagger+\frac{a_2}{a_1}\omega_3\right) & -2x\omega_3^\dagger & x^2-\omega_3^2-(\omega_3^\dagger)^2+2\beta_{12} & -\omega_3^2-(\omega_2^\dagger)^2-(\omega_3^\dagger)^2+2\beta_{12} & \omega_2^\dagger\omega_3^\dagger\frac{a_2}{a_1} & 0 \\
2x\omega_2^\dagger & 2x\omega_2^\dagger & \omega_2^\dagger\omega_3^\dagger & 2\omega_2^\dagger\omega_3\frac{a_2}{a_1} & \frac{a_2}{a_1}[x^2-\omega_3^2-(\omega_2^\dagger)^2+2\beta_{13}] & -2x\omega_3\frac{a_2}{a_1} \\
2\omega_2^\dagger\omega_3^\dagger & \omega_2^\dagger\omega_3^\dagger & 0 & -2x\omega_2^\dagger & 2x\omega_3 & x^2-(\omega_2^\dagger)^2
\end{bmatrix} = 0\,. \tag{166}
$$

where

$$
(1,1) = -\left(\frac{a_2}{a_1}+\frac{a_1}{a_2}\right)[x^2-2\omega_3^2-2(\omega_3^\dagger)^2-2\beta_{12}]+\frac{a_1}{a_2}[2(\omega_2^\dagger)^2-12\beta_{11}],
$$

$$
(1,2) = (2,1) = -\frac{a_1}{a_2}[x^2-2\omega_3^2-2(\omega_2^\dagger)^2-2(\omega_3^\dagger)^2+6\beta_{11}-2\beta_{12}], \tag{167}
$$

and

$$
(2,2) = -\frac{a_1}{a_2}[x^2+2\omega_3^2-2(\omega_2^\dagger)^2-2(\omega_3^\dagger)^2+6\beta_{11}-2\beta_{13}].
$$

On expanding the determinant (165) and making use of the special forms of ω_3^2 and $(\omega_3^\dagger)^2$ (see eqs. [161]), we find that equation (165) provides the two characteristic roots

$$x^2 = -\omega_3^2 \quad \text{and} \quad x^2 = -\frac{3}{a_2^2}(4a_1^2 - a_2^2)\omega_3^2 . \tag{168}$$

Accordingly, the second of these two roots makes all the disklike ellipsoids of type II (for which $a_2 \geq 2a_1$) unstable. But the determination of the stability of the analogous ellipsoids of type I requires a consideration of the roots of equation (166) (see § XI below).

XI. THE DOMAINS OF STABILITY WITH RESPECT TO THE OSCILLATIONS BELONGING TO THE SECOND HARMONICS IN THE $(a_2/a_1, a_3/a_1)$-PLANE

The characteristic equation (134) has been solved for its roots for some hundred ellipsoids in their domains of occupancy; and by interpolation among the roots so obtained, the loci of the marginally stable configurations in the $(a_2/a_1, a_3/a_1)$-plane were determined. The results of the calculations, as they pertain to the loci of marginal stability (rather *marginal overstability*, as it happens), are summarized in Tables 6 and 7; and in Tables 8 and 9 we enumerate the characteristic frequencies of oscillation of the ellipsoids whose properties have been listed in § VII.

TABLE 6a

THE PROPERTIES OF THE MARGINALLY OVERSTABLE RIEMANN ELLIPSOIDS OF TYPE I*
(Along the Locus $O_2X_2^{(\text{I})}$)

a_2/a_1...	1.0000	1.0526	1.1111	1.1765	1.2500	1.3333	1.4286	1.5385	1.6722
a_3/a_1...	0.3033	0.3712	0.4230	0.4560	0.4703	0.4676	0.4474	0.4053	0.3278
Ω_2....	0	+0.1283	+0.2153	+0.2942	+0.3639	+0.4269	+0.4877	+0.5550	+0.7107
Ω_3....	+0.7073	+0.7176	+0.7098	+0.6901	+0.6633	+0.6329	+0.5999	+0.5635	+0.5142
ζ_2....	0	−1.5014	−1.8984	−2.1276	−2.2794	−2.3842	−2.4626	−2.5307	−2.4011
ζ_3....	−2.7417	−2.5977	−2.3978	−2.1787	−1.9637	−1.7621	−1.5752	−1.3984	−1.1673
$\Omega_2^\dagger$....	0	+0.4898	+0.6812	+0.8032	+0.8778	+0.9150	+0.9178	+0.8807	+0.7107
$\Omega_3^\dagger$....	+1.3708	+1.2972	+1.1922	+1.0751	+0.9579	+0.8458	+0.7400	+0.6390	+0.5142
$\zeta_2^\dagger$....	0	−0.3931	−0.6000	−0.7794	−0.9450	−1.1125	−1.3082	−1.5937	−2.4011
$\zeta_3^\dagger$....	−1.4147	−1.4371	−1.4275	−1.3984	−1.3599	−1.3186	−1.2768	−1.2330	−1.1673

* The angular velocities and the vorticities are expressed in the unit $(\pi G\rho)^{1/2}$.

TABLE 6b

THE PROPERTIES OF THE MARGINALLY OVERSTABLE RIEMANN ELLIPSOIDS OF TYPE I*
(Along the Locus D_2Q)

a_2/a_1........	1.1582	1.1846	1.2124	1.2418	1.2727	1.3050	1.3707
a_3/a_1........	0.1411	0.1238	0.1057	0.0866	0.0666	0.0455	ϵ
Ω_2..........	+0.0618	+0.0558	+0.0480	+0.0389	+0.0286	+0.0176	$+2.1492\ \epsilon^{3/2}$
Ω_3..........	+0.5209	+0.4903	+0.4554	+0.4146	+0.3658	+0.3044	$+1.4485\ \epsilon^{1/2}$
ζ_2..........	−4.1927	−4.7796	−5.4901	−6.4045	−7.6880	−9.7572	$-2.2795\ \epsilon^{-1/2}$
ζ_3..........	−1.8047	−1.6695	−1.5236	−1.3629	−1.1810	−0.9654	$-4.4390\ \epsilon^{1/2}$
$\Omega_2^\dagger$..........	+0.5802	+0.5829	+0.5737	+0.5506	+0.5098	+0.4436	$+2.2795\ \epsilon^{1/2}$
$\Omega_3^\dagger$..........	+0.8927	+0.8229	+0.7479	+0.6658	+0.5737	+0.4661	$+2.1135\ \epsilon^{1/2}$
$\zeta_2^\dagger$..........	−0.4469	−0.4573	−0.4598	−0.4523	−0.4308	−0.3873	$-2.1492\ \epsilon^{1/2}$
$\zeta_3^\dagger$..........	−1.0530	−0.9947	−0.9277	−0.8488	−0.7529	−0.6305	$-3.0422\ \epsilon^{1/2}$

* The angular velocities and the vorticities are expressed in the unit $(\pi G\rho)^{1/2}$.

TABLE 6c

THE PROPERTIES OF THE MARGINALLY OVERSTABLE RIEMANN ELLIPSOIDS OF TYPE I*
(Along the Locus QR_{I})

a_2/a_1	1.2907	1.4954	1.6417	1.7679	1.8651
a_3/a_1	0.1573	0.1563	0.1431	0.1233	0.0976
Ω_2	+0.0979	+0.1427	+0.1769	+0.2211	+0.2856
Ω_3	+0.5082	+0.4633	+0.4219	+0.3784	+0.3310
ζ_2	−4.6947	−5.1812	−5.5580	−6.0132	−6.7416
ζ_3	−1.6093	−1.3221	−1.1381	−0.9802	−0.8341
$\Omega_2\dagger$	+0.7206	+0.7908	+0.7788	+0.7280	+0.6299
$\Omega_3\dagger$	+0.7791	+0.6109	+0.5056	+0.4200	+0.3471
$\zeta_2\dagger$	−0.6376	−0.9355	−1.2612	−1.8137	−2.8546
$\zeta_3\dagger$	−1.0498	−1.0027	−0.9496	−0.8829	−0.7942

* The angular velocities and the vorticities are expressed in the unit $(\pi G\rho)^{1/2}$.

TABLE 7

THE PROPERTIES OF THE MARGINALLY OVERSTABLE RIEMANN ELLIPSOIDS OF TYPE III*
(Along the Locus $X_2^{(III)}O'$)

a_2/a_1	4.0000	4.4141	4.9777	5.3909
a_3/a_1	1.7210	1.6000	1.4933	1.4000
Ω_2	+0.4966	+0.5410	+0.5567	+0.5497
Ω_3	+0.3281	+0.2553	+0.1968	+0.1657
ζ_2	+0.5283	+0.3190	+0.1992	+0.1430
ζ_3	−1.9954	−2.1577	−2.2582	−2.2936
$\Omega_2\dagger$	+0.2198	+0.1440	+0.0914	+0.0676
$\Omega_3\dagger$	+0.4787	+0.4641	+0.4359	+0.4113
$\zeta_2\dagger$	+1.0946	+1.2088	+1.1951	+1.1610
$\zeta_3\dagger$	−1.4219	−1.1826	−1.0190	−0.9242

* The angular velocities and the vorticities are expressed in the unit $(\pi G\rho)^{1/2}$.

TABLE 8

THE SQUARES OF THE CHARACTERISTIC FREQUENCIES OF OSCILLATIONS OF SOME TYPICAL RIEMANN ELLIPSOIDS OF TYPE I*

$\Omega_2^2+\Omega_3^2$	0.55867	0.59683	0.68570	0.78661	0.16888	0.17311	0.08871	0.08956
$\Omega_2\dagger^2+\Omega_3\dagger^2$	1.95795	1.66888	1.28261	0.78661	0.75034	0.73533	0.42328	0.42713
σ_1^2	3.95684	2.90595	2.42382	1.30300	1.98060	0.76906	1.11840	1.24447
σ_2^2	3.10667	2.14020	1.41447	1.08486	1.13927	0.48044	0.59519	0.49385
σ_3^2	1.41497	1.06182	0.79359	0.60901	0.31730	0.18666	0.14508	0.06994
σ_4^2	6.75010	7.99131	8.22190	6.47555	0.62319	2.36704	0.20346	0.17370

* The squares of the characteristic frequencies are expressed in the unit $\pi G\rho$.

The loci of the marginally overstable configurations are delineated in Figures 1 and 3; and the properties of these configurations are further exhibited in Figures 4 and 5.

It will be observed that, among the ellipsoids of type I, there are two disconnected domains of stability with respect to the oscillations considered. The existence of the stable domain, adjoining the stable Maclaurin spheroids along SO_2 is, of course, to be expected. But the existence of the second domain, bounded by the segment D_2R_{I} of the a_2-axis, is unexpected. The point (see Table 6*b*),

$$a_2/a_1 = 1.3707\,, \tag{169}$$

limiting the stable disklike ellipsoids of type I was determined with the aid of equation (166). In Table 10 we list the asymptotic forms of the characteristic frequencies of oscillation, together with some of the other properties, of these disklike ellipsoids.

The calculations show that *all ellipsoids of type II are unstable.* As we have already remarked in § X, their instability along the a_2-axis follows directly from equations (168).

TABLE 9

THE SQUARES OF THE CHARACTERISTIC FREQUENCIES OF OSCILLATIONS OF SOME TYPICAL RIEMANN ELLIPSOIDS OF TYPES II AND III*

	Ellipsoids of Type II			Ellipsoids of Type III			
$\Omega_2^2+\Omega_3^2$	+0.52061	+1.31413	+0.70393	+0.39458	+0.35751	+0.23651	+0.24718
$\Omega_2^{\dagger 2}+\Omega_3^{\dagger 2}$	+0.06631	+0.14392	+0.04312	+0.39458	+0.35751	+0.23192	+0.19297
σ_1^2	+0.11530	+0.21425	+0.07331	+0.50590 }	+0.46352 }	+0.70381	+0.55228
σ_2^2	−0.10399	−0.07905	−0.09417	±0.28039*i* }	±0.20708*i* }	+0.15463	+0.20340
σ_3^2	+0.51325	+1.25723	+0.68147	+1.27177	+1.79119	+1.74081	+1.18118
σ_4^2	+1.44145	+3.61524	+2.04314	+1.39306	+1.87486	+2.65271	+3.06709

* The squares of the characteristic frequencies are expressed in the unit $\pi G\rho$.

TABLE 10

THE ASYMPTOTIC PROPERTIES AND THE SQUARES OF THE CHARACTERISTIC FREQUENCIES OF OSCILLATION OF THE DISKLIKE ELLIPSOIDS OF TYPE I

$\theta=\sec^{-1}(a_2/a_1)$	42°	43°	44°	45°	50°	58°
a_2/a_1	1.34563	1.36733	1.39016	1.41421	1.55572	1.88708
ω_2	+ 2.05429	+ 2.13629	+ 2.22576	+ 2.32434	+ 3.05912	+11.27045
ω_3	+ 1.46610	+ 1.45084	+ 1.43516	+ 1.41906	+ 1.33207	+ 1.16871
ζ_2	− 2.24870	− 2.27560	− 2.30081	− 2.32434	− 2.41628	− 2.47347
ζ_3	− 4.55156	− 4.45371	− 4.35556	− 4.25717	− 3.76491	− 2.99380
$\omega_2^\dagger$	+ 2.24870	+ 2.27560	+ 2.30081	+ 2.32434	+ 2.41628	+ 2.47347
$\omega_3^\dagger$	+ 2.17905	+ 2.12215	+ 2.06473	+ 2.00685	+ 1.71248	+ 1.23864
$\zeta_2^\dagger$	− 2.05429	− 2.13629	− 2.22576	− 2.32434	− 3.05912	−11.27045
$\zeta_3^\dagger$	− 3.06236	− 3.04484	− 3.02747	− 3.01027	− 2.92858	− 2.82477
ω_3^2	2.14945	2.10493	2.05968	2.01372	1.77442	1.36588
$(\omega_2^\dagger)^2+(\omega_3^\dagger)^2$	9.80493	9.68185	9.55686	9.43002	8.77100	7.65228
$3(4a_1^2-a_2^2)\omega_3^2/a_2^2$	7.79644	7.19577	6.61032	6.04117	3.47451	0.50507
σ_1^2	19.86078 ± }	19.55939 ± }	17.53812	16.45406	12.98259	9.39418
σ_2^2	2.05833*i* }	0.73302*i* }	20.96929	21.43377	21.70035	19.90914
σ_3^2	2.58994	2.55633	2.52123	2.48458	2.27657	1.84879

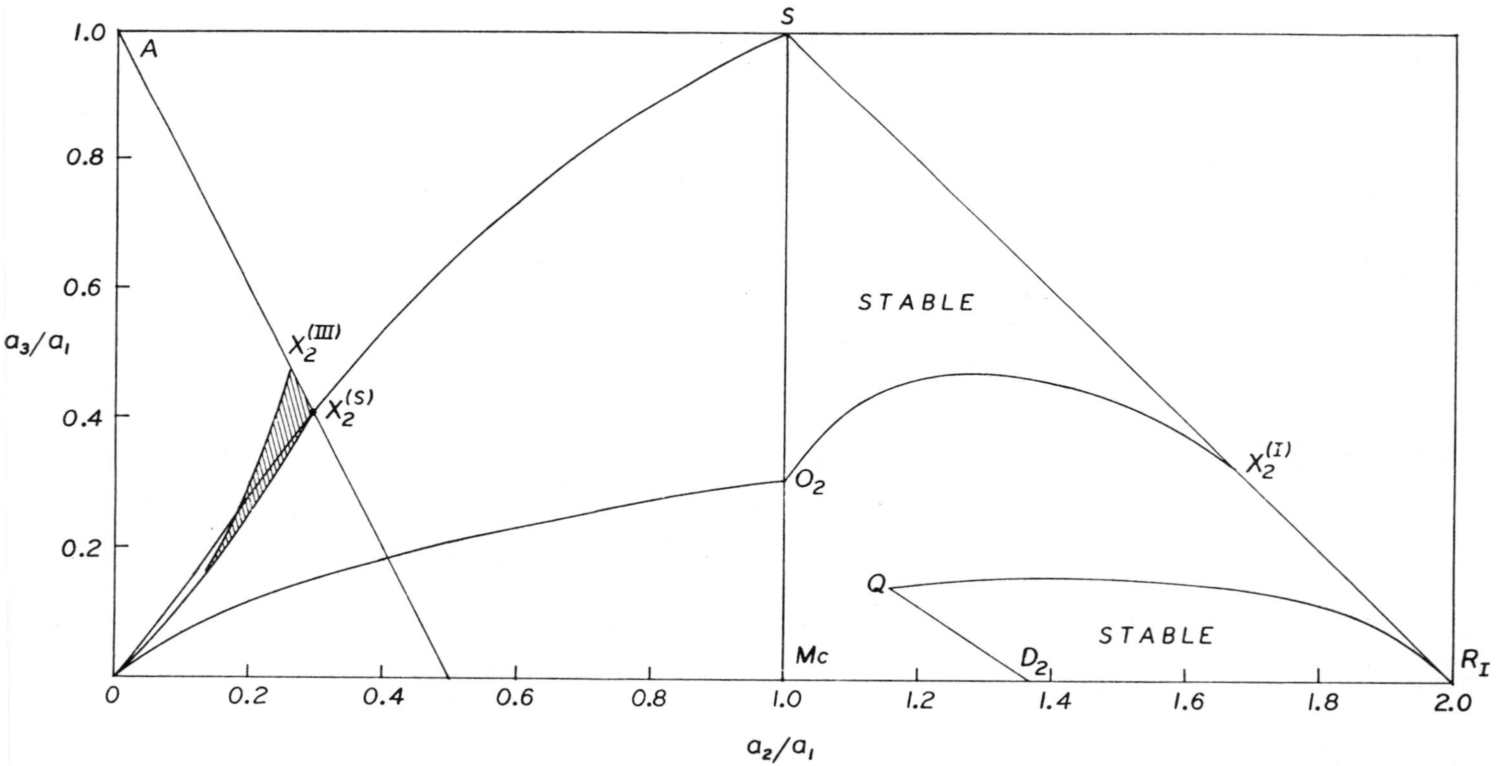

FIG. 3.—The loci of marginally stable configurations in the $(a_2/a_1, a_3/a_1)$-plane. The type S ellipsoids are bounded by two self-adjoint sequences (SO and O_2O) and the stable part of the Maclaurin sequence represented by SO_2. Along the arc $X_2^{(S)}O$ the type S ellipsoids become unstable by a mode of oscillation belonging to the second hormonics; and along this same arc the stability passes to the type III ellipsoids whose domain of occupancy is $AX_2^{(S)}O$. The shaded region included between $X_2^{(III)}O$ and $X_2^{(S)}O$ represents the domain of stability for type III ellipsoids with respect to oscillations belonging to the second harmonics.

The type I ellipsoids occupy the triangle SM_cR_I; and the region of the stable members is included in the two domains marked "stable." The domain $SO_2X_2^{(I)}$ of stable ellipsoids adjoining the stable Maclaurin spheroids is to be expected; but the domain D_2QR_I including disklike ellipsoids along D_2R_I is unexpected.

Among the ellipsoids of type III, there is a fringe of stable configurations (stable, that is, with respect to the oscillations considered) bordering on the boundary $X_2'O'$ of their domain of occupancy. As we have shown in § VI, by interchanging the roles of the indices 1 and 2 (so that a_1 becomes the longest axis, as it is among the ellipsoids of type S) the locus $X_2'O'$ is transformed to the locus X_2O of the marginally unstable ellipsoids of type S (see Fig. 3). One should expect under these circumstances that the stability passes from the ellipsoids of type S to the ellipsoids of type III along their common curve of bifurcation; and this is exactly what happens. However, since the ellipsoids of type S become

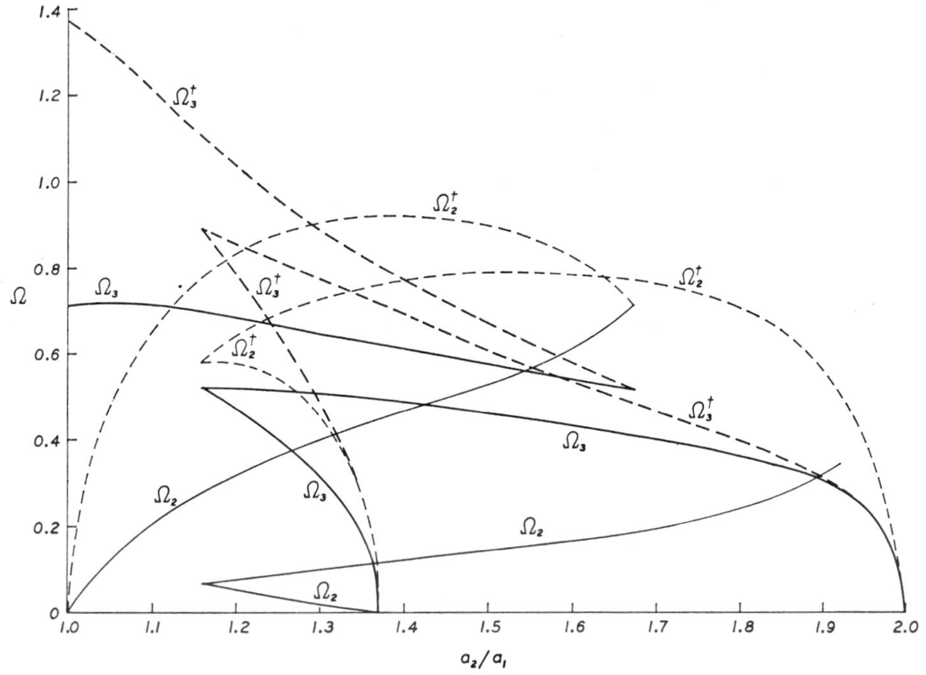

FIG. 4.—The variation of the components of the angular velocity along the marginally overstable ellipsoids delineated in Fig. 3. The curves distinguished by Ω and Ω† are appropriate for the adjoint configurations having the same figure.

unstable with respect to an oscillation belonging to the *third* harmonics, *prior* to the onset of instability by an odd mode of oscillation belonging to the second harmonics, *it is very likely that ellipsoids of type III are all unstable with respect to a third-harmonic oscillation.*

From the foregoing account it would appear that only among the ellipsoids of types I and S do stable ones occur.

In his paper, Riemann considers the stability of his ellipsoidal figures by an energy criterion. But most of the conclusions he derives from his criterion (with the notable exception of those pertaining to the Maclaurin spheroid) are false. His criterion is clearly in error; the origin of this error (which mars an otherwise most remarkable paper) is clarified by Lebovitz (1966) in the paper following this one.

XII. CONCLUDING REMARKS

The present paper completes the series of investigations initiated some six years ago with a view toward completing and consolidating the classical work on the ellipsoidal figures of equilibrium of homogeneous masses. As the investigations proceeded, several misconceptions in the earlier work (e.g., the Roche ellipsoids become "unstable" at the Roche limit, or that the bifurcation of the Jacobian from the Maclaurin sequence is a "unique" phenomenon) became apparent; and these have been eliminated.

In many ways, the most curious aspect of the subject has been the almost total neglect of the fundamental papers of Dirichlet, Dedekind, and Riemann (all published in 1860). Nevertheless, the completion of Riemann's work has been essential to a comprehensive view of the subject. The fruitful exploration of these classical avenues is by

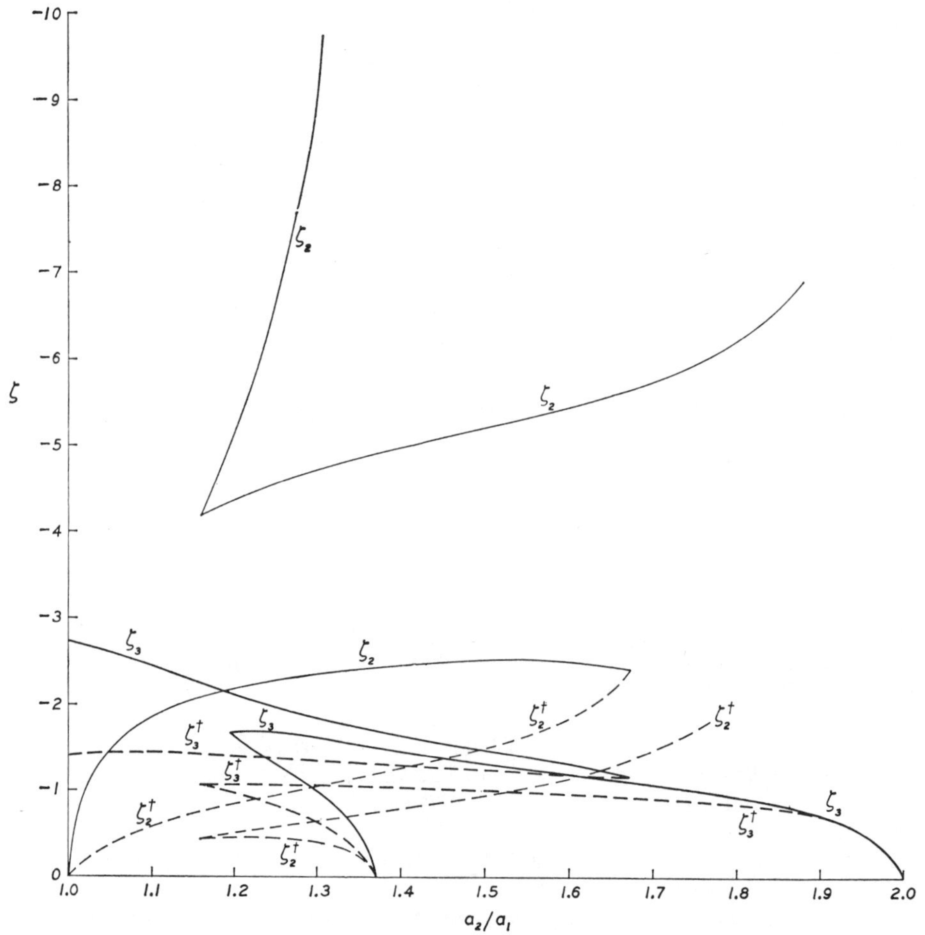

FIG. 5.—The variation of the components of the vorticity along the marginally overstable ellipsoids delineated in Fig. 3. The curves distinguished by ζ and $\zeta\dagger$ are appropriate for the adjoint configurations having the same figure.

no means ended: the continuation of Dirichlet's work on the non-linear finite amplitude oscillations of ellipsoidal figures appears to hold rich promise. These further areas of research are, however, beyond the scope of the present series.

I am grateful to Dr. M. Clement whose careful scrutiny of the analysis helped the elimination of a number of oversights and obscurities; he also generously programmed for machine calculations,[1] the characteristic equation (both in its finite and asymptotic forms) for the determination of its roots. I am equally grateful to Dr. N. R. Lebovitz for many clarifying discussions and particularly for his examination of Riemann's criterion for determining the stability of his figures and demonstration of the place where he erred. I am also indebted to Miss Donna Elbert for her continued patience in assisting with these investigations; in this particular instance it was most essential.

The research reported in this paper has in part been supported by the Office of Naval Research under contract Nonr-2121(24) with the University of Chicago.

REFERENCES

Basset, A. B. 1888, *A Treatise on Hydrodynamics* (Cambridge, Eng.: Deighton Bell & Co.; reprinted 1961, Dover Publications, New York), Vol. **2**.

Chandrasekhar, S. 1964, *Lectures in Theoretical Physics*, ed. W. E. Brittin and W. R. Chappell (Boulder: University of Colorado Press), p. 1.

———. 1965*a*, *Ap. J.*, **141**, 1043.

———. 1965*b*, *ibid.*, **142**, 890.

Chandrasekhar, S., and Lebovitz, N. R. 1962, *Ap. J.*, **136**, 1037.

Hicks, W. M. 1882, *Reports to the British Association*, pp. 57–61.

Lebovitz, N. R. 1961, *Ap. J.*, **134**, 500.

———. 1966, *ibid.*, **145**, 878.

Riemann, B. 1860, *Abh. d. Königl. Gesell. der Wiss. zur Göttingen*, **9**, 3; see also 1892, *Gesammelte mathematische Werke* (Leipzig: Verlag von B. G. Teubner; reprinted 1953, Dover Publications, New York), p. 182.

[1] The machine calculations were performed with an IBM 1620 computer at the Yerkes Observatory.

THE VIRIAL EQUATIONS OF THE FOURTH ORDER

S. CHANDRASEKHAR
University of Chicago
Received August 17, 1967

ABSTRACT

The virial equations of the fourth order are derived; and the results of certain formal developments needed for their practical usefulness are also given. The equations are then used to locate the neutral points, along the Maclaurin and the Jacobian sequences, that belong to the fourth harmonics.

I. INTRODUCTION

In papers published in the *Astrophysical Journal* (Volumes **134–148**) during the past few years, the tensor virial equations of the second and the third orders have been made the basis of a systematic development of the theory of the equilibrium and the stability of the ellipsoidal figures which arise in various gravitational contexts. In particular, the first variations of the virial equations governing equilibrium were used to isolate the neutral points, belonging to the second and the third harmonics, along the different sequences. More recently, in connection with the construction of sequences, analogous to those of Maclaurin and Jacobi, in the post-Newtonian framework of general relativity, a further neutral point, along each of the two sequences and belonging to the fourth harmonics, was isolated (Chandrasekhar 1967*a*, *b*). However, along the Maclaurin sequence there must exist neutral points, besides the one already located, belonging to the fourth harmonics. To find them and, as well, for the sake of completing the general theory, we shall obtain in this paper the relevant virial equations of the fourth order and illustrate their usefulness.

II. THE FOURTH-ORDER VIRIAL EQUATIONS IN A ROTATING FRAME OF REFERENCE

For purposes of later applications (and also for greater generality) we shall obtain the basic equations in a frame of reference rotating uniformly with an angular velocity Ω about the x_3-axis (say). The appropriate form of the equation of motion is

$$\rho \frac{du_i}{dt} = -\frac{\partial p}{\partial x_i} + \rho \frac{\partial}{\partial x_i}[\mathfrak{B} + \tfrac{1}{2}\Omega^2(x_1^2 + x_2^2)] + 2\rho\Omega\epsilon_{im3}u_m , \tag{1}$$

where the various symbols have their usual meanings.

Multiplying equation (1) by $x_jx_kx_l$ and integrating over the volume V occupied by the fluid, we obtain by transformations familiar in this theory,

$$\begin{aligned}\frac{d}{dt}\int_V \rho u_i x_j x_k x_l d\mathbf{x} = {} & 2(\mathfrak{T}_{ij;kl} + \mathfrak{T}_{ik;lj} + \mathfrak{T}_{il;jk}) \\ & + \Omega^2(I_{ijkl} - \delta_{i3}I_{3jkl}) + \int_V \rho \frac{\partial \mathfrak{B}}{\partial x_i} x_j x_k x_l d\mathbf{x} \\ & + 2\Omega\epsilon_{im3}\int_V \rho u_m x_j x_k x_l d\mathbf{x} + \delta_{ij}\Pi_{kl} + \delta_{ik}\Pi_{lj} + \delta_{il}\Pi_{jk} ,\end{aligned} \tag{2}$$

where

$$I_{ijkl} = \int_V \rho x_i x_j x_k x_l d\mathbf{x} , \qquad \Pi_{jk} = \int_V p x_j x_k d\mathbf{x} ,$$

and

$$\mathfrak{T}_{ij;kl} = \tfrac{1}{2}\int_V \rho u_i u_j x_k x_l d\mathbf{x} . \tag{3}$$

The term involving $\partial\mathfrak{B}/\partial x_i$ can be transformed as follows. First writing

$$\int_V \rho \frac{\partial \mathfrak{B}}{\partial x_i} x_j x_k x_l d\boldsymbol{x} = -G\int_V\int_V \rho(\boldsymbol{x})\rho(\boldsymbol{x}') \frac{x_j x_k x_l (x_i - x_i')}{|\boldsymbol{x}-\boldsymbol{x}'|^3} d\boldsymbol{x} d\boldsymbol{x}'$$

$$= -\tfrac{1}{2}G\int_V\int_V \rho(\boldsymbol{x})\rho(\boldsymbol{x}') \frac{(x_j x_k x_l - x_j' x_k' x_l')(x_i - x_i')}{|\boldsymbol{x}-\boldsymbol{x}'|^3} d\boldsymbol{x} d\boldsymbol{x}', \tag{4}$$

we replace

$$(x_j x_k x_l - x_j' x_k' x_l')(x_i - x_i') \tag{5}$$

by

$$\begin{aligned}\tfrac{1}{3}[&(x_j - x_j')x_k x_l + x_j'(x_k - x_k')x_l + x_j' x_k'(x_l - x_l')\\ &+(x_k - x_k')x_l x_j + x_k'(x_l - x_l')x_j + x_k' x_l'(x_j - x_j')\\ &+(x_l - x_l')x_j x_k + x_l'(x_j - x_j')x_k + x_l' x_j'(x_k - x_k')]\,,\end{aligned} \tag{6}$$

and obtain

$$\int_V \rho \frac{\partial \mathfrak{B}}{\partial x_i} x_j x_k x_l d\boldsymbol{x} = -\tfrac{1}{6}\int_V \rho(2x_k x_l \mathfrak{B}_{ij} + 2x_l x_j \mathfrak{B}_{ik} + 2x_j x_k \mathfrak{B}_{il} + x_l \mathfrak{D}_{ik;j} + x_j \mathfrak{D}_{il;k} + x_k \mathfrak{D}_{ij;l})d\boldsymbol{x}\,, \tag{7}$$

where

$$\mathfrak{B}_{ij}(\boldsymbol{x}) = G\int_V \rho(\boldsymbol{x}') \frac{(x_i - x_i')(x_j - x_j')}{|\boldsymbol{x}-\boldsymbol{x}'|^3} d\boldsymbol{x}' \tag{8}$$

is the Newtonian tensor potential as usually defined and

$$\mathfrak{D}_{ij;k}(\boldsymbol{x}) = G\int_V \rho(\boldsymbol{x}') x_k' \frac{(x_i - x_i')(x_j - x_j')}{|\boldsymbol{x}-\boldsymbol{x}'|^3} d\boldsymbol{x}' \tag{9}$$

is the tensor potential due to a distribution of "density" ρx_k. In terms of the Newtonian potentials $\mathfrak{D}_i$ and $\mathfrak{D}_{ij}$, due to the distributions ρx_i and $\rho x_i x_j$, we can express $\mathfrak{D}_{ij;k}$ in the manner (cf. Chandrasekhar and Lebovitz 1962, eq. [71])

$$\mathfrak{D}_{ij;k} = -x_i \frac{\partial \mathfrak{D}_k}{\partial x_j} + \frac{\partial \mathfrak{D}_{ik}}{\partial x_j}. \tag{10}$$

Now define the tensors

$$\mathfrak{W}_{ij;kl} = -\tfrac{1}{2}\int_V \rho \mathfrak{B}_{ij} x_k x_l d\boldsymbol{x} \tag{11}$$

and

$$\mathfrak{W}_{ij;k;l} = -\tfrac{1}{2}\int_V \rho \mathfrak{D}_{ij;k} x_l d\boldsymbol{x} = -\tfrac{1}{2}\int_V \rho \mathfrak{D}_{ij;l} x_k d\boldsymbol{x}\,, \tag{12}$$

which (like the tensors $\mathfrak{T}_{ij;kl}$ and $\delta_{ij}\Pi_{kl}$) are symmetric in the indices (i,j) and (k,l), separately. In terms of these tensors, we can now write equation (2) in the form

$$\begin{aligned}\frac{d}{dt}\int_V \rho u_i x_j x_k x_l d\boldsymbol{x} = {} & 2(\mathfrak{T}_{ij;kl} + \mathfrak{T}_{ik;lj} + \mathfrak{T}_{il;jk}) + \Omega^2(I_{ijkl} - \delta_{i3} I_{3jkl})\\ & + \tfrac{1}{3}(2\mathfrak{W}_{ij;kl} + 2\mathfrak{W}_{ik;lj} + 2\mathfrak{W}_{il;jk} + \mathfrak{W}_{ij;k;l} + \mathfrak{W}_{ik;l;j} + \mathfrak{W}_{il;j;k})\\ & + \delta_{ij}\Pi_{kl} + \delta_{ik}\Pi_{lj} + \delta_{il}\Pi_{jk} + 2\Omega\epsilon_{im3}\int_V \rho u_m x_j x_k x_l d\boldsymbol{x}).\end{aligned} \tag{13}$$

Equation (13) is the desired form of the virial equation of the fourth order. It provides a set of thirty moment equations in addition to the nine and the eighteen provided by the second- and the third-order equations, respectively.

Under conditions when no internal motions are present (in the rotating frame considered) and a stationary state prevails, equation (13) gives

$$\tfrac{1}{3}(2\mathfrak{W}_{ij;kl}+2\mathfrak{W}_{ik;lj}+2\mathfrak{W}_{il;jk}+\mathfrak{W}_{ij;k;l}+\mathfrak{W}_{ik;l;j}+\mathfrak{W}_{il;j;k}) \\ +\Omega^2(I_{ijkl}-\delta_{i3}I_{3jkl})=-(\delta_{ij}\Pi_{kl}+\delta_{ik}\Pi_{lj}+\delta_{il}\Pi_{jk}). \tag{14}$$

a) The First Variations of $\mathfrak{W}_{ij;kl}$ *and* $\mathfrak{W}_{ij;k;l}$

For treating small departures from equilibrium by the virial equation (13), it is necessary that we know the first-order changes in $\mathfrak{W}_{ij;kl}$ and $\mathfrak{W}_{ij;k;l}$ that result from a deformation caused by the different elments having suffered different displacements. If $\boldsymbol{\xi}(\boldsymbol{x},t)$ denotes the corresponding Lagrangian displacement, then the resulting first variations in $\mathfrak{W}_{ij;kl}$ and $\mathfrak{W}_{ij;k;l}$ can be readily evaluated by standard procedures. Thus,

$$\begin{aligned}-2\delta\mathfrak{W}_{ij;kl}&=\delta\int_V\rho\mathfrak{B}_{ij}x_kx_l\,d\boldsymbol{x}\\&=\int_V\rho\mathfrak{B}_{ij}\xi_m\frac{\partial}{\partial x_m}(x_kx_l)\,d\boldsymbol{x}\\&+G\int_V\int_V\rho(\boldsymbol{x})\rho(\boldsymbol{x}')x_kx_l\left[\xi_m(\boldsymbol{x})\frac{\partial}{\partial x_m}+\xi_m(\boldsymbol{x}')\frac{\partial}{\partial x_m'}\right]\frac{(x_i-x_i')(x_j-x_j')}{|\boldsymbol{x}-\boldsymbol{x}'|^3}\,d\boldsymbol{x}\,d\boldsymbol{x}'\\&=\int_V\rho\left[\mathfrak{B}_{ij}\xi_m\frac{\partial}{\partial x_m}(x_kx_l)+x_kx_l\xi_m\frac{\partial\mathfrak{B}_{ij}}{\partial x_m}\right]d\boldsymbol{x}\\&\quad+\int_V d\boldsymbol{x}'\rho(\boldsymbol{x}')\xi_m(\boldsymbol{x}')\frac{\partial}{\partial x_m'}\int_V d\boldsymbol{x}\,\rho(\boldsymbol{x})x_kx_l\frac{(x_i-x_i')(x_j-x_j')}{|\boldsymbol{x}'-\boldsymbol{x}|^3},\end{aligned}\tag{15}$$

or, equivalently,

$$-2\delta\mathfrak{W}_{ij;kl}=\int\rho\xi_m\frac{\partial}{\partial x_m}(x_kx_l\mathfrak{B}_{ij}+\mathfrak{D}_{ij;kl})\,d\boldsymbol{x}, \tag{16}$$

where $\mathfrak{D}_{ij;kl}$ is the "$\mathfrak{B}_{ij}$" induced by the distribution ρx_kx_l and is determined by the Newtonian potentials $\mathfrak{D}_{kl}$ and $\mathfrak{D}_{ikl}$ due to the distributions ρx_kx_l and $\rho x_ix_kx_l$ by (cf. eq. [10])

$$\mathfrak{D}_{ij;kl}=-x_i\frac{\partial\mathfrak{D}_{kl}}{\partial x_j}+\frac{\partial\mathfrak{D}_{ikl}}{\partial x_j}. \tag{17}$$

Similarly,

$$\begin{aligned}-2\delta\mathfrak{W}_{ij;k;l}&=\delta\int_V\rho x_k\mathfrak{D}_{ij;l}\,d\boldsymbol{x}\\&=\int_V\rho\xi_k\mathfrak{D}_{ij;l}\,d\boldsymbol{x}\\&+G\int_V\int_V\rho(\boldsymbol{x})\rho(\boldsymbol{x}')x_k\left[\xi_m(\boldsymbol{x})\frac{\partial}{\partial x_m}+\xi_m(\boldsymbol{x}')\frac{\partial}{\partial x_m'}\right]\frac{x_l'(x_i-x_i')(x_j-x_j')}{|\boldsymbol{x}-\boldsymbol{x}'|^3}\,d\boldsymbol{x}\,d\boldsymbol{x}'\\&=\int_V\rho\left(\xi_k\mathfrak{D}_{ij;l}+\xi_l\mathfrak{D}_{ij;k}+x_k\xi_m\frac{\partial\mathfrak{D}_{ij;l}}{\partial x_m}+x_l\xi_m\frac{\partial\mathfrak{D}_{ij;k}}{\partial x_m}\right)d\boldsymbol{x}\end{aligned}\tag{18}$$

or, equivalently,

$$-2\,\delta\mathfrak{W}_{ij;k;l}=\int_V \rho\,\xi_m \frac{\partial}{\partial x_m}(x_k\mathfrak{D}_{ij;l}+x_l\mathfrak{D}_{ij;k})\,dx\,. \tag{19}$$

b) Explicit Expressions for $\delta\mathfrak{W}_{ij;kl}$ and $\delta\mathfrak{W}_{ij;k;l}$ for Homogeneous Ellipsoids

For homogeneous ellipsoids, the potentials $\mathfrak{D}_{ij}$ and $\mathfrak{D}_{ijk}$ are explicitly known (cf. Chandrasekhar and Lebovitz 1963*a*, eqs. [107]–[118] and Chandrasekhar 1967*a*, eq. [63]; in the former reference explicit expressions for $\mathfrak{D}_{ij;k}$ are also given). Making use of them, the formulae (16) and (19) can be directly evaluated and expressed in terms of the symmetrized second-order and fourth-order virials:

$$V_{ij}=\delta I_{ij}=\int_V \rho\,\xi_m\frac{\partial}{\partial x_m}(x_ix_j)\,dx$$

and

$$V_{ijkl}=\delta I_{ijkl}=\int_V \rho\,\xi_m\frac{\partial}{\partial x_m}(x_ix_jx_kx_l)\,dx\,. \tag{20}$$

We find, distinguishing the various cases,[1]

$$\begin{aligned}
-2\delta\mathfrak{W}_{ii;ii} &= a_i^2(-\tfrac{15}{2}a_i^2B_{iii}+2A_i)V_{ii}-\tfrac{3}{2}a_i^4B_{iij}V_{jj}-\tfrac{3}{2}a_i^4B_{iik}V_{kk}\\
&\quad+(\tfrac{35}{4}a_i^4B_{iiii}+4B_{ii}-2A_i)V_{iiii}+\tfrac{3}{4}a_i^4B_{iijj}V_{jjjj}+\tfrac{3}{4}a_i^4B_{iikk}V_{kkkk}\\
&\quad+a_i^2(\tfrac{15}{2}a_i^2B_{iiij}-2A_{ij})V_{iijj}+a_i^2(\tfrac{15}{2}a_i^2B_{iiik}-2A_{ik})V_{iikk}+\tfrac{3}{2}a_i^4B_{iijk}V_{jjkk}\,,\\
-2\delta\mathfrak{W}_{ii;jj} &= a_j^2(-\tfrac{3}{2}a_i^2B_{iij}+B_{ij})V_{ii}+a_i^2(-\tfrac{1}{2}a_j^2B_{ijj}+a_j^4A_{ijj}+A_i)V_{jj}-\tfrac{1}{2}a_i^2a_j^2B_{ijk}V_{kk}\\
&\quad+a_j^2(\tfrac{5}{4}a_i^2B_{iiij}-B_{iij})V_{iiii}+a_i^2(\tfrac{1}{4}a_j^2B_{ijjj}-a_j^4A_{ijjj}-A_{ij})V_{jjjj}+\tfrac{1}{4}a_i^2a_j^2B_{ijkk}V_{kkkk}\\
&\quad+a_j^2(\tfrac{3}{2}a_i^2B_{iijk}-B_{ijk})V_{iikk}+a_i^2(\tfrac{1}{2}a_j^2B_{ijjk}-a_j^4A_{ijjk}-A_{ik})V_{jjkk}\\
&\quad+[3a_j^2(a_j^2+\tfrac{1}{2}a_i^2)B_{iijj}+3B_{ii}+B_{ij}-2A_i]V_{iijj}\,,\\
-2\delta\mathfrak{W}_{ij;ij} &= -\tfrac{3}{2}a_i^2a_j^2B_{iij}V_{ii}-\tfrac{3}{2}a_i^2a_j^2B_{ijj}V_{jj}-\tfrac{1}{2}a_i^2a_j^2B_{ijk}V_{kk}\\
&\quad+\tfrac{5}{4}a_i^2a_j^2B_{iiij}V_{iiii}+\tfrac{5}{4}a_i^2a_j^2B_{ijjj}V_{jjjj}+\tfrac{1}{4}a_i^2a_j^2B_{ijkk}V_{kkkk}\\
&\quad+(\tfrac{9}{2}a_i^2a_j^2B_{iijj}+2B_{ij})V_{iijj}+\tfrac{3}{2}a_i^2a_j^2B_{iijk}V_{iikk}+\tfrac{3}{2}a_i^2a_j^2B_{ijjk}V_{jjkk}\,,\\
-2\delta\mathfrak{W}_{ij;ik} &= -a_i^2a_k^2B_{ijk}V_{jk}+(3a_i^2a_k^2B_{iijk}+2B_{ij})V_{iijk}\\
&\quad+a_i^2a_k^2B_{ijjk}V_{jjjk}+a_i^2a_k^2B_{ijkk}V_{jkkk}\,,\\
-2\delta\mathfrak{W}_{ij;kk} &= a_k^2C_{ijk}V_{ij}+(3a_k^4B_{ijkk}-a_k^2B_{ijk}+2B_{ij})V_{ijkk}\\
&\quad-a_k^2C_{iijk}V_{iiij}-a_k^2C_{ijjk}V_{ijjj}\,,\\
-2\delta\mathfrak{W}_{ij;ii} &= a_i^2(-3a_i^2B_{iij}+B_{ij})V_{ij}+(5a_i^4B_{iiij}-a_i^2B_{iij}+2B_{ij})V_{iiij}\\
&\quad+a_i^2(3a_i^2B_{iijj}-B_{ijj})V_{ijjj}+a_i^2(3a_i^2B_{iijk}-B_{ijk})V_{ijkk}\,,\\
-2\delta\mathfrak{W}_{ii;jk} &= a_i^2(a_j^2a_k^2A_{ijk}+A_i)V_{jk}+(2B_{ii}-3a_i^2a_j^2a_k^2A_{iijk}+2a_j^2a_k^2A_{ijk}-a_i^2A_{ii})V_{iijk}\\
&\quad-a_i^2(a_j^2a_k^2A_{ijjk}+A_{ij})V_{jjjk}-a_i^2(a_j^2a_k^2A_{ijkk}+A_{ik})V_{jkkk}\,,\\
-2\delta\mathfrak{W}_{ii;ij} &= a_i^2(-a_j^2B_{iij}+B_{ij}+2a_i^2a_j^2A_{iij})V_{ij}+a_i^2(a_j^2B_{iijj}-B_{ijj}-2a_i^2a_j^2A_{iijj})V_{ijjj}\\
&+a_i^2(a_j^2B_{iijk}-B_{ijk}-2a_i^2a_j^2A_{iijk})V_{ijkk}+(5a_i^2a_j^2B_{iiij}+2B_{ii}-a_i^2a_j^2A_{iij}-a_i^2A_{ii})V_{iiij}\,,
\end{aligned} \tag{21}$$

[1] The formulae (21) and (22) were also derived independently by Dr. C. E. Rosenkilde. I am grateful to him for providing a check on their correctness.

and

$$
\begin{aligned}
-2\delta\mathfrak{W}_{ii;i;i} &= 2a_i^2[(-2B_{ii}+a_i^2A_{ii})V_{ii}+(4B_{iii}-a_i^2A_{iii})V_{iiii}\\
&\quad +(2B_{iij}-a_i^2A_{iij})V_{iijj}+(2B_{iik}-a_i^2A_{iik})V_{iikk}]\,,\\
-2\delta\mathfrak{W}_{ii;j;j} &= 2a_j^2[a_i^2A_{ij}V_{jj}+(2B_{iij}-a_i^2A_{iij})V_{iijj}-a_i^2A_{ijj}V_{jjjj}-a_i^2A_{ijk}V_{jjkk}]\,,\\
-2\delta\mathfrak{W}_{ij;i;j} &= -a_j^2B_{ij}V_{ii}-a_i^2B_{ij}V_{jj}+a_j^2B_{iij}V_{iiii}+a_i^2B_{ijj}V_{jjjj}\\
&\quad +a_i^2B_{ijk}V_{jjkk}+a_j^2B_{ijk}V_{iikk}+3(a_i^2B_{iij}+a_j^2B_{ijj})V_{iijj}\,,\\
-2\delta\mathfrak{W}_{ij;i;k} &= -a_i^2B_{ij}V_{jk}+(3a_i^2B_{iij}+2a_k^2B_{ijk})V_{iijk}+a_i^2B_{ijj}V_{jjjk}+a_i^2B_{ijk}V_{jkkk}\,,\\
-2\delta\mathfrak{W}_{ij;k;k} &= 4a_k^2B_{ijk}V_{ijkk},\\
-2\delta\mathfrak{W}_{ij;i;i} &= 2a_i^2(-B_{ij}V_{ij}+B_{ijj}V_{ijjj}+3B_{iij}V_{iiij}+B_{ijk}V_{ijkk})\,,\\
-2\delta\mathfrak{W}_{ii;j;k} &= a_i^2(a_j^2A_{ij}+a_k^2A_{ik})V_{jk}+(2a_j^2B_{iij}+2a_k^2B_{iik}-a_i^2a_j^2A_{iij}-a_i^2a_k^2A_{iik})V_{iijk}\\
&\quad -a_i^2(a_j^2A_{ijj}+a_k^2A_{ijk})V_{jjjk}-a_i^2(a_k^2A_{ikk}+a_j^2A_{ijk})V_{jkkk}\,,\\
-2\delta\mathfrak{W}_{ii;i;j} &= a_i^2(-2B_{ii}+a_i^2A_{ii}+a_j^2A_{ij})V_{ij}+(5a_i^2B_{iii}+2a_j^2B_{iij}-a_i^2a_j^2A_{iij}-a_i^2A_{ii})V_{iiij}\\
&\quad +a_i^2(3B_{iij}+B_{ijj}-2A_{ij})V_{ijjj}+a_i^2(3B_{iik}+B_{ijk}-2A_{ik})V_{ijkk}\,.
\end{aligned}
\tag{22}
$$

(In eqs. [21] and [22] the summation convention has been suspended; also $i \neq j \neq k$ represent distinct indices.)

In equations (21) and (22) the index symbols $A_{ijk\ldots}$ and $B_{ijk\ldots}$ are so normalized that $A_1 + A_2 + A_3 = 2$; and the symbol $C_{ijk\ldots}$ is related to the symbol $B_{ijk\ldots}$ in the manner defined in Chandrasekhar (1967*b*), equation (67). Also, a common factor $\pi G\rho$ in the expressions for the potentials has been suppressed; and, consistently, Ω^2 will be measured in the unit $\pi G\rho$.

c) The Divergence Conditions on V_{ijkl} in Case of Incompressibility

When we are dealing with an incompressible ellipsoid, as we shall in §§ IV and V, the Lagrangian displacement considered in § II*b* should be divergence-free. This requirement leads to certain restrictions on the symmetrized virials V_{ijkl} even as there are restrictions on the second- and the third-order virials (cf. Lebovitz 1961, eq. [83]; Chandrasekhar and Lebovitz 1963*a*, § VI); and they can be similarly found. Thus, by an integration by parts we find

$$
\begin{aligned}
&\int_V \rho\boldsymbol{\xi}\cdot\text{grad}\left[x_i x_j\left(\sum_{k=1}^{3}\frac{x_k^2}{a_k^2}-1\right)\right]d\boldsymbol{x}\\
&\quad = \int_S \rho x_i x_j\left(\sum_{k=1}^{3}\frac{x_k^2}{a_k^2}-1\right)\boldsymbol{\xi}\cdot d\boldsymbol{S} - \int_V \rho x_i x_j\left(\sum_{k=1}^{3}\frac{x_k^2}{a_k^2}-1\right)\text{div}\,\boldsymbol{\xi}\,d\boldsymbol{x} = 0\,,
\end{aligned}
\tag{23}
$$

the surface integral (over S) vanishing on account of the equation satisfied by the ellipsoidal boundary and the volume integral (over V) vanishing on account of the solenoidal character of $\boldsymbol{\xi}$; and we obtain, in view of the definitions (20),

$$
V_{ij} = \sum_{k=1}^{3}\frac{V_{ijkk}}{a_k^2}\,. \tag{24}
$$

(Note that we have suspended the summation convection over repeated indices. It will not be adopted in the rest of this paper. Summation will be indicated if required.)

Equation (24) provides six relations. On the other hand, since the divergence condition on the second-order virials requires

$$\sum_{k=1}^{3} \frac{V_{kk}}{a_k^2} = 0\,, \tag{25}$$

we have the further relation

$$\sum_{k=1}^{3} \frac{V_{kkkk}}{a_k^4} + \sum_{i\neq j} \frac{V_{iijj}}{a_i^2 a_j^2} = 0\,. \tag{26}$$

III. THE CONDITIONS TO BE SATISFIED AT A NEUTRAL POINT

At a point of bifurcation, or more generally at any neutral point, where the configuration allows a neutral mode of oscillation for some non-trivial displacement $\boldsymbol{\xi}$, the corresponding first variations of the equilibrium conditions given by the virial equations of the various orders must also be necessarily satisfied.

We shall now consider the conditions that follow from equation (14).

Equation (14) provides a total of thirty conditions to be satisfied. These conditions can be divided into four groups: a group of nine equations which are even in all three indices 1, 2, and 3; and three groups of seven equations each which are even in a particular index k and odd in the remaining two.

The nine even equations are, in turn, of two types: three equations of the form

$$2\delta\mathfrak{W}_{ii;ii} + \delta\mathfrak{W}_{ii;i;i} + \Omega^2(1-\delta_{i3})V_{iiii} = -3\delta\Pi_{ii} \qquad (i = 1,2,3)\,; \tag{27}$$

and six equations obtained by letting i and j $(i \neq j)$ represent an *ordered pair* of indices in the equation

$$4\delta\mathfrak{W}_{ij;ij} + 2\delta\mathfrak{W}_{jj;ii} + 2\delta\mathfrak{W}_{ij;i;j} + \delta\mathfrak{W}_{jj;i;i} + 3\Omega^2(1-\delta_{j3})V_{iijj} = -3\delta\Pi_{ii}\,. \tag{28}$$

(Note that eq. [28] is not symmetric in i and j and that the equations obtained by interchanging i and j are different.)

The seven equations, odd in a selected pair of indices i and j $(i \neq j)$ and even in the index k $(\neq i \neq j)$, are of four types. The equations

$$2\delta\mathfrak{W}_{ij;jj} + \delta\mathfrak{W}_{ij;j;j} + \Omega^2(1-\delta_{i3})V_{ijjj} = 0\,, \tag{29}$$

$$4\delta\mathfrak{W}_{ik;jk} + 2\delta\mathfrak{W}_{ij;kk} + 2\delta\mathfrak{W}_{ik;j;k} + \delta\mathfrak{W}_{ij;k;k} + 3\Omega^2(1-\delta_{i3})V_{ijkk} = 0\,, \tag{30}$$

and

$$4\delta\mathfrak{W}_{ii;ij} + 2\delta\mathfrak{W}_{ij;ii} + 2\delta\mathfrak{W}_{ii;i;j} + \delta\mathfrak{W}_{ij;i;i} + 3\Omega^2(1-\delta_{i3})V_{iiij} = -6\delta\Pi_{ij}\,, \tag{31}$$

together with the equations obtained from these by interchanging i and j, and the further equation

$$2\delta\mathfrak{W}_{kk;ij} + 2\delta\mathfrak{W}_{jk;ki} + 2\delta\mathfrak{W}_{ki;jk} + \delta\mathfrak{W}_{kk;i;j} + \delta\mathfrak{W}_{jk;k;i} + \delta\mathfrak{W}_{ki;j;k} + 3\Omega^2(1-\delta_{k3})V_{ijkk} = -3\delta\Pi_{ij} \tag{32}$$

(symmetric in i and j) provide the seven equations.

When dealing with an incompressible configuration (and sometimes even quite generally) it is convenient to eliminate the six $\delta\Pi_{ij}$'s from equations (27)–(32). After the elimination, we shall be left with twenty-four equations. In the case of incompressible ellipsoids, the twenty-four equations must be supplemented by the divergence conditions (24)–(26); and in the general case, six equations including the different $\delta\Pi_{ij}$'s must be considered and the $\delta\Pi_{ij}$'s must be evaluated by some additional assumption, such as, that the perturbed motions take place adiabatically (cf. Chandrasekhar and Lebovitz 1963*c*, § VI).

In the case of the even equations, the six equations which remain after the elimination of the three $\delta\Pi_{ii}$'s are of two types:

$$2\delta\mathfrak{W}_{ii;ii} + \delta\mathfrak{W}_{ii;i;i} - (4\delta\mathfrak{W}_{ij;ij} + 2\delta\mathfrak{W}_{jj;ii} + 2\delta\mathfrak{W}_{ij;i;j} + \delta\mathfrak{W}_{jj;i;i}) + \Omega^2(1-\delta_{i3})V_{iiii} - 3\Omega^2(1-\delta_{j3})V_{iijj} = 0 \qquad (i \neq j)\,, \tag{33}$$

and

$$+(4\delta\mathfrak{W}_{ij;ij} + 2\delta\mathfrak{W}_{jj;ii} + 2\delta\mathfrak{W}_{ij;i;j} + \delta\mathfrak{W}_{jj;i;i}) + 3\Omega^2(1-\delta_{j3})V_{iijj} - (4\delta\mathfrak{W}_{ik;ik} + 2\delta\mathfrak{W}_{kk;ii} + 2\delta\mathfrak{W}_{ik;i;k} + \delta\mathfrak{W}_{kk;i;i}) - 3\Omega^2(1-\delta_{k3})V_{iikk} = 0 \qquad (i \neq j \neq k)\,. \tag{34}$$

In an Appendix, we list, for the case of homogeneous ellipsoids, the explicit expressions for the various combinations of $\delta\mathfrak{W}_{ij;kl}$ and $\delta\mathfrak{W}_{ij;k;l}$ that occur in equations (29)–(34).

IV. THE NEUTRAL POINTS, BELONGING TO THE FOURTH HARMONICS, ALONG THE MACLAURIN SEQUENCE

One can readily convince oneself[2] that the neutral points, along the Maclaurin and the Jacobian sequences that belong to the fourth harmonics, are derived from displacements whose only non-vanishing fourth-order symmetrized virials are those that are even in the index 3. Accordingly, we need to consider only the nine equations which are even in all three indices and the seven equations which are odd in the indices 1 and 2.

In the case of the Maclaurin sequence, when $a_1 = a_2$ and the indices 1 and 2 can be identified in all the index symbols, the various equations simplify considerably.

Considering first the even equations (33) and (34) and using the formulae (A.1) and (A.2) given in the Appendix, we find that a non-trivial satisfaction of these equations can be accomplished in only one of two ways. *Either*

$$V_{1111} = V_{2222} \qquad \text{and} \qquad V_{1133} = V_{2233}\,, \tag{35}$$

and the fourth-order virials, after the elimination of the second-order virials with the aid of equation (24), satisfy the equations

$$(\Omega^2 - 2A_1 + 2a_1^6 A_{1111})(V_{1111} - 3V_{1122}) = 0\,, \tag{36}$$

$$\left(4.5a_1^2a_3^2B_{1113} - 1.5a_3^4B_{1133} - 6a_3^2B_{113} + 3\frac{a_3^4}{a_1^2}B_{133}\right)(V_{1111}+V_{1122}) - \left(6a_1^2C_{1133} - 24a_3^4B_{1133} - 6\frac{a_3^2}{a_1^2}B_{33}\right)V_{1133} + (3.75a_1^2a_3^2B_{1333} - 8.75a_3^4B_{3333} - 1.5a_1^2B_{133} + 2.5a_3^2B_{333} - 2B_{33})V_{3333} = 0\,, \tag{37}$$

$$(15a_1^4B_{1111} - 9a_1^2a_3^2B_{1113} - 12a_1^2B_{111} + 6a_3^2B_{113})V_{1111} + (27a_1^4B_{1111} - 9a_1^2a_3^2B_{1113} + 6a_3^2B_{113} + 12B_{11} - 6\Omega^2)V_{1122} + 6\left[3\frac{a_1^6}{a_3^2}B_{1113} - 6a_1^2a_3^2B_{1133} + \frac{a_1^2}{a_3^2}(a_3^2 - 2a_1^2)B_{113} + B_{13} - 3B_{33}\right]V_{1133} + \left(1.5a_1^4B_{1133} - 7.5a_1^2a_3^2B_{1333} + 9a_1^2B_{133} - 3\frac{a_1^4}{a_3^2}B_{113}\right)V_{3333} = 0\,, \tag{38}$$

and

$$\frac{2V_{1111}}{a_1^4} + \frac{2V_{1122}}{a_1^4} + \frac{4V_{1133}}{a_1^2a_3^2} + \frac{V_{3333}}{a_3^4} = 0\,, \tag{39}$$

[2] I am grateful to Dr. N. Lebovitz for discussions relating to this point.

(where the last eq. [39] is the divergence condition [26] appropriate for this case), *or*

$$V_{1111} = -V_{2222}, \; V_{1133} = -V_{2233}, \quad \text{and} \quad V_{1122} = V_{3333} = 0\,, \tag{40}$$

and the non-vanishing virials satisfy the equations

$$(\Omega^2 - 2B_{11} - a_1^2 C_{1113}) V_{1111} + 6\frac{a_1^4}{a_3^2} C_{1113} V_{1133} = 0 \tag{41}$$

and

$$a_3^2 C_{1113} V_{1111} + (\Omega^2 - 2B_{11} - 6a_1^2 C_{1113}) V_{1133} = 0\,. \tag{42}$$

It is manifest that equations (36)–(39) can, in turn, be satisfied non-trivially in one of two ways: *either* by letting

$$\Omega^2 = 2A_1 - 2a_1^6 A_{1111}\,, \tag{43}$$

and determining V_{1111}, V_{1122}, V_{1133}, and V_{3333}, apart from an arbitrary constant of proportionality, with the aid of equations (37)–(39), *or* by letting

$$V_{1122} = \tfrac{1}{3} V_{1111} \tag{44}$$

and requiring that the determinant of the system of equations

$$\begin{aligned} &\left(6a_1^2 a_3^2 B_{1113} - 2a_3^4 B_{1133} - 8a_3^2 B_{113} + 4\frac{a_3^4}{a_1^2} B_{133}\right) V_{1111} \\ &\quad - 6\left(a_1^2 C_{1133} - 4a_3^4 B_{1133} - \frac{a_3^2}{a_1^2} B_{33}\right) V_{1133} \\ &+ (3.75a_1^2 a_3^2 B_{1333} - 8.75a_3^4 B_{3333} - 1.5a_1^2 B_{133} + 2.5a_3^2 B_{333} - 2B_{33}) V_{3333} = 0\,, \end{aligned} \tag{45}$$

$$\begin{aligned} &(24a_1^4 B_{1111} - 12a_1^2 a_3^2 B_{1113} + 8a_3^2 B_{113} - 12a_1^2 B_{111} + 4B_{11} - 2\Omega^2) V_{1111} \\ &\quad + 6\left[3\frac{a_1^6}{a_3^2} B_{1113} - 6a_1^2 a_3^2 B_{1133} + \frac{a_1^2}{a_3^2}(a_3^2 - 2a_1^2) B_{113} + B_{13} - 3B_{33}\right] V_{1133} \\ &\quad + \left(1.5a_1^4 B_{1133} - 7.5a_1^2 a_3^2 B_{1333} + 9a_1^2 B_{133} - 3\frac{a_1^4}{a_3^2} B_{113}\right) V_{3333} = 0\,, \end{aligned} \tag{46}$$

and

$$\frac{8V_{1111}}{3a_1^4} + \frac{4V_{1133}}{a_1^2 a_3^2} + \frac{V_{3333}}{a_3^4} = 0 \tag{47}$$

(which are the appropriate forms of eqs. [37]–[39] when the relation [44] obtains) vanish. The two neutral points, which follow in this fashion, occur for eccentricities of the Maclaurin spheroid given by

$$e_1^{(4)} = 0.93275 \qquad (\text{where } \Omega^2 = 2A_1 - a_1^6 A_{1111})\,, \tag{48}$$

and

$$e_2^{(4)} = 0.98531 \text{ (where the determinant of the system of eqs. [45]–[47] vanishes).} \tag{49}$$

The latter point at $e_2^{(4)}$ agrees with the point located earlier (Chandrasekhar 1967*a*) by a different method.[3]

An alternative form of the condition (43) is

$$\Omega^2 = 2(B_{11} + a_1^2 B_{111} + a_1^4 B_{1111})\,. \tag{50}$$

[3] The value 0.98526 determined earlier differs slightly from the value (49). But it is believed that the present determination is the more accurate of the two.

It is of interest to contrast this condition, for the occurrence of a neutral mode belonging to the fourth harmonics, with the conditions

$$\Omega^2 = 2B_{11} \quad \text{and} \quad \Omega^2 = 2(B_{11} + a_1^2 B_{111}) , \tag{51}$$ [4]

for the occurrence of neutral modes belonging to the second and the third harmonics. In view of these results, one may perhaps conjecture that a neutral mode belonging to the nth harmonics occurs where

$$\Omega^2 = 2(B_{11} + a_1^2 B_{11} + \ldots + a_1^{2(n-1)} B_{(n)}) . \tag{52}$$

Considering next equations (41) and (42), we infer that a neutral point occurs when the determinant of this system vanishes; and this condition gives

$$(\Omega^2 - 2B_{11})(\Omega^2 - 2B_{11} - 7a_1^2 C_{1113}) = 0 . \tag{53}$$

We must, therefore, have

$$\Omega^2 = 2B_{11} , \tag{54}$$

or

$$\Omega^2 = 2B_{11} + 7a_1^2 C_{1113} . \tag{55}$$

The occurrence of the "Jacobi point" (54) in this context is a consequence of the fact that neutral points belonging to the second harmonics are automatically included in the present analysis for the same reason that the *equilibrium* conditions (14) determine the same Maclaurin and Jacobi figures as the equations

$$\mathfrak{W}_{11} + \Omega^2 I_{11} = \mathfrak{W}_{22} + \Omega^2 I_{22} = \mathfrak{W}_{33} \tag{56}$$

provided by the second-order virial theorem.

The point determined by the condition (55) is, however, new and belongs, genuinely, to the fourth harmonics. It is found that the condition is satisfied for a Maclaurin spheroid with the eccentricity

$$e_3^{(4)} = 0.98097 \qquad (\text{where } \Omega^2 = 2B_{11} + 7a_1^2 C_{1113}) . \tag{57}$$

This completes the discussion of the even equations.

We turn now to a consideration of the seven equations odd in the indices 1 and 2. On inserting from equations (A.3)–(A.6) in equations (29)–(32), we find that the six equations remaining after the elimination of $\delta\Pi_{12}$ and V_{12} (the latter with the aid of eq. [24]) can be reduced to the following four equations:

$$(\Omega^2 - 2A_1 + 2a_1^6 A_{1111})(V_{1211} - V_{1222}) = 0 , \tag{58}$$

$$(\Omega^2 - 2B_{11} - a_1^2 C_{1113})(V_{1211} + V_{1222}) + 6\frac{a_1^4}{a_3^2} C_{1113} V_{1233} = 0 , \tag{59}$$

$$a_3^2 C_{1113}(V_{1211} + V_{1222}) + (\Omega^2 - 2B_{11} - 6a_1^2 C_{1113}) V_{1233} = 0 , \tag{60}$$

and

$$\begin{aligned} &\tfrac{1}{2}(5a_1^2 C_{1113} + 2B_{11} - \Omega^2)(V_{1211} + V_{1222}) \\ &-\tfrac{1}{2}(\Omega^2 - 2A_1 + 2a_1^6 A_{1111})(V_{1211} - V_{1222}) \\ &\qquad -\frac{a_1^2}{a_3^2}(4B_{11} + 15a_1^2 C_{1113} - 2\Omega^2) V_{1233} = 0 , \end{aligned} \tag{61}$$

[4] The first of these two points is, of course, the point of bifurcation where the Jacobian sequence branches off from the Maclaurin sequence; and for the location of the second point belonging to the third harmonics see Chandrasekhar and Lebovitz (1963*b*) and also Chandrasekhar (1963).

where it may be noted that equations (59) and (60) are each repeated twice among the original six equations.

It is clear that equations (58)–(60) can be satisfied non-trivially in one of two ways: *either* by requiring

$$V_{1211} = -V_{1222} \qquad \text{and} \qquad V_{1233} = 0 \tag{62}$$

or by requiring

$$V_{1211} = V_{1222} \qquad \text{and} \qquad V_{1233} \neq 0\,. \tag{63}$$

In the former case, we are led to the same condition (43) that followed from equation (36); and in the latter case, we are led to the same characteristic equation (53) for Ω^2 that followed from equations (41) and (42). It remains to verify that equation (61) is satisfied under both these requirements. It is clearly satisfied under (62); that it is also satisfied under (63) can be established as follows. When the determinant of equations (59) and (60) vanishes (i.e., when eq. [53] holds)

$$\frac{V_{1211}}{V_{1233}} = -\frac{a_1^4}{a_3^2}\frac{3C_{1113}}{\Omega^2 - 2B_{11} - a_1^2C_{1113}} = -\frac{\Omega^2 - 2B_{11} - 6a_1^2C_{1113}}{2a_3^2C_{1113}}\,. \tag{64}$$

In particular,

$$\begin{aligned} \frac{V_{1211}}{V_{1233}} &= \frac{3a_1^2}{a_3^2} \qquad \text{when} \qquad \Omega^2 = 2B_{11} \\ &= -\frac{a_1^2}{2a_3^2} \qquad \text{when} \qquad \Omega^2 = 2B_{11} + 7a_1^2C_{1113}\,; \end{aligned} \tag{65}$$

and we verify that under these circumstances equation (61) is indeed satisfied.

The neutral points $e_1^{(4)}$ and $e_3^{(4)}$ are, therefore, repeated as solutions of the odd equations. We conclude that the corresponding neutral states are degenerate. On the other hand, the point $e_2^{(4)}$ occurs only once as the solution of the even equations; and this indicates the non-degeneracy of the neutral state—a fact manifest from the uniqueness of the proper solution determined for this point in the earlier paper (Chandrasekhar 1967*a;* see particularly § IX).

V. THE NEUTRAL POINT ALONG THE JACOBIAN SEQUENCE BELONGING TO THE FOURTH HARMONICS

The six even equations provided by equations (33) and (34), after the elimination of the second-order virials with the aid of equation (24), represent a set of six homogeneous equations for the six fourth-order virials V_{1111}, V_{2222}, V_{3333}, V_{1122}, V_{2233}, and V_{3311}. However, only five of these six equations are linearly independent when they are supplemented by the further condition (26) that must also hold. Therefore, including equation (26), we have only six linear homogeneous equations between the six virials; and the vanishing of the determinant of these equations is the condition for the occurrence of a neutral point. It is found that the condition is met for a Jacobi ellipsoid for which

$$\cos^{-1}(a_3/a_1) = 75^\circ\!.068\,; \tag{66}$$

it is the same point that was determined earlier (Chandrasekhar 1967*b*) by a different method.[5] The uniqueness of this neutral point indicates the non-degeneracy of this neutral state—a fact that is in agreement with the unique proper solution that was explicitly determined for this point in the earlier paper.

Turning next to the equations that are odd in the indices 1 and 2, we find that they

[5] The value $\cos^{-1}(a_3/a_1) = 75^\circ\!.081$ found earlier differs slightly from the value (66); but the difference is not outside the limits of accuracy of the numerical evaluation.

provide no additional information: the six equations governing V_{1211}, V_{1222}, and V_{1233} (remaining after the elimination of $\delta\Pi_{12}$ and V_{12}) are all satisfied *identically* by the substitution

$$V_{1211} : V_{1222} : V_{1233} = 3a_1^2 : 3a_2^2 : a_3^2 \,, \tag{67}$$

by virtue of the properties of the Jacobi ellipsoid, namely, that

$$\Omega^2 = 2B_{12} \qquad \text{and} \qquad a_1^2 a_2^2 A_{12} = a_3^2 A_3 \,, \tag{68}$$

along the Jacobian sequence. The reason for this behavior is that the neutral mode of oscillation that characterizes the Maclaurin spheroid at its point of bifurcation persists through the entire Jacobian sequence.

I am greatly indebted to Miss Donna D. Elbert for her assistance with the reduction of the formulae given in the Appendix and for the numerical location of the various neutral points.

The research reported in this paper has in part been supported by the Office of Naval Research under contract Nonr-2121(24) with the University of Chicago.

APPENDIX

The particular combinations of $\delta\mathfrak{W}_{ij;kl}$ and $\delta\mathfrak{W}_{ij;k;l}$ that occur in equations (29)–(34) can be written down by making use of the general formulae (21) and (22). In most cases, the resulting expressions can be simplified substantially if proper use is made of the various identities that relate the different index symbols. The expressions derived in this fashion are listed below.

$$\begin{aligned}
&2\delta\mathfrak{W}_{ii;ii} + \delta\mathfrak{W}_{ii;i;i} - (4\delta\mathfrak{W}_{ij;ij} + 2\delta\mathfrak{W}_{jj;ii} + 2\delta\mathfrak{W}_{ij;i;j} + \delta\mathfrak{W}_{jj;i;i})\\
&\quad = 1.5a_i^2[5a_i^2B_{iii} - (2a_i^2 + a_j^2)B_{iij}]V_{ii}\\
&\qquad +1.5a_i^2(a_i^2B_{iij} - 3a_j^2B_{ijj})V_{jj} + 1.5a_i^2(a_i^2B_{iik} - a_j^2B_{ijk})V_{kk}\\
&\qquad +[2B_{ij} - 4B_{ii} + 2a_j^2B_{iij} - 4a_i^2B_{iii} + a_i^2(a_i^2 + 2.75a_j^2)B_{iiij} - 8.75a_i^4B_{iiii}]V_{iiii}\\
&\qquad +0.75a_i^2(5a_j^2B_{ijjj} - a_i^2B_{iijj})V_{jjjj} - 0.75a_i^2(a_i^2B_{iikk} - a_j^2B_{ijkk})V_{kkkk}\\
&\qquad -1.5a_i^2[5a_i^2B_{iiik} - (2a_i^2 + a_j^2)B_{iijk}]V_{iikk} - 1.5a_i^2(a_i^2B_{iijk} - 3a_j^2B_{ijjk})V_{jjkk}\\
&\qquad +1.5\,[2B_{ij} + 2B_{jj} + 2(a_i^2 + a_j^2)B_{ijj} + a_i^2(2a_i^2 + 7a_j^2)B_{iijj} - 5a_i^4B_{iiij}]V_{iijj}\,,
\end{aligned} \tag{A·1}$$

$$\begin{aligned}
&+ (4\delta\mathfrak{W}_{ij;ij} + 2\delta\mathfrak{W}_{jj;ii} + 2\delta\mathfrak{W}_{ij;i;j} + \delta\mathfrak{W}_{jj;i;i})\\
&\qquad -(4\delta\mathfrak{W}_{ik;ik} + 2\delta\mathfrak{W}_{kk;ii} + 2\delta\mathfrak{W}_{ik;i;k} + \delta\mathfrak{W}_{kk;i;i})\\
&\quad = 1.5a_i^2(a_j^2 - a_k^2)(B_{ijk} - 3a_i^2B_{iijk})V_{ii}\\
&\qquad -1.5a_i^2(a_k^2B_{ijk} - 3a_j^2B_{ijj})V_{jj} + 1.5a_i^2(a_j^2B_{ijk} - 3a_k^2B_{ikk})V_{kk}\\
&\qquad +0.75a_i^2(a_j^2 - a_k^2)(5a_i^2B_{iiijk} - B_{iijk})V_{iiii}\\
&\qquad -0.75a_i^2(5a_j^2B_{ijjj} - a_k^2B_{ijjk})V_{jjjj} + 0.75a_i^2(5a_k^2B_{ikkk} - a_j^2B_{ijkk})V_{kkkk}\\
&\qquad -4.5a_i^2(a_j^2 - a_k^2)C_{ijjkk}V_{jjkk}\\
&\qquad +[2B_{jk} - 5B_{ij} - 3B_{jj} + (a_i^2 + 2a_k^2)B_{ijk} - 3a_i^2B_{iij} - 3(a_i^2 + a_j^2)B_{ijj}\\
&\qquad +4.5a_i^2a_k^2B_{iijk} - 3a_i^2(a_i^2 + 3.5a_j^2)B_{iijj}]V_{iijj}\\
&\qquad -[2B_{jk} - 5B_{ik} - 3B_{kk} + (a_i^2 + 2a_j^2)B_{ijk} - 3a_i^2B_{iik} - 3(a_i^2 + a_k^2)B_{ikk}\\
&\qquad +4.5a_i^2a_j^2B_{iijk} - 3a_i^2(a_i^2 + 3.5a_k^2)B_{iikk}]V_{iikk}\,,
\end{aligned} \tag{A·2}$$

$$-(2\delta\mathfrak{W}_{ij;jj} + \delta\mathfrak{W}_{ij;j;j}) = 3a_j^4(-B_{ijj}V_{ij} + B_{iijj}V_{iiij} + B_{ijjk}V_{ijkk}) + (2B_{ij} + 2a_j^2B_{ijj} + 5a_j^4B_{ijjj})V_{ijjj} \, , \tag{A.3}$$

$$-(4\delta\mathfrak{W}_{ik;jk} + 2\delta\mathfrak{W}_{ij;kk} + 2\delta\mathfrak{W}_{ik;j;k} + \delta\mathfrak{W}_{ij;k;k}) = 3a_j^2a_k^2(-B_{ijk}V_{ij} + B_{iijk}V_{iiij} + B_{ijjk}V_{ijjj}) + [2B_{ij} + 4B_{ik} + (2a_j^2 + a_k^2)B_{ijk} + 3a_k^2B_{ikk} + 3a_k^2(2a_j^2 + a_k^2)B_{ijkk}]V_{ijkk} \, , \tag{A.4}$$

$$-(4\delta\mathfrak{W}_{ii;ij} + 2\delta\mathfrak{W}_{ij;ii} + 2\delta\mathfrak{W}_{ii;i;j} + \delta\mathfrak{W}_{ij;i;i}) = 3a_i^2a_j^2[-(A_{ij} - 3a_i^2A_{iij})V_{ij} + (A_{ijj} - 3a_i^2A_{iijj})V_{ijjj} + (A_{ijk} - 3a_i^2A_{iijk})V_{ijkk}] + 3(2A_i + a_i^2a_j^2A_{iij} - 5a_i^4a_j^2A_{iiij})V_{iiij} \, , \tag{A.5}$$

$$-(2\delta\mathfrak{W}_{kk;ij} + 2\delta\mathfrak{W}_{ik;jk} + 2\delta\mathfrak{W}_{jk;ik} + \delta\mathfrak{W}_{kk;i;j} + \delta\mathfrak{W}_{ik;j;k} + \delta\mathfrak{W}_{jk;i;k}) = 3a_i^2a_j^2a_k^2(A_{ijk}V_{ij} - A_{iijk}V_{ijii} - A_{ijjk}V_{ijjj}) + (6A_k - 9a_i^2a_j^2a_k^2A_{ijkk})V_{ijkk} \, . \tag{A.6}$$

REFERENÇES

Chandrasekhar, S. 1963, *Ap. J.*, **137**, 1185.
———. 1967*a*, *ibid.*, **147**, 334.
———. 1967*b*, *ibid.*, **148**, 621.
Chandrasekhar, S., and Lebovitz, N. R. 1962, *Ap. J.*, **136**, 1037.
———. 1963*a*, *ibid.*, **137**, 1142.
———. 1963*b*, *ibid.*, p. 1162.
———. 1963*c*, *ibid.*, p. 1185.
Lebovitz, N. R. 1961, *Ap. J.*, **134**, 500.

A TENSOR VIRIAL-EQUATION FOR STELLAR DYNAMICS

S. Chandrasekhar and Edward P. Lee

(Received 1967 November 6)

Summary

A tensor virial-equation is derived for a system consisting of equal mass-points, appropriately for stellar dynamics, by starting with the Liouville equation governing an ensemble of such systems in the six N-dimensional phase space.

1. *Introduction.* The tensor form of the second-order virial equation has, in recent years, found numerous applications to a wide variety of problems (for a general account of these topics, see Chandrasekhar 1964). In the gravitational context (to which we shall restrict ourselves) there are two limiting classes of systems for which the virial equation has been explicitly written down: (1) systems consisting of a number of discrete mass points (or particles) $m^{(\alpha)}$ ($\alpha = 1, 2, \ldots,$) under their mutual attractions and (2) systems described in terms of a density ρ and an isotropic pressure p and governed by the usual hydrodynamic equations. For a system belonging to the first class, the equation takes the form

$$\frac{1}{2}\frac{d^2}{dt^2}\sum_{\alpha} m^{(\alpha)}x_i^{(\alpha)}x_j^{(\alpha)} = \sum_{\alpha} m^{(\alpha)}u_i^{(\alpha)}u_j^{(\alpha)} - \frac{1}{2}G\sum_{\alpha}\sum_{\alpha\neq\beta} m^{(\alpha)}m^{(\beta)}\frac{[x_i^{(\alpha)}-x_i^{(\beta)}][x_j^{(\alpha)}-x_j^{(\beta)}]}{|\mathbf{x}^{(\alpha)}-\mathbf{x}^{(\beta)}|^3}, \tag{1}$$

where the Greek superscripts distinguish the different particles and the Latin subscripts distinguish the different Cartesian components of the position $\mathbf{x}$ and the velocity $\mathbf{u}$ of the particle α. And for a system belonging to the second class, the equation takes the form

$$\frac{1}{2}\frac{d^2}{dt^2}\int_V \rho x_i x_j\, d\mathbf{x} = 2\mathfrak{T}_{ij} + \Pi\,\delta_{ij} + \mathfrak{W}_{ij}, \tag{2}$$

where

$$\mathfrak{T}_{ij} = \frac{1}{2}\int_V \rho u_i u_j\, d\mathbf{x} \tag{3}$$

and

$$\mathfrak{W}_{ij} = -\frac{1}{2}\int_V \rho \mathfrak{V}_{ij}\, d\mathbf{x} = -\frac{1}{2}G\int_V\int_V \rho(\mathbf{x})\rho(\mathbf{x}')\frac{(x_i-x_i')(x_j-x_j')}{|\mathbf{x}-\mathbf{x}'|^3}\, d\mathbf{x}\, d\mathbf{x}' \tag{4}$$

are the kinetic-energy and the potential-energy tensors; also

$$\Pi = \int_V p\, d\mathbf{x}. \tag{5}$$

(In equations (2)–(5) the integrations are effected over the entire volume V occupied by the fluid.)

While the forms of the two equations (1) and (2) are very similar, the essential difference between them is reflected in the separation, in equation (2), of the contributions to the kinetic-energy tensor by the macroscopic (or mean) and the microscopic (or molecular) motions: the former is included in $\mathfrak{T}_{ij}$ and the latter is manifested by the pressure term $\Pi\,\delta_{ij}$. But neither of these two equations is suitable under the normal circumstances of stellar dynamics. For in the context of stellar dynamics, the idealization in terms of the strict N-body problem (which does not permit a separation between the 'mean' and the 'peculiar' motions) or in terms of a fully relaxed hydrodynamic system (which does permit a unique separation of the two) are both unrealistic.

A recent discussion by Camm (1967) has suggested that it might be worthwhile to draw attention to a form of the tensor virial-equation that is 'exact' in the context of stellar dynamics in that it follows, rigorously, without any assumptions or approximations, from the six N-dimensional Liouville equation governing an ensemble of systems.

2. *The six N-dimensional Liouville equation and its integrated form.* We consider an ensemble of systems consisting of a (large) number N of discrete mass points (or particles) under their mutual gravitational attractions. For the sake of simplicity, we shall suppose that the masses of all the particles are the same so that any statistical property of the ensemble in which we may be interested, may be assumed to be symmetric in all the particles. And this assumption of symmetry will be made in this paper.

In the six N-dimensional phase space of the system, we describe an ensemble by the density function

$$f^{(N)} \equiv f^{(N)}(\mathbf{x}^{(1)}, \mathbf{x}^{(2)}, \ldots, \mathbf{x}^{(N)}; \mathbf{u}^{(1)}, \mathbf{u}^{(2)}, \ldots, \mathbf{u}^{(N)}; t), \tag{6}$$

where the superscripts (for which we shall use Greek letters) distinguish the different particles; and in accordance with our earlier remark, $f^{(N)}$ will be assumed to be symmetric in the variables of the different particles.

The Liouville equation governing $f^{(N)}$ is

$$\frac{\partial f^{(N)}}{\partial t} + \sum_{\alpha} u_k^{(\alpha)} \frac{\partial f^{(N)}}{\partial x_k^{(\alpha)}} + Gm \sum_{\alpha} \sum_{\beta \neq \alpha} \frac{x_k^{(\beta)} - x_k^{(\alpha)}}{|\mathbf{x}^{(\beta)} - \mathbf{x}^{(\alpha)}|^3} \frac{\partial f^{(N)}}{\partial u_k^{(\alpha)}} = 0; \tag{7}$$

and on integrating this equation over the coordinates and the velocities of all the particles except one, say the first, we obtain

$$\frac{\partial f^{(1)}}{\partial t} + u_k^{(1)} \frac{\partial f^{(1)}}{\partial x_k^{(1)}} + (N-1)Gm \int\int \frac{x_k^{(2)} - x_k^{(1)}}{|\mathbf{x}^{(2)} - \mathbf{x}^{(1)}|^3} \frac{\partial f^{(2)}}{\partial u_k^{(1)}} \, d\mathbf{x}^{(2)} \, d\mathbf{u}^{(2)} = 0, \tag{8}$$

where

$$f^{(2)} \equiv f^{(2)}(\mathbf{x}^{(1)}, \mathbf{x}^{(2)}; \mathbf{u}^{(1)}, \mathbf{u}^{(2)}; t)$$
$$= \int \ldots \int f^{(N)} \, d\mathbf{x}^{(3)} \, d\mathbf{x}^{(4)} \ldots d\mathbf{x}^{(N)} \, d\mathbf{u}^{(3)} \, d\mathbf{u}^{(4)} \ldots d\mathbf{u}^{(N)}, \tag{9}$$

and

$$f^{(1)} \equiv f^{(1)}(\mathbf{x}^{(1)}; \mathbf{u}^{(1)}; t) = \int\int f^{(2)} \, d\mathbf{x}^{(2)} \, d\mathbf{u}^{(2)}$$
$$= \int \ldots \int f^{(N)} \, d\mathbf{x}^{(2)} \, d\mathbf{x}^{(3)} \ldots d\mathbf{x}^{(N)} \, d\mathbf{u}^{(2)} \, d\mathbf{u}^{(3)} \ldots d\mathbf{u}^{(N)}. \tag{10}$$

The factor $(N-1)$ in the last term in equation (8) originates in the assumed symmetry of $f^{(N)}$ in the variables of the different particles: the result of integrating the term in the summand involving the pair of particles (α, β), with α fixed and β variable, is independent of β.

3. *The hydrodynamic equations.* By integrating equation (8) over the velocities of the particle 1, as well, we obtain (cf. the corresponding derivation based on the six-dimensional 'collisionless' Boltzmann equation in Chandrasekhar (1942, see pp. 185, 186))

$$\frac{\partial n^{(1)}}{\partial t}+\frac{\partial}{\partial x_k^{(1)}}[n^{(1)}\langle u_k^{(1)}\rangle] = 0, \tag{11}$$

where

$$n^{(1)} \equiv n^{(1)}(\mathbf{x}^{(1)}, t) = \int f^{(1)}\, d\mathbf{u}^{(1)}, \tag{12}$$

and $\langle u_k^{(1)}\rangle$ is the average 1-particle velocity defined by

$$n^{(1)}\langle u_k^{(1)}\rangle = \int f^{(1)}u_k^{(1)}\, d\mathbf{u}^{(1)}. \tag{13}$$

It should be noted that here (and in the sequel) 'averages' (indicated by angular brackets) are averages over an ensemble.

Next, multiplying equation (8) by $u_i^{(1)}$ and integrating over the velocities of the particle 1, we obtain (cf. once again, the corresponding derivation based on the six-dimensional Liouville equation in Chandrasekhar (1942))

$$\frac{\partial}{\partial t}[n^{(1)}\langle u_i^{(1)}\rangle]+\frac{\partial}{\partial x_k^{(1)}}[n^{(1)}\langle u_i^{(1)}u_k^{(1)}\rangle] \\ -(N-1)Gm\int\int\int f^{(2)}\frac{x_i^{(2)}-x_i^{(1)}}{|\mathbf{x}^{(2)}-\mathbf{x}^{(1)}|^3}\, d\mathbf{x}^{(2)}\, d\mathbf{u}^{(2)}\, d\mathbf{u}^{(1)} = 0, \tag{14}$$

where in reducing the last term on the right-hand side of equation (8), an integration by parts has been effected and an assumption, appropriate to ignoring the integrated part, has been made.

In terms of the two-particle function,

$$n^{(1,2)} \equiv n^{(1,2)}(\mathbf{x}^{(1)}, \mathbf{x}^{(2)}, t) = \int\int f^{(2)}\, d\mathbf{u}^{(1)}\, d\mathbf{u}^{(2)}, \tag{15}$$

we can rewrite equation (14) in the form

$$\frac{\partial}{\partial t}[n^{(1)}\langle u_i^{(1)}\rangle]+\frac{\partial}{\partial x_k^{(1)}}[n^{(1)}\langle u_i^{(1)}u_k^{(1)}\rangle] \\ -(N-1)Gm\int n^{(1,2)}(\mathbf{x}^{(1)}, \mathbf{x}^{(2)}, t)\frac{x_i^{(2)}-x_i^{(1)}}{|\mathbf{x}^{(2)}-\mathbf{x}^{(1)}|^3}\, d\mathbf{x}^{(2)} = 0, \tag{16}$$

where it should be noted that $n^{(1,2)}$ (on our original assumption concerning $f^{(N)}$) is symmetric in $\mathbf{x}^{(1)}$ and $\mathbf{x}^{(2)}$.

We can now separate the 'mean' ($\mathbf{U}^{(1)}$) from the 'peculiar' ($\mathbf{v}^{(1)}$) velocities and define a *pressure-tensor* $P_{ik}^{(1)}$ by means of the relations

$$u_i^{(1)} = v_i^{(1)}+\langle u_i^{(1)}\rangle = v_i^{(1)}+U_i^{(1)}, \tag{17}$$

and

$$n^{(1)}\langle u_i^{(1)}u_k^{(1)}\rangle = n^{(1)}\langle v_i^{(1)}v_k^{(1)}\rangle + n^{(1)}U_i^{(1)}U_k^{(1)}$$
$$= P_{ik}^{(1)} + n^{(1)}U_i^{(1)}U_k^{(1)}. \tag{18}$$

With these definitions, equations (11) and (16) take the forms

$$\frac{\partial}{\partial t}[n^{(1)}U_i^{(1)}] + \frac{\partial}{\partial x_k^{(1)}}[n^{(1)}U_i^{(1)}U_k^{(1)}] = -\frac{\partial P_{ik}^{(1)}}{\partial x_k^{(1)}} + (N-1)Gm\int n^{(1,2)}\frac{x_i^{(2)}-x_i^{(1)}}{|\mathbf{x}^{(2)}-\mathbf{x}^{(1)}|^3}\,d\mathbf{x}^{(2)} \tag{19}$$

and

$$\frac{\partial n^{(1)}}{\partial t} + \frac{\partial}{\partial x_k^{(1)}}[n^{(1)}U_k^{(1)}] = 0. \tag{20}$$

4. *The tensor virial-equation.* From equation (19) we may derive a 'virial equation' by multiplying it by $x_j^{(1)}$ and integrating over the coordinates of the particle 1. The left-hand side of the equation gives, after an integration by parts,

$$\frac{d}{dt}\int n^{(1)}U_i^{(1)}x_j^{(1)}\,d\mathbf{x}^{(1)} - 2\mathfrak{T}_{ij}, \tag{21}$$

where

$$\mathfrak{T}_{ij} = \tfrac{1}{2}\int n^{(1)}U_i^{(1)}U_j^{(1)}\,d\mathbf{x}^{(1)}. \tag{22}$$

And the terms on the right-hand side give

$$-\int x_j^{(1)}\frac{\partial P_{ik}^{(1)}}{\partial x_k^{(1)}}\,d\mathbf{x}^{(1)} = \int P_{ij}^{(1)}\,d\mathbf{x}^{(1)} = \Pi_{ij}\ \text{(say)}, \tag{23}$$

and

$$(N-1)Gm\int\int n^{(1,2)}(\mathbf{x}^{(1)},\mathbf{x}^{(2)},t)\frac{[x_i^{(2)}-x_i^{(1)}]x_j^{(1)}}{|\mathbf{x}^{(2)}-\mathbf{x}^{(1)}|^3}\,d\mathbf{x}^{(1)}\,d\mathbf{x}^{(2)}$$
$$= -\tfrac{1}{2}(N-1)Gm\int\int n^{(1,2)}(\mathbf{x}^{(1)},\mathbf{x}^{(2)},t)\frac{[x_i^{(2)}-x_i^{(1)}][x_j^{(2)}-x_j^{(1)}]}{|\mathbf{x}^{(2)}-\mathbf{x}^{(1)}|^3}\,d\mathbf{x}^{(1)}\,d\mathbf{x}^{(2)}$$
$$= \mathfrak{W}_{ij}\ \text{(say)}, \tag{24}$$

where in integrating by parts in equation (23), it has been assumed that $P_{ik}^{(1)}$ vanishes on the boundary of the system; and, further, in passing from the first to the second line in equation (24), the symmetry of $n^{(1,2)}$ in $\mathbf{x}^{(1)}$ and $\mathbf{x}^{(2)}$ has been used and the expression for $\mathfrak{W}_{ij}$ is the average of the original form and the one obtained from it by interchanging the variables of integration $\mathbf{x}^{(1)}$ and $\mathbf{x}^{(2)}$. Combining the results of these reductions, we obtain

$$\frac{d}{dt}\int n^{(1)}U_i^{(1)}x_j^{(1)}\,d\mathbf{x}^{(1)} = 2\mathfrak{T}_{ij} + \Pi_{ij} + \mathfrak{W}_{ij}. \tag{25}$$

The tensors on the right-hand side of equation (25) are manifestly symmetric in i and j. Therefore

$$\frac{d}{dt}\int n^{(1)}[U_i^{(1)}x_j^{(1)} - U_j^{(1)}x_i^{(1)}]\,d\mathbf{x}^{(1)} = 0; \tag{26}$$

and this equation represents the conservation of the angular momentum. And taking the symmetric part of the term on the left-hand side of equation (25), we have

$$\frac{1}{2}\frac{d}{dt}\int n^{(1)}[U_i^{(1)}x_j^{(1)} + U_j^{(1)}x_i^{(1)}]\, d\mathbf{x}^{(1)}$$

$$= \frac{1}{2}\frac{d^2}{dt^2}\int n^{(1)}x_i^{(1)}x_j^{(1)}\, d\mathbf{x}^{(1)} = \frac{1}{2}\frac{d^2I_{ij}}{dt^2} \text{ (say).} \quad (27)$$

Thus, we finally obtain

$$\frac{1}{2}\frac{d^2I_{ij}}{dt^2} = 2\mathfrak{T}_{ij} + \Pi_{ij} + \mathfrak{W}_{ij}. \quad (28)$$

While equation (28) is similar in form to the tensor virial-equation (2) for fluids, there are two important differences: *first*, the averages that have been taken before arriving at equation (28) are averages over an ensemble and *second*, $\mathfrak{W}_{ij}$ is *not* the potential-energy tensor (4) as defined in hydrodynamics: it is now defined in terms of the symmetric two-particle function $n^{(1,\,2)}(\mathbf{x}^{(1)}, \mathbf{x}^{(2)}, t)$ and not in terms of the product $n^{(1)}(\mathbf{x}^{(1)}, t)\, n^{(2)}(\mathbf{x}^{(2)}, t)$ of two one-particle functions.

Acknowledgment. The research reported in this paper has in part been supported by the Office of Naval Research under contract Nonr-2121(24) with the University of Chicago.

University of Chicago,
Chicago,
Illinois.
1967 *November.*

References

Camm, G. L., 1967. *Bull. astr.*, **2**, 141.
Chandrasekhar, S., 1942. *Principles of Stellar Dynamics*, University of Chicago Press.
Chandrasekhar, S., 1964. *Lectures in Theoretical Physics*, ed. by W. E. Brittin, Vol. 6, p. 1, University of Colorado Press, Boulder.

SOME ELEMENTARY APPLICATIONS OF THE VIRIAL THEOREM TO STELLAR DYNAMICS

S. Chandrasekhar and Donna D. Elbert

(Received 1971 September 1)

SUMMARY

The dynamical evolution of spherical and spheroidal systems of mass points is examined with the aid of the scalar and the tensor forms of the virial theorem. Spherical systems with positive total energy tend to disperse to infinity while those with negative total energy execute periodic oscillations of finite amplitude. Spheroidal systems with positive total energy exhibit expansion like the spherical systems; but they also become less oblate (and sometimes actually become prolate) if they are initially oblate, and less prolate if they are initially prolate. Spheroidal systems with negative total energy collapse to smaller volumes while enhancing their initial oblateness or prolateness.

1. INTRODUCTION

The most common (if not the sole) application of the virial theorem to stellar dynamics has been for the purposes of estimating the kinetic energy, $\mathfrak{T}$, of a group (' cluster ') of stars (or galaxies) from their dimensions. The basis of this estimation is, of course, the relation

$$2\mathfrak{T}+\mathfrak{W} = 0 \tag{1}$$

provided by the scalar form of the virial theorem under stationary conditions. The gravitational potential-energy $\mathfrak{W}$ that occurs in equation (1) is, in practice, estimated from a dimensional relation of the form

$$\mathfrak{W} = -\frac{1}{2}\frac{GM^2}{\bar{R}}, \tag{2}$$

where $\bar{R}$ is some average measure of the linear dimension of the system and M is its mass.

In this paper we shall outline some other elementary applications of the non-stationary form of the virial theorem (in both its scalar and tensor forms) to suggest the scope of its applications that remains to be exploited.

2. THE VIRIAL THEOREM

For a system consisting of a number (N) of discrete mass-points (or particles) under their mutual gravitational attractions, the tensor form of the virial theorem takes the standard form (cf. Chandrasekhar 1964)

$$\frac{1}{2}\frac{d^2I_{ij}}{dt^2} = 2\mathfrak{T}_{ij}+\mathfrak{W}_{ij}, \tag{3}$$

Figures 3 and 5 have been replaced to correct oversights. I am grateful to Dr. R. Miller for pointing out the errors and providing corrected figures—S. Chandrasekhar.

where I_{ij}, $\mathfrak{T}_{ij}$, and $\mathfrak{W}_{ij}$ are, respectively, the moment of inertia, the kinetic-energy, and the potential-energy tensors given by

$$I_{ij} = \sum_{\alpha} m^{(\alpha)} x_i^{(\alpha)} x_j^{(\alpha)}, \qquad \mathfrak{T}_{ij} = \tfrac{1}{2} \sum_{\alpha} m^{(\alpha)} u_i^{(\alpha)} u_j^{(\alpha)}, \tag{4}$$

and

$$\mathfrak{W}_{ij} = -\tfrac{1}{2} G \sum_{\alpha} \sum_{\alpha \neq \beta} m^{(\alpha)} m^{(\beta)} \frac{[x_i^{(\alpha)} - x_i^{(\beta)}][x_j^{(\alpha)} - x_j^{(\beta)}]}{|\mathbf{x}^{(\alpha)} - \mathbf{x}^{(\beta)}|^3}. \tag{5}$$

In equations (4) and (5) the Greek superscripts distinguish the different mass points while the Latin subscripts distinguish the different Cartesian components of the position $\mathbf{x}$ and the velocity $\mathbf{u}$ of the different particles.

The contracted version of equation (3) provides the usual scalar form of the virial equation, namely,

$$\frac{1}{2} \frac{d^2 I}{dt^2} = 2\mathfrak{T} + \mathfrak{W} = 2\mathfrak{E} - \mathfrak{W}, \tag{6}$$

where

$$\mathfrak{E} = \mathfrak{T} + \mathfrak{W} \tag{7}$$

is the total energy of the system that remains constant.

In applying the virial theorem to a system of particles described statistically in terms of a density distribution $\rho(x)$, the common procedure is to use the same equations (3) and (6) altering only the definitions of I_{ij}, $\mathfrak{T}_{ij}$, and $\mathfrak{W}_{ij}$ as sums to integrals; thus,

$$I_{ij} = \int \rho x_i x_j \, d\mathbf{x}, \qquad \mathfrak{T}_{ij} = \tfrac{1}{2} \int \rho u_i u_j \, d\mathbf{x}, \tag{8}$$

and

$$\mathfrak{W}_{ij} = -\tfrac{1}{2} G \iint \rho(\mathbf{x}) \rho(\mathbf{x}') \frac{(x_i - x_i')(x_j - x_j')}{|\mathbf{x} - \mathbf{x}'|^3} \, d\mathbf{x} \, d\mathbf{x}'. \tag{9}$$

But this procedure is not strictly justified. For, if one wishes to describe a stellar system in terms of distribution functions, then one must start with an ensemble of systems described by a density function

$$f^{(N)}(\mathbf{x}^{(1)}, \mathbf{x}^{(2)}, \ldots, \mathbf{x}^{(N)};\ \mathbf{u}^{(1)}, \mathbf{u}^{(2)}, \ldots, \mathbf{u}^{(N)};\ t) \tag{10}$$

defined in the six N-dimensional phase space of the system and consider the hierarchy of equations that one obtains by integrating the Liouville equation governing $f^{(N)}$ over the coordinates and velocities of all particles except a particular one, of all particles except a particular pair, and so on. And whatever integral relation one uses must be a consequence of this hierarchy of equations. Equation (3) with the definitions (8) and (9) is not such a consequence. However, the correct generalization of equation (3) to stellar dynamics has been derived in an earlier paper (Chandrasekhar & Lee 1968). Thus, by separating the 'mean' ($\mathbf{U}^{(1)}$) from the 'peculiar' ($\mathbf{v}^{(1)}$) velocities in the manner

$$u_i^{(1)} = v_i^{(1)} + \langle u_i^{(1)} \rangle = v_i^{(1)} + U_i^{(1)}, \tag{11}$$

and defining a *pressure tensor* $P_{ij}^{(1)}$ by the relation

$$\begin{aligned} n^{(1)} \langle u_i^{(1)} u_j^{(1)} \rangle &= n^{(1)} \langle v_i^{(1)} v_j^{(1)} \rangle + n^{(1)} U_i^{(1)} U_j^{(1)} \\ &= P_{ij}^{(1)} + n^{(1)} U_i^{(1)} U_j^{(1)}, \end{aligned} \tag{12}$$

(where $n^{(1)}(\mathbf{x}^{(1)}; t)$ is the one-particle distribution and the angular brackets denote ensemble averages), the following integral relation was derived:

$$\frac{1}{2}\frac{d^2 I_{ij}}{dt^2} = 2\mathfrak{T}_{ij} + \Pi_{ij} + \mathfrak{W}_{ij}, \tag{13}$$

where

$$I_{ij} = m\int n^{(1)} x_i^{(1)} x_j^{(1)}\, d\mathbf{x}^{(1)},$$

$$\mathfrak{T}_{ij} = \tfrac{1}{2}m\int n^{(1)} U_i^{(1)} U_j^{(1)}\, d\mathbf{x}^{(1)}, \tag{14}$$

$$\Pi_{ij} = \int P_{ij}^{(1)}\, d\mathbf{x}^{(1)},$$

$$\mathfrak{W}_{ij} = -\tfrac{1}{2}GNm^2 \iint n^{(1,\,2)}(\mathbf{x}^{(1)}, \mathbf{x}^{(2)}; t)\frac{[x_i^{(1)} - x_i^{(2)}][x_j^{(1)} - x_j^{(2)}]}{|\mathbf{x}^{(1)} - \mathbf{x}^{(2)}|^3}\, d\mathbf{x}^{(1)}\, d\mathbf{x}^{(2)}, \tag{15}$$

and $n^{(1,\,2)}(\mathbf{x}^{(1)}, \mathbf{x}^{(2)}; t)$ is the two-particle distribution defined by

$$n^{(1,\,2)} = \int\cdots\int f^{(N)}\, d\mathbf{x}^{(3)}\, d\mathbf{x}^{(4)}\ldots d\mathbf{x}^{(N)}\, d\mathbf{u}^{(1)}\, d\mathbf{u}^{(2)}\ldots d\mathbf{u}^{(N)}$$

$$= \iint f^{(2)}(\mathbf{x}^{(1)}, \mathbf{x}^{(2)}; \mathbf{u}^{(1)}, \mathbf{u}^{(2)}; t)\, d\mathbf{u}^{(1)}\, d\mathbf{u}^{(2)}. \tag{16}$$

In deriving the foregoing equations the further simplifying assumption has been made that all the particles have the same mass m.

From equation (15) it is clear that it is only when we ignore the two-particle correlations and assume that

$$n^{(1,\,2)} = \frac{1}{N}\, n^{(1)}(\mathbf{x}^{(1)}; t)\, n^{(1)}(\mathbf{x}^{(2)}; t), \tag{17}$$

does this new definition of $\mathfrak{W}_{ij}$ coincide with the common definition (9).

When there are no mass motions and $\mathbf{U}^{(1)} = 0$, equation (13) takes the more familiar form

$$\frac{1}{2}\frac{d^2 I_{ij}}{dt^2} = \Pi_{ij} + \mathfrak{W}_{ij}, \tag{18}$$

where

$$\Pi_{ij} = \int \rho^{(1)}(\mathbf{x}^{(1)}, t)\langle u_i^{(1)} u_j^{(1)}\rangle\, d\mathbf{x}^{(1)}, \tag{19}$$

and

$$\mathfrak{W}_{ij} = -\tfrac{1}{2}G \iint \rho^{(1,\,2)}(\mathbf{x}^{(1)}, \mathbf{x}^{(2)}; t)\frac{[x_i^{(1)} - x_i^{(2)}][x_j^{(1)} - x_j^{(2)}]}{|\mathbf{x}^{(1)} - \mathbf{x}^{(2)}|^3}\, d\mathbf{x}^{(1)}\, d\mathbf{x}^{(2)}. \tag{20}$$

And these last equations are equivalent to equations (3), (8) and (9) only if we ignore two-point correlations and assume that

$$\rho^{(1,\,2)}(\mathbf{x}^{(1)}, \mathbf{x}^{(2)}; t) = \rho^{(1)}(\mathbf{x}^{(1)}, t)\, \rho^{(1)}(\mathbf{x}^{(2)}, t). \tag{21}$$

In Sections 3 and 4 which follow, the two-point correlations will be ignored and the analysis will be based on equations (3) and (6) together with the definitions (8) and (9). In Section 5 a first attempt will be made to include the two-point correlations.

3. THE DYNAMICAL EVOLUTION OF A SPHERICAL SYSTEM

Suppose that at some initial instant, a cluster is characterized by a certain moment of inertia and kinetic and potential energies. The evolution of such a cluster, while it must be consistent with equation (6), it cannot, of course, be uniquely determined by equation (6). On the other hand, if we choose to describe the cluster ' grossly ' by a certain linear size, then it would be proper to inquire how this size changes with time. Such a description can, at best, give one only the crudest indication of what might happen. The following analysis based on equation (6) is undertaken with the hope that even crude inferences concerning the evolution of isolated clusters (of stars or galaxies) are of some interest.

In the spirit of the foregoing remarks, it should not also matter (within limits) how we choose to describe the internal distribution of density in the cluster. Accordingly, to be specific, we shall choose a spherically symmetric cluster of N equal mass-points with a gaussian distribution of density

$$\rho(r) = \frac{Nm}{a^3\pi^{3/2}} \exp(-r^2/a^2), \tag{22}$$

where $a = a(t)$ is a measure of the radius of the cluster. For clusters in which the distribution (22) continues to be maintained (by assumption!) equation (6) leads to the following differential equation for a:

$$\frac{3}{4} Nm \frac{d^2a^2}{dt^2} = 2\mathfrak{E} + \frac{GN^2m^2}{(2\pi)^{1/2}} \frac{1}{a}. \tag{23}$$

It should be noted here that if instead of a gaussian distribution, we had assumed a homogeneous distribution inside a sphere of radius a, we should have obtained an exactly similar equation with only somewhat different numerical coefficients; and the equation will be *identically* the same in the non-dimensional form given below (equation (25)). The particular assumption (22) is therefore not an essential one so long as we assume that the evolution proceeds homologously.

Letting

$$a = a_0 z \tag{24}$$

where a_0 is the value of a at some time to be specified later and measuring time in the unit

$$t_0 = \left[\left(\frac{9\pi}{8}\right)^{1/2} \frac{a_0^3}{NmG}\right]^{1/2}, \tag{25}$$

we find that equation (23) becomes

$$\frac{d^2z^2}{dt^2} = 2(\operatorname{sgn} \mathfrak{E})\, Q + \frac{1}{z}, \tag{26}$$

where sgn $\mathfrak{E}$ denotes the sign (+ or −) of the constant total energy of the system,

$$Q = \frac{|\mathfrak{E}|}{|\mathfrak{W}_0|}, \tag{27}$$

and $\mathfrak{W}_0$ is the potential energy when $a = a_0$. It is seen that equation (26) admits the first integral

$$z^2 \left(\frac{dz}{dt}\right)^2 = (\text{sgn } \mathfrak{E})\, Qz^2 + z + \text{constant}. \tag{28}$$

At this point it is convenient to distinguish systems with positive or negative total energies.

(a) *Systems with positive total energies*

In this case, suppose that

$$a = a_0, \quad z = 1 \quad \text{and} \quad \frac{dz}{dt} = 0 \quad \text{at time} \quad t = 0. \tag{29}$$

Then equation (28) can be written alternatively in the form

$$z^2 \left(\frac{dz}{dt}\right)^2 = (z-1)(Qz+Q+1). \tag{30}$$

The required solution of this equation is

$$t = \frac{1}{Q}[(z-1)(Qz+Q+1)]^{1/2} - \frac{1}{Q^{3/2}} \tanh^{-1} \left[\frac{Q(z-1)}{Qz+Q+1}\right]^{1/2}; \tag{31}$$

and for the particular case $Q = 0$, the solution is

$$t = \tfrac{2}{3}(z+2)(z-1)^{1/2}. \tag{32}$$

The variations of z with t that follow from equations (31) and (32) are illustrated in Fig. 1 for various initially assigned values of Q. As might be expected, the solutions show that systems with positive total energy are eventually dispersed.

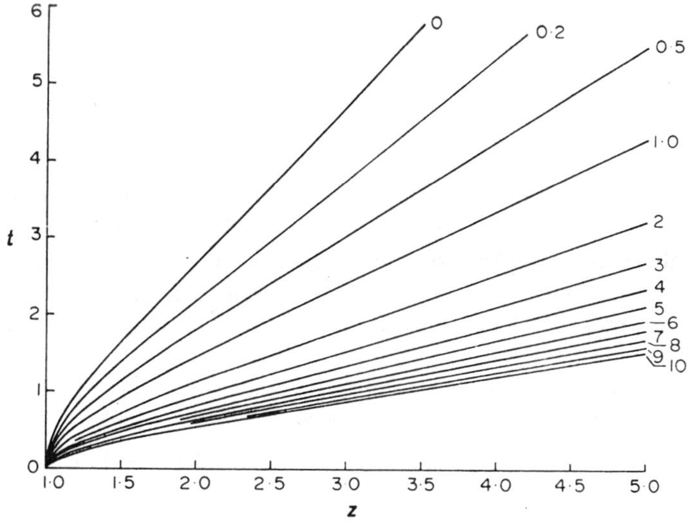

FIG. 1. *The evolution of a spherical cluster with a gaussian distribution of density; the case of positive total energy. The abscissa* (z) *measures the mean radius* (a) *of the cluster in terms of its initial value* (a_0). *The ordinate measures t in the units specified in equation* (25). *The curves are labelled with the values of* Q ($= |\mathfrak{E}|/|\mathfrak{W}|$) *to which they belong.*

(b) *Systems with negative total energies*

In this case, it is convenient to rewrite equation (28) in the form

$$z^2\left(\frac{dz}{dt}\right)^2 = \Lambda^2 - Q\left(z-\frac{1}{2Q}\right)^2, \tag{33}$$

where Λ^2 is now the constant of integration. The solution of equation (33) is

$$t = \frac{1}{Q^{1/2}}\left[\Lambda^2-\left(z-\frac{1}{2Q}\right)^2\right]^{1/2}+\frac{1}{2Q^{3/2}}\cos^{-1}\left[\frac{1}{\Lambda}\left(z-\frac{1}{2Q}\right)\right], \tag{34}$$

if we ignore an inessential arbitrary constant defining the origin of t.

The solution (34) makes it manifest that when the total energy of the system is negative, it executes periodic oscillations of finite amplitude. The period and the amplitude of the oscillation are given by

$$\text{period: } \pi/Q^{3/2} \tag{35}$$

and

$$\text{amplitude of oscillation: } +\Lambda+\frac{1}{2Q} \geqslant z \geqslant -\Lambda+\frac{1}{2Q}. \tag{36}$$

The requirement that $z \geqslant 0$ (by definition) implies that

$$\Lambda \leqslant \frac{1}{2Q}. \tag{37}$$

It can be verified that the critical value $\Lambda = 1/2Q$ corresponds to the system having a vanishing kinetic energy when $t = 0$ and $z = 1/Q$; that under these circumstances the system should collapse to a point—as it does after a time $t = \pi/2Q^{3/2}$—is to be expected.

The variation of z ($= a/a_0$) during a half-period of oscillation is illustrated in Fig. 2 for a number of selected values of Q and Λ.

4. THE DYNAMICAL EVOLUTION OF A SPHEROIDAL SYSTEM

The analysis of Section 3 applies to the case when a single linear dimension provides an adequate description of it. We now turn to the slightly more general case when the cluster departs so much from even a gross spherical symmetry that at least two linear dimensions are required to describe it. Under these circumstances it might suffice to idealize the system as being homogeneous and spheroidal (either oblate or prolate).

We shall suppose then that we are given a system that is initially homogeneous and spheroidal with semi-axes a_1, a_2 ($= a_1$), and a_3 (which is not necessarily the least axis) and a given total energy $\mathfrak{E}$. We shall further suppose that the distribution of velocities is (locally) isotropic so that

$$\mathfrak{T}_{11} = \mathfrak{T}_{22} = \mathfrak{T}_{33} = \tfrac{1}{3}\mathfrak{T}. \tag{38}$$

On these assumptions

$$\mathfrak{E} = \mathfrak{T}+2\mathfrak{W}_{11}+\mathfrak{W}_{33} = \text{constant} \tag{39}$$

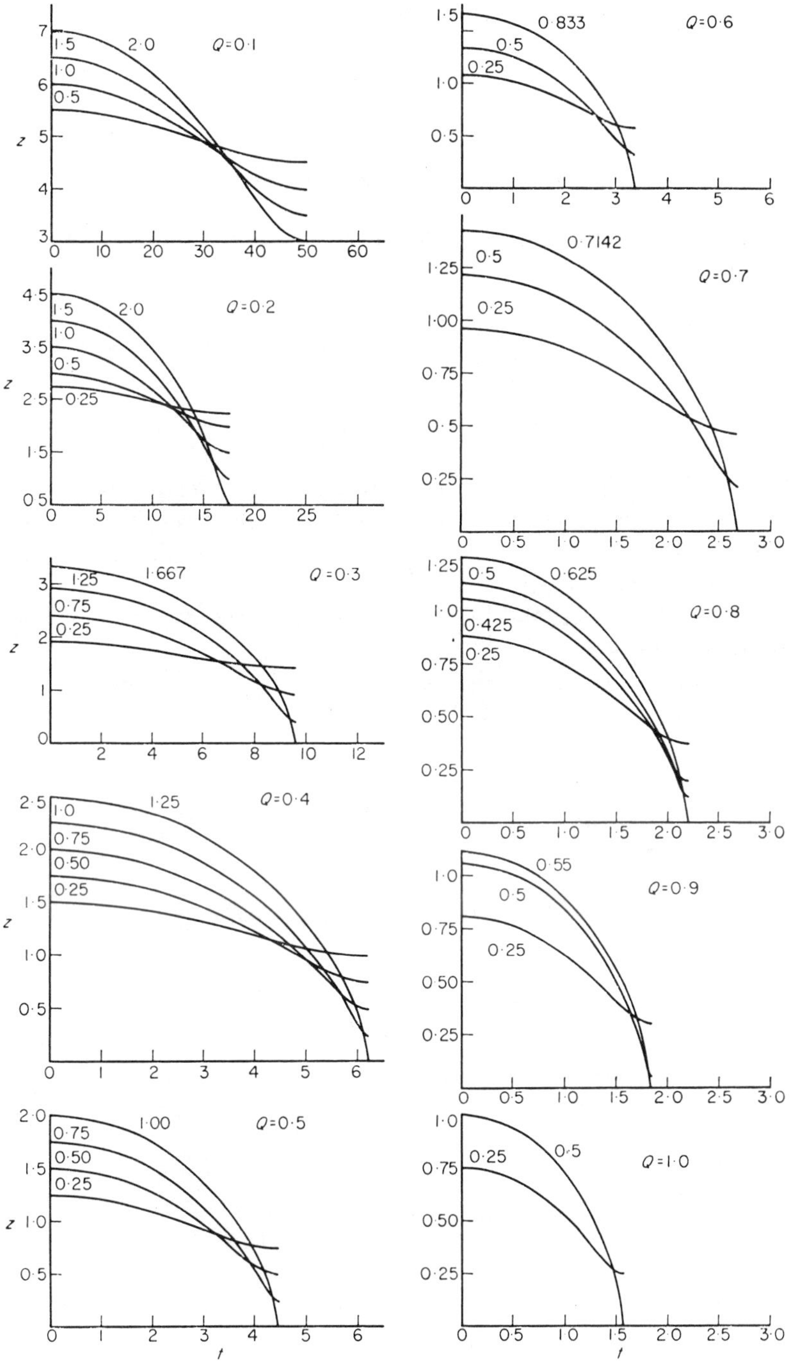

FIG. 2. *The finite amplitude oscillations of a spherical cluster with a gaussian distribution of density; the case of negative total energy. Each panel illustrates the oscillations for a particular value of* Q $(= |\mathfrak{E}|/|\mathfrak{W}|)$*; and in each panel the curves are labelled by the values of* Λ *to which they belong. In each panel the ordinate measures the amplitude* ($z = a/a_0$) *while the abscissa measures the time in the unit specified in equation* (25). *The period and the amplitude of these oscillations are given by* $\pi Q^{-3/2}$ *and* 2Λ, *respectively.*

and the tensor virial-theorem provides the pair of equations

$$\frac{1}{2}\frac{d^2I_{11}}{dt^2} = \tfrac{2}{3}\mathfrak{E} - \tfrac{1}{3}\mathfrak{W}_{11} - \tfrac{2}{3}\mathfrak{W}_{33} \tag{40}$$

and

$$\frac{1}{2}\frac{d^2I_{33}}{dt^2} = \tfrac{2}{3}\mathfrak{E} - \tfrac{4}{3}\mathfrak{W}_{11} + \tfrac{1}{3}\mathfrak{W}_{33}. \tag{41}$$

The assumption that the cluster is homogeneous and spheroidal provides the further relations

$$\mathfrak{W}_{11} = -2\pi G\rho A_1 I_{11} = -\frac{3}{10}\frac{GM^2}{a_3}A_1 \tag{42}$$

and

$$\mathfrak{W}_{33} = -2\pi G\rho A_3 I_{33} = -\frac{3}{10}\frac{GM^2}{a_1}\frac{a_3}{a_1}A_3, \tag{43}$$

where A_1 and A_3 are the 'index symbols' defined in the theory of the gravitational potential of ellipsoids (cf. Chandrasekhar 1969, Section 21). The values of A_1 and A_3 are explicitly known in the case of spheroids (cf. Chandrasekhar 1969, p. 43, equations (36) and (38)).

With the values of $\mathfrak{W}_{11}$ and $\mathfrak{W}_{33}$ given by equations (42) and (43), equations (40) and (41) reduce to

$$\frac{d^2a_1^2}{dt^2} = \frac{20\mathfrak{E}}{3M} + \frac{GM}{a_1}\left(\frac{a_1}{a_3}A_1 + 2\frac{a_3}{a_1}A_3\right) \tag{44}$$

and

$$\frac{d^2a_3^2}{dt^2} = \frac{20\mathfrak{E}}{3M} + \frac{GM}{a_1}\left(4\frac{a_1}{a_3}A_1 - \frac{a_3}{a_1}A_3\right). \tag{45}$$

Making the substitutions

$$a_1^2 = a_0^2 z, \qquad a_3^2 = a_1^2(1-y) = a_0^2 z(1-y), \tag{46}$$

and measuring time in the unit

$$t_0 = \left(\frac{3Ma_0^2}{20|\mathfrak{E}|}\right)^{1/2}, \tag{47}$$

we find that equations (44) and (45) can be reduced to the forms

$$\frac{\mathfrak{E}}{|\mathfrak{E}|}\frac{d^2z}{dt^2} = 1 + \frac{Q}{\sqrt{z}}f_1(y) \tag{48}$$

and

$$\frac{\mathfrak{E}}{|\mathfrak{E}|}z\frac{d^2y}{dt^2} + 2\frac{dz}{dt}\frac{dy}{dt} = -y + \frac{Q}{\sqrt{z}}f_2(y), \tag{49}$$

where

$$Q = \frac{3}{20}\frac{GM^2}{a_0\mathfrak{E}}, \tag{50}$$

and

$$f_1(y) = \frac{3}{y}(1-y)^{1/2} + \frac{4y-3}{y}S(y), \tag{51}$$

$$f_2(y) = \left(\frac{9}{y} - 3\right)(1-y)^{1/2} - \frac{4y^2-9y+9}{y}S(y), \tag{52}$$

and

$$S(y) = \frac{\sin^{-1}\sqrt{y}}{\sqrt{y}} \qquad (y \geqslant 0)$$
$$= \frac{\sinh^{-1}\sqrt{-y}}{\sqrt{-y}} \qquad (y < 0). \tag{53}$$

Equations (48) and (49) have been integrated for various values of Q with the initial conditions

$$z = 1,\ y = y_0 \text{ (assigned), and } \frac{dz}{dt} = \frac{dy}{dt} = 0 \text{ at } t = 0. \tag{54}$$

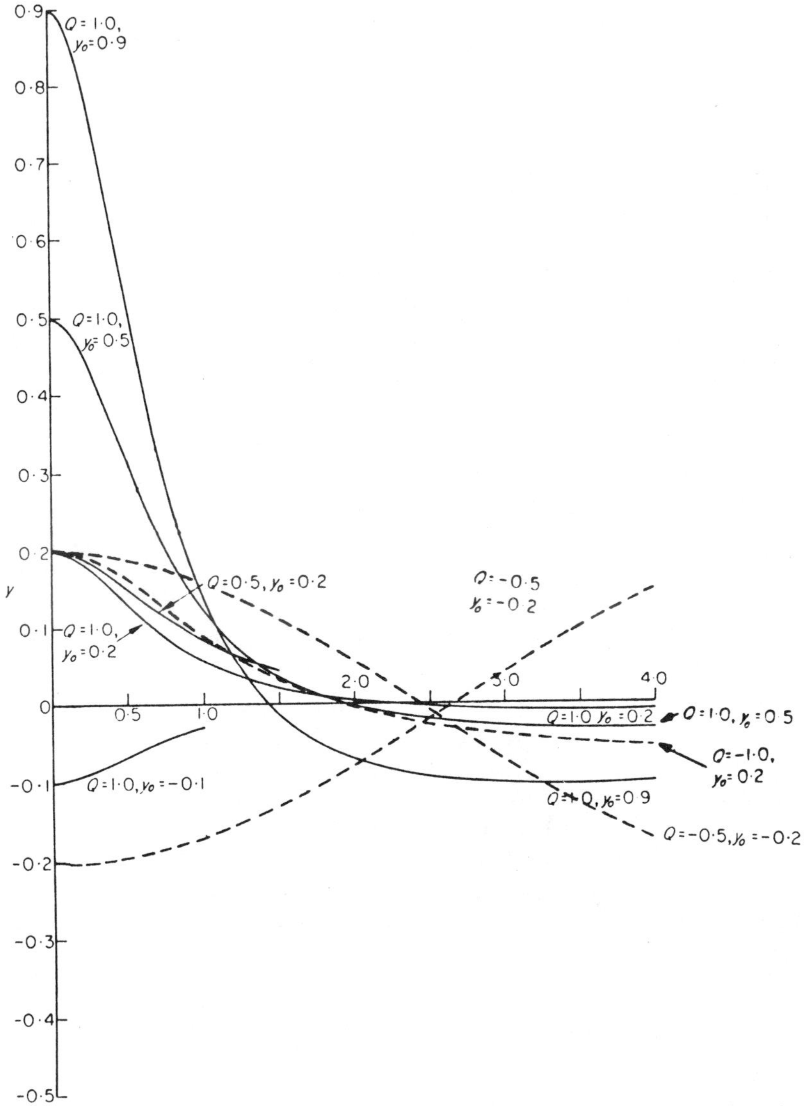

FIG. 3. *The evolution of homogeneous spheroidal clusters for various initially assigned values of Q and y_0. The curves are labelled by the values of Q and y_0 to which they belong. The ordinate y measures the square of the eccentricity (e^2) positive for oblate spheroids and negative for prolate spheroids. The abscissa measures the time in the unit specified in equation (47). The full-line curves refer to clusters of positive total energy while the dashed curves refer to clusters of negative total energy.*

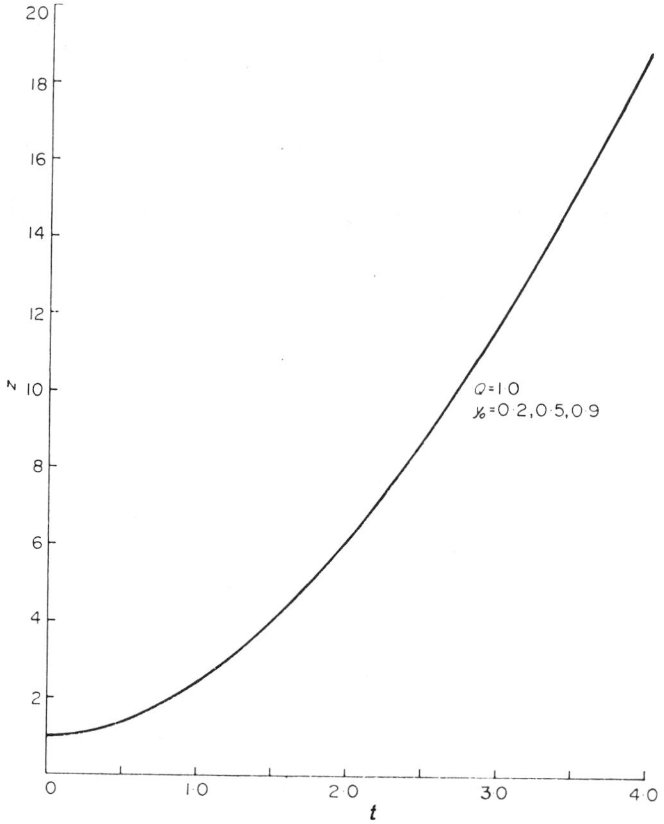

FIG. 4. *The evolution of homogeneous oblate spheroidal clusters. The ordinate* z ($= a_1^2/a_0^2$) *is a measure of the area of the circular equatorial section; and the abscissa measures the time in the same unit as in Fig.* 3. *The curve applies to* $Q = 1\cdot0$ *and* $y_0 = 0\cdot2$, $0\cdot5$, *and* $0\cdot9$. (*The three cases cannot be distinguished in the scale of the figure.*)

The derived solution curves are exhibited in Figs 3, 4 and 5 where it should be noted that $y > 0$ corresponds to oblate spheroids while $y < 0$ corresponds to prolate spheroids. An examination of these solution curves reveals the following features.

Consider first the evolution of systems with positive total energy. The series of integrations for $Q = 1\cdot0$ and $y_0 = 0\cdot9$, $0\cdot5$, and $0\cdot2$ shows that these objects which are initially oblate eventually become prolate (see Fig. 3); and while the eccentricity in all these cases appears to tend to a finite limit the object itself grows considerably in size (see Fig. 4). It will be observed that in all these three cases, the variations of z ($= a_1^2/a_0^2$) with time are so nearly the same that they cannot be distinguished in the scale of Fig. 4. For the same value of Q ($= 1\cdot0$), if the object is initially prolate ($y_0 = -0\cdot2$) it evolves towards a less prolate and a more spherical shape as it grows in size (see Figs 3 and 5).

For the case $Q = 0\cdot5$, $y_0 = 0\cdot2$, the behaviour is essentially the same as for $Q = 1\cdot0$, $y_0 = 0\cdot2$ except that the object continues to remain oblate though of considerably reduced eccentricity (see Figs 3 and 5).

Considering next the integrations for systems with negative total energy ($Q = -0\cdot5$, $y_0 = \pm0\cdot2$ and $Q = -1\cdot0$, $y_0 = +0\cdot2$) we observe (see Fig. 3) that an object which is initially oblate becomes increasingly more oblate while

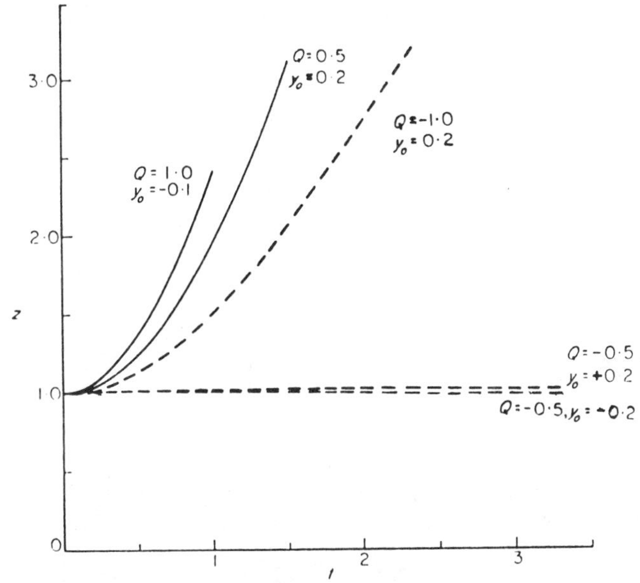

FIG. 5. *Same as Fig. 4 but for other initially assigned values of Q and y_0. The curves are labelled appropriately. The full-line curves refer to clusters of positive total energy while the dashed curves refer to clusters of negative total energy.*

an object which is initially prolate becomes increasingly more prolate. This last result is in agreement with an earlier investigation of Lin, Mestel & Shu (1965) in which the collapse of an object initially at rest with zero kinetic-energy is considered. For the case $Q = -0{\cdot}5$ the increase in the oblateness or prolateness is not accompanied by any substantial changes in the length of the a_1 ($= a_2$)-axis. But in the case $Q = -1{\cdot}0$, $y_0 = +0{\cdot}2$, the increase in the oblateness is accompanied by a substantial decrease in a_1 ($= a_2$).

5. A SEMI-EMPIRICAL ALLOWANCE FOR TWO-POINT CORRELATIONS

As we have noted in Section 2, the allowance for two-point correlations in the present context modifies only the expression for the potential energy. And we shall try to allow for it in a semi-empirical fashion by assuming that

$$\rho^{(1,\,2)}(\mathbf{x}^{(1)}, \mathbf{x}^{(2)}) = \rho^{(1)}(\mathbf{x}^{(1)})\, \rho^{(1)}(\mathbf{x}^{(2)})[1+\kappa \exp(-k|\mathbf{x}^{(1)}-\mathbf{x}^{(2)}|)], \qquad (55)$$

where κ is a constant and k is a measure of the *correlation length*. On this assumption, we may define a *correlation potential* (of the Yukawa form)

$$\phi(\mathbf{x}) = \kappa G \int \rho(\mathbf{x}') \frac{\exp(-k|\mathbf{x}-\mathbf{x}'|)}{|\mathbf{x}-\mathbf{x}'|}\, d\mathbf{x}' \qquad (56)$$

and a *correlation energy*

$$\mathfrak{W}_{\text{COR.}} = -\tfrac{1}{2}\int \rho\phi\, d\mathbf{x} \qquad (57)$$

that must be added to the Newtonian potential energy $\mathfrak{W}_N$ to obtain the true potential energy.

For spherically symmetric systems it can be readily shown that

$$\phi(r) = \frac{2\pi G\kappa}{k^2}\frac{1}{r}\left(\exp(-r)\int_0^r \exp(-z)\,\rho z\,dz\right.$$
$$\left.-\exp(+r)\int_0^r \exp(-z)\,\rho z\,dz + 2\sinh r\int_0^\infty \exp(-z)\,\rho z\,dz\right), \qquad (58)$$

where r is measured in the unit $1/k$. For the two cases

$$\text{(i)}\ \rho = \rho_0 = \text{constant and (ii)}\ \rho = \rho_0 \exp(-r^2/a^2) \qquad (59)$$

we find that

$$\phi(r) = \frac{4\pi G\kappa\rho_0}{k^2}\left[1 - \frac{\sinh r}{r}(R+1)\exp(-R)\right] \qquad \text{(case i)} \quad (60)$$

and

$$\phi(r) = \frac{2\pi G\kappa\rho_0}{k^2}\left(\frac{\sqrt{\pi}}{4}a^3\exp(a^2/4)\right)\frac{1}{r}\left[\Phi\left(\frac{r}{a}-\frac{a}{2}\right)\exp(-r)\right.$$
$$\left.+\Phi\left(\frac{r}{a}+\frac{a}{2}\right)\exp(+r) - 2\sinh r\right] \qquad \text{(case ii)}, \quad (61)$$

where r, a, and R (the radius of the sphere in case i) are all measured in the unit $1/k$ and

$$\Phi(x) = \frac{2}{\sqrt{\pi}}\int_0^x \exp(-z^2)\,dz \qquad (62)$$

is the error function. The corresponding expressions for the correlation energy are

$$\frac{\mathfrak{W}_{\text{COR.}}}{\mathfrak{W}_N} = \kappa\frac{9}{R^5}\left[\tfrac{1}{3}R^3 - (R\cosh R - \sinh R)\exp(-R)(R+1)\right] \qquad \text{(case i)} \quad (63)$$

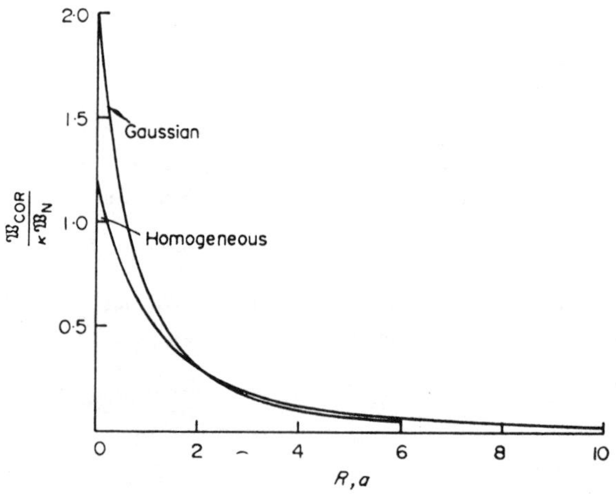

FIG. 6. *The correlation energy for clusters with homogeneous and gaussian distributions of density for a Yukawa-type potential. The abscissa measures, in units of the correlation length, the radius R (in the case of a homogeneous cluster) and the mean radius a (in the case of a cluster of gaussian distribution). The ordinate gives the ratio of the correlation energy to the Newtonian potential energy.*

and

$$\frac{\mathfrak{W}_{\text{COR.}}}{\mathfrak{W}_N} = \kappa\left(\frac{\pi}{2}\right)^{1/2}\exp(a^2/2)\int_0^\infty [1+\Phi(x-\tfrac{1}{2}a)] \times[1-\Phi(x+\tfrac{1}{2}a)]\,dx \quad \text{(case ii)}, \quad (64)$$

where $\mathfrak{W}_N$ denotes the Newtonian potential energy (without correlations) in the two cases. Fig. 6 exhibits the functions $\mathfrak{W}_{\text{COR.}}/\mathfrak{W}_N$ defined by the foregoing equations.

The effect of the correlation energy on the finite amplitude oscillations of spherical systems considered in 3(b) has been examined on the assumptions that κ in equation (55) is 0·1 and that a and $1/k$ are equal and remain equal during the oscillation. The results of these new calculations are contrasted with those of Section 3(b) in Fig. 7. It will be observed that the qualitative features of the solution are not affected.

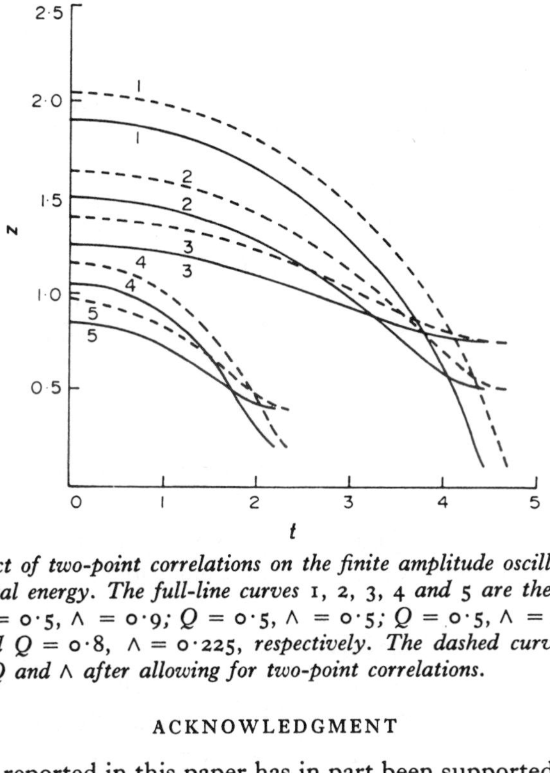

FIG. 7. *The effect of two-point correlations on the finite amplitude oscillations of clusters with negative total energy. The full-line curves 1, 2, 3, 4 and 5 are the cases illustrated in Fig. 3 for Q = 0·5, Λ = 0·9; Q = 0·5, Λ = 0·5; Q = 0·5, Λ = 0·25; Q = 0·8, Λ = 0·425; and Q = 0·8, Λ = 0·225, respectively. The dashed curves refer to these same values of Q and Λ after allowing for two-point correlations.*

ACKNOWLEDGMENT

The research reported in this paper has in part been supported by the National Science Foundation under contract GP–28342 with the University of Chicago.

University of Chicago, Chicago, Illinois, U.S.A.

REFERENCES

Chandrasekhar, S., 1964. *Lectures in Theoretical Physics*, ed. W. E. Brittin, Vol. 6, p. 1. University of Colorado Press, Boulder.
Chandrasekhar, S., 1969. *Ellipsoidal Figures of Equilibrium*, Yale University Press.
Chandrasekhar, S. & Lee, E. P., 1968. *Mon. Not. R. astr. Soc.*, **139**, 135.
Lin, C. C., Mestel, L. & Shu, F. H., 1965. *Astrophys. J.*, **142**, 1431.

The stability of a rotating liquid drop

By S. Chandrasekhar, F.R.S.
University of Chicago

(*Received* 18 *September* 1964)

In this paper, the stability of a rotating drop held together by surface tension is investigated by an appropriate extension of the method of the tensor virial. Consideration is restricted to axisymmetric figures of equilibrium which enclose the origin. These figures form a one-parameter sequence; and a convenient parameter for distinguishing the members of the sequence is $\Sigma = \rho\Omega^2a^3/8T$, where Ω is the angular velocity of rotation, a is the equatorial radius of the drop, ρ is its density, and T is the interfacial surface tension. It is shown that $\Sigma \leqslant 2{\cdot}32911$ (*not* $1+\surd 2$ as is sometimes supposed) if the drop is to enclose the origin. It is further shown that with respect to stability, the axisymmetric sequence of rotating drops bears a remarkable similarity to the Maclaurin sequence of rotating liquid masses held together by their own gravitation. Thus, at a point along the sequence (where $\Sigma = 0{\cdot}4587$) a neutral mode of oscillation occurs *without* instability setting in at that point (i.e. provided no dissipative mechanism is present); and the instability actually sets in at a subsequent point (where $\Sigma = 0{\cdot}8440$) by overstable oscillations with a frequency Ω. The dependence on Σ of the six characteristic frequencies, belonging to the second harmonics, is determined (tables 3 and 4) and exhibited (figures 3 and 4).

1. Introduction

The interpretation of phenomena related to nuclear fission has revived interest in the classical problem of the equilibrium and the stability of a rotating liquid drop held together by surface tension. This problem was the subject of a remarkable series of experiments by Plateau (1843–69); and certain aspects of the associated theoretical problem of the allowed figures of equilibrium were investigated by Rayleigh (1914; see also Poincaré 1895 and Appell 1932). A brief account of the equilibrium theory, which carries the analytical treatment somewhat further than Rayleigh's, is given in the Appendix.

If one restricts oneself to axisymmetric forms and uniform density, the figure of equilibrium depends on the value of the non-dimensional parameter

$$\Sigma = \rho\Omega^2a^3/8T, \tag{1}$$

where Ω denotes the angular velocity of rotation, a the equatorial radius of the drop, ρ its density, and T the interfacial surface tension. A sequence of equilibrium figures derived on the basis of the formulae given in the Appendix is illustrated in figure 1. When $\Sigma \to 0$, the figure tends to an oblate spheroid with an ellipticity ϵ given by (see equation (48) below)

$$\epsilon = \Sigma \quad (\Sigma \to 0). \tag{2}$$

However, as Σ increases, the figure rapidly departs from the spheroidal form: at $\Sigma = 1$, the polar regions are flat; and as Σ increases beyond 1, the drop develops a dimple that deepens quickly and reduces the polar axis to evanescence when $\Sigma = 2{\cdot}3291$; and for $\Sigma > 2{\cdot}3291$, the equilibrium figures no longer enclose the origin and are tori. The stability of these axisymmetric figures has never been investigated properly. Indeed, as we shall see, the supposed onset of instability at $\Sigma = 1$ (cf. Sperber 1962) is incorrect.

In this paper, the stability of the rotating liquid drop will be investigated. It will appear that the results are remarkably similar to those along the Maclaurin sequence of rotating incompressible masses in gravitational equilibrium. Thus, at a point along the sequence (where $\Sigma = 0{\cdot}4587$) a neutral mode of oscillation occurs *without* instability setting in at that point (i.e. provided no dissipative mechanism is present); and the instability actually sets in at a subsequent point (where $\Sigma = 0{\cdot}8440$) by overstable oscillations with a frequency equal to the angular

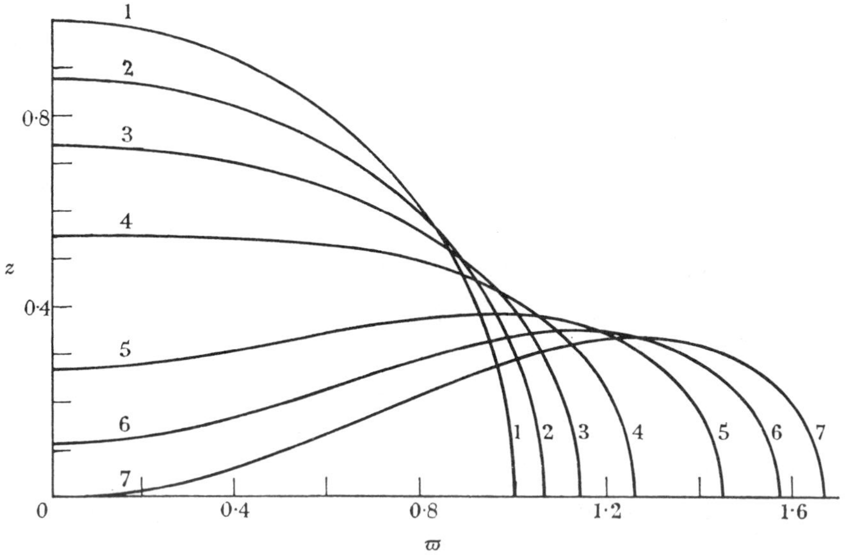

FIGURE 1. The figures of equilibrium of a rotating liquid drop for various values of the parameter Σ. The curves labelled 1, 2, ..., 7 are for $\Sigma = 0$, 0·210478, 0·510790, 1·0, 1·72598, 2·09589, and 2·32911 ($= \Sigma_{max.}$), respectively. The radial co-ordinates have been normalized in such a way that the volumes of all the figures are the same. The curve labelled 1 is, of course, a quadrant of a circle; the curve labelled 4 (for $\Sigma = 1$) is flat at the poles; and the curve labelled 7 for $\Sigma_{max.}$ is the limiting figure enclosing the origin.

velocity of rotation. (Along the Maclaurin sequence, which is a sequence of oblate spheroids, the corresponding points occur at eccentricities of the spheroid equal to $e = 0{\cdot}8127$ and 0·9529, respectively.) At its first neutral point (at $\Sigma = 0{\cdot}4587$) the rotating drop must be 'secularly' unstable (i.e. unstable if some dissipative mechanism is present), and one may anticipate that at this point a stable sequence of triaxial forms branches off in a manner analogous to the way the Jacobian sequence of ellipsoids branches off from the Maclaurin sequence of spheroids.

In many ways it is remarkable that even though Plateau's original investigations were explicitly motivated by 'cosmogonical' considerations, the almost complete similarity of Plateau's problem to the one, then, under active theoretical consideration by Jacobi, Riemann, Poincaré, Bryan, and Darwin (cf. Jeans 1919) seems to have been overlooked. Apparently, the emergence of a sequence of toroidal figures of equilibrium (for $\Sigma > 2{\cdot}3291$) and its striking similarity (in appearance!) with systems such as Saturn's rings succeeded in drawing attention away from the real significance of Plateau's problem to the astronomical problem of the time.

In concluding this introductory section, it may be stated that, on the analytical side, the method that will be developed in this paper is an extension, to allow for capillary forces, of the *method of the virial* (Chandrasekhar 1960, 1961) which has found applications to a wide variety of problems in the theory of homogeneous masses in gravitational equilibrium (see the various papers by Chandrasekhar and by Chandrasekhar & Lebovitz in the *Astrophysical Journal* of the past four years; and for a summary of the principal ideas see Chandrasekhar 1964.)

2. The virial theorem; the surface-energy tensor

Consider a fluid confined to a volume V bounded by a *closed surface* $\mathbf{S}$ on which surface tension (of magnitude T) is operative. To keep the present development at its simplest level, we shall suppose that the external pressure on $\mathbf{S}$ is zero. In virtue of surface tension, the pressure in the interior, immediately adjacent to $\mathbf{S}$, is given by

$$p = T \operatorname{div} \mathbf{n} \quad (\text{on } \mathbf{S}_{-}), \tag{3}$$

where $\mathbf{n}$ is the unit outward normal on $\mathbf{S}$. Interior to $\mathbf{S}$ and in the volume V the equation of motion governing the fluid is

$$\rho \frac{\mathrm{d}u_i}{\mathrm{d}t} = -\frac{\partial p}{\partial x_i} \quad (\text{inside } V), \tag{4}$$

where u_i denotes the velocity and p is the pressure.

To obtain the form of the virial theorem appropriate to the problem on hand, we multiply equation (4) by x_j and integrate over the volume V. By familiar transformations, we obtain

$$\frac{\mathrm{d}}{\mathrm{d}t}\int_V \rho u_i x_j \,\mathrm{d}\mathbf{x} = 2\mathfrak{T}_{ij} - \int_V x_j \frac{\partial p}{\partial x_i}\,\mathrm{d}\mathbf{x}, \tag{5}$$

where

$$\mathfrak{T}_{ij} = \frac{1}{2}\int_V \rho u_i u_j \,\mathrm{d}\mathbf{x} \tag{6}$$

is the kinetic-energy tensor. The second term on the right-hand side of equation (5), after an integration by parts, gives

$$\int_V x_j \frac{\partial p}{\partial x_i}\,\mathrm{d}\mathbf{x} = \int_{\mathbf{S}} x_j p \,\mathrm{d}S_i - \delta_{ij}\int_V p\,\mathrm{d}\mathbf{x}, \tag{7}$$

where

$$\mathrm{d}S_i = n_i \,\mathrm{d}S, \tag{8}$$

$\mathrm{d}S$ being an element of area of $\mathbf{S}$. By making use of the condition (3) obtaining on $\mathbf{S}$, we have

$$\int_V x_j \frac{\partial p}{\partial x_i}\,\mathrm{d}\mathbf{x} = T\int_{\mathbf{S}} x_j \operatorname{div}\mathbf{n}\,\mathrm{d}S_i - \delta_{ij}\int_V p\,\mathrm{d}\mathbf{x}. \tag{9}$$

Equation (5) can now be written in the form

$$\frac{\mathrm{d}}{\mathrm{d}t}\int_V \rho u_i x_j \,\mathrm{d}\mathbf{x} = 2\mathfrak{T}_{ij} + \mathfrak{S}_{ij} + \delta_{ij}\Pi, \tag{10}$$

where
$$\Pi = \int_V p \, d\mathbf{x} \tag{11}$$
and
$$\mathfrak{S}_{ij} = -T \int_{\mathbf{S}} x_j \operatorname{div} \mathbf{n} \, dS_i. \tag{12}$$

We shall call $\mathfrak{S}_{ij}$ *the surface-energy tensor.*

From equation (10) it is apparent that *the conservation of the angular momentum of the fluid requires the surface-energy tensor $\mathfrak{S}_{ij}$ to be symmetric.* While $\mathfrak{S}_{ij}$, as defined by equation (12), is not manifestly symmetric in its indices, its symmetry, nevertheless, can be demonstrated. I am indebted to Professor G. Wentzel for the following demonstration.

LEMMA. *Let* **S** *be a surface bounded by a simple closed contour* C. *Then*
$$\mathbf{L} = \oint_C \mathbf{x} \times (d\mathbf{x} \times \mathbf{n}) = -\int_{\mathbf{S}} (\mathbf{x} \times \mathbf{n}) \operatorname{div} \mathbf{n} \, dS. \tag{13)*$$

Proof. Let (i, j, k) represent the indices $(1, 2, 3)$ in some cyclic order. Then, suspending the summation convention throughout this proof, we have for the ith component of **L**
$$L_i = \oint_C [dx_i(n_j x_j + n_k x_k) + dx_j(-n_i x_j) + dx_k(-n_i x_k)]. \tag{14}$$

Transforming this line integral around C to a surface integral over **S** by Stokes's theorem, we obtain
$$\begin{aligned} L_i = \int_{\mathbf{S}} \Bigg\{ & n_i \left[\frac{\partial}{\partial x_j}(-n_i x_k) - \frac{\partial}{\partial x_k}(-n_i x_j)\right] \\ & + n_j \left[\frac{\partial}{\partial x_k}(n_j x_j + n_k x_k) - \frac{\partial}{\partial x_i}(-n_i x_k)\right] \\ & + n_k \left[\frac{\partial}{\partial x_i}(-n_i x_j) - \frac{\partial}{\partial x_j}(n_j x_j + n_k x_k)\right]\Bigg\} dS. \end{aligned} \tag{15}$$

On carrying out the requisite differentiations, we find after some rearrangement,
$$\left.\begin{aligned} L_i &= \int_{\mathbf{S}} \bigg[-\tfrac{1}{2} x_k \frac{\partial}{\partial x_j}(n_i^2 + n_k^2) + \tfrac{1}{2} x_j \frac{\partial}{\partial x_k}(n_i^2 + n_j^2) \\ &\qquad + n_j x_k \left(\frac{\partial n_i}{\partial x_i} + \frac{\partial n_k}{\partial x_k}\right) - n_k x_j \left(\frac{\partial n_i}{\partial x_i} + \frac{\partial n_j}{\partial x_j}\right)\bigg] dS, \\ &= \int_{\mathbf{S}} \left[\tfrac{1}{2}\left(x_j \frac{\partial}{\partial x_k} - x_k \frac{\partial}{\partial x_j}\right) |\mathbf{n}|^2 - (x_j n_k - x_k n_j) \operatorname{div} \mathbf{n}\right] dS; \end{aligned}\right\} \tag{16}$$

or, since $|\mathbf{n}|^2 = 1$,
$$L_i = -\int_{\mathbf{S}} (x_j n_k - x_k n_j) \operatorname{div} \mathbf{n} \, dS; \tag{17}$$

and this establishes the lemma.

* Contrary to the definition of **S** that has been used up to the present (and that will be used subsequently) we are, in the statement of this lemma and during the course of its proof, letting **S** be an open surface bounded by a finite closed contour C.

Now let the surface $\mathbf{S}$ in the statement of the lemma tend to become closed. Then the contour C will tend to evanescence and L_i, defined as the line integral in equation (14), will tend to zero; and from its equivalent definition as a surface integral in equation (17) it follows that

$$\int_{\mathbf{S}} x_j \operatorname{div} \mathbf{n} \, \mathrm{d}S_k = \int_{\mathbf{S}} x_k \operatorname{div} \mathbf{n} \, \mathrm{d}S_j \quad (\mathbf{S} \text{ a closed surface}); \tag{18}$$

and this establishes the symmetry (in fact) of the surface-energy tensor.

Returning to equation (10), we observe that under conditions of a steady state

$$2\mathfrak{T}_{ij} + \mathfrak{S}_{ij} + \delta_{ij} \Pi = 0; \tag{19}$$

and if, in addition, no macroscopic motions are present

$$\mathfrak{S}_{ij} = -\delta_{ij} \Pi. \tag{20}$$

The contracted versions of equations (19) and (20) are

$$2\mathfrak{T} + \mathfrak{S} + 3\Pi = 0 \quad \text{and} \quad \mathfrak{S} = -3\Pi, \tag{21}$$

where $\mathfrak{S} = \mathfrak{S}_{ii}$ is the surface energy and $\mathfrak{T} = \mathfrak{T}_{ii}$ is the kinetic energy.

3. The virial theorem for a rotating drop

In a frame of reference rotating uniformly with an angular velocity Ω about the x_3 axis, the equations of motion take the form

$$\rho \frac{\mathrm{d}u_i}{\mathrm{d}t} = -\frac{\partial p}{\partial x_i} + \rho \Omega^2 (x_i - \delta_{i3} x_3) + 2\rho \Omega \epsilon_{il3} u_l. \tag{22}$$

By multiplying equation (22) by x_j and integrating over the volume V occupied by the fluid, we obtain, by the same sequence of transformations that led to equation (10),

$$\frac{\mathrm{d}}{\mathrm{d}t} \int_V \rho u_i x_j \, \mathrm{d}\mathbf{x} = 2\mathfrak{T}_{ij} + \mathfrak{S}_{ij} + \Omega^2 (I_{ij} - \delta_{i3} I_{3j}) + \delta_{ij} \Pi + 2\Omega \epsilon_{il3} \int_V \rho u_l x_j \, \mathrm{d}\mathbf{x}, \tag{23}$$

where

$$I_{ij} = \int_V \rho x_i x_j \, \mathrm{d}\mathbf{x} \tag{24}$$

is the moment of inertia tensor.

When no relative motions are present in the chosen frame of reference and a state of hydrostatic equilibrium prevails, equation (23) gives

$$\mathfrak{S}_{ij} + \Omega^2 (I_{ij} - \delta_{i3} I_{3j}) = -\delta_{ij} \Pi. \tag{25}$$

Writing out equation (25) explicitly for the different components, we have

$$\mathfrak{S}_{11} + \Omega^2 I_{11} = \mathfrak{S}_{22} + \Omega^2 I_{22} = \mathfrak{S}_{33} = -\Pi, \tag{26}$$

$$\mathfrak{S}_{12} + \Omega^2 I_{12} = \mathfrak{S}_{21} + \Omega^2 I_{21} = 0, \tag{27}$$

$$\mathfrak{S}_{13} + \Omega^2 I_{13} = \mathfrak{S}_{31} = 0, \quad \text{and} \quad \mathfrak{S}_{23} + \Omega^2 I_{23} = \mathfrak{S}_{32} = 0. \tag{28}$$

From the symmetry of the tensors $\mathfrak{S}_{ij}$ and I_{ij}, it follows from equations (28) that

$$\mathfrak{S}_{13} = \mathfrak{S}_{23} = 0 \quad \text{and} \quad I_{13} = I_{23} = 0. \tag{29}$$

But it is not required that $\mathfrak{S}_{12}$ and I_{12} also vanish; all that is required is

$$\mathfrak{S}_{12} = -\Omega^2 I_{12}. \tag{30}$$

It is often convenient to eliminate Π from equations (26) and obtain two integral relations that are in no way dependent on any constitutive relations that may exist. Two such integral relations are

$$\mathfrak{S}_{11} - \mathfrak{S}_{22} + \Omega^2(I_{11} - I_{22}) = 0 \tag{31}$$

and

$$\mathfrak{S}_{11} + \mathfrak{S}_{22} - 2\mathfrak{S}_{33} + \Omega^2(I_{11} + I_{22}) = 0. \tag{32}$$

It is particularly to be noted that *equations* (30) *to* (32) *do not require* $I_{11} = I_{22}$ *and* $\mathfrak{S}_{11} = \mathfrak{S}_{22}$ *as necessary conditions of equilibrium; and neither is it required that* $I_{12} = 0$ *and* $\mathfrak{S}_{12} = 0$. The most that can be deduced about the tensors $\mathfrak{S}_{ij}$ and I_{ij} (apart from their known symmetry) is that they are reducible to the forms

$$\mathfrak{S}_{ij} = \begin{pmatrix} \mathfrak{S}_{11} & \mathfrak{S}_{12} & 0 \\ \mathfrak{S}_{21} & \mathfrak{S}_{22} & 0 \\ 0 & 0 & \mathfrak{S}_{33} \end{pmatrix} \quad \text{and} \quad I_{ij} = \begin{pmatrix} I_{11} & I_{12} & 0 \\ I_{21} & I_{22} & 0 \\ 0 & 0 & I_{33} \end{pmatrix}. \tag{33}$$

It is important to observe that the virial equations provide no substance to the common expectation that symmetry about the rotational axis should be associated, necessarily, with any form produced by an axisymmetric rotational field. Indeed, we may expect that along a sequence of axisymmetric forms *a point of bifurcation* occurs where a sequence of genuine triaxial forms branches off.

4. Relations obtaining for axisymmetric figures of equilibrium

An axisymmetric figure of equilibrium can be uniquely specified by giving the equation of its meridional section. Let this equation, in the positive hemisphere ($x_3 > 0$) be

$$z = x_3 = f(\varpi), \tag{34}$$

where

$$\varpi = \sqrt{(x_1^2 + x_2^2)} \tag{35}$$

is the perpendicular distance from the axis of rotation.

The equatorial plane of a rotating configuration may in all cases be expected to be a plane of symmetry.* Accordingly, to be valid over the whole sphere, we should write

$$x_3 = \pm f(\varpi). \tag{36}$$

However, to avoid ambiguities in sign which will arise if equation (36) is explicitly used, we shall (when it is necessary to use this equation) restrict our attention to the positive hemisphere and use equation (34). And to avoid misunderstanding,

* For configurations in equilibrium under their own gravitation, a proof of this symmetry about the equatorial plane has been given by Lichtenstein (1933); an analogous proof, valid for configurations in equilibrium under capillary forces, may be constructed.

we shall distinguish integrals extended (under the imposed restriction) over the positive hemisphere by specifying the domains of integrations as V^+ or $\mathbf{S}^+$ as the case may be.

Letting

$$\phi = \mathrm{d}f/\mathrm{d}\varpi, \tag{37}$$

we can write (on $\mathbf{S}^+$)

$$n_1 = -\frac{\phi}{\sqrt{(1+\phi^2)}}\frac{x_1}{\varpi}, \quad n_2 = -\frac{\phi}{\sqrt{(1+\phi^2)}}\frac{x_2}{\varpi}, \quad \text{and} \quad n_3 = \frac{1}{\sqrt{(1+\phi^2)}}; \tag{38}$$

and correspondingly (also on $\mathbf{S}^+$)

$$\mathrm{d}S_1 = -\phi x_1 \,\mathrm{d}\varpi\,\mathrm{d}\theta, \quad \mathrm{d}S_2 = -\phi x_2\,\mathrm{d}\varpi\,\mathrm{d}\theta, \quad \text{and} \quad \mathrm{d}S_3 = \varpi\,\mathrm{d}\varpi\,\mathrm{d}\theta, \tag{39}$$

where $\mathrm{d}\theta$ is an element of the cylindrical polar angle θ.

From the expressions for the components of $\mathbf{n}$ given in (38), we readily verify that

$$\operatorname{div}\mathbf{n} = -\frac{\phi}{\varpi(1+\phi^2)^{\frac{1}{2}}} - \frac{\phi'}{(1+\phi^2)^{\frac{3}{2}}} = -\frac{1}{\varpi}\frac{\mathrm{d}}{\mathrm{d}\varpi}\frac{\varpi\phi}{(1+\phi^2)^{\frac{1}{2}}}, \tag{40}$$

where a dash denotes differentiation with respect to ϖ.

With the aid of equations (38) and (39) we find (cf. equation (12))

$$\begin{aligned}\mathfrak{S}_{11} &= -2T\int_{\mathbf{S}^+} x_1 \operatorname{div}\mathbf{n}\,\mathrm{d}S_1 = 2T\int_0^{2\pi}\int_0^a \phi x_1^2 \operatorname{div}\mathbf{n}\,\mathrm{d}\varpi\,\mathrm{d}\theta \\ &= 2\pi T\int_0^a \phi\varpi^2 \operatorname{div}\mathbf{n}\,\mathrm{d}\varpi;\end{aligned} \tag{41}$$

and $\mathfrak{S}_{22}$ has this same value. On the other hand,

$$\begin{aligned}\mathfrak{S}_{33} &= -2T\int_{\mathbf{S}^+} x_3 \operatorname{div}\mathbf{n}\,\mathrm{d}S_3 = -2T\int_0^{2\pi}\int_0^a f(\varpi)\operatorname{div}\mathbf{n}\varpi\,\mathrm{d}\varpi\,\mathrm{d}\theta \\ &= -4\pi T\int_0^a f(\varpi)\,\varpi\operatorname{div}\mathbf{n}\,\mathrm{d}\varpi.\end{aligned} \tag{42}$$

Hence

$$\begin{aligned}\mathfrak{S}_{33}-\mathfrak{S}_{11} &= -2\pi T\int_0^a\left(\varpi^2\frac{\mathrm{d}f}{\mathrm{d}\varpi}+2\varpi f\right)\operatorname{div}\mathbf{n}\,\mathrm{d}\varpi \\ &= -2\pi T\int_0^a(\varpi^2 f)'\operatorname{div}\mathbf{n}\,\mathrm{d}\varpi.\end{aligned} \tag{43}$$

And since

$$\begin{aligned}I_{11} &= 2\rho\int_{V^+} x_1^2\,\mathrm{d}\mathbf{x} = 2\pi\rho\int_0^a \mathrm{d}\varpi\varpi^3\int_0^{f(\varpi)}\mathrm{d}x_3 \\ &= 2\pi\rho\int_0^a \varpi^3 f(\varpi)\,\mathrm{d}\varpi = -\pi\rho\int_0^a(\varpi^2 f)'\varpi^2\,\mathrm{d}\varpi,\end{aligned} \tag{44}$$

we have by the virial theorem

$$\Omega^2 = \frac{\mathfrak{S}_{33}-\mathfrak{S}_{11}}{I_{11}} = -\frac{T}{\rho}\frac{\displaystyle\int_0^a(\varpi^2 f)'\operatorname{div}\mathbf{n}\,\mathrm{d}\varpi}{\displaystyle\int_0^a \varpi^3 f\,\mathrm{d}\varpi}. \tag{45}$$

If ϖ is measured in units of the equatorial radius a, the foregoing relation takes the form

$$8\Sigma = -\frac{\int_0^1 (\varpi^2 f)' \operatorname{div} \mathbf{n} \, d\varpi}{\int_0^1 \varpi^3 f \, d\varpi} = \frac{\int_0^1 (\varpi^2 f)' \operatorname{div} \mathbf{n} \, d\varpi}{\frac{1}{2}\int_0^1 (\varpi^2 f)' \varpi^2 \, d\varpi}, \tag{46}$$

where Σ is the non-dimensional parameter defined in equation (1).

As an illustration of equation (46) consider a slightly oblate spheroid with an ellipticity ϵ. Then, in the units in which equation (46) is expressed,

$$\operatorname{div} \mathbf{n} \doteq 2 + 4\epsilon\varpi^2; \tag{47}$$

and the second equality in (46) gives

$$\Sigma = \epsilon, \tag{48}$$

a known result.

5. The virial equations for small oscillations about equilibrium

We shall suppose that a uniformly rotating drop, initially in a state of equilibrium, is slightly perturbed; and further that the ensuing motions are described by a Langrangian displacement of the form

$$\boldsymbol{\xi}(x)\, e^{\lambda t}, \tag{49}$$

where λ is a parameter whose characteristic values are to be determined. To the first-order in $\boldsymbol{\xi}$ the general time-dependent equation (23) gives

$$\lambda^2 V_{i;j} - 2\lambda\Omega\epsilon_{il3} V_{l;j} = \delta\mathfrak{S}_{ij} + \Omega^2(\delta I_{ij} - \delta_{i3}\delta I_{3j}) + \delta_{ij}\delta\Pi, \tag{50}$$

where

$$V_{i;j} = \int_V \rho\xi_i x_j \, d\mathbf{x} \tag{51}$$

is the second-order unsymmetrized virial, and $\delta\mathfrak{S}_{ij}$, δI_{ij}, and $\delta\Pi$ are the first variations in the respective quantities due to the deformations caused by the displacement $\boldsymbol{\xi}$. (Note that the index after the semi-colon indicates that a moment with respect to the associated space co-ordinate is involved.)

The first variation of I_{ij} is readily written down. We have

$$\delta I_{ij} = \delta \int_V \rho x_i x_j \, d\mathbf{x} = \int_V \rho(\xi_i x_j + x_i \xi_j) \, d\mathbf{x}$$
$$= V_{i;j} + V_{j;i}. \tag{52}$$

We shall write

$$\delta I_{ij} = V_{ij}, \tag{53}$$

where V_{ij} without the semi-colon is the symmetrized virial.*

Turning next to $\delta\mathfrak{S}_{ij}$, we have, by definition,

$$\delta\mathfrak{S}_{ij} = -T\delta \int_{\mathbf{S}} x_j \operatorname{div} \mathbf{n} \, d\mathbf{S}_i; \tag{54}$$

* Note that $V_{11} = 2V_{1;1}$, etc.

and since the operations of displacement and integration are permutable, we can write

$$\delta\mathfrak{S}_{ij} = -T\left[\int_{\mathbf{S}} \xi_j \operatorname{div}\mathbf{n}\, \mathrm{d}S_i + \int_{\mathbf{S}} x_j\, \delta(\operatorname{div}\mathbf{n})\, \mathrm{d}S_i + \int_{\mathbf{S}} x_j \operatorname{div}\mathbf{n}\, \delta(\mathrm{d}S_i)\right]; \tag{55}$$

and we require to express $\delta(\operatorname{div}\mathbf{n})$ and $\delta(\mathrm{d}S_1)$ in terms of $\boldsymbol{\xi}$. We proceed as follows.

First, we observe that if F is any explicitly known function of the position, then (cf. Chandrasekhar & Lebovitz 1964, equation (A 3))

$$\delta\left(\frac{\partial F}{\partial x_i}\right) = \frac{\partial}{\partial x_i}\,\delta F - \frac{\partial F}{\partial x_k}\frac{\partial \xi_k}{\partial x_i}. \tag{56}$$

Accordingly,

$$\delta(\operatorname{div}\mathbf{n}) = \operatorname{div}(\delta\mathbf{n}) - \frac{\partial n_i}{\partial x_k}\frac{\partial \xi_k}{\partial x_i}, \tag{57}$$

where $\delta\mathbf{n}$ is the Lagrangian change in $\mathbf{n}$ as the surface $\mathbf{S}$ gets displaced by the application of $\boldsymbol{\xi}$. It is known that

$$\delta n_i = \left(n_j n_k \frac{\partial \xi_k}{\partial x_j}\right) n_i - n_k \frac{\partial \xi_k}{\partial x_i}. \tag{58}$$

(Note that, as required, $\delta\mathbf{n}$ is orthogonal to $\mathbf{n}$.) It remains to determine $\delta(\mathrm{d}S_i)$.

Let F be any arbitrary continuous function defined in V and on $\mathbf{S}$. Then

$$\delta\int_{\mathbf{S}} F\, \mathrm{d}S_i = \int_{\mathbf{S}} \delta F\, \mathrm{d}S_i + \int_{\mathbf{S}} F\, \delta(\mathrm{d}S_i). \tag{59}$$

On the other hand, we can also write

$$\delta\int_{\mathbf{S}} F\, \mathrm{d}S_i = \delta\int_V \frac{\partial F}{\partial x_i}\,\mathrm{d}\mathbf{x} = \int_V \delta\left(\frac{\partial F}{\partial x_i}\right)\mathrm{d}\mathbf{x} + \int_V \frac{\partial F}{\partial x_i}\,\delta(\mathrm{d}\mathbf{x}). \tag{60}$$

But it is known that

$$\delta(\mathrm{d}\mathbf{x}) = \operatorname{div}\boldsymbol{\xi}\,\mathrm{d}\mathbf{x}. \tag{61}$$

Making use of this result and of equation (56), we have

$$\delta\int_{\mathbf{S}} F\, \mathrm{d}S_i = \int_V \frac{\partial}{\partial x_i}\,\delta F\,\mathrm{d}\mathbf{x} - \int_V \frac{\partial F}{\partial x_k}\frac{\partial \xi_k}{\partial x_i}\,\mathrm{d}\mathbf{x} + \int \frac{\partial F}{\partial x_i}\operatorname{div}\boldsymbol{\xi}\,\mathrm{d}\mathbf{x}. \tag{62}$$

The first of the three integrals on the right-hand side of equation (62) is clearly expressible as a surface integral over $\mathbf{S}$; and integrating by parts the other two, we have

$$\delta\int_{\mathbf{S}} F\, \mathrm{d}S_i = \int_{\mathbf{S}} \delta F\, \mathrm{d}S_i - \int_{\mathbf{S}} F \frac{\partial \xi_k}{\partial x_i}\,\mathrm{d}S_k + \int_V F \frac{\partial}{\partial x_i}(\operatorname{div}\boldsymbol{\xi})\,\mathrm{d}\mathbf{x}$$
$$+ \int_{\mathbf{S}} F \operatorname{div}\boldsymbol{\xi}\,\mathrm{d}S_i - \int_V F\frac{\partial}{\partial x_i}(\operatorname{div}\boldsymbol{\xi})\,\mathrm{d}\mathbf{x}, \tag{63}$$

or

$$\delta\int_{\mathbf{S}} F\, \mathrm{d}S_i = \int_{\mathbf{S}} \delta F\, \mathrm{d}S_i + \int_{\mathbf{S}} F\left(\operatorname{div}\boldsymbol{\xi}\,\mathrm{d}S_i - \frac{\partial \xi_k}{\partial x_i}\,\mathrm{d}S_k\right). \tag{64}$$

Comparing equations (59) and (64), and remembering that F is arbitrary, we conclude that

$$\delta(\mathrm{d}S_i) = \operatorname{div}\boldsymbol{\xi}\,\mathrm{d}S_i - \frac{\partial \xi_k}{\partial x_i}\,\mathrm{d}S_k. \tag{65}$$

In particular, if $\boldsymbol{\xi}$ is required to be divergence-free, as in the case we are presently considering, then

$$\delta(\mathrm{d}S_i) = -\frac{\partial \xi_k}{\partial x_i}\,\mathrm{d}S_k \quad (\operatorname{div}\boldsymbol{\xi} = 0). \tag{66}$$

Inserting this last expression for $\delta(\mathrm{d}S_i)$ in equation (55), we finally have

$$\delta\mathfrak{S}_{ij} = -T\left[\int_S \xi_j \operatorname{div}\mathbf{n}\,\mathrm{d}S_i + \int_S x_j\,\delta(\operatorname{div}\mathbf{n})\,\mathrm{d}S_i - \int_S x_j \operatorname{div}\mathbf{n}\frac{\partial \xi_k}{\partial x_i}\,\mathrm{d}S_k\right], \tag{67}$$

where, it may be recalled that

$$\delta(\operatorname{div}\mathbf{n}) = \operatorname{div}(\delta\mathbf{n}) - \frac{\partial n_l}{\partial x_k}\frac{\partial \xi_k}{\partial x_l} \tag{68}$$

and

$$\delta n_i = \left(n_k n_l \frac{\partial \xi_k}{\partial x_l}\right) n_i - n_k \frac{\partial \xi_k}{\partial x_i}. \tag{69}$$

Returning to equation (50), and replacing δI_{ij} by V_{ij} in accordance with equation (53), we have

$$\lambda^2 V_{i;j} - 2\lambda\Omega\epsilon_{il3}V_{l;j} = \delta\mathfrak{S}_{ij} + \Omega^2(V_{ij} - \delta_{i3}V_{3j}) + \delta_{ij}\,\delta\Pi. \tag{70}$$

The nine equations which equation (70) represents fall into two non-combining groups of five and four equations distinguished respectively by their parity (i.e. evenness or oddness) with respect to the index 3. It is convenient to have these equations written out explicitly. The equations of even parity are

$$\lambda^2 V_{3;3} = \delta\mathfrak{S}_{33} + \delta\Pi, \tag{71}$$

$$\lambda^2 V_{1;1} - 2\lambda\Omega V_{2;1} = \delta\mathfrak{S}_{11} + \Omega^2 V_{11} + \delta\Pi, \tag{72}$$

$$\lambda^2 V_{2;2} + 2\lambda\Omega V_{1;2} = \delta\mathfrak{S}_{22} + \Omega^2 V_{22} + \delta\Pi, \tag{73}$$

$$\lambda^2 V_{1;2} - 2\lambda\Omega V_{2;2} = \delta\mathfrak{S}_{12} + \Omega^2 V_{12}, \tag{74}$$

$$\lambda^2 V_{2;1} + 2\lambda\Omega V_{1;1} = \delta\mathfrak{S}_{12} + \Omega^2 V_{12}; \tag{75}$$

and the equations of odd parity are

$$\lambda^2 V_{3;1} = \delta\mathfrak{S}_{31}, \tag{76}$$

$$\lambda^2 V_{3;2} = \delta\mathfrak{S}_{32}, \tag{77}$$

$$\lambda^2 V_{1;3} - 2\lambda\Omega V_{2;3} = \delta\mathfrak{S}_{13} + \Omega^2 V_{13}, \tag{78}$$

$$\lambda^2 V_{2;3} + 2\lambda\Omega V_{1;3} = \delta\mathfrak{S}_{23} + \Omega^2 V_{23}. \tag{79}$$

It will be observed that while the even equations (71) to (75) involve $\delta\Pi$, the odd equations (76) to (79) do not. In general, the specification of $\delta\Pi$ in terms of $\boldsymbol{\xi}$ will require some supplementary assumption concerning the physical nature of the oscillations, such as for example that they are taking place adiabatically. However, for an incompressible fluid, the Lagrangian displacement is required to be solenoidal, and we can dispense with the evaluation of $\delta\Pi$ by eliminating it from

(71) to (75) and considering the eliminant system. It will appear subsequently that it is appropriate to associate the equation, obtained by subtracting (73) from (72), with the equation obtained by adding (74) and (75); and similarly to associate the equation, obtained by subtracting twice equation (71) from the sum of (72) and (73), with the equation obtained by differencing (74) and (75). Thus, the eliminant system of equations which we shall consider are the two pairs of equations

$$\tfrac{1}{2}\lambda^2(V_{11}-V_{22})-2\lambda\Omega V_{12} = \delta\mathfrak{S}_{11}-\delta\mathfrak{S}_{22}+\Omega^2(V_{11}-V_{22}), \tag{80}$$

$$\lambda^2 V_{12}+\lambda\Omega(V_{11}-V_{22}) = 2\delta\mathfrak{S}_{12}+2\Omega^2 V_{12}, \tag{81}$$

and

$$\tfrac{1}{2}\lambda^2(V_{11}+V_{22})+2\lambda\Omega(V_{1;2}-V_{2;1})-\lambda^2 V_{33} = \delta\mathfrak{S}_{11}+\delta\mathfrak{S}_{22}-2\delta\mathfrak{S}_{33}+\Omega^2(V_{11}+V_{22}), \tag{82}$$

$$\lambda^2(V_{1;2}-V_{2;1})-\lambda\Omega(V_{11}+V_{22}) = 0. \tag{83}$$

And as already stated, these equations must be supplemented by the solenoidal condition on $\boldsymbol{\xi}$.

We shall see in detail in § 6 below how equations (76) to (83) enable us to determine all the characteristic frequencies of oscillation belonging to the second harmonics.

6. The characteristic equations governing the oscillations belonging to the second harmonics

It can be shown quite generally that the proper solutions belonging to the different characteristic frequencies of oscillation of a rotating incompressible mass in gravitational equilibrium are polynomials of different degrees in the co-ordinates (cf. Cartan 1922; also Lyttleton 1953). But this is not necessarily the case in the problem we are presently considering. However, we may expect that an adequate approximation for the lowest modes will be provided for the problem on hand (particularly for $\Sigma \leqslant 1$ when the equilibrium figures do not depart appreciably from spheroidal forms) by a form for $\boldsymbol{\xi}$ which is linear in the co-ordinates (cf Chandrasekhar & Lebovitz 1962 for the analogous treatment of the oscillations of rotating compressible masses). On this approximation the *eight* virial equations (76) to (83) will suffice to determine all the characteristic frequencies belonging to the second harmonics. For, the *nine* constants L_{ij} $(i, j = 1, 2, 3)$ in the most general linear form for $\boldsymbol{\xi}$, namely

$$\xi_i = L_{ij} x_j \tag{84}$$

are reduced to *eight* by the solenoidal condition on $\boldsymbol{\xi}$ which requires the trace of L_{ij} to vanish:

$$\operatorname{div} \boldsymbol{\xi} = L_{ii} = 0. \tag{85}$$

In actual practice it is not necessary to insert the general form (84) for $\boldsymbol{\xi}$ in the virial equations and carry out the reductions on that basis, since the equations separate into three non-combining groups characterized by the requirements that only certain of the virials, or certain specific linear combinations of them, do not vanish. The three non-combining groups of equations are, in fact, the equations

(76) to (79), the equations (80) and (81), and the equations (82) and (83). The requirements appropriate to each of these three groups are as follows:

(A) equations (76) to (79): only the 'odd' virials $V_{1;3}$, $V_{3;1}$, $V_{2;3}$, and $V_{3;2}$ are non-vanishing;

(B) equations (80) and (81): all the odd virials vanish, $V_{1;2} = V_{2;1}$, $V_{11} = -V_{22}$, and $V_{33} = 0$;

(C) equations (82) and (83): all the odd virials vanish, $V_{1;2} = -V_{2;1}$, $V_{11} = V_{22}$, and $V_{33} \neq 0$.

This same separation of the eight virial equations into three non-combining groups happens when one is treating the oscillations of a Maclaurin spheroid (cf. Lebovitz 1961); and as in the case of the Maclaurin spheroids, we shall call the different modes of oscillations that are obtained as proper solutions of the three groups of equations, which we have distinguished by (A), (B), and (C), as the *transverse-shear*, the *toroidal*, and the *pulsation* modes, respectively. Of these three types of oscillations, that we have called toroidal is the most important, since it is by one of these mòdes that the liquid drop manifests both 'secular' and 'ordinary' instability. On this account, we shall treat the toroidal modes of oscillation first.

(*a*) *The toroidal modes*

The form of the Lagrangian displacement appropriate for these modes is

$$\xi_1 = \alpha x_1 + \beta x_2, \quad \xi_2 = \beta x_1 - \alpha x_2, \quad \text{and} \quad \xi_3 = 0, \tag{86}$$

where α and β are constants unspecified for the present; for, consistent with the requirements for these modes, the only non-vanishing virials that the displacement (86) gives are

$$V_{1;2} = V_{2;1} = 2\pi\rho\beta \int_0^a \varpi^3 f(\varpi)\, d\varpi \tag{87}$$

and

$$V_{11} = -V_{22} = 4\pi\rho\alpha \int_0^a \varpi^3 f(\varpi)\, d\varpi. \tag{88}$$

From equations (68) and (69), we find that for the chosen displacement

$$\delta n_1 = [\alpha(n_1^2 - n_2^2) + 2\beta n_1 n_2]n_1 - (\alpha n_1 + \beta n_2), \tag{89}$$

$$\delta n_2 = [\alpha(n_1^2 - n_2^2) + 2\beta n_1 n_2]n_2 - (\beta n_1 - \alpha n_2), \tag{90}$$

$$\delta n_3 = [\alpha(n_1^2 - n_2^2) + 2\beta n_1 n_2]n_3, \tag{91}$$

and

$$\delta(\operatorname{div} \mathbf{n}) = \alpha \left[(n_1^2 - n_2^2) \operatorname{div} \mathbf{n} + \mathbf{n} . \operatorname{grad} (n_1^2 - n_2^2) - 2 \left(\frac{\partial n_1}{\partial x_1} - \frac{\partial n_2}{\partial x_2} \right) \right] + 2\beta \left[n_1 n_2 \operatorname{div} \mathbf{n} + \mathbf{n} . \operatorname{grad} (n_1 n_2) - \left(\frac{\partial n_1}{\partial x_2} + \frac{\partial n_2}{\partial x_1} \right) \right]. \tag{92}$$

Inserting in equation (92) the explicit forms for n_i given in equation (38), we find

$$\delta(\operatorname{div} \mathbf{n}) = -[\alpha(x_1^2 - x_2^2) + 2\beta x_1 x_2] Q(\varpi), \tag{93}$$

where

$$Q(\varpi) = \frac{1}{\varpi^2} \left[\frac{\phi(2 + 3\phi^2)}{\varpi(1 + \phi^2)^{\frac{3}{2}}} + \frac{\phi'(\phi^2 - 2)}{(1 + \phi^2)^{\frac{5}{2}}} \right]. \tag{94}$$

Next, we verify that for the displacement we are presently considering

$$\delta\mathfrak{S}_{12} = -2T\int_{\mathbf{S}^+} x_2\,\delta(\operatorname{div}\mathbf{n})\,\mathrm{d}S_1 = -2T\int_{\mathbf{S}^+} x_1\,\delta(\operatorname{div}\mathbf{n})\,\mathrm{d}S_2 \tag{95}$$

and
$$\delta\mathfrak{S}_{11}-\delta\mathfrak{S}_{22} = -2T\int_{\mathbf{S}^+}(x_1\,\mathrm{d}S_1-x_2\,\mathrm{d}S_2)\,\delta(\operatorname{div}\mathbf{n}). \tag{96}$$

Now making use of equations (39) and (93), we readily find that

$$\delta\mathfrak{S}_{12} = -\pi T\beta\int_0^a \phi\varpi^4 Q(\varpi)\,\mathrm{d}\varpi \tag{97}$$

and
$$\delta\mathfrak{S}_{11}-\delta\mathfrak{S}_{22} = -2\pi T\alpha\int_0^a \phi\varpi^4 Q(\varpi)\,\mathrm{d}\varpi. \tag{98}$$

Also, from (87) and (88) we have

$$V_{12} = 4\pi\rho\beta\int_0^a \varpi^3 f(\varpi)\,\mathrm{d}\varpi \tag{99}$$

and
$$V_{11}-V_{22} = 8\pi\rho\alpha\int_0^a \varpi^3 f(\varpi)\,\mathrm{d}\varpi. \tag{100}$$

Inserting the foregoing expressions for $\delta\mathfrak{S}_{12}$, $\delta\mathfrak{S}_{11}-\delta\mathfrak{S}_{22}$, V_{12}, and $V_{11}-V_{22}$ in (80) and (81), we obtain the pair of equations

$$\lambda^2\alpha-2\lambda\Omega\beta = -2\mathfrak{J}\alpha+2\Omega^2\alpha \tag{101}$$

and
$$\lambda^2\beta+2\lambda\Omega\alpha = -2\mathfrak{J}\beta+2\Omega^2\beta, \tag{102}$$

where
$$2\mathfrak{J} = \frac{T}{2\rho}\,\frac{\displaystyle\int_0^a \phi\varpi^4 Q(\varpi)\,\mathrm{d}\varpi}{\displaystyle\int_0^a \varpi^3 f(\varpi)\,\mathrm{d}\varpi}. \tag{103}$$

Equations (101) and (102) can be rewritten in the forms

$$[\lambda^2+2(\mathfrak{J}-\Omega^2)]\alpha-2\lambda\Omega\beta = 0 \tag{104}$$

and
$$[\lambda^2+2(\mathfrak{J}-\Omega^2)]\beta+2\lambda\Omega\alpha = 0; \tag{105}$$

and these equations lead to the characteristic equation

$$[\lambda^2+2(\mathfrak{J}-\Omega^2)]^2+4\lambda^2\Omega^2 = 0. \tag{106}$$

Setting $\lambda^2 = -\sigma^2$ (so that a real σ is a necessary and a sufficient condition for the stability of these modes), we can factorize (106) to give

$$\sigma^2-2\sigma\Omega-2(\mathfrak{J}-\Omega^2) = 0 \tag{107}$$

and a similar equation with $-\Omega$ in place of Ω. The roots of (107) are

$$\sigma = \Omega\pm\sqrt{(2\mathfrak{J}-\Omega^2)}. \tag{108}$$

Equation (108) for the frequencies of the toroidal modes predicts the occurrence of a neutral mode of oscillation at the point where

$$\Omega^2 = \mathfrak{J} = \frac{T}{4\rho} \frac{\int_0^a \phi \varpi^4 Q(\varpi)\,\mathrm{d}\varpi}{\int_0^a \varpi^3 f(\varpi)\,\mathrm{d}\varpi}; \tag{109}$$

but since σ is real in the vicinity of this point, on either side, instability does not occur here.* However, instability does occur at the point where

$$\Omega^2 = 2\mathfrak{J}; \tag{110}$$

for $\Omega^2 > 2\mathfrak{J}$ *the rotating drop is unstable by overstable oscillations with a frequency* Ω.

If we measure ϖ in the unit a, and σ and Ω in the unit $(8T/\rho a^3)^{\frac{1}{2}}$, the expression (108) for σ takes the form

$$\sigma = \sqrt{\Sigma} \pm \sqrt{(2J - \Sigma)}, \tag{111}$$

where Σ is the non-dimensional parameter defined in equation (1) and

$$J = \frac{\int_0^1 \phi \varpi^4 Q(\varpi)\,\mathrm{d}\varpi}{32 \int_0^1 \varpi^3 f(\varpi)\,\mathrm{d}\varpi}. \tag{112}$$

(*b*) *The pulsation mode*

The form of the Lagrangian displacement appropriate for this mode—we are using the singular since, as we shall see presently, there is only one non-trivial mode of this kind—is

$$\xi_1 = \alpha x_1 + \beta x_2, \quad \xi_2 = -\beta x_1 + \alpha x_2, \quad \text{and} \quad \xi_3 = -2\alpha x_3, \tag{113}$$

where α and β are constants unspecified for the present. As defined, the displacement is clearly solenoidal; and, moreover, consistent with the requirements for these modes, the only non-vanishing virials which the displacement gives are

$$V_{11} = V_{22} = 2\alpha I_{11}, \quad V_{33} = -4\alpha I_{33}, \quad \text{and} \quad V_{1;2} = -V_{2;1} = \beta I_{11}, \tag{114}$$

where
$$I_{11} = 2\pi\rho \int_0^a \varpi^3 f(\varpi)\,\mathrm{d}\varpi \quad \text{and} \quad I_{33} = \tfrac{4}{3}\pi\rho \int_0^a \varpi f^3(\varpi)\,\mathrm{d}\varpi \tag{115}$$

are the two distinct components of the moment of inertia tensor.

From (68) and (69) we now find

$$\delta(\operatorname{div}\mathbf{n}) = \alpha\left[(n_1^2 + n_2^2 - 2n_3^2)\operatorname{div}\mathbf{n} + \mathbf{n}.\operatorname{grad}(n_1^2 + n_2^2 - 2n_3^2) - 2\left(\frac{\partial n_1}{\partial x_1} + \frac{\partial n_2}{\partial x_2}\right)\right]. \tag{116}$$

And on inserting the explicit forms for n_i given in (38), we obtain

$$\delta(\operatorname{div}\mathbf{n}) = \alpha R(\varpi), \tag{117}$$

where
$$R(\varpi) = \frac{\phi(4+\phi^2)}{\varpi(1+\phi^2)^{\frac{3}{2}}} + \frac{\phi'(4-5\phi^2)}{(1+\phi^2)^{\frac{5}{2}}}. \tag{118}$$

* It should be qualified that the non-occurrence of instability at this point depends on the assumed absence of any dissipative mechanism.

We next verify that for the displacement we are considering at present

$$\delta\mathfrak{S}_{11} = \delta\mathfrak{S}_{22} = 2\pi T\alpha\int_0^a \phi\varpi^2 R(\varpi)\,d\varpi \tag{119}$$

and

$$\delta\mathfrak{S}_{33} = -4\pi T\alpha\int_0^a \varpi f R(\varpi)\,d\varpi. \tag{120}$$

Hence

$$\delta\mathfrak{S}_{11}+\delta\mathfrak{S}_{22}-2\delta\mathfrak{S}_{33} = 4\pi T\alpha\int_0^a (\varpi^2 f)'\,R(\varpi)\,d\varpi. \tag{121}$$

In view of equations (114) and (121, equations (82) and (83) become

$$2\alpha\lambda^2 I_{11}+4\beta\lambda\Omega I_{11}+4\alpha\lambda^2 I_{33} = 4\pi T\alpha\int_0^a (\varpi^2 f)'\,R(\varpi)\,d\varpi+4\alpha\Omega^2 I_{11} \tag{122}$$

and

$$2\beta\lambda^2 I_{11} = 4\alpha\lambda\Omega I_{11}. \tag{123}$$

If we ignore the possibility that λ may be zero—we return to this possibility later—(123) gives

$$\beta\lambda = 2\alpha\Omega. \tag{124}$$

From equation (122) it now follows, after some rearrangements, that

$$\sigma^2\left(1+2\frac{I_{33}}{I_{11}}\right) = 2\Omega^2-\frac{T}{\rho}\frac{\int_0^a (\varpi^2 f)'\,R(\varpi)\,d\varpi}{\int_0^a \varpi^3 f\,d\varpi}, \tag{125}$$

where we have written $\sigma^2 = -\lambda^2$.

If we measure ϖ in the unit a, and σ and Ω in the unit $(8T/\rho a^3)^{\frac{1}{2}}$, the expression (125) for σ^2 becomes

$$\sigma^2\left(1+\frac{4}{3}\frac{\int_0^1 \varpi f^3\,d\varpi}{\int_0^1 \varpi^3 f\,d\varpi}\right) = 2\Sigma-\frac{\int_0^1 (\varpi^2 f)'\,R(\varpi)\,d\varpi}{8\int_0^1 \varpi^3 f\,d\varpi}. \tag{126}$$

Returning to the possibility that λ may be zero, we observe that (122) and (123) do allow this possibility (without requiring α and β to vanish) provided that

$$\Omega^2 I_{11} = -\pi T\int_0^a (\varpi^2 f)'\,R(\varpi)\,d\varpi, \tag{127}$$

or in the units used in expressing σ^2 as in equation (126),

$$\Sigma = -\frac{\int_0^1 (\varpi^2 f)'\,R(\varpi)\,d\varpi}{16\int_0^1 \varpi^3 f\,d\varpi}. \tag{128}$$

The occurrence of a neutral mode, should condition (127) be satisfied, corresponds to the possible existence of an equilibrium configuration, along the axisymmetric sequence, whose figure is invariant to a suitable infinitesimal deformation. The simplest form of such a deformation is represented by the displacement,

$$\xi_1 = \alpha x_1, \quad \xi_2 = \alpha x_2, \quad \text{and} \quad \xi_3 = -2\alpha x_3, \tag{129}$$

since it manifestly preserves the axisymmetry and conserves the volume. Of the three integral relations (30), (31), and (32), which must always obtain under conditions of equilibrium, the first two are clearly invariant to the displacement (129), while the invariance of the last requires

$$\delta\mathfrak{S}_{11}+\delta\mathfrak{S}_{22}-2\delta\mathfrak{S}_{33}+\Omega^2(V_{11}+V_{22}) = 0; \tag{130}$$

it is exactly this last requirement that leads to the condition (127) (cf. (114) and (121)).

Along the sequence of the axisymmetric figures of equilibrium of a rotating drop, the condition (128) is *not* met; therefore, a shape-preserving deformation along the sequence does not exist. However, it may be recalled that, along the Maclaurin sequence, the condition analogous to (130) *is* met where Ω^2 attains its maximum value.

(c) The transverse-shear modes

Finally, turning to the odd equations (76) to (79), we have for the form of the Lagrangian displacement, appropriate to a proper solution of these equations,

$$\xi_1 = \alpha x_3, \quad \xi_2 = \beta x_3, \quad \text{and} \quad \xi_3 = \gamma x_1+\delta x_2. \tag{131}$$

where α, β, γ, and δ are constants unspecified for the present. For this displacement, the only non-vanishing virials (as required) are $V_{1;3}$, $V_{3;1}$, $V_{2;3}$, and $V_{3;2}$; and these are given by

$$V_{1;3} = \alpha I_{33}, \quad V_{2;3} = \beta I_{33}, \quad V_{3;1} = \gamma I_{11}, \quad \text{and} \quad V_{3;2} = \delta I_{11}, \tag{132}$$

where I_{11} and I_{33} have the same meanings as in equations (115).

From equations (68) and (69) we find that for the displacement (131),

$$\begin{aligned}\delta(\operatorname{div}\mathbf{n}) = {} & (\alpha+\gamma)\,(n_1 n_3 \operatorname{div}\mathbf{n}+\mathbf{n}\,.\operatorname{grad} n_1 n_3 - \partial n_3/\partial x_1) \\ & +(\beta+\delta)\,(n_2 n_3 \operatorname{div}\mathbf{n}+\mathbf{n}\,.\operatorname{grad} n_2 n_3 - \partial n_3/\partial x_2);\end{aligned} \tag{133}$$

and when the known expressions for n_i are inserted we find

$$\delta(\operatorname{div}\mathbf{n}) = [(\alpha+\gamma)x_1+(\beta+\delta)x_2]P(\varpi), \tag{134}$$

where

$$\begin{aligned}P(\varpi) &= \frac{\phi}{\varpi(1+\phi^2)^{\frac{3}{2}}}\left(\frac{\phi}{\varpi}+3\frac{\phi'}{1+\phi^2}\right) \\ &= \frac{1}{\varpi^2(1+\phi^2)^{\frac{1}{2}}}-\frac{1}{\varpi^2}\frac{\mathrm{d}}{\mathrm{d}\varpi}\frac{\varpi}{(1+\phi^2)^{\frac{3}{2}}}.\end{aligned} \tag{135}$$

From (67) we find that for the displacement we are now considering

$$\delta\mathfrak{S}_{13} = 2\pi T\gamma\int_0^a(\varpi^2 f)'\operatorname{div}\mathbf{n}\,\mathrm{d}\varpi+2\pi T(\alpha+\gamma)\int_0^a \phi f\varpi^2 P(\varpi)\,\mathrm{d}\varpi; \tag{136}$$

or in view of the identity (45), we can write

$$\delta\mathfrak{S}_{13} = -\gamma\Omega^2 I_{11}-(\alpha+\gamma)L, \tag{137}$$

where

$$L = -2\pi T\int_0^a \phi f\varpi^2 P(\varpi)\,\mathrm{d}\varpi. \tag{138}$$

By the same procedure we also find

$$\delta\mathfrak{S}_{31} = +\alpha\Omega^2 I_{11} - (\alpha+\gamma)\,M, \tag{139}$$

where
$$M = 2\pi T \int_0^a \varpi^3 P(\varpi)\,\mathrm{d}\varpi. \tag{140}$$

The expressions (137) and (139) for $\delta\mathfrak{S}_{13}$ and $\delta\mathfrak{S}_{31}$ look very different; but the known symmetry of $\mathfrak{S}_{ij}$ requires that the two expressions, in fact, be the same; and this equality requires

$$\Omega^2 I_{11} = -2\pi T \int_0^a (\varpi^2 f)' \operatorname{div}\mathbf{n}\,\mathrm{d}\varpi = M - L$$

$$= 2\pi T \int_0^a (\varpi + \phi f)\,\varpi^2 P(\varpi)\,\mathrm{d}\varpi, \tag{141}$$

an integral relation which can be established.*

Analogous to (137) and (139) we have

$$\delta\mathfrak{S}_{23} = -\delta\Omega^2 I_{11} - (\beta+\delta)\,L \tag{142}$$

and
$$\delta\mathfrak{S}_{32} = +\beta\Omega^2 I_{11} - (\beta+\delta)\,M. \tag{143}$$

Returning to (76) to (79) and inserting for the various first variations the expressions we have found for them, we obtain, after some minor simplifications, the set of equations

$$\alpha(M - \Omega^2 I_{11}) + \gamma(M + \lambda^2 I_{11}) = 0, \tag{144}$$

$$\beta(M - \Omega^2 I_{11}) + \delta(M + \lambda^2 I_{11}) = 0, \tag{145}$$

$$\alpha(\lambda^2 I_{33} + L - \Omega^2 I_{33}) + \gamma L = +2\beta\lambda\Omega I_{33}, \tag{146}$$

$$\beta(\lambda^2 I_{33} + L - \Omega^2 I_{33}) + \delta L = -2\alpha\lambda\Omega I_{33}. \tag{147}$$

From equations (144) and (145) it is apparent that

$$\beta = q\alpha \quad \text{and} \quad \delta = q\gamma, \tag{148}$$

where q is some constant; from equations (146) and (147) it then follows that

$$q = \pm \mathrm{i}. \tag{149}$$

* The required equality follows from the relations

$$\int_0^a \varpi^3 P(\varpi)\,\mathrm{d}\varpi = \int_0^a \frac{2+\phi^2}{(1+\phi^2)^{\frac{3}{2}}}\,\varpi\,\mathrm{d}\varpi$$

and
$$\int_0^a \phi f \varpi^2 P(\varpi)\,\mathrm{d}\varpi = -\int_0^a \frac{\phi^4}{(1+\phi^2)^{\frac{3}{2}}}\,\varpi\,\mathrm{d}\varpi_1$$

which can be derived, after two or more integrations by parts, from the second form of $P\varpi)$) given in (135), and use is made of the further relation

$$\int_0^a (\varpi^2 f)' \operatorname{div}\mathbf{n}\,\mathrm{d}\varpi = -\int_0^a \frac{2-\phi^2}{(1+\phi^2)^{\frac{1}{2}}}\,\varpi\,\mathrm{d}\varpi$$

which can be similarly derived when use is made of the expression for div $\mathbf{n}$ given in (40).

Choosing the lower sign in equation (149) for the sake of definiteness and writing $i\lambda = \sigma$, we obtain the pair of equations

$$\alpha(M-\Omega^2 I_{11})+\gamma(M-\sigma^2 I_{11}) = 0 \tag{150}$$

and

$$\alpha(\sigma^2 I_{33}-2\sigma\Omega I_{33}+\Omega^2 I_{33}-L)-\gamma L = 0. \tag{151}$$

Equations (150) and (151) lead to the following characteristic equation for σ:

$$(\sigma^2 I_{11}-M)(\sigma^2 I_{33}-2\sigma\Omega I_{33}+\Omega^2 I_{33}-L)+L(\Omega^2 I_{11}-M) = 0. \tag{152}$$

On simplification (after expansion and division by $I_{11}I_{33}$) we find that equation (152) can be brought to the form

$$(\sigma-\Omega)[\sigma^2(\sigma-\Omega)-K_1(\sigma-\Omega)-K_3(\sigma+\Omega)] = 0, \tag{153}$$

where

$$K_1 = M/I_{11} \quad \text{and} \quad K_3 = L/I_{33}. \tag{154}$$

Thus $\sigma = \Omega$ is an allowed root; and the three remaining roots are determined by the cubic equation

$$\sigma^3-\sigma^2\Omega-\sigma(K_1+K_3)+\Omega(K_1-K_3) = 0. \tag{155}$$

We observe the remarkable similarity of (155) with the equation one obtains for the corresponding modes of a rotating *compressible* mass on an approximative treatment based on the virial equations (Chandrasekhar & Lebovitz 1962, equations (54) and (55)). In the analogous *exact* treatment of the Maclaurin spheroid, one initially obtains equations similar to equations (153) and (155); but the coefficients which replace K_1 and K_3 in that treatment happen to be equal; consequently, a root $\sigma = 0$ is obtained and the remaining non-zero roots are given by a quadratic equation.

If we measure ϖ in the unit a, and σ and Ω in the unit $(8T/\rho a^3)^{\frac{1}{2}}$, the equation for σ takes the form

$$\sigma^3-\sigma^2\sqrt{\Sigma}-\sigma(\kappa_1+\kappa_3)+(\kappa_1-\kappa_3)\sqrt{\Sigma} = 0, \tag{156}$$

where

$$\kappa_1 = \frac{\int_0^1 \varpi^3 P(\varpi)\,d\varpi}{8\int_0^1 \varpi^3 f\,d\varpi} \quad \text{and} \quad \kappa_3 = -\frac{\int_0^1 \phi f \varpi^2 P(\varpi)\,d\varpi}{\frac{16}{3}\int_0^1 \varpi f^3\,d\varpi}. \tag{157}$$

Also, it may be recalled that the relation (141), in the present units requires

$$\int_0^1 \varpi^3 P(\varpi)\,d\varpi+\int_0^1 \phi f\varpi^2 P(\varpi)\,d\varpi = 8\Sigma\int_0^1 \varpi^3 f\,d\varpi. \tag{158}$$

7. Numerical results

The characteristic equations (111), (126), and (156), governing the modes of oscillation of a rotating drop belonging to the second harmonics, depend on certain integral properties of the equilibrium figure. The values of the requisite integrals (besides $\int\varpi^3 f\,d\varpi$, already included in table A 1 of the Appendix) are listed in table 1 for a few values of Σ; with their aid, the characteristic frequencies can be evaluated.

Considering first the discriminant $2J-\Sigma$ of the toroidal modes, we find by interpolation among the values listed in table 1, and by subsequent verification and refinement, that along the sequence $\Sigma = J$ (the condition for the occurrence of

TABLE 1. VALUES OF THE INTEGRALS WHICH OCCUR IN THE CHARACTERISTIC EQUATIONS (111), (126), AND (156)

$\sin^{-1}k$	Σ	$\int_0^1 \phi\varpi^4 Q(\varpi)\,d\varpi$	$2J-\Sigma$	$\frac{2}{3}\int_0^1 f^3\varpi\,d\varpi$	$-\int_0^1 (\varpi^2 f)' R(\varpi)\,d\varpi$	$-\int_0^1 \phi f\varpi^2 P(\varpi)\,d\varpi$	$\int_0^1 \varpi^3 P(\varpi)\,d\varpi$
78°	0·275621	1·61355	+0·67736	0·064978	3·03696	0·37839	0·61173
77°	0·366754	1·48750	+0·56871	0·052915	2·98386	0·34364	0·63523
76°	0·510790	1·31766	+0·39636	0·039003	2·90118	0·29896	0·66993
75°	1·00000	0·91779	−0·18180	0·015234	2·65144	0·20395	0·76483
76°	1·72598	0·62650	−0·95377	0·0040096	2·45867	0·13684	0·83698
77°·7509	2·32911 ($=\Sigma_{\max.}$)	0·57456	−1·40324	—	—	—	—

TABLE 2. SOME PROPERTIES OF THE EQUILIBRIUM FIGURES AT NEUTRAL STABILITY AND AT OVERSTABILITY

(z_0 is the polar radius in units of the equatorial radius; $\bar{a}_1$ is the equatorial radius in units of the radius of a sphere of the same volume; the integrals $\int_0^1 \varpi f\,d\varpi$ and $\int_0^1 \varpi^3 f\,d\varpi$ are simply related to the volume and the moment of inertia I_{11} of the drop; and $A = \Sigma^{-1}\sqrt{(1+2\Sigma)}$.)

	at point of neutral stability	at point of overstability
Σ	0·458706	0·84404
A	3·0187	1·9426
$2\int_0^1 \varpi f\,d\varpi$	0·46259	0·3646
$\int_0^1 \varpi^3 f\,d\varpi$	0·093704	0·07569
z_0	0·67139	0·4924
$\bar{a}_1$	1·1295	1·2228
$\int_0^1 \phi\varpi^4 Q(\varpi)\,d\varpi$	1·375444	1·02215
$2J-\Sigma$	0·458706	0

a neutral mode) and $\Sigma = 2J$ (the condition for the occurrence of two coincident roots and the onset of overstability) at the points

$$\Sigma = 0{\cdot}458706 \quad \text{and} \quad \Sigma = 0{\cdot}84404, \tag{159}$$

respectively. The equilibrium figures for these two values of Σ are illustrated in figure 2; and some of their properties are listed in table 2. In table 3 the characteristic frequencies of the toroidal modes, derived in accordance with equation (111),

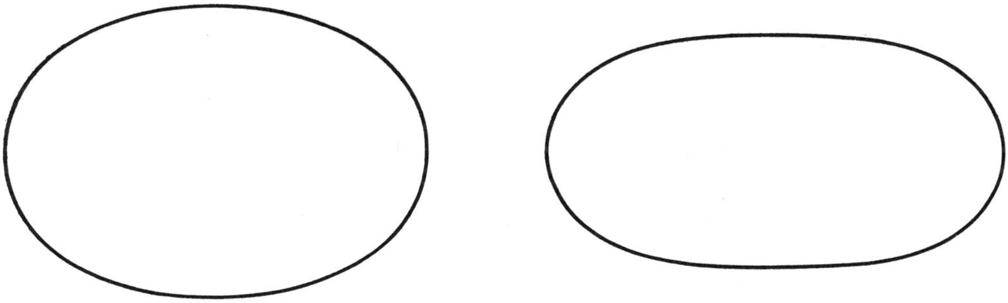

FIGURE 2. The equilibrium figures for $\Sigma = 0{\cdot}458706$ (the one on the left) and $\Sigma = 0{\cdot}84404$ (the one on the right). The figure on the left allows a neutral mode of oscillation; and the one on the right is on the verge of overstability.

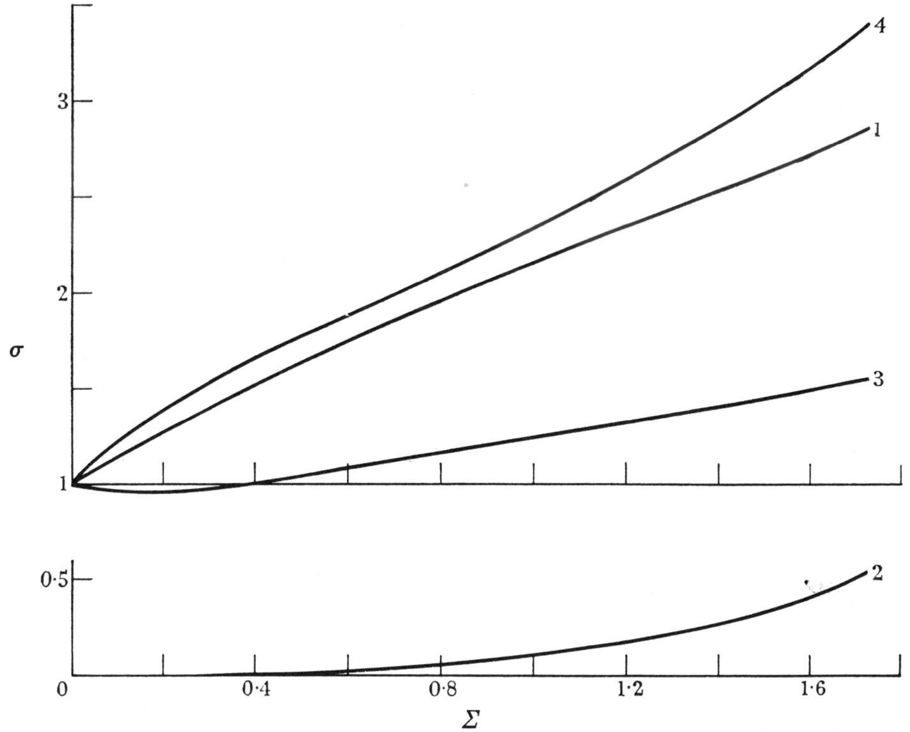

FIGURE 3. The dependence on Σ of the characteristic frequencies of the two toroidal modes of oscillation of a rotating liquid drop. At B ($\Sigma = 0{\cdot}45871$) the drop allows a neutral mode but does not pass through instability in the absence of any dissipative mechanism; at S ($\Sigma = 0{\cdot}84404$) the drop becomes overstable; and the curves labelled $\mathscr{R}(\sigma)$ and $\mathscr{I}(\sigma)$ are the real and the imaginary parts of the frequency for larger values of Σ.

TABLE 3. THE FREQUENCIES OF THE TOROIDAL MODES OF OSCILLATION EXPRESSED IN THE UNIT $(8T/\rho a^3)^{\frac{1}{2}}$

(The values in parentheses were obtained by interpolation applied to the values of $2J - \Sigma$ given in tables 1 and 2.)

Σ	σ_1	σ_2	Σ	$\sigma_{1,2}$
0	1	1	1·0	1·0000 ± 0·4264i
0·05	(1·1944)	(0·7472)	1·2	(1·0954 ± 0·6394i)
0·10	(1·2566)	(0·6242)	1·4	(1·1832 ± 0·7915i)
0·20	(1·3230)	(0·4286)	1·6	(1·2649 ± 0·9121i)
0·275621	1·3480	0·2980	1·725982	1·3138 ± 0·9766i
0·366754	1·3597	0·1485	2·329113	1·5261 ± 1·1846i
0·458706	1·3546	0		
0·510790	1·3443	0·0851		
0·75	(1·1993)	(0·5328)		
0·84404	0·9187	0·9187		

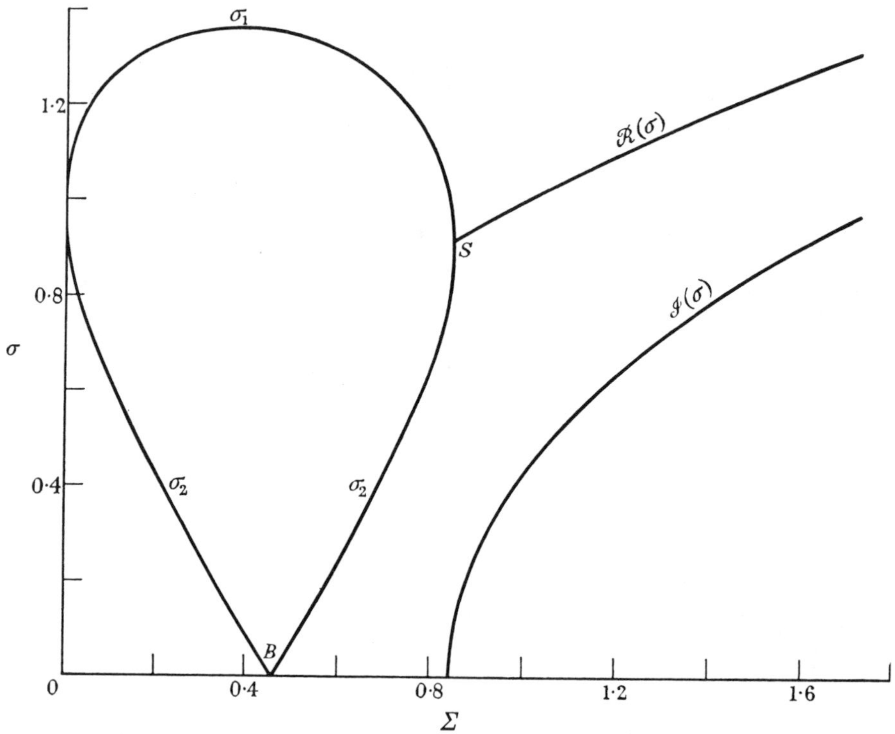

FIGURE 4. The dependence on Σ of the characteristic frequencies of the pulsation mode (curve labelled 1) and the three transverse-shear modes (curves labelled 2, 3 ,and 4).

are listed; and their dependence on Σ is further exhibited in figure 3. It is of interest to compare the behaviour exhibited in figure 3 with the corresponding behaviour along the Maclaurin sequence (cf. Lebovitz 1961, figure 2) and notice their remarkable similarity.

Finally, in table 4 and figure 4 the results for the pulsation and the transverse shear modes are presented.

TABLE 4. THE FREQUENCIES OF THE PULSATION AND THE TRANSVERSE-SHEAR MODES EXPRESSED IN THE UNIT $(8T/\rho a^3)^{\frac{1}{2}}$

	pulsation mode	transverse-shear modes		
Σ	σ_3	σ_4	σ_5	σ_6
0·275621	1·3629	1·4959	0·9690	0·00193
0·366754	1·4740	1·6094	0·9990	0·00482
0·510790	1·6426	1·7795	1·0512	0·01364
1·000000	2·1655	2·3491	1·2429	0·10613
1·725982	2·8660	3·4152	1·5573	0·54413

8. CONCLUDING REMARKS

It is clear that the present investigation requires to be extended along several directions. On the side of hydrodynamics, the extension to the case where the medium surrounding the drop is not a vacuum is particularly important: since only then can the results of the theory be compared with experiments (such as Plateau's) in which the drop is freed of the effects of gravity by its suspension in a liquid of the same density. And on the side of nuclear physics (for applications to phenomena related to nuclear fission), the extension to include the effects of volume charges would appear to be most important. Extensions of the present analysis along both these directions are now being considered.

In conclusion, I wish to record my indebtedness to Professor G. Wentzel for helpful discussions and especially for his proof of the symmetry of the surface-energy tensor $\mathfrak{S}_{ij}$; to Mr M. J. Clement for checking the analysis and providing tables of the elliptic integrals for the three critical values of Σ (namely, 2·32911, 0·8440, and 0·45871); and to Miss Donna D. Elbert for carrying out the necessary numerical calculations.

The research reported in this paper has in part been supported by the Office of Naval Research under contract Nonr–2121(24) with the University of Chicago.

APPENDIX. THE AXISYMMETRIC FIGURES OF EQUILIBRIUM OF A ROTATING DROP

The axisymmetric figures of equilibrium of a rotating drop have been considered by Poincaré (1895), Rayleigh (1914), Appell (1932), and Sperber (1962). But the extant analytic treatments are not as complete as it is needed for the purposes of this paper. We therefore give a brief account of the theory in the context of the present paper and assemble the basic formulae of the subject.

The pressure in the interior of the drop is clearly given by

$$p = p_0 + \tfrac{1}{2}\rho\Omega^2\varpi^2, \tag{A 1}$$

where p_0 is a constant. Applying equation (A 1) to the boundary $z = f(\varpi)$, we have, by equations (3) and (40),

$$p_0 + \tfrac{1}{2}\rho\Omega^2\varpi^2 = T\,\mathrm{div}\,\mathbf{n} = -\frac{T}{\varpi}\frac{\mathrm{d}}{\mathrm{d}\varpi}\frac{\varpi\phi}{\sqrt{(1+\phi^2)}}, \tag{A 2}$$

where $\phi = \mathrm{d}f/\mathrm{d}\varpi$. Equation (A 2) directly integrates to give

$$\tfrac{1}{2}p_0\varpi^2 + \tfrac{1}{8}\rho\Omega^2\varpi^4 = -T\frac{\varpi\phi}{\sqrt{(1+\phi^2)}} + C, \tag{A 3}$$

where C is a constant.

We shall suppose that the drop encloses the origin. Then $\varpi = 0$ is attained; and at this point ϕ must vanish by symmetry. From equation (A 3) it follows that under these circumstances the constant C must be set equal to zero; and we have

$$\frac{p_0}{2T}\varpi + \frac{\rho\Omega^2}{8T}\varpi^3 = -\frac{\phi}{\sqrt{(1+\phi^2)}}. \tag{A 4}$$

At the equator (where $\varpi = a$) $\phi \to -\infty$ (cf. the remarks following equation (36) with respect to the convention of restricting ourselves to the positive hemisphere when ambiguities in sign arise) and equation (A 4) gives

$$\frac{p_0 a}{2T} + \frac{\rho\Omega^2 a^3}{8T} = 1. \tag{A 5}$$

Hence with Σ defined as in equation (1)

$$\frac{p_0 a}{2T} = 1 - \Sigma. \tag{A 6}$$

If we now measure ϖ and z in the unit a, we can rewrite equation (A 4) in the form

$$\frac{\phi}{\sqrt{(1+\phi^2)}} = -\varpi(1-\Sigma+\Sigma\varpi^2), \tag{A 7}$$

or

$$\phi = \frac{\mathrm{d}f}{\mathrm{d}\varpi} = -\frac{\varpi(1-\Sigma+\Sigma\varpi^2)}{[1-\varpi^2(1-\Sigma+\Sigma\varpi^2)^2]^{\frac{1}{2}}}. \tag{A 8}$$

From this last equation it follows that

$$z = f(\varpi) = \int_{\varpi}^{1} \frac{w(1-\Sigma+\Sigma w^2)}{[1-w^2(1-\Sigma+\Sigma w^2)^2]^{\frac{1}{2}}}\,\mathrm{d}w. \tag{A 9}$$

By the transformations

$$w^2 = 1 - A\tan^2\tfrac{1}{2}\theta, \tag{A 10}$$

where

$$A = \frac{1}{\Sigma}\sqrt{(1+2\Sigma)}, \tag{A 11}$$

the integral defining z can be reduced to the form

$$z = \frac{1}{2\Sigma\sqrt{A}}\int_0^{\varphi}\left[(1+A\Sigma) - \frac{2A\Sigma}{1+\cos\theta}\right]\frac{\mathrm{d}\theta}{\sqrt{(1-k^2\sin^2\theta)}}, \tag{A 12}$$

where

$$k^2 = \frac{1}{2}\left[1 + \frac{1}{A}\left(\frac{1}{2}+\frac{1}{\Sigma}\right)\right], \tag{A 13}$$

and φ is related to the value of ϖ at the boundary (where z has the value (A 12)) by

$$\varpi = \sqrt{(1 - A\tan^2\tfrac{1}{2}\varphi)}. \tag{A 14}$$

[23]

The integrals appearing in equation (A 12) are expressible in terms of the standard elliptic integrals

$$F(k,\varphi) = \int_0^{\varphi} \frac{d\theta}{\sqrt{(1-k^2\sin^2\theta)}} \quad \text{and} \quad E(k,\varphi) = \int_0^{\varphi} d\theta \sqrt{(1-k^2\sin^2\theta)} \qquad \text{(A 15)}$$

of the two kinds. We have

$$z = \frac{1}{2\Sigma\sqrt{A}}\left\{(1-A\Sigma)\,F(k,\varphi) + 2A\Sigma\left[E(k,\varphi) - \frac{\sin\varphi\sqrt{(1-k^2\sin^2\varphi)}}{1+\cos\varphi}\right]\right\}. \qquad \text{(A 16)}$$

Equations (A 14) and (A 16) define the equation of the boundary of the drop, parametrically, for a given Σ. In particular

$$\varpi = 0 \quad \text{when} \quad \varphi = \varphi_{\text{max.}} = 2\tan^{-1} A^{-\frac{1}{2}}, \qquad \text{(A 17)}$$

and the corresponding value of z ($=z_0$, say) at the poles follows from (A 16). Since the maximum allowed value of φ is $\frac{1}{2}\pi$, it follows from equation (A 17) that a *necessary* (but, as we shall presently see, not a *sufficient*) condition for the equilibrium figure to enclose the origin is

$$A \geqslant 1. \qquad \text{(A 18)}$$

From equation (A 11) we find that this lower bound on A requires that

$$\Sigma \leqslant 1+\sqrt{2}. \qquad \text{(A 19)}$$

But it should not be supposed, as Sperber (1962) does, that if $\Sigma < 1+\sqrt{2}$ the equilibrium figure necessarily encloses the origin. Clearly, the maximum value of Σ, compatible with the requirement that the origin be enclosed, is determined by the condition that z as given by equation (A 16) vanishes *simultaneously* with ϖ, i.e. when $\varphi = \varphi_{\text{max.}} = 2\tan^{-1}[\Sigma^{\frac{1}{2}}/(1+2\Sigma)^{\frac{1}{4}}]$; and this condition is more stringent than (A 19): it gives

$$\Sigma_{\text{max.}} = 2{\cdot}32911. \qquad \text{(A 20)*}$$

Most of the integrals encountered in § 6, in the treatment of the stability, can be reduced to elliptic integrals. We shall state only the results:

$$z_0 = (1-\Sigma)\,J_0 + \Sigma J_1, \qquad \text{(A 21)}$$

$$\int_0^{\varphi_{\text{max.}}} \varpi f\, d\varpi = \tfrac{1}{2}[(1-\Sigma)\,J_1 + \Sigma J_2], \qquad \text{(A 22)}$$

$$\int_0^{\varphi_{\text{max.}}} \varpi^3 f\, d\varpi = \tfrac{1}{4}[(1-\Sigma)\,J_2 + \Sigma J_3], \qquad \text{(A 23)}$$

* According to equations (A 11) and (A 13)

$$\sqrt{(1+2\Sigma)} = 2(2k^2-1) \pm [4(2k^2-1)^2-3]^{\frac{1}{2}}. \qquad \text{(i)}$$

The reality of the right-hand side requires that

$$k^2 \geqslant \tfrac{1}{4}(2+\sqrt{3}) \quad \text{or} \quad \vartheta = \sin^{-1}k \geqslant 75^\circ. \qquad \text{(ii)}$$

For $\vartheta = 75^\circ$, the two values of Σ given by (i) coincide and are equal to 1. For $75^\circ \leqslant \vartheta \leqslant 90^\circ$, the smaller of the two values of Σ given by (i) is less than 1 and is an 'allowed' root; but the present requirement that $\Sigma \leqslant 2{\cdot}32911$ implies that *both* the roots of (i) are allowed only for ϑ in the range $75^\circ \leqslant \vartheta \leqslant 77^\circ{\cdot}7509$ (see table A 1).

$$\int_0^{\varphi_{\max.}} \phi\varpi^4 Q(\varpi)\,\mathrm{d}\varpi = 4(1-\Sigma)\,[(1-\Sigma)^3-\Sigma]\,J_2$$
$$+[22\Sigma(1-\Sigma)^3-4\Sigma^2]\,J_3+42\Sigma^2(1-\Sigma)^2J_4+34\Sigma^3(1-\Sigma)\,J_5+10\Sigma^4J_6, \quad \text{(A 24)}$$

$$\int_0^{\varphi_{\max.}} \varpi^3 P(\varpi)\,\mathrm{d}\varpi = 2J_0-3(1-\Sigma)^2J_1+(1-\Sigma)\,[(1-\Sigma)^3-6\Sigma]\,J_2$$
$$+[4\Sigma(1-\Sigma)^3-3\Sigma^2]\,J_3+6\Sigma^2(1-\Sigma)^2J_4+4\Sigma^3(1-\Sigma)\,J_5+\Sigma^4J_6, \quad \text{(A 25)}$$

$$-\int_0^{\varphi_{\max.}} (\varpi^2 f)'\,R(\varpi)\,\mathrm{d}\varpi = 2(1-\Sigma)\,[3(1-\Sigma)^3-4\Sigma]\,J_2$$
$$+[42\Sigma(1-\Sigma)^3-8\Sigma^2]\,J_3+90\Sigma^2(1-\Sigma)^2J_4+78\Sigma^3(1-\Sigma)\,J_5+24\Sigma^4J_6, \quad \text{(A 26)}$$

where
$$J_n = \int_0^1 \frac{\varpi^{2n+1}\,\mathrm{d}\varpi}{[1-\varpi^2(1-\Sigma+\Sigma\varpi^2)^2]^{\frac{1}{2}}}. \quad \text{(A 27)}$$

By the same transformations that were used in the reduction of the integral (A 12) it can be shown that

$$J_n = \frac{1}{2\Sigma\sqrt{A}} \sum_{m=0}^{n} \frac{n!}{m!\,(n-m)!}\,(1+A)^{n-m}\,(-2A)^m\,I_m(\varphi_{\max.}), \quad \text{(A 28)}$$

where
$$I_m(\varphi) = \int_0^{\varphi} \frac{\mathrm{d}\theta}{(1+\cos\theta)^m\sqrt{(1-k^2\sin^2\theta)}}. \quad \text{(A 29)}$$

The integrals I_0 and I_1 are given by (cf. (A 15) and (A 16))

$$I_0 = F(k,\varphi) \quad \text{and} \quad I_1 = F(k,\varphi)-E(k,\varphi)+\frac{\sin\varphi\sqrt{(1-k^2\sin^2\varphi)}}{1+\cos\varphi}, \quad \text{(A 30)}$$

and the integrals of the higher orders can be reduced to I_0 and I_1 by means of the recurrence relation

$$I_{m+1} = \frac{1}{2m+1}\Bigg[(4k^2+1)\,m\,I_m-2(2m-1)\,k^2\,I_{m-1}+(m-1)\,k^2 I_{m-2}$$
$$+\frac{\sin\varphi\sqrt{(1-k^2\sin^2\varphi)}}{(1+\cos\varphi)^{m+1}}\Bigg] \quad (m \geqslant 1). \quad \text{(A 31)}$$

TABLE A 1. CONSTANTS DESCRIBING THE AXISYMMETRIC FIGURES OF EQUILIBRIUM OF A ROTATING DROP

(z_0 is the polar radius in units of the equatorial radius; $\bar{a}_1$ is the equatorial radius in units of the radius of a sphere of the same volume; and the integrals $\int_0^1 \varpi f\,\mathrm{d}\varpi$ and $\int_0^1 \varpi^3 f\,\mathrm{d}\varpi$ are simply related to the volume and the moment of inertia I_{11} of the drop.)

$\sin^{-1} k$	k^2	Σ	A	$\varphi_{\max.}$	$2\int_0^1 \varpi f\,\mathrm{d}\varpi$	$\int_0^1 \varpi^3 f\,\mathrm{d}\varpi$	z_0	$\bar{a}_1$
80°	0·969846	0·161182	7·13442	41°·0505	0·57605	0·11542	0·86008	1·0499
79°	0·963592	0·210478	5·66348	45°·5845	0·55351	0·11105	0·82391	1·0640
78°	0·956773	0·275621	4·51885	50°·3865	0·52641	0·10582	0·77961	1·0819
77°	0·949397	0·366754	3·58995	55°·6487	0·49268	0·099384	0·72314	1·1061
76°	0·941474	0·510790	2·78358	61°·8748	0·44699	0·090784	0·64400	1·1425
75°	0·933013	1·000000	1·73205	74°·4577	0·33333	0·070109	0·43118	1·2599
76°	0·941474	1·72598	1·22247	84°·2570	0·21935	0·050706	0·18698	1·4485
77°	0·949397	2·09589	1·08715	87°·6069	0·17173	0·043139	0·07281	1·5716
77°·75085	0·954987	2·32911	1·02129	89°·3965	0·14312	0·038785	0	1·6701

In the particular case $\Sigma = 1$, the integrals J_n can be reduced to Γ-functions; thus

$$J_n = \frac{\sqrt{\pi}}{6} \frac{\Gamma(\frac{1}{3} + \frac{1}{3}n)}{\Gamma(\frac{5}{6} + \frac{1}{3}n)}. \tag{A 32}$$

In table A 1 we collect together a number of constants describing the equilibrium figures for nine values of Σ (including $\Sigma_{\text{max.}} = 2{\cdot}32911$); they were evaluated with the aid of the foregoing formulae.

References

Appell, P. 1932 *Trâité de Mecanique Rationelle*, vol. 4, p. 295. Paris: Gauthier Villars.

Cartan, E. 1922 *Bull. Sci. Math.* **46**, 317, 356.

Chandrasekhar, S. 1960 *J. Math. Anal. Appl.* **1**, 240.

Chandrasekhar, S. 1961 *Hydrodynamic and hydromagnetic stability*, §§117, 118. Oxford: Clarendon Press.

Chandrasekhar, S. 1964 The higher order virial equations and their applications to the equilibrium and the stability of rotating configurations in *Lectures in theoretical physics*, vol. 6. Boulder: University of Colorado Press.

Chandrasekhar, S. & Lebovitz, N. R. 1962 *Astrophys. J.* **135**, 248.

Chandrasekhar, S. & Lebovitz, N. R. 1964 *Astrophys. J.* **140**, 1517.

Jeans, J. H. 1919 *Problems of cosmogony and stellar dynamics*, p. 40. Cambridge University Press.

Lebovitz, N. R. 1961 *Astrophys. J.* **134**, 500.

Lichtenstein, I. 1933 *Gleichgewichtsfigüren Rotierender Flüssigkeiten*. Berlin: Springer Verlag.

Lyttleton, R. A. 1953 *The stability of rotating liquid masses*. Cambridge University Press.

Plateau, J. 1843–69 *Mém. Acad. Roy. Sci. Lett. Belg.*, vol. 16–37; various papers with the common title 'Réchérches Experimentales et Theorique sur les Figures d'Equilibre d'une Masse Liquide sans Tesanteur'; these papers are available in English translation in *The Annual Report of the Board of Regents of the Smithsonian Institution*, 1863 (pp. 207–285), 1864 (pp. 286–369), 1865 (pp. 411–435), and 1866 (pp. 255–289) under the common title 'Experimental and theoretical researches on the figures of equilibrium of a liquid mass withdrawn from the action of gravity'. Washington: Government Printing Office.

Poincaré, H. 1895 *Capillarité*, p. 118. Paris: George Garre, Editeur.

Rayleigh, Lord 1914 *Phil. Mag.* **28**, 161; also *Scientific papers*, vol. vi, p. 257.

Sperber, D. 1962 *Phys. Rev.* **130**, 468.

Acknowledgements

The author and publisher are grateful to the following societies and publishers for their permission to reprint in this volume papers that originally appeared in print under their auspices.

Academic Press (*Annals of Physics, Journal of Mathematical Analysis and Applications*)
American Mathematical Society
Journal of the American Academy of Arts and Sciences
Journal of Mathematics and Mechanics (formerly *Journal of Rational Mechanics and Analysis*)
Mathematics Association of America (*The American Mathematical Monthly*)
Oxford University Press (*Quarterly Journal of Mechanics and Applied Mathematics*)
Stanford University Press
The Colorado Associated University Press
The Royal Astronomical Society
The Royal Society
VEB Deutscher Verlag der Wissenschaften